STUDE[illegible] AND SOLUTIONS MANUAL TO ACCOMPANY

TERNAY'S

CONTEMPORARY ORGANIC CHEMISTRY

Second Edition

ROBERT F. FRANCIS
Department of Chemistry
University of Texas, Arlington
Arlington, Texas

SAUNDERS GOLDEN SUNBURST SERIES

1979

W.B. SAUNDERS COMPANY • Philadelphia • London • Toronto

W. B. Saunders Company: West Washington Square
Philadelphia, PA 19105

1 St. Anne's Road
Eastbourne, East Sussex BN21 3UN, England

1 Goldthorne Avenue
Toronto, Ontario M8Z 5T9, Canada

Student Guide and Solutions Manual to Accompany
Ternay's CONTEMPORARY ORGANIC CHEMISTRY ISBN 0-7216-3834-1

Last digit is the print number: 9 8 7 6 5 4 3 2 1

Preface

Working problems will be the "secret of success" for most of you who are enrolled in an organic chemistry course. By working problems, you will be able to test your understanding of the information presented in lectures and the textbook. After you work a problem or group of problems, having additional input from other students, your instructor, or from a written source is generally desirable, and this book is intended to serve you in that capacity.

The chapters which introduce the major types of compounds (the major functional groups) contain:

1. learning objectives
2. answers to problems
3. reaction summary
4. reaction review exercises

Learning objectives outline the topics of each chapter, and each objective is correlated with the appropriate questions in the text. The reactions of each chapter are presented in general form in the reaction summary. Reaction review exercises are also found in most chapters. These are presented in a question and answer format and are generally divided into two parts. In part A, only the reactions of the chapter are considered. Part B also reviews reactions from previous chapters.

A periodic table and tables which contain IR and NMR data are found in the appendix.

CONTENTS

Chapter 1
THE ATOM.. 1

Chapter 2
FROM BONDS TO SMALL MOLECULES............................. 4

Chapter 3
ALKANES... 16

Chapter 4
STEREOISOMERISM IN ALPHATIC COMPOUNDS..................... 30

Chapter 5
ALKYL HALIDES - NUCLEOPHILIC SUBSTITUTION REACTIONS........ 48

Chapter 6
ALKYL HALIDES - ELIMINATION REACTIONS AND GRIGNARD REAGENTS. 72

Chapter 7
THE STEREOCHEMISTRY OF RING SYSTEMS....................... 94

Chapter 8
ALKENES... 118

Chapter 9
ALKYNES... 179

Chapter 10
ALCOHOLS.. 209

Chapter 11
ETHERS, EPOXIDES AND DIOLS................................ 253

Chapter 12
ORGANIC SYNTHESIS... 280

Chapter 13
DIENES AND TERPENES....................................... 300

Chapter 14
ELECTROCYCLIC AND CYCLOADDITION REACTIONS.................. 331

Chapter 15
AROMATICITY... 338

Chapter 16
ELECTROPHILIC AROMATIC SUBSTITUTION........................ 352

Chapter 17
SYNTHESIS AND SIMPLE ADDITION REACTIONS
OF ALDEHYDES AND KETONES.................................... 397

Chapter 18
ALDEHYDES AND KETONES - THEIR CARBANIONS AND SPECTRA....... 442

Chapter 19
CARBOXYLIC ACIDS AND THEIR DERIVATIVES...................... 494

Chapter 20
SYNTHESIS OF CARBON- CARBON BONDS........................... 537

Chapter 21
AMINES.. 572

Chapter 22
ARYL NITROGEN COMPOUNDS..................................... 619

Chapter 23
PHENOLS AND QUINONES.. 640

Chapter 24
ORGANIC COMPOUNDS OF SULFUR AND PHOSPHORUS.................. 670

Chapter 25
AMINO ACIDS, PEPTIDES AND PROTEINS.......................... 693

Chapter 26 A
CARBOHYDRATES - MONOSACCHARIDES............................. 712

Chapter 26 B
OLIGOSACCHARIDES AND POLYSACCHARIDES........................ 726

Chapter 27
NUCLEIC ACIDS... 736

Chapter 28
INFRARED SPECTROSCOPY, ULTRAVIOLET SPECTROSCOPY
AND MASS SPECTROMETRY....................................... 747

Chapter 29
NUCLEAR MAGNETIC RESONANCE SPECTROSCOPY..................... 757

APPENDICES.. 774

CHAPTER 1, THE ATOM

ANSWERS TO QUESTIONS

1. (a) The atomic number represents the number of protons in the nucleus of an atom. The mass number is the sum of the number of protons and neutrons in the nucleus.

   (b) The atomic weight of an element is the average weight of the atoms of an element (derived from the contribution made by all isotopes of the element) relative to an atom of carbon-12. The atomic number is defined in 1a.

   (c) Main and principal quantum numbers are the same and give the size of a main energy level (the general distance from the nucleus to the electron).

   (d) Electronegativity is a measure of the ability of an atom to attract electrons. Electron affinity is related to the energy released when an atom gains an electron. Both give information concerning the "tendency" of atoms to gain electrons.

   (e) A cation is a particle (atom or group of atoms) possessing a positive charge; an anion is a particle possessing a negative charge.

   (f) Both are isotopes of hydrogen. The deuterium nucleus contains 1 proton and 1 neutron (mass no. = 2); the tritium nucleus contains 1 proton and 2 neutrons (mass no. = 3).

   (g) A 2s orbital is symmetrical about the nucleus and contains a nodal sphere. A 2p orbital is "dumbbell" shaped.

   (h) An orbit is the pathway followed by an electron in the Bohr atom. An orbital is an energy region about the nucleus described by three quantum numbers (n, l, m).

2. (a) Examples mentioned in the text include ^{3}H, ^{14}C, and ^{90}Sr. The Handbook of Chemistry and Physics reveals that all elements with atomic number less than 92 have one or more radioactive isotopes. However, for some elements, the radioactive isotopes occur only in trace amounts.

   (b) All elements with atomic number less than 92 have one or more isotopes which are not naturally occurring.

3. (a) Pauli: Proposed that no two electrons within an atom may have the same four quantum numbers (Pauli exclusion principle).

   (b) de Broglie: Proposed the quantitative relationship between the mass and wavelength of a particle.

   (c) Bohr: Proposed the concept of quantized energy levels based on atomic emission spectra.

   (d) Rutherford: Established the presence of a nucleus in an atom.

(continued next page)

(e) Stern and Gerlach: Demonstrated that the electrons possess a magnetic moment. Introduction of the spin quantum number.

(f) Thomson: Discovered the electron. Proposed an early model of the atom as a homogeneous sphere of positive charge which contained the negatively charged electrons.

(g) Schrödinger: Introduced a wave equation which describes the electron in terms of its wave properties.

(h) Pauling: Devised an electronegativity scale for the elements.

4. (a) 1s does not have a nodal surface.

(b) 2_p has a nodal plane.

5. Carbon (at. no. 6) has a total of 6 electrons; silicon (at. no. 14) has a total of 14 electrons. Both carbon and silicon have 4 valence electrons (both are group IV elements). Since the valence electrons of silicon are in the third energy level, the atomic radius is larger, the ionization potential is less and the electronegativity is lower.

6. See Table 1-2 for isotope masses and abundance.

(a) O: 15.995(0.9976) + 16.999(0.00037) + 17.999(0.002) = 16.00

(b) B: 10.013(0.196) + 11.009(0.804) = 10.8

(c) N: 14.003(0.9963) + 15.00(0.0037) = 14.01

7. Yes. One process is the reverse of the other. The energy involved in adding an electron to an atom to form an anion has the same value as the energy required to remove an electron from the anion to form the atom.

8. Only LiBr (electronegativity difference of 1.8)

9. The most loosely held electron (generally that corresponding to the highest quantum level) is lost first as shown by ionization potential data. The aufbau process is an artificial method of predicting and relating electronic configurations of atoms but cannot predict the properties of atoms.

10. The s orbital is more tightly held by the nucleus because of greater penetration (see Figure 1.2).

11. (a) 17 protons (at. no.), 18 neutrons (mass no.-at. no.)

(b) 17 protons, 20 neutrons

12. The increase in nuclear charge (going down in the group) is more than compensated for by the screening effect of the inner electrons and the increase in distance from the nucleus to the outer electron.

13. Generally in formulas of compounds as well as polyatomic ions, the most electronegative element is written last (notable exceptions are $OH^{\ominus}$ and NH_3). This is also true of complex anions.

CHAPTER 2, FROM BONDS TO SMALL MOLECULES

LEARNING OBJECTIVES

When you have completed this chapter, you should be able to:

1. compare the properties of bonds formed between charged species (problem 1);
2. predict the stabilities of ions and molecules based on their M.O. diagrams (problems 2, 3);
3. predict ionic and covalent bonding (problems 7, 32);
4. draw Lewis (electron-dot) structures (problems 12, 13, 16, 20, 28, 35, 39, 40, 41, 43);
5. predict the polarity of bonds and molecules (problems 8, 9, 19);
6. determine the hybridization of atoms in molecules (problems 11, 18, 36, 38);
7. identify Lewis acids and bases (problems 12, 13, 16, 20, 28, 35, 39, 40, 41, 43);
8. draw resonance structures and predict their relative stabilities (problems 14, 23, 24, 25, 26, 27, 42);
9. assign the formal charge to atoms (problem 31);
10. identify functional groups in organic molecules (problems 15, 19);
11. show how hydrogen bonding can account for certain properties (problems 21, 22, 34, 37).

ANSWERS TO QUESTIONS

1. All of these bonds are a result of the attraction between species which have opposite charges or partial charges. All of these types of bonds lower the energy of the systems in which they are found (all systems are stabilized by bonds). All of these bonds are nondirectional. They differ in their strength (ionic > ion-dipole > dipole-dipole) and in the types of substances involved in the bonding process. An <u>ionic bond</u> is formed between oppositely charged <u>ions</u> (such as $Na^{\oplus}$ and $Cl^{\ominus}$), an <u>ion-dipole</u> bond is formed between an <u>ion</u> and a <u>polar bond or molecule</u> (such as $Na^{\oplus}$ and H_2O), and a <u>dipole-dipole bond</u> is formed between oppositely charged ends of dipoles (such as the association of water molecules).

2. $He_2^{\oplus}$ which has one less electron in an antibonding orbital is more stable than He_2. He, which has no electrons in an antibonding orbital is more stable than $He_2^{\oplus}$.

3. NO contains an unpaired electron (in π^*); CO does not contain an unpaired electron.

4.

element		(a) valence electrons	(b) electrons in kernel
(i)	B	3	2
(ii)	C	4	2
(iii)	N	5	2
(iv)	O	6	2
(v)	F	7	2
(vi)	Cl	7	10

5. Recall that the <u>total number of electrons</u> which can be used in drawing a Lewis structure is equal to the sum of the valence electrons of the atoms in the structure. For <u>ions</u>, one electron is added for each unit negative charge, and one electron is subtracted for each unit positive charge. All atoms should have eight electrons except hydrogen which can have only two electrons.

```
         H                               H H
         |                               | |
(a)    H-Si-H                 (c)      H-C-C-H
         |                               | |
         H                               H H

        ..
       :Cl:                              H H
     ..  |  ..                           | |
(b)  :Cl-Si-Cl:               (d)      H-C-N:
     ..  |  ..                           | |
       :Cl:                              H H
        ..
```

6. See general rules for drawing Lewis structures in question 5.

```
      ..  ..                           ..  ..
(a)  :Br-Br:                  (d)     .S=C=S.
      ..  ..

                                         ..    ..
                              (e)      H-O-C=O:
(b)  H-C≡C-H                             ..  |
                                             H

      ..    ..                         H H H                H   H
(c)  :Cl-C≡C-Cl:                       | | |                 \ /
      ..    ..                                                C
                              (f)    H-C-C=C-H    and     H  / \  H
                                       |                   C - C
                                       H                  H     H
```

7. The ions can be identified by determining the net charge for each structure. This is accomplished by taking the sum of the formal charges of the atoms in each structure.

```
       ..
(a)  H-O:⊖    (ion)                   H
       ..                             | ⊕
       ⊕                      (c)   H-C-Ö-H      (ion)
(b)  H-Ö-H    (ion)                   | |
       |                              H H
       H
```

(continued next page)

(d) H—C(=Ö:)—Ö-H

(e) H—C(=Ö-H⊕)—Ö-H (ion)

(f) :C̈l-C≡N:

(g) H-Ö-N̈=Ö.

(h) H-P(H)-Ö-H

(i) H-S̈-H

(j) [Ö=N(2+)-Ö:⊖]⊕ (ion)

(k) :C̈l-Al(:C̈l:)(:C̈l:)⊖-C̈l: (ion)

8. The individual bond moments cancel each other. For example, in BF_3 the resultant dipole (c) of bonds a and b cancels the dipole of bond d.

⟵+ F —d— B (a/ F, b\ F) +c⟶

9. The H-F bond is more polar than the bonds of other hydrogen halides due to the greater electronegativity of fluorine.

10. (b) and (c)

11.

	sp^3	sp^2	sp
Designation	tetrahedral	trigonal (planar)	digonal (linear)
No. of bonds to carbon	4	4	4
No. of σ bonds	4	3	2
No. of π bonds	0	1	2
No. of substituents	4	3	2
Bond Angle	109.5°	120°	180°
% s character	25	33	50

12. Lewis bases. These elements possess nonbonding electron pairs.

13. In the reaction of magnesium with fluorine each magnesium atom loses one electron to each of two fluorine atoms. A Lewis base donates two electrons to each atom it reacts with.

14. (a) 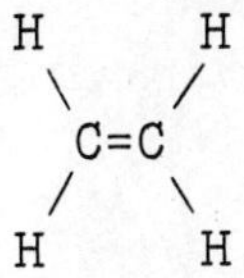
```
H     H
 \   /
  C=C
 /   \
H     H
```
(b)
```
    H
    |
  H-C-H
    |
    H
```
(c)
```
H     H
 \   /
  C=C
 /   \
H     :F:
       ..
```

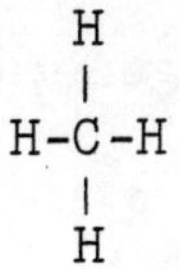

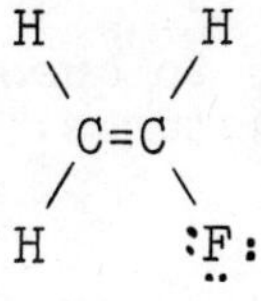

The structures above are better approximations of the hybrid because they have no charge separation.

(d) 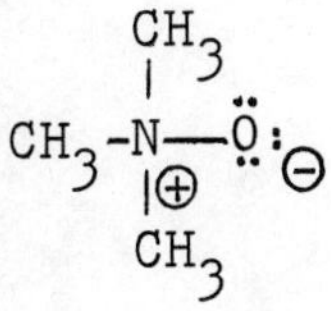
```
       CH3
       |   ..
  CH3-N---O:
       |⊕  ..  ⊖
       CH3
```
(the other structure has 10 electrons around N)

15. (a) $-CH_3$ methyl
$-OH$ hydroxy
$-CO_2H$ carboxy

(b) 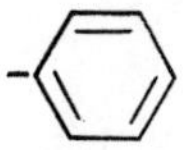 $-C_6H_5$ (benzene ring) phenyl
$-CH=CH_2$ vinyl

(c) $-Br$ bromo 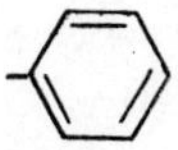
(benzene ring) phenyl
$-NH_2$ amino

(d) $-CH_3$ methyl
$-OH$ hydroxy
$-CN$ cyano

(e) C=O carbonyl (keto)
$-CH_3$ methyl
$-CH_2CH_3$ ethyl

(f) $-C_6H_5$ phenyl
-N=N- azo

16.

(a) 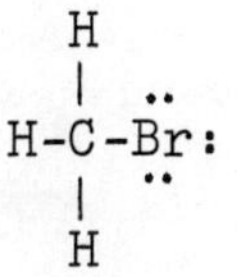
```
    H
    |  ..
  H-C-Br:
    |  ..
    H
```
(b)
```
    H
    | ..  ..
  H-C-S-Cl:
    | ..  ..
    H
```
(c)
```
  ..
 :Cl-C≡N:
  ..
```
(d)
```
   .. ..
  H-N=N-H
```
(e) 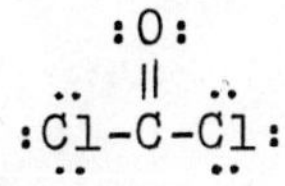
```
       :O:
  ..   ||   ..
 :Cl - C - Cl:
  ..        ..
```
(f)
```
    H   :O:   H
    | ..  || .. |
  H-C-O-C-O-C-H
    | ..    .. |
    H          H
```
(g)
```
         :S:
   H  ..  ||  .. H
    \ N - C - N /
   /           \
  H             H
```
(h)
```
    H
    | .. ..  H
  H-C-N-N  /
    | |    \
    H H     H
```

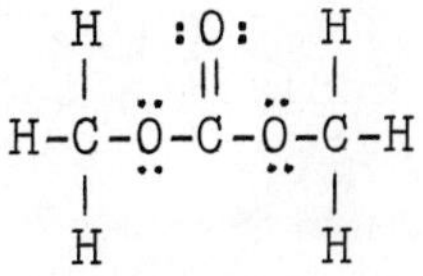

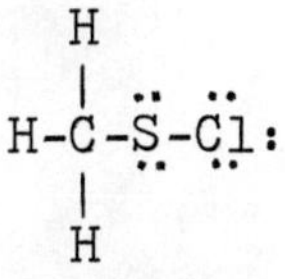

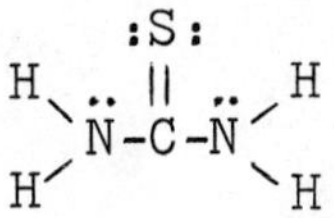

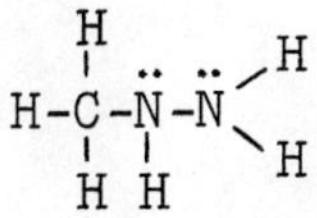

17. See answer to question 5 for general rules used in drawing Lewis structures.

```
          H
          |  ..
(a)     H-C-O-H
          |  ..
          H
```

```
          H H
          | |   ..
(b)     H-C-C-O-H
          | |   ..
          H H
```

```
          H
          |   .
          |  .O:
(c)     H-C-N=
          |  ⊕\ .O:
          H      . . ⊖
```

```
          .O.
          ||  ..
(d)     H-C-O-H
              ..
```

```
                ..
                Cl:
         ..  ..,  ..
(e)     :Cl-P
         ..   \ ..
                Cl:
                ..
```

```
                     H
                     |
              H    H-C-H
              |      |⊕       ..
(f)         H-C------P------O:⊖
              |      |        ..
              H    H-C-H
                     |
                     H
```

```
                         ..
                       =O:(2-)
(g)    Ca(2+)     :O-C
                   ..  \ ..
                        :O:
                         ..
```

```
          H
           \        .
(h)         C=C=O:
           /        .
          H
```

Structures c,f each contain a coordinate covalent bond. Only g has an ionic bond.

18. (b) $:N\equiv N:$ 2 π bonds 2p-2p
σ 2 sp-2sp

```
          ..
         :Cl
          ..\   ..
(c)          C=O:
          ../   ..
         :Cl
          ..
```

C-Cl σ $2sp^2$-3p
C=O π 2p-2p
σ $2sp^2$-$2sp^2$

(continued next page)

(d) $\ominus$:Ö-C(=Ö:)-Ö:$\ominus$ *

C=O σ $2sp^2$-$2sp^2$
π 2p-2p

* carbonate ion, CO_3^{2-}, is a hybrid of the <u>three</u> resonance structures:

[$\ominus$:Ö-C(=O:)-Ö:$\ominus$ ⟷ O=C(-Ö:$\ominus$)-Ö:$\ominus$ ⟷ $\ominus$:Ö-C(-Ö:$\ominus$)=O:]

the three carbon-oxygen bonds are equivalent with C and O sp^2.

(e) H_2C=N̈-H

H-C σ $1s$-$2sp^2$

C=N σ $2sp^2$-$2sp^2$
π 2p-2p

N-H σ $2sp^2$-1s

(f) H_2C=C(C≡C-H)(C≡N:)

H-C σ 1s-$2sp^2$

C=C σ $2sp^2$-$2sp^2$
π 2p-2p

C-C σ $2sp^2$-2sp

C≡C π 2p-2p (two)
σ 2sp-2sp

C≡N – same as C≡C

C-H σ 2sp-1s

(g) :S̈=C=S̈:

C-S π 2p-3p
σ 2sp-$3sp^2$

19. (a) [structure: $CHCl_3$] (b) [structure: CH_2Cl_2] (c) [structure: CH_2ClBr]

(d) [structure: CCl_3F] (e) [structure: CCl_4] (f) [structure: $H_2C{=}CCl_2$]

nonpolar

(g) [structure: $H_2C{=}CHCl$] (h) H-C≡C-C≡C-H (i) Br-C≡N

nonpolar

(j) [structure: cis-BrHC=CHBr] and [structure: trans-BrHC=CHBr] (k) $(CH_3)_2C{=}O$

nonpolar

(Text S.2.6)

20. $H_2\ddot{O}:$, $:NH_3$, $:CH_3^{\ominus}$, $:H^{\ominus}$, $:\ddot{Cl}:^{\ominus}$ all possess nonbonding electron pairs and are Lewis bases. $H^{\oplus}$ and $AlCl_3$ (six electrons around Al) are electron deficient and are Lewis acids. If the definition of a Lewis acid -- electron-pair acceptor -- is extended to include substances which accept an electron pair while also losing an electron pair, then H_2O and NH_3 may be considered Lewis acids as in the reactions below.

$$H_3N: + H{-}\ddot{O}{-}H \longrightarrow NH_4^{\oplus} + :\ddot{O}H^{\ominus}$$

$$H:^{\ominus} + H{-}\ddot{N}H_2 \longrightarrow H_2 + \ddot{N}H_2^{\ominus}$$

(Text S. 2.5)

21. As the temperature is lowered, water molecules become more highly associated with one another and are less "available" for solvating other species.

22. Dimethyl sulfoxide is better able to associate with water molecules through hydrogen bonding between water and the polar $S^{\oplus}-O^{\ominus}$ bond. Dimethyl sulfide, which does not contain a dipolar functional group, does not show the same degree of hydrogen bonding with water.

23. (a) 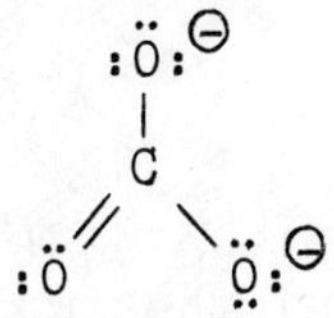 important-same energy 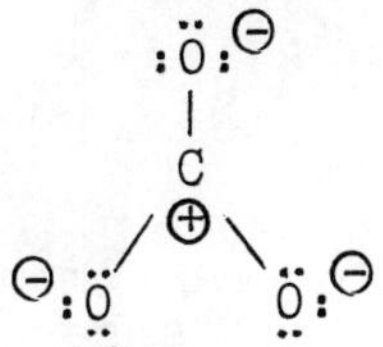 | less important-separation of charge

(b) 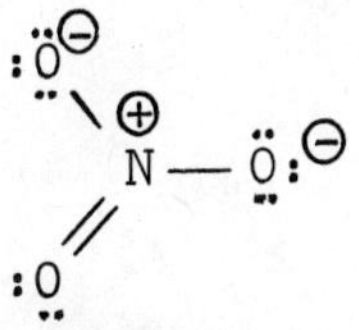 important-same energy 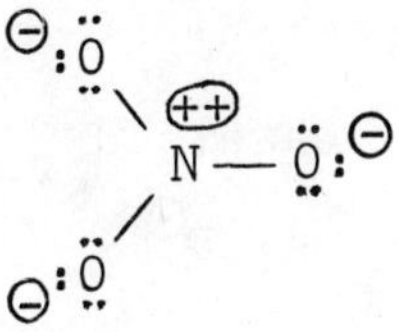 | less important-greater separation of charge

(c)  important-same energy 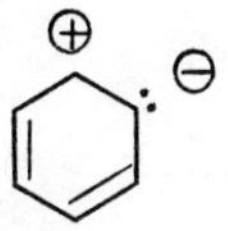 | less important-separation of charge

(Text S. 2.10)

24. (a) 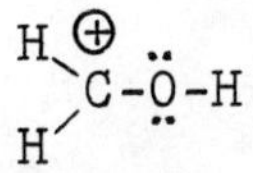 $H_2C^{\oplus}-\ddot{O}-H$

(d) 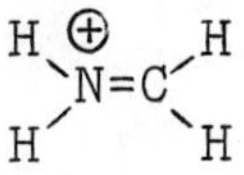 $H_2N^{\oplus}=CH_2$

(b) 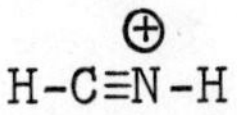 $H-C\equiv N^{\oplus}-H$

(e) $H_2C=N^{\oplus}=\ddot{N}^{\ominus}$

(c) $H_2C=C(H)-\overset{\oplus}{C}HCH_3$

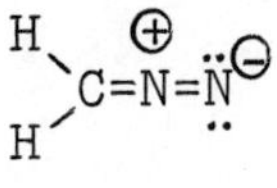

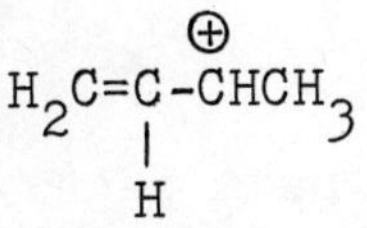

25. (a) each atom (except H) has an octet of electrons, greater number of bonds.

(b) each atom has an octet of electrons, greater number of bonds, no charges

(c) no charges

(d) each atom has an octet of electrons, greater number of bonds.

26. The more important structure has the negative charge on the more electronegative atom and the positive charge on the more electropositive atom.

27. (b), (d), (e), (g)

In b, d, and g, a rearrangement of atoms is involved (not allowed in drawing resonance structures). In (e) the two structures have different numbers of atoms.

28. (a) acid: $H^{\oplus}$; base: $H{-}\overset{O}{\overset{\|}{C}}{-}O{-}H$

(b) acid: $AlCl_3$; base: $Cl^{\ominus}$

(c) acid: BF_3 ; base: $F^{\ominus}$

(d) acid: $^{\oplus}CH_3$; base: CH_3OCH_3

(e) acid: $H^{\oplus}$; base: $H^{\ominus}$

29. (a) $-CH_2-$ methylene

$-CH_2-\ddot{\underset{..}{O}}-CH_2-$ ether

(b) $>C{=}O$ carbonyl

(c) CH_3CH_2- ethyl

$-\overset{O}{\overset{\|}{C}}-OH$ carboxy (a carboxylic acid)

(d) $-CH_2-$ methylene

$-NH_2$ amino (an amine)

$-CO_2H$ carboxy (a carboxylic acid)

(e) 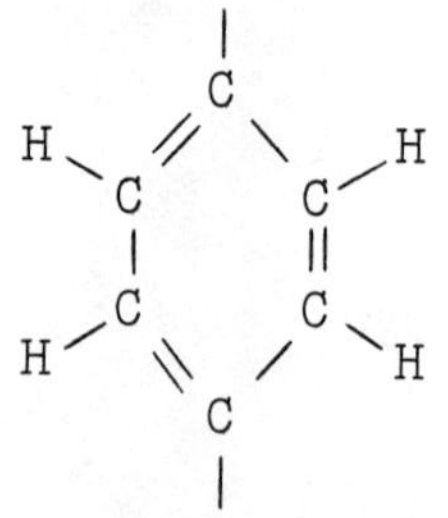 $-C_6H_4-$ (benzene ring with H on four carbons, two open bonds at opposite carbons) phenyl

(continued next page)

$-NO_2$ nitro

-OH hydroxy (an alcohol)

```
    O
    ||
 -N-C-   carboxamido
  |      (an amide)
  H
```

-Cl chloro

(Text S. 2.12)

30. The nitrogen atom has ten electrons around it.

31.

(a) +2	(b) +1	(c) 0	(d) -1	(e) +1
(f) 0	(g) -1	(h) 0	(i) +1	(j) 0
(k) -1	(l) 0	(m) -1	(n) +1	(o) 0
(p) 0	(q) 0	(r) +1	(s) +2	(t) 0
(u) -1				

(Text S. 2.5)

32. $\chi A - \chi B$ = (a) 0.2, (b) 1.0, (c) 3.0

(a) $\% = 16(0.2) + 3.5(0.2)^2 = 3.34$

(b) $\% = 16(1) + 3.5(1)^2 = 19.5$

(c) $\% = 16(3) + 3.5(3)^2 = 79.5$

33. Water molecules solvate cations and anions from an ionic salt in preference to less polar organic molecules. Thus addition of a salt effectively displaces from solution slightly soluble organic compounds.

Methanol is soluble in water and, therefore, unsuitable as an extraction solvent.

(Text S. 2.5)

34. Both ethyl ether and n-butyl alcohol can form hydrogen bonds with water and have about the same solubility.

A sample of n-butyl alcohol consists of molecules associated by hydrogen bonding. Since hydrogen bonding among ethyl ether molecules is not possible, n-butyl alcohol has a higher boiling point.

35. $CH_3-\overset{\oplus}{\ddot{O}}(H)-CH_3$ and $:\ddot{F}:^{\ominus}$. Water functions as a base to remove a proton from $CH_3-\overset{\oplus}{O}(H)-CH_3$.

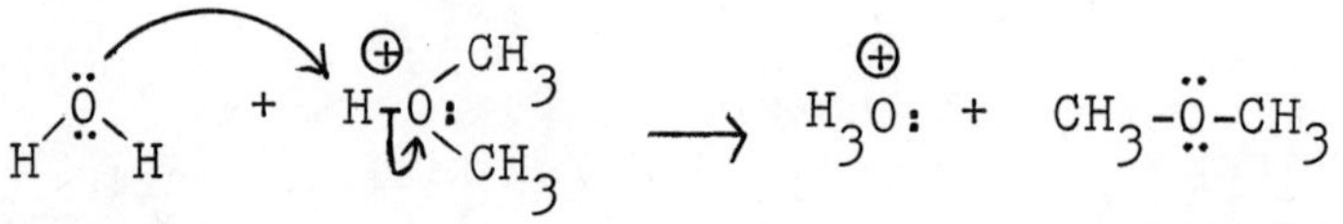

$$H-\ddot{O}-H + H-\overset{\oplus}{O}(CH_3)_2 \longrightarrow H_3\overset{\oplus}{O}: + CH_3-\ddot{O}-CH_3$$

36. $1s^2 2s^2 2p^6 3s^2 3p^6 3d^{10} 4s^2 4p^6 4d^{10} 4f^{14} 5s^2 5p^6 5d^{10} 6sp 6sp$

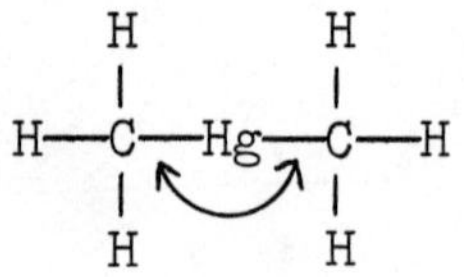

$H_3C-Hg-CH_3$ (180°)

H-C: $1s-2sp^3$

C-Hg: $2sp^3-6sp$

(Text S.2.8)

37. Molecules of 2-methylpyrrolidine are associated with one another by hydrogen bonding. Additional energy is required to break these bonds. Since N-methylpyrrolidine lacks an N-H bond, association by hydrogen bonding is not possible.

38. (a) carbons from sp^2 to sp^3

(b) carbons from sp to sp^2

(c) carbons from sp to $CH_3-\overset{O}{\overset{\|}{C}}-H$ (CH_3: sp^3; C=O carbon: sp^2)

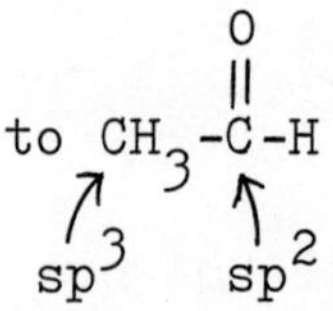

(d) carbon from sp^2 to sp^3

oxygen from sp^2 to sp^3 (continued next page)

(e) center carbon from sp^2 to sp^3

(f) 
```
 H H            H       H
 | |             \     /
H-C-C-O~~~  -->   C=C
 | |             /     \
 H H            H       H
```

(carbons sp^3) (carbons sp^2)

39. $CH_3^{\oplus}$ + $CH_2{=}CH_2$ $\longrightarrow$ $CH_3CH_2CH_2^{\oplus}$

acid base $sp^3 sp^3 sp^2$

sp^2-C sp^2-C's

(Text: Hybridization, S. 2.8)

40. Because $:CH_2$ is electron deficient (lacking an octet of electrons). If it functioned as a base, it would not complete its octet.

41. To complete its octet, $\cdot CH_3$ can accept only <u>one</u> electron. A Lewis acid is an electron-<u>pair</u> acceptor. $\cdot CH_3$ has sp^2 carbon. The formal charge of carbon is 0.
(Text S. 2.10)

42. Since a single bond between two atoms is longer than a double bond between the same two atoms, the compound in this problem should show only one kind of carbon-oxygen bond if it is a hybrid of these structures. Bond equivalency (bond lengths) can be determined by the x-ray diffraction method.
(Text S. 2.8)

43. The <u>strongest acid has the weakest conjugate base</u>. Relative strengths of bases may be determined by considering electronegativity and size factors (greater electronegativity and larger size stabilize a negative charge).

Relative strengths of the conjugate bases are:

(a) $:CH_3^{\ominus} > :\ddot{N}H_2^{\ominus} > :\ddot{O}H^{\ominus} > :\ddot{\underset{..}{F}}:^{\ominus}$ and

(b) $:\ddot{S}\text{-}H^{\ominus} > :\ddot{\underset{..}{Cl}}:^{\ominus}$

(continued next page)

Strengths of the conjugate bases increase as the electronegativity of the central atoms decrease. All are sppproximately the same size.

CHAPTER 3, ALKANES

LEARNING OBJECTIVES

When you have completed this chapter, you should be able to:

1. relate the terms conformation and transition state to energy profiles for C-C bond rotations in a molecule (problems 1, 2, 3, 4);
2. write IUPAC names for structures (problems 5, 13, 14, 15);
3. identify common alkyl groups (problems 5, 11);
4. draw the isomers of a general formula (problems 9, 16);
5. identify primary, secondary and tertiary hydrogens and carbons in structures (problem 12);
6. write the steps in the free-radical halogenation of an alkane (problems 17, 18);
7. calculate ΔH for free-radical halogenation reactions (problems 18, 20);
8. calculate the percents of monohaloalkanes formed in free-radical halogenation (problem 19);
9. show how lithium dialkylcuprates can be used in synthesis of alkanes (problems 10, 24);
10. calculate the equilibrium constant in a chemical reaction (problems 6, 21, 22);
11. perform calculations which involve the determination of molecular weights by use of cryoscopic constants (problems 27, 28);
12. determine an empirical formula from combustion data (problem 29).

ANSWERS TO QUESTIONS

1. An infinite number since each point on the curve represents a conformation.

2. (a) All energy maxima are transition states.

   (b) They are of equal energy since they have identical structures.

3. C and E are mirror images of one another (same energy, similar geometry). D is the transition state between them.

4. Increasing temperatures make energy barriers less significant and lead eventually to the same concentrations of all conformations. However, temperatures necessary for this usually are beyond decomposition temperatures.

5. (a) (i) n-butane (a butane)
(ii) 2,3-dimethylbutane (a hexane)
(iii) 4-ethyl-2,6-dimethylheptane (an undecane)
(iv) 2,2,3,3-tetramethylpentane (a nonane)
(v) 3,3,4,4-tetramethylhexane (a decane)
(vi) 2,3,4,5-tetramethylhexane (a decane)

(b) (i) those compounds containing ethyl groups are:

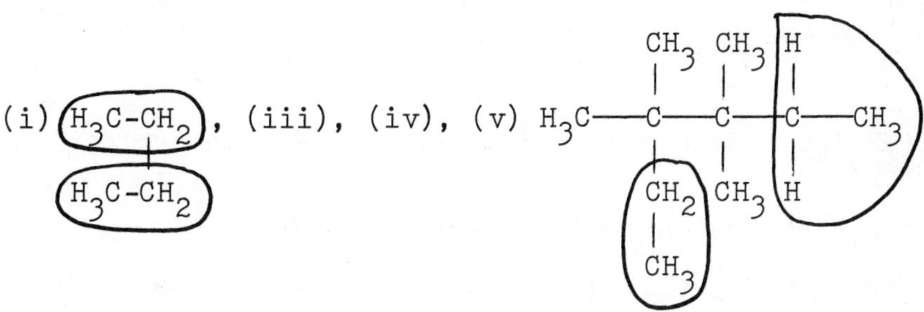

(ii) those containing isopropyl groups are:

(ii) $H_3C-C(H)(CH_3)-C(CH_3)(CH_3)-H$, (iii), (vi)

6. $\Delta G = -RT \ln K_{eq}$, Assume $\Delta H = \Delta G$.

For $CH_{4(g)} + 2O_{2(g)} \longrightarrow CO_{2(g)} + H_2O_{(l)}$ $\Delta H = -213$ kcal.

$$-\ln K_{eq} = \frac{-213 \text{ kcal/mole}}{\left(\frac{1.99 \times 10^{-3} \text{ kcal}}{\text{mole - }^{\circ}K}\right)\left(298^{\circ}K\right)} = -360; \quad K_{eq} = 10^{156}$$

7. Ions will attract one another and greater energy is required to keep ions apart compared to the energy required to keep chlorine radicals (atoms) apart.

8. N_2 and varying amounts of:

(continued next page)

(a) $\left(CH_3\text{-}\underset{CH_3}{\overset{CN}{C}}\text{—}\right)_2$, (b) $CH_3\text{-}\overset{CN}{C}=CH_2$ and (c) $CH_3\text{-}\overset{CN}{CH}\text{-}CH_3$

These products are derived from the process:

$$CH_3\text{—}\underset{CH_3}{\overset{CN}{C}}\text{—}\ddot{N}=\ddot{N}\text{—}\underset{CH_3}{\overset{CN}{C}}\text{—}CH_3 \longrightarrow :N\equiv N: + CH_3\text{—}\underset{CH_3}{\overset{CN}{C}}\cdot$$

9. (a) $(CH_3)_4C$ (MW = 72) forms only one monochloro derivative.

   (b) Two dichloro derivatives can be formed,
   $(CH_3)_3CCHCl_2$ (1,1-dichloro-2,2-dimethylpropane) and
   $(CH_3)_2C(CH_2Cl)_2$ (1,3-dichloro-2,2-dimethylpropane).

10. (a) CH_3X

    (b) CH_3CH_2X

    (c) $CH_3CH_2CH_2X$

    (d) $(CH_3)_2CHCH_2X$ or $CH_3CH_2\underset{CH_3}{CH}X$

    (e) $CH_3CH_2CH_2C(CH_3)_2CH_2X$ or $CH_3CH_2C(CH_3)_2CH_2CH_2X$

    (f) $(CH_3CH_2CH_2)_2CHCH_2CH_2X$

11. (a) methyl

    (b) ethyl

    (c) isopropyl

    (d) <u>sec</u>-butyl

    (e) isoheptyl

    (f) isodecyl

    (g) <u>tert</u>-pentyl (1,1-dimethylpropyl)

    (h) <u>t</u>-butyl

(Text S. 3.4, Table 3.3)

12. (a) $CH_3(CH_2)_5CH_3$

1° 2° 1°

(b) $(CH_3)_3CCH_2CH(CH_3)_2$

1° 2° 3° 1°

(c) $(CH_3)_4C$

1°

(d) $(CH_3)_2CH(CH_2)_4CH_3$

1° 3° 2° 1°

(Text S. 3.7)

13. (a) ethane
(b) 2-methylbutane
(c) 2,2-dimethylbutane
(d) 2,2,3-trimethylbutane
(e) 4-methylheptane
(f) 2,4,4,6-tetramethyloctane
(g) 2,5-dimethylhexane
(h) 2,2,3,3-tetramethylbutane
(i) 3,3-dimethylpentane

(Text S. 3.4)

14. (a) The number has been omitted, should be 2-methylbutane.
(b) Longest chain is eight carbons (4-methyloctane).
(c) Incorrect numbering, should be 2-methylpentane.
(d) The longest chain contains 7 carbons; the name is 2,3-dimethylheptane.

(Text S. 3.4)

15. (a) chloromethane
(b) dichloromethane
(c) bromochloromethane
(d) chloroethane
(e) 1,1-dichloroethane
(f) 1,2-dichloroethane
(g) 1,2-difluoroethane
(h) 2,3-difluorobutane
(i) 1,2,3-trifluorobutane
(j) 1,3,4-triiodohexane
(k) 1-bromo-2-fluorobutane
(l) 2-bromo-1-fluorobutane
(m) 2,3-difluoro-4-methylpentane
(n) 2,4-difluoro-3-methylpentane

(Text S. 3.4)

16. (a) $CH_3CH_2CH_2CH_2CH_2CH_2CH_3$, heptane

$(CH_3)_2CHCH_2CH_2CH_2CH_3$, 2-methylhexane

$CH_3CH_2CH(CH_3)CH_2CH_2CH_3$, 3-methylhexane

$(CH_3)_3CCH_2CH_2CH_3$, 2,2-dimethylpentane

$CH_3CH_2\text{-}C(CH_3)_2\text{-}CH_2CH_3$, 3,3-dimethylpentane

$CH_3\text{—}CH(CH_3)\text{—}CH(CH_3)\text{—}CH_2CH_3$, 2,3-dimethylpentane

$CH_3CH(CH_3)CH_2CH(CH_3)CH_3$, 2,4-dimethylpentane

$(CH_3)_3C\text{-}CH(CH_3)_2$, 2,2,3-trimethylbutane

$CH_3CH_2CH(CH_2CH_3)_2$, 3-ethylpentane

(b) $CH_3CH_2CH_2CH_2CH_2Cl$ 1-chloropentane

$CH_3CH_2CH_2CHClCH_3$ 2-chloropentane

$CH_3CH_2CHClCH_2CH_3$ 3-chloropentane

$(CH_3)_2CHCH_2CH_2Cl$ 1-chloro-3-methylbutane

$(CH_3)_2CHCHClCH_3$ 2-chloro-3-methylbutane

$(CH_3)_2CClCH_2CH_3$ 2-chloro-2-methylbutane

$ClCH_2CH(CH_3)CH_2CH_3$ 1-chloro-2-methylbutane

(continued next page)

$(CH_3)_3CCH_2Cl$
1-chloro-2,2-dimethylpropane

(c) $CH_3CH_2CH_2CH_2CH_2CHCl_2$
1,1-dichlorohexane

$CH_3CH_2CH_2CH_2CCl_2CH_3$
2,2-dichlorohexane

$CH_3CH_2CH_2CCl_2CH_2CH_3$
3,3-dichlorohexane

$CH_3CH_2CH_2CH_2CHClCH_2Cl$
1,2-dichlorohexane

$CH_3CH_2CH_2CHClCH_2CH_2Cl$
1,3-dichlorohexane

$CH_3CH_2CHClCH_2CH_2CH_2Cl$
1,4-dichlorohexane

$CH_3CHClCH_2CH_2CH_2CH_2Cl$
1,5-dichlorohexane

$ClCH_2CH_2CH_2CH_2CH_2CH_2Cl$
1,6-dichlorohexane

$CH_3CHClCHClCH_2CH_2CH_3$
2,3-dichlorohexane

$CH_3CHClCH_2CHClCH_2CH_3$
2,4-dichlorohexane

$CH_3CHClCH_2CH_2CHClCH_3$
2,5-dichlorohexane

$CH_3CH_2CHClCHClCH_2CH_3$
3,4-dichlorohexane

$CH_3CHClCH(CH_2Cl)CH_2CH_3$
2-chloro-3-(chloromethyl)pentane

$(CH_3)_2CHCH_2CH_2CHCl_2$
1,1-dichloro-4-methylpentane

$(CH_3)_2CHCH_2CCl_2CH_3$
2,2-dichloro-4-methylpentane

$(CH_3)_2CHCCl_2CH_2CH_3$
3,3-dichloro-2-methylpentane

$Cl_2CHCH(CH_3)CH_2CH_2CH_3$
1,1-dichloro-2-methylpentane

$(CH_3)_2CHCH_2CHClCH_2Cl$
1,2-dichloro-4-methylpentane

$(CH_3)_2CHCHClCH_2CH_2Cl$
1,3-dichloro-4-methylpentane

$CH_3CCl(CH_2Cl)CH_2CH_2CH_3$

2-chloro-2-(chloromethyl)pentane

$CHCl_2CH_2CH(CH_3)CH_2CH_3$

1,1-dichloro-3-methylpentane

$CH_3CHClCCl(CH_3)CH_2CH_3$

2,3-dichloro-3-methylpentane

$CH_2ClCHClCH(CH_3)CH_2CH_3$

1,2-dichloro-3-methylpentane

$CH_3CHClCH(CH_3)CHClCH_3$

2,4-dichloro-3-methylpentane

$CH_2ClCH_2CCl(CH_3)CH_2CH_3$

1,3-dichloro-3-methylpentane

$CH_3CHClCH(CH_2Cl)CH_2CH_3$

2-chloro-3(chloromethyl)-pentane

$CH_2ClCH_2CH(CH_3)CHClCH_3$

1,4-dichloro-3-methylpentane

$CHCl_2CH(CH_3)CH(CH_3)CH_3$

1,1-dichloro-2,3-dimethyl-pentane

$CH_2ClCH_2CH(CH_3)CH_2CH_2Cl$

1,5-dichloro-3-methylpentane

$CH_2ClCCl(CH_3)CH(CH_3)CH_3$

1,2-dichloro-2,3-dimethyl-pentane

$CH_2ClCH_2CH(CH_2Cl)CH_2CH_3$

1-chloro-3-(chloromethyl)-pentane

$CH_2ClCH(CH_3)CCl(CH_3)CH_3$

1,3-dichloro-2,3-dimethyl-pentane

$CH_3CCl_2CH(CH_3)CH_2CH_3$

2,2-dichloro-3-methylpentane

$CH_2ClCH(CH_2Cl)CH(CH_3)CH_3$

1-chloro-2-(chloromethyl)-3-methylpentane

$CH_2ClCH(CH_3)CH(CH_2Cl)CH_3$

1-chloro-3-(chloromethyl)-2-methylpentane

$CH_3CCl(CH_2Cl)CH(CH_3)CH_3$

2-chloro-2-(chloromethyl)-3-methylpentane

$(CH_3)_2CClCH_2CH_2CH_2Cl$
1,4-dichloro-4-methylpentane

$ClCH_2CH(CH_3)CHCH_2CH_2Cl$
1,5-dichloro-2-methylpentane

$(CH_3)_2CHCHClCHClCH_3$
2,3-dichloro-4-methylpentane

$(CH_3)_2CClCH_2CHClCH_3$
2,4-dichloro-2-methylpentane

$ClCH_2CH(CH_3)CH_2CHClCH_3$
1,4-dichloro-2-methylpentane

$(CH_3)_2CClCHClCH_2CH_3$
2,3-dichloro-2-methylpentane

$ClCH_2CH(CH_3)CHClCH_2CH_3$
1,3-dichloro-2-methylpentane

$ClCH_2CCl(CH_3)CH_2CH_2CH_3$
1,2-dichloro-2-methylpentane

$(CH_3)_2CClCCl(CH_3)_2$
2,3-dichloro-2,3-dimethylbutane

$ClCH_2CH(CH_3)CH(CH_3)CH_2Cl$
1,4-dichloro-2,3-dimethylbutane

$(CH_3)_3CCH_2CHCl_2$
1,1-dichloro-3,3-dimethylbutane

$(CH_3)_3CCl_2CH_3$
2,2-dichloro-3,3-dimethylbutane

$Cl_2CHC(CH_3)_2CH_2CH_3$
1,1-dichloro-2,2-dimethylbutane

$(CH_3)_3CCHClCH_2Cl$
1,2-dichloro-3,3-dimethylbutane

$ClCH_2C(CH_3)_2CH_2CH_2Cl$
1,4-dichloro-2,2-dimethylbutane

$ClCH_2C(CH_3)_2CHClCH_3$
1,3-dichloro-2,2-dimethylbutane

$(ClCH_2)_2C(CH_3)CH_2CH_3$
1-chloro-2-chloromethyl-2-methylbutane

(Text S. 3.4)

17. (a) Product Analysis: The reaction mechanism proposed in this problem predicts the formation of H_2, HCl and Cl_2 from the combination of hydrogen and chlorine atoms (H· and :C̤̈l·). Experimentally, however, H_2 is not a

product in the chlorination of methane.

(b) Energetics: The 1st propagation step in the mechanism of the reaction proposed in this problem is less favorable (ΔH = +21 kcal*) than for the same step in the mechanism proposed in the text (ΔH = -1 kcal*).

*Bond dissociation energies: CH_4=102, CH_3Cl=81, HCl=103 kcal/mole

(Text S. 3.7)

18. Consult Table 2-3 for Bond Dissociation Energies

(a) $CH_3CH_2CH_3 + Br_2 \xrightarrow{h\nu} CH_3CHBrCH_3 + HBr$

(2°C-H = 95) (46) (68) (87)

total energy required to break bonds = 141 kcal

total energy released in bond formation = 155 kcal

ΔH = -14 kcal/mole (exothermic)

(b) and (c)

	Reaction	ΔH (kcal/mole)
(i)	:Br-Br: $\xrightarrow{h\nu}$ 2 :Br· (46)	+ 46
(ii)	:Br· + $CH_3CH_2CH_3 \longrightarrow$ HBr + $CH_3\dot{C}HCH_3$ (95) (87)	+ 8
(iii)	$CH_3\dot{C}HCH_3$ + :Br-Br: $\longrightarrow$ $CH_3CH(Br)CH_3$ + :Br· (46) (68)	- 22
(iv)	2:Br· $\longrightarrow$ Br_2 (46)	- 46
net:	$Br_2 + CH_3CH_2CH_3 \longrightarrow CH_3CH(Br)CH_3$ + HBr	- 14

(Text S. 3.7)

19. $CH_3CH_2CH_2CH_2Cl$ (from $CH_3CH_2CH_2CH_3$ containing 6 H's which can give $CH_3CH_2CH_2CH_2Cl$)

$$\frac{6 \times 1}{(6 \times 1) + (4 \times 4)} \times 100 = 27.3\%$$

$CH_3CH_2CHClCH_3$ (from $CH_3CH_2CH_2CH_3$ containing 4 H's which can give $CH_3CH_2CHClCH_3$)

$$\frac{(4 \times 4)}{(6 \times 1) + (4 \times 4)} \times 100 = 72.7\%$$

(Text S. 3.7)

20. The iodination of methane is endothermic (ΔH = 14 kcal/mole). Entropy changes in free-radical halogenation reactions are negligible, and the reactions are enthalpy controlled.

(Text S. 3.7)

21. $\Delta G = (32 \text{ kcal/mole}) - (298^\circ K \times 30 \times 10^{-3} \text{ kcal/deg-mole} =$

23 kcal/mole

$$\Delta G = -RT \ln K, \; -\ln K = \frac{-23 \text{ kcal/mole}}{\left(\frac{1.99 \times 10^{-3} \text{ kcal}}{\text{mole} - {}^\circ K}\right)\left(298^\circ K\right)} = -38.98;$$

$$K = 1.2 \times 10^{-17}$$

(Text S. 3.5)

22. n-pentane $\rightleftharpoons$ isopentane; $\Delta G = G_{f(isopentane)} - G_{f(pentane)}$

$$\Delta G = (-3.50) - (-2.00) = -1.50 \text{ kcal/mole}$$

$$\Delta G = -RT \ln K; \ -\ln K = \frac{-1.50 \text{ kcal/mole}}{\left(\frac{1.99 \times 10^{-3} \text{ kcal}}{\text{mole} - {}^{o}\text{K}}\right)\left(298^{o}\text{K}\right)} = -2.54;$$

$K = 12.6$

(Text S. 3.5)

23. Methane is produced from methyl free radicals reacting with hydrogen.

$$CH_3\text{-}\ddot{\underset{..}{Cl}}: + Na\cdot \longrightarrow Na^{\oplus} :\ddot{\underset{..}{Cl}}:^{\ominus} + \cdot CH_3$$

$$H_3C\cdot + H\text{-}H \longrightarrow CH_4 + H\cdot \quad \text{etc.}$$

Iodine can also react with the methyl free radicals generated from methyl chloride and sodium.

$$H_3C\cdot + :\ddot{\underset{..}{I}}\text{-}\ddot{\underset{..}{I}}: \longrightarrow CH_3I + :\ddot{\underset{..}{I}}\cdot$$

(Text S. 3.7)

24. A variety of combinations are shown; however, recall that best yields of alkanes are obtained from 1° alkyl halides and 1° or 2° lithium dialkylcuprates.

(a) $Li(CH_3)_2Cu + CH_3I$

(b) $Li(CH_3)_2Cu + CH_3CH_2Br$

or $Li(CH_3CH_2)_2Cu + CH_3I$

(c) $Li(CH_3)_2Cu + CH_3CH_2CH_2Cl$

or $Li(CH_3CH_2)_2Cu + CH_3CH_2\text{-}Br$

or $Li(CH_3CH_2CH_2)_2Cu + CH_3I$

(continued next page)

(d) $Li(CH_3)_2Cu + (CH_3)_2CHCl$

or $Li[(CH_3)_2CH]Cu + CH_3I$

(e) $Li(CH_3)_2Cu + (CH_3)_2CHCH_2Cl$

or $Li(CH_3CH_2)_2Cu + (CH_3)_2CHCl$

or $Li[(CH_3)_2CHCH_2]_2Cu + CH_3I$

or $Li(CH_3)_2Cu + CH_3CH_2CH(CH_3)\text{-}Cl$

or $Li(CH_3CH_2CH(CH_3))_2Cu + CH_3I$

(f) $Li(CH_3)_2Cu + (CH_3CH_2)_2CHCH_2Cl$

or $Li[(CH_3CH_2)_2CHCH_2]Cu + CH_3I$

or $Li(CH_3CH_2)_2Cu + (CH_3CH_2)_2CHCl$

(g) $Li(CH_3)_2Cu + (CH_3)_2CHCH(CH_3)Cl$

or $Li[(CH_3)_2CH]Cu + (CH_3)_2CHCl$

(Text S. 3.8)

25. All components of both reactions are identical and have the same energies except for butane and isobutane, i.e., both reactions are $2C_4H_{10} + 13O_2 \longrightarrow 8CO_2 + 10H_2O$. Therefore, the greater ΔH for the combustion of butane reflects the greater energy content of butane.

26. $k = Ae^{-(Ea/RT)} = 10^{12}\ e^{-(37/(1.98 \times 10^{-3})(373)}$

$k = (10^{12})(e^{-50}) = 1.9 \times 10^{-10}\ \text{sec}^{-1}$

(Text, Appendix B)

27. $MW = K_f \dfrac{1000\ w_2}{\Delta T\ w_1}$; $K_f = \dfrac{(300)(0.1)(2000)}{(1000)(30)} = 2$

(Text, Appendix A)

28. K_f benzene = 5.12

$w_2 = \dfrac{(72)(1.58)(15)}{(1000)(5.12)} = 0.333g$

(Text, Appendix A)

29. $0.88g\ CO_2 \times \dfrac{12g\ C}{44g\ CO_2} = 0.24g\ C$

$0.55g\ H_2O \times \dfrac{2g\ H}{18g\ H_2O} = 0.06g\ H$

Weight of oxygen = 0.614 - 0.300 = 0.314g

$0.24g\ C \times \dfrac{1\ \text{mole C atoms}}{12g} = 0.02\ \text{mole C}$

$0.06g\ H \times \dfrac{1\ \text{mole H atoms}}{1g\ H} = 0.06\ \text{mole H}$

$0.314g\ O \times \dfrac{1\ \text{mole O atoms}}{16g\ O} = 0.02\ \text{mole O}$

simple ratio CH_3O

(Text, Appendix A)

SUMMARY OF REACTIONS

A. Free-radical Halogenation

general: $R\text{-}H + X_2 \xrightarrow{\text{heat or light}(h\nu)} R\text{-}X + HX$

reactivity of C-H bonds: $3^\circ > 2^\circ > 1^\circ$

reactivity of X_2: $F_2 > Cl_2 > Br_2 > I_2$

examples: $CH_3CH_2CH_3 + Br_2 \xrightarrow{hv} CH_3CHBrCH_3 + CH_3CH_2CH_2Br$

(major)

$(CH_3)_3CH + Cl_2 \xrightarrow{hv} (CH_3)_3CCl + (CH_3)_2CHCH_2Cl$

(major)

B. Lithium Dialkylcuprates

1. Preparation

general: 1). $RX + 2Li \xrightarrow{ether} RLi + LiX$

2). $2RLi + CuI \longrightarrow Li[R_2Cu] + LiX$

(or may be written as: $RX \xrightarrow{Li} \xrightarrow{CuI} Li[R_2Cu]$)

example: 1). $CH_3Cl + 2Li \xrightarrow{ether} CH_3Li + LiCl$

2). $2CH_3Li + CuI \longrightarrow Li[(CH_3)_2Cu] + LiI$

(or: $CH_3Cl \xrightarrow{Li} \xrightarrow{CuI} Li[(CH_3)_2Cu]$)

2. Synthesis of Alkanes from Lithium Dialkylcuprates

general: $LiR_2Cu + R'X \longrightarrow R\text{-}R' + RCu + LiX$

(best results when R is 1° or 2° and R' is 1°)

example: $Li(CH_3)_2Cu + (CH_3)_2CHCH_2Cl \longrightarrow (CH_3)_2CHCH_2CH_3 +$

$CH_3Cu + LiCl$

CHAPTER 4, STEREOISOMERISM IN ALIPHATIC COMPOUNDS

LEARNING OBJECTIVES

When you have completed this chapter, you should be able to:

1. identify chiral objects or structures which are capable of existing as enantiomers (problems 1, 3, 4, 23, 37);
2. locate planes, points and simple axes of symmetry in objects or structures (problems 2, 4, 27, 43);
3. interconvert three-dimensional drawings and Fischer projections (problems 6, 7);
4. assign priorities to groups and absolute configurations to structures (problems 8, 9, 10, 11, 12, 13, 20, 24, 25, 26, 41);
5. define common terms associated with the stereochemistry of compounds (glossary, problem 19);
6. associate structures with the terms: identical, enantiomers, diastereomers, meso, d, l, threo, erythro, or optically active (problems 14, 15, 17, 18, 20, 21, 22, 32, 35, 36);
7. predict the number of stereoisomers possible for a general structure from the number of chiral centers (problem 30);
8. identify enantiotopic and prochiral groups in structures (problems 16, 29, 31);
9. perform calculations which involve specific rotation (problems 5, 33, 34, 38, 40);
10. identify properties which differentiate and allow separation of enantiomers (problems 28, 39).

ANSWERS TO QUESTIONS

1. (a) a point at the center, an axis through the center, any plane through the axis, a plane perpendicular to the axis bisecting the tire.

   (b) a vertical plane bisecting the stem.

   (c) same as (a)

   (d) a plane through the center of the edge of the roll.

   (e) a point in the center lengthwise, an axis through the center (from top to bottom), a plane through the center (long or short direction).

   (f) an axis through the center of the top and bottom, any plane along that axis.

   (g) an axis through the center of the top and bottom, any plane along that axis.

   (h) a point at the center, an axis perpendicular to the rings, any plane along that axis, a plane through the rings.

2. The dissymmetric objects are capable of existing as pairs of enantiomers. There are: a, c, d, f, g, h, i and k.

3. (a) No, because the planet possesses a plane of symmetry which passes through the center of the red spot. (b) The earth is chiral. Considering surface features, no plane or point of symmetry exists.

4. Structures capable of existing as enantiomers are b, d, and i. Each has a nonsuperimposable minor image.

   Each of the remaining structures contains a plane of symmetry and is drawn below so that the page represents the plane of symmetry.

   (a) 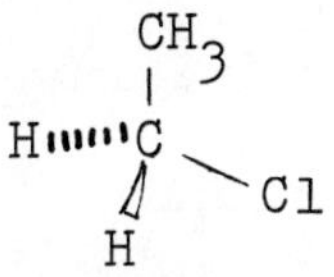 CH_3, H, Cl, H on C

   (c) Br, CH_3, F, CH_3 on C

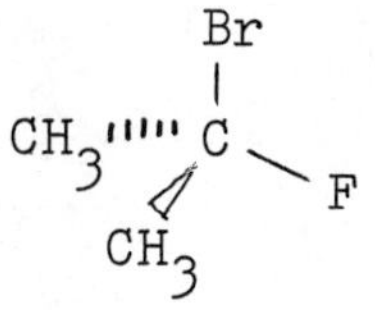

   (e) 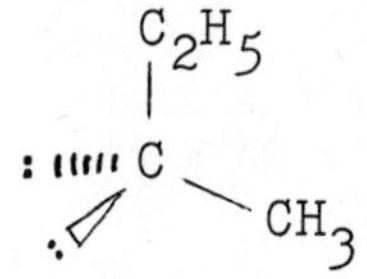 C_2H_5, :, CH_3, : on C

   (f) 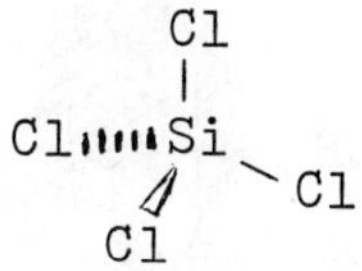 Cl, Cl, Cl, Cl on Si

   (g) 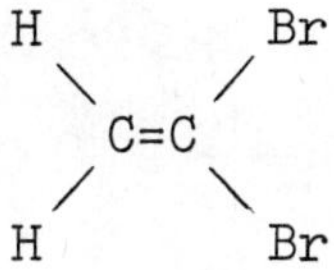 H, Br, H, Br on C=C (top view) or $H_2C=CBr_2$ (side view)

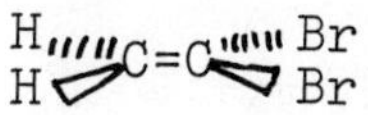

   (h) 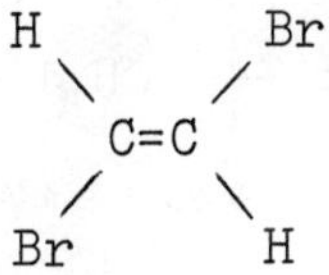 H, Br, Br, H on C=C

   (j) 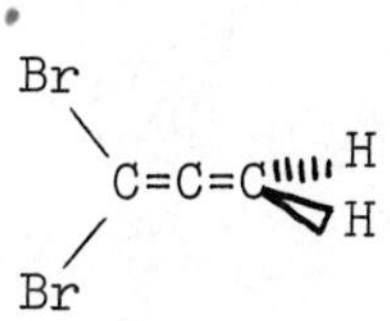 Br, Br on C=C=C, H, H

5. (a) $+120^{o} = \dfrac{0.10^{o}}{1\text{ dm x g/100 ml}}$; g = 0.083g

   (b) for a 2 dm tube, 0.041g

6. Remember the convention: the structure is flattened with the <u>horizontal</u> "bow tie" <u>facing</u> you. In (a) all possible (9) orientations are shown. In (b) through (h) only one of the possible orientations is shown.

(a) 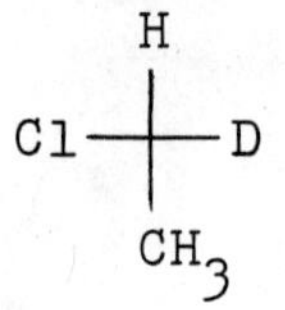 Fischer projections:

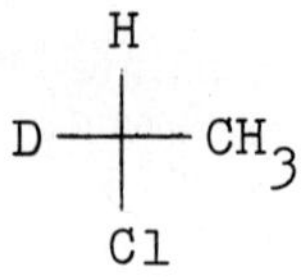

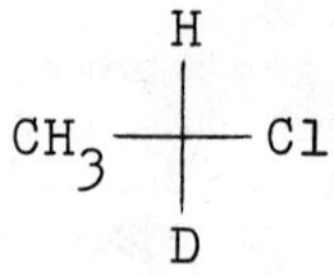

Top	Left	Right	Bottom
H	Cl	D	CH_3
H	D	CH_3	Cl
H	CH_3	Cl	D
Cl	CH_3	D	H
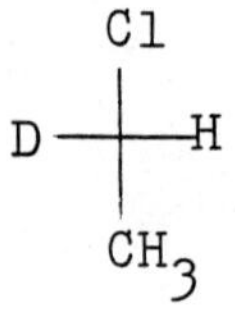 Cl	D	H	CH_3
Cl	H	CH_3	D
CH_3	D	Cl	H
CH_3	Cl	H	D
CH_3	H	D	Cl
D	H	Cl	CH_3
D	Cl	CH_3	H
D	CH_3	H	Cl

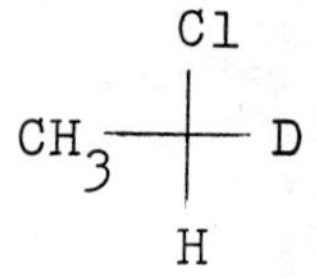

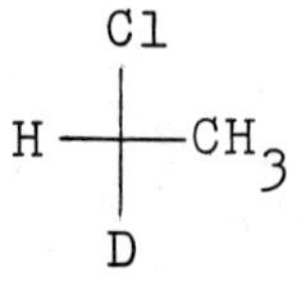

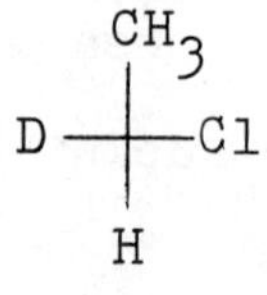

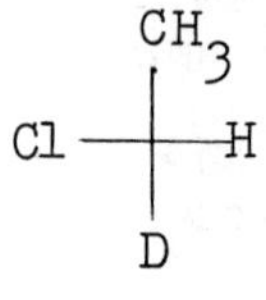

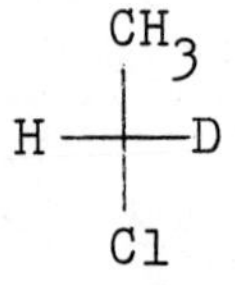

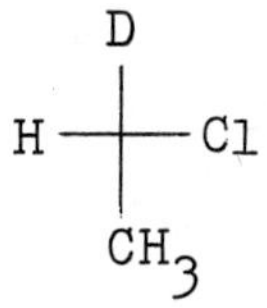

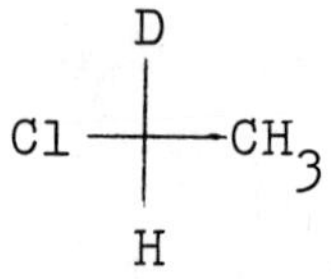

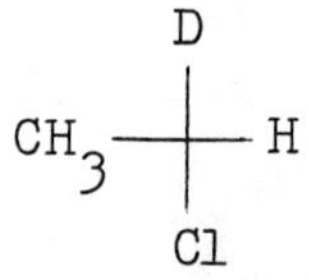

Part	Top	Left	Right	Bottom
(b)	H	D	Cl	CH_3
(c)	CH_3	Cl	D	H
(d)	H	CH_3	Cl	Br
(e)	H	Br	CH_3	D
(f)	CH_3	H	D	Br
(g)	H	Br	CH_3	D
(h)	NH_2	CH_3	CO_2H	H

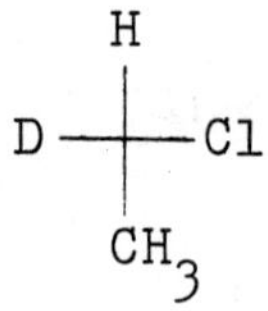

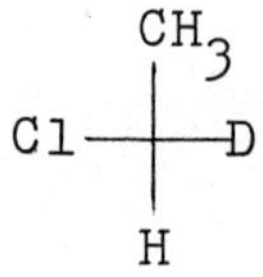

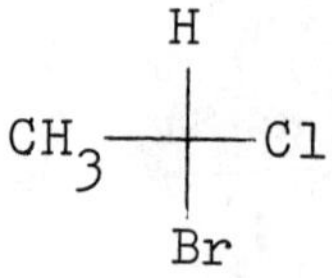

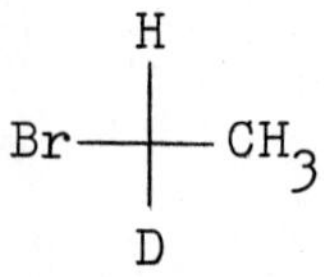

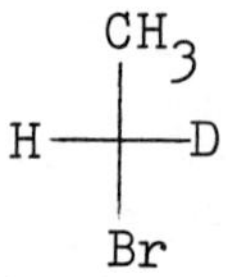

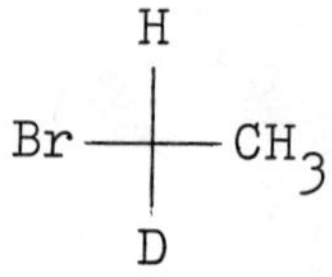

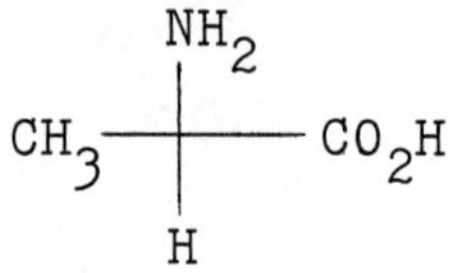

7. (i) (a) 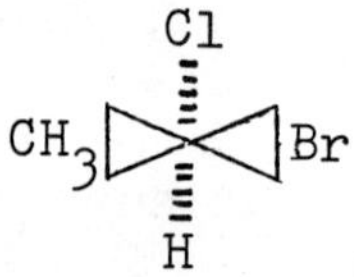 CH_3, Cl, Br, H (wedge-dash structure) or H, C, Br, Cl, CH_3 (wedge-dash structure) etc.

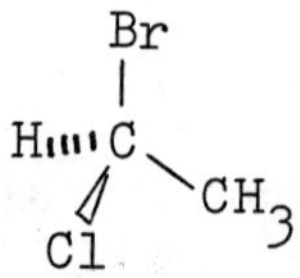

(continued next page)

(b) 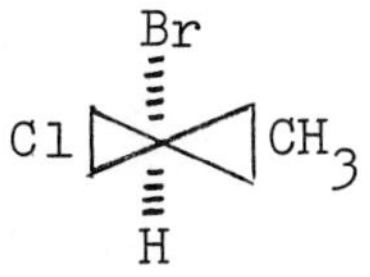 [structure: Br top, Cl left, CH_3 right, H bottom] or [structure: CH_3, H, Br, Cl around C] etc.

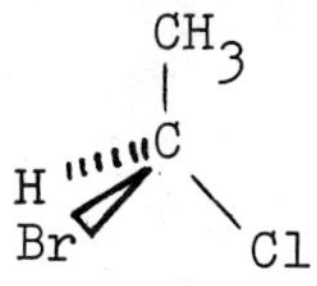

(c) 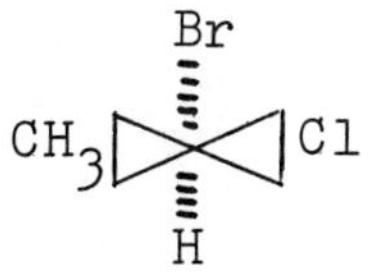 [structure: Br top, CH_3 left, Cl right, H bottom] (d) 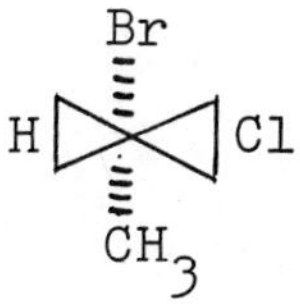 [structure: Br top, H left, Cl right, CH_3 bottom] (e) 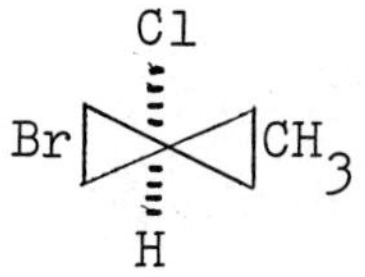 [structure: Cl top, Br left, CH_3 right, H bottom]

(f) 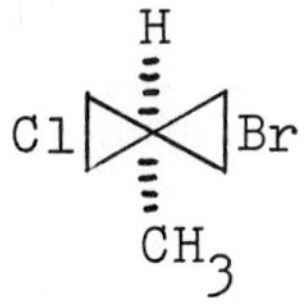 [structure: H top, Cl left, Br right, CH_3 bottom]

(ii) a, b, d and f are identical; c and e are identical

8. (a) Br > Cl > H (d) Sn > Br > H > lone e^- pair

(b) I > S > N (e) I > Br > Cl > F

(c) $^3H > ^2H > ^1H$

9. (a) $-C(CH_3)_3 > -CH(CH_3)_2 > -C_2H_5 > -CH_3 > -H$

(b) $-CH(CH_2CH_3)_2 > -CH_2CH(CH_3)_2 > -CH_2CH_2CH_3$

(c) $-C(CH_3)_2CH_2CH_2CH_3 > -C(CH_3)_2CH_2CH_3 > -CH(CH_3)CH_2CH_3$

10. (a) $-CH(CH_3)=CH_2 > -CH=CHCH_3 > -CH=CH_2$

(b) $-C\equiv CH > -C(CH_3)_3 > -CH_2CH_2CH_2C(CH_3)_3$

(c) $-OCH_3 > -CH_2SH > -CH_2OCH_3 > -CH_2OH$

(continued next page)

(d) 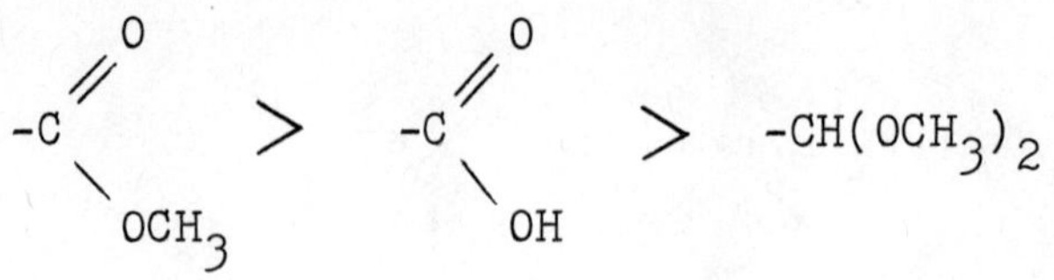 $-C(=O)OCH_3$ > $-C(=O)OH$ > $-CH(OCH_3)_2$

11. (a) 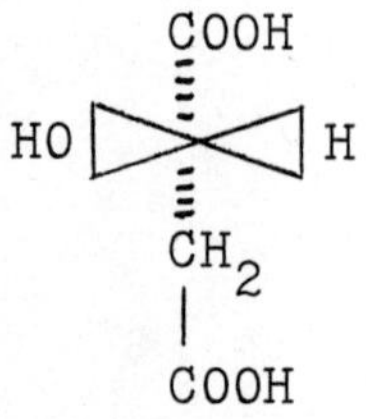 HO, H, COOH, CH_2–COOH (wedge structure)

(b) the sign of rotation is negative (c) the structure has an S configuration: (S)-malic acid

12. Problem 4.6: (a) S, (b) R, (c) R, (d) S, (e) R, (f) S, (g) R, (h) R.

Problem 4.7: (a) S, (b) S, (c) R, (d) S, (e) R, (f) S.

13. (a) R (d) R (g) S
(b) S (e) S (h) S
(c) S (f) R (i) S

14. (a) a d,l form (b) meso (c) a d,l form
(d) erythro (e) threo (f) erythro

15. Those which will exhibit optical activity are: b, c, d, e, h and i (compounds in (i) are stereoisomers).

16. (a) hydrogens
(b) methyl groups

(continued next page)

(c) ethyl groups

(d), (e) hydrogens of either methylene group

17. Isomers are shown for selected examples.

(a) 2; (b) 4: $CH_3CH_2CHBr_2$, (R) and (S)-$CH_3CHBrCH_2Br$, $CH_3CBr_2CH_3$ and $CH_2BrCH_2CH_2Br$; (c) 7: $CH_3CH_2CH_2CHBr_2$, $CH_3CH_2CBr_2CH_3$, (R) and (S)-$CH_3CH_2CHBrCH_2Br$, (R) and (S)-$CH_3CHBrCH_2CH_2Br$, $CH_2BrCH_2CH_2CH_2Br$, meso-, (2R, 3R)- and (2S, 3S) CH_3-CHBr-CHBr-CH_3; (d) 3; (e) 2; (f) 7; (g) 11.

18. (a) 2 (d) 3 (g) 17
(b) 5 (e) 2
(c) 10 (f) 9

19. In addition to examples, a summary of definitions is included.

(a) A chiral molecule is not superimposable upon its mirror image. It may be dissymmetric or asymmetric. Example:

CH_3 | H''''C — CH_2CH_3, Cl (wedge)

(b) An achiral molecule is superimposable upon its mirror image. It is symmetric. Example:

H | CH_3''''C — Cl, CH_3 (wedge)

(page represents plane of symmetry)

(continued next page)

(c) A <u>prochiral molecule</u> contains at least one prochiral center which has two identical and two nonidentical substituents. Replacing one of the identical substituents with an achiral substituent produces a chiral center. Example:

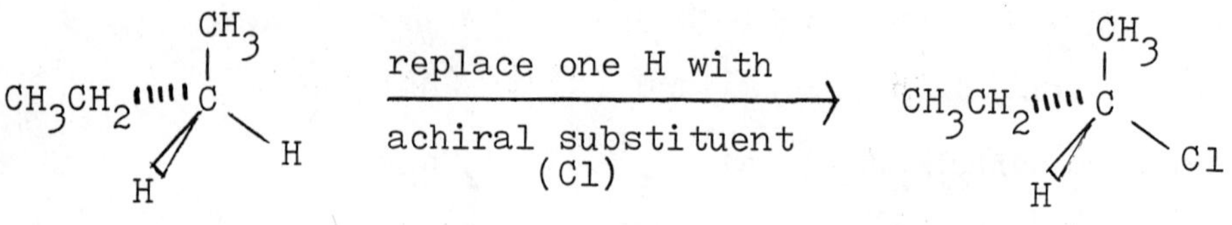

CH_3CH_2–C(CH_3)(H)(H) —replace one H with achiral substituent (Cl)→ CH_3CH_2–C(CH_3)(H)(Cl)

achiral chiral

(d) A <u>pair of enantiomers</u>. Stereoisomers which are nonsuperimposable mirror images. Example:

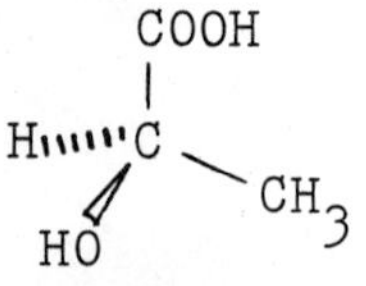

COOH, H, C, CH_3, HO and COOH, C, H, CH_3, OH

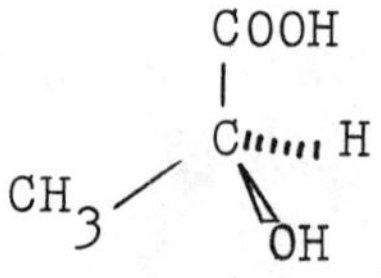

R S

(e) <u>Diastereomers</u> are stereoisomers which are <u>not</u> mirror images and <u>not</u> superimposable; i.e., nonenantiomeric stereoisomers. Example:

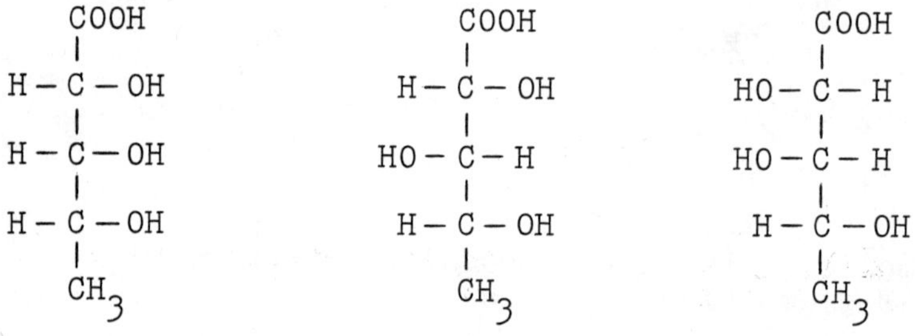

COOH	COOH	COOH
H – C – OH	H – C – OH	HO – C – H
H – C – OH	HO – C – H	HO – C – H
H – C – OH	H – C – OH	H – C – OH
CH_3	CH_3	CH_3

(f) A <u>threo form</u> will have groups a, b, d and a, b, e on adjacent chiral centers. Only two of the like groups (a, a or b, b) can be superimposed in an eclipsed conformation. Example:

(continued next page)

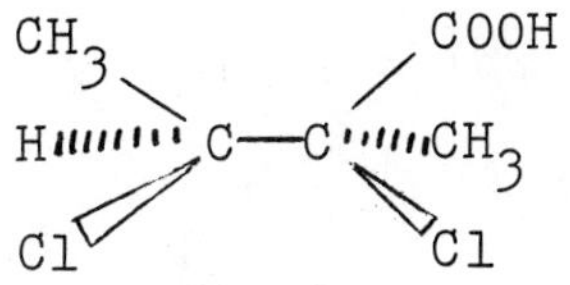

(g) An <u>erythro form</u> will have groups a, b, d and a, b, e on adjacent chiral centers. The like groups (a, a and b, b) can be superimposed in an eclipsed conformation. Example:

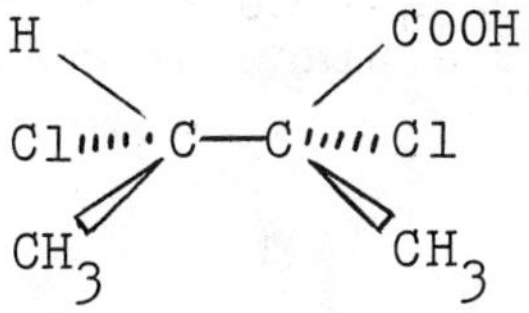

(h) A <u>racemization reaction</u> produces both enantiomers (a <u>racemic modification</u>) from one of the enantiomers. Example:

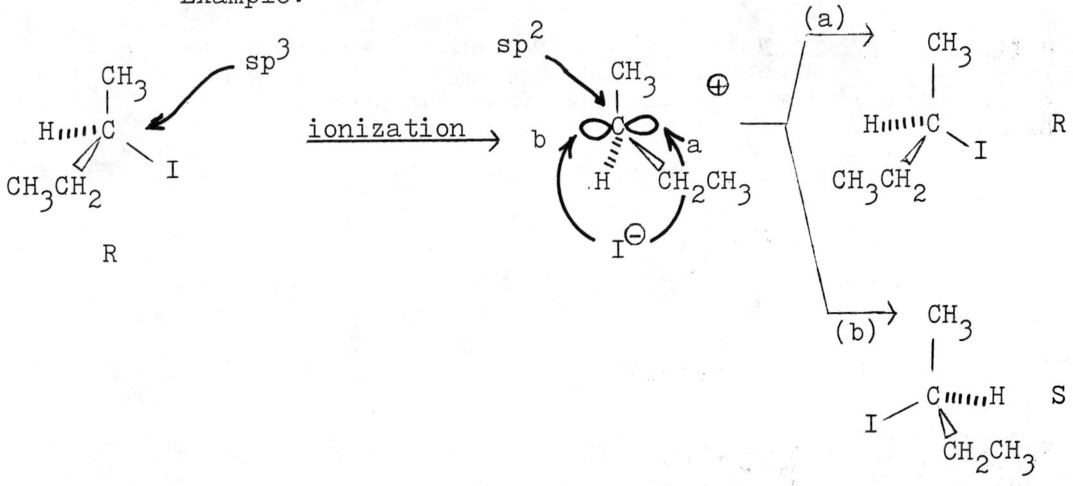

The mechanism of this reaction will be discussed in Ch. 5.

(i) A <u>meso form</u> is the diastereomer with chiral centers which is achiral. Example:

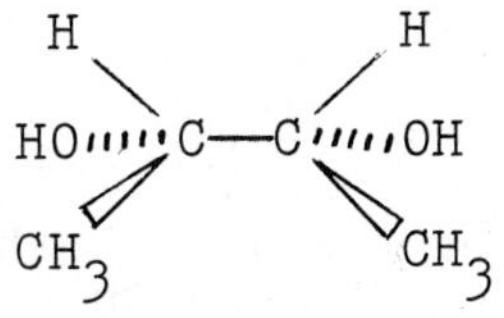

(continued next page)

(j) A d,l form is the diastereomer which exists in enantiomeric forms. Example:

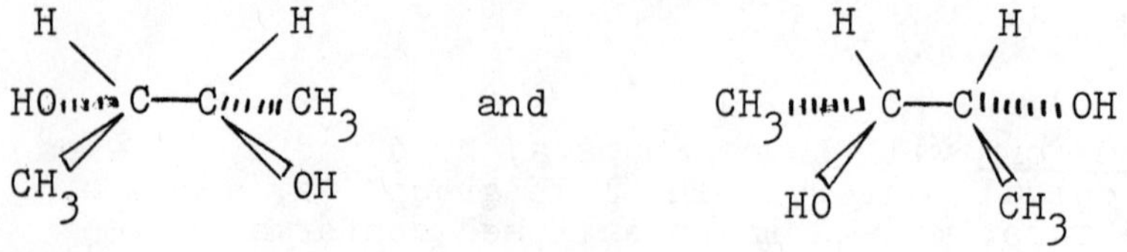

H, HO, C—C, CH$_3$, CH$_3$, H, OH

and

H, H, CH$_3$, C—C, OH, HO, CH$_3$

(k) A plane of symmetry bisects an object so that the halves are mirror images. This page is a plane of symmetry for:

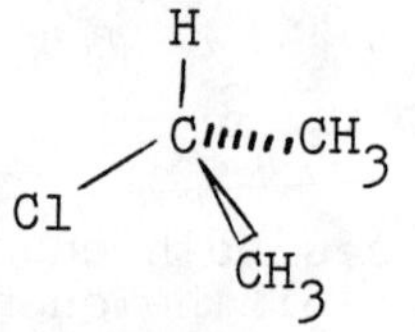

H, C, CH$_3$, Cl, CH$_3$

(l) A point of symmetry is a point from which two lines can be directed at an angle of 180° and encounter the same environment at the same distance from the point. Example:

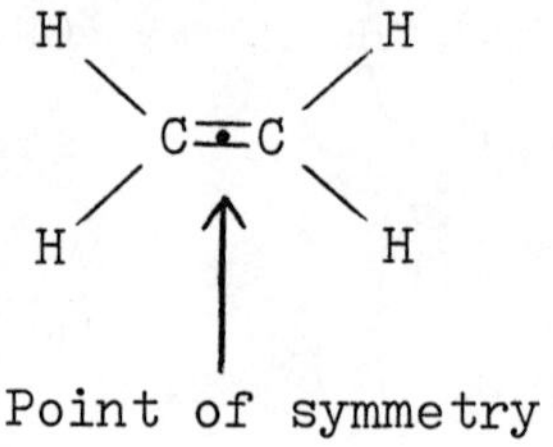

H, H, C=C, H, H

Point of symmetry

(m) Enantiotopic protons when replaced each in turn by an achiral group produce enantiomers. Example:

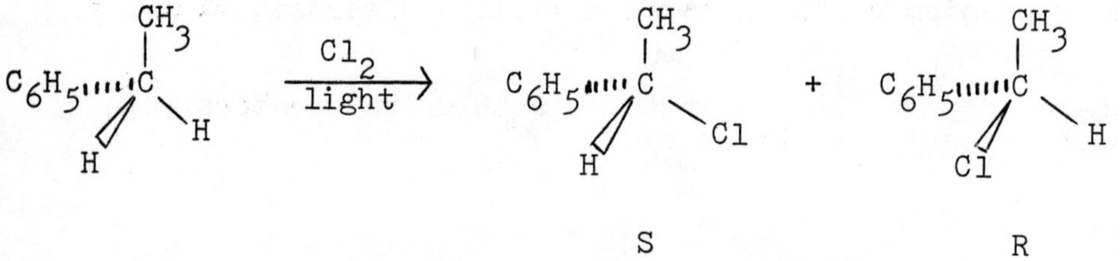

$$C_6H_5CH(H)(H)CH_3 \xrightarrow[\text{light}]{Cl_2} C_6H_5C(H)(Cl)CH_3 + C_6H_5C(Cl)(H)CH_3$$

S R

(n) Diastereotopic protons exist in a group CHHab with a or b a chiral center somewhere in its structure. Replacement of protons in CHHab in turn by an achiral center produces diastereomers. Example:

(continued next page)

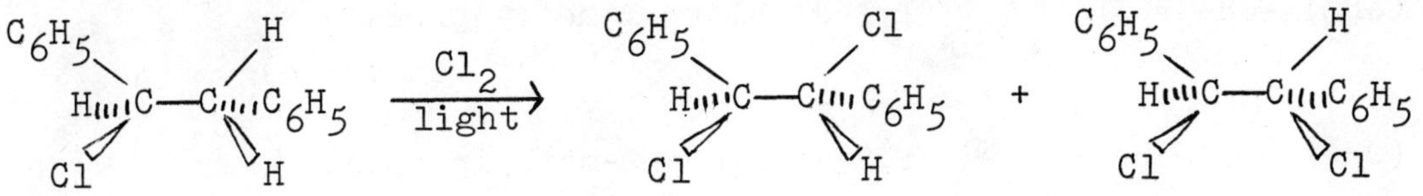

C_6H_5, H, H···C—C···C_6H_5, Cl, H $\xrightarrow[\text{light}]{Cl_2}$ C_6H_5, Cl, H···C—C···C_6H_5, Cl, H + C_6H_5, H, H···C—C···C_6H_5, Cl, Cl

diastereomers

20. Assigning R and S configurations is a convenient way of identifying identical structures.

(a) S: 1,5; R: 2,3,4 (d) S: 3; R: 1,2

(b) S: 1,2,4; R: 3,5 (e) S,S: 1; R,R: 2,3; R,S: 4

(c) S: 1,2,3; R: 4

(Text S. 4.4)

21. (a) identical (b) enantiomers (c) diastereomers

(d) identical (e) enantiomers (f) identical

(Text S. 4.2, 4.5)

22. Chiral centers are starred.

(a) CH_3-CH_2-CH_2-CH_2-Cl 1-chlorobutane (inactive)

CH_3-CH_2-C*H(Cl)-CH_3 (R)- and (S)-2-chlorobutane

$(CH_3)_2CH$-CH_2Cl 1-chloro-2-methylpropane (inactive)

$(CH_3)_3C$-Cl 2-chloro-2-methylpropane (inactive)

(b) CH_3-CH_2-CH_2-$CHCl_2$ 1,1-dichlorobutane (inactive)

CH_3-CH_2-C*HCl-CH_2Cl (R)- and (S)-1,2-dichlorobutane

CH_3-CH_2-CCl_2-CH_3 2,2-dichlorobutane (inactive)

CH_3-C*HCl-C*HCl-CH_3 (2R,3S)-, (2R,3R)-, and (2S,3S)-2,3-dichlorobutane
(2R,3S- is <u>meso</u>)

$ClCH_2$-CH_2-CH_2-CH_2Cl 1,4-dichlorobutane (inactive)

$(CH_3)_2$-CH-$CHCl_2$	1,1-dichloro-2-methylpropane (inactive)
$(CH_3)_2$-CCl-CH_2Cl	1,2-dichloro-2-methylpropane (inactive)
$ClCH_2$-CH(CH_3)-CH_2Cl	1,3-dichloro-2-methylpropane (inactive)
$CH_3\overset{*}{C}HClCH_2CH_2Cl$	(R)- and (S)-1,3-dichlorobutane

(c)

CH_3-CH_2-CH_2-$\overset{*}{C}HBrCl$	(R)- and (S)-1-bromo-1-chloro-butane
CH_3-CH_2-$\overset{*}{C}HBr$-CH_2Cl	(R)- and (S)-2-bromo-1-chloro-butane
CH_3-$\overset{*}{C}HBr$-CH_2-CH_2Cl	(R)- and (S)-3-bromo-1-chloro-butane
$BrCH_2$-CH_2-CH_2-CH_2Cl	1-bromo-4-chlorobutane (inactive)
$BrCH_2$-$\overset{*}{C}HCl$-CH_2-CH_3	(R)- and (S)-1-bromo-2-chloro-butane
$BrCH_2$-CH_2-$\overset{*}{C}HCl$-CH_3	(R)- and (S)-1-bromo-3-chloro-butane
CH_3-CH_2-$\overset{*}{C}ClBr$-CH_3	(R)- and (S)-2-bromo-2-chloro-butane
CH_3-$\overset{*}{C}HCl$-$\overset{*}{C}HBr$-CH_3	(2R,3S)-, (2S,3R)-, (2R,3R)-, (2S,3S)-2-bromo-3-chlorobutane
$CH_3CH(CH_3)\overset{*}{C}HBrCl$	(R)- and (S)-1-bromo-1-chloro-2-methylpropane
$BrCH_2\overset{*}{C}H(CH_3)CH_2Cl$	(R)- and (S)-1-bromo-3-chloro-2-methylpropane
$(CH_3)_2CClCH_2Br$	1-bromo-2-chloro-2-methylpropane
$(CH_3)_2CBrCH_2Cl$	2-bromo-1-chloro-2-methylpropane

(Text: enantiomers, S. 4.2; diastereomers, S. 4.5)

23. The product, 2-methylbutane, does not have a chiral center.

24. (a) (2R,3R)-CH_3-CH(Br)-CH(Cl)-CH_3 and (2S,3S)-CH_3-CH(Br)-CH(Cl)-CH_3

H / CH_3—C—Br (R) / Cl—C—CH_3 (R) / H   and   H / CH_3—C—Br (R) / CH_3—C—Cl (S) / H

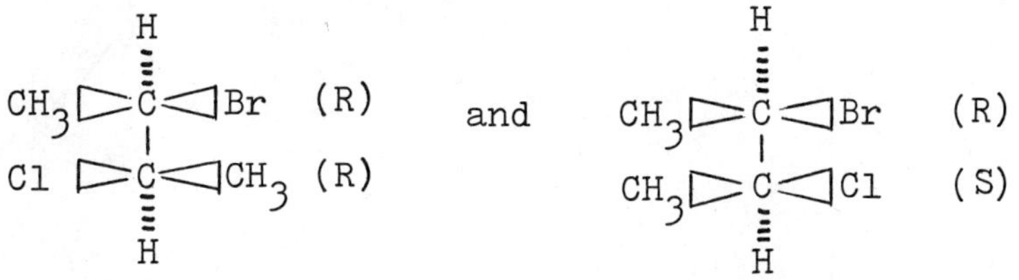

(b) The products are diastereomers and are formed through diastereomeric activated complexes which have unequal energies.

(Text: halogenation S. 3.7, 4.6)

25. (a) -Br > $-CF_3$ > $-CH_2CH_3$ > $-CD_3$ > CH_3

(b) $-CHClCH_2CH_3$ > $-CH_2CH_2CCl_3$ > $-CH_2CH_2CH_3$

(c) -Cl > -OCl > $-CCl_3$ > $-CH_2Cl$

(d) -C≡N > -C≡C-CH_3 > -C≡CH > -CH=CH_2

(e) $-N(CH_3)_2$ > $-NHCH_3$ > $-NH_2$

(f) $-CHClCH_3$ > $-CH_2CCl_3$ > $-CH_2(CH_2)_7CH_2Cl$ > $-CH_2(CH_2)_8CCl_3$

(Text S. 4.4)

26. (a) S (b) R (c) R (d) R (e) S

(f) S,R (g) S,S (h) 2R,4R (i) no chiral center

(j) S (k) R(N), S(C) (l) S

(Text S. 4.4)

27. All possess a plane of symmetry. Structures are drawn so that page represents plane of symmetry. Points of symmetry are indicated by dots.

(continued next page)

(a) 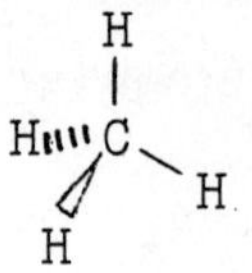 H, H, C, H, H (b) H, H, C=C=C, H, H (c) H, H, C≡C, H, H, Point

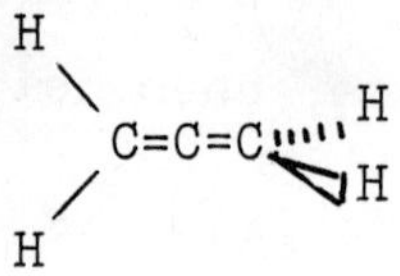

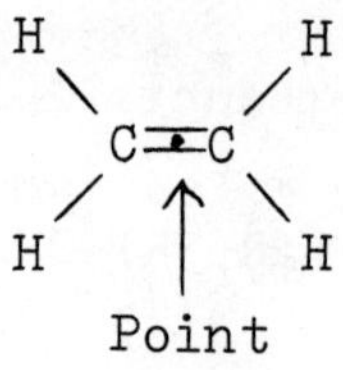

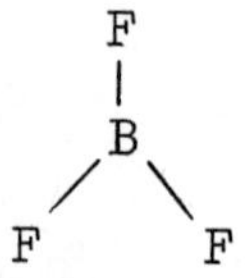

(d) 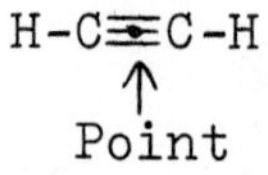 H-C≡C-H, Point (e) F, B, F, F (f) O, C, H, H

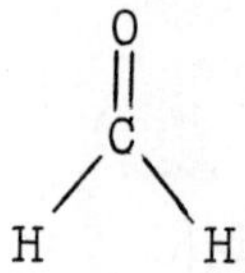

(g) 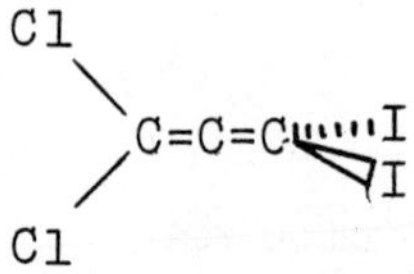 Cl, Cl, C=C=C, I, I (h) H, H, C, Cl, H 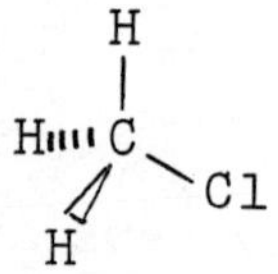 (i) 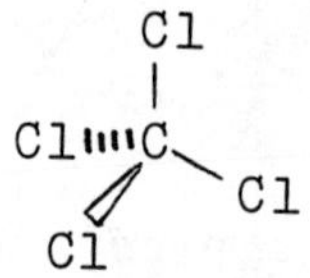 Cl, Cl, C, Cl, Cl

Structures b, c, f, g and i also have planes of symmetry perpendicular to the page. Structure h has two additional planes passing through the other two hydrogens.

(Text S. 4.2, glossary)

28. Enantiomers can be differentiated with all chiral "probes". Thus enantiomers will differ in principle in c, e, h, i, k and l.

(Text S. 4.3)

29. (a) Prochiral hydrogens are those on carbons 2, 4, and 5.

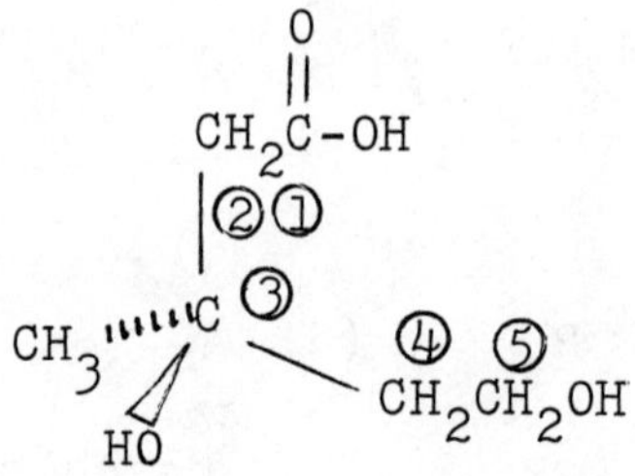

$$\begin{array}{c} O \\ \| \\ CH_2C\text{-}OH \\ ②① \\ CH_3\text{-}C③ \quad ④\ ⑤ \\ HO \quad CH_2CH_2OH \end{array}$$

(b) Replacement of hydrogens on carbon 4.

(Text S. 4.5)

30. (a) Chiral centers are starred. $HOCH_2-\overset{*}{C}H(OH)-\overset{*}{C}H(OH)-\overset{*}{C}H(OH)-\overset{*}{C}H(OH)-CHO$

16 isomers possible.

(b) This is one of the 16 possible isomers of the general structure in (a). Thus, there are 15 possible stereoisomers.

(c) 1 chiral center ($CONH_2$ on $-\overset{*}{C}H-$)

(d) 2 chiral centers ($-\overset{*}{C}HOH-\overset{*}{C}H(CH_2OH)-$)

(e) 1 chiral center ($-\overset{*}{C}HCl-$)

(f) 1 chiral center ($-\overset{*}{C}HOH-$)

(g) 1 chiral center ($-\overset{*}{C}HCH_3-$) Number of possible stereoisomers: Two for (c), (e), (f) and (g); Four for (d).

(Text: enantiomers, S. 4.2; diastereomers, S. 4.5)

31. (a) H's of $-CH_2-$ groups, $-ONO_2$ groups on carbons 1 and 3, $-CH_2ONO_2$ groups; (b) H's of CH_2 groups, $-OCH_2CH_3$ groups; (c) H's of $-CH_2-$ groups; (d) H's of CH_2 group; (e) H's of $-CH_2-$ groups, H's of $-NH_2$ (nitrogen inversion leads to racemization); (f) H's of $-CH_2-$ groups, phenyl (C_6H_5-) groups.
(Text S. 4.5)

32. 2-aminobutane, consisting of R and S forms, reacts with (R)-4-chloropentanoic acid to give two salts which are diastereomers.

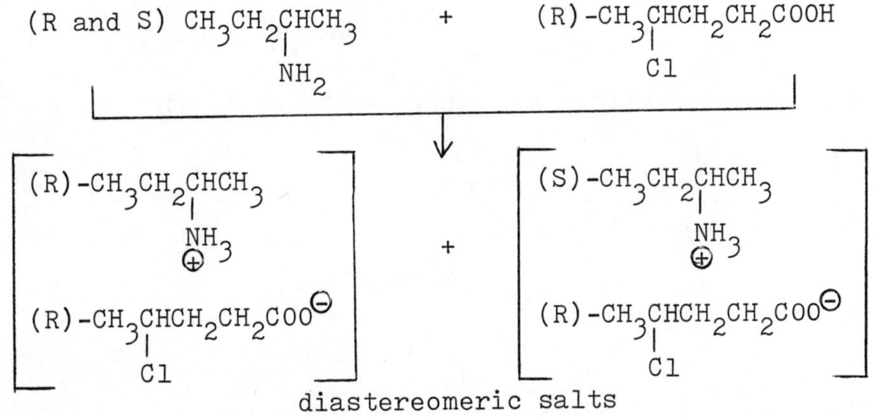

diastereomeric salts

33. (a) $[\alpha]_{\lambda}^{20^\circ} = \dfrac{18.88^\circ}{(1\text{ dm})\ \frac{5.678\text{g}}{20\text{cc}}} = 66.5^\circ$

(b) concentration $= \dfrac{10.75^\circ}{(66.5^\circ)(2\text{ dm})} = 0.0808\text{g/cc}.$

(c) No, the molecular rotation must be known.

(Text S. 4.3)

34. (a) $[\alpha]_{\lambda}^{20^\circ} = \dfrac{9.44^\circ}{(2\text{ dm})(0.8\text{g/cc})} = 5.9^\circ$

(b) $[\alpha]_{\lambda}^{20^\circ} = \dfrac{3.56^\circ}{(4\text{ dm})(0.8\text{g/cc})} = 1.11^\circ$

% optically active alcohol $= \dfrac{1.11^\circ}{5.9^\circ} \times 100 = 18.9\%$

(Text S. 4.3)

35. Replacement of enantiotopic hydrogens produces enantiomers.

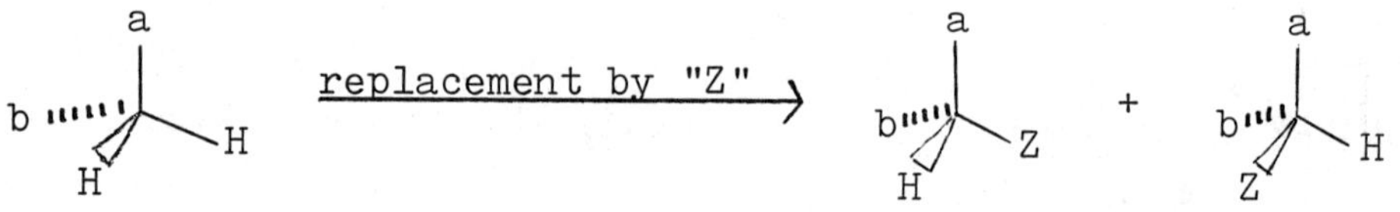

replacement by "Z"

enantiomers

Replacement of diastereotopic hydrogens produces diastereomers.

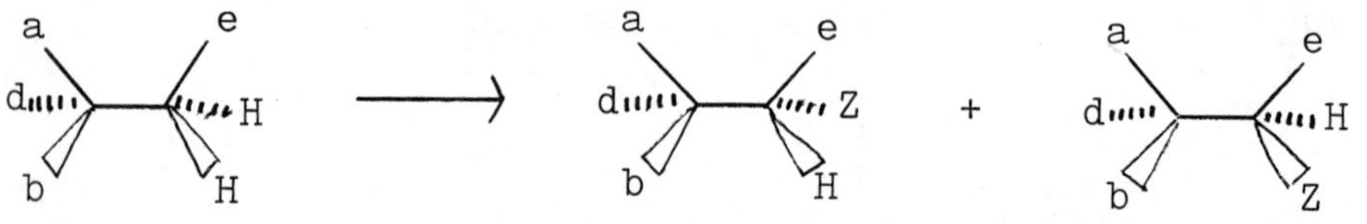

diastereomers

36. (a) A molecule containing a single chiral center is chiral. However, molecules containing more than one chiral center may be achiral.

An example of an achiral molecule containing two chiral centers is meso-2,3-dichlorobutane:

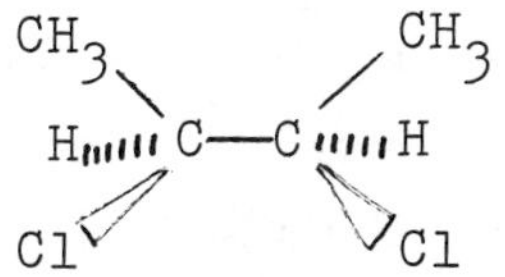

(b) A molecule which possesses a plane of symmetry can not have a chiral center in that plane. The groups on either side of the plane must have the same composition; thus, a carbon in the plane will have two like groups attached.

37. (a) 6,6′-difluoro-2,2′-diphenic acid is an example of a compound which can exist as a pair of enantiomers because of restricted rotation about the carbon-carbon single bond. This occurs because at room temperature the large ortho-substituted groups cannot pass each other. These rotational isomers shown below are non-superimposable mirror images.

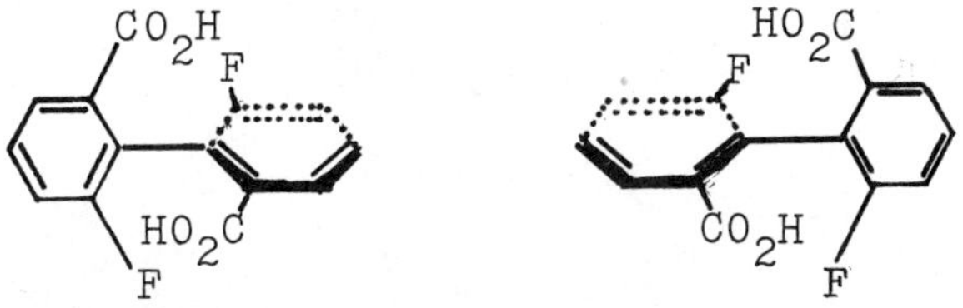

(b) Heating an enantiomer can provide the energy needed to overcome the barrier to rotation.

(c) The transition state for racemization of both A and B requires two H and I interactions. In isomer B the van der Waals interactions of the adjacent iodine atoms (which pushes them apart) causes the H and I interactions to be more severe.

(d) No, the structure is symmetric (the page is the plane of symmetry in the drawing below).

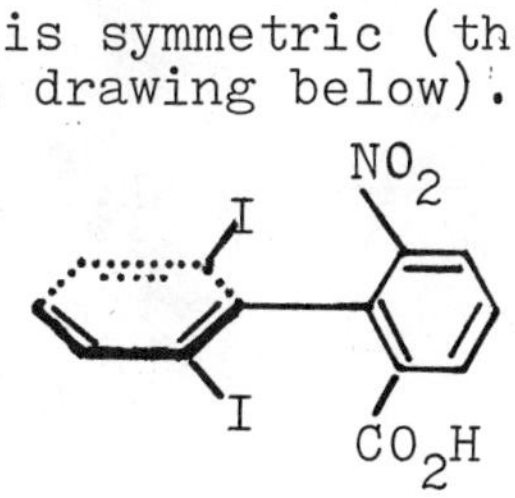

38. (a) $[\alpha] = \dfrac{+30}{1 \text{ dm} \times 1\text{g/cc}} = +30$

(b) specific rotation is still +30

$\alpha = (+30)(1 \text{ dm})(1.05\text{g/ml}) = 31.5$

(c) $\alpha = (+30)(0.5\ dm)(1.00 g/ml) = 15$

(d) Optical rotation is directly proportional to concentration; therefore, determining the rotation at different concentrations will allow you to distinguish between +30 and -330°. For instance, if the solutions are diluted to 0.1 of their original concentrations, the rotations will be +3° if the specific rotation is 30°, but -33° if the specific rotation is -330°.

(e) Yes. The same procedure described in (d) can be used. Dilution of the solution to 0.1 of their original concentrations will give rotations of +3.0° and +21.0° respectively. Thus, this method can be used to distinguish between rotations which differ by 360° (d) and by 180° (e).

39. Racemic modifications of carboxylic acids may be resolved using an optically active (+ or -) amine (RNH_2). The reaction involves proton exchange.

$$\underset{\text{acid}}{RCOOH} + \underset{\text{amine}}{RNH_2} \longrightarrow \underset{\text{salt}}{RNH_3^{\oplus}\ RCOO^{\ominus}}$$

$$(\pm)HO_2C-C(CH_3)_2-CH_2-CH(CH_3)-CH_2CH_3 \xrightarrow{(+)R\ddot{N}H_2} \begin{bmatrix} (+)RNH_3^{\oplus}\ (+)RCO_2^{\ominus} \\ (+)RNH_3^{\oplus}\ (-)RCO_2^{\ominus} \end{bmatrix}$$

(represented as (±) RCO_2H) diastereomers

separated by crystallization

$$(+)RNH_3^{\oplus}Cl^{\ominus} + (+)RCO_2H^{*} \xleftarrow{HCl(aq)} (+)RNH_3^{\oplus}\ (+)RCO_2^{\ominus}$$

$$(+)RNH_3^{\oplus}Cl^{\ominus} + (-)RCO_2H^{*} \xleftarrow{HCl(aq)} (+)RNH_3^{\oplus}\ (-)RCO_2^{\ominus}$$

*Amine salt is soluble in water. The optically active acid is extracted with an organic solvent and recovered by evaporation of the solvent.

(Text S. 4.7)

40. This is sufficient data. The optical purity is:

$$\frac{9^\circ}{11^\circ} \times 100 = 81.8\%$$

(Text S. 4.3)

41. Their absolute configurations differ because of a different order of group priorities.

For L-cysteine: $NH_2 > CH_2SH > CO_2H > H$ and

for L-serine: $NH_2 > CO_2H > CH_2OH > H$.

42. (2R,3S)-tartaric acid or (2S,3R)-tartaric acid (numbering from either end is correct).

43. The axis of symmetry passes through the center atom at an angle of 45° relative to the planes defined by groups a, b, d and e. This is perhaps best viewed from a front view (drawing B).

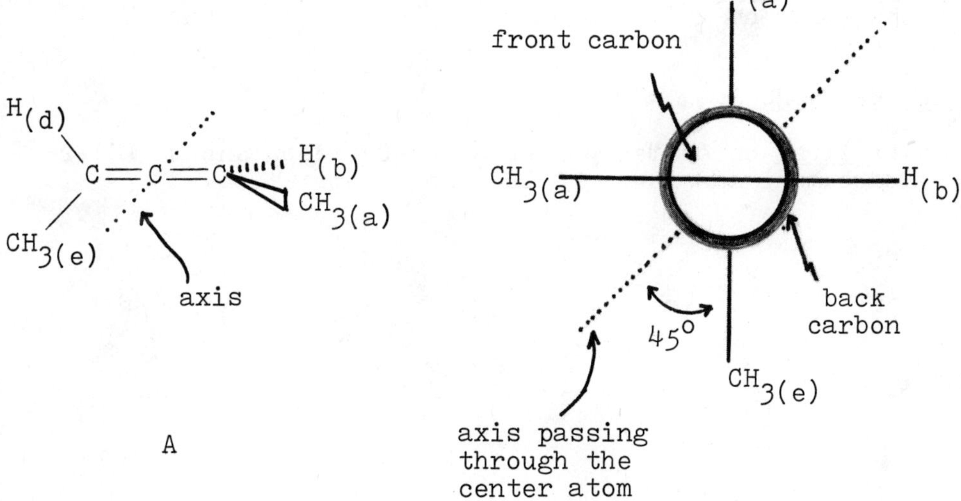

CHAPTER 5. ALKYL HALIDES-NUCLEOPHILIC SUBSTITUTION REACTIONS

LEARNING OBJECTIVES

When you have completed this chapter, you should be able to:

1. explain the stereochemistry of the S_N2 reaction (problems 1, 2, 3, 4, 39);
2. explain the reactivity and compare reactivity rates of substrates in S_N2 reactions (problems 5, 6, 7, 35, 43, 45);
3. predict reactants (products given) or products (reactants given) in S_N2 reactions (problems 8, 9, 10, 11, 24, 25, 28, 30, 32);
4. write equations which illustrate S_N2 mechanisms (problems 33, 34, 37, 46);
5. explain the stereochemistry of S_N1 reactions (problems 13, 42);
6. write equations which illustrate S_N1 mechanisms (problems 12, 26, 37, 38, 46);
7. predict the products of S_N1 reactions (problem 14);
8. draw resonance structures for carbocations (problems 16, 17, 18, 19, 20);
9. define terms associated with nucleophilic substitution reactions (glossary, problem 23);
10. write equations which illustrate how nucleophilic substitution reactions can be used in multistep syntheses (problems 16, 22, 31);
11. compare the general characteristics of S_N1 and S_N2 reactions (problem 27).

ANSWERS TO QUESTIONS

1. (a) The <u>first</u> encounter converts (R)-2-iodobutane to (S)-2-iodobutane and does <u>not</u> lead to racemization.

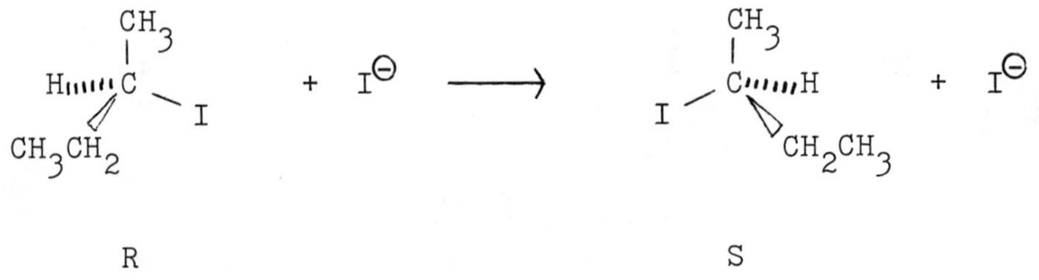

CH_3, H, CH_3CH_2, I on C (R) + $I^{\ominus}$ ⟶ CH_3, H, CH_2CH_3, I on C (S) + $I^{\ominus}$

R S

(b) The first successful encounter converts one (R)-molecule to (S) and results in racemization (one S and one R are present).

(c) In general, the assumption is that there are enough molecules in the sample that probability favors attainment of nearly equal populations of enantiomers. (If one assumes that exactly equal concentrations of enantiomers are present after racemization, an even number of molecules must be present initially).

2. (R)-chloride $\xrightarrow{I^{\ominus}}$ (S)-iodide $\xrightarrow{Cl^{\ominus}}$ (S)-iodide $\xrightarrow{Cl^{\ominus}}$ (R)-chloride
   (R)-chloride   (R)-chloride   (S)-chloride   (S)-chloride

3. Product of first encounter:

   (2R,3R)-$CH_3\underset{\displaystyle Cl}{\underset{|}{C}}HCH(CH_3)CH_2CH_3$ $\xrightarrow{Cl^{\ominus}}$ (2S,3R)-$CH_3\underset{\displaystyle Cl}{\underset{|}{C}}HCH(CH_3)CH_2CH_3$

   erythro   threo

   Reaction between a large amount of (2R,3R)-2-chloro-3-methylpentane and chloride ion can not lead to racemization since inversion occurs only at carbon-2. The configuration at carbon-3 remains unchanged, and the product is, therefore, optically active. The reactant and product above are diastereomers.

4. (a) 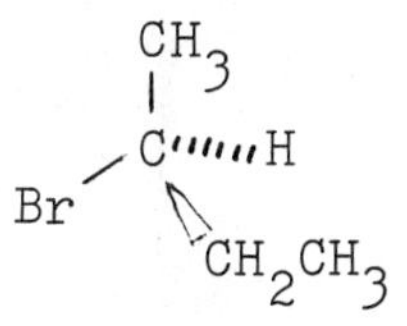 C with CH_3 (up), H (hashed), Br, CH_2CH_3 (wedge) — (S)-2-bromobutane

   (b) 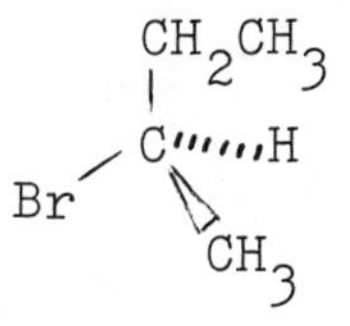 C with CH_2CH_3 (up), H (hashed), Br, CH_3 (wedge) — (R)-2-bromobutane

   (c) 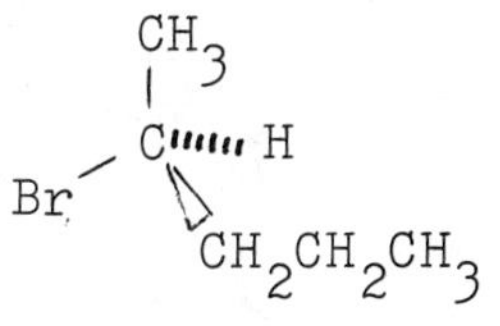 C with CH_3 (up), H (hashed), Br, $CH_2CH_2CH_3$ (wedge) — (S)-2-bromopentane

   (d) 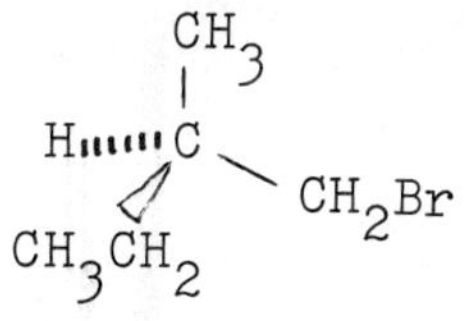 C with CH_3 (up), H (hashed), CH_2Br, CH_3CH_2 (wedge) — (S)-1-bromo-2-methylbutane

5. Both 6-chloro-1-hexene and 1-chlorohexane react normally as primary alkyl halides; however, the reaction of allyl chloride is accelerated because of the stabilization afforded by the π system to the activated complex in an S_N2 process.

6. These reactions are S_N2, and decreased reactivity parallels increased steric hindrance to backside attack of the nucleophile. In this series steric hindrance increases (S_N2 reactivity decreases) as G becomes larger.

7. Because of steric hindrance (the two methyl groups on the carbon adjacent to the C-I bond) an S_N2 process is prevented.

8. (a) (±) $CH_3CH(Cl)(CH_2)_5CH_3$

(b) (±)-$CH_3CH(I)(CH_2)_5CH_3$

(c) (±)-$CH_3CH(SH)(CH_2)_5CH_3$

(d) (±)-$CH_3CH(NH_3^{\oplus})(CH_2)_5CH_3$ $I^{\ominus}$

(e) (±)-$CH_3CH(CH_2C(=O)CH_3)(CH_2)_5CH_3$

(f) 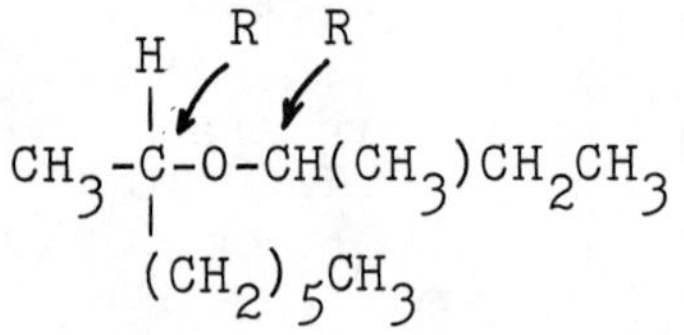 $CH_3-C(H)((CH_2)_5CH_3)-O-CH(CH_3)CH_2CH_3$ (R, R) and $CH_3-C(H)(O-CH(CH_3)CH_2CH_3)-(CH_2)_5CH_3$ (S, R)

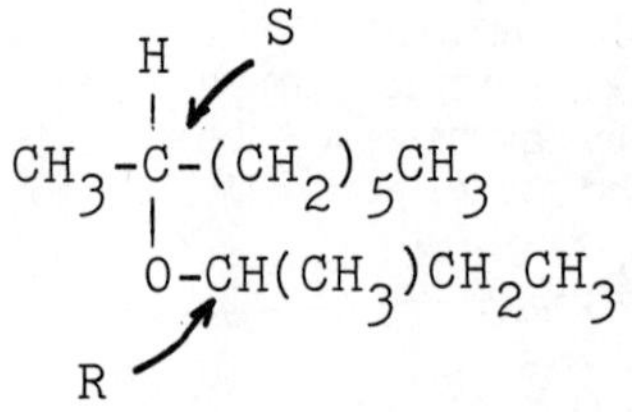

(g) 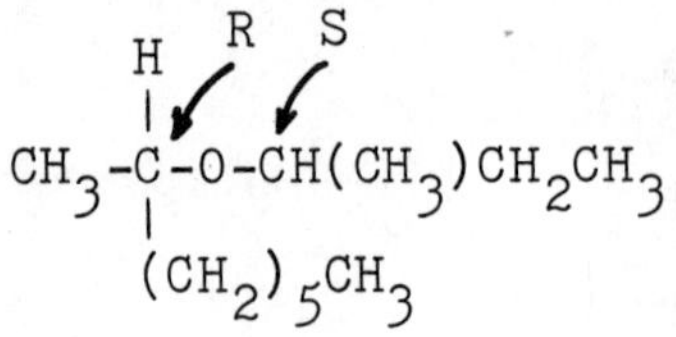 $CH_3-C(H)((CH_2)_5CH_3)-O-CH(CH_3)CH_2CH_3$ (R, S) and $CH_3-C(H)(OCH(CH_3)CH_2CH_3)-(CH_2)_5CH_3$ (S, S)

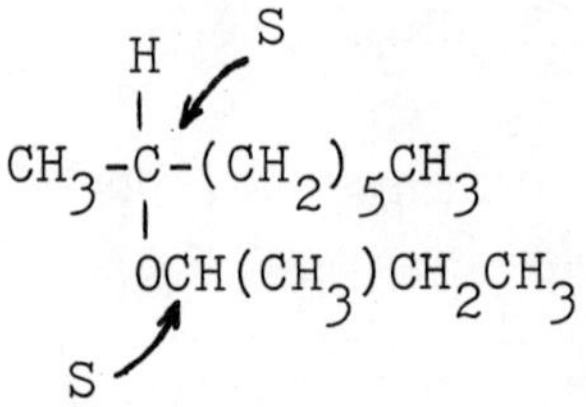

(h) (R,R)-, (R,S)-, (S,S)-, and (S,R)-$CH_3\overset{*}{C}H(O\underset{*}{C}H(CH_3)CH_2CH_3)(CH_2)_5CH_3$

9. Neutral groups leave as anions: R-L → $:L^{\ominus}$

Positive groups leave as neutral molecules: $R-L^{\oplus}$ → :L

(a) $-OS(O)_2CH_3$ leaves as $^{\ominus}:OS(O)_2CH_3$

(continued next page)

(b) $-OS(O)_2OCH_3$ leaves as $^{\ominus}:OS(O)_2OCH_3$. This leaving group can react again:

$$Nu:^{\ominus} + CH_3{-}OSO_3^{\ominus} \longrightarrow NuCH_3 + SO_4^{2-}$$

(c) $-\overset{\oplus}{O}H_2$ leaves as $H_2O:$; (d) nucleophilic substitution may occur in two ways: $-\overset{\oplus}{O}(CH_3)_2$ may leave as $:O(CH_3)_2$; $-\overset{\oplus}{O}(CH_3)CH_2CH_2CH(CH_3)_2$ may leave as $:O(CH_3)CH_2CH_2CH(CH_3)_2$;

(e) nucleophilic substitution may occur in three ways: $-\overset{\oplus}{O}(CH_3)C_2H_5$ may leave as $:O(CH_3)C_2H_5$; $-\overset{\oplus}{O}(CH_3)CH_2CH_2C(CH_3)_3$ may leave as $:O(CH_3)CH_2CH_2C(CH_3)_3$; $-\overset{\oplus}{O}(C_2H_5)CH_2CH_2C(CH_3)_3$ may leave as $:O(C_2H_5)CH_2CH_2C(CH_3)_3$; (f) $-\overset{\oplus}{N}(CH_3)_3$ leaves as $:N(CH_3)_3$.

10. (a) $CH_3CH_2CH_2I$; (b) CH_3I; (c) CH_3CH_2I; (d) CH_3I (major product. Substitution occurs mainly at the methyl carbon. Lesser amount of $(CH_3)_2CHCH_2CH_2I$ formed.); (e) CH_3I ($(CH_3)_3CCH_2CH_2I$ and CH_3CH_2I are minor products); (f) CH_3I

11. (a) $CH_3I + CH_3O^{\ominus}$
(b) $CH_3Br + Cl^{\ominus}$
(c) $CH_3I + HS^{\ominus}$
(d) $CH_3Br + H_2N^{\ominus}$
(e) $CH_3Cl + HO^{\ominus}$
(f) $CH_3CH_2Br + H_2P^{\ominus}$
(g) $CH_3I + CH_3CH_2O^{\ominus}$ (or $CH_3CH_2Br + CH_3O^{\ominus}$)
(h) $CH_3I + (CH_3)_3CO^{\ominus}$
(i) $CH_3CH_2Cl + CN^{\ominus}$
(j) $CH_3I + (CH_3)_3CCH_2S^{\ominus}$
(k) $CH_3I + (CH_3)_2N^{\ominus}$
(l) $CH_3CH_2Cl + HC{\equiv}C^{\ominus}$

(continued next page)

(m) $C_6H_5CH_2Cl + C_6H_5\overset{O}{\overset{\|}{C}}O^{\ominus}$

(n) (S)-$CH_3\underset{Cl}{\underset{|}{C}}HCH_2CH_3 + CH_3O^{\ominus}$

(or CH_3I + (R)-$CH_3CH_2CH(CH_3)O^{\ominus}$)

(o) $CH_3Cl + (CH_3)_3N{:}$

(p) $CH_2{=}CH{-}CH_2{-}Br + CH_2{=}CH{-}CH_2^{\ominus}$

(q) $Cl{-}CH_2CH_2CH_2CH_2{-}O^{\ominus}$ (intramolecular S_N2)

(r) $2(Cl{-}CH_2CH_2{-}O^{\ominus})$

(all nucleophiles have one or more nonbonding electron pairs; e.g, $CH_3\ddot{\underset{..}{O}}{:}^{\ominus}$, $:\ddot{\underset{..}{Cl}}{:}^{\ominus}$, $H{-}\ddot{\underset{..}{S}}{:}^{\ominus}$, $H_2\ddot{N}{:}^{\ominus}$ etc.) See Table 5-3, text, for similar examples.

12. (a) $(CH_3)_3COH + HCl \longrightarrow (CH_3)_3CCl + H_2O$

(b)(1) $(CH_3)_3C{-}\ddot{\underset{..}{O}}{-}H \quad H{-}\ddot{\underset{..}{Cl}}{:} \rightleftharpoons (CH_3)_3C{-}\overset{\oplus}{O}{:}\!\!\begin{smallmatrix}H\\H\end{smallmatrix} \quad :\ddot{\underset{..}{Cl}}{:}^{\ominus}$

(2) $(CH_3)_3C{-}\overset{\oplus}{O}{:}\!\!\begin{smallmatrix}H\\H\end{smallmatrix} \rightleftharpoons (CH_3)_3C^{\oplus} + H_2O$

(3) $(CH_3)_3C^{\oplus} + :\ddot{\underset{..}{Cl}}{:}^{\ominus} \rightleftharpoons (CH_3)_3C{-}\ddot{\underset{..}{Cl}}{:}$

(continued next page)

(c)

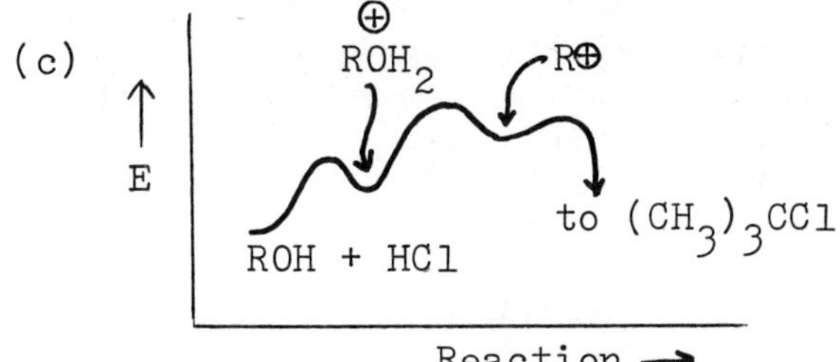

(d) ionization to form the carbocation

$$(CH_3)_3C\text{-}\overset{\oplus}{O}H_2 \longrightarrow (CH_3)_3C^{\oplus} + H_2O$$

(e) $R=k\left[(CH_3)_3CO\overset{\oplus}{H}_2\right] = k^1\left[((CH_3)_3COH)\,(HCl)\right]$

(f) No. The rate is dependent only on the ionization of $(CH_3)_3CO\overset{\oplus}{H}_2$; therefore, the rate is dependent only on the concentration of $(CH_3)_3CO\overset{\oplus}{H}_2$.

13. The products are diastereomers which have the same configuration at C-4 and opposite configurations at C-3. The mixture would likely be optically active even if equal amounts of isomers were present since the direction and magnitude of rotation of the two isomers cannot be correlated. The diastereomers have different energies and are formed at different rates. They have different physical properties such as boiling points and specific rotations.

14. All products are alcohols from the reaction

$$R^{\oplus} + H_2O \longrightarrow ROH_2^{\oplus} \xrightarrow{-H^{\oplus}} R\text{-}OH$$

(a) $(CH_3)_3COH$ (b) $(CH_3)_3COH$

(c) (R)-and (S)- $CH_3CH_2C(CH_3)(OH)\text{-}CH_2CH_2CH_3$ (d) same as (c)

(e) (3R,5S) and (3S,5S)-$CH_3CH_2C(CH_3)(OH)CH_2CH(CH_3)CH_2CH_3$

15. In this case "quite stable" refers to the reaction of limestone with atmospheric constituents while "extremely reactive" refers to limestone in a different reaction. A substance may be quite unreactive to one set of conditions and quite reactive to another.

16. Allyl chloride has a covalent bond (C-Cl) which the ionic form lacks. Energy is required to keep the charged species $[(CH_2 \cdots CH \cdots CH_2)^{\oplus}$ and $Cl^{\ominus}]$ apart in the ionic form.

17. (a) [resonance structures of benzyl cation: C_6H_5-$\overset{\oplus}{C}H_2$ ↔ ring(⊕ ortho)=CH_2 ↔ ⊕ring=CH_2 ↔ ring(⊕ para)=CH_2]

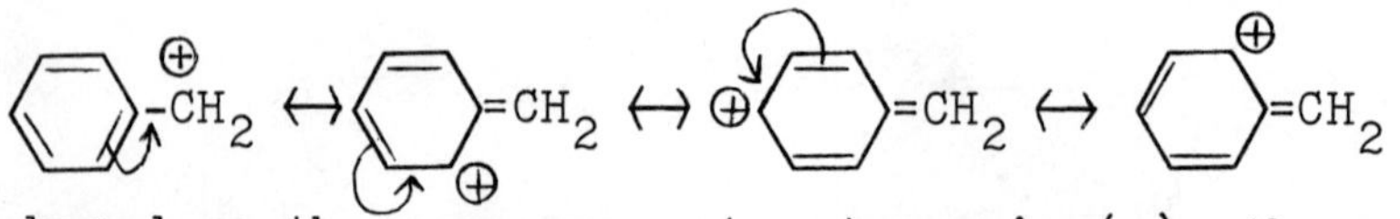

(b) based on the resonance structures in (a), the possible products are:

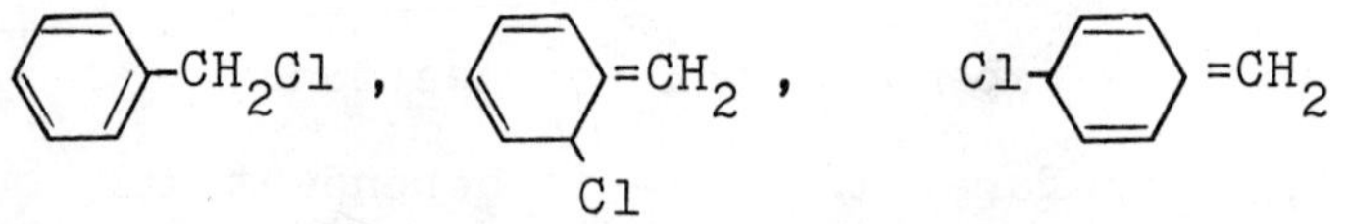

C_6H_5-CH_2Cl , (ring with Cl)=CH_2 , Cl-(ring)=CH_2

18. (a) Moving an electron pair to nitrogen in order to place the charge on the phenyl group gives 10 electrons around nitrogen. Elements of the second period of the periodic table cannot accommodate more than eight electrons.

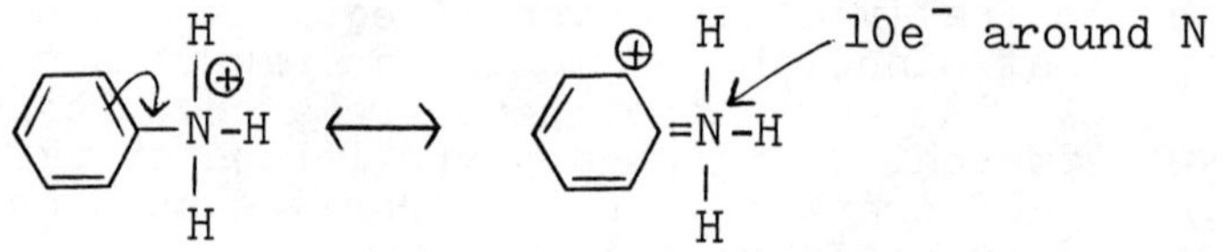

C_6H_5-$\overset{\oplus}{N}H_3$ ⟷ ⊕ring=NH_3 (10e⁻ around N)

(b) No. Attack on N by $Cl^{\ominus}$ would also give N ten electrons.

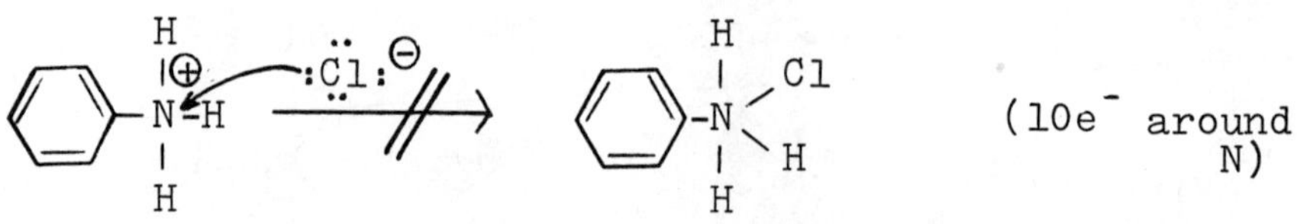

C_6H_5-$\overset{\oplus}{N}H_3$ + :Cl:$^{\ominus}$ ⟶/⟶ C_6H_5-NH_3Cl (10e⁻ around N)

19. The order of stability of carbocations is allyl $> 3^o > 2^o > 1^o$. Rearrangements are predicted on the basis of this order, i.e., generally a less stable carbocation rearranges to a more stable one.

(a) $(CH_3)_2\overset{\oplus}{C}CH_2CH_3$ (b) $(CH_3)_2\overset{\oplus}{C}CH(CH_3)_2$

(c) $CH_3(CH_2)_3\overset{\oplus}{C}C(CH_3)_3$ with CH_3 on the cationic carbon

(continued next page)

(d) $\left[\overset{\oplus}{C}H_2\text{=CH-}\overset{\oplus}{C}HCH_2CH_2CH_3 \leftrightarrow \overset{\oplus}{C}H_2\text{-CH=CHCH}_2CH_2CH_3\right]$

(e) $\left[CH_3\text{-}\ddot{O}\text{-}\overset{\oplus}{C}H\text{-CH}_2C(CH_3)_3 \leftrightarrow CH_3\text{-}\overset{\oplus}{O}\text{=CH-CH}_2C(CH_3)_3\right]$

(f) 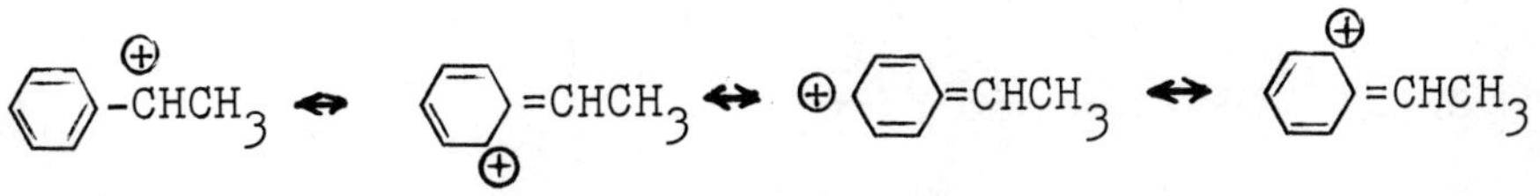 $C_6H_5\text{-}\overset{\oplus}{C}HCH_3 \leftrightarrow$ (ring)=CHCH$_3$ (⊕ ortho) $\leftrightarrow$ ⊕(ring)=CHCH$_3$ (⊕ para) $\leftrightarrow$ (ring)=CHCH$_3$ (⊕ ortho)

(g) 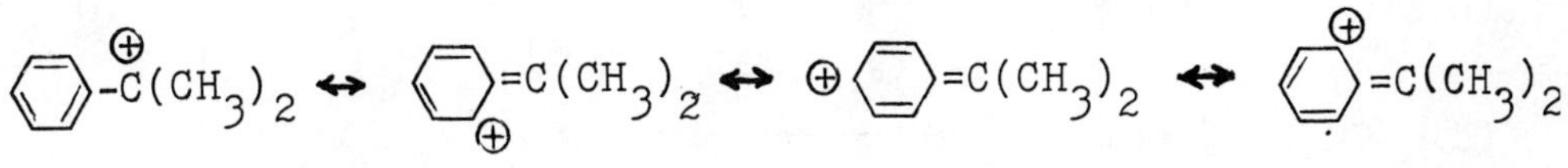 $C_6H_5\text{-}\overset{\oplus}{C}(CH_3)_2 \leftrightarrow$ (ring)=C(CH$_3$)$_2$ (⊕ ortho) $\leftrightarrow$ ⊕(ring)=C(CH$_3$)$_2$ (⊕ para) $\leftrightarrow$ (ring)=C(CH$_3$)$_2$ (⊕ ortho)

(h) 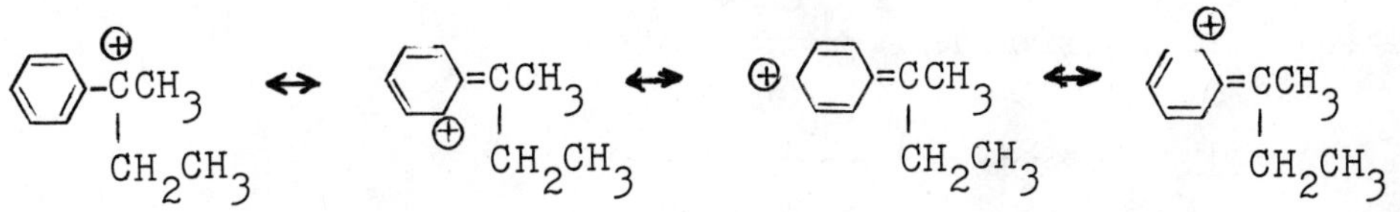 $C_6H_5\text{-}\overset{\oplus}{C}CH_3(CH_2CH_3) \leftrightarrow$ (ring)=CCH$_3$(CH$_2$CH$_3$) (⊕ ortho) $\leftrightarrow$ ⊕(ring)=CCH$_3$(CH$_2$CH$_3$) (⊕ para) $\leftrightarrow$ (ring)=CCH$_3$(CH$_2$CH$_3$) (⊕ ortho)

20. Resonance structures for the cation are:

$$CH_3\text{-}\ddot{\underset{..}{O}}\text{-}\overset{\oplus}{C}H_2 \leftrightarrow CH_3\text{-}\overset{\oplus}{\underset{..}{O}}\text{=}CH_2$$

Reaction of a negatively charged nucleophile at the positive oxygen gives a structure with a negative charge on carbon and a positive charge on oxygen.

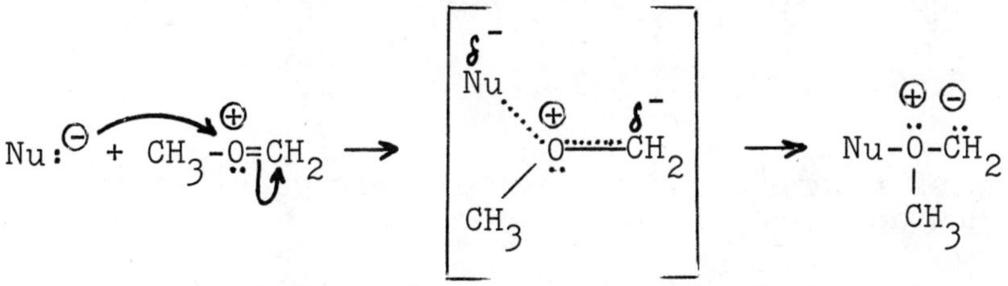

$$Nu{:}^{\ominus} + CH_3\text{-}\overset{\oplus}{\underset{..}{O}}\text{=}CH_2 \longrightarrow \left[\overset{\delta^-}{Nu}\cdots\overset{\oplus}{O}(CH_3)\cdots\overset{\delta^-}{C}H_2\right] \longrightarrow Nu\text{-}\overset{\oplus}{\ddot{O}}(CH_3)\text{-}\overset{\ominus}{\ddot{C}}H_2$$

Attack of a nucleophile at carbon leads to a lower energy activated complex with less separation of charge.

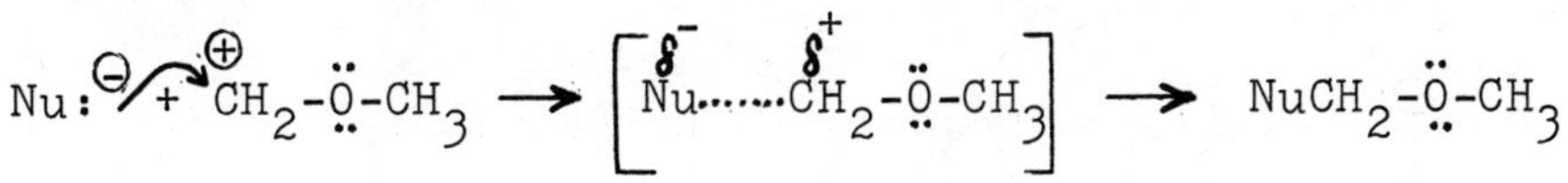

$$Nu{:}^{\ominus} + \overset{\oplus}{C}H_2\text{-}\ddot{\underset{..}{O}}\text{-}CH_3 \longrightarrow \left[\overset{\delta^-}{Nu}\cdots\cdots\overset{\delta^+}{C}H_2\text{-}\ddot{\underset{..}{O}}\text{-}CH_3\right] \longrightarrow NuCH_2\text{-}\ddot{\underset{..}{O}}\text{-}CH_3$$

21. Ethanol has a much lower dielectric constant than water and formic acid, and ethanol is less effective in solvating (stabilizing) the charged species formed in an S_N1 solvent.

22. (a) $CH_4 \xrightarrow{Cl_2/h\nu}$

(b) CH_3Cl (from a) $\xrightarrow{Na^{\oplus}OH^{\ominus}}$

(c) CH_3Cl (from a) $\xrightarrow{Na^{\oplus}SH^{\ominus}}$

(d) CH_3Cl (from a) $\xrightarrow{Na^{\oplus}SCH_3^{\ominus}}$

(e) CH_3Cl (from a) $\xrightarrow{Na^{\oplus}I^{\ominus}}$

(f) CH_3Cl (from a) $\xrightarrow{Na^{\oplus}OCH_3^{\ominus}}$

(g) CH_3Cl (from a) $\xrightarrow{Na^{\oplus}CH_3^{\ominus}}$

(h) CH_3CH_3 (from g) $\xrightarrow{Cl_2/h\nu}$

(i) CH_3CH_2Cl (from h) $\xrightarrow{Na^{\oplus}OCH_2CH_3^{\ominus}}$

Reactions shown in b, c, d, e, f, g and i are nucleophilic displacements (S_N2) which involve displacement of the leaving group Cl by the nucleophiles:

$:\ddot{O}H^{\ominus}$, $\ddot{O}R^{\ominus}$, $:\ddot{S}H^{\ominus}$, $:\ddot{S}R^{\ominus}$, $:\ddot{X}:^{\ominus}$ and $:CH_3^{\ominus}$.

See text S. 5.2, 5.3 and Table 5-3.

23. Specific examples related to the terms in this question are given here. How do these fit <u>your</u> definitions?

(a) $(CH_3)_3N:$ (b) $(CH_3)_3C^{\oplus}$ (c) $R = k[A]^2[B]$ is 3rd order overall

(d) $CH_3CH_2\ddot{O}:^{\ominus} + CH_3\ddot{I}: \longrightarrow CH_3CH_2\overset{\delta\ominus}{\ddot{O}}\cdots CH_3\cdots\overset{\delta\ominus}{\ddot{I}}: \longrightarrow CH_3CH_2\ddot{O}CH_3 + :\ddot{I}:^{\ominus}$

This reaction is bimolecular since two species are associated in the activated complex.

(e) example in (d)

(continued next page)

(f) $(CH_3)_3CCl \xrightarrow{slow} (CH_3)_3C^{\oplus}Cl^{\ominus} \xrightarrow[fast]{H_2O} (CH_3)_3C\overset{\oplus}{O}H_2\ Cl^{\ominus}$

The rate is affected only by concentration of $(CH_3)_3CCl$.

(g) DMF, DMSO etc. which effectively solvate cations but not anions. Thus, anions are made more reactive.

24. The key to answering this question is the realization that $:NH_3$, $CH_3\ddot{S}:^{\ominus}$ and $:CN:^{\ominus}$ are nucleophiles in nucleophilic substitution reactions while Cl_2 and Br_2 with light are reactants in free-radical substitution reactions.

(a) CH_3NH_2 $(CH_3\overset{\oplus}{N}H_3Cl^{\ominus} + NH_3 \rightarrow CH_3NH_2 + NH_4^{\oplus}Cl^{\ominus})$

(b) $(CH_3)_4\overset{\oplus}{N}\ Cl^{\ominus}$ (c) $(CH_3)_3CCH_2CH_2SCH_3$

(d) (R) and (S)-$CH_3CH_2CClBrCH_3$

(e) (R) and (S)-$CH_3CH_2CClBrCH_3$ (f) $NCCH_2CH_2CH_2CH_2CN$

(Text: nucleophilic substitution (S_N2), S. 5.2, 5.3; free-radical substitution, S. 3.7)

25. (a) Lewis acid: A substance which forms a covalent bond by accepting a nonbonding electron pair from a base (in short, an electron-pair acceptor).

Lewis base: A substance which forms a covalent bond by donating a nonbonding electron pair to an acid (in short, an electron-pair donor).

(b) $(CH_3)_3CBr + H_2O \longrightarrow (CH_3)_3COH + HBr$

$(CH_3)_3CBr \xrightarrow{H_2O} (CH_3)_3C^{\oplus}\ :Br^{\ominus}$

(continued next page)

$$(CH_3)_3C^{\oplus} + H\text{-}\ddot{O}\text{-}H \longrightarrow (CH_3)_3C\text{-}\overset{\oplus}{O}{:}H_2$$

acid base

$$(CH_3)_3C\text{-}\overset{\oplus}{O}{:}H_2 + {:}\ddot{Br}{:}^{\ominus} \longrightarrow (CH_3)_3COH + HBr$$

acid base

(Text S. 5.4)

26. a = 7 , b = 3 , c = 20 , d = 19 , e = 1 , f = 8 ,
g = 2 , h = 11 , i = 9 , j = 17 , k = 16 , l = 4 ,
m = 10 , n = 18 , o = 5 , p = 12 , q = 13 , r = 6 ,
s = 14 , t = 15

27. (Mechanisms are shown for selected examples.)

	Nucleophile	Electrophile	Leaving Group
(a)	$I^{\ominus}$	CH_3 of CH_3Cl	-Cl of CH_3Cl
(b)	$Br^{\ominus}$	CH_3 of CH_3I	-I of CH_3I

$$({:}\ddot{Br}{:}^{\ominus} + CH_3\text{-}\ddot{I}{:} \longrightarrow CH_3\text{-}\ddot{Br}{:} + {:}\ddot{I}{:}^{\ominus})$$

	Nucleophile	Electrophile	Leaving Group
(c)	$OH^{\ominus}$	$NH_4^{\oplus}$	NH_3

$$(H\text{-}\ddot{O}{:}^{\ominus} + H\text{-}\overset{\oplus}{N}H_3 \longrightarrow H\text{-}\ddot{O}\text{-}H + {:}NH_3)$$

	Nucleophile	Electrophile	Leaving Group
(d)	NH_3	S of $ClS(O_2)CH_3$	-Cl
(e)	-F of $CH_3CHFCH_2CH_3$	$Ag^{\oplus}$	-F
(f)	$H{:}^{\ominus}$	S of CH_3-S-S-CH_3	-SCH_3
(g)	S of $CH_3SCH_2CH_2Cl$	CH_2 of CH_2-Cl	-Cl

$$\left(CH_3\text{-}\ddot{S}\text{-}CH_2CH_2\text{-}\ddot{Cl}{:} \longrightarrow CH_3\text{-}\overset{\oplus}{S}(CH_2CH_2) + {:}\ddot{Cl}{:}^{\ominus} \right)$$

(Text S_N2 S. 5.2)

28. (a) the phenoxide ion is a lower energy (more stable) anion because of resonance stabilization.

$C_6H_5\ddot{O}:^{\ominus} \leftrightarrow \leftrightarrow \leftrightarrow$

(b) the CF_3 groups are electron withdrawing because of the high electronegativity of fluorine; therefore, N of $(CF_3)_3N$ has a lower electron density compared to the N of $(CH_3)_3N$.

(c) alcohols are relatively poor nucleophiles but do react with substrates when the reaction is catalyzed by $Ag^{\oplus}$ which aids in removing the leaving group.

(Text: Leaving groups S. 5.3, electrophilic catalysis S. 5.7)

29. (a) $:\ddot{Cl}:^{\ominus}$ (b) $CH_3CH_2CH_2\ddot{C}H_2^{\ominus}$ (c) $(CH_3CH_2CH_2CH_2)_3N:$

(d) $CH_3\ddot{O}:^{\ominus}$ (e) $(CH_3)_2CH:^{\ominus}$ (f) $CH_3C(=\ddot{O}:)\ddot{O}:^{\ominus}$

(g) $:C{\equiv}\overset{\ominus}{N}:$

(Text S. 5.2, 5.3)

30. (a) the reaction is S_N2 and the product is $CH_3CH_2CH_2I$.

(b) the conditions (NaI/acetone) are S_N2 and S_N2 cannot occur with a 3° substrate.

(c) OH is a poor leaving group and is not displaced by $CH_3O^{\ominus}$.

(d) only the -Cl is a leaving group. The product is $CH_3CH_2CH_2I$.

(e) the substrate and nucleophile each contain one carbon. The product is CH_3OCH_3.

(f) CH_3CH_3 does not have a leaving group. No reaction occurs.

31. (a) CH_3CH_3

(continued next page)

(b) CH_3OCH_3

(c) $CH_3CH{=}CHCH_2OH$ + $CH_3CH(OH)CH{=}CH_2$

(S_N1. Intermediate cation is $CH_3\overset{\delta+}{C}H{\cdots}CH{\cdots}\overset{\delta+}{C}H_2$

(d) CH_3OCH_2I

(e) $CH_3-\overset{\oplus}{N}(CH_3)_2-CH_2CH_3 \quad Cl^{\ominus}$

32. These transformations involve nucleophilic substitution and free-radical halogenation (Chapter 3). Review the fundamentals of these reactions before attempting this problem.

(a) Cl_2, light (b) $Na^{\oplus}I^{\ominus}$, DMF (c) $OH^{\ominus}$, H_2O

(d) excess NH_3 (e) Cl_2, light (f) CH_3OH

(g) (1) Cl_2, light; (2) $Na^{\oplus}CN^{\ominus}$, DMSO

(h) Cl_2, light (i) (1) Cl_2, light; (2) H_2O, acetone

(DMF, DMSO, HMPT, H_2O and acetone are some of the solvents commonly used in nucleophilic substitution reactions.)

(Text: free-radical chlorination S. 3.7, nucleophilic substitution S. 5.2, 5.3)

33. Each reaction is conducted in an appropriate solvent.

(a) $CH_3I + H\ddot{O}:^{\ominus}$

(b) $CH_3CH_2Br + :\ddot{Cl}:^{\ominus}$

(c) $CH_3CH(Cl)CH_3 + :\ddot{I}:^{\ominus}$

(d) $CH_3I + CH_3\ddot{O}:^{\ominus}$

(e) $CH_3I + CH_3CH_2\ddot{O}:^{\ominus}$

(or $CH_3CH_2Br + CH_3\ddot{O}:^{\ominus}$)

(f) $CH_3I + (CH_3)_3C\ddot{O}:^{\ominus}$

(<u>not</u> $(CH_3)_3CX + CH_3\ddot{O}:^{\ominus}$)

(g) $CH_3Br + :\overset{\ominus}{C}N:$

(h) $CH_3Cl + H_2\ddot{N}:^{\ominus}$

(i) $CH_3I + (CH_3)_2\ddot{N}:^{\ominus}$

(j) $CH_3Br + CH_3\overset{O}{\overset{\|}{C}}-\ddot{O}:^{\ominus}$

(continued next page)

(k) $CH_3CH_2Cl + :N_3^{\ominus}$ ($:\overset{\ominus}{N}=\overset{\oplus}{N}=\overset{\ominus}{N}:$)

(l) $CH_3Br + (CH_3)_2\ddot{S}:$

(m) $CH_2Cl_2 + 2CH_3\ddot{O}:^{\ominus}$

(n) $CH_3I + CH_3C{\equiv}C:^{\ominus}$

(or $2CH_3I + (:C{\equiv}C:)^{2-}$)

(o) $CH_3Br + (CH_3)_3CC{\equiv}C:^{\ominus}$

(not $(CH_3)_3CBr + CH_3C{\equiv}C:^{\ominus}$)

(p) $CH_3Cl + CH_3\ddot{S}:^{\ominus}$

(q) CH_3CHCH_3 (with Cl on the central C) $+ :\overset{\ominus}{C}N:$

(r) $CH_3CH_2CH_2Cl + :\overset{\ominus}{C}N:$

(s) (+)-or(-)-$CH_3CHICH_2CH_3$

$+ I^{\ominus}$ /heat

34. (a) $H_3N: + H{-}\ddot{Cl}: \longrightarrow H_3\overset{\delta\oplus}{N}\cdots H\cdots\overset{\delta\ominus}{\ddot{Cl}}: \longrightarrow NH_4^{\oplus} + :\ddot{Cl}:^{\ominus}$

$H_2\ddot{O}: + H{-}\ddot{Cl}: \longrightarrow H_2\overset{\delta\oplus}{\ddot{O}}\cdots H\cdots\overset{\delta\ominus}{\ddot{Cl}}: \longrightarrow H_3O:^{\oplus} + :\ddot{Cl}:^{\ominus}$

(b) The criterion of an S_N2 reaction which can not be detected in each of the reactions is inversion of configurations at the atom attacked by the nucleophile.

(c) Ionization of hydrogen chloride requires a polar solvent to solvate the ions formed.

(Text S_N2 S. 5.2)

35. $(CH_3)_3CCH_2MgBr$ is a source of the nucleophile $(CH_3)_3CCH_2:^{\ominus}$. The reaction is S_N2 on iodine.

$(CH_3)_3CCH_2:^{\ominus} + :\ddot{I}{-}\ddot{I}: \longrightarrow (CH_3)_3CCH_2{-}I + :\ddot{I}:^{\ominus}$

(Text: S_N2 S. 5.2)

36. Increasing reactivity to S_N2 parallels decreasing steric hindrance in the substrates.

$CH_3(CH_2)_4I$ ($I^{\ominus}$ is a better leaving group than $Cl^{\ominus}$) >

$CH_3(CH_2)_3CH_2Cl > (CH_3)_3CCH_2Cl$ $CH_3CH_2C(CH_3)_2$ (with Cl on that C)

(Text S_N2 S. 5.2)

37. 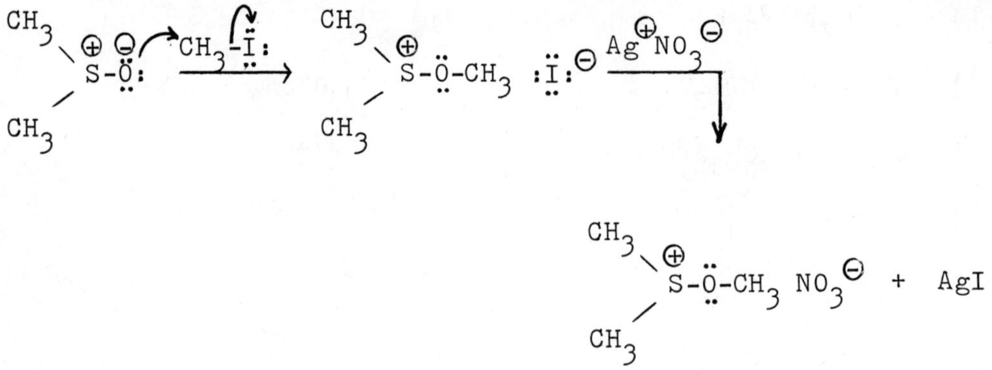 $(CH_3)_2\overset{\oplus}{S}-\ddot{O}:^{\ominus}$ + $CH_3-\ddot{I}:$ ⟶ $(CH_3)_2\overset{\oplus}{S}-\ddot{O}-CH_3$ $:\ddot{I}:^{\ominus}$ $\xrightarrow{Ag^{\oplus}NO_3^{\ominus}}$ $(CH_3)_2\overset{\oplus}{S}-\ddot{O}-CH_3$ $NO_3^{\ominus}$ + AgI

38. (a) any two of: [three resonance structures of $NO_3^{\ominus}$ drawn with ⟷]

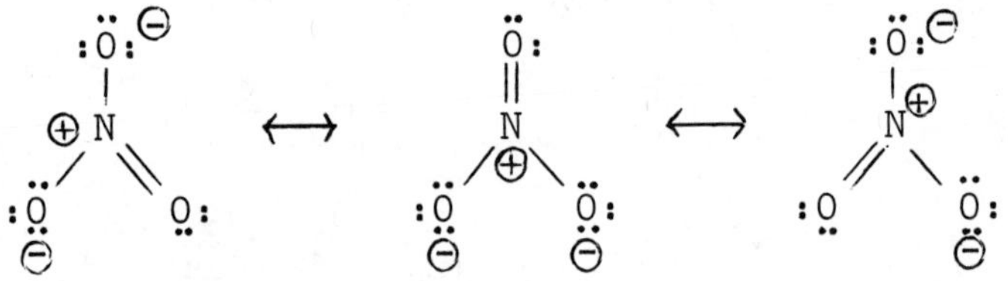

Sulfur and phosphorus can accommodate more than eight electrons. Resonance structures having less separation of charge are shown first.

(b) 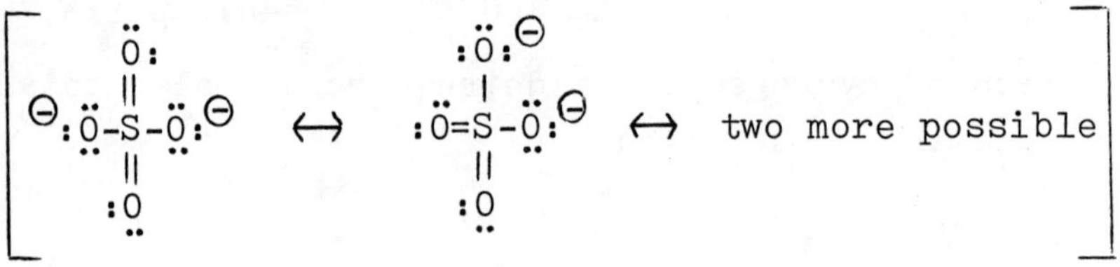 [$^{\ominus}:\ddot{O}-S(=O)_2-\ddot{O}:^{\ominus}$ ⟷ $:\ddot{O}=S(=O)(-\ddot{O}:^{\ominus})-\ddot{O}:^{\ominus}$ ⟷ two more possible]

less important are:

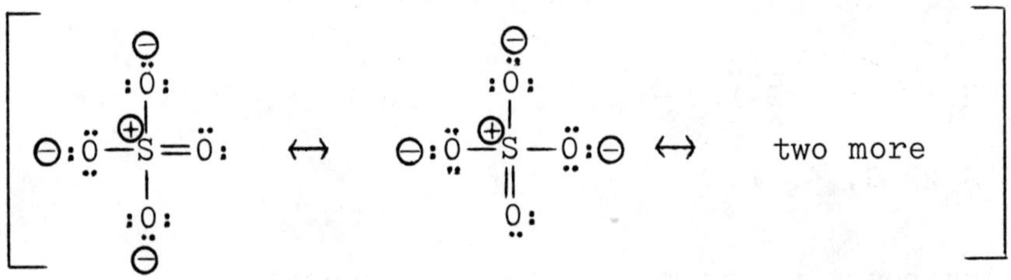

[$\ominus:\ddot{O}-\overset{\oplus}{S}(=\ddot{O}:)(-\ddot{O}:^{\ominus})-\ddot{O}:^{\ominus}$ ⟷ $\ominus:\ddot{O}-\overset{\oplus}{S}(=O:)(-\ddot{O}:^{\ominus})-\ddot{O}:\ominus$ ⟷ two more]

and $\ominus:\ddot{O}-\overset{++}{S}(-\ddot{O}:^{\ominus})(-\ddot{O}:^{\ominus})-\ddot{O}:\ominus$

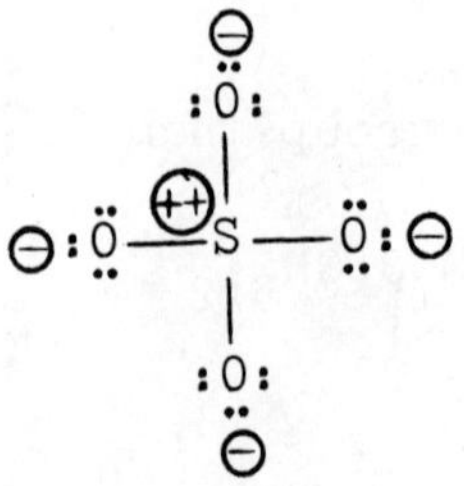

(continued next page)

(c) $\left[{}^{\ominus}\!:\ddot{O}-P(=O)(-O^{\ominus})-\ddot{O}:^{\ominus} \leftrightarrow :\ddot{O}=P(-O^{\ominus})(-O^{\ominus})-\ddot{O}:^{\ominus} \leftrightarrow \text{two more} \right]$

less important is: $^{\ominus}:\ddot{O}-P^{\oplus}(-O^{\ominus})(-O^{\ominus})-\ddot{O}:^{\ominus}$

(Text: resonance S. 2.9)

39. (a) $CH_3-\ddot{O}-CH_3 \xrightarrow{H-I} CH_3-\overset{H}{\underset{\oplus}{O}}-CH_3 \quad :\ddot{I}:^{\ominus} \xrightarrow{S_N2} CH_3OH + CH_3I$

$I:^{\ominus} + CH_3-\ddot{O}-CH_3 \not\longrightarrow CH_3I + CH_3\ddot{O}:^{\ominus}$ (does not occur)

$-\overset{H}{\underset{\oplus}{O}}-CH_3$ is a good leaving group, $-\ddot{O}-CH_3$ is not.

(b) $(CH_3)_3C-\ddot{O}-CH_3 \xrightarrow{H-I} (CH_3)_3C-\overset{H}{\underset{\oplus}{O}}-CH_3 \quad :I^{\ominus}$

$\downarrow$

$(CH_3)_3CI \xleftarrow[S_N1]{I:^{\ominus}} (CH_3)_3C^{\oplus} + CH_3OH$

40. $H\text{-}C(CH_3)(Et)\text{-}\ddot{O}H \xrightarrow{H^{\oplus}} H\text{-}C(CH_3)(Et)\text{-}\overset{\oplus}{O}H_2 \xrightarrow{-H_2\ddot{O}:} \left[C^{\oplus}(CH_3)(H)(Et) \right]^{\oplus}$

R   R   achiral

(continued next page)

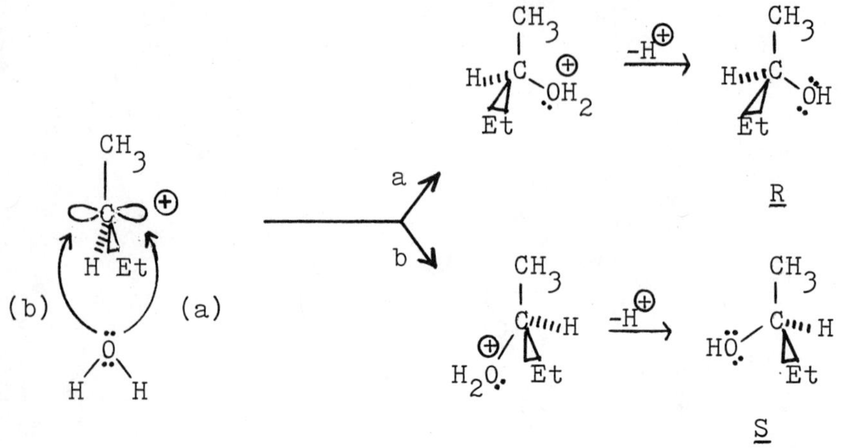

(Reaction scheme: planar carbocation attacked by H_2O at face (a) or (b); path a gives protonated alcohol, then $-H^{\oplus}$, giving R; path b gives protonated alcohol, then $-H^{\oplus}$, giving S.)

CH3 H C ⊕ OH2 Et → -H⊕ → CH3 H C OH Et — R

CH3 C ⊕ H Et (b) (a) O H H — a / b

CH3 C H ⊕ H2O Et → -H⊕ → CH3 HO C H Et — S

Attack by H_2O occurs at either face of the planar carbocation. (Text: stereochemistry of the S_N1 reaction S. 5.4)

41. From meso-2,3-dichlorobutane:

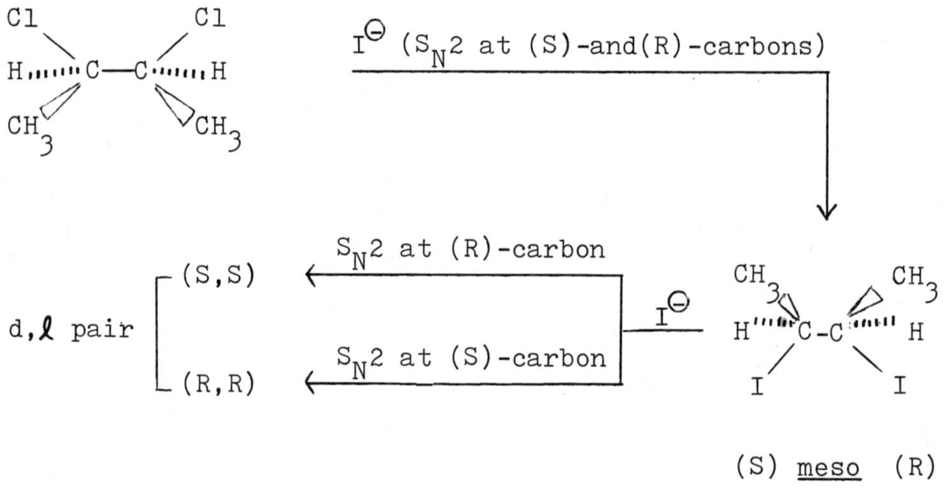

Cl Cl H C—C H CH3 CH3 — $I^{\ominus}$ (S_N2 at (S)-and(R)-carbons) →

d,ℓ pair [(S,S) ← S_N2 at (R)-carbon; (R,R) ← S_N2 at (S)-carbon] ← $I^{\ominus}$ — CH3 CH3 H C-C H I I

(S) meso (R)

From d,ℓ-2,3-dichlorobutane. An S_N2 reaction of iodide ion at each chiral center of the (d) and (ℓ) forms gives (d,ℓ)-diiodobutane. An additional S_N2 reaction at either chiral center of (d)- or (ℓ)-2,3-diiodobutane gives meso-diiodobutane. The meso and d, ℓ are of unequal energy and, therefore, are formed in unequal amounts.

(Text: stereochemistry of the S_N2 reaction S. 5.2)

42. The observations are explained by a cyclic bromonium ion intermediate. The bridging occurs simultaneously with loss of the leaving group.

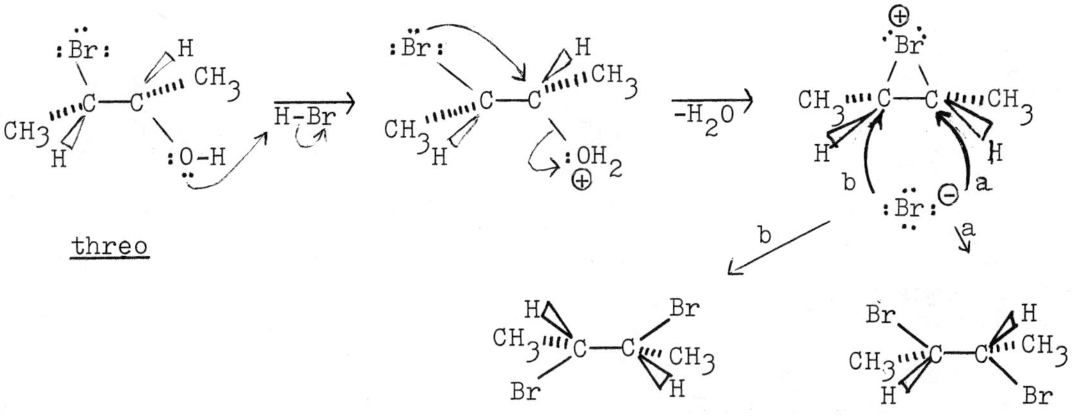

:Br: H CH$_3$ CH$_3$ H :O-H —H-Br→ :Br: H CH$_3$ CH$_3$ H :OH$_2^{\oplus}$ —$-H_2O$→ Br$^{\oplus}$ CH$_3$ C—C CH$_3$ H H b a :Br:$^{\ominus}$

threo

b ↙ a ↘

H CH$_3$ C—C Br Br CH$_3$ H   Br CH$_3$ C—C H H CH$_3$ Br

enantiomers

A similar intermediate accounts for the formation of meso-2,3-dibromobutane from (+) or (-)-erythro-3-bromo-2-butanol.

43. Products are derived from carbocations A and B shown below:

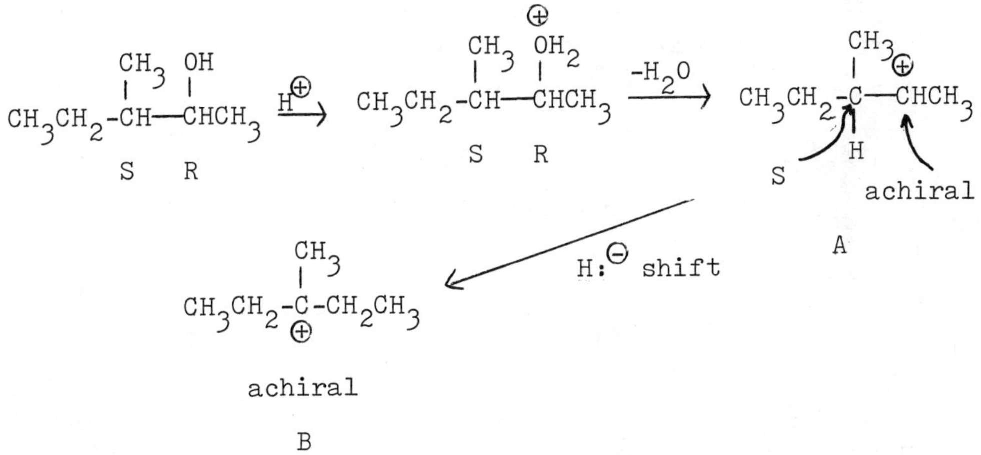

$CH_3CH_2\text{-}CH(CH_3)\text{—}CH(OH)CH_3$ (S, R) $\xrightarrow{H^{\oplus}}$ $CH_3CH_2\text{-}CH(CH_3)\text{—}CH(OH_2^{\oplus})CH_3$ (S, R) $\xrightarrow{-H_2O}$ $CH_3CH_2\text{-}C(CH_3)(H)\text{—}\overset{\oplus}{C}HCH_3$

S   achiral

A

H:$^{\ominus}$ shift

$CH_3CH_2\text{-}\overset{\oplus}{C}(CH_3)\text{-}CH_2CH_3$

achiral

B

From A: (2S,3S)- and (2R,3S)-$CH_3CH_2CH(CH_3)CHBrCH_3$

From B: $CH_3CH_2CBr(CH_3)CH_2CH_3$ (inactive)

(Text: carbocation rearrangements, S. 5.6)

44. If racemization is defined as the conversion of one enantiomer into a racemic modification, the rate of racemization is twice the rate of label incorporation since the reaction of $I^{*\ominus}$ with two molecules of RI brings about racemization but introduces only one I*.

For example:

$$\begin{matrix} (+)\text{-R-I} \\ (+)\text{-R-I} \end{matrix} \xrightarrow{I^{*\ominus}} \begin{matrix} (-)\text{R-I}^* \\ (+)\text{R-I} \end{matrix}$$

racemic modification

Defining racemization as the rate of interconversion of enantiomers requires a minimum of two steps -- the same number of steps required for label incorporation. Thus, the rates are equal.

For example:

$$\begin{matrix} (+)\text{-R-I} \\ (-)\text{-R-I} \end{matrix} \xrightarrow{I^{*\ominus}} \begin{matrix} (-)\text{-R-I}^* \\ (-)\text{R-I} \end{matrix} \xrightarrow{I^{*\ominus}} \begin{matrix} (-)\text{-R-I}^* \\ (+)\text{-R-I}^* \end{matrix}$$

(see related problem 53)

45. The electron-withdrawing C=O group makes the carbon of $-CH_2-Cl$ more positive and more susceptible to nucleophilic attack. In addition, the π bond of the C=O may stabilize the activated complex in this S_N2 process.

An S_N1 process requires formation of $CH_3-\overset{\overset{O}{\|}}{C}-\overset{\oplus}{C}H_2$. This ion is destabilized by the electron-withdrawing C=O which intensifies the positive charge.

46. $(CH_3CH_2)_2\ddot{N}-CH_2-CH(\ddot{Cl}:)-CH_2CH_3 \xrightarrow{-Cl^{\ominus}}$ $Et_2\overset{\oplus}{N}$ (three-membered ring with $CH-CH_2CH_3$ and CH_2) $\xrightarrow{:\ddot{O}H^{\ominus}}$ $Et_2N-CHCH_2CH_3$ with $CH_2\ddot{O}H$ on the CH

(Text: neighboring group participation S. 5.7)

47. $R_3N: + R'\text{-}X \longrightarrow \overset{\delta\oplus}{R_3N}\cdots\cdots R'\cdots\cdots\overset{\delta\ominus}{X} \longrightarrow \overset{\oplus}{R_3NR'}\ X^{\ominus}$

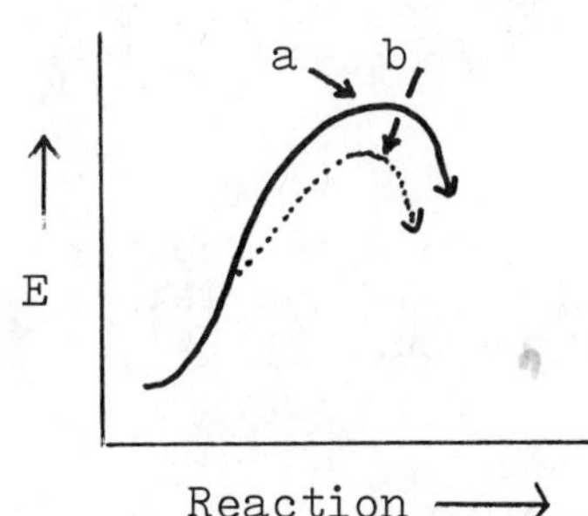

a. activated complex in less polar solvent. Little stabilization afforded to charged complex.

b. activated complex in more polar solvent. Greater stabilization afforded charged complex. System has lower energy.

(Text: solvent effects in S_N2 S. 5.2)

48. (a) In the presence of sodium ethoxide, the reaction is S_N2.

$$C_2H_5\ddot{O}:^{\ominus} + CH_3\underset{\displaystyle Cl}{\underset{|}{C}}HCH{=}CH_2 \longrightarrow CH_3\underset{\displaystyle OC_2H_5}{\underset{|}{C}}HCH{=}CH_2 + Cl^{\ominus}$$

(b) In the absence of sodium ethoxide, the reaction is S_N1.

$$CH_3\underset{\displaystyle :\ddot{C}l:}{\underset{|}{C}}HCH{=}CH_2 \xrightarrow{C_2H_5OH} \left[CH_3\text{-}\overset{\oplus}{C}H\text{-}CH{=}CH_2 \leftrightarrow CH_3\text{-}CH{=}CH\text{-}\overset{\oplus}{C}H_2\right]^{*} :\ddot{C}l:^{\ominus}$$

$\downarrow$ $C_2H_5\ddot{O}H$ (nucleophile)

$$CH_3\text{-}CH\text{-}CH{=}CH_2 \text{ with } \overset{\oplus}{O}(C_2H_5)(H) \quad \xrightarrow{H\ddot{O}C_2H_5} \quad CH_3\underset{\displaystyle C_2H_5\ddot{O}:}{\underset{|}{C}}HCH{=}CH_2$$

$$CH_3CH{=}CHCH_2\text{-}\overset{\oplus}{O}:(C_2H_5)(H) \quad \xrightarrow{C_2H_5\ddot{O}H} \quad CH_3CH{=}CHCH_2\ddot{O}C_2H_5$$

*(equivalent to $CH_3\overset{\delta\oplus}{C}H\text{===}CH\text{===}\overset{\delta\oplus}{C}H_2$)

(Text: S_N2, S 5.2 and allylic cations, S 5.5)

49. Carbon, like other second-period elements, has valence electrons in s and p orbitals of the second level. Phosphorus, on the other hand, has its valence electrons in the third level where d orbitals are available allowing phosphorus to expand its valence shell (accommodate more than just eight electrons).

50. The loss of optical activity can be attributed to the conversion of some (+)-1-phenylethyl chloride to (-)-1-phenylethyl chloride in the presence of the Lewis acid mercury(II) chloride.

$$(+)\text{-}C_6H_5CHClCH_3 + HgCl_2 \rightleftharpoons \underset{\text{achiral}}{C_6H_5\overset{\oplus}{C}HCH_3}\ HgCl_3^{\ominus} \rightleftharpoons (-)\text{-}C_6H_5CHClCH_3 + HgCl_2$$

51. Forming a 1:1 ratio of enantiomers in an S_N1 process requires, in step two of the mechanism, that the attacking nucleophile have equal access to both faces of the carbocation. The more stable carbocations exist for a longer period of time which permits the initially formed ion pair to become separated by solvent molecules. Thus, the carbocation becomes symmetrically solvated, and reaction by the nucleophile may occur with equal probability at either face of the carbocation.

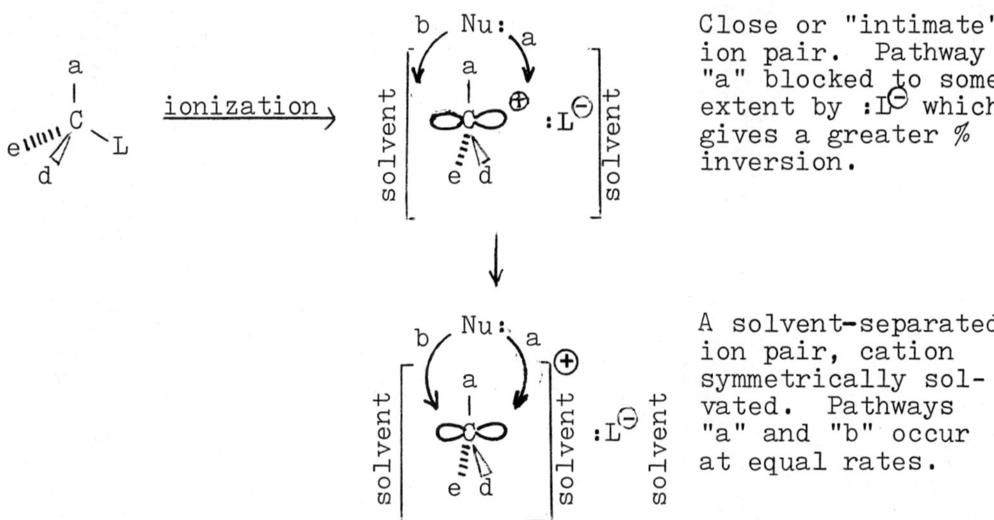
a
C
e
d
L
ionization
b
Nu:
a
solvent
:L⊖
solvent
Close or "intimate" ion pair. Pathway "a" blocked to some extent by :L⊖ which gives a greater % inversion.
A solvent-separated ion pair, cation symmetrically solvated. Pathways "a" and "b" occur at equal rates.

52. In the gas phase and in aprotic solvents, increasing nucleophilicity of halide ions parallels increasing basicity (which increases as the electronegativity increases) and decreasing polarizability (which increases as the size increases). <u>In protic solvents</u>, the more electronegative and more basic the anion, the more strongly it hydrogen bonds to the solvent and the <u>less</u> nucleophilic it is.

53. The rate of racemization is twice the rate of label incorporation since the reaction of $^{131}I^{\ominus}$ with two molecules of (+)-2-iodobutane brings about racemization but introduces only one $^{131}I^{\ominus}$.

$$\begin{matrix} (+)\text{-}CH_3CHICH_2CH_3 \\ (+)\text{-}CH_3CHICH_2CH_3 \end{matrix} \xrightarrow{^{131}I^{\ominus}} \begin{matrix} (-)\text{-}CH_3CH(^{131}I)CH_2CH_3 \\ (+)\text{-}CH_3CHICH_2CH_3 \end{matrix} \quad \text{racemic}$$

(see related problem 44)

SUMMARY OF REACTIONS

Nucleophilic Substitution

1. General: $-\overset{|}{\underset{|}{C}}-L + :Nu^{\ominus} \longrightarrow -\overset{|}{\underset{|}{C}}-Nu + :L^{\ominus}$
(-L usually -X or -OTs)

example: $CH_3CH_2-\ddot{B}r: + :\ddot{O}H^{\ominus} \longrightarrow CH_3CH_2-\ddot{O}H + :\ddot{B}r:^{\ominus}$

2. General: $-\overset{|}{\underset{|}{C}}-L + :NuH \longrightarrow -\overset{|}{\underset{|}{C}}-\overset{\oplus}{N}uH\ :L^{\ominus} \longrightarrow -\overset{|}{\underset{|}{C}}-Nu + HL$
(-L usually -X, -OTs; :NuH usually HOH, ROH)

example: $(CH_3)_3C-\ddot{B}r: + H_2\ddot{O}: \longrightarrow (CH_3)_3C-\overset{\oplus}{O}{<}^{H}_{H}\ :\ddot{B}r:^{\ominus} \longrightarrow$
$(CH_3)_3C\ddot{O}H + HBr$

3. General: $-\overset{|}{\underset{|}{C}}-\overset{\oplus}{L} + :Nu^{\ominus} \longrightarrow -\overset{|}{\underset{|}{C}}-Nu + :L$
($-L^{\oplus}$ usually $-\overset{\oplus}{O}{<}^{H}_{H}$; $:Nu^{\ominus}$ usually $:\ddot{X}:^{\ominus}$)

example: $CH_3CH_2\ddot{O}H + H\ddot{B}r: \longrightarrow CH_3CH_2\overset{\oplus}{\ddot{O}}H_2\ :\ddot{B}r:^{\ominus} \longrightarrow CH_3CH_2-\ddot{B}r:$
$+ H_2\ddot{O}:$

REACTION REVIEW

The following exercises are designed to test your knowledge of the reactions covered in this and previous chapters.

In part A, a general reaction from this chapter is followed by specific examples. The reactants are given on the "question" side and the products appear on the "answer" side. These questions and answers can be effectively used to make a set of reaction cards by pasting the question to one side of a 3 x 5" index card and the answer to the opposite side. Alternatively, you can simply cover the answer side with a blank sheet of paper while writing the answers. Part B consists of exercises which use reactions from this and previous chapters.

A. Nucleophilic Substitution

Questions

1. $-\overset{|}{\underset{|}{C}}-L + :Nu^{\ominus} \longrightarrow$?

Predict major substitution products in the following:

(a) $CH_3CH_2Br \xrightarrow{:\ddot{O}H^{\ominus}/H_2O}$?

(b) CH_3CHCH_3 (with Cl on the middle carbon) $\xrightarrow[DMF]{Na^{\oplus}\ :CN^{\ominus}}$?

(c) $CH_3CH_2OTs \xrightarrow[NH_3(\ell)]{Na^{\oplus}\ :C{\equiv}CCH_3^{\ominus}}$?

(d) (R)-$CH_3CHCH_2CH_3$ (with I on the second carbon) $\xrightarrow[CH_3OH]{Na^{\oplus}\ :\ddot{O}CH_3^{\ominus}}$?

(e) (S)-$CH_3CHCH_2CH_2Cl$ (with OCH_3 on the second carbon) $\xrightarrow[C_2H_5OH]{Na^{\oplus}:\ddot{O}C_2H_5^{\ominus}}$?

Answers

1. $-\overset{|}{\underset{|}{C}}-Nu + :L^{\ominus}$

(a) CH_3CH_2OH

(b) CH_3CHCH_3 (with CN on the middle carbon)

(c) $CH_3CH_2C{\equiv}CCH_3$

(d) (S)-$CH_3CHCH_2CH_3$ (with OCH_3 on the second carbon)

(e) (S)-$CH_3CHCH_2CH_2OC_2H_5$ (with OCH_3 on the second carbon)

(chiral center not affected)

Questions

2. $-\overset{|}{\underset{|}{C}}-\overset{\oplus}{L} + :Nu^{\ominus} \longrightarrow$?

Predict the major substitution products in the following:

(a) $CH_3CH_2CH_2OH \xrightarrow{HBr}$?

(b) (R)-$(CH_3)_2CHCHCH_3$ (with OH on the third carbon) $\xrightarrow{HCl}$?

Answers

2. $-\overset{|}{\underset{|}{C}}-Nu + :L$

(a) $CH_3CH_2CH_2Br$

(b) $(CH_3)_2CCH_2CH_3$ (with OH on the second carbon) (achiral)

(product via rearrangement)

(continued next page)

(c) $CH_3CH{=}CHCH_2OH \xrightarrow{HBr}$?	(c) $CH_3CH{=}CHCH_2Br$ and $CH_3CH(Br)CH{=}CH_2$ (allylic rearrangement)
3. $-\overset{\vert}{\underset{\vert}{C}}-L$ + :NuH $\longrightarrow$? Predict the major substitution products in the following: (a) $(CH_3)_3C\ddot{Cl}: \xrightarrow{H_2\ddot{O}:}$? (b) (R)-$CH_3CH(:\ddot{Br}:)CH_2CH_3 \xrightarrow{H_2\ddot{O}:}$?	3. $\left[-\overset{\vert}{\underset{\vert}{C}}-\overset{\oplus}{Nu}H\ :L^{\ominus}\right] \longrightarrow -\overset{\vert}{\underset{\vert}{C}}-Nu + HL$ (a) $(CH_3)_3COH$ (b) (R)- and (S)-$CH_3CH(OH)CH_2CH_3$ (inverted product predominates)
B. Write structures for all letters 1. $CH_3CH_3 \xrightarrow{Cl_2/h\nu} A \xrightarrow{OH^{\ominus}/H_2O} B$	1. A: CH_3CH_2Cl, B: CH_3CH_2OH
2. $CH_3CH_2Cl \xrightarrow{Li(CH_3CH_2)_2Cu} A \xrightarrow{Br_2/h\nu} B \xrightarrow[DMSO]{Na^{\oplus}\ CN^{\ominus}} C$	2. A: $CH_3CH_2CH_2CH_3$, B: $CH_3CH(Br)CH_2CH_3$, C: $CH_3CH(CN)CH_2CH_3$
3. $(CH_3)_3CH \xrightarrow{Br_2/h\nu} A \xrightarrow{H_2O} B \xrightarrow{HI} C$	3. A: $(CH_3)_3CBr$, B: $(CH_3)_3COH$, C: $(CH_3)_3CI$

CHAPTER 6, ALKYL HALIDES - ELIMINATION REACTIONS AND GRIGNARD REAGENTS

LEARNING OBJECTIVES

When you have completed this chapter, you should be able to:

1. identify α, β, γ and δ elimination reactions and compare the stability of products in some of these reactions (problems 1, 2);
2. predict or account for products formed in α-elimination reactions (problems 11, 35);
3. write rate expressions for substitution and elimination reactions (problem 3, related problem 28);
4. predict the stereochemistry of alkenes formed in E2 reactions (problems 4, 7, 37);
5. identify and predict alkenes formed in elimination reactions according to Saytzeff's and Hofmann's rules (problems 5, 9, 10, 20, 26);
6. choose reactants which will give specified alkenes in E2 reactions (problem 8);
7. predict the stereochemistry of alkenes formed in Elcb reactions (problems 15, 30);
8. write equations which illustrate the preparation and reactions of organometallic compounds (problems 13, 14, 16, 17, 24);
9. predict whether substitution or elimination reactions predominate under specified conditions (problems 18, 19, 23, 34);
10. write equations for the synthesis of alkenes and other compounds by combining reactions from this chapter with those from previous chapters (problems 21, 24);
11. write equations which account for the formation of products in E1, E2 and related mechanisms (problems 4, 7, 15, 25, 29, 33, 36);
12. write equations which illustrate the formation of substrates used in Hofmann elimination reactions (problem 27).

ANSWERS TO QUESTIONS

1. (a) β (d) δ

   (b) β (e) α

   (c) γ (f) β

2. More bonds are formed in the β-elimination (alkene formation). Assuming the energies of the activated complexes are related to the energies of the products, the energy of the activated complex leading to the alkene should be lower than that for α- elimination.

3. (a) $R = k\left[CH_3O^{\ominus}\right]\left[(CH_3)_2CHBr\right]$

(b) $R = k'\left[CH_3O^{\ominus}\right]\left[(CH_3)_2CHBr\right]$

These rate expressions differ only in k and k'.

(c) Doubling the concentration of methoxide ion (doubling the number of methoxide ions per unit volume) doubles the rate of each reaction because the number of effective collisions is doubled.

4. These reactions are <u>anti</u>-eliminations. The term "<u>anti</u>-elimination" refers to <u>the orientation in the activated complex</u> of the groups being eliminated (H and L of H–C–C–L). As a matter of convenience, the groups being eliminated may be aligned in the substrate as they will appear in the activated complex.

$CH_3(Br)C(H)–C(Br)(CH_3)(H) + C_2H_5O^{\ominus} \rightarrow$ [activated complex, with $\delta^{\ominus}Br$ and $\delta^{\ominus}OC_2H_5$] $\rightarrow$ $Br(CH_3)C{=}C(CH_3)(H) + :Br:^{\ominus} + C_2H_5OH$

(or mirror image) activated complex

$CH_3(Br)C(H)–C(Br)(H)(CH_3) + C_2H_5O^{\ominus} \rightarrow$ [activated complex, with $\delta^{\ominus}Br$ and $\delta^{\ominus}OC_2H_5$] $\rightarrow$ $Br(CH_3)C{=}C(H)(CH_3) + :Br:^{\ominus} + C_2H_5OH$

meso activated complex

(continued next page)

(Note: If two hydrogens are attached to the carbon adjacent to the leaving group (as in $CH_3CH_2CHBrCH_3$), an E2 reaction gives both cis- and trans-alkenes since either hydrogen may be aligned anti to the leaving group.)

5. In an elimination reaction which may give more than one alkene, the alkene having the C=C with the most alkyl groups attached (the more highly substituted C=C) is called the Saytzeff product. The alkene having the C=C with the fewest alkyl groups attached is called the Hofmann product.

   (a), (c), and (d) are formed according to Hofmann's rule (Hofmann products); (b) and (e) are formed according to Saytzeff's rule (Saytzeff products).

6. The $-\overset{\oplus}{O}H_2$ group reacts with a base by donating a proton to the base. For example, the reaction between $CH_3CH_2CH_2\overset{\oplus}{O}H_2$ and $OH^{\ominus}$ is shown below.

$$CH_3CH_2CH_2-\overset{\oplus}{\ddot{O}}(H)H \quad + \quad :\ddot{O}H^{\ominus} \longrightarrow CH_3CH_2CH_2\ddot{O}-H \quad + \quad H-\ddot{O}-H$$

   This reaction is analogous to that between hydronium ion ($H_3O^{\oplus}$) and $OH^{\ominus}$.

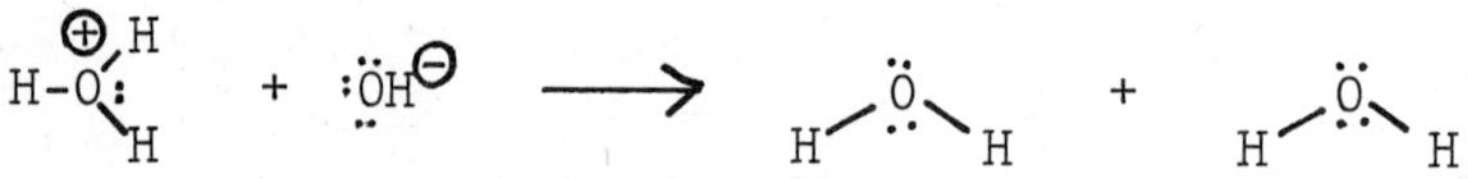

$$H-\overset{\oplus}{\ddot{O}}(H)H \quad + \quad :\ddot{O}H^{\ominus} \longrightarrow H-\ddot{O}-H \quad + \quad H-\ddot{O}-H$$

7. Review problem 4. The reaction is E2 and an anti- elimination occurs.

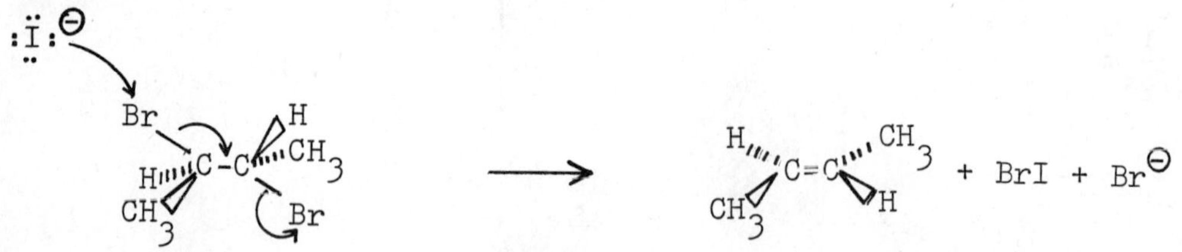

$:\ddot{I}:^{\ominus}$ + Br–C(H)(CH$_3$)–C(H)(CH$_3$)–Br ⟶ H(CH$_3$)C=C(CH$_3$)H + BrI + $Br^{\ominus}$

meso ⟶ trans

(continued next page)

(d,ℓ)-2,3-dibromobutane $\rightarrow$ cis-2-butene

8. (a) $CH_3CH_2Br + OH^{\ominus}$ (KOH/alcohol/heat is a standard combination associated with E2 reactions)

(b) $CH_3CH_2CH_2Br$ (or $CH_3CHBrCH_3$) + $NH_2^{\ominus}$($Na^{\oplus}NH_2^{\ominus}$/liq. NH_3)

(c) $(CH_3)_2CHCH_2OTs + OC_2H_5^{\ominus}$ ($Na^{\oplus}OC_2H_5^{\ominus}/C_2H_5OH$/heat)

(d) $CH_3CH_2CHClCH_2CH_3 + OH^{\ominus}$ (KOH/DMSO/heat)

(e) $CH_3CH_2CHClCH_3 + OH^{\ominus}$

(f) $(CH_3)_2CHCH(CH_3)CH_2\overset{\oplus}{S}(CH_3)_2 + OH^{\ominus}$

The base-solvent combinations in examples (a) $\rightarrow$ (f) are, in general, interchangeable.

(g) $(CH_3)_3\overset{\oplus}{N}CH_2(CH_2)_3CH_3$ $OH^{\ominus}$ heat (see also problem 5d this chapter)

9. Increasing the size of (steric hindrance in) the base, i.e., changing from smaller ethoxide to larger t-butoxide, increases the amount of Hofmann product at the expense of Saytzeff product. The larger base reacts to a greater extent with the more accessible hydrogens.

10. Saytzeff's rule is used in predicting the major alkene product (more highly substituted alkene) when an alcohol is dehydrated in acidic solution.

(a) $CH_3CH{=}CH_2$

(b, c) both give $CH_3CH{=}CHCH_3$ and $CH_3CH_2CH{=}CH_2$

(major, trans > cis)

(d) $CH_2{=}C(CH_3)_2$

(e) $CH_3CH{=}C(CH_3)_2$ + $CH_3CH_2C(CH_3){=}CH_2$

(major)

(f) $(CH_3)_2C{=}C(CH_3)_2$ and $(CH_3)_2CHC(CH_3){=}CH_2$

(major)

11. a concerted mechanism:

$$>C(-\ddot{O}-H)(-\ddot{X}:) + :B^{\ominus} \longrightarrow >C=O + HB + :\ddot{X}:^{\ominus}$$

or a two-step mechanism:

$$>C(-\ddot{O}-H)(-X:) + :B^{\ominus} \underset{HB}{\rightleftharpoons} >C(-\ddot{O}:^{\ominus})(-\ddot{X}:) \longrightarrow >C=O + :\ddot{X}:^{\ominus}$$

12. 1,2-Dimethoxyethane is higher boiling (desirable if a higher reaction temperature is required) and can coordinate with magnesium using both oxygens.

13. (a) the hydrocarbon is $C_6H_5-CH_2-CH_2-C_6H_5$ and

(b) is formed as shown below.

$$C_6H_5-CH_2-Br \xrightarrow[\text{ether}]{Mg} C_6H_5-CH_2MgBr \xrightarrow[S_N2]{C_6H_5-CH_2-Br} MgBr_2 + C_6H_5-CH_2-CH_2-C_6H_5$$

14. (a) $CH_4 \xrightarrow[h\nu]{Cl_2} CH_3Cl \xrightarrow[\text{ether}]{Mg} CH_3MgCl \xrightarrow{D_2O} CH_3D$

$+HCl$ $\qquad$ $+ MgClOD$

products in (b) and (c) are prepared by a similar series of reactions summarized below.

(b) $CH_3CH_3 \xrightarrow[h\nu]{Cl_2} \xrightarrow[\text{ether}]{Mg} \xrightarrow{D_2O} CH_3CH_2D$

(c) $(CH_3)_3CH \xrightarrow[h\nu]{Br_2} \xrightarrow[\text{ether}]{Mg} \xrightarrow{D_2O} (CH_3)_3CD$

15. Treating dehalogenation as an E2 elimination allows one to predict the geometry of the major product.

(+)- and (-)-2,3-dibromobutane give cis-2-butene. Meso-2,3-dibromobutane gives trans-2-butene. These reactions are Elcb. The reaction for meso-2,3-dibromobutane is illustrated below.

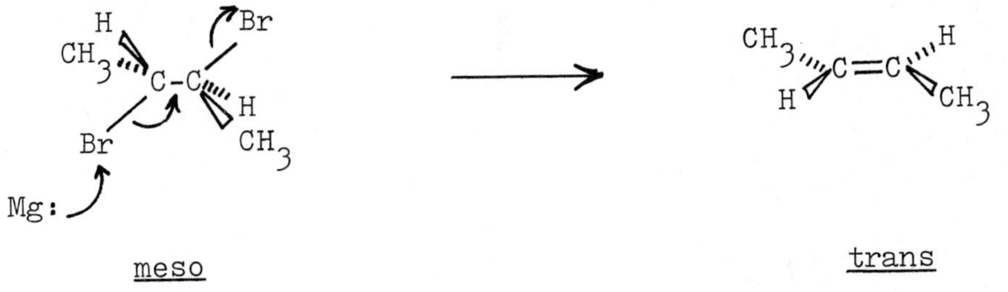

H, CH_3, Br C-C Br, H, CH_3; Mg: $\longrightarrow$ CH_3, H C=C H, CH_3

meso   trans

16. (complete reactions are shown in the first two examples)

(a) $2CH_3MgCl + HgCl_2 \xrightarrow{\text{ether}} (CH_3)_2Hg + 2MgCl_2$

(b) $4CH_3MgCl + SnCl_4 \xrightarrow{\text{ether}} (CH_3)_4Sn + 4MgCl_2$

(c) $4(CH_3)_2CHMgCl + SnCl_4$

(d) $2(CH_3)_3CCH_2CH_2MgCl + CdCl_2$

17. (a) $CH_4 \xrightarrow{Cl_2/h\nu}$

(b) $CH_3CH_3 \xrightarrow{Cl_2/h\nu}$

(c) CH_3Cl (from a) $\xrightarrow{2Li}$

(d) CH_3CH_2Cl (from b) $\xrightarrow{2Li}$

(e) $2CH_3Li$ (from c) + CuCl

(or stepwise $CH_3Li \xrightarrow{CuCl} CH_3Cu \xrightarrow{CH_3Li}$)

(f) $2CH_3CH_2Li$ (from d) + CuCl

(or $CH_3CH_2Li \xrightarrow{CuCl} CH_3CH_2Cu \xrightarrow{CH_3CH_2Li}$)

(g) CH_3CH_2Cl (from b) $\xrightarrow{Li(CH_3)_2Cu \text{ (from e)}}$

(or CH_3Cl (from a) $\xrightarrow{Li(CH_3CH_2)_2Cu \text{ (from f)}}$)

(h) CH_3CH_2Cl (from b) $\xrightarrow{Li(CH_3CH_2)_2Cu \text{ (from f)}}$

(i) $CH_3CH_2CH_3 \xrightarrow{Cl_2/h\nu}$

(j) $CH_3CHClCH_3$ (from i) $\xrightarrow{Li(CH_3)_2Cu \text{ (from e)}}$

(k) CH_3Li (from c) $\xrightarrow{D_2O}$

(l) CH_3CH_2Li (from d) $\xrightarrow{D_2O}$

(m) CH_3Li (from c) $\xrightarrow{O_2}$ $\xrightarrow{H_3O^{\oplus}}$

18. The optically active compounds with the formula $C_5H_{11}Cl$ and their reactions with (a) alcoholic KOH, (b) aqueous KOH, and (c) sodium amide are shown below.

A knowledge of the kind of reaction generally obtained with each reagent listed is essential. Alcoholic KOH and $NaNH_2$ tend to give elimination; aqueous KOH tends to give substitution (unless prohibited by steric hindrance).

$CH_3CH_2CH_2CHClCH_3$ (Cl on C-2) $\xrightarrow{\text{KOH, alcohol}}$ or $\xrightarrow{Na^{\oplus}NH_2^{\ominus}, \text{ liq. } NH_3}$ $CH_3CH_2CH{=}CHCH_3$ (*cis* and *trans*)

$\xrightarrow{KOH,\ H_2O}$ $CH_3CH_2CH_2CH(OH)CH_3$

(continued next page)

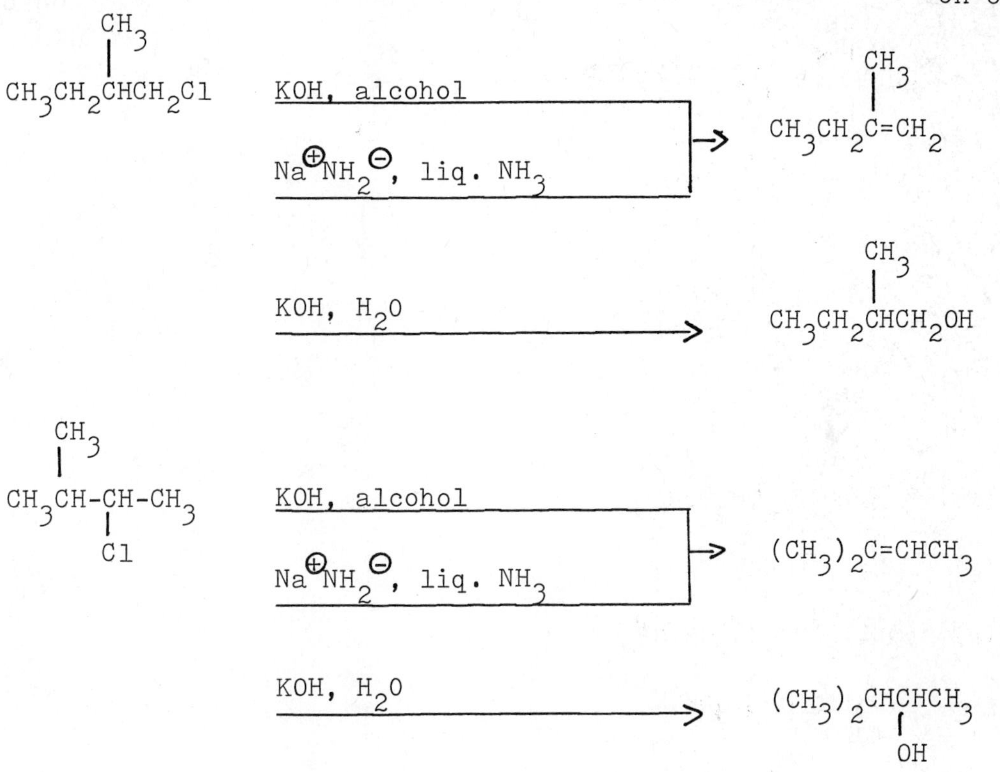

(Text: Substitution and Elimination S. 6.6)

19. (a) $CH_2=CH_2$ (from $BrMg-CH_2-CH_2-SCH_3$ —— $CH_2=CH_2$ + $MgBrSCH_3$)

(b) $:CH_2$ (α- elimination)

(c) $CH_3CH_2O:^{\ominus}\ Na^{\oplus}$ formed in very small quantity

(d) Little, if any reaction. E2 not possible because no β-hydrogen is present. $(CH_3)_3CCH_2OH$ by S_N2 would be formed slowly because of steric hindrance. S_N1 and E1 products are possible but would be formed slowly since C_2H_5OH is not a good ionizing solvent and the substrate is primary. An α- elimination is possible under forcing conditions.

(e) $CH_3CH_2C(OH)(CH_3)_2$ from rearrangement.

(continued next page)

$$CH_3\text{-}\overset{\displaystyle CH_3}{\underset{\displaystyle CH_3}{C}}\text{-}CH_2\text{-}Br \xrightarrow{\text{KOH/water}} CH_3\text{-}\overset{\oplus}{\underset{\displaystyle CH_3}{C}}\text{-}CH_2CH_3\ Br^{\ominus}$$

$$\downarrow OH^{\ominus}$$

$$CH_3\text{-}\overset{\displaystyle OH}{\underset{\displaystyle CH_3}{C}}\text{-}CH_2CH_3$$

(f) $(CH_3)_3CCH_2CH_2SCH_3$

(g) $(CH_3)_3C(CH_2)_4SCH_3$

(h) $(CH_3)_3CCH_2CH_2CH{=}CH_2$

(i) $(CH_3)_3C(CH_2)_4OH$

(j) $CH_3CH{=}CHCH_3$ (<u>trans</u> > <u>cis</u>)

(k) $CH_3CH{=}CHCH_3$ (<u>trans</u> > <u>cis</u>)

(l) $CH_3\overset{\oplus}{N}H_2\text{-}CH_2CH_2CH_2CH_2\text{-}\overset{\oplus}{N}H_2CH_3\ (CH_3SO_3^{\ominus})_2$

(Text: Substitution and Elimination, S. 6.6)

20. The type of elimination product, Saytzeff or Hofmann, is indicated (by Say or Hof).

(a) $CH_3CH_2CH{=}C(CH_3)_2$ (Say)

(b) $CH_3CH_2CH_2\underset{\displaystyle CH_3}{C}{=}CH_2$ (Hof)

(c) $CH_3CH{=}CHCH_3$, <u>trans</u> > <u>cis</u> (Say)

(d) $CH_3CH_2\underset{\displaystyle CH_3}{C}{=}CH_2$ (Hof)

(e) $CH_3CH_2CH_2CH{=}CH_2$ (Hof. product due to large base)

(f) $(CH_3)_2C{=}C(CH_3)_2$ (Say) (Cationic rearrangement precedes Saytzeff elimination)

(Text S. 6.2-6.6)

21. (a) $CH_4 \xrightarrow{Br_2,\ light} CH_3Br \xrightarrow{Mg,\ ether} CH_3MgBr \xrightarrow{D_2O} CH_3D$

(Note: In this method of showing a reaction sequence, only the organic products of interest are shown. The student should be aware of the other products, i.e., in step 1, HBr; in step 3, MgODBr.)

(b) $CH_3CH_3 \xrightarrow{Cl_2,\ light} CH_3CH_2Cl \xrightarrow{Mg,\ ether} CH_3CH_2MgCl \xrightarrow{D_2O}$ CH_3CH_2D

(c) CH_3CH_2Cl (from b) $\xrightarrow{KOH,\ alcohol,\ heat} CH_2{=}CH_2$

(d) $CD_3CD_3 \xrightarrow{Cl_2,\ light} CD_3CD_2Cl \xrightarrow{KOH,\ alcohol,\ heat} CD_2{=}CD_2$

(e) $CH_2{=}CHCH_2Br \xrightarrow{Mg,\ ether} CH_2{=}CHCH_2MgBr \xrightarrow{CH_2=CHCH_2Br}$

$CH_2{=}CHCH_2CH_2CH{=}CH_2$

22. x moles of CH_3OH (32g/mole) and of C_2H_5OH (46g/mole)

x(32) + x(46) = 100g; x = 1.28 moles each

2.56 moles $CH_3OH + C_2H_5OH$ gives 2.56 mole CH_4.

2.56 moles CH_4 x 16g/mole = 41.0 grams CH_4.

2.56 moles x 22.4 ℓ/mole = 57.3 liters at STP.

23. (a) $CH_3CH_2CH_2C(Br)(CH_3)CH_2CH_3$. Substitution product is formed by S_N1. -Br is a better leaving group than -Cl.

(b) $CH_3CH_2CH_2C(Cl)(CH_3)_2$. Substitution product is formed by S_N1.

(c) $(CH_3)_2CHCH_2Cl$. Substitution product is formed by S_N2. Less steric hindrance at leaving group.

(d) 10% KOH in H_2O. Substitution product is formed by S_N2. More polar solvents favor S_N2 over E2.

(Text S. 6.6)

24. (a) $CH_4 \xrightarrow{Cl_2/h\nu}$

(b) CH_3Cl (from a) $\xrightarrow{Li} CH_3Li \xrightarrow{D_2O}$

(c) $CH_3CH_3 \xrightarrow{Br_2/h\nu}$

(d) CH_3CH_2Br (from c) $\xrightarrow{Li} CH_3CH_2Li \xrightarrow{CuCl} Li\left[(CH_3CH_2)_2Cu\right]$

$\xrightarrow{CH_3CH_2Br}$

(e) $CH_3CH_2CH_3 \xrightarrow{Cl_2/h\nu} CH_3CH_2CH_2Cl$ $\xrightarrow{\text{* } Li\left[(CH_3CH_2)_2Cu\right] \text{ (from d)}}$

*(separate from mixture of1- and 2-chloropropane)

(f) $CH_3CH_2CH_2Cl$ (from e) $\xrightarrow{Li} CH_3CH_2CH_2Li \xrightarrow{CuCl}$

$Li\left[(CH_3CH_2CH_2)_2Cu\right] \xrightarrow{CH_3CH_2CH_2Cl}$

(g) CH_3Cl (from a) $\xrightarrow{Mg/ether} CH_3MgCl \xrightarrow{ZnCl_2}$

(h) CH_3CH_2Br (from c) $\xrightarrow{KOH/alcohol}$

(i) $CH_3CH_2CH_3 \xrightarrow{Cl_2/h\nu}$

(j) $(CH_3)_2CHCl$ (from i) $\xrightarrow{Li\ (CH_3)_2Cu\ *}$

*(from CH_3Cl $\xrightarrow{Li}$ $\xrightarrow{CuCl}$)

(k) $(CH_3)_3CH \xrightarrow{Br_2/h\nu} \xrightarrow{Li}$

(l) $(CH_3)_3CLi$ (from k) $\xrightarrow{D_2O}$

(m) $CH_3CH_2CH_2CH_3$ (from d) $\xrightarrow{Br_2/h\nu} \xrightarrow[heat]{KOH/alcohol}$

(n) CH_3CH_2Br (from c) $\xrightarrow{Li}$ $\xrightarrow{CuCl}$

(o) $CH_3CH_2CH_2Cl$ (from f) $\xrightarrow{OH^{\ominus}/H_2O}$

25. 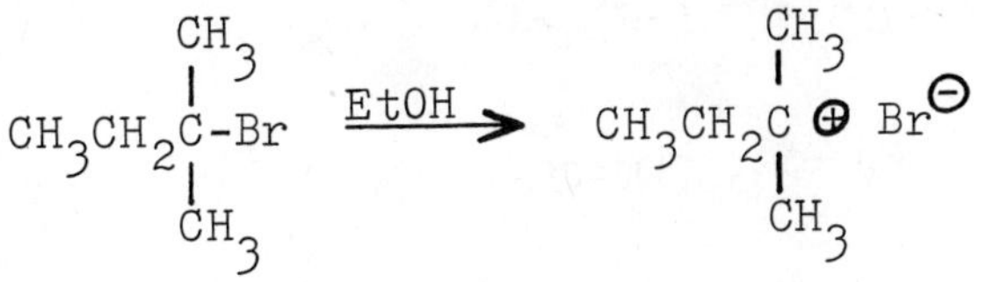 $CH_3CH_2C(CH_3)_2\text{-}Br \xrightarrow{EtOH} CH_3CH_2C^{\oplus}(CH_3)_2 \; Br^{\ominus}$ (all products are derived from this cation)

(a) 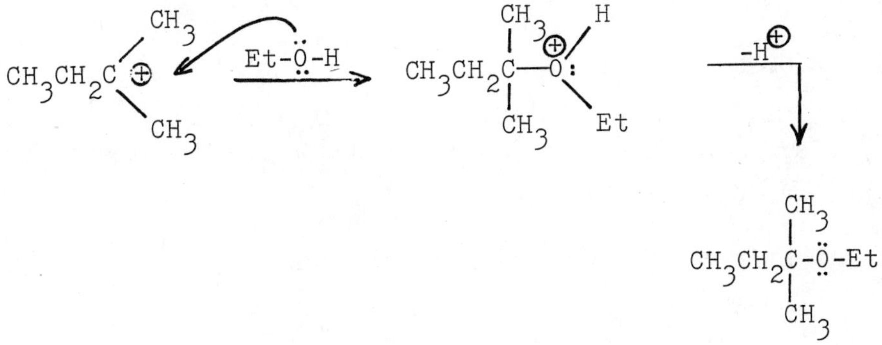 $CH_3CH_2C^{\oplus}(CH_3)_2 \xrightarrow{Et\text{-}\ddot{O}\text{-}H} CH_3CH_2C(CH_3)_2\text{-}\overset{\oplus}{O}(H)(Et) \xrightarrow{-H^{\oplus}} CH_3CH_2C(CH_3)_2\text{-}\ddot{O}\text{-}Et$

(b) 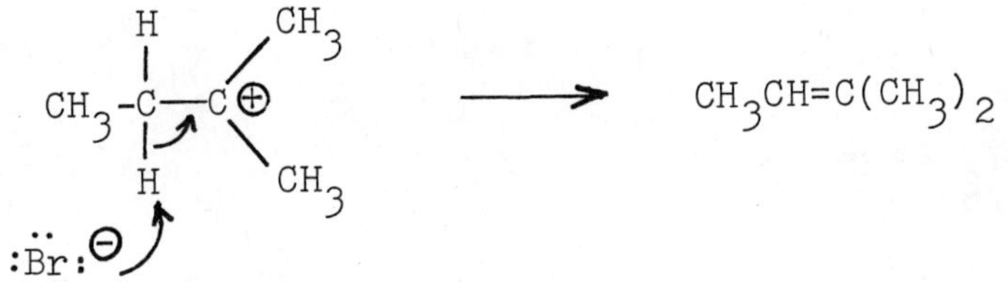 $CH_3\text{-}CH_2\text{-}C^{\oplus}(CH_3)_2$ (H removed by $:\ddot{Br}:^{\ominus}$) $\longrightarrow CH_3CH{=}C(CH_3)_2$

(c) $CH_3CH_2\text{-}C^{\oplus}(CH_3)(CH_2\text{-}H)$ (H removed by $:\ddot{Br}:^{\ominus}$) $\longrightarrow CH_3CH_2\text{-}C(CH_3){=}CH_2$

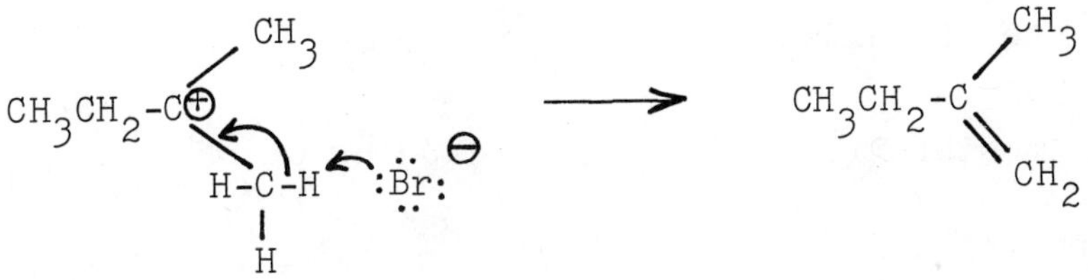

26. (a) $CH_3CH_2CH{=}CH_2$

(b) $CH_3CH{=}CHCH_3$ (trans > cis) and $CH_3CH_2CH{=}CH_2$

(c) $(CH_3)_2C{=}CH_2$ and $CH_2{=}CH_2$ (major)

from: $(CH_3)_2CH\text{-}CH_2\text{-}\overset{\oplus}{N}(CH_3)_2\text{-}CH_2\text{-}CH_2\text{-}H$ + $HO^{\ominus}$ $\xrightarrow{a} CH_2{=}CH_2$; $\xrightarrow{b} (CH_3)_2C{=}CH_2$

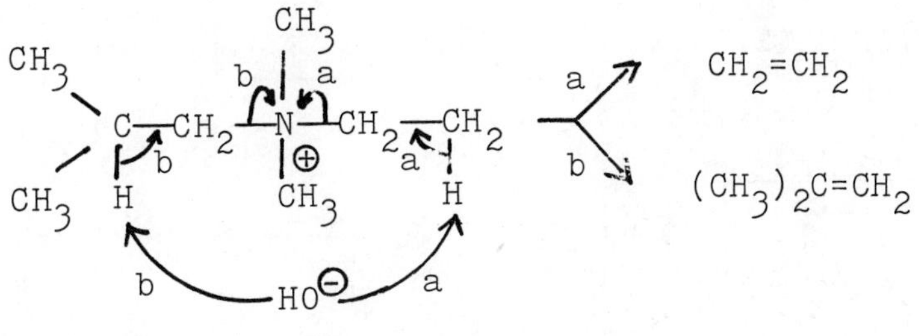

(continued next page)

(d) $(CH_3)_2C{=}CH_2$ and $CH_3CH{=}CH_2$ (major)

(e) $(CH_3)_2C{=}CH_2$ and $CH_3CH_2CH{=}CH_2$ (major)

(Text: Hofmann elimination S. 6.3)

27. (b) $CH_3CH_2CH(CH_3)SCH_3$ + CH_3I

(less acceptable is $(CH_3)_2S$ + $CH_3CH_2CH(CH_3)$-Cl because of greater steric hindrance in the halide)

(c) $(CH_3)_2CHCH_2N(CH_3)_2$ + CH_3CH_2Br <u>or</u>

$(CH_3)_2CHCH_2N(CH_3)CH_2CH_3$ + CH_3I <u>or</u>

$CH_3CH_2N(CH_3)_2$ + $(CH_3)_2CHCH_2Br$

(d) same possibilities as in (c) but using $CH_3CH_2CH_2$- in place of CH_3CH_2-

(e) same possibilities as in (c) but using $CH_3CH_2CH_2CH_2$- instead of CH_3-

28. (a) The elimination and substitution products in either reaction are derived from the same cation. Ionization of <u>either</u> substrate (rate determining step of an S_N1 or E1 process) gives the <u>t</u>-butyl cation.

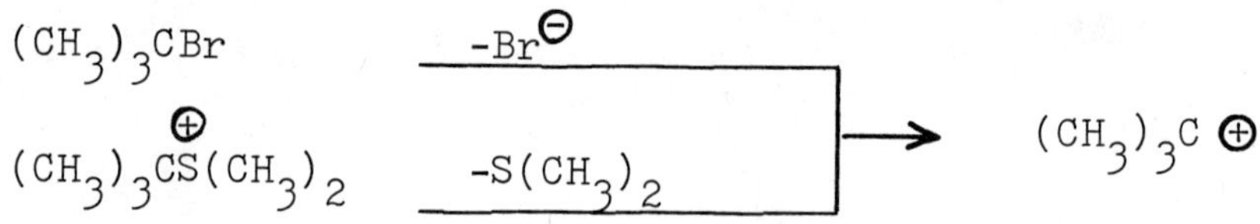

<u>elimination</u>:

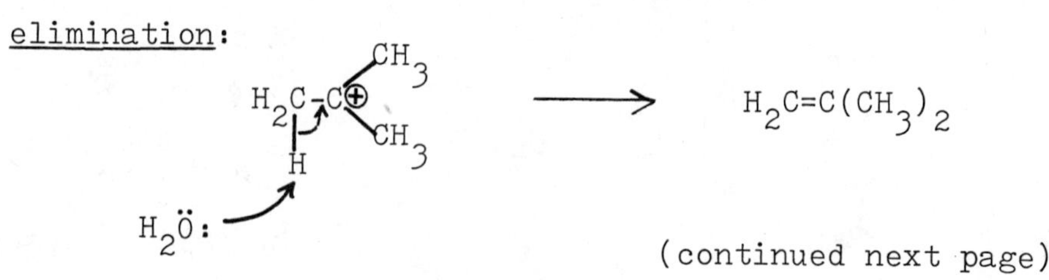

(continued next page)

(b) Substitution products expected are $(CH_3)_3COH$ and $(CH_3)_3COC_2H_5$.

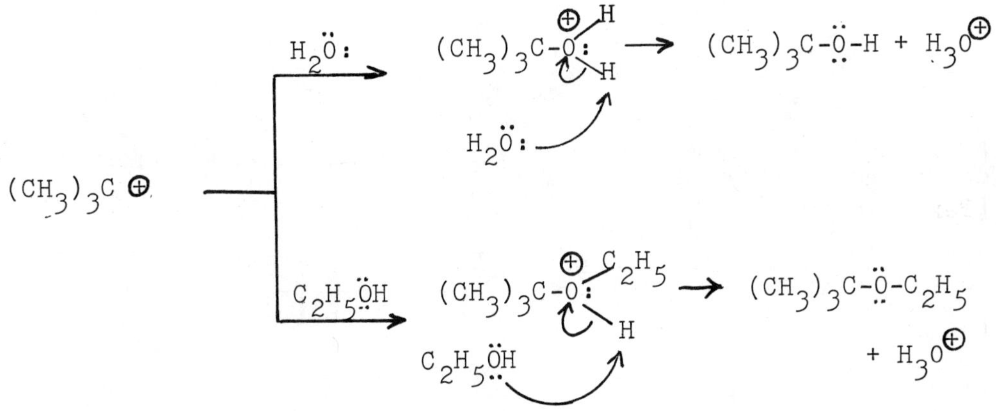

$(CH_3)_3C^{\oplus}$ —$H_2\ddot{O}:$→ $(CH_3)_3C-\overset{\oplus}{O}(H)_2$ → $(CH_3)_3C-\ddot{O}-H + H_3O^{\oplus}$ ($H_2\ddot{O}:$ removes H)

$(CH_3)_3C^{\oplus}$ —$C_2H_5\ddot{O}H$→ $(CH_3)_3C-\overset{\oplus}{O}(C_2H_5)H$ → $(CH_3)_3C-\ddot{O}-C_2H_5 + H_3O^{\oplus}$ ($C_2H_5\ddot{O}H$ removes H)

(Text: S_N1, S. 5.4; E1, S. 6.4)

29. (a) E1. Since the leaving group is relatively large, an E2 process should give a higher percentage of Hofmann elimination.

(b) To test the mechanism, use $(CH_3)_2C(OI)CD_2CH_3$. If the mechanism is E2, it will show a kinetic isotope effect since deuterium must be removed in the rate-determining step. However, in an E1 mechanism, deuterium is removed after the rate-determining step; therefore, no isotope effect will be observed.

30. These reactions are probably E1cb; however, the reactions follow the E2 geometry (anti-elimination).

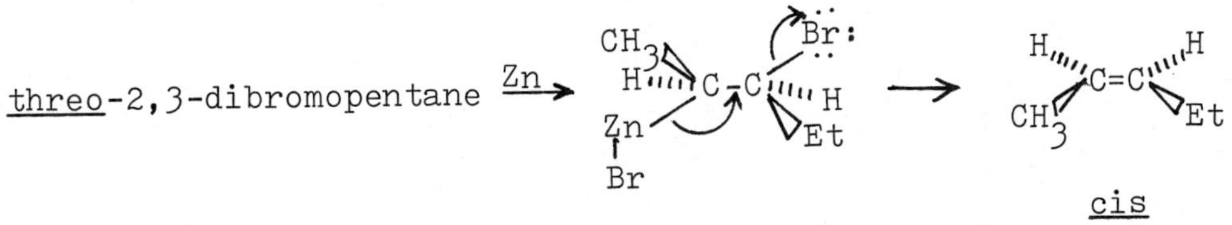

threo-2,3-dibromopentane $\xrightarrow{Zn}$ [CH_3, H, Zn(Br) on C; Br, H, Et on C] → [H, CH_3 C=C H, Et] cis erythro-2,3-dibromopentane $\xrightarrow{Zn}$ CH_3, H, Zn–Br, C–C, :Br:, Et, H $\longrightarrow$ H, Et, C=C, CH_3, H

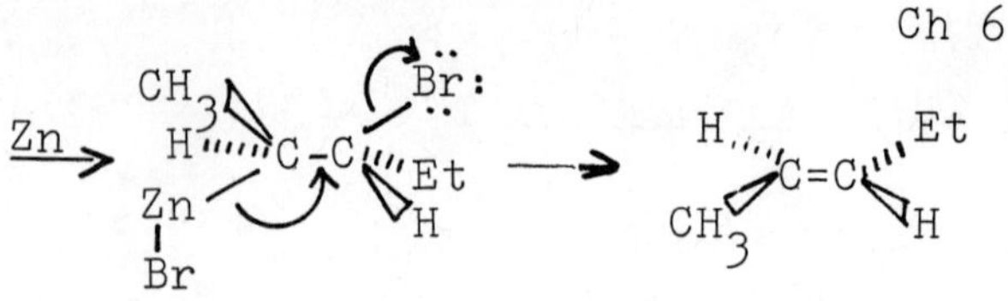

trans

(Text S. 6.5, also see problems 7 and 15).

31. A carbanion (B) may be formed allowing for incorporation of deuterium (B→C); however, the elimination could involve A (or C) in an E2 process.

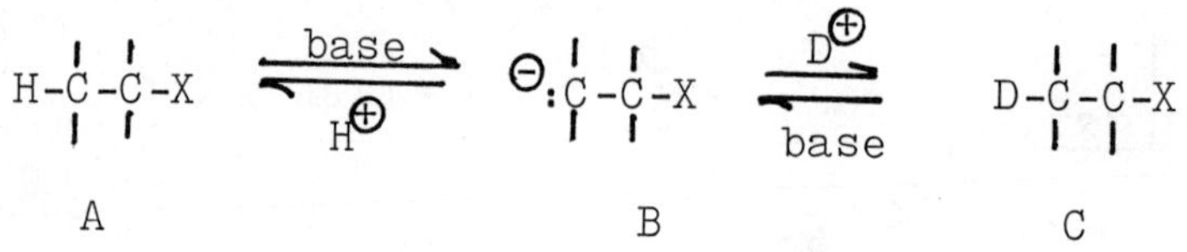

$$\underset{A}{H\text{-}\overset{|}{\underset{|}{C}}\text{-}\overset{|}{\underset{|}{C}}\text{-}X} \underset{H^{\oplus}}{\overset{base}{\rightleftharpoons}} \underset{B}{\ominus:\overset{|}{\underset{|}{C}}\text{-}\overset{|}{\underset{|}{C}}\text{-}X} \underset{base}{\overset{D^{\oplus}}{\rightleftharpoons}} \underset{C}{D\text{-}\overset{|}{\underset{|}{C}}\text{-}\overset{|}{\underset{|}{C}}\text{-}X}$$

32. $(CH_3)_3CH_2\text{-}MgBr + :\ddot{I}\text{-}\ddot{I}: \xrightarrow{S_N2} (CH_3)_3CH_2\ddot{I}: + MgBr\ddot{I}:$

(Text S. 6.7)

33. (a) 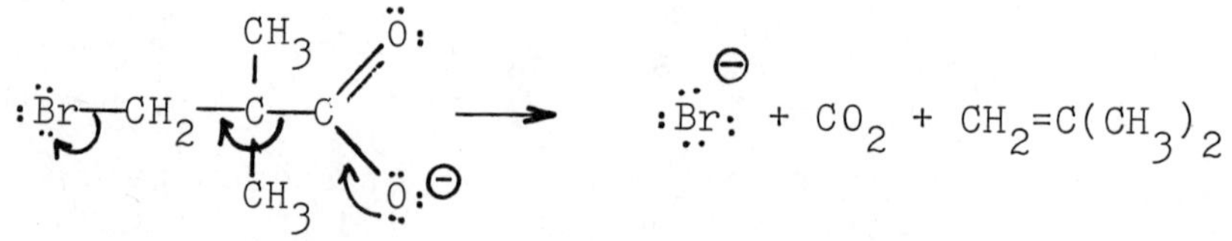 $:\ddot{Br}\text{-}CH_2\text{-}C(CH_3)_2\text{-}C(=\ddot{O}:)\text{-}\ddot{O}:^{\ominus} \longrightarrow :\ddot{Br}:^{\ominus} + CO_2 + CH_2\text{=}C(CH_3)_2$

(b) 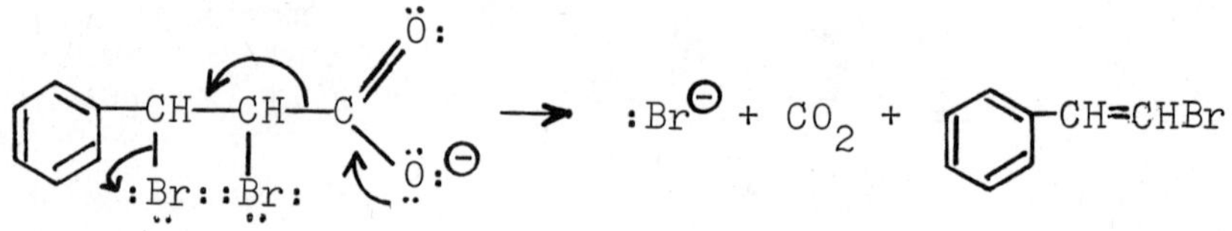 $C_6H_5\text{-}CH(:\ddot{Br}:)\text{-}CH(:\ddot{Br}:)\text{-}C(=\ddot{O}:)\text{-}\ddot{O}:^{\ominus} \longrightarrow :Br^{\ominus} + CO_2 + C_6H_5\text{-}CH\text{=}CHBr$

34. (a) Route 1 yields the ether and route 2 the alkene.

(b) The ether is formed by an S_N2 reaction requiring attack of the alkoxide, $RO^{\ominus}$, at the carbon of the C-L bond. In route 2, the substrate is 3° and, because of steric hindrance, an S_N2 reaction is not possible.

Substrate reactivity in S_N2: $1^{o} > 2^{o} > 3^{o}$

35. (a) $CH_3-\overset{O}{\overset{\|}{C}}-CH_3$ (from $CH_3-\overset{:\ddot{O}:^{\ominus}}{\overset{|}{C}}(-OCH_3)(CH_3)$ $\longrightarrow$ $CH_3\overset{:O:}{\overset{\|}{C}}CH_3 + CH_3O^{\ominus}$)

(b) $CH_3\overset{O}{\overset{\|}{C}}CH_3$

(c) $CH_3\overset{O}{\overset{\|}{C}}OCH_3 + CH_3\overset{O}{\overset{\|}{C}}OCH_2CH_3$

(d) $CH_3\overset{O}{\overset{\|}{C}}OCH_3$ (-Cl better leaving group)

(e) $CH_3\overset{O}{\overset{\|}{C}}-NH_2$ (-Cl better leaving group)

(f) $CH_3\overset{O}{\overset{\|}{C}}-NH_2$ ($-O\overset{O}{\overset{\|}{C}}CH_3$ better leaving group)

(g) $CH_3\overset{O}{\overset{\|}{C}}CH_3$ ($-O\overset{O}{\overset{\|}{C}}CH_3$ better leaving group)

(b) These reactions are most similar to the Elcb process.

36. (a) acidic solution:

$$CH_3-CH_2-\underset{O-H}{\overset{:\ddot{O}-H}{C}}-CH_2-CH_3 \xrightarrow{H^{\oplus}} CH_3-CH_2-\underset{O-H}{\overset{H\overset{\oplus}{O}H}{C}}-CH_2-CH_3 \searrow$$

$$H_3O^{\oplus} + CH_3-CH_2-\underset{O}{\underset{\|}{C}}-CH_2-CH_3 \longleftarrow CH_3-CH_2-\underset{OH}{\overset{\oplus}{C}}-CH_2-CH_3 + H_2O \quad (:\ddot{O}H_2)$$

(continued next page)

36. (b) basic solution:

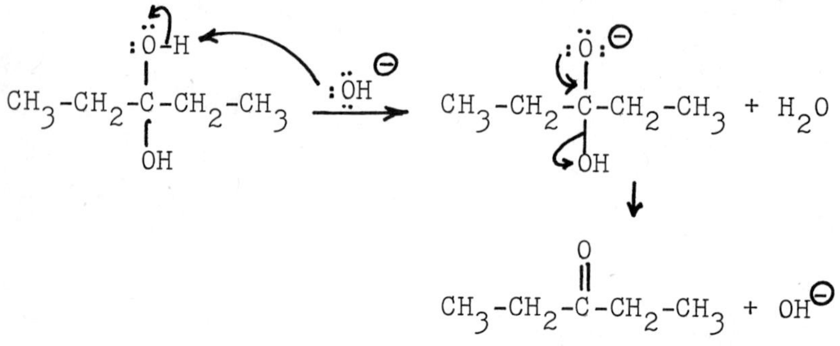

$$CH_3\text{-}CH_2\text{-}\underset{OH}{\overset{:\ddot{O}\text{-}H}{C}}\text{-}CH_2\text{-}CH_3 \xrightarrow{:\ddot{O}H^{\ominus}} CH_3\text{-}CH_2\text{-}\underset{OH}{\overset{:\ddot{O}:^{\ominus}}{C}}\text{-}CH_2\text{-}CH_3 + H_2O$$

$$\downarrow$$

$$CH_3\text{-}CH_2\text{-}\overset{O}{\overset{\|}{C}}\text{-}CH_2\text{-}CH_3 + OH^{\ominus}$$

37. These statements are correct. For example:

(a) enantiotopic hydrogens a and b give the same product.

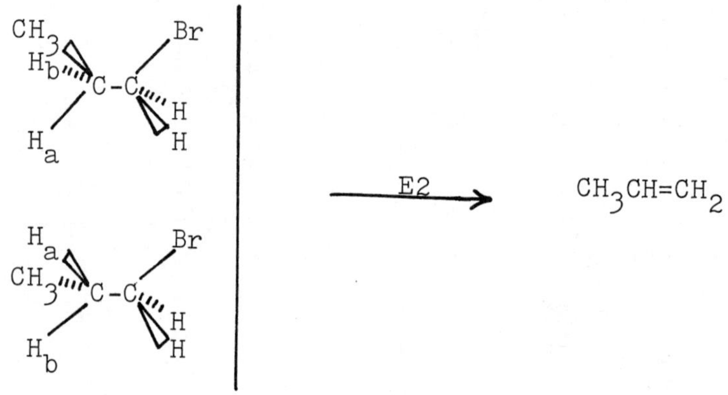

CH_3, H_b, H_a C-C Br, H, H

H_a, CH_3, H_b C-C Br, H, H

$\xrightarrow{E2}$ $CH_3CH{=}CH_2$

(b) diastereotopic hydrogens d and e give diastereomers.

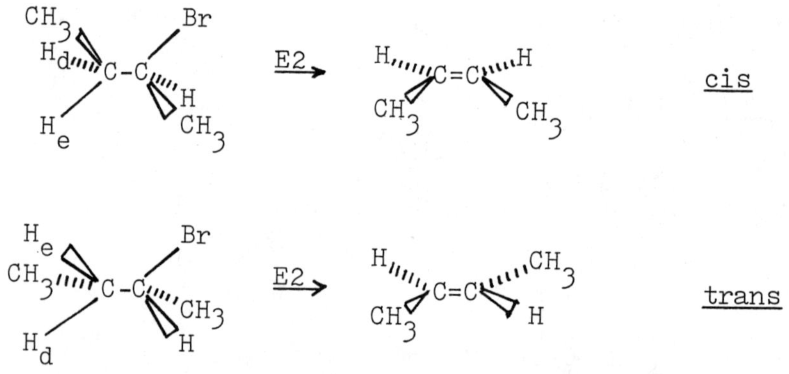

CH_3, H_d, H_e C-C Br, H, CH_3 $\xrightarrow{E2}$ H, CH_3 C=C H, CH_3 cis

H_e, CH_3, H_d C-C Br, CH_3, H $\xrightarrow{E2}$ H, CH_3 C=C CH_3, H trans

(Text S. 6.3, review the answer to question 4)

SUMMARY OF REACTIONS

A. Elimination Reactions

1. Dehydrohalogenation (major alkene predicted by Saytzeff's rule)

general: $-\underset{H}{\overset{|}{C}}-\overset{L}{\underset{|}{C}}- \; + \; :B^{\ominus} \longrightarrow \; >C=C< \; + \; HB \; + \; :L^{\ominus}$

$-L$ is $-X$; $:B^{\ominus}$ is $OH^{\ominus}$ (KOH/alcohol), $NH_2^{\ominus}$ ($Na^{\oplus}NH_2^{\ominus}$ /$NH_{3(\ell)}$), $OR^{\ominus}$ ($Na^{\oplus}OR^{\ominus}$ /ROH), etc.

example: $CH_3CHBrCH_2CH_3 \xrightarrow[\text{heat}]{\text{KOH/alcohol}} CH_3CH{=}CHCH_3 + H_2O + KBr$

major (trans > cis)

2. Hofmann Elimination (major alkene predicted by Hofmann's rule)

general: $-\underset{H}{\overset{|}{C}}-\overset{L^{\oplus}}{\underset{|}{C}}- \; + \; :B^{\ominus} \longrightarrow \; >C=C< \; + \; HB \; + \; :L$

$-L^{\oplus}$ is $-\overset{\oplus}{N}(CH_3)_3$ or $-\overset{\oplus}{S}(CH_3)_2$; $:B^{\ominus}$ usually $OH^{\ominus}$

example: $CH_3\underset{\oplus N(CH_3)_3}{CH}CH_2CH_3 \;\; OH^{\ominus} \xrightarrow{\text{heat}} CH_2{=}CHCH_2CH_3 + H_2O + (CH_3)_3N:$

major

3. Dehydration of Alcohols (major alkene predicted by Saytzeff's rule)

general: $-\underset{H}{\overset{|}{C}}-\overset{OH}{\underset{|}{C}}- \; + \; H_2SO_4 \text{ (or } H_3PO_4) \xrightarrow{\text{heat}} \; >C=C< \; + \; H_2O$

example: $CH_3\underset{OH}{CH}CH_2CH_3 \xrightarrow[\text{heat}]{H_2SO_4} CH_3CH{=}CHCH_3 + H_2O$

major

(trans > cis)

B. Synthesis of Organometallic Compounds

1. Grignard Reagents

general: $RX + Mg \xrightarrow{ether} RMgX$

example: $CH_3CH_2Br + Mg \xrightarrow{ether} CH_3CH_2MgBr$

2. Organolithium Compounds

general: $R\text{-}X + 2Li \longrightarrow RLi + LiX$

3. Organometallic Compounds Prepared from Grignard Reagents

general: (a) $2RMgX + MX_2 \longrightarrow R_2M + 2MgX_2$

(b) $4RMgX + MX_4 \longrightarrow R_4M + 4MgX_2$

example: (a) $2CH_3MgCl + HgCl_2 \longrightarrow (CH_3)_2Hg + 2MgCl_2$

(b) $4CH_3MgCl + SnCl_4 \longrightarrow (CH_3)_4Sn + 4MgCl_2$

4. Lithium Dialkylcuprates from Organolithium Compounds

general: $2RLi + CuX \longrightarrow Li\left[R_2Cu\right] + LiX$

example: $2CH_3CH_2Li + CuCl \longrightarrow Li\left[(CH_3CH_2)_2Cu\right] + LiCl$

C. Grignard Reagents as Bases

general: $R\text{-}MgX + H\text{-}Z \longrightarrow R\text{-}H + MgXZ$

(H-Z: H bonded to an electronegative atom such as O, X, N, S)

examples: $CH_3MgI + HOH \longrightarrow CH_4 + MgIOH$

$CH_3CH_2MgCl + NH_3 \longrightarrow CH_3CH_3 + MgClNH_2$

REACTION REVIEW

A. Reactions, Chapter 6

Questions	Answers
1. $-\overset{\vert}{\underset{H}{C}}-\overset{L}{\underset{\vert}{C}}-$ + :$B^{\ominus}$ $\longrightarrow$?	1. $>C=C<$ + HB + :$L^{\ominus}$
Write the structure for the major elimination products in the following:	
(a) $(CH_3)_2CClCH_2CH_3$ $\xrightarrow[\text{heat}]{\text{KOH/alcohol}}$?	(a) $(CH_3)_2C=CHCH_3$
(b) $CH_3CH_2CHBrCH_3$ $\xrightarrow[NH_{3(\ell)}]{NaNH_2}$?	(b) $CH_3CH=CHCH_3$ (<u>trans</u> > <u>cis</u>)
(c) $C_6H_5-CH_2CHCl-C_6H_5$ $\xrightarrow[\text{heat}]{\text{KOH/alcohol}}$?	(c) $C_6H_5-CH=CH-C_6H_5$ (<u>trans</u> > <u>cis</u>)
(d) $C_6H_5-CH(CH_3)CHCl-C_6H_5$ $\xrightarrow[\text{heat}]{\text{KOH/alcohol}}$? (<u>threo</u>)	(d) $(C_6H_5)(CH_3)C=CH(C_6H_5)$ (<u>trans</u> only)
2. $-\overset{\vert}{\underset{H}{C}}-\overset{L^{\oplus}}{\underset{\vert}{C}}-$:$B^{\ominus}$ $\xrightarrow{\text{heat}}$?	2. $>C=C<$ + HB + :L
Write the structure for the <u>major</u> elimination product in the following:	
(a) $(CH_3)_2CHCH(\overset{\oplus}{N}(CH_3)_3)CH_3$ $OH^{\ominus}$ $\xrightarrow{\text{heat}}$?	(a) $(CH_3)_2CHCH=CH_2$
3. $-\overset{\vert}{\underset{H}{C}}-\overset{OH}{\underset{\vert}{C}}-$ $\xrightarrow[\text{heat}]{H_2SO_4 \text{ (or } H_3PO_4)}$?	3. $>C=C<$ + H_2O
	(continued next page)

Write the structure for the major elimination product when

$$\begin{array}{cc} & OH \\ | & | \\ -C- & C- \\ | & | \\ H & \end{array}$$

is:

(a) $CH_3CH_2CH_2CH_2OH$

(b) $CH_3CHCH_2CH_3$ (OH on C-2)

(c) $(CH_3)_3CCHCH_3$ (OH on CH)

Answers:

(a) $CH_3CH_2CH{=}CH_2$

(b) $CH_3CH{=}CHCH_3$
(trans > cis)

(c) $(CH_3)_2C{=}C(CH_3)_2$
(rearrangement via E1)

---

4. $R\text{-}X \xrightarrow[\text{ether}]{Mg} ?$

Write the structure of the product when RX is:

(a) CH_3CH_2Br

(b) $CH_3CHCH_2CH_3$ (Cl on CH)

Answers:

4. $RMgX$

(a) CH_3CH_2MgBr

(b) $CH_3CHCH_2CH_3$ (MgCl on CH)

---

5. $R\text{-}X \xrightarrow[\text{ether}]{Li} ?$

Write the structure of the product when RX is:

(a) CH_3Cl

(b) $(CH_3)_3CBr$

Answers:

5. $RLi + LiX$

(a) CH_3Li

(b) $(CH_3)_3CLi$

---

6. (a) $RMgX + MX_2 \longrightarrow ?$

(b) $RMgX + MX_4 \longrightarrow ?$

Write the structure of the product in each of the following:

(c) $CH_3CH_2MgCl \xrightarrow{HgCl_2} ?$

Answers:

6. (a) $R_2M + MgX_2$

(b) $R_4M + MgX_2$

(c) $(CH_3CH_2)_2Hg$

(continued next page)

Question	Answer
(d) $(CH_3)_2CHMgCl \xrightarrow{SnCl_4}$?	(d) $[(CH_3)_2CH]_4Sn$
7. $RMgX + HZ \longrightarrow$? (Z=O, X, N, etc.) Write the structures of the products in each of the following: (a) $CH_3MgI \xrightarrow{H_2O}$? (b) $CH_3CH_2MgCl \xrightarrow{D_2O}$? (c) $CH_3MgI \xrightarrow{C_2H_5OH}$? (d) $CH_3CH_2MgBr \xrightarrow{NH_3}$?	7. $RH + MgXZ$ (a) $CH_4 + MgIOH$ (b) $CH_3CH_2D + MgClOD$ (c) $CH_4 + MgIOC_2H_5$ (d) $CH_3CH_3 + MgBrNH_2$

B. Write structures for all letters.

Question	Answer
1. $CH_3CH_2CH_3 \xrightarrow[h\nu]{Cl_2} A \xrightarrow{KOH/alcohol} B$	1. A: $CH_3CHClCH_3 + CH_3CH_2CH_2Cl$ B: $CH_3CH{=}CH_2$
2. $CH_3CH_2Cl \xrightarrow{Li[(CH_3CH_2)_2Cu]} A$ $\xrightarrow[h\nu]{Br_2} B \xrightarrow[ether]{Mg} C \xrightarrow{D_2O} D$	2. A: $CH_3CH_2CH_2CH_3$ B: $CH_3CH_2CHBrCH_3$ (major) C: $CH_3CH_2CH(MgBr)CH_3$ D: $CH_3CH_2CHDCH_3$
3. $(CH_3)_2CHCH_2Cl \xrightarrow{OH^{\ominus}/H_2O} A$ $\xrightarrow[heat]{H_2SO_4} B$	3. A: $(CH_3)_2CHCH_2OH$ B: $(CH_3)_2C{=}CH_2$
4. $CH_3CH_2Br \xrightarrow{Li} A \xrightarrow{CuCl} B$	4. A: CH_3CH_2Li B: $Li[(CH_3CH_2)_2Cu]$

CH_3CH_2Cl → C $\xrightarrow[h\nu]{Br_2}$ D $\xrightarrow[NH_{3(\ell)}]{NaNH_2}$ E	C: $CH_3CH_2CH_2CH_3$ D: $CH_3CHBrCH_2CH_3$ (major) E: $CH_3CH=CHCH_3$ (trans > cis)
5. $(CH_3)_3COH \xrightarrow{HBr}$ A $\xrightarrow[ether]{Mg}$ B $\xrightarrow{NH_3}$ C	5. A: $(CH_3)_3CBr$ B: $(CH_3)_3CMgBr$ C: $(CH_3)_3CH$

CHAPTER 7, THE STEREOCHEMISTRY OF RING SYSTEMS

LEARNING OBJECTIVES

When you have completed this chapter, you should be able to:

1. draw and name cyclic compounds (problems 1, 7);
2. classify cyclic structures by the use of the terms: identical, _meso_, _d,l_ forms, enantiomers, diastereomers, _cis_, _trans_ and optically active (problems 2, 6, 13, 20, 28, 29);
3. draw structures for the stereoisomers formed in the free-radical halogenation of cyclic compounds (problems 13, 28, 31, 32, 50);
4. establish the _cis_ and _trans_ relationship of groups in cyclic systems (problems 5, 26);
5. draw _cis_ and _trans_ structures by use of the dot convention (problem 14);
6. draw conformations of cyclohexane and substituted cyclohexanes (problem 17 and related problems 16 and 38);
7. identify chiral centers in cyclic structures (problems 15, 18, 30);
8. predict and account for the stability of cyclic compounds (problems 3, 19, 34, 35, 37, 49);
9. predict the products of nucleophilic substitution reactions (problems 21, 51);
10. predict the products of elimination reactions (problems 22, 23, 24, 25, 51);
11. write equations which illustrate the synthesis of cyclic compounds by use of reactions from previous chapters (problem 39);
12. account for and predict conformational preferences of

groups in substituted cyclohexanes (problems 42, 43, 52 and related problem 44);

13. account for the reactivity of cyclic compounds in substitution and elimination reactions (problems 40, 45, 46, 47, 48);
14. explain the properties of heterocyclic compounds which are associated with the presence of the heteroatom(s) (problems 8, 9, 10, 11, 13, 36);
15. correlate the acidity of C-H bonds with different hybridizations of carbon (problem 12).

ANSWERS TO QUESTIONS

1. (a)

```
 H H
 | |
H-C-C-H
 | |
H-C-C-H
 | |
 H H
```

monocyclic

(b)

```
    H   H   H
     \ /    |
 H    C   H C     H
 |  /  \  |/|\   /
H-C      C  H  C-H
  |      |      \
H-C------C-H     H
  |      |
  H      H
```

monocyclic

(c)

```
  H H H H H
  | | | | |
H-C-C-C-C-C-H
  | | | | |
  H H H H H
```

acyclic

(d)

```
      H   H
       \ /
   H    C    H
   |  /  \   |
 H-C       C
   |       |  O
 H-C       C
   |  \  /  |
   H    C    H
       / \
      H   H
```

heterocyclic, bicyclic

(e)

```
      H   H     H   H
       \ /       \ /
   H    C    H    C
   |  /  \   |   /  \
 H-C       C-C-H     H
   |       |   \    H
 H-C       C-H  C
   |  \  / |    | \
   H    C  H    H  H
       / \
      H   H
```

monocyclic

(f)

```
      H   H H   H
       \ /   \ /
   H    C     C
   |  /  \   /  \      H
 H-C       C------C
   |       |        \
 H-C-------C-H       H
   |       |
   H       H
```

spiro ring system

(continued next page)

(g)

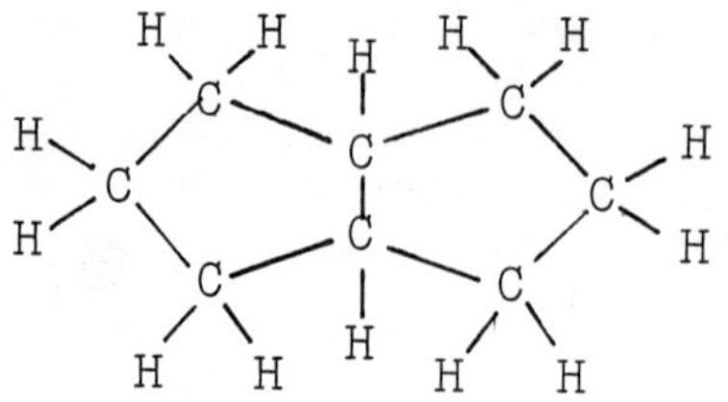

bicyclic

(j)

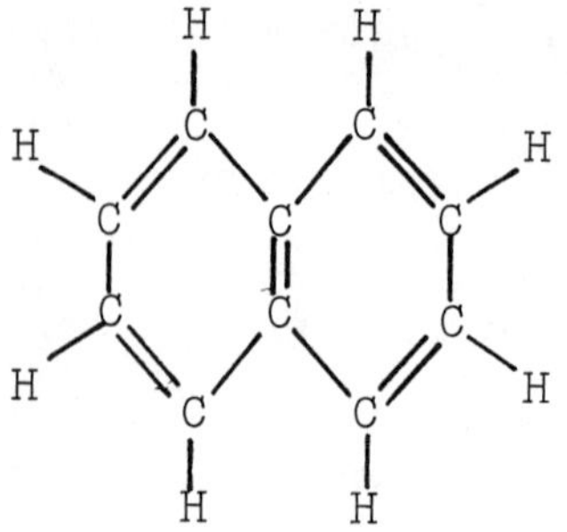

bicyclic

(h)

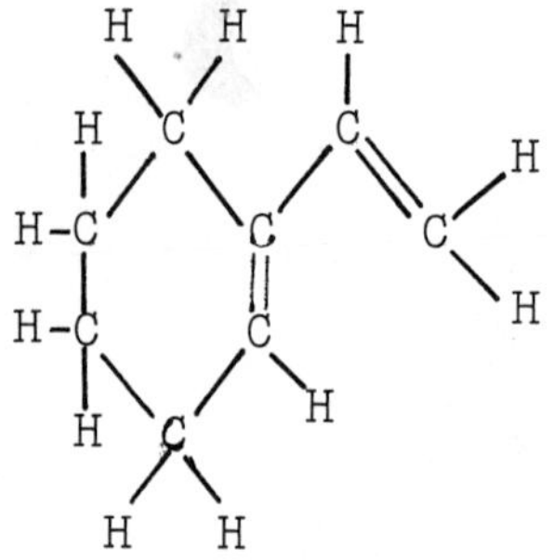

monocyclic, alkene

(k)

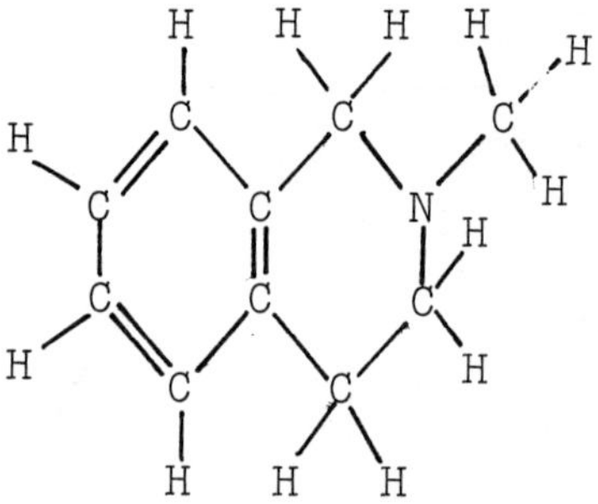

bicyclic, heterocyclic

(i)

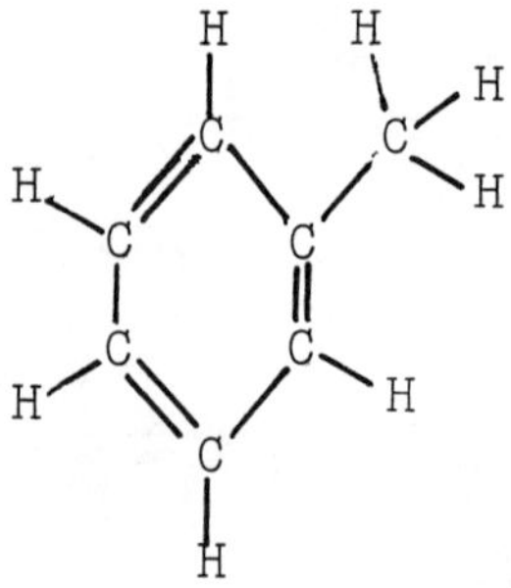

monocyclic

2. (i) (a) member of d,l pair   (c) member of d,l pair

(b) meso   (d) member of d,l pair

(ii) these structures do not have chiral centers.

(iii) threo and erythro designations

3. There are 8 pairs of eclipsed hydrogens in planar cyclobutane. Puckered cyclobutane has no eclipsed hydrogens (build a model).

4. (a)  (b) 

The page is a plane of symmetry for (a) and (b). In addition, (b) has a plane of symmetry perpendicular to the page.

(c)  (d) 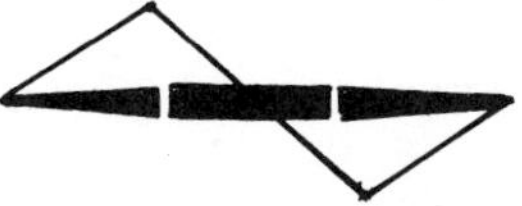

no plane of symmetry   no plane of symmetry

5. (a) 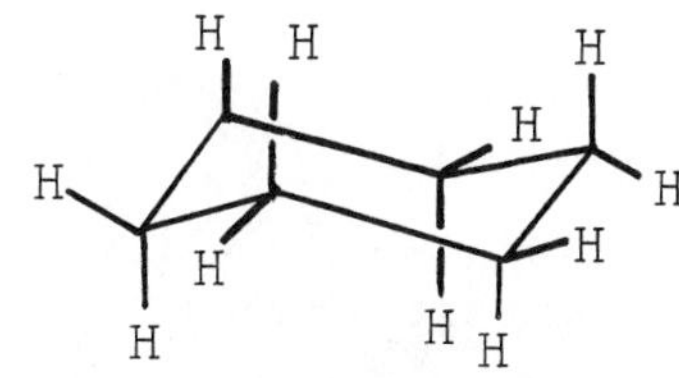

(b) Only one axial chlorocyclohexane is possible. All structures below are the same (superimposable).

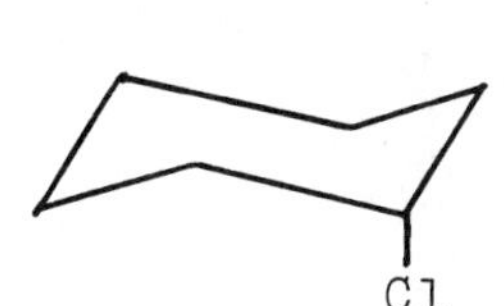

≡ Cl ≡
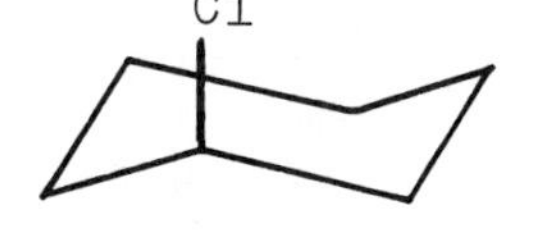

etc.

Build a model of axial chlorocyclohexane and move the Cl from one axial position to another. Note that each axial chlorocyclohexane has a plane of symmetry.

(c) Only one equatorial chlorocyclohexane is possible. Build a model and move the Cl from one equatorial position to another. Note that each equatorial chlorocyclohexane has a plane of symmetry.

(d) 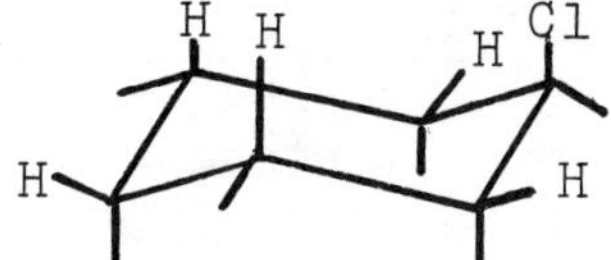

the hydrogens shown are all pointing up (on the same side as the chlorine) and are cis to the chlorine.

(e) The hydrogens pointing in the direction opposite that of the axial chlorine are trans to the chlorine; e.g.,

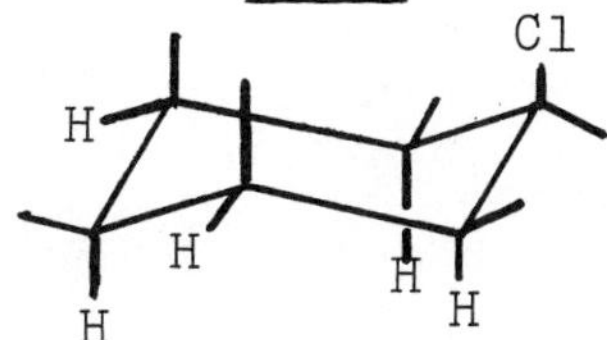

(continued next page)

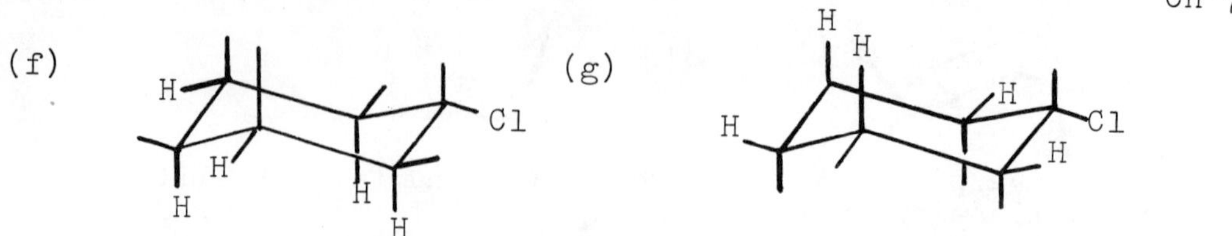

6. All pairs are identical (superimposable) except the pair in j.

7. (a) 2-methyltetrahydrofuran
(2-methyloxacyclopentane)

(b) 2,2-dimethyltetrahydrofuran
(2,2-dimethyloxacyclopentane)

(c) 2,3,3-trimethyltetrahydropyran
(2,3,3-trimethyloxacyclohexane)

(d) 3-ethyl-3-methylmorpholine
(3-ethyl-3-methyl-1-oxa-4-azacyclohexane)

(e) 4,4-dichloro-3-isopropyl-3-methylpiperidine
(4,4-dichloro-3-isopropyl-3-methylazacyclohexane)

8. The interconversion can be accounted for by inversion at N (configurational instability) which leads to:

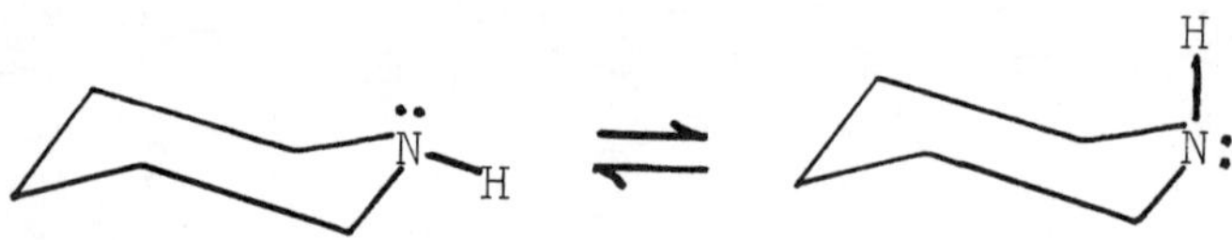

9. Three dioxanes are possible - the 1,2-, 1,3- and 1,4-dioxane. Because of repulsion of nonbonding electron pairs on adjacent oxygens, the 1,2-dioxane should be much less stable (the O-O bond energy is 35 kcal/mole while the C-O bond energy is 85 kcal/mole. See Table 2-3).

10. Solubility in water: 1,4-dioxane > tetrahydropyran > thiane

1,4-dioxane has two oxygen atoms which may participate in hydrogen bonding with water. Tetrahydropyran hydrogen bonds with water more effectively than thiane since it possesses the more electronegative atom.

11. Piperidine is a relatively good proton donor and reacts with Grignard reagents as shown below.

12. Acidity may be related to the stability of the conjugate base (a strong acid has a weak conjugate base). The electronegativity of an orbital increases as its s character increases. Thus the order of stability observed for the conjugate base is:

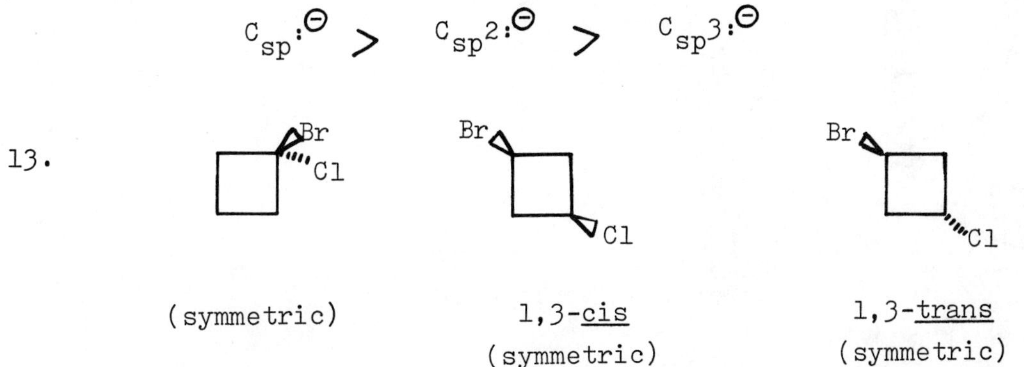

Capable of existing as enantiomers are:

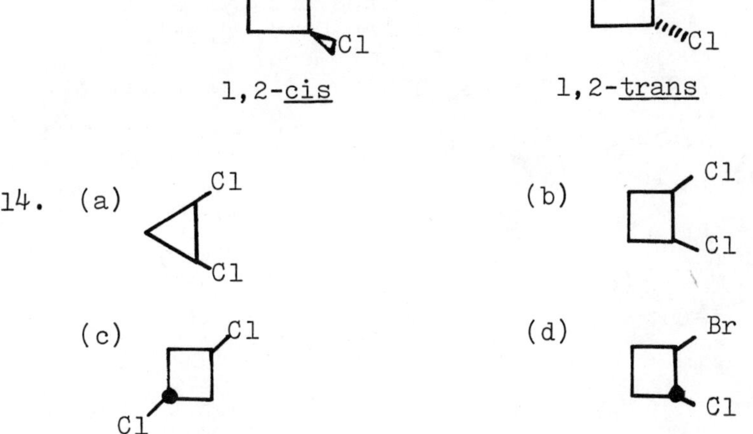

15. (a) All except 1,1-dimethylcyclopentane

(b) and (c)

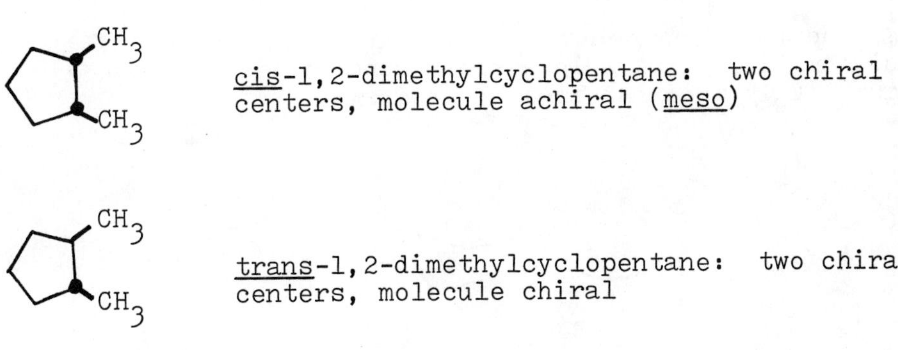

<u>cis</u>-1,2-dimethylcyclopentane: two chiral centers, molecule achiral (<u>meso</u>)

<u>trans</u>-1,2-dimethylcyclopentane: two chiral centers, molecule chiral

(continued next page)

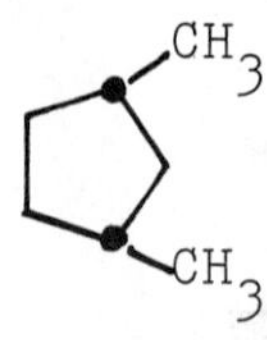

cis-1,3-dimethylcyclopentane: two chiral centers, molecule achiral (meso)

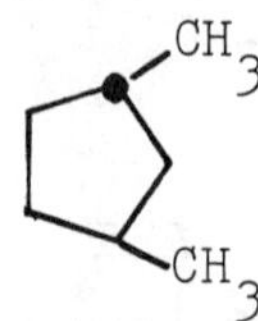

trans-1,3-dimethylcyclopentane: two chiral centers, molecule chiral.

16. (a) trans-1,3-Di-t-butylcyclohexane

(b) trans-1,3-Di-t-butylcyclohexane must have one t-butyl group in an axial position if in a chair conformation. (Thermodynamic studies have shown it to exist in a twist-boat conformation with both t-butyl groups occupying equatorial-like positions.) On the other hand, cis-1,3-di-t-butylcyclohexane can exist in a chair conformation with both t-butyl groups occupying equatorial positions.

17. Compounds (a), (c) and (d) can all exist in chair conformations with t-butyl groups occupying equatorial positions.

(a)

(c)

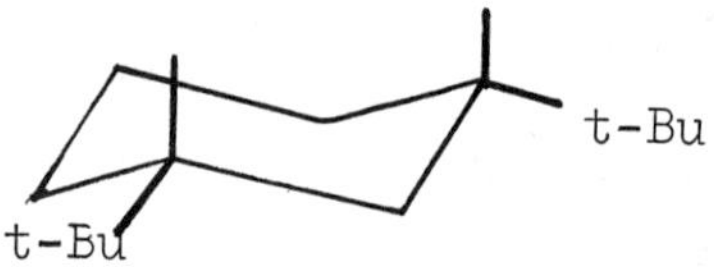

(d)

Compound (b), however, must have one t-butyl group in an axial position if it exists in a chair conformation. Thus, (b) is least likely to exist in a chair form.

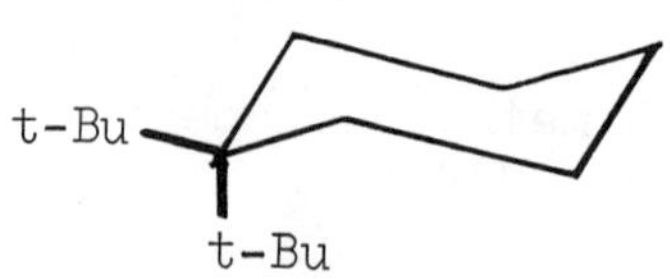

18. (a)

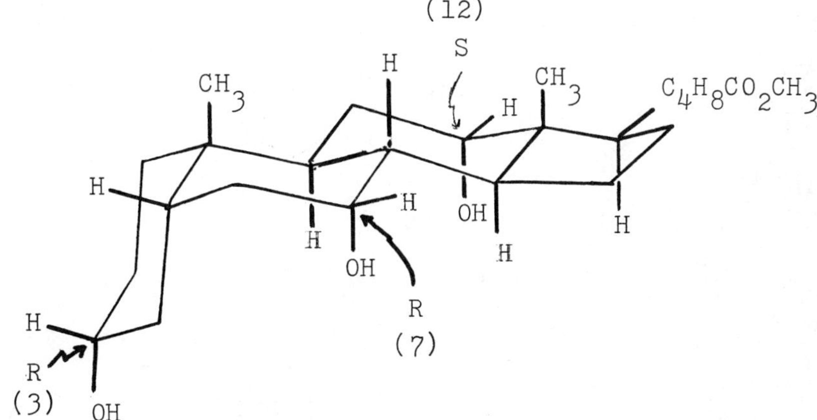

(b) methyl cholate is 5-β

19. The 5α-steroid is more stable. The 5β-steroid below is destabilized by skew-butane and axial group interactions because of the A/B <u>cis</u> fusion.

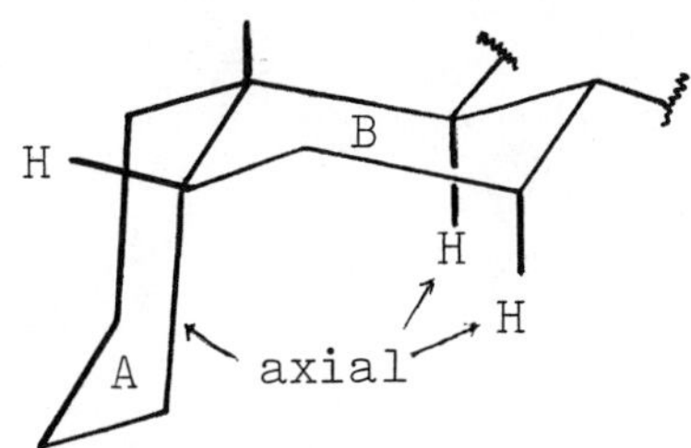

20. (a) "<u>Exo</u>" and "<u>endo</u>" refer to the orientation of a group relative to the bridge (smaller ring). In this example all rings contain the same number of carbons.

(b) A and B can be compared by the R and S designation: A is S, B is R.

(c) A and C are enantiomers.

(d) A and D are enantiomers.

21. (a)

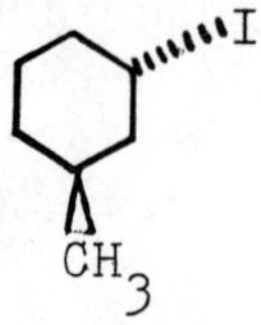

CH_3, I

trans

(b)

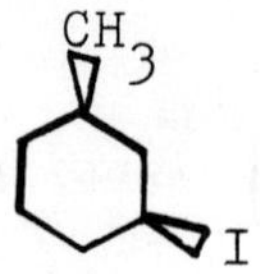

CH_3, I

cis

(c)

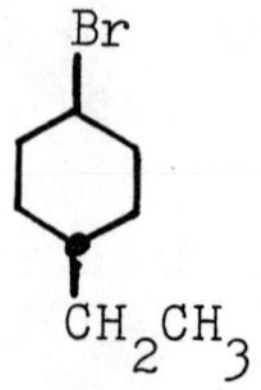

Br, CH_2CH_3

trans

(d)

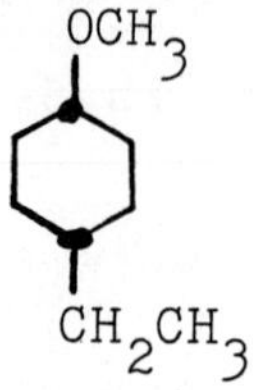

OCH_3, CH_2CH_3 + CH_3I

cis

(e)

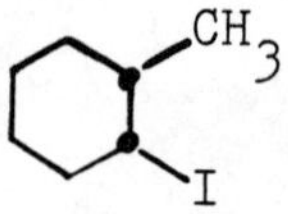

CH_3, I

cis

22. Recall that the stereoelectronic requirement for an E2 process is that a hydrogen and the leaving group on adjacent carbons be <u>anti</u> (in a cyclohexane this means the groups are <u>trans</u> and <u>axial</u>).

Neomenthyl chloride exists mainly in conformation A which has the necessary stereochemistry (either H_A or H_B may be removed along with the axial Cl). Conformation B is less stable since the larger isopropyl group must occupy an axial position.

The more stable conformation of menthyl chloride (C) does not have the Cl in an axial position.

neomenthyl chloride: Cl $(CH_3)_2CH$ CH_3 H_a A H_b ⇄ $CH(CH_3)_2$ Cl CH_3 B

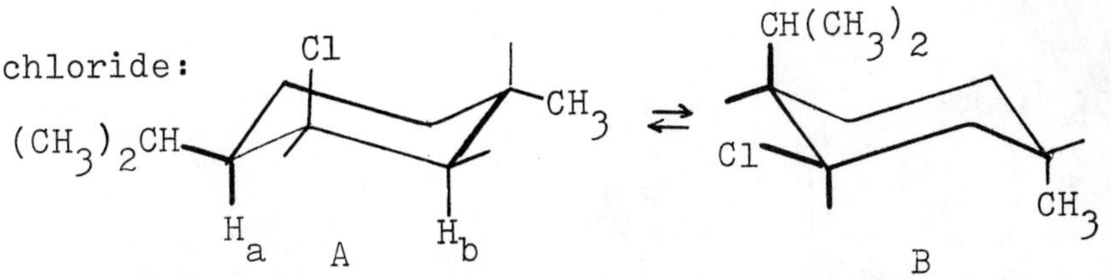

menthyl chloride: H $(CH_3)_2CH$ Cl CH_3 H C ⇄ $CH(CH_3)_2$ H Cl D CH_3

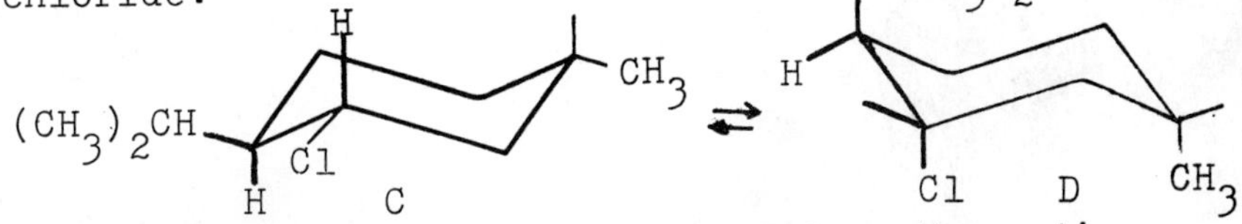

23. The reaction is E1. Both alkenes are formed from the cation:

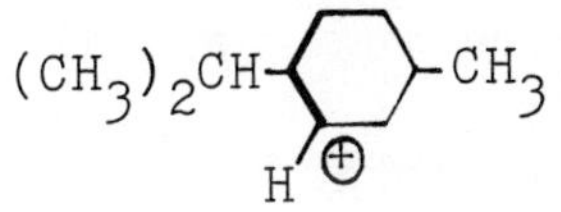

$(CH_3)_2CH$–⬡–CH_3 H ⊕

The major product, 3-methene, is the more stable alkene (the Saytzeff product).

24. Although 1-phenylcyclopentene and 3-phenylcyclopentene are possible products, 1-phenylcyclopentene is more stable because the C=C of the alkene is conjugated with the benzene ring.

The cis-isomer is more reactive because it possesses the necessary stereoelectronic requirement for an E2 reaction. (see below)

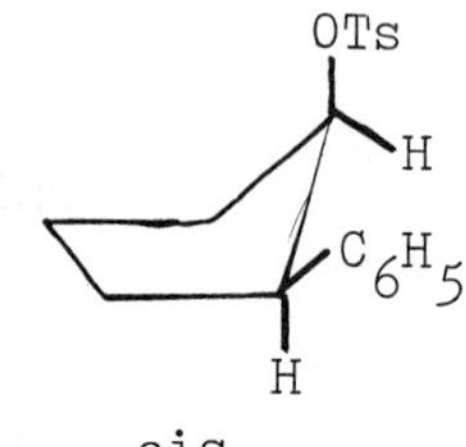

cis

25. (a)

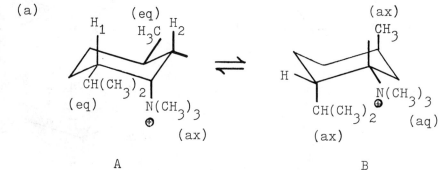

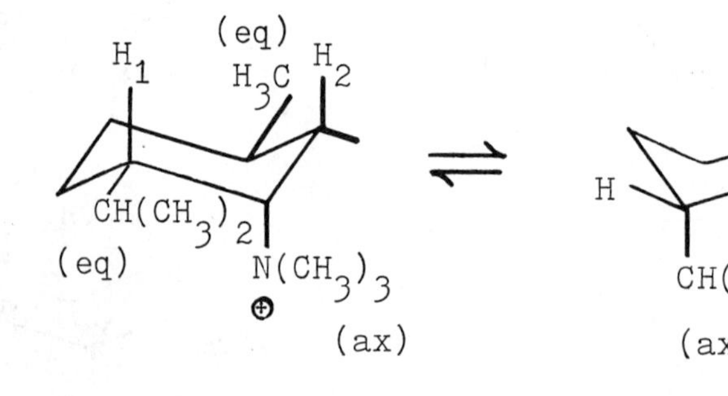

(b) Conformation A, with the leaving group axial, is involved in the elimination.

(c) The loss of H_1 and $:N(CH_3)_3$ leads to 3-menthene.

(d) The loss of H_2 and $:N(CH_3)_3$ leads to 2-menthene.

26. cis to H_3: H_2, H_5, H_7, H_9

trans to H_3: H_1, H_6, H_8, H_{10}

(Text: meaning of cis and trans, S. 7.3)

27. (a) identical (b) diastereotopic (c) enantiotopic

(d) diastereotopic (e) enantiotopic (f) diastereotopic

(g) diastereotopic (h) enantiotopic

(Text: definitions of enantiotopic and diastereotopic, S. 4.5, 4.6)

28. (a)

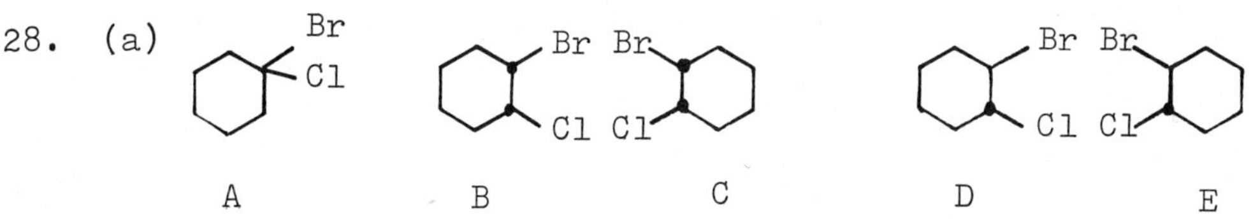

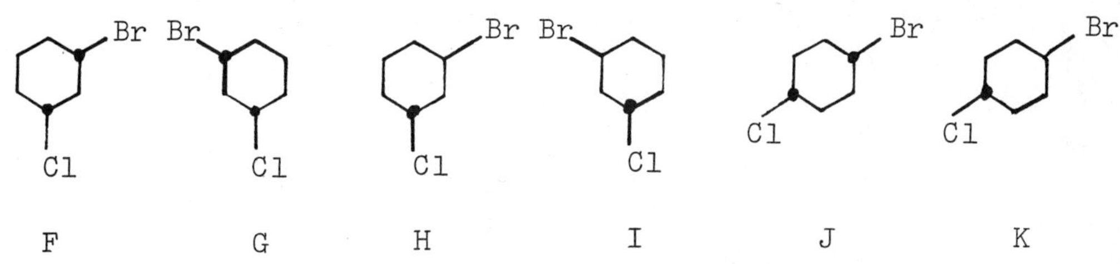

(b) A, J and K are symmetric

(c) All nonenantiomeric stereoisomers have different boiling points (enantiomeric pairs are: B and C, D and E, F and G, H and I).

29. (a)

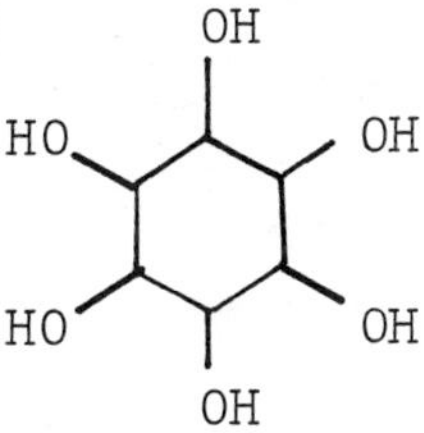

A 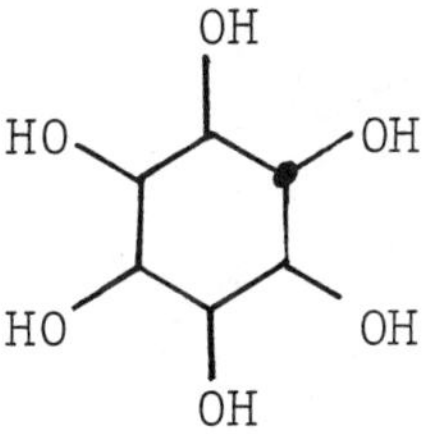 B 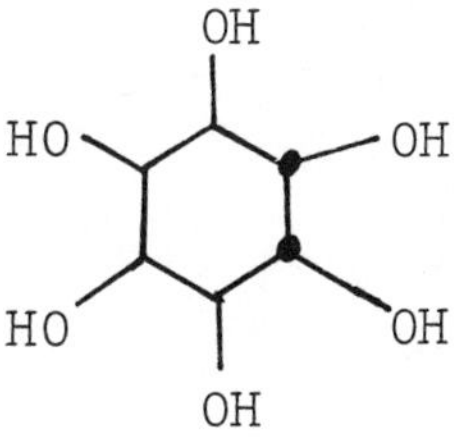 C

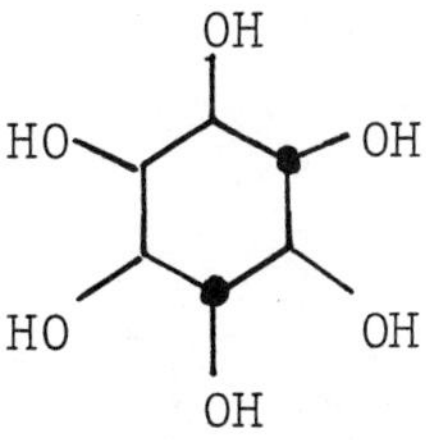

D 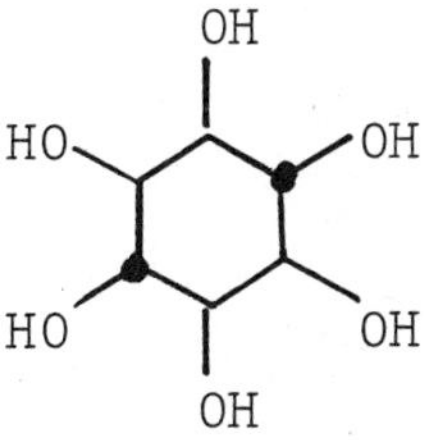 E 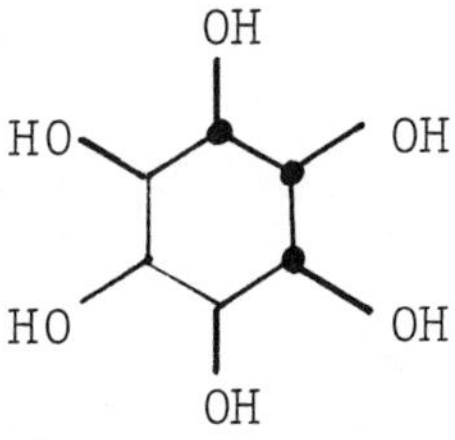 F

(continued next page)

G H

(b) All are symmetric and, therefore, incapable of optical activity except G which can exist as enantiomers.

30.

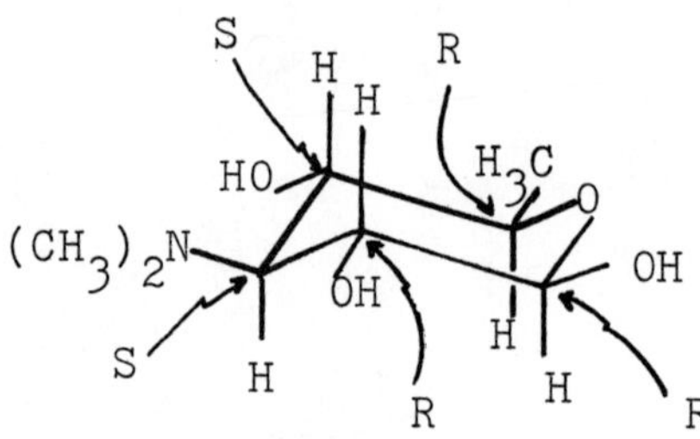

(Text: absolute configuration, S. 4.4)

31.

inactive active inactive

active active active

inactive active active

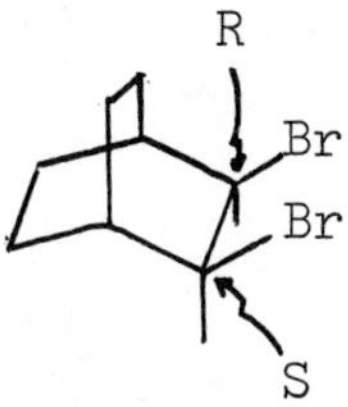

R Br Br S

inactive

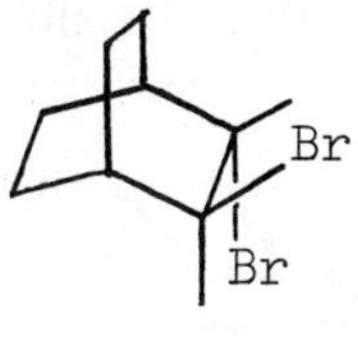

Br Br

active

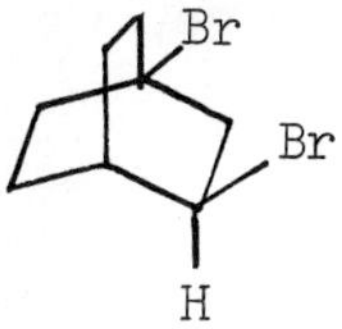

Br Br H

active

32. Cl Br | Br Cl | Cl Br | Br Cl

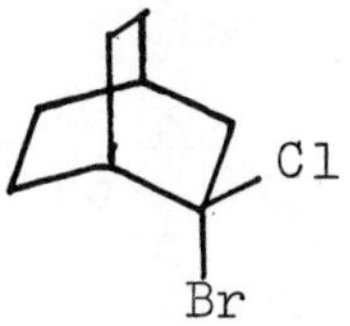

(±) A 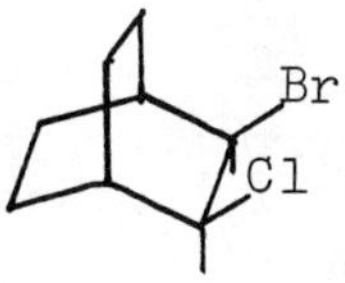 (±) B 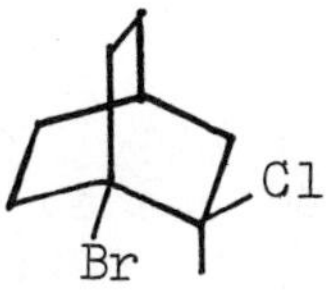 (±) C  (±) D

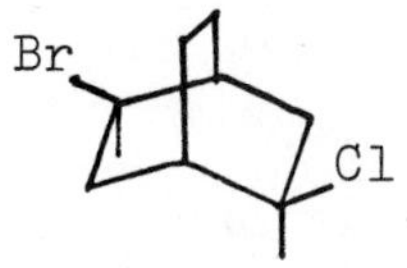

Br Cl | Cl Br | Cl Br | Br Cl

(±) E 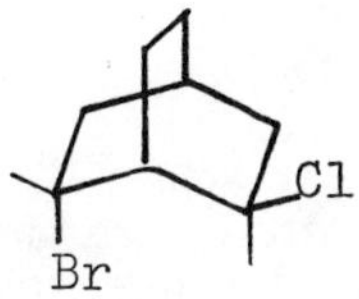 (±) F 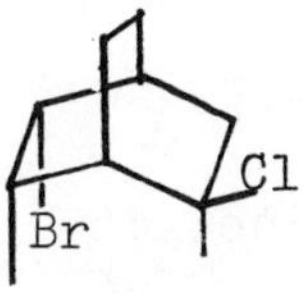 (±) G 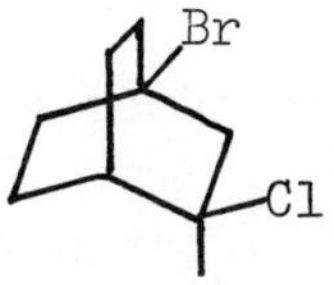 (±) H

Cl Br | Br Cl | Br Cl | Br Cl | Br Cl

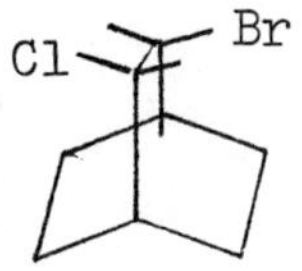

(±) I 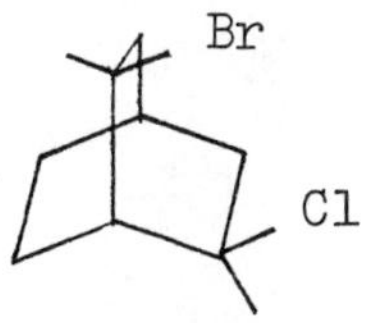 (±) J 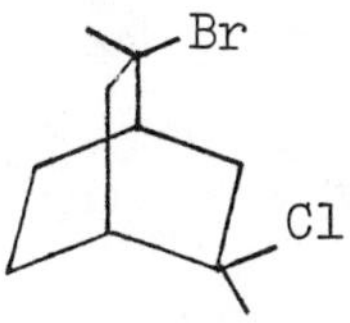 (±) K 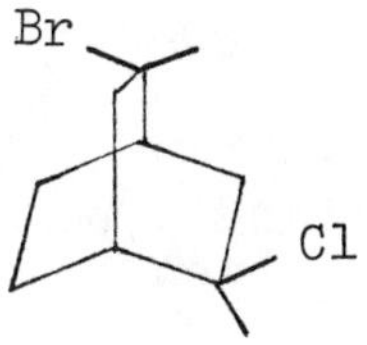 (±) L 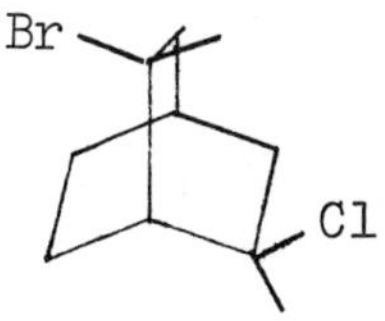 (±) M

All are resolvable into two enantiomers.

33. Possible conformations are:

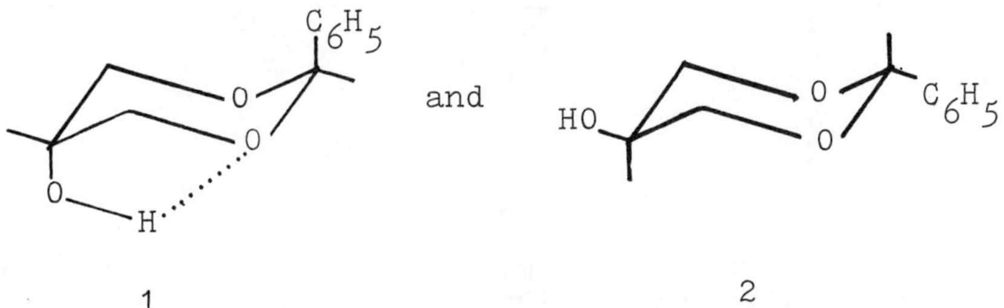

Conformation 1 has the OH group axial and hydrogen bonding with a ring O is possible.

34. Torsional strain (bond opposition strain) due to eclipsing of hydrogens, van der Waals strain due to "flagpole" bonds.

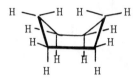

(Text: types of strain, S. 7.5)

35. Torsional strain has a value of approximately 1 kcal/mole for two adjacent, eclipsed bonds. Thus for planar cyclopentane with 10 C-H bond interactions, the torsional strain is approximately 10 kcal/mole. In going from the planar to puckered form there are 8 fewer C-H interactions, and the torsional strain is approximately 2 kcal/mole in puckered cyclopentane.

Torsional strain in planar cyclobutane = 8 kcal/mole, planar cyclohexane = 12 kcal/mole.

(Text: types of strain, S. 7.5)

36. This is reasonable since the chair conformation of dioxane is the lowest energy conformation and does not have a dipole moment.

(A molecule having a point of symmetry can not have a permanent dipole moment.)

37. (a) In the envelope (puckered) form of cyclopentane, the two methyl groups of <u>cis</u>-1,3-dimethylcyclopentane occupy equatorial-like positions (A). In <u>trans</u>-1,3-dimethylcyclopentane one methyl group must occupy an axial-like position (B).

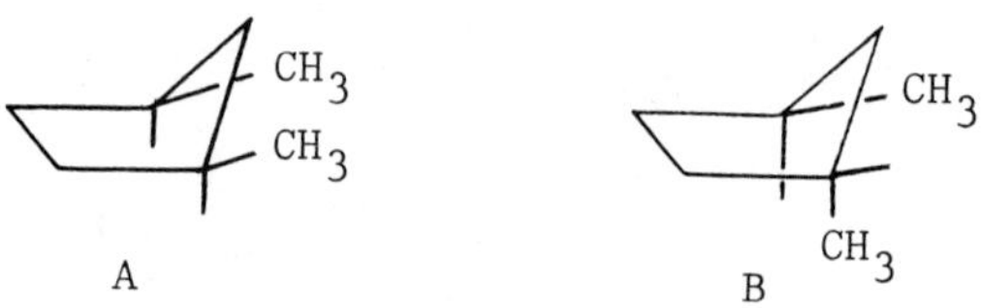

Building the models is helpful.

(b) Yes. A gives seven monobrominated derivatives separable by gas chromatographic analysis; B gives only six.

38. The central ring in each of these structures is a shallow boat. Build a model.

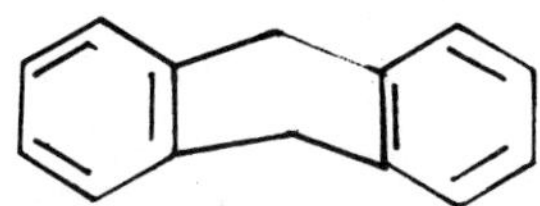

The preferred conformation for VI is that in which the ethyl group is away from the boat.

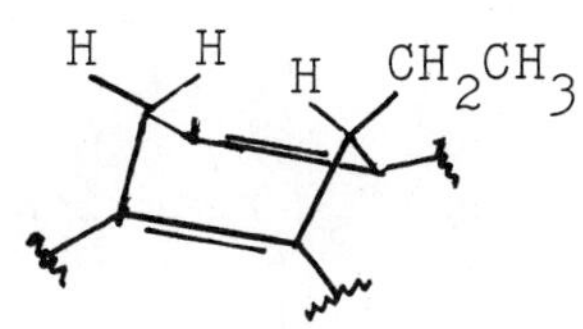

39. Other answers are possible in some parts.

(a) $Br_2/h\nu$

(b) 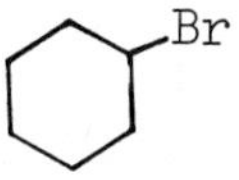 Br (from a) $\xrightarrow{Li(CH_3)_2Cu}$

(c) 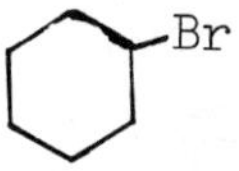 Br (from a) $\xrightarrow{OH^{\ominus}/H_2O}$

(d) 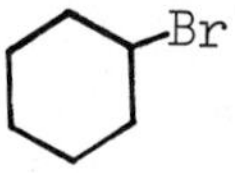 Br (from a) $\xrightarrow[\text{heat}]{KOH/alcohol}$ OR

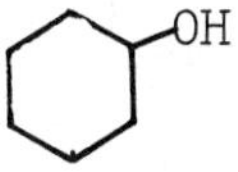

OH (from c) $\xrightarrow[\text{heat}]{H_2SO_4}$

(e) $\xrightarrow[\text{heat}]{KOH/alcohol}$

(f) $\xrightarrow[\text{acetone}]{Na^{\oplus}\ I^{\ominus}}$  I $\xrightarrow[\text{heat}]{KOH/alcohol}$

(g) $\xrightarrow{Na^{\oplus}CN^{\ominus}}$

(h) $\xrightarrow{Li}$ (or $\xrightarrow[\text{ether}]{Mg}$) $\xrightarrow{H_2O}$

(continued next page)

(i) 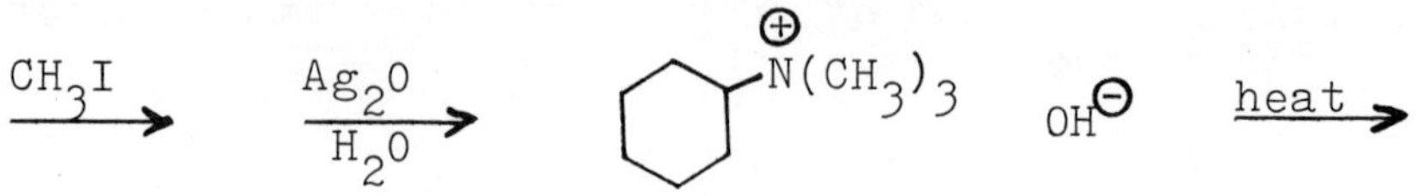 $\xrightarrow{CH_3I}$ $\xrightarrow[H_2O]{Ag_2O}$ cyclohexyl-$\overset{\oplus}{N}(CH_3)_3$ $OH^{\ominus}$ $\xrightarrow{heat}$

(j) 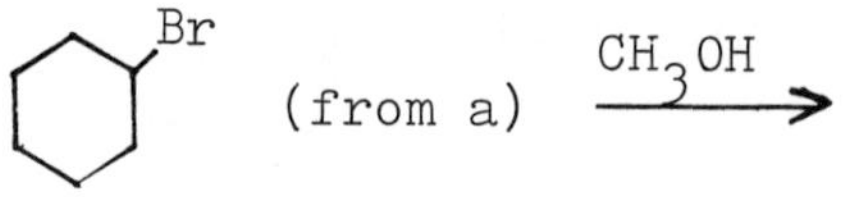 bromocyclohexane (from a) $\xrightarrow{CH_3OH}$

Text: review free-radical halogenation (Ch. 3), nucleophilic substitution (Ch. 5) and elimination reactions (Ch. 6).

40. The isomer shown below.

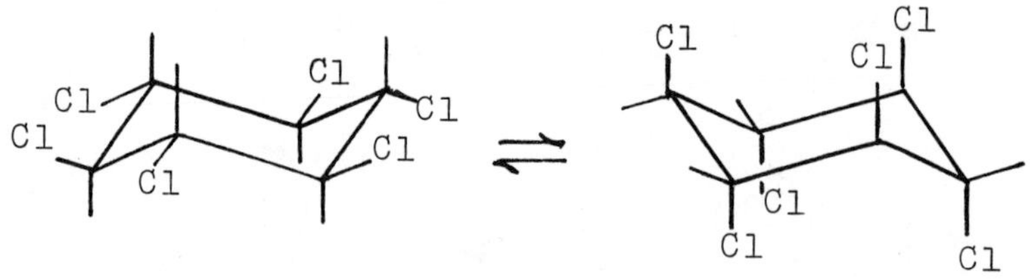

Chlorine and an adjacent hydrogen must be axial for an E2 reaction to occur. This orientation of H and Cl does not exist in either conformation.

41. The formation of the trans-product from the trans-substrate eliminates the possibility of a simple carbocation being formed since it would react with acetic acid to give cis- and trans-products. The exclusive formation of trans-4-methylcyclohexyl acetate suggests an intermediate in which acetic acid is prohibited from attacking from the side which leads to cis-product. Such an intermediate is provided by neighboring group participation but requires a boat conformation.

This mechanism could be tested using deuterium-labeled substrate. Products A and B should be formed from this intermediate.

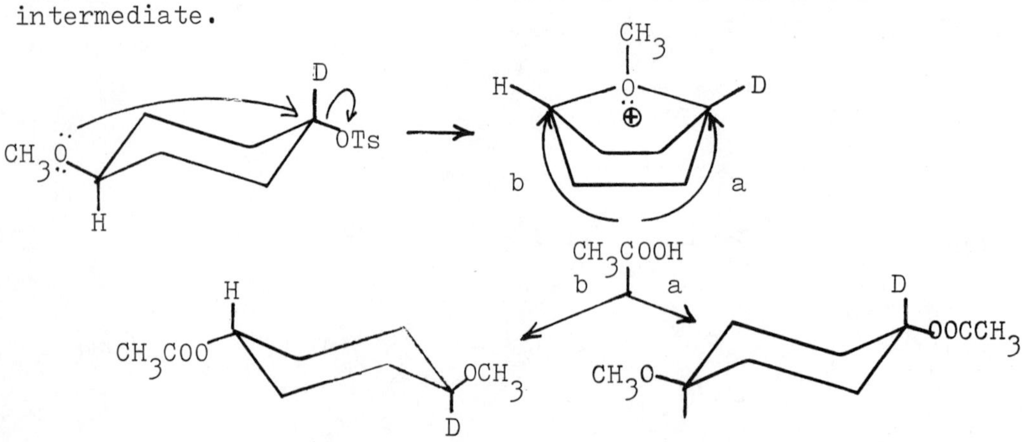

42. Bond length as well as group size must be considered in determining the net steric interaction of a group with other axial groups. Because the carbon-iodine bond is longer than the carbon-chlorine bond, the larger iodine atom is greater distance from the axial hydrogens on the same side of the ring.

43. The equatorial conformations of each have comparable energies. The axial conformations have nearly identical energies since each alkyl group can assume an orientation with a hydrogen directed toward other axial hydrogens.

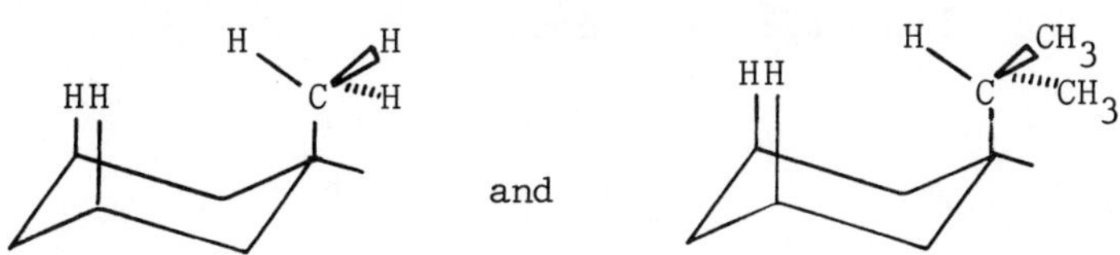

44. (a) $\Delta G_{CH_3} = -1.7$, $-1.7 = -(1.987 \times 10^{-3})(298)\ln K$

$\ln k = 2.87$, $K = \text{antilog}\,\frac{2.87}{2.30} = 17.8$

Let x = fraction of equatorial, 1-x = fraction of axial

$K = \frac{eq}{a} = 17.8 = \frac{x}{1-x};\quad x = 17.8 - 17.8x$

x = 0.947. % equatorial = 94.7

(b) % equatorial = 97.2

(c) % equatorial = 58.3

(d) % equatorial = 66.2

45. The carbocation formed in an S_N1 reaction prefers a planar configuration with bond angles of 120°.

The reactivity observed for these compounds parallels their ability to form cations with the preferred configuration.

stability of cations: 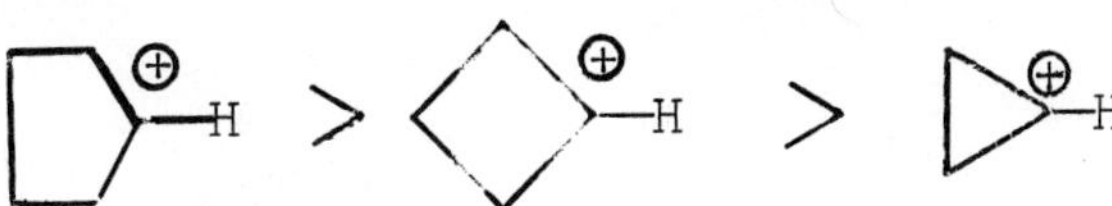

46. The activated complex is the same for both reactions; however, the ground state energy is higher for <u>cis</u>-4-<u>t</u>-butylcyclohexyl iodide (iodine in an axial position), thus the energy of activation is less for the reaction of the <u>cis</u>-isomer with iodide ion.

(Text: S_N2 reactivity, S. 7.9)

47. These reactions are S_N2. The activated complex in the reaction of cyclopropyl bromide with a halide ion has greater steric hindrance than the activated complex in the same reaction with cyclopentyl bromide.

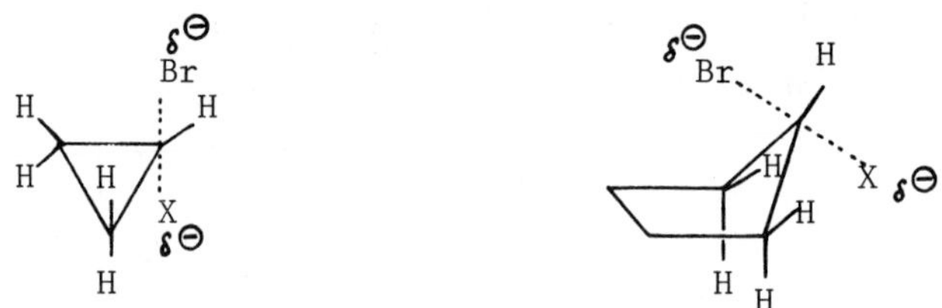

48. E undergoes the Elcb reaction shown below.

Br

MgBr

+ $MgBr_2$

D does not react in a similar manner since to do so would require formation of a double bond at the bridgehead (Bredt's rule). Thus, D forms a stable Grignard reagent.

(Text: E reactions and Bredt's rule, S. 7.9)

49. <u>9-methyldecalins</u>. Skew-butane interactions with the 9-methyl group are marked "a".

interactions with the methyl group = 4

interactions with the methyl group = 2

Recall that <u>cis</u>-decalin has three more skew-butane interactions than <u>trans</u>-decalin.
(Text S. 7.7)

The net difference is one more skew-butane interaction in the <u>cis</u>-isomer. This suggests than an equilibrium mixture of the 9-methyldecalins should contain more of the <u>cis</u>-isomer than decalin itself.

<u>9,10-dimethyldecalins</u>. Skew-butane interactions with methyl groups a and b are marked "a" and "b" respectively.

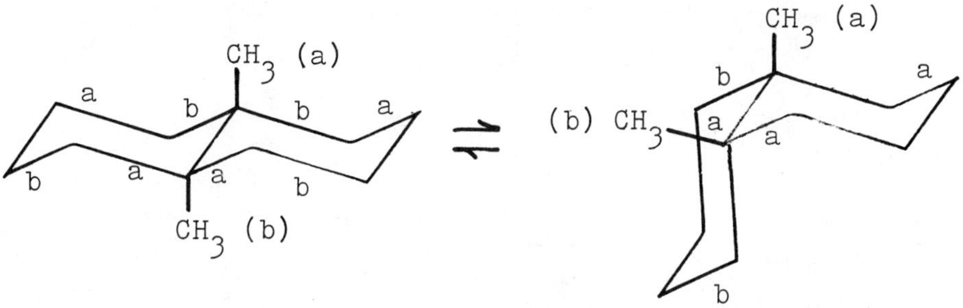

CH_3 (a) ... CH_3 (b) $\rightleftharpoons$ (b) CH_3 ... CH_3 (a)

interactions with methyl groups = 8 | interactions with methyl groups = 5

50. (a) Mechanism for formation of Cl–⟨C₆H₁₀⟩–⟨C₆H₁₀⟩–Cl

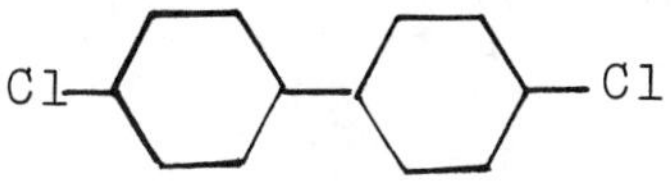

$Cl_2 \xrightarrow{\text{light}} 2Cl\cdot$

$Cl\cdot$ + $H_2C_6H_9Cl$ ⟶ HCl + H–$\dot{C}_6H_9$–Cl

2 H–$\dot{C}_6H_9$–Cl ⟶ Cl–C_6H_{10}–C_6H_{10}–Cl

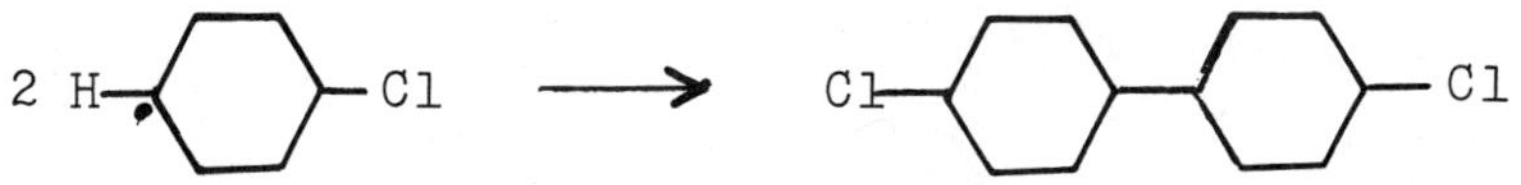

(b) Although six structures can be drawn using flat rings and the dot convention, rotation about the carbon-carbon single bond makes some structures identical with others, and the number of stereoisomers reduces to three. For example, rotation of the ring on the left in structure A converts it to B.

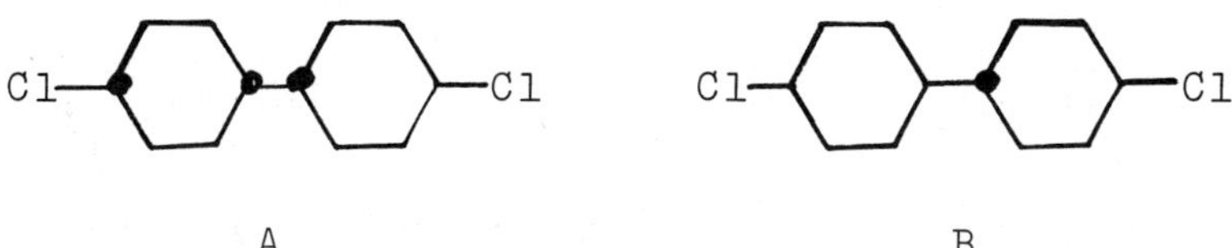

(c) The three stereoisomers in their preferred conformations are shown below.

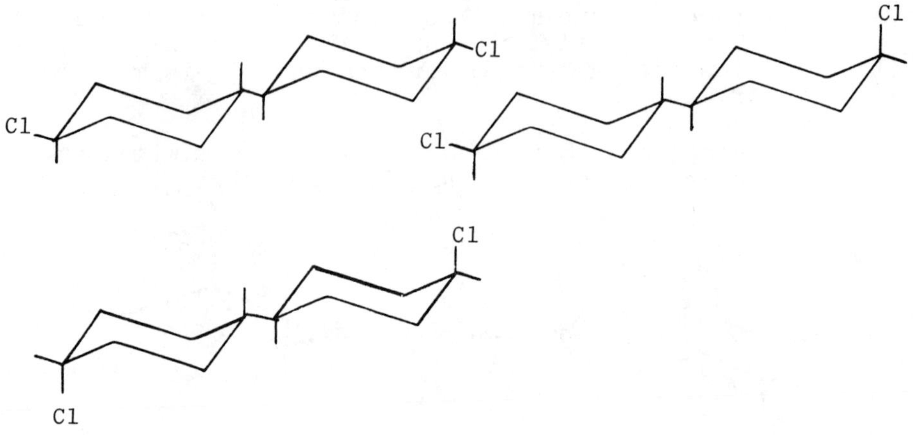

Cl Cl Cl Cl

Cl

Cl

51. A: (piperidinium, N with H and CH_3) $I^{\ominus}$ B: (N-methylpiperidine, N–CH_3)

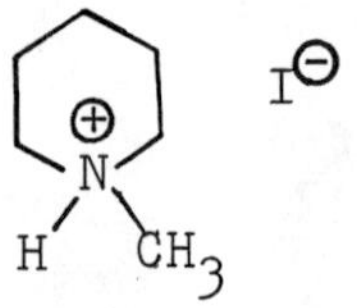

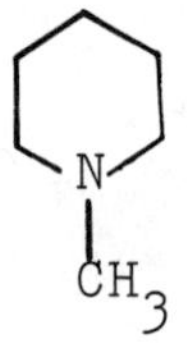

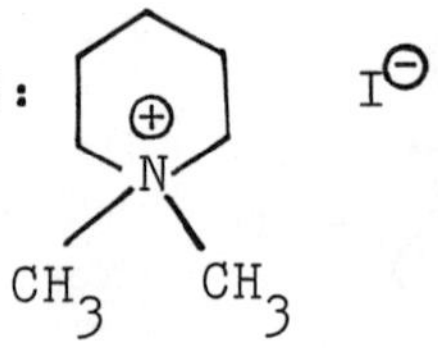

C: (piperidinium, N with CH_3 and CH_3) $I^{\ominus}$ D: (piperidinium, N with CH_3 and CH_3) $OH^{\ominus}$

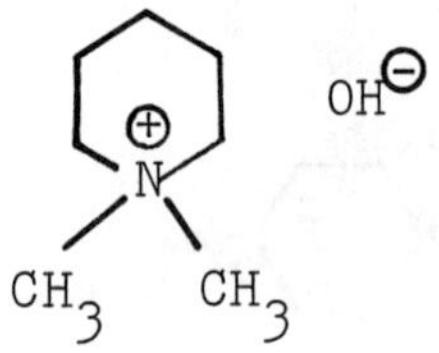

E: $CH_2{=}CH{-}CH_2CH_2CH_2{-}N(CH_3)_2$

F: $CH_2{=}CHCH_2CH_2CH_2\overset{\oplus}{N}(CH_3)_3\ I^{\ominus}$

G: $CH_2{=}CHCH_2CH_2CH_2\overset{\oplus}{N}(CH_3)_3\ OH^{\ominus}$

52. (a) Yes. They have the same configurations at all carbons except one.

(b) The ring-inverted conformation of β-D-glucopyranose is shown by structure 1 and that of α-D-glucopyranose by structure 2.

(continued next page)

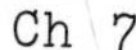

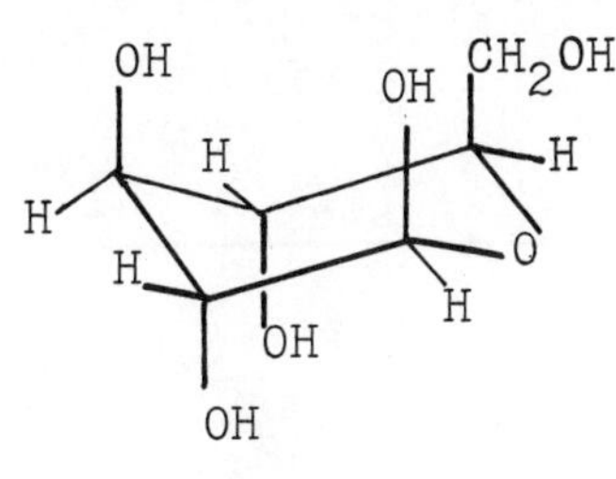

1

2

(c) In each set, the conformation with the greatest number of the larger groups (OH and CH_2OH) equatorial is the more stable (i.e. those shown in the question). These possess fewer destabilizing interactions (skew butane and axial).

SUMMARY OF REACTIONS

Review:

1. free-radical halogenation of alkanes, Chapter 3
2. synthesis of alkanes from lithium dialkyl cuprates, Chapter 3
3. nucleophilic substitution reactions, Chapter 5
4. elimination reactions, Chapter 6
5. synthesis of organometallic compounds, Chapter 6

REACTION REVIEW

A. Predict the major product in each of the following. In some examples the mechanism is indicated.

(continued next page)

Question	Answer
1. 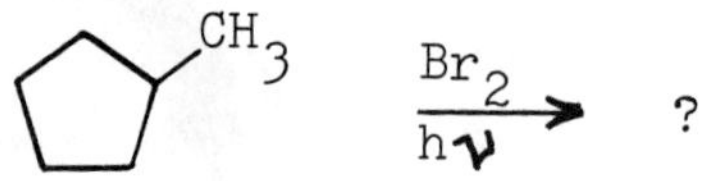 (methylcyclopentane) $\xrightarrow[h\nu]{Br_2}$?	1. (1-bromo-1-methylcyclopentane: CH_3, Br) 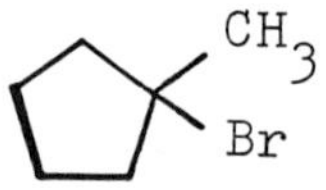
2. (Cl, CH_3 cyclopentane) $\xrightarrow[(S_N2)]{OH^{\ominus}/H_2O}$? 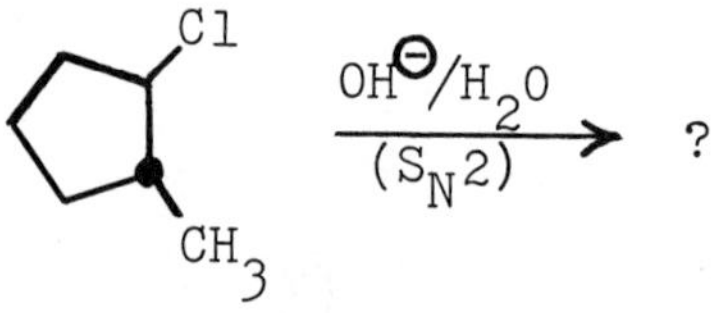	2. (OH, CH_3 cyclopentane) 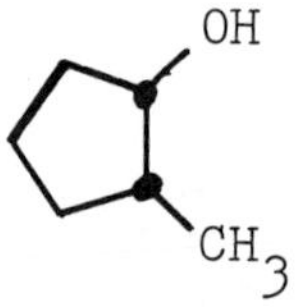
3. (cyclohexane: CH_3, OH, H, CH_3) $\xrightarrow{HBr}$? 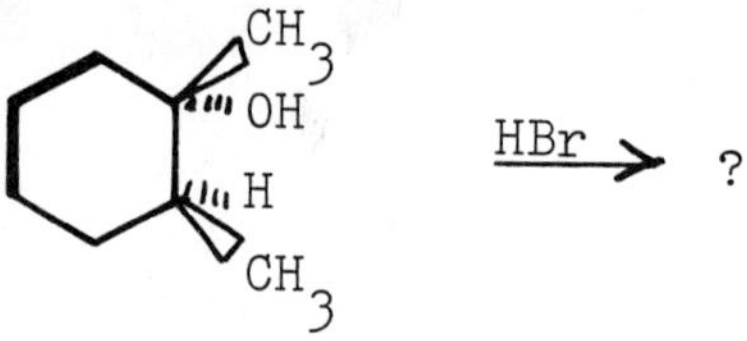	3. (cyclohexane: CH_3, Br, H, CH_3) + (cyclohexane: Br, CH_3, H, CH_3) 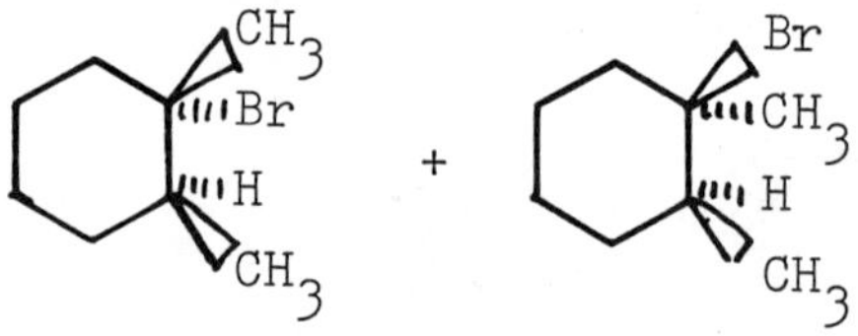
4. 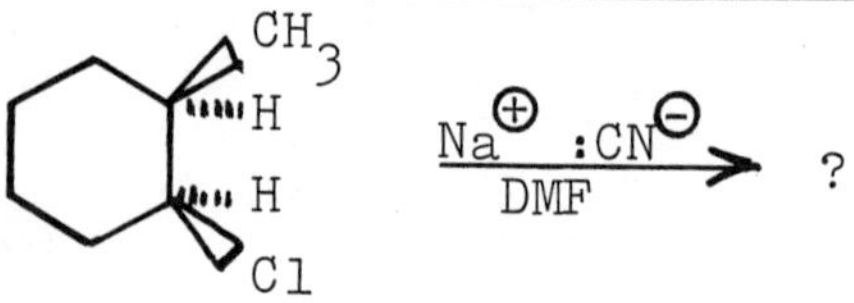 (cyclohexane: CH_3, H, H, Cl) $\xrightarrow[DMF]{Na^{\oplus}\ :CN^{\ominus}}$?	4. 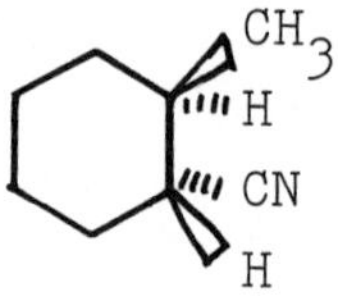 (cyclohexane: CH_3, H, CN, H) (S_N2)
5. 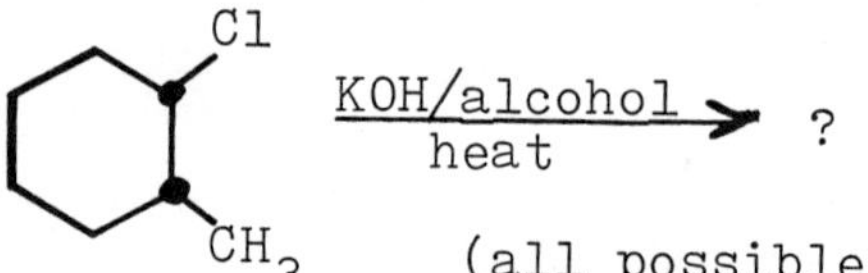 (Cl, CH_3 cyclohexane) $\xrightarrow[heat]{KOH/alcohol}$? (all possible E2 products)	5. 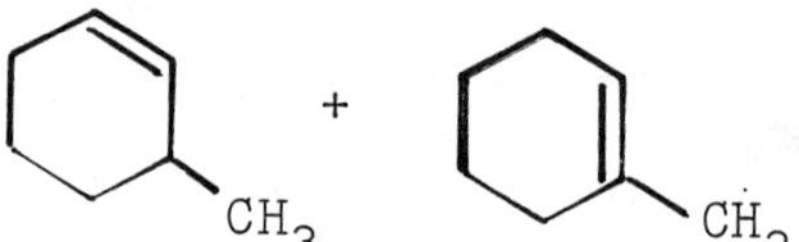 (3-methylcyclohexene, CH_3) + (1-methylcyclohexene, CH_3) (major)
6. 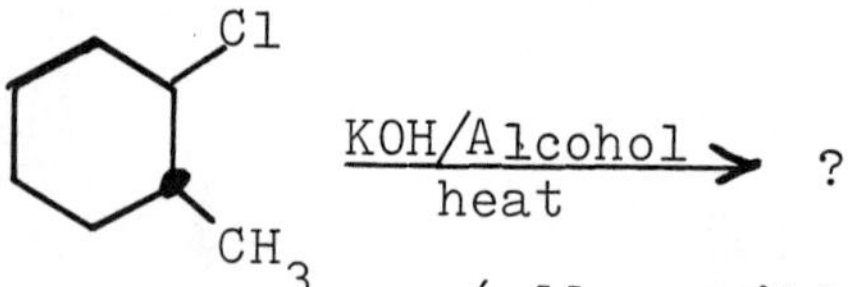 (Cl, CH_3 cyclohexane) $\xrightarrow[heat]{KOH/Alcohol}$? (all possible E2 products)	6. only 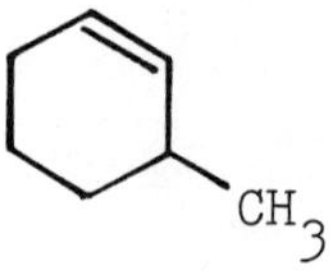 (3-methylcyclohexene, CH_3)

(continued next page)

7. CH_3 OH CH_3 $\xrightarrow[\text{heat}]{H_2SO_4}$ (all possible E products)	7. CH_3 CH_3 (major) + CH_2 CH_3 + CH_3 CH_3
8. CH_3 CH_3 OH $\xrightarrow{HBr}$?	8. CH_3 Br CH_3 (rearrangement via S_N1) (cis and trans)

B. Write structures for all letters.

1. $\xrightarrow[h\nu]{Cl_2}$ A $\xrightarrow{Li\ (CH_3)_2Cu}$ B $\xrightarrow[h\nu]{Br}$ C	1. A: Cl B: CH_3 C: CH_3 Br
2. CH_3 $\xrightarrow[h\nu]{Br_2}$ A $\xrightarrow[\text{heat}]{\text{KOH/alcohol}}$ B	2. A: CH_3 Br B: CH_3
3. OH $\xrightarrow{HBr}$ A $\xrightarrow[\text{ether}]{Mg}$ B $\xrightarrow{D_2O}$ C	3. A: Br B: MgBr C: D
4. CH_3 Cl $\xrightarrow{OH^{\ominus}/H_2O}$ A $\xrightarrow[\text{heat}]{H_2SO_4}$ B	4. A: CH_3 OH B: CH_3

CHAPTER 8. ALKENES

LEARNING OBJECTIVES

When you have completed this chapter, you should be able to:

1. name and specify the E or Z configuration of alkenes (problems 2, 33, 34 and related problem 46);
2. predict or explain the stability of alkenes on the basis of heats of hydrogenation data (problems 3, 55);
3. determine the oxidation number of carbon in various functional groups (problem 4, related problem 5);
4. predict the reactants or products (or explain the formation of products) in the catalytic hydrogenation of alkenes (problems 6, 7, 8, 35, 58);
5. suggest experiments which support the syn-addition of hydrogen in the reduction of alkenes (problems 6, 9, 11 and related problem 10);
6. account for products formed in the diborane reduction of alkenes (problems 52, 54);
7. predict and explain the formation of products in the ionic addition of halogens to alkenes (problems 12, 13, 14, 15, 16, 25, 44, 60);
8. predict the products formed in the addition of acids to alkenes (problems 17, 18, 19, 20, 35);
9. account for the formation of products in the addition of acids to alkenes (problems 21, 23, 40, 41, 42, 43, 45, 47, 57, 60, 64);
10. compare the intermediates formed in ionic and free-radical addition reactions (problem 22);
11. predict the products or reactants in the oxidation of alkenes (problems 26, 27, 28, 29, 35, 39 and related problem 49);
12. predict the reactants or products (or account for the formation of products) in the reactions of alkenes with methylene or dichlorocarbene (problems 31, 32, 35, 50, 59);
13. predict or explain the formation of products in the reactions of alkanes with methylene (problem 30);
14. write equations for the synthesis of compounds by use of reactions from this chapter and those from previous chapters (problems 36, 37 and related problem 38);
15. explain the stereospecificity of radical addition reactions (problem 38);
16. suggest chemical and spectral methods for distinguishing structures (problems 65, 66);
17. write a mechanism for the disproportionation of alkyl radicals (problem 24);
18. compare the intermediates in free-radical addition and substitution reaction (problem 53).

ANSWERS TO PROBLEMS

1. 1,2-propadiene , $CH_2=\overset{sp}{C}=CH_2$

(continued next page)

2. For structures which may be designated as E or Z, assign priorities to the groups on one carbon of C=C, then do the same for the other carbon of the C=C. If the groups of highest priority are on the same side, the structure is Z, if on opposite sides, the structure is E.

(a) (E)-1,1,1-triflouro-2,3-dimethyl-2-pentene

(b) 1-chloro-2-methylcyclohexene

(c) 1-isopropyl-2,3,3-trimethylcyclohexene

(d) (Z)-3-methyl-1,3,5-hexatriene

(e) isopropylidenecyclohexane

3. When heats of hydrogenation are used to compare the relative energies of two alkenes, the alkenes must give the same product. Cyclopentene and cyclohexene do not give the same cycloalkane on reduction.

4. Summary of oxidation number rules:

(1) an electronegative atom makes a +1 contribution to the O.N. of the carbon to which it is bonded.

(2) a hydrogen makes a -1 contribution.

(3) another carbon makes a 0 contribution.

<u>All hydrogens are +1 in the examples below</u>.

(a) CH_3-CH=CH_2
-3 -1 -2

(b) CH_3-CH_2-CH=CH_2
-3 -2 -1 -2

(c) CH_3-CH=CH-CH_3
-3 -1 -1 -3

(d) CH_3-CH=CH-CH_2-Cl
-3 -1 -1 -1 -1

(e) CH_3 (-3) and Cl (-1) on C (+1); =C (-1) bearing CH_3 (-3) and H:
(CH₃)(Cl)C=C(CH₃)(H) → CH_3(-3), Cl(-1), C(+1), C(-1), CH_3(-3)

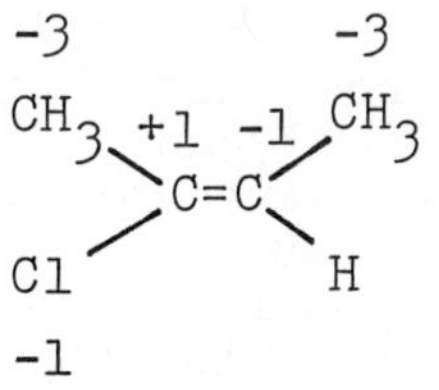

(f) CH_3-CH_2-CH(Cl)-Cl
-3 -2 +1, Cl -1, Cl -1

(g) CH_3—C(=O)—O—H
-3 +3 -2, =O -2

(h) CH_3-S-H
-2 -2

5. The atom being oxidized is the carbon bonded to oxygen.

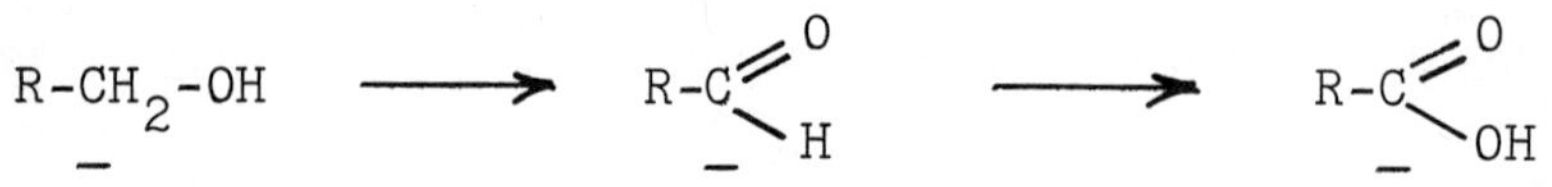

$$R\text{-}\underline{C}H_2\text{-}OH \longrightarrow R\text{-}\underline{C}(=O)H \longrightarrow R\text{-}\underline{C}(=O)OH$$

6. Dehydrogenation as well as hydrogenation may occur in the metal catalyst surface.

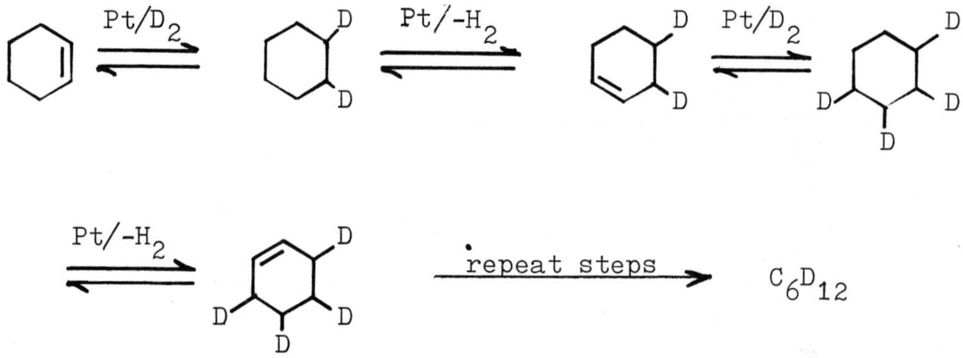

Pt/D_2 ⇌ Pt/$-H_2$ ⇌ Pt/D_2 ⇌

D D D D D D

D

Pt/$-H_2$ ⇌ repeat steps → C_6D_{12}

D D D D

7. Recall that catalytic hydrogenation gives syn-addition of hydrogen.

	Product with H_2	Product with D_2
(a)	$CH_3CH_2CH_2CH_3$	$CH_3CH_2CHDCH_2D$
(b)	$CH_3CH_2CH_2CH_3$	(d,ℓ) $CH_3CHDCHDCH_3$
(c)	$CH_3CH_2CH_2CH_3$	meso-$CH_3CHDCHDCH_3$*

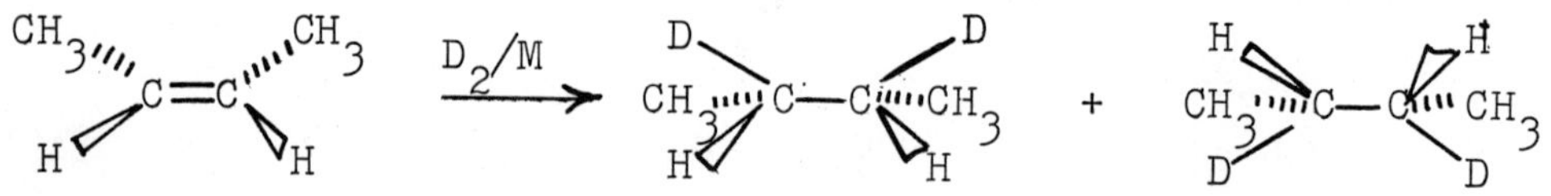

* CH_3 CH_3 H H C=C D_2/M → D D CH_3 C—C CH_3 H H + H H CH_3 C—C CH_3 D D

superimposable

meso

(continued next page)

(d) $CH_3CH_2CH_2CH_2CH_3$ (d,l)-$CH_3CHDCHDCH_2CH_3$ (threo)

(e) $CH_3CH_2CH_2CH_2CH_3$ (d,l)-$CH_3CHDCHDCH_2CH_3$ (erythro)

(f)  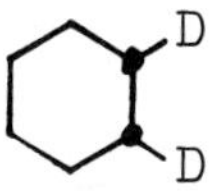 [cyclohexane] [cyclohexane-1,2-d_2: D, D]

8. (a) $CH_3CH=CH_2$

(b) $CH_3CH=CHCH_3$ or $CH_3CH_2CH=CH_2$

(cis and trans)

(c) $(CH_3)_2CHCH=CH_2$ or $(CH_3)_2C=CHCH_3$

(d) $(CH_3)_3CCH_2CH=CH_2$ or $(CH_3)_3CCH=CHCH_3$

(e) cis-$CH_3CD=CDCH_3$

(f) trans-$CH_3CD=CDCH_3$

(g) 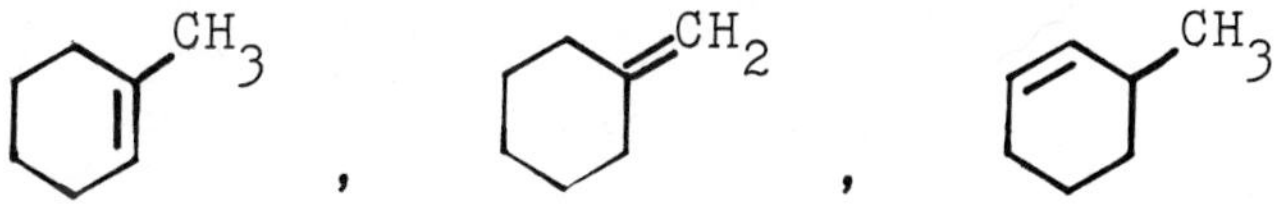 [1-methylcyclohexene, CH_3], [methylenecyclohexane, CH_2], [3-methylcyclohexene, CH_3] etc.

(h) 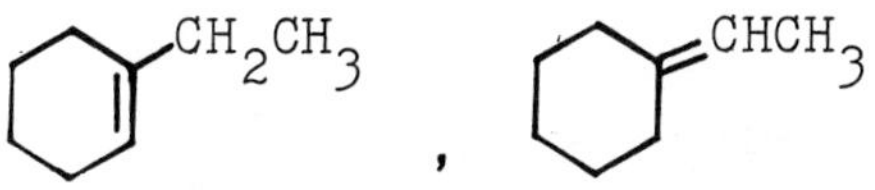 [1-ethylcyclohexene, CH_2CH_3], [ethylidenecyclohexane, $CHCH_3$] etc.

9. Reduce with diimide dimethylcyclohexene or any other alkene capable of demonstrating the stereospecifity of the reaction (the syn-addition of two hydrogens from diimide). For example:

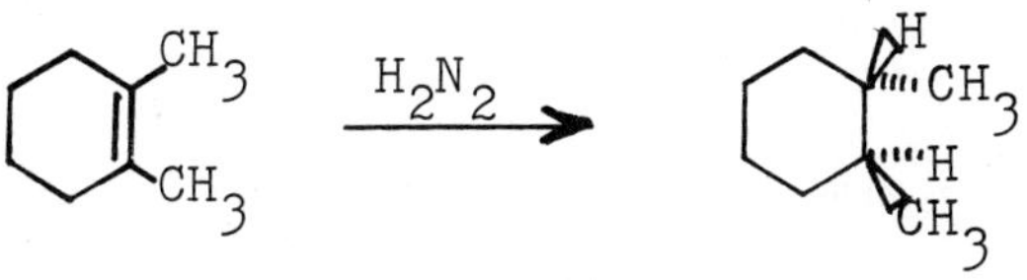

[1,2-dimethylcyclohexene: CH_3, CH_3] $\xrightarrow{H_2N_2}$ [cis-1,2-dimethylcyclohexane: H, CH_3, H, CH_3]

(continued next page)

A reaction involving two molecules of diimide would likely result in anti-addition, although both syn- and anti-addition may occur.

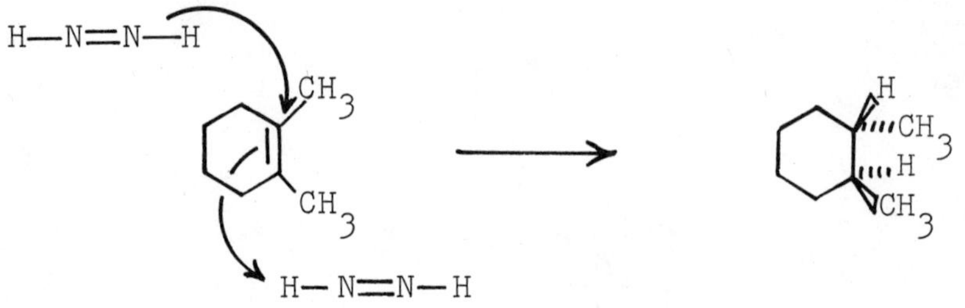

H—N=N—H

CH_3

CH_3

H—N=N—H

H, CH_3, H, CH_3

10. π (2p-2p) σ $(2sp^2\text{-}2sp^2)$

H, N, N, H

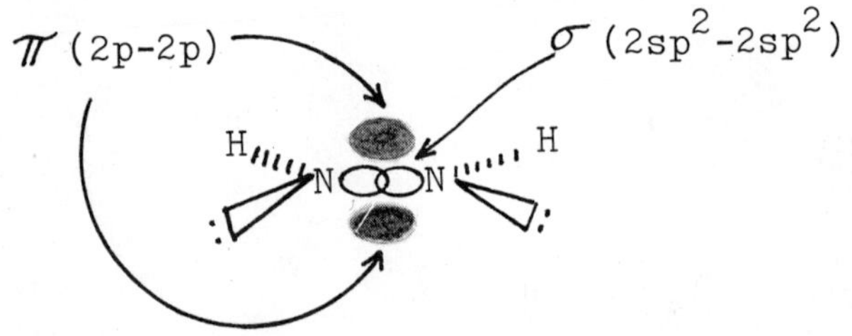

N is sp^2

Dimide can exist in syn- and anti- forms.

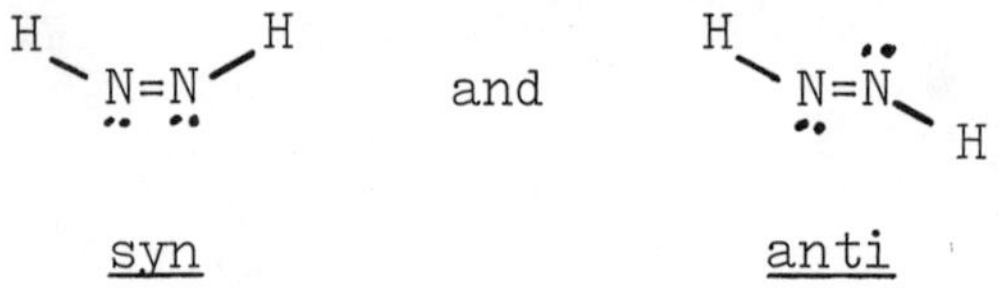

H–N=N–H	and	H–N=N–H
syn		anti

Only the syn-isomer can participate in syn-addition of hydrogens to the double bond.

11. React an alkene with BD_3/RCOOH or BH_3/RCOOD.

For example: $\text{C=C} \xrightarrow{BH_3} \text{-C-C-}$ (H, B-) $\xrightarrow{RCOOD} \text{-C-C-}$ (H, D)

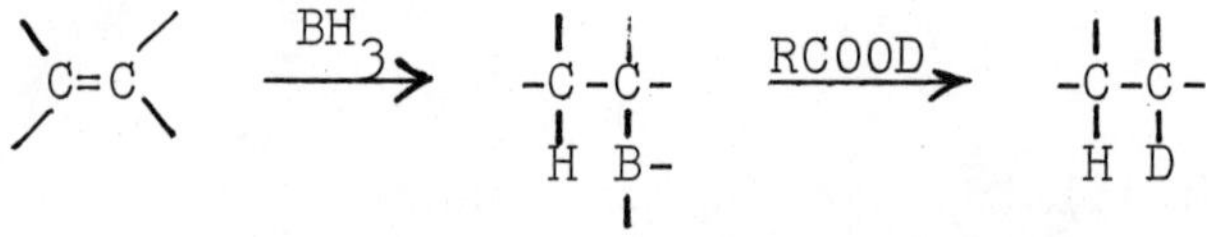

12. Closure of the σ-complex to the cyclic bromonium ion is comparable to the second step of an S_N1 process in which a nucleophile adds to a carbocation.

(continued next page)

bromonium ion: $-\overset{|}{\underset{|}{C}}-\overset{|}{\underset{\oplus}{C}}-$ with :Br: on the first carbon attacking the cation $\longrightarrow$ cyclic bromonium ion $-\overset{|}{C}-\overset{|}{C}-$ bridged by :Br: with $\oplus$ on Br

S_N1: $-\overset{|}{\underset{|}{C}}-\overset{|}{\underset{\oplus}{C}}-$ + Nu:$^\ominus$ $\longrightarrow$ $-\overset{|}{\underset{|}{C}}-\overset{|}{\underset{Nu}{C}}-$

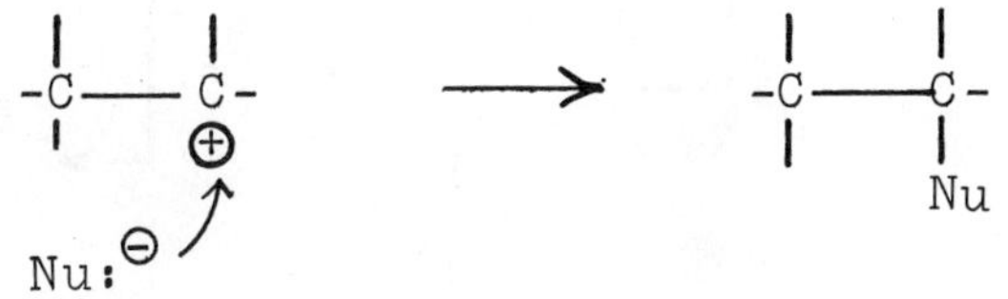

(Text: S_N1 mechanism, S. 5.4)

13. (a) Bromonium ion A is more reactive because of greater angle strain in the three-membered ring.

 (b) Products are $BrCH_2CH_2Br$ from A and $BrCH_2CH_2CH_2Br$ from B.

(Text: ring stability, S. 7.5)

14. No. Initial attack by the nucleophile, $Cl^\ominus$, followed by reaction of the resultant anion with bromine could lead to the same product.

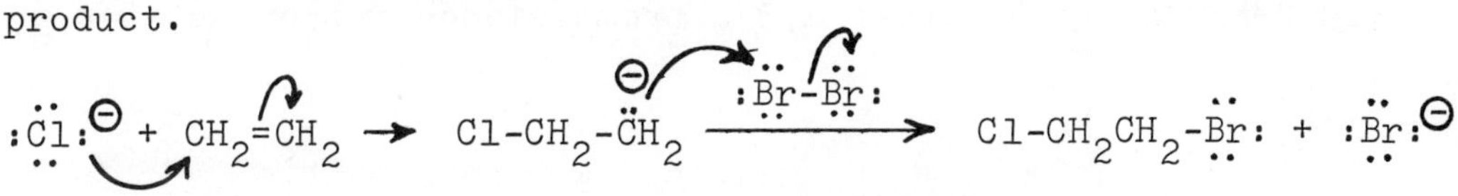

$$:\ddot{Cl}:^\ominus + CH_2{=}CH_2 \rightarrow Cl\text{-}CH_2\text{-}\ddot{C}H_2^\ominus \xrightarrow{:\ddot{Br}\text{-}\ddot{Br}:} Cl\text{-}CH_2CH_2\text{-}\ddot{Br}: + :\ddot{Br}:^\ominus$$

It is also possible that the reaction below occurs and that Cl-Br reacts with the alkene by electrophilic addition.

$$Cl^\ominus + Br_2 \longrightarrow Cl\text{-}Br + Br^\ominus$$

15. trans-2-Butene gives only meso-2,3-dibromobutane if only anti-addition of bromine occurs.

(continued next page)

cis-2-Butene gives (d,ℓ)-2,3-dibromobutane if bromine adds anti.

16.

	symmetric bromonium ion	unsymmetric bromonium ion	open carbocation
a. free rotation about C-C bond	no	no	yes
b,c. extent of positive charge on carbon vs. bromine	little or no charge on carbon, mostly on bromine	shared by the more highly substituted carbon and bromine	entirely on carbon
d. possibility that they lead to syn addition	anti-addition only	anti-addition only	syn and anti-addition may occur
e. distance between bromine and both carbons	bromine equidistance from carbons	bromine nearer the less highly substituted carbon	bonded to only one carbon

(Text S. 8.6)

17. (a) 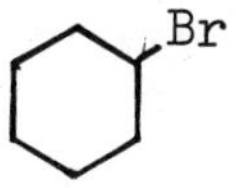 Bromocyclohexane (cyclohexane ring with Br)

(b) 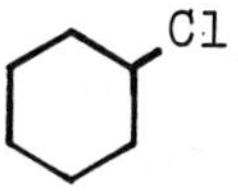 Chlorocyclohexane (cyclohexane ring with Cl)

(c) $(CH_3)_2CHCBr(CH_3)_2$

(d) $CH_3CH_2CHClCH_3$

(e) $CH_3CH_2CHICH_3$

(f) 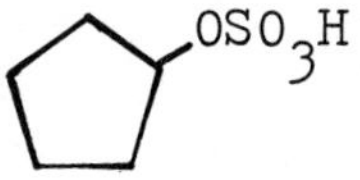 Cyclopentane ring with OSO_3H

18. (a) CH_3CHICH_3

(b) 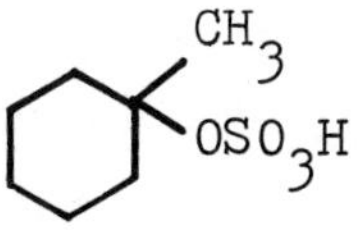 Cyclohexane ring bearing CH_3 and OSO_3H on the same carbon

(c) $CH_3CH_2CHClCH_2CH_3$ + $CH_3CH_2CH_2CHClCH_3$

(d) $(CH_3)_3CBr$

(e) 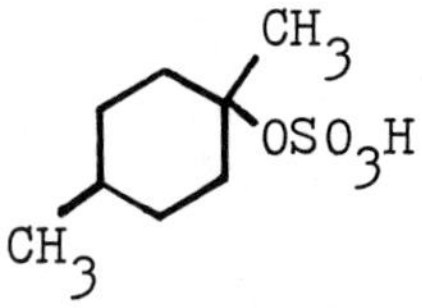 Cyclohexane ring bearing CH_3 and OSO_3H on one carbon, CH_3 at the 4-position

19. (a) 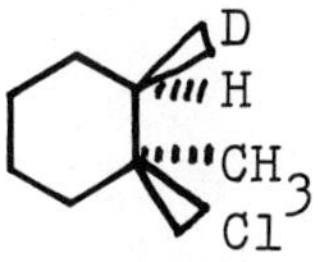 Cyclohexane ring with adjacent carbons bearing D, H and CH_3, Cl (wedge/hash stereochemistry)

(b) $CH_3CH(OCH_3)CH_2Br$ + $CH_3CHBrCH_2Br$

(c) 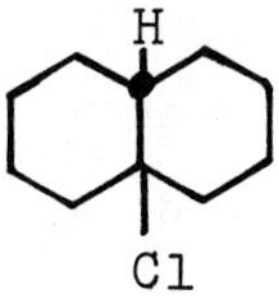 Decalin with H at one ring-fusion carbon and Cl at the other

20. General reaction: $RCH{=}CH_2 \xrightarrow{H_3O^{\oplus}} RCH(OH)CH_3$ (Markownikoff addition)

(a) $CH_3CH(OH)CH_2CH_3$

(b) $CH_3CH(OH)CH_2CH_3$

(c) $C_6H_5CH(OH)CH_3$

(d) 1-methylcyclohexanol (cyclohexane ring bearing CH_3 and OH on the same carbon)

(e) 1-ethylcyclohexanol (cyclohexane ring bearing CH_2CH_3 and OH on the same carbon)

(Text S. 8.6)

21. The reaction of A with $H^{\oplus}$ can in theory give carbocations 1 and 2; however since no D is formed, it is likely that less stable cation 1 does not form.

$$(CH_3)_3CCH{=}CH_2 \xrightarrow{H^{\oplus}} (CH_3)_3CCH_2\overset{\oplus}{C}H_2 \quad (1) \text{ (does not form)}$$

$$(CH_3)_3CCH{=}CH_2 \xrightarrow{H^{\oplus}} CH_3-\underset{CH_3}{\overset{CH_3}{C}}-\overset{\oplus}{C}H-CH_3 \quad (2)$$

$$\downarrow$$

$$CH_3-\overset{CH_3}{\underset{\oplus}{C}}-\underset{CH_3}{CH}-CH_3 \quad (3)$$

Cation 2 (2°) can rearrange to the more stable cation 3 (3°). These cations are the source of products B and C as shown below.

$$\underset{2}{(CH_3)_3C\overset{\oplus}{C}HCH_3} \xrightarrow{H_2\ddot{O}:} (CH_3)_3CCH(\overset{\oplus}{O}H_2)CH_3 \xrightarrow{:\ddot{O}H_2} \underset{C}{(CH_3)_3CCH(\ddot{O}H)CH_3}$$

(continued next page)

$$(CH_3)_2\overset{\oplus}{C}CH_2CH_3 \xrightarrow{H_2\ddot{O}:} (CH_3)_2C(\overset{\oplus}{O}H_2)CH_2CH_3 \xrightarrow{:\ddot{O}H_2} (CH_3)_2C(\ddot{O}H)CH_2CH_3$$

3 B

22. In each the more stable (lower energy) of two possible intermediates is formed. For example:

(a) cationic Markownikoff addition

$$RCH{=}CH_2 \xrightarrow{H^{\oplus}} R CH_2\overset{\oplus}{C}H_2 \text{ (not formed)}; \quad R\overset{\oplus}{C}HCH_3 \text{ (more stable)}$$

(b) cationic anti-Markownikoff addition

$$R\text{-}\overset{\overset{\delta\ominus}{O}}{\overset{\|}{C}}{}^{\delta\oplus}CH{=}CH_2 \xrightarrow{H^{\oplus}} RC(=O)\overset{\oplus}{C}HCH_3 \text{ (not formed)}; \quad RC(=O)CH_2\overset{\oplus}{C}H_2 \text{ (more stable)}$$

(c) free-radical anti-Markownikoff addition

$$RCH{=}CH_2 \xrightarrow{X\cdot} RCHBr\dot{C}H_2 \text{ (not formed)}; \quad R\dot{C}HCH_2Br \text{ (more stable)}$$

23. The alkenes, C_8H_{16}, are $(CH_3)_3CCH{=}C(CH_3)_2$ (A) and $(CH_3)_3CCH_2C(CH_3){=}CH_2$ (B) formed by the sequence:

(continued next page)

$$(CH_3)_2C{=}CH_2 \xrightarrow{H^{\oplus}} (CH_3)_2\overset{\oplus}{C}{-}CH_3 \xrightarrow{(CH_3)_2C=CH_2} (CH_3)_3C{-}CH_2{-}\overset{\oplus}{C}(CH_3)_2$$

3° 3°

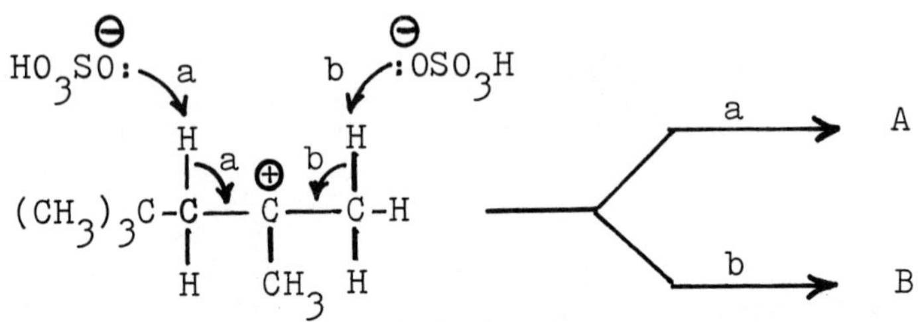

24. $$CH_3{-}CH_2{-}\dot{C}H_2 \;+\; \dot{C}H_2{-}CH(H){-}CH_3 \longrightarrow CH_3CH_2CH_3 + CH_2{=}CHCH_3$$

25. The cyclic bromonium ion formed from bromine and ethene has a structure analogous to that of neutral ethylene oxide. The bromonium ion and ethylene oxide react with nucleophiles <u>via</u> S_N2.

$$\underset{\text{(cation)}}{CH_2{-}CH_2\ (\overset{\oplus}{:Br:})} \xrightarrow{:\ddot{Br}:^{\ominus}} \underset{\text{(neutral)}}{:\ddot{Br}{-}CH_2CH_2{-}\ddot{Br}:}$$

$$\underset{\text{(neutral)}}{CH_2{-}CH_2\ (:O:)} \xrightarrow{:\ddot{O}{-}H^{\ominus}} \underset{\text{(anion)}}{{}^{\ominus}:\ddot{O}{-}CH_2CH_2{-}\ddot{O}{-}H}$$

26. Both OsO_4 and $MnO_4^{\ominus}$ (mild conditions) give the same product with an alkene resulting from syn-addition of hydroxyl groups.

(a) $CH_3CH(OH)CH_2OH$

(b) $CH_3CH_2CH(OH)CH_2OH$

(c) (2R,3R)- and (2S,3R)-$CH_3CHClCH(OH)CH_2OH$

(d) (2S,3S)- and (2R,3S)-$CH_3CHClCH(OH)CH_2OH$

(e) 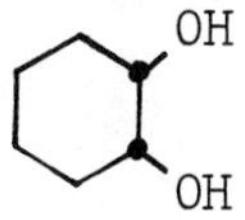 OH OH (meso)

(f) 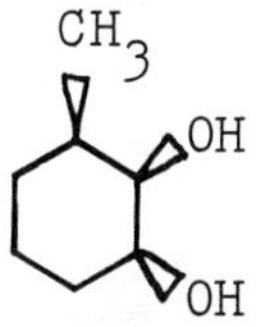 CH_3 (R) OH (R) OH (S) and CH_3 (R) OH (S) OH (R)

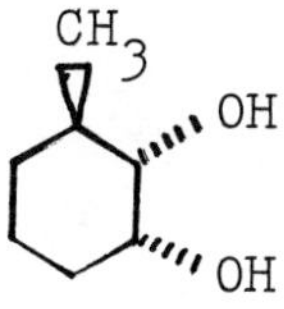

(g) 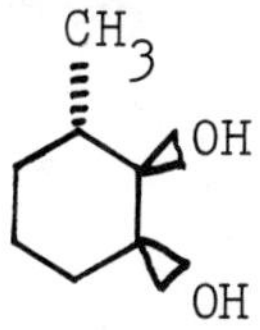 CH_3 (S) OH (R) OH (S) and CH_3 (S) OH (S) OH (R)

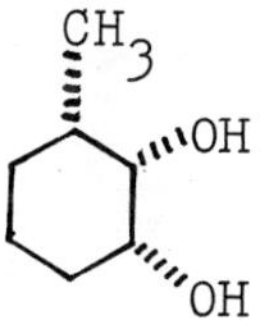

27. Review in the Summary of Functional Group Transformations the correlation between reactants and products in this reaction.

(a) $CH_3COOH + CO_2$

(b) $CH_3CH_2COOH + CO_2$

(c) (R)-$CH_3CHClCOOH + CO_2$

(d) (S)-$CH_3CHClCOOH + CO_2$

(e) $HOOCCH_2CH_2CH_2CH_2COOH$

(continued next page)

(f) (R)-$CH_3CHCH_2CH_2CH_2COOH$ (with COOH on the CH)

(g) (S)-$CH_3CHCH_2CH_2CH_2COOH$ (with COOH on the CH)

28. Review in the Summary of Functional Group Transformations the correlation between reactants and products in this reaction.

(a) $HC(=O)CH_2CH_2CH_2CH_2C(=O)H$

(b) $CH_3C(=O)CH_2CH_2CH_2CH_2C(=O)H$

(c) $HC(=O)CH_2CH_2C(=O)H$ and $CH_3C(=O)-C(=O)CH_3$

29. (a) $CH_3CH{=}CHCH_3$ (E or Z)

(b) $(CH_3)_2C{=}C(CH_3)_2$

(c) $CH_3CH{=}CHCH_2CH_3$ (E or Z)

(d) $(CH_3)_2C{=}CHCH_3$

(e) $CH_3CH{=}CH(CH_2)_4CH{=}CHCH_3$

30. The reactions of ethane, propane and triplet carbene will produce methyl, ethyl, n-propyl and isopropyl alkyl free radicals. All possible combinations of these radicals give:

CH_3CH_3

$CH_3CH_2CH_2CH_3$

$CH_3CH_2CH_2CH_2CH_2CH_3$

$(CH_3)_2CHCH(CH_3)_2$

$CH_3CH_2CH_3$

$CH_3CH_2CH_2CH_3$

$CH_3CH(CH_3)_2$

$CH_3CH_2CH_2CH_2CH_3$

$CH_3CH_2CH(CH_3)_2$

$CH_3CH_2CH_2CH(CH_3)_2$

31. Because of restricted rotation in the intermediate (C) only <u>syn</u> addition is possible.

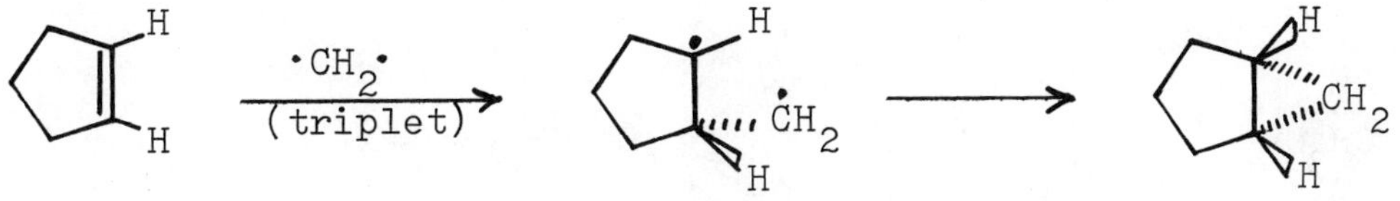

$\cdot CH_2\cdot$ (triplet)

C

restricted rotation

32. (a) $CH_2=CH_2$ + ICH_2ZnI (or $CH_2N_2/h\nu$)

(b)  cyclohexene + ICH_2ZnI (or $CH_2N_2/h\nu$)

(c) 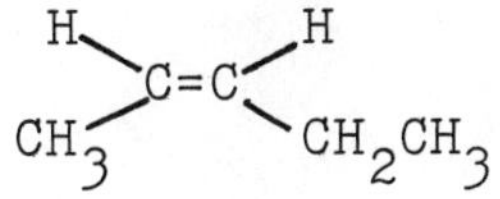 (H)(CH$_3$)C=C(H)(CH$_2$CH$_3$) + ICH_2ZnI (or $CH_2N_2/h\nu$/solvent)

(d) $CH_3CH=CH_2$ + $CHCl_3$/base or $CH_3CH=CCl_2$ + $CH_2N_2/h\nu$

(e) $CH_2=CH_2$ + $CHCl_2CCl_3$/base or $CH_2=C(CCl_3)Cl$ + $CH_2N_2/h\nu$

33. $CH_3CH_2CH_2CH_2CH=CH_2$

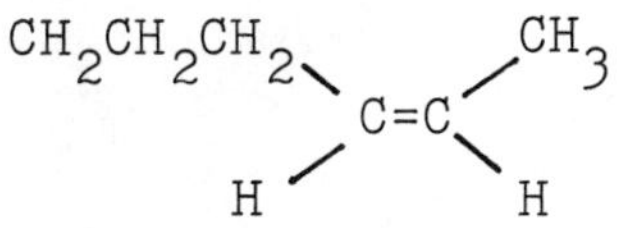

1-hexene

(CH$_3$CH$_2$CH$_2$)(H)C=C(CH$_3$)(H)

(*Z*)-2-hexene

(continued next page)

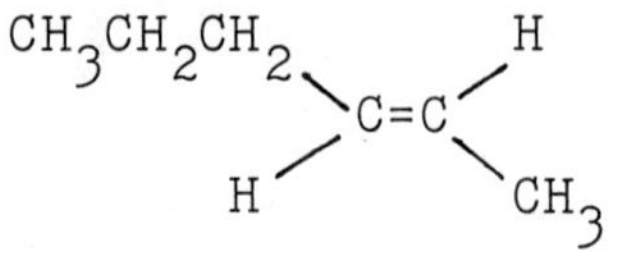

$CH_3CH_2CH_2$(H)C=C(H)(CH_3)

(<u>E</u>)-2-hexene

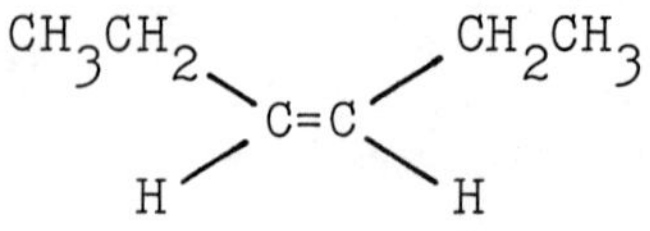

CH_3CH_2(H)C=C(CH_2CH_3)(H)

(<u>Z</u>)-3-hexene

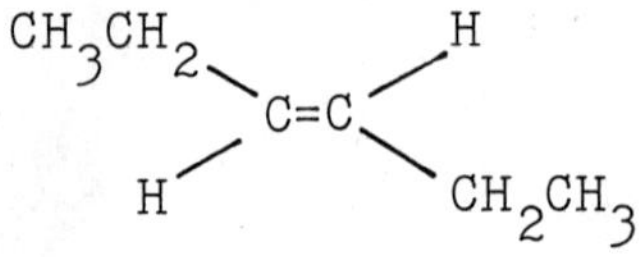

CH_3CH_2(H)C=C(H)(CH_2CH_3)

(<u>E</u>)-3-hexene

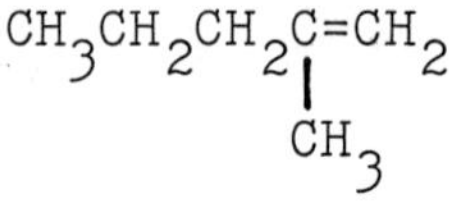

$CH_3CH_2CH_2C(CH_3)=CH_2$

2-methyl-1-pentene

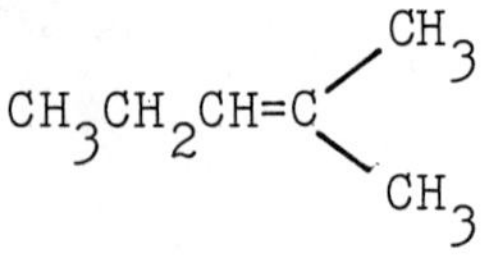

$CH_3CH_2CH=C(CH_3)_2$

2-methyl-2-pentene

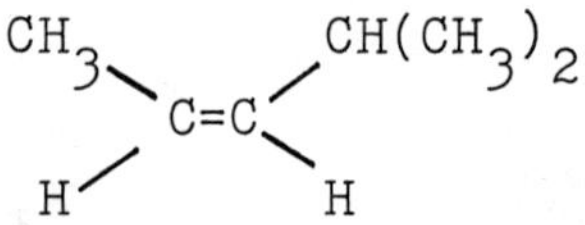

CH_3(H)C=C($CH(CH_3)_2$)(H)

(<u>Z</u>)-4-methyl-2-pentene

CH_3(H)C=C(H)($CH(CH_3)_2$)

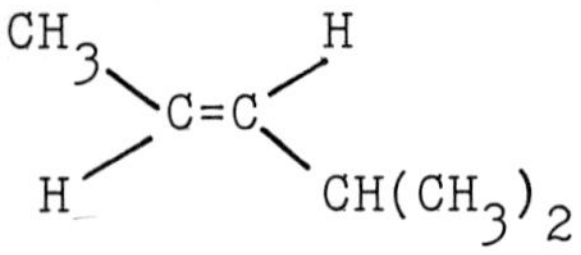

(<u>E</u>)-4-methyl-2-pentene

$CH_2=CHCH_2CH(CH_3)_2$

4-methyl-1-pentene

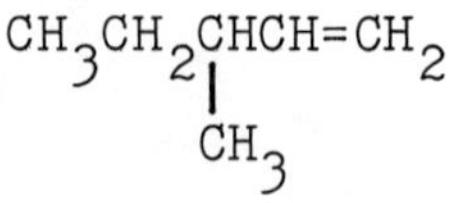

$CH_3CH_2CH(CH_3)CH=CH_2$

(R) and (S)-3-methyl-1-pentene

(continued next page)

$CH_3CH_2(CH_3)C{=}CH(CH_3)$ (CH₃CH₂ and CH₃ on one carbon; CH₃ cis to CH₃CH₂, H cis to CH₃)

(Z)-3-methyl-2-pentene

$CH_3CH_2(CH_3)C{=}CH(CH_3)$

(E)-3-methyl-2-pentene

$CH_2{=}C(CH_3){-}CH(CH_3){-}CH_3$

2,3-dimethyl-1-butene

$(CH_3)_2C{=}C(CH_3)_2$

2,3-dimethyl-2-butene

$CH_2{=}CH{-}C(CH_3)_2{-}CH_3$

3,3-dimethyl-1-butene

$CH_2{=}C(CH_2CH_3){-}CH_2CH_3$

2-ethyl-1-butene

(Text: nomenclature S. 8.2, E, Z system, S. 8.3)

34. (a) E (b) Z (c) Z (d) Z (e) Z (f) E

(g) Z

(Text S. 8.3)

35.

	Product with cis-2-butene	Product with trans-2-butene
(a)	(R)- & (S)-$CH_3CH_2CH(OH)CH_3$	(R)- & (S)-$CH_3CH_2CH(OH)CH_3$
(b)	(R)- & (S)-$CH_3CH_2CH(OCH_3)CH_3$	(R)- & (S)-$CH_3CH_2CH(OCH_3)CH_3$

(continued next page)

	Products with cis 2-butene	Products with trans-2-butene
(c)	(R,R)- & (S,S)-$CH_3CHClCHICH_3$ (*threo*)	(S,R)- & (R,S)-$CH_3CHClCHICH_3$ (*erythro*)
(d)	(R,R)- & (S,S)- $CH_3CHClCHBrCH_3$ (*threo*)	(S,R)- & (R,S)- $CH_3CHClCHBrCH_3$ (*erythro*)
(e)	$CH_3C(=O)-OH$	$CH_3-C(=O)-OH$
(f)	*meso*-$CH_3CH(OH)CH(OH)CH_3$	(S,R)- & (R,S)- $CH_3CH(OH)CH(OH)CH_3$
(g)	*meso*-$CH_3CH(OH)CH(OH)CH_3$	(S,R)- & (R,S)- $CH_3CH(OH)CH(OH)CH_3$
(h)	(R,R)- & (S,S)- $CH_3CHBrCHBrCH_3$ and (R,R) and (S,S)- $CH_3CHBrCHClCH_3$ (*threo*)	*meso*-$CH_3CHBrCHBrCH_3$ and (S,R)- & (R,S)- $CH_3CHBrCHClCH_3$ (*erythro*)
(i)	$CH_3CH_2CH_2CH_3$	$CH_3CH_2CH_2CH_3$
(j)	(R)- & (S)-$CH_3CH_2CH_2CH_2D$	(R)- & (S)-$CH_3CH_2CH_2CH_2D$
(k)	(R)- & (S)-$CH_3CH_2CHDCH_3$	(R)- & (S)-$CH_3CH_2CHDCH_3$
(l)	(R)- & (S)-$CH_3CH_2CH(OSO_3H)CH_3$	(R)- & (S)-$CH_3CH_2CH(OSO_3H)CH_3$
(m)	(R,R)- & (S,S)- $CH_3CHClCH(OH)CH_3$ (*threo*)	(R,S)- & (S,R)- $CH_3CHClCH(OH)CH_3$ (*erythro*)
(n)	*meso*-$CH_3CHDCHDCH_3$	(R,R) and (S,S)-$CH_3CHDCHDCH_3$
(o)	(R,R)-, (S,S)-, (R,S)-, (S,R)- $CH_3CHDCHClCH_3$ (*threo* and *erythro*)	(R,R)-, (S,S)-, (R,S)-, (S,R)- $CH_3CHDCHClCH_3$ (*threo* and *erythro*)

(continued next page)

	Products with cis-2-butene	Products with trans-2-butene
(p) 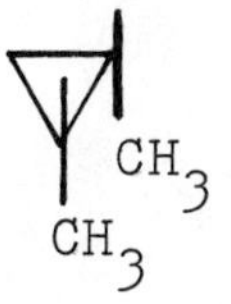	CH_3 CH_3	CH_3 CH_3 & CH_3 CH_3
(q) 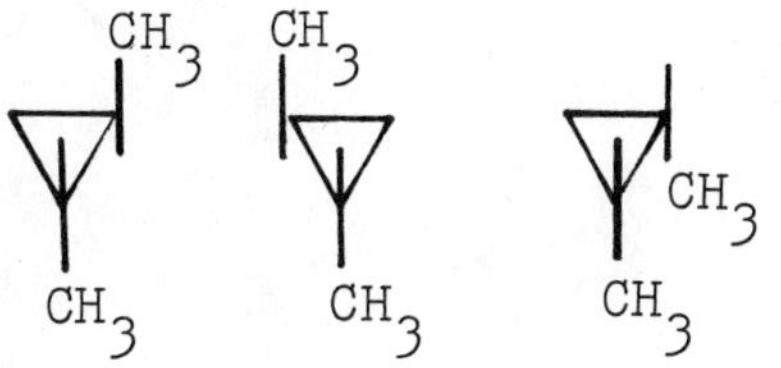	CH_3 CH_3 CH_3 CH_3 CH_3 CH_3 major	same
(r) 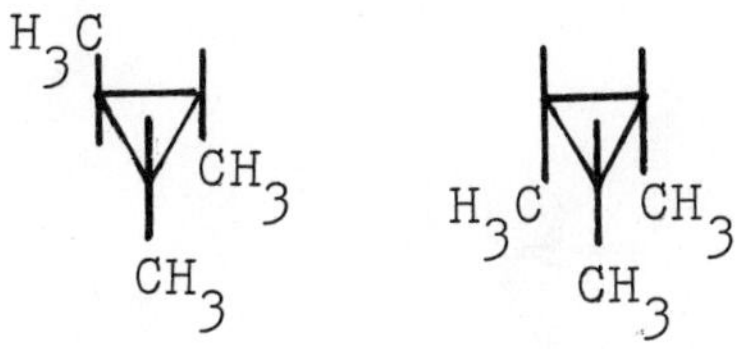	H_3C CH_3 CH_3 H_3C CH_3 CH_3	H_3C CH_3 CH_3
(s) 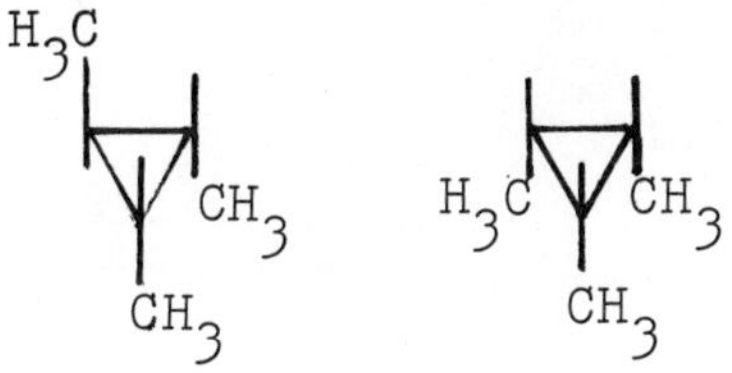	H_3C CH_3 CH_3 H_3C CH_3 CH_3	same

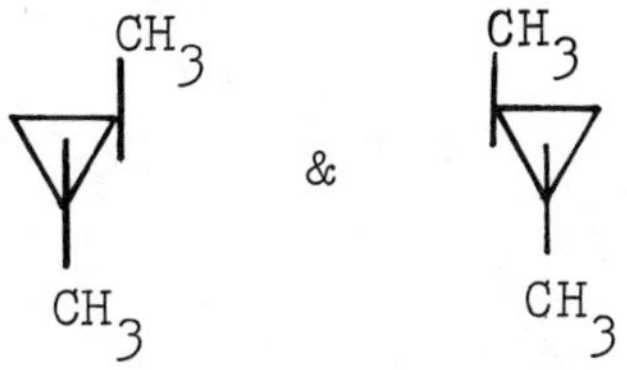

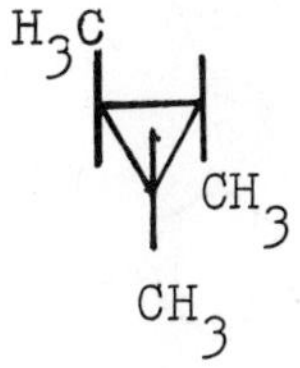

(Ref: Summary of Functional Group Transformations)

36. (a) (1) H_2/metal (metal=Ni, Pt or Pd) <u>or</u>

(2) $H_2/[(C_6H_5)_3P]_3RhCl$ <u>or</u> (3) BH_3 followed by CH_3CO_2H

(b) D_2/metal (Ni, Pt or Pd)

(continued next page)

(c) (1) hot $KMnO_4/H^{\oplus}$ or (2) O_3 followed by $H_2O_2/H^{\oplus}$

(d) O_3 followed by $Zn/H^{\oplus}$

(e) Br_2/CCl_4

(f) (1) $Br_2/h\nu$ or (2) NBS/peroxide

(g) $Cl_2/h\nu$

(h) HCl

(i) (1) 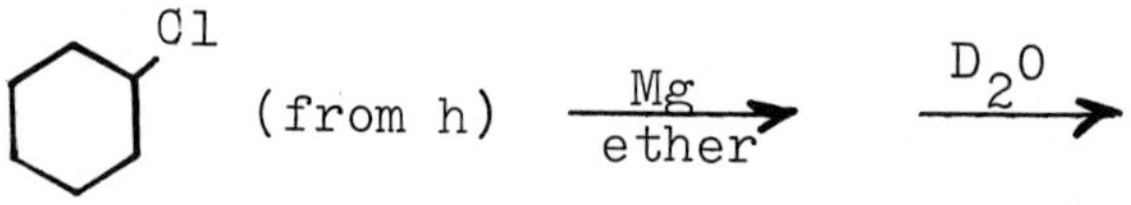 chlorocyclohexane (Cl) (from h) $\xrightarrow[\text{ether}]{\text{Mg}}$ $\xrightarrow{D_2O}$ or

(2) 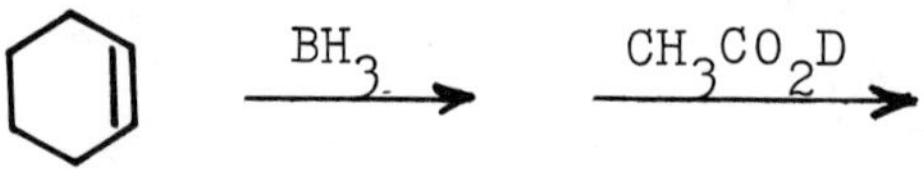 cyclohexene $\xrightarrow{BH_3}$ $\xrightarrow{CH_3CO_2D}$

(j) cold H_2SO_4

(k) (1) $H_2O/H^{\oplus}$ or (2) 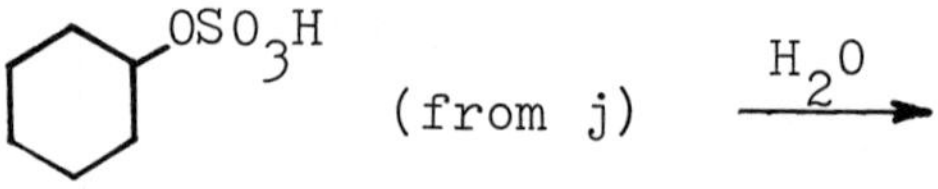 cyclohexyl-OSO_3H (from j) $\xrightarrow{H_2O}$

(l) 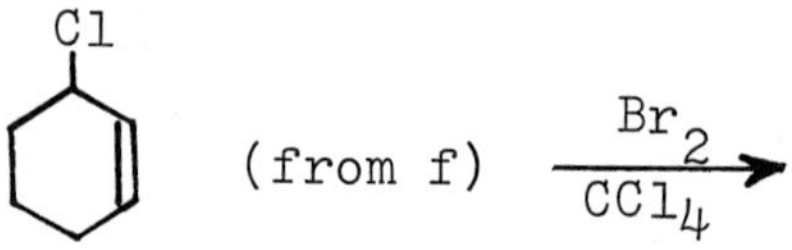 3-chlorocyclohexene (Cl) (from f) $\xrightarrow[CCl_4]{Br_2}$

(m) 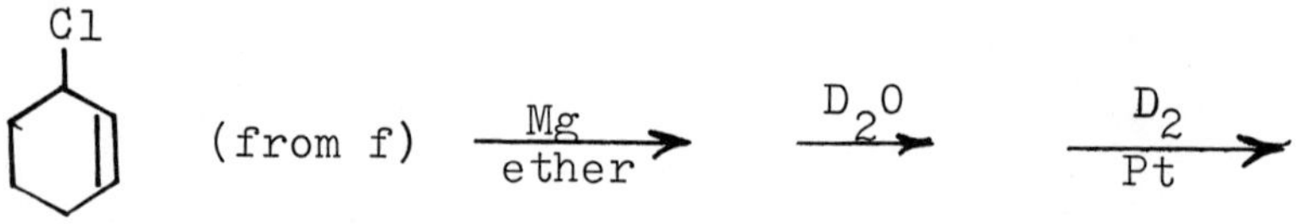 3-chlorocyclohexene (Cl) (from f) $\xrightarrow[\text{ether}]{\text{Mg}}$ $\xrightarrow{D_2O}$ $\xrightarrow[\text{Pt}]{D_2}$

37. Syntheses other than those shown are possible in some cases.

(a) $\xrightarrow{Br_2/\text{light}}$ $CH_3CHBrCH_3$, 2-bromopropane

(continued next page)

(b) $CH_3CHBrCH_3$ (from a) $\xrightarrow{Na^{\oplus}I^{\ominus}/acetone}$ CH_3CHICH_3

2-iodopropane

(c) $CH_3CHBrCH_3$ (from a) $\xrightarrow{KOH/alcohol/heat}$ $CH_3CH{=}CH_2$

propene

(d) $CH_3CH{=}CH_2$ (from c) $\xrightarrow{HBr/peroxide}$ $CH_3CH_2CH_2Br$

1-bromopropane

(e) $CH_3CH{=}CH_2$ (from c) $\xrightarrow{Br_2/CCl_4}$ $CH_3CHBrCH_2Br$

1,2-dibromopropane

(f) $CH_3CH_2CH_2Br$ (from d) $\xrightarrow[acetone]{Na^{\oplus}Cl^{\ominus}}$ $CH_3CH_2CH_2Cl$

1-chloropropane

(g) $CH_3CH{=}CH_2$ (from c) $\xrightarrow{Cl_2/300^{\circ}}$ $ClCH_2CH{=}CH_2$

3-chloropropene

(h) $ClCH_2CH{=}CH_2$ (from c) $\xrightarrow{Mg/ether}$ $ClMgCH_2CH{=}CH_2$ $\xrightarrow{D_2O}$

$DCH_2CH{=}CH_2$, 3-deuteriopropene

(i) $CH_3CH{=}CH_2$ (from c) $\xrightarrow{D_2/Pt}$ CH_3CHDCH_2D

1,2-dideuteriopropane

(j) $DCH_2CH{=}CH_2$ (from h) $\xrightarrow{D_2/Pt}$ DCH_2CHDCH_2D

1,2,3-trideuteriopropane

(continued next page)

(k) $CH_3CH_2CH_2Cl$ (from f) $\xrightarrow[\text{ether}]{\text{Mg}}$ $\xrightarrow{D_2O}$ $CH_3CH_2CH_2D$

1-deuteriopropane

(l) $CH_3CH_2CH_2Cl$ (from f) $\xrightarrow{Na^{\oplus}CN^{\ominus}}$ $CH_3CH_2CH_2CN$

butanenitrile
(n-propyl cyanide)

(m) $CH_3CHBrCH_3$ (from a) $\xrightarrow[\text{ether}]{\text{Mg}}$ $\xrightarrow{D_2O}$ CH_3CHDCH_3

2-deuteriopropane

(n) $CH_3CH_2CH_2Cl$ (from f) $\xrightarrow{OH^{\ominus}/H_2O}$ $CH_3CH_2CH_2OH$

1-propanol
(n-propyl alcohol)

(o) $CH_2{=}CHCH_2Cl$ (from g) $\xrightarrow{OH^{\ominus}/H_2O}$ $CH_2{=}CHCH_2OH$

2-propen-1-ol

(p) $CH_3CHBrCH_3$ (from a) $\xrightarrow{CN^{\ominus}}$ $CH_3CH(CN)CH_3$

2-methylpropanenitrile
(isopropyl cyanide)

(q) $CH_3CH{=}CH_2$ (from c) $\xrightarrow{O_3}$ $\xrightarrow{Zn/CH_3COOH}$ $CH_3C(=O)H$

acetaldehyde

(r) $CH_3CH{=}CH_2$ (from c) $\xrightarrow{\text{hot } KMnO_4,\ H^{\oplus}}$ $CH_3C(=O)OH$ (acetic acid) + CO_2

(continued next page)

(s) $CH_3CH{=}CH_2$ (from c) $\xrightarrow[\text{room temp.}]{\text{dil. }KMnO_4}$ $CH_3CH(OH)CH_2OH$

1,2-propanediol

38. (a) Difficulty in controlling the fluorination reaction makes it impractical. Chlorination or bromination is preferred.

(b) The presence of water with magnesium in step two will prevent formation of the Grignard reagent. Use magnesium with dry ether as the solvent. Also, CD_4 is not strong enough an acid to supply deuterium. Use D_2O.

(c) A tetrasubstituted double bond is difficult to reduce with the Rh reagent. H_2 and Pt or Pd is more suitable. Also, reaction of the dibromide with Mg (step 3) gives the mono-Grignard which undergoes internal elimination (Elcb).

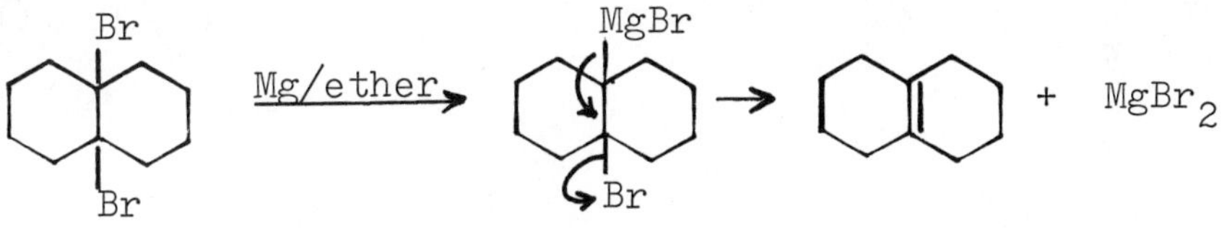

Br, Br (dibromide) $\xrightarrow{\text{Mg/ether}}$ MgBr, Br $\rightarrow$ (alkene) + $MgBr_2$

Use:

(diene) $\xrightarrow[\text{benzene}]{(Ph_3P)_3RhCl}$ (alkene) $\xrightarrow[\text{Pt}]{D_2}$ D, D

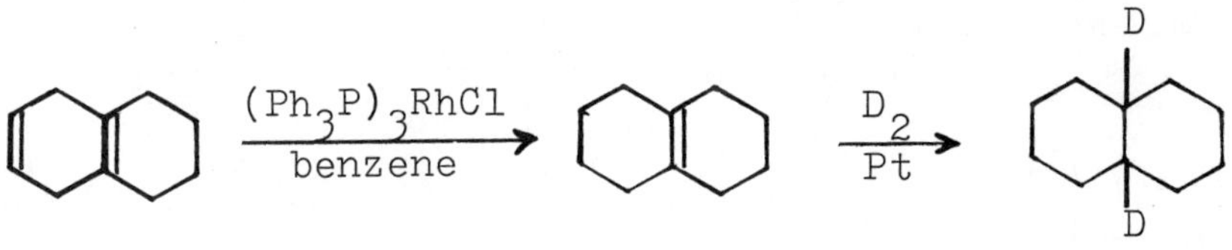

(d) Rearrangement is likely in step 1.

(S)-$CH_2{=}CH{-}CH(CH_3){-}(CH_2)_4CH_3$ $\xrightarrow{\text{HCl}}$ (R)- and (S)-$CH_3CH_2CCl(CH_3)(CH_2)_4CH_3$

(continued next page)

Even if $CH_3CH{=}\overset{CH_3}{\overset{|}{C}}(CH_2)_4CH_3$ were formed after step two, dimide reduction of this product gives both (R)- and (S)-$CH_3CH_2\underset{(CH_2)_4CH_3}{\underset{|}{C}H}CH_3$ since syn-addition of hydrogen occurs from both faces of the C=C.

Catalytic hydrogenation of the initial reactant gives the desired product (the R,S priorities change).

39.

Name	Reductive Work Up	Oxidative Work Up
a. methylpropene	$CH_3\overset{O}{\overset{\|}{C}}CH_3$, $H\text{-}\overset{O}{\overset{\|}{C}}\text{-}H$	$CH_3\overset{O}{\overset{\|}{C}}CH_3$, $H\text{-}\overset{O}{\overset{\|}{C}}\text{-}OH$
b. 2-methyl-2-butene	$CH_3\overset{O}{\overset{\|}{C}}CH_3$, $CH_3\overset{O}{\overset{\|}{C}}H$	$CH_3\overset{O}{\overset{\|}{C}}CH_3$, $CH_3\overset{O}{\overset{\|}{C}}OH$
c. 2,3-dimethyl-2-butene	$CH_3\overset{O}{\overset{\|}{C}}CH_3$	same
d. (2E,4E)-2,4-hexadiene	$CH_3\overset{O}{\overset{\|}{C}}H$, $H\overset{O}{\overset{\|}{C}}\text{-}\overset{O}{\overset{\|}{C}}H$	$CH_3\overset{O}{\overset{\|}{C}}OH$, $HO\overset{O}{\overset{\|}{C}}\text{-}\overset{O}{\overset{\|}{C}}OH$
e. (2E,4Z)-2,4-hexadiene	$CH_3\overset{O}{\overset{\|}{C}}H$, $H\overset{O}{\overset{\|}{C}}\text{-}\overset{O}{\overset{\|}{C}}H$	$CH_3\overset{O}{\overset{\|}{C}}OH$, $HO\overset{O}{\overset{\|}{C}}\text{-}\overset{O}{\overset{\|}{C}}OH$
f. (2Z,4Z)-2,4-hexadiene	$CH_3\overset{O}{\overset{\|}{C}}H$, $H\overset{O}{\overset{\|}{C}}\text{-}\overset{O}{\overset{\|}{C}}H$	$CH_3\overset{O}{\overset{\|}{C}}OH$, $HO\overset{O}{\overset{\|}{C}}\text{-}\overset{O}{\overset{\|}{C}}OH$
g. (2E,4E)-3-methyl-2,4-hexadiene	$CH_3\overset{O}{\overset{\|}{C}}H$, $CH_3\overset{O}{\overset{\|}{C}}\text{-}\overset{O}{\overset{\|}{C}}H$	$CH_3\overset{O}{\overset{\|}{C}}OH$, $CH_3\overset{O}{\overset{\|}{C}}\text{-}\overset{O}{\overset{\|}{C}}OH$
h.	$H\overset{O}{\overset{\|}{C}}(CH_2)_4\overset{O}{\overset{\|}{C}}H$	$HO\overset{O}{\overset{\|}{C}}(CH_2)_4\overset{O}{\overset{\|}{C}}OH$

(continued next page)

i. $HC(=O)-C(=O)-(CH_2)_4C(=O)(CH_2)_2CH(=O)$ $HOC(=O)-C(=O)(CH_2)_4C(=O)(CH_2)_2COH(=O)$

j. $HC(=O)CH_2CH(=O)$, 2-oxocyclohexane-C(=O)-H $HOC(=O)CH_2C(=O)OH$, 2-oxocyclohexane-C(=O)OH

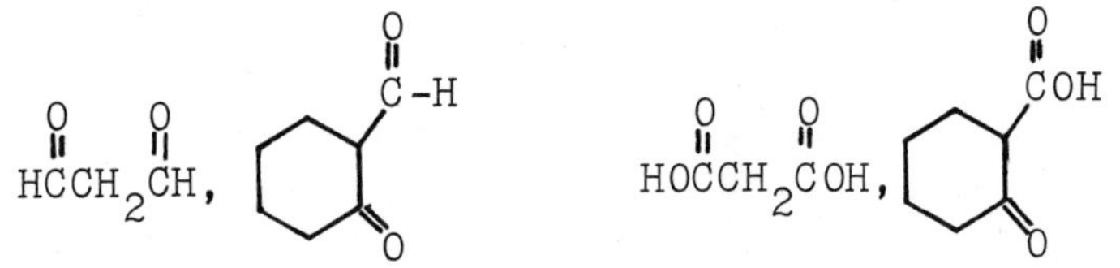

k. cyclopentane-1,3-di-CH(=O)   cyclopentane-1,3-di-C(=O)OH

l. cyclohexane with C(=O)H and $CH_2CH_2C(=O)H$ cyclohexane with C(=O)OH and $CH_2CH_2C(=O)OH$

(Text: ozonalysis S. 8.9)

40. Compare the relative stabilities of the carbocations formed in reaction of $CH_3CH{=}CHBr$ and $H^{\oplus}$. The more stable cation is formed faster and accounts for the observed product, $CH_3CH_2CHBrCl$.

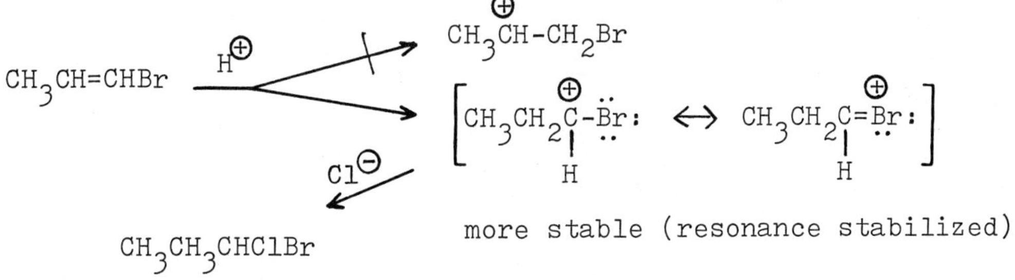

$$CH_3CH{=}CHBr \xrightarrow{H^{\oplus}} \not\to CH_3\overset{\oplus}{C}H-CH_2Br$$

$$\to \left[CH_3CH_2\overset{\oplus}{C}(H)-\ddot{B}r: \leftrightarrow CH_3CH_2C(H)=\overset{\oplus}{B}r:\right] \xrightarrow{Cl^{\ominus}} CH_3CH_3CHClBr$$

more stable (resonance stabilized)

$CH_3CH_3CHClBr$
(major product)

(Text: resonance stabilization of the type $\left[\overset{\oplus}{C}-\ddot{Z} \leftrightarrow -C=\overset{\oplus}{Z}\right]$ S 5.7)

41. Compare the relative stabilities of the carbocations formed in the reaction between $CH_2{=}CH{-}CO_2H$ and $H^{\oplus}$. The more stable cation is formed faster and accounts for the observed product, A.

$$CH_2{=}CH{-}\overset{O}{\overset{\|}{C}}{-}OH \xrightarrow{H^{\oplus}} \begin{cases} CH_3\overset{\oplus}{C}H \rightarrow \overset{\delta\ominus\ O}{\overset{\|}{C}}{-}OH\ (\delta\oplus) & \text{(not formed)} \\ \overset{\oplus}{C}H_2CH_2 \rightarrow \overset{\delta\ominus\ O}{\overset{\|}{C}}{-}OH\ (\delta\oplus) \xrightarrow{Cl^{\ominus}} ClCH_2CH_2\overset{O}{\overset{\|}{C}}{-}OH \end{cases}$$

(more stable)

Because oxygen is strongly electronegative, the carbon of the -COOH group has a positive ($\delta^{\oplus}$) character; thus, the cation with adjacent positive charges is destabilized.

(Text: anti-Markownikoff addition of HX, S. 8.6)

42. The carbocation intermediates in this reaction are 1 and 2 below.

Methylenecyclobutane ($=CH_2$) $\xrightarrow{H^{\oplus}}$ cyclobutyl cation with CH_3 and $\oplus$ on ring carbon, ring angle 90° (1); cyclobutane ring carbon bearing H and $\overset{\oplus}{C}H_2$ (2)

(continued next page)

Although cation 1 is 3°, it can not attain the stable, planar configuration (bond angles 120°) that 1° cation 2 can, thus in this example, the 1° cation is formed faster.

Since either cation derived from methylenecyclohexane can achieve the desired configuration, the normal order of cation stability is observed; i.e., the 3° cation forms faster.

(Text: S. 8.6)

43. This ring expansion requires a 1,2-shift occurring simultaneously with loss of the leaving group.

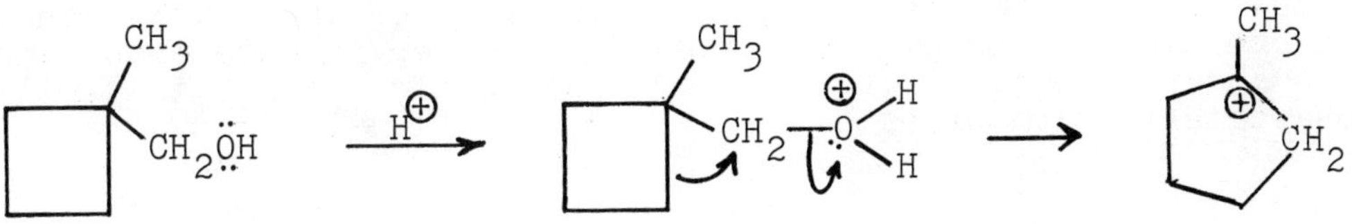

CH_3 $CH_2\ddot{O}H$ $\xrightarrow{H^{\oplus}}$ CH_3 CH_2 $\overset{\oplus}{O}H_2$ $\longrightarrow$ CH_3 $\oplus$ CH_2

(3° cation)

CH_3 $\oplus$ H H $:OSO_3H^{\ominus}$ $\xrightarrow{E1}$ CH_3

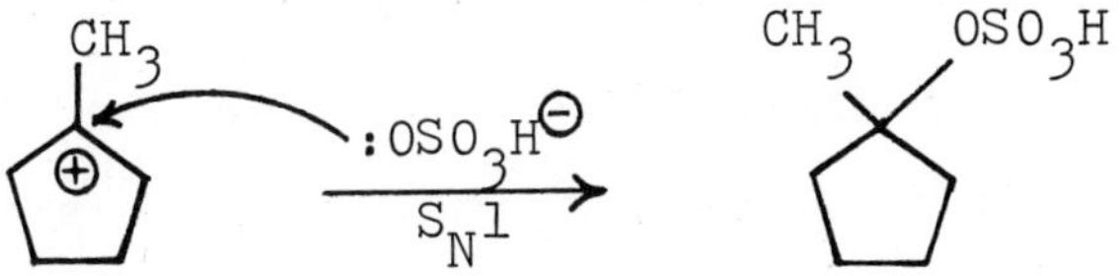

CH_3 $\oplus$ $:OSO_3H^{\ominus}$ $\xrightarrow{S_N1}$ CH_3 OSO_3H

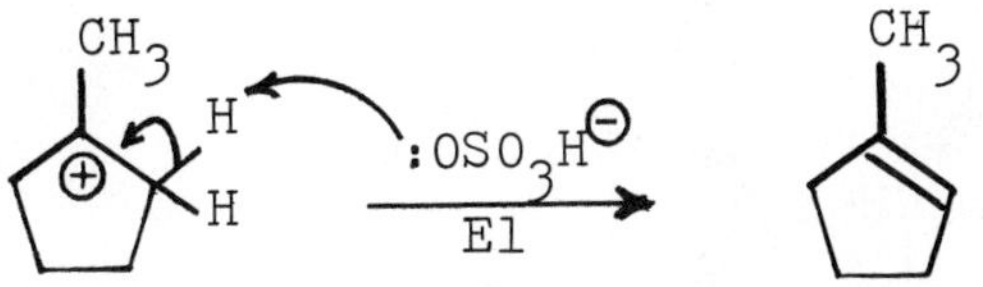

44. The increasing reaction rates parallel the increasing polarity of the bonds (electronegativity: Cl > Br > I).

45. If the reaction involves one HCl per alkene unit, a kinetic study will show $R = k[\text{alkene}][\text{HCl}]$; if two HCl molecules per alkene unit are required, $R = k[\text{alkene}][\text{HCl}]^2$.

46. trans-Cyclooctene can exist in enantiomeric forms.

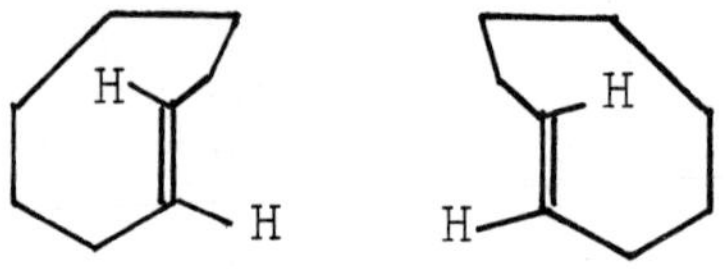

Chirality results from "twisting" about the carbon-carbon double bond. The carbon chain is above the plane of the double bond on one side and below the plane on the other. These enantiomers are interconvertible, but the process is slow enough to permit characterization of each.

47. (1) $CH_3CH{=}CH_2 + D^{\oplus} \underset{k_{-1}}{\overset{k_1}{\rightleftharpoons}} CH_3\overset{\oplus}{C}HCH_2D$

(2) $CH_3\overset{\oplus}{C}HCH_2D + Cl^{\ominus} \underset{k_{-2}}{\overset{k_2}{\rightleftharpoons}} CH_3CH(Cl)CH_2D$

The reverse of reaction 1 would be expected to provide some $CH_3CH{=}CHD$ since a C-D bond is more difficult to break than a C-H bond. Since no $CH_3CH{=}CHD$ is formed, k_2 must be much greater than k_{-1}.

48. 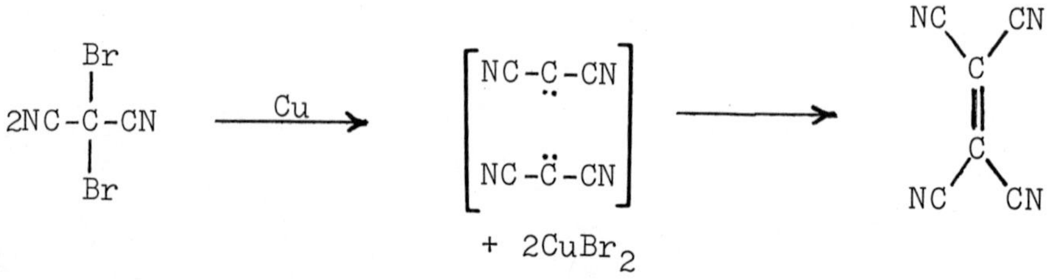 $2NC\text{-}CBr_2\text{-}CN \xrightarrow{Cu} [NC\text{-}\ddot{C}\text{-}CN \quad NC\text{-}\ddot{C}\text{-}CN] + 2CuBr_2 \longrightarrow (NC)_2C{=}C(CN)_2$

49. (a) Both compounds give $HO\overset{O}{\overset{\|}{C}}CH_2CH_2\overset{O}{\overset{\|}{C}}OH$; however, 1,5-hexadiene also gives carbon dioxide which the chemist failed to detect.

(b) Conduct the reaction under reductive conditions ($Zn/H^{\oplus}$) which gives formaldehyde instead of carbon dioxide.

50. Methylene is generated in the reaction between tetramethylammonium chloride and phenylsodium.

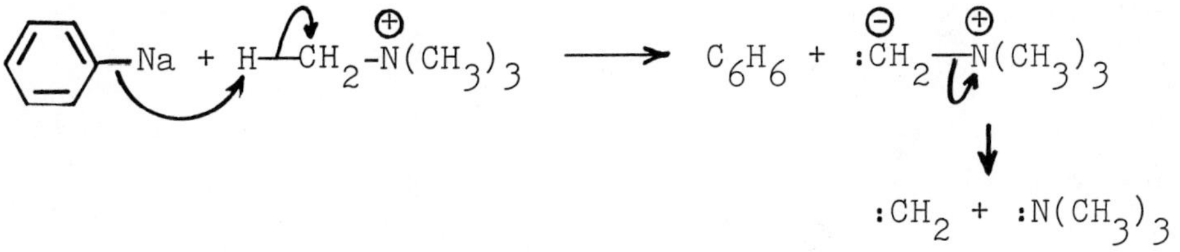

$$C_6H_5Na + H-CH_2-\overset{\oplus}{N}(CH_3)_3 \longrightarrow C_6H_6 + :\overset{\ominus}{C}H_2-\overset{\oplus}{N}(CH_3)_3 \longrightarrow :CH_2 + :N(CH_3)_3$$

The reaction of methylene with cyclohexene generates norcarane.

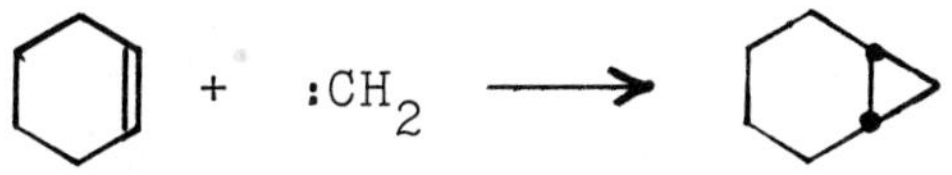

$$\text{cyclohexene} + :CH_2 \longrightarrow \text{norcarane}$$

51. London dispersion forces. A plasticizer may reduce these forces preventing "bonding" between polymer layers.

52. (a) $R\text{-}\underset{H}{CH}\text{—}\underset{B<}{CH}\text{-}CH_2\text{-}CH_3 \rightleftharpoons RCH_2\text{-}HC{\cdots}CHCH_3\ (\text{—B}\cdots H) \rightleftharpoons RCH_2CH{=}CHCH_3$

$$\overset{BH_3}{\rightleftharpoons} RCH_2HC{\cdots}CHCH_3\ (H\cdots B\text{—}) \rightleftharpoons RCH_2\underset{H}{CH}\text{—}\underset{B-}{CHCH_3} \rightleftharpoons RCH_2CH_2HC{\cdots}CH_2\ (\text{—B}\cdots H)$$

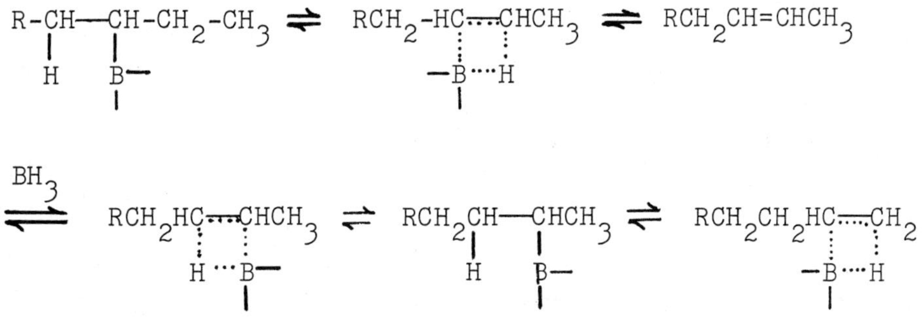

(continued next page)

$$\rightleftharpoons RCH_2CH_2CH{=}CH_2 \xrightleftharpoons{BH_3} RCH_2CH_2C{\cdots}CH_2 \ (\text{H}\cdots\text{B}) \rightleftharpoons RCH_2CH_2C(H){-}CH_2(B{-})$$

(b) $RCH_2CH_2CH_2CH_2B\langle$

(c) Reduce with RCO_2D. The product will be $RCH_2CH_2CH_2CH_2D$.

53. (a) The initial step of bromine <u>addition</u> gives an intermediate (A) higher in energy than the intermediate (B) obtained in <u>free-radical substitution</u>. Thus the formation of B is favored, and at low bromine concentration, cyclohexane is eventually converted to 3-bromocyclohexene.

A (2-bromocyclohexyl radical) $\xleftarrow{Br\cdot}$ cyclohexene $\xrightarrow{Br\cdot}$ [allylic radical ↔ allylic radical] + HBr

A — B resonance stabilized

(b) The function of NBS is to provide bromine which is produced by the reaction between NBS and traces of acid present.

N-bromosuccinimide (N-Br) + HBr ⟶ succinimide (N-H) + Br_2

$Br_2 + RO\cdot \longrightarrow Br\cdot + ROBr$

$Br\cdot$ + cyclohexene ⟶ HBr + [allylic radical ↔ allylic radical]

(continued next page)

$$\left[\text{cyclohexenyl radical}\right]^{\cdot} + Br_2 \longrightarrow \text{3-bromocyclohexene} + Br\cdot$$

etc.

54. (a) $(CH_3CH_2)_3B$, $\left[(CH_3)_2CHCH(CH_3)\right]_2BH$, $(CH_3)_3C\text{-}CH(BH_2)\text{-}CH(CH_3)_2$

(b) The extent of alkylation of diborane is related to the branching at the C=C of the alkene and, therefore, to the steric hindrance at the boron of the initially formed alkylborane. For example, the monoalkylborane from 2,4,4-trimethyl-2-pentene is too crowded about the boron to permit a second alkene molecule to react.

(Text S. 8.5)

55. While the compounds of set B each have carbon-carbon double bonds which are trisubstituted, one compound in set A has a disubstituted double bond and the other compound has a trisubstituted double bond. Thus the difference in heats of hydrogenation of the compounds in set A may in fact reflect the extent of substitution at the carbon-carbon double bond rather than endocyclic vs. exocyclic stability.

56. The stereochemistry of the reaction is <u>anti</u>-addition. The observed experimental results can be explained by assuming that a cyclic intermediate free radical is formed between the alkene and a bromine atom.

(continued next page)

$$\text{D-Br} + \text{R}\cdot \longrightarrow \text{R-D} + \text{Br}\cdot$$

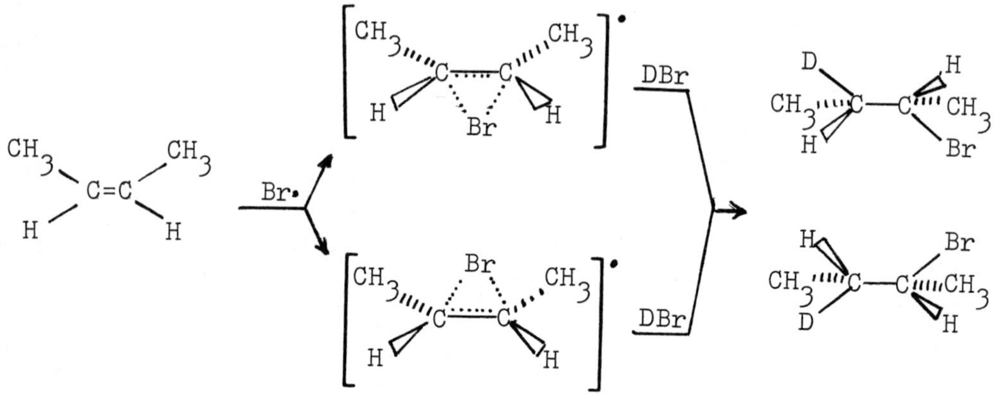

CH3 C=C CH3 / H H —Br•→ [bridged radicals] —DBr→

enantiomers
(threo)

CH3 C=C H / H CH3 —Br•→ [bridged radicals] —DBr→

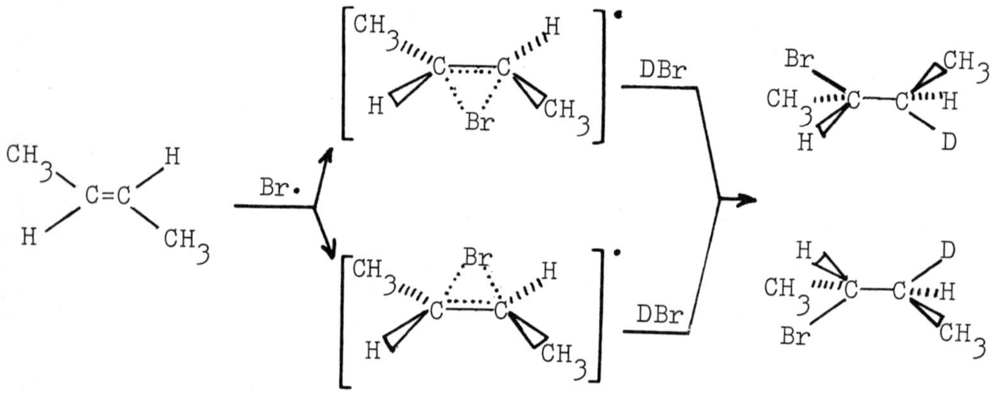

enantiomers
(erythro)

Only one resonance form is shown above for clarity in illustrating bond changes.

57. 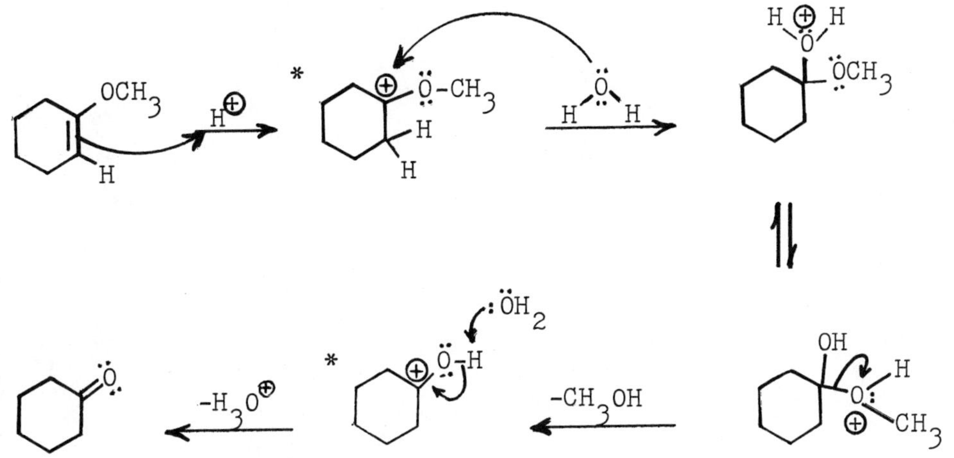 [Mechanism: 1-methoxycyclohexene + $H^{\oplus}$ → * cation; + H_2O → oxonium adduct ⇌ protonated hemiacetal; $-CH_3OH$ → * cation; $-H_3O^{\oplus}$ → cyclohexanone]

*These cations are resonance stabilized.

$$\left[-\overset{\oplus}{\underset{|}{C}}-\ddot{\underset{\cdot\cdot}{O}}- \longleftrightarrow -\underset{|}{C}=\overset{\oplus}{O}- \right]$$

58. (a) Subject a sample of either pure cis- or trans-1,2-dimethylcyclohexane to catalytic reduction conditions. Either should give a mixture containing both cis- and trans-1,2-dimethylcyclohexane if isomerization occurs on the catalyst surface.

(b) [cis-1,2-dimethylcyclohexane $\underset{}{\overset{-H_2}{\rightleftharpoons}}$ [1,2-dimethylcyclohexene or 2,3-dimethylcyclohexene or 1,6-dimethylcyclohexene] $\overset{+H_2}{\rightleftharpoons}$ trans-1,2-dimethylcyclohexane]

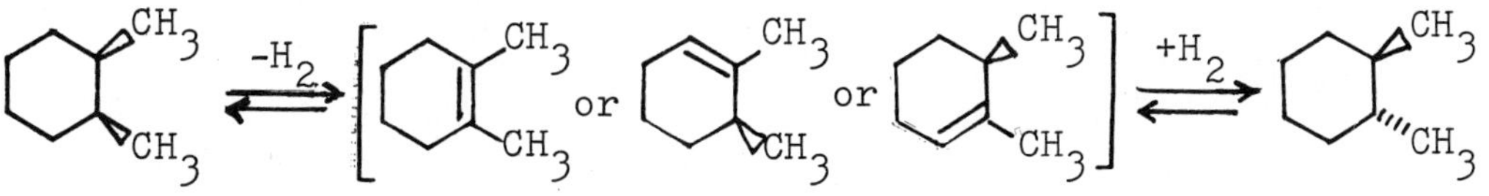

(continued next page)

(c)

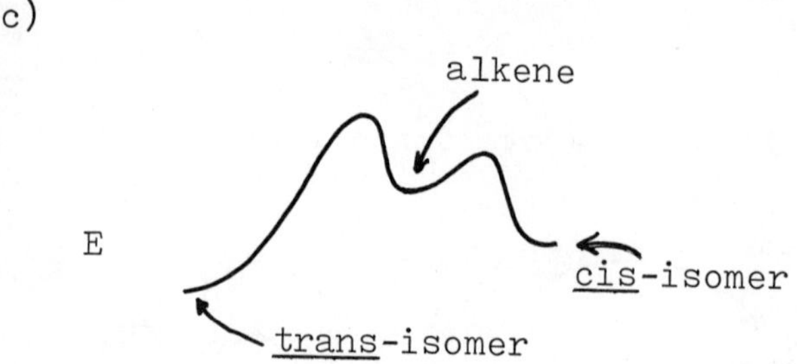

Reaction Coordinate

(Text S. 8.5)

59. (a) $\rightarrow$ (stereospecific)

singlet

(minor)

d,ℓ pair (major)

(continued next page)

(b) Cyclobutene:

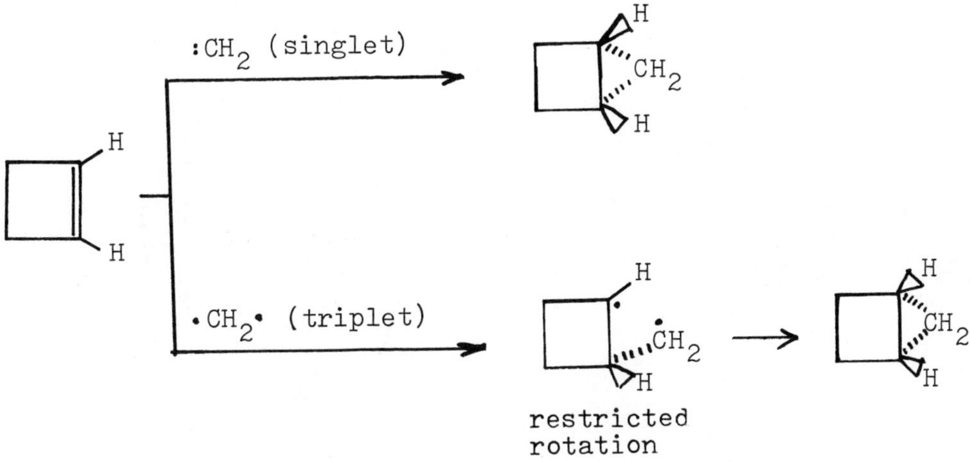

Cyclodecene:

H
H
:CH_2 (singlet)
H
CH_2
H
•CH_2• (triplet)
H
CH_2
H
H
•CH_2
H
CH_2
H
H
H

(Text S 8.10)

60. (a) Consider the relative stabilities of the intermediate carbocations.

$$CH_2{=}CH{-}\ddot{F}: \xrightarrow{H-F} \overset{\oplus}{C}H_2{-}CH_2F$$

$$CH_2{=}CH{-}\ddot{F}: \xrightarrow{H-F} \left[CH_3{-}\overset{\oplus}{C}H{-}\ddot{F}: \longleftrightarrow CH_3{-}CH{=}\overset{\oplus}{F}: \right]$$

more stable

$$\downarrow F^{\ominus}$$

$$CH_3CHF_2$$

(b) Radicals are produced in the reaction of HBr + $h\nu$. The species which initially reacts with the alkene is Br• which adds to the carbon-carbon double bond to give the more stable alkyl free radical.

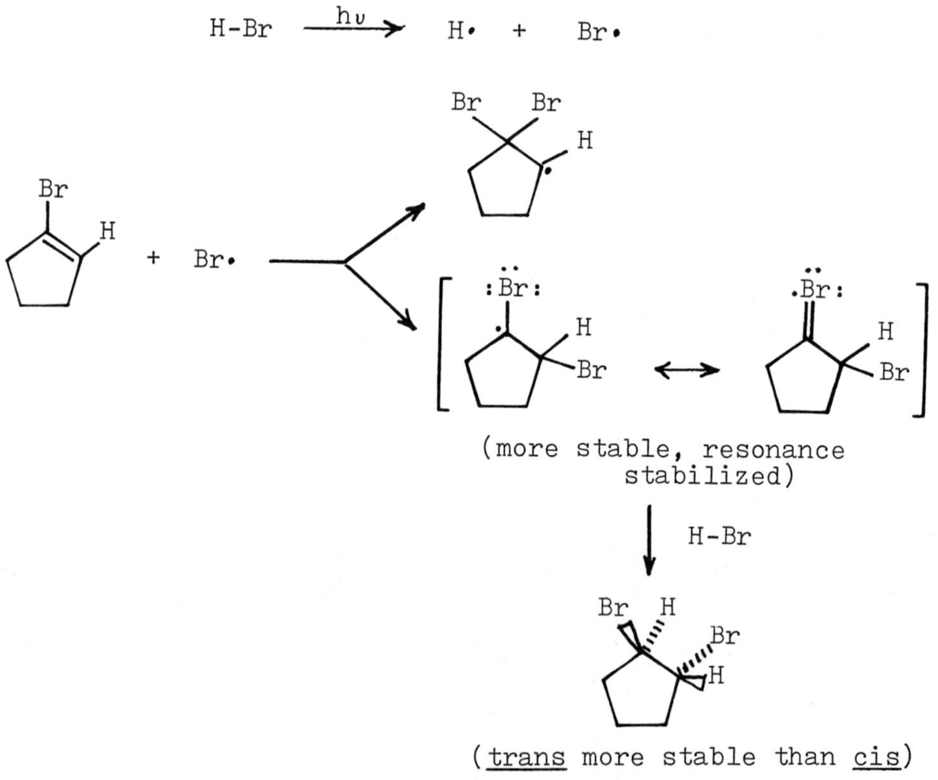

(<u>trans</u> more stable than <u>cis</u>)

(continued next page)

(c) The cation formed by reaction of 2-pentene and chlorine can react with the nucleophiles Cl^- and CH_3OH (two ways).

$$CH_3CH{=}CHCH_2CH_3 \xrightarrow{Cl_2} CH_3\overset{}{CH}\text{—}CHCH_2CH_3 \;(\text{bridged by } Cl^{\oplus})\quad Cl^{\ominus}$$

<u>reaction with $Cl^{\ominus}$</u>

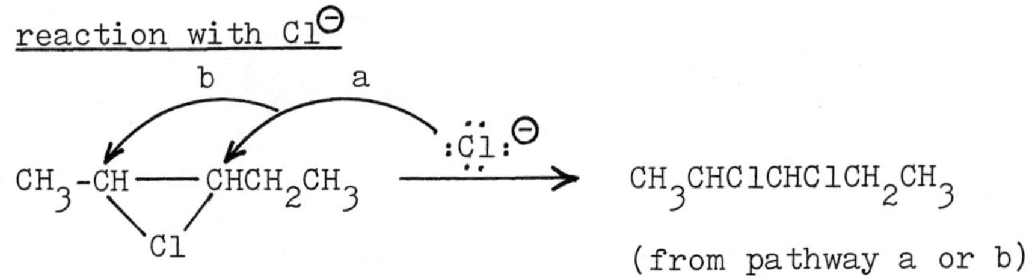

<u>reaction with CH_3OH</u>

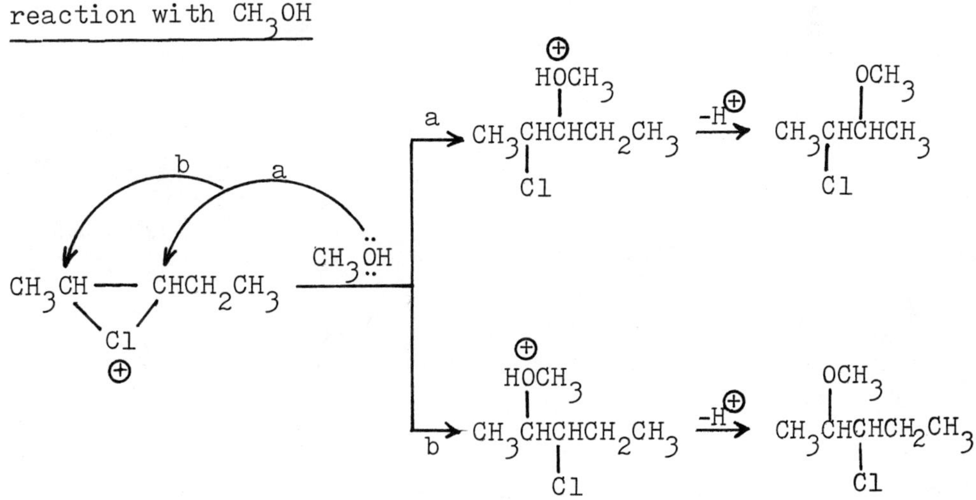

(d) $(CH_3)_2C{=}CH_2 \xrightarrow{H^{\oplus}} (CH_3)_3C^{\oplus} \xrightarrow{CH_2=CH_2} (CH_3)_3CCH_2\overset{\oplus}{C}H_2$

(more reactive than $CH_2{=}CH_2$; reacts first)

(3° cation)

$(CH_3)_3CCH_2\overset{\oplus}{C}H_2 \xrightarrow{:\ddot{Cl}:^{\ominus}} (CH_3)_3CCH_2CH_2Cl$

(continued next page)

(e) SbF_6 is a Lewis acid.

$$:\ddot{Br}-CH_2CH_2-\ddot{F}: + SbF_5 \longrightarrow :\ddot{Br}-CH_2-CH_2-\overset{\oplus}{\ddot{F}}-\overset{\ominus}{SbF_5}$$

$$\downarrow$$

$$\underset{:\overset{\oplus}{Br}:}{CH_2-CH_2} + SbF_6^{\ominus}$$

(f) Chlorine is more electronegative than iodine.

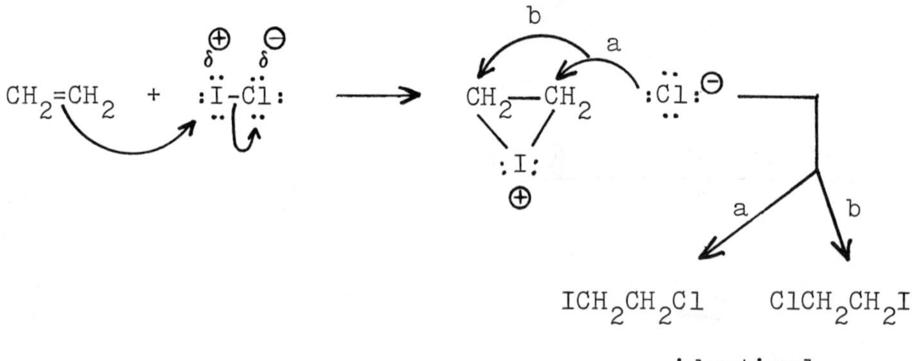

61. In addition polymerization, π bonds are converted to σ bonds which stabilize (lower the energy) of the system and energy is released.

62. It is possible to develop an R,S description. At any carbon other than the one at the exact center, one part of the polymer chain is longer than the other; therefore, the carbon is chiral and can be assigned an R or S configuration. Even a carbon at the center of the molecule may be chiral if the two end groups are not the same.

63. Oxygen has an unpaired electron and may initiate a free-radical polymerization.

(Text: See the MO diagram for oxygen, Ch. 2)

64. (a) C and D are shown below along with equations illustrating their formation.

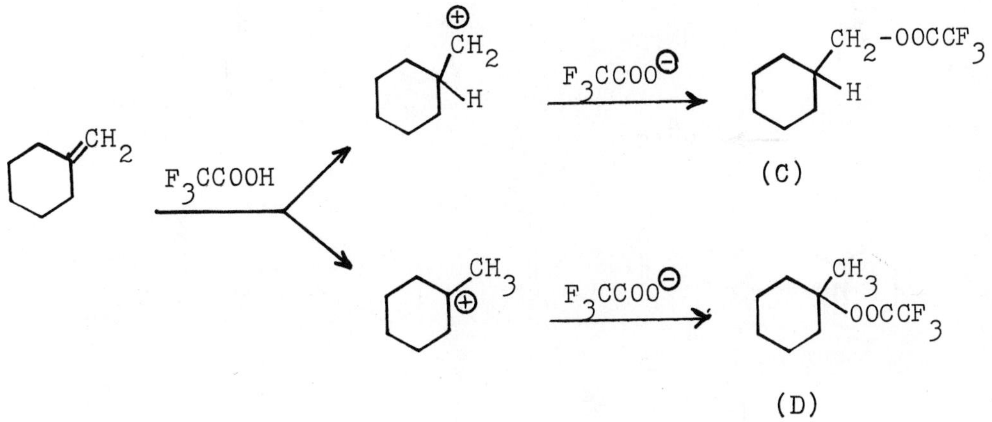

F_3CCOOH ; $F_3CCOO^{\ominus}$; CH_2-$OOCCF_3$, H (C) ; CH_3, $OOCCF_3$ (D)

(b) D is the major product (from the more stable carbocation). The 3H singlet in the nmr is due to the methyl group in D.

(Text: Ionic Addition S. 8.6)

65. (a) Chemical: different ozonolysis products for the first compound (reductive conditions are used to illustrate).

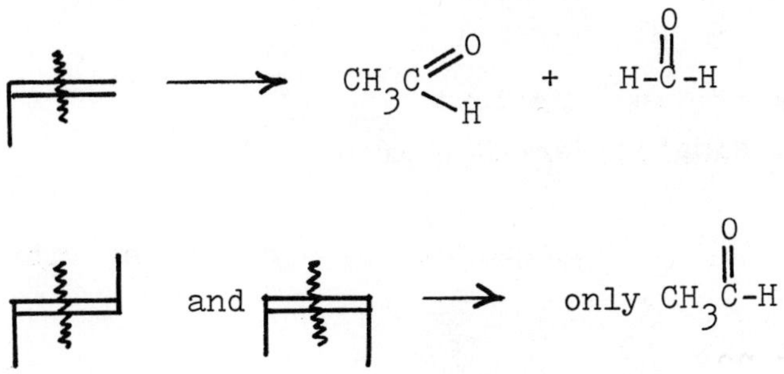

$\longrightarrow$ CH_3CHO + H-C(=O)-H

and $\longrightarrow$ only CH_3C(=O)-H

ir (C-H out-of-plane bending vibrations)

(continued next page)

990 and 910 cm^{-1}, 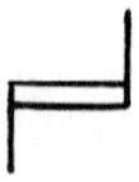 965 cm^{-1}, 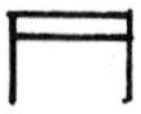 670-735 cm^{-1}

(b) Chemical: each gives different ozonolysis products (reductive conditions are used to illustrate).

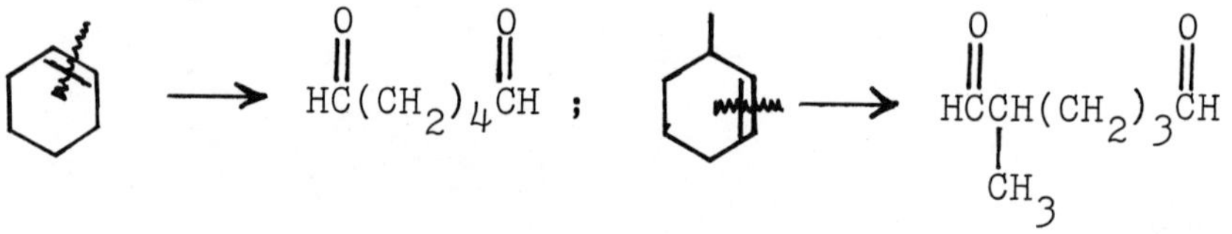

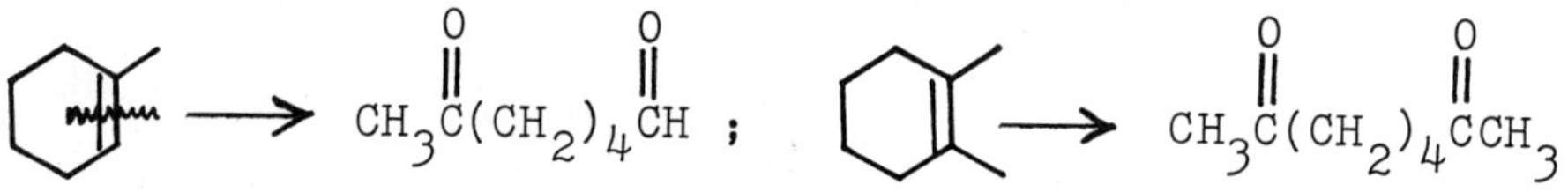

$\rightarrow$ $HC(=O)(CH_2)_4CH(=O)$; $\rightarrow$ $HC(=O)CH(CH_3)(CH_2)_3CH(=O)$

$\rightarrow$ $CH_3C(=O)(CH_2)_4CH(=O)$; $\rightarrow$ $CH_3C(=O)(CH_2)_4C(=O)CH_3$

nmr:

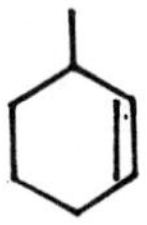

2 vinyl protons ~6 δ
doublet for CH_3 ~1 δ

2 vinyl protons ~6 δ
no CH_3 signal

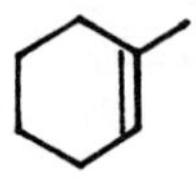

1 vinyl proton ~ 6 δ
singlet for CH_3 ~1.5 δ

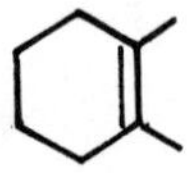

no vinyl protons
singlet for $2CH_3$ groups

(these can also be distinguished by C-H bending vibrations in the ir)

(continued next page)

(c) Chemical: no ozonolysis with cyclooctane; different ozonolysis products with the cyclooctadienes.

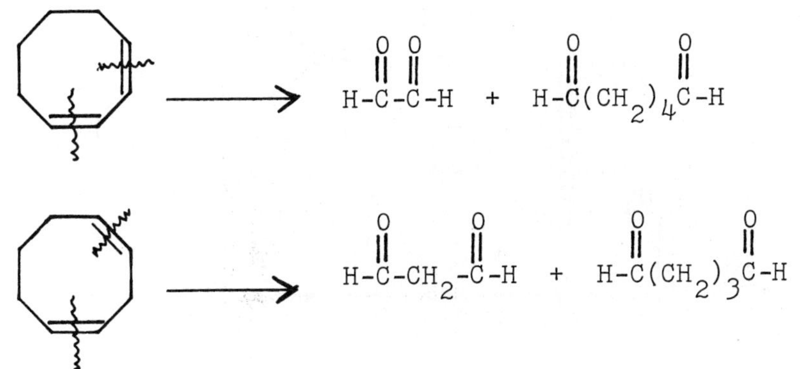

uv: no uv absorption for cyclooctane

1,3-cyclooctadiene, a conjugated diene, will absorb at longer wavelength than the non-conjugated diene.

(d) Chemical: different ozonolysis products or measure uptake of hydrogen on catalytic hydrogenation:

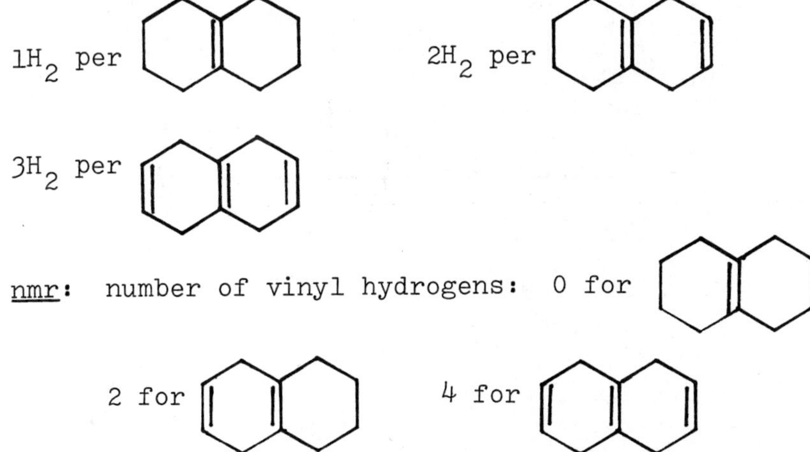

nmr: number of vinyl hydrogens: 0 for

2 for 4 for

66.

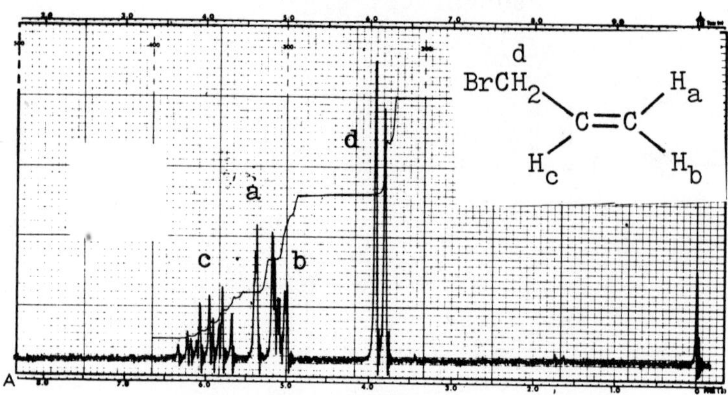
d
$BrCH_2$
H_a
C=C
H_c
H_b
a
b
c
d
A

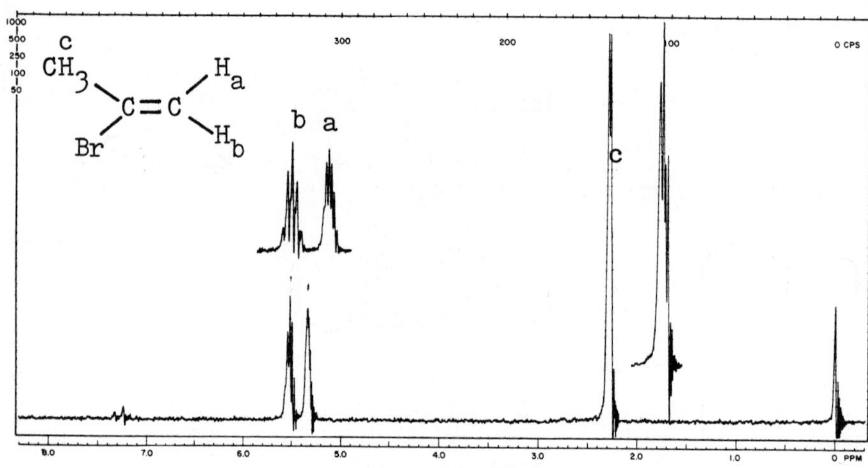
c
CH_3
H_a
C=C
Br
H_b
b a
c
300
200
100
0 CPS
8.0
7.0
6.0
5.0
4.0
3.0
2.0
1.0
0 PPM

SUMMARY OF REACTIONS

Specific examples of the general reactions below are provided in the Reaction Review section.

A. Reduction of Alkenes (S. 8.5)

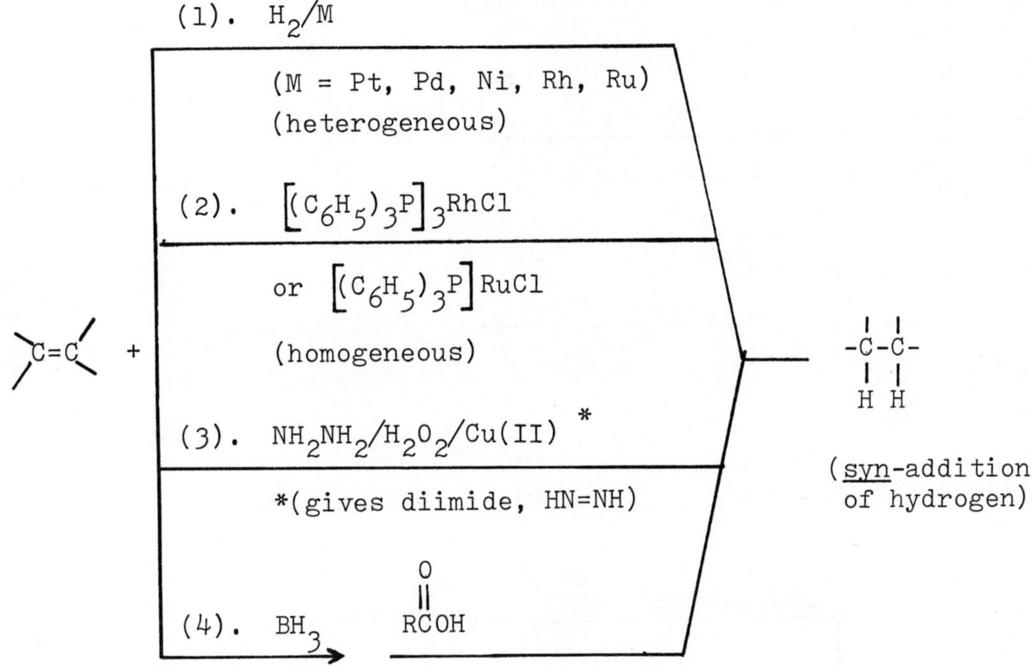

B. Electrophilic Addition Reactions of Alkenes (S. 8.6)

(reactivity of alkenes: $R_2C{=}CH_2 > RHC{=}CH_2 > CH_2{=}CH_2$)

(continued next page)

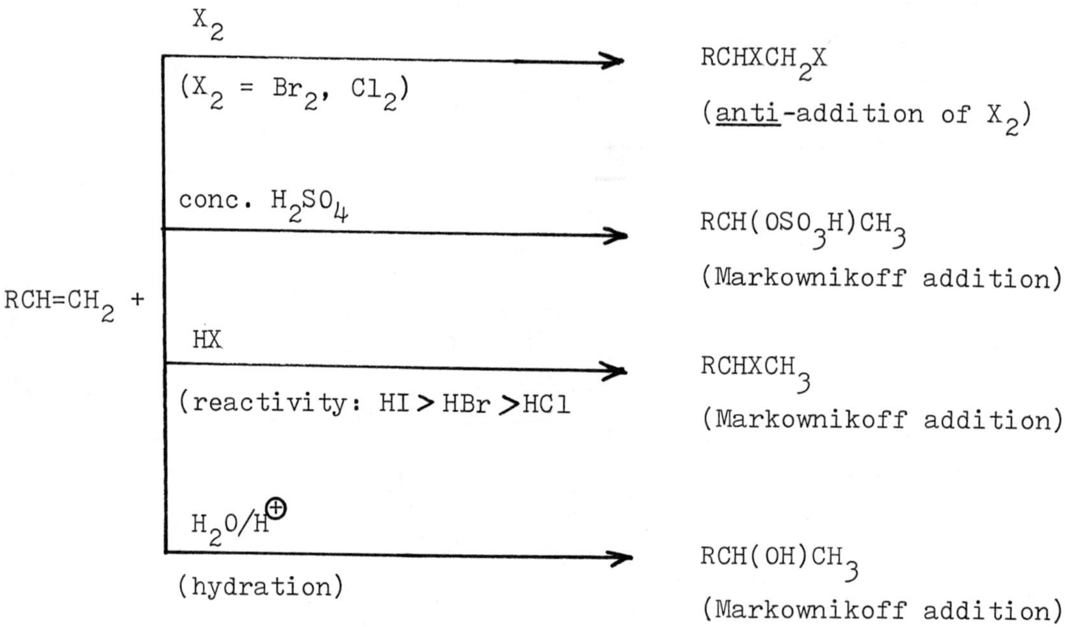

C. Free-radical Addition Reactions of Alkenes (S. 8.7)

$RCH=CH_2$ + HBr/peroxide (R·) → RCH_2CH_2Br (anti-Markownikoff addition)

$RCH=CH_2$ + $X_2/h\nu$ → $RCHXCH_2X$

(other free-radical additions see p. , text)

D. Free-radical Substitution Reactions of Alkenes (S. 8.7)

(continued next page)

Reaction scheme (allylic substitution):

$$-\overset{|}{\underset{\underset{H}{|}}{C}}-\overset{|}{C}=\overset{|}{C}- \quad + \quad \xrightarrow[(X_2 = Br_2 \text{ or } Cl_2)]{X_2/\text{heat}} \quad -\overset{|}{\underset{\underset{X}{|}}{C}}-\overset{|}{C}=\overset{|}{C}- \quad + \quad HX$$

(allylic substitution)

$$\xrightarrow[\text{(N-bromosuccinimide (NBS)}]{\text{(succinimide ring, C=O at both sides, N-Br)}/\text{peroxide}} \quad -\overset{|}{\underset{\underset{X}{|}}{C}}-\overset{|}{C}=\overset{|}{C}-$$

(allylic substitution)

E. Polymerization of Alkenes (S. 8.8)

$$nRCH{=}CH_2 \xrightarrow{\text{catalyst}} \sim\!\!(\!-CHRCH_2-\!)_n\!\sim$$

catalysts: (1). acids: $H^{\oplus}$, BF_3, $AlCl_3$ etc.

(2). bases: $NH_2^{\ominus}$ etc.

(3). free radicals: O_2 (air), peroxide, etc.

(4). coordination: $Al(C_2H_5)_3TiCl_4$

(Ziegler-Natta)

F. Oxidation of Alkenes (S. 8.9)

$>C=C<$ reacts with:

- $R-\overset{O}{\overset{\|}{C}}OOH$ (a peracid) $\longrightarrow$ $-\overset{|}{C}-\overset{|}{C}-$ bridged by O (epoxidation)

- $MnO_4^{\ominus}$ (aq) (or OsO_4) $\longrightarrow$ $-\overset{|}{\underset{HO}{C}}-\overset{|}{\underset{OH}{C}}-$ (<u>vic</u>-diol formation) (<u>syn</u>-addition of OH groups)

- $MnO_4^{\ominus}/H^{\oplus}$/heat $\longrightarrow$ $2\ >C=O$ *

  * $=CH_2 \longrightarrow [H_2C=O \longrightarrow (HO)_2C=O] \longrightarrow CO_2 + H_2O$

  $=CHR \longrightarrow [HRC=O] \longrightarrow R-C(=O)OH$ (an acid)

  $=CR_2 \longrightarrow R_2C=O$ (a ketone)

- $O_3 \longrightarrow$ $H_2O_2/H^{\oplus}$ $\longrightarrow$ $2\ >C=O$ ** (ozonolysis, oxidative conditions)

  ** $=CH_2 \longrightarrow [H_2C=O] \longrightarrow H-C(=O)OH$ (formic acid)

  $=CHR \longrightarrow HRC=O \longrightarrow R-C(=O)OH$ (an acid)

  $=CR_2 \longrightarrow R_2C=O$

- $O_3 \longrightarrow$ $Zn/H^{\oplus}$ $\longrightarrow$ $2\ >C=O$ # (ozonolysis, reductive conditions)

  # $=CH_2 \longrightarrow H-C(=O)H$ (formaldehyde)

  $=CHR \longrightarrow R-C(=O)H$ (an aldehyde)

  $=CR_2 \longrightarrow R_2C=O$ (a ketone)

- $CrO_3/H^{\oplus}$ $\longrightarrow$ $2\ >C=O$

  (same products obtained as from oxidative ozonolysis)

G. Carbene Reactions (S. 8.11)

1. Methylene (CH_2) and Dichloromethylene (CCl_2) Additions

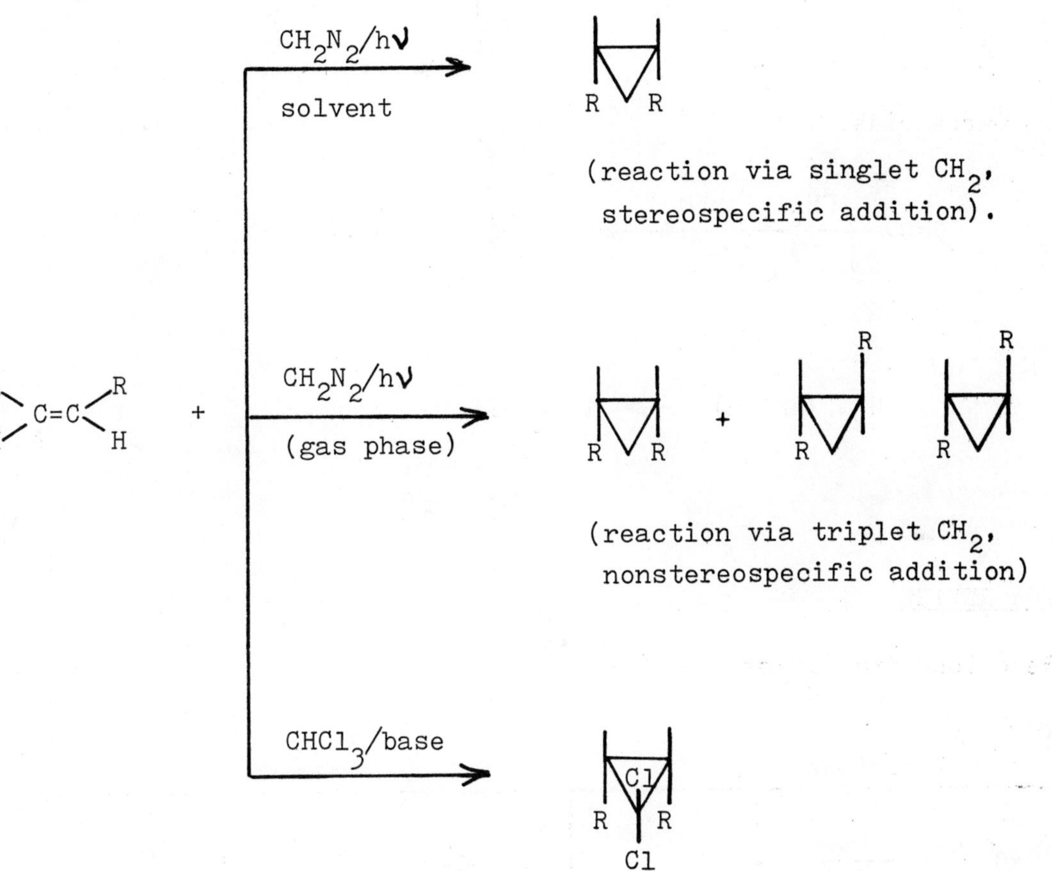

(reaction via singlet CH_2, stereospecific addition).

(reaction via triplet CH_2, nonstereospecific addition)

(reaction via singlet CCl_2, stereospecific addition)

2. Methylene Insertion (singlet CH_2)

$$R\text{-}H \xrightarrow{CH_2N_2/h\nu} RCH_3$$

(continued next page)

3. Hydrogen Abstraction (triplet CH_2)

$$R\text{-}H \xrightarrow{CH_2N_2/h\nu} [R\cdot + \cdot CH_3] \longrightarrow CH_3CH_3 + RR + RCH_3$$

4. Carbenoids

$$>C=C< \xrightarrow{CH_2I_2/Zn(Cu)} -\overset{|}{C}-\overset{|}{C}- \text{ (cyclopropane ring with } CH_2\text{)}$$

$$(CH_2I_2/Zn(Cu) \longrightarrow ICH_2ZnI)$$

REACTION REVIEW

A. Reactions from Chapter 8.

Questions	Answers
1. $>C=C< \xrightarrow[\text{catalyst}]{H_2}$?	1. $-\underset{H}{\overset{|}{C}}-\underset{H}{\overset{|}{C}}-$ *syn*-addition of hydrogen
catalysts are ?	catalysts: Pt, Pd, Ni etc. (heterogeneous) $[(C_6H_5)_3P]_3RhCl$ (homogeneous)

(continued next page)

2. Give the stereochemistry of the products in: a). $CH_3(D)C{=}C(CH_3)(D)$ $\xrightarrow{H_2/Pt}$? b). 1-methyl-(R)-3-ethylcyclohexene (ring with CH_3 and CH_2CH_3; (R)) $\xrightarrow[[(C_6H_5)_3P]_3RhCl]{H_2}$?	2. a). $CH_3-\underset{H}{\overset{D}{C}}-\underset{H}{\overset{D}{C}}-CH_3$ (meso) b). (R)-CH_3, (R)-CH_2CH_3 cyclohexane + (S)-CH_3, (R)-CH_2CH_3 cyclohexane (diastereomers)
3. $>C{=}C<$ $\xrightarrow[H_2O_2/Cu(II)]{NH_2NH_2}$? Write the structures of the major product(s) when $>C{=}C<$ is: a. $CH_3CH{=}CH_2$ b. 1,2-dimethylcyclopentene (CH_3, CH_3)	3. $-\underset{H}{\overset{\vert}{C}}-\underset{H}{\overset{\vert}{C}}-$ *syn*-addition of hydrogen ($NH_2NH_2/H_2O_2/Cu(II)$ → HN=NH) diimide a. $CH_3CH_2CH_3$ b. cis-1,2-dimethylcyclopentane (CH_3, CH_3) (meso)

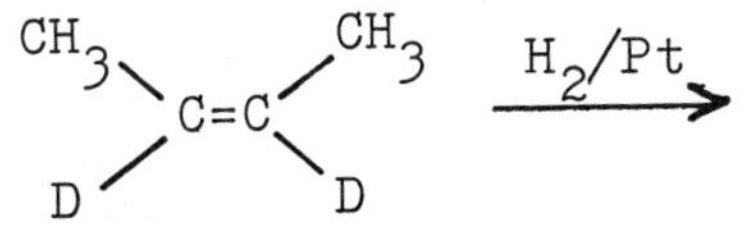

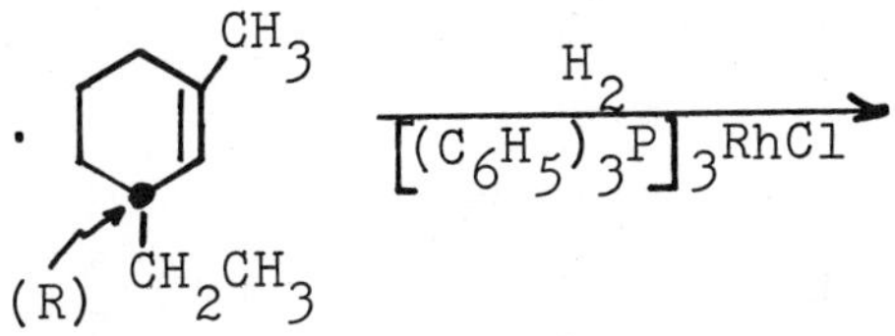

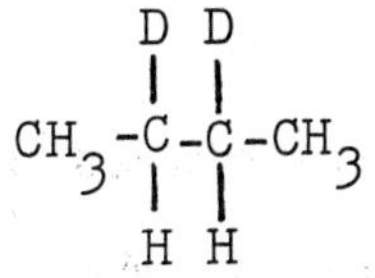

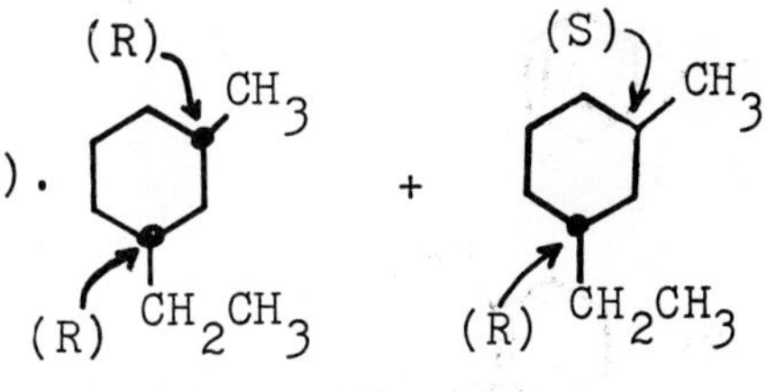

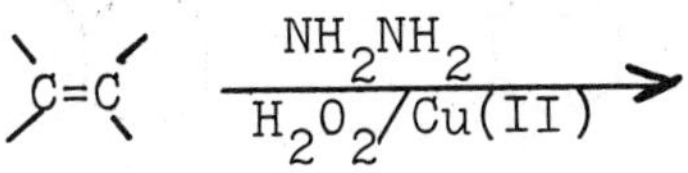

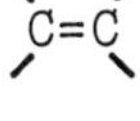

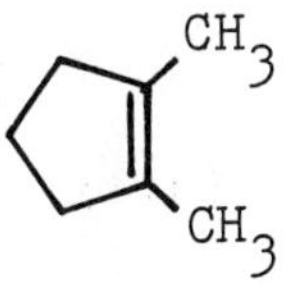

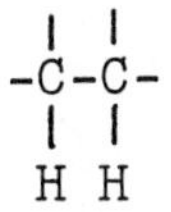

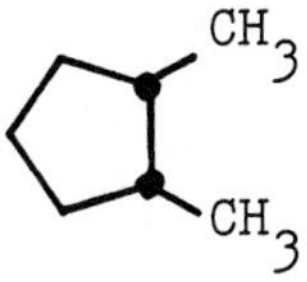

4. $\rangle C=C\langle$ $\xrightarrow{BH_3}$ $\xrightarrow{RCO_2H}$?

Write the structures of the major product(s) when $\rangle C=C\langle$ is:

a.  (cyclohexene)

b. 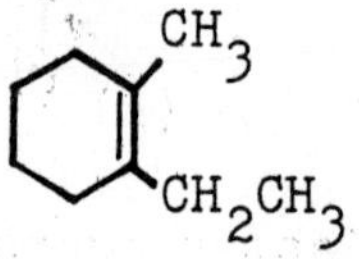 (1-methyl-2-ethylcyclohexene: CH_3, CH_2CH_3)

4. $-\overset{|}{\underset{H}{C}}-\overset{|}{\underset{H}{C}}-$ *syn*-addition of hydrogen

a.  (cyclohexane)

b. 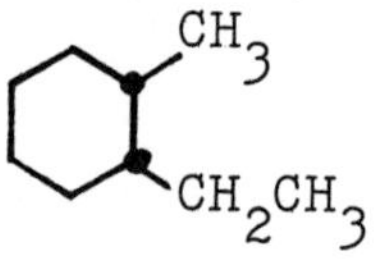 (1-methyl-2-ethylcyclohexane: CH_3, CH_2CH_3) (d,l) *

*(i.e. (CH_3, CH_2CH_3) + (CH_3, CH_2CH_3))

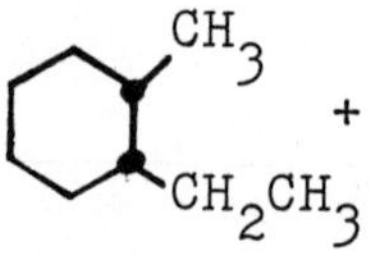

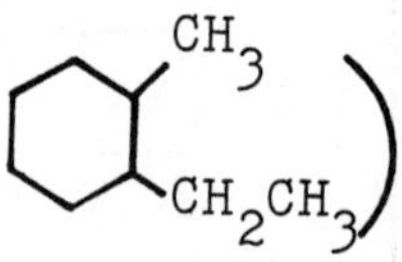

enantiomers

5. $RCH=CH_2$ $\xrightarrow{X_2}$?

Write the structure of the major product(s) when X_2 is Br_2 and $RCH=CH_2$ is:

a.  (cyclohexene)

5. $RCHXCH_2X$ (*anti*-addition of X_2)

a. 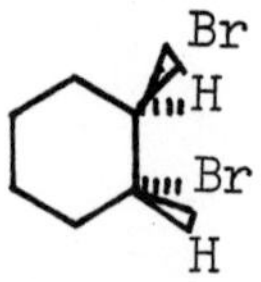 (trans-1,2-dibromocyclohexane: Br, H, Br, H) (d,l)

(continued next page)

b. ($\underline{E}$)-$CH_3CH{=}CHCH_3$	b. 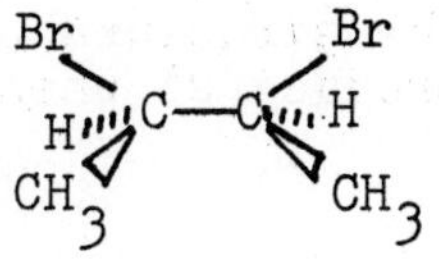 Br(H)(CH$_3$)C—C(H)(CH$_3$)Br (meso)
6. $RCH{=}CH_2 \xrightarrow{X_2/H_2O}$?	6. $RCH(OH)CH_2X$
Write the structure for the major product(s) for Br_2/H_2O when $RCH{=}CH_2$ is:	
a. $CH_3CH{=}CH_2$	a. $CH_3CH(OH)CH_2Br$ (d,l)
b. 1-methylcyclopentene (ring with CH_3)	b. cyclopentane ring with Br, H on one carbon and OH, CH_3 on the adjacent carbon (d,l)
7. $RCH{=}CH_2 \xrightarrow{H\text{-}X}$?	7. $RCHXCH_3$ Markownikoff addition
Write the structure of the major product in the reactions:	
a. $(CH_3)_2C{=}CH_2 \xrightarrow{HBr}$?	a. $(CH_3)_2CBrCH_3$
b. 1-methylcyclohexene (ring with CH_3) $\xrightarrow{H\text{-}I}$?	b. cyclohexane ring with CH_3 and I on the same carbon
8. $RCH{=}CH_2 \xrightarrow{\text{conc. } H_2SO_4}$?	8. $RCH(OSO_3H)CH_3$ Markownikoff addition

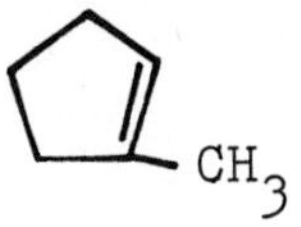

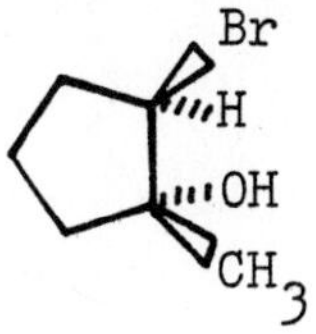

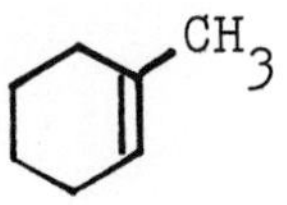

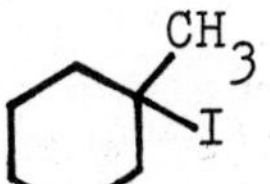

(continued next page)

Write the structure of the major product(s) when $RCH=CH_2$ is:

a. $CH_3CH=CH_2 \xrightarrow{H_2SO_4}$

a. $CH_3CH(OSO_3H)CH_3$

b. $(CH_3)_2C=CHCH_3 \xrightarrow{H_2SO_4}$

b. $(CH_3)_2C(OSO_3H)CH_2CH_3$

9. $RCH=CH_2 \xrightarrow{H_2O/H^{\oplus}}$?

9. $RCH(OH)CH_3$ Markownikoff addition

Write the structure of the major product(s) when $RCH=CH_2$ is:

a. 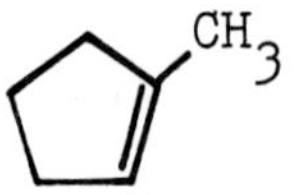 1-methylcyclopentene (cyclopentene ring with CH_3 on the double-bond carbon)

a. 1-methylcyclopentanol (cyclopentane ring carbon bearing CH_3 and OH)

b. $(CH_3)_2CHCH=CHCH_3$

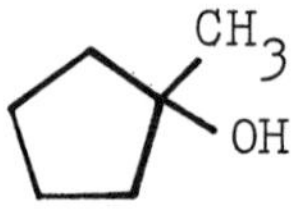

b. $(CH_3)_2CHCH(OH)CH_2CH_3$

and

$(CH_3)_2CHCH_2CH(OH)CH_3$

10. $RCH=CH_2 \xrightarrow[\text{peroxide}]{HBr}$?

10. RCH_2CH_2Br anti-Markownikoff addition

Write the structure of the major product(s) when $RCH=CH_2$ is:

a. $CH_3CH=CH_2$

a. $CH_3CH_2CH_2Br$

(continued next page)

b. (cyclohexene ring with CH_3 on the double-bond carbon)	b. (cyclohexane ring with Br and adjacent CH_3) (cis and trans)
11. $-\overset{\mid}{\underset{H}{C}}-\overset{\mid}{C}=C<$ $\xrightarrow[\text{heat}]{X_2}$ Write the structure of all possible product(s) with Cl_2/heat when $-\overset{\mid}{\underset{H}{C}}-\overset{\mid}{C}=C<$ is: a. $CH_3CH_2CH=CH_2$ b. (cyclohexene)	11. $-\overset{\mid}{\underset{X}{C}}-\overset{\mid}{C}=C<$ + HX (allylic halogenation) a. $CH_3CHClCH=CH_2$ + $CH_3CH=CHCH_2Cl$ b. (3-chlorocyclohexene: Cl on ring carbon adjacent to the double bond)
12. $-\overset{\mid}{\underset{H}{C}}-\overset{\mid}{C}=C<$ $\xrightarrow[\text{peroxide}]{\text{NBS}}$ (N-Br (NBS): N-bromosuccinimide) Write the structure of the product(s) when $-\overset{\mid}{\underset{H}{C}}-\overset{\mid}{C}=C<$ is:	12. $-\overset{\mid}{\underset{Br}{C}}-\overset{\mid}{C}=C<$ (allylic bromination) (continued next page)

a. cyclopentene (structure) b. $(CH_3)_2CHCH{=}CH_2$	a. 3-bromocyclopentene (structure, Br) b. $(CH_3)_2CBrCH{=}CH_2$ + $(CH_3)_2C{=}CHCH_2Br$
13. $>C{=}C<$ $\xrightarrow{R-\overset{O}{\overset{\parallel}{C}}-OOH}$? Write the structure of the major product when $>C{=}C<$ is: a. $CH_3(H)C{=}C(H)CH_3$ (trans) b. cyclohexene (structure)	13. $-\overset{\vert}{C}-\overset{\vert}{C}-$ with O bridging (epoxide) a. trans-epoxide: CH_3, H on one C; H, CH_3 on other C; O b. cyclohexene oxide (H, O, H)
14. epoxide $\xrightarrow[H^{\oplus} \text{ or } OH^{\ominus}]{H_2O}$ Write the structure of the major product in:	14. $-\overset{\vert}{C}(OH)-\overset{OH}{\overset{\vert}{C}}-$ *anti*-orientation of OH groups (continued next page)

a. [cis-2,3-dimethyloxirane: CH$_3$, H on each C, epoxide O] $\xrightarrow{H_2O/H^{\oplus}}$?	a. CH_3–C(H)(HO)–C(OH)(H)–CH_3 (meso)
b. [cyclohexene] $\xrightarrow{C_6H_5\overset{O}{\overset{\parallel}{C}}OOH}$ $\xrightarrow{H_2O/OH^{\ominus}}$?	b. [trans-1,2-cyclohexanediol: OH, H, OH, H] (d,l)
15. >C=C< $\xrightarrow[\text{(or } O_sO_4)]{MnO_4^{\ominus}\text{ (aq)}}$? Write the structure of the major product when >C=C< is: a. [trans: CH_3(H)C=C(H)CH_3] b. [cyclohexene]	15. –C(OH)–C(OH)– <u>syn</u>-orientation of OH groups a. H(CH$_3$)C(HO)–C(OH)(H)CH$_3$ (d,l) b. [cis-1,2-cyclohexanediol: OH, OH] (meso)
16. >C=C< $\xrightarrow[\text{heat}]{MnO_4^{\ominus}/H^{\oplus}}$? Write the structure of the products obtained when >C=C< is:	16. >C=O + O=C< (H attached to C=O is oxidized to OH) (continued next page)

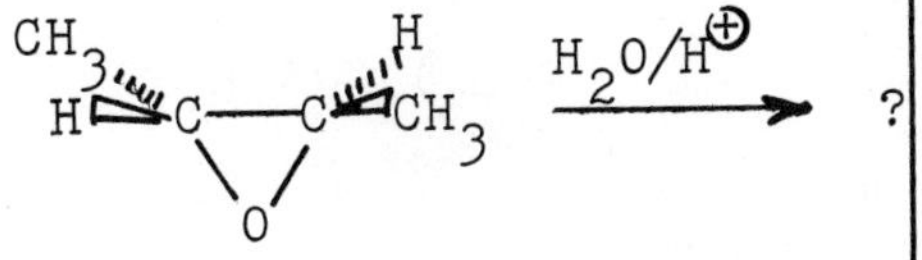

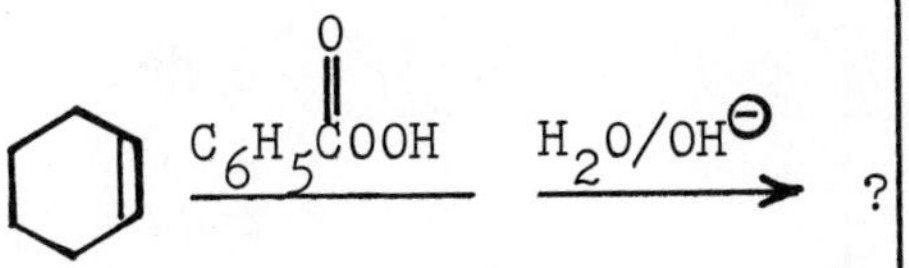

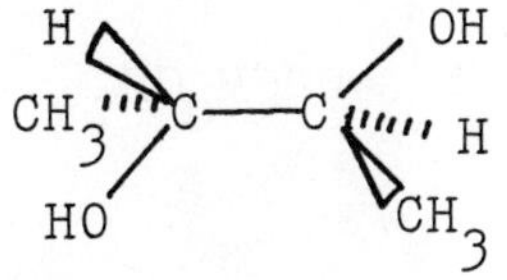

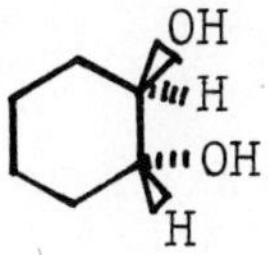

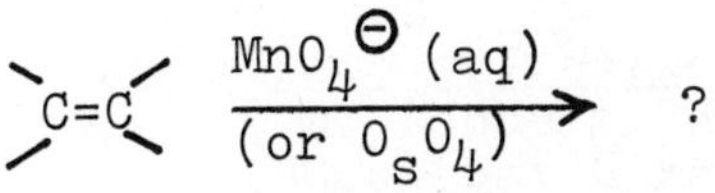

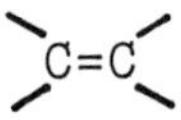

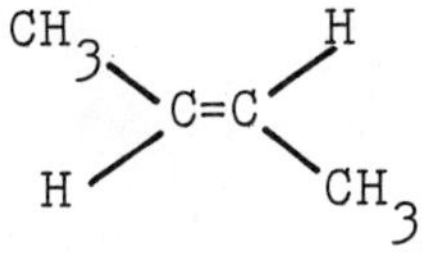

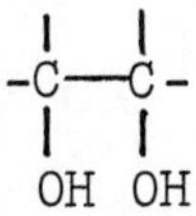

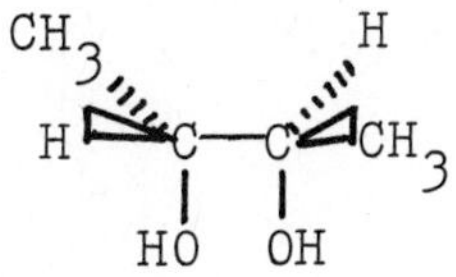

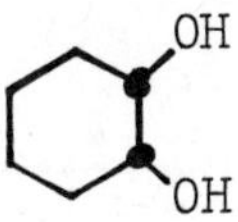

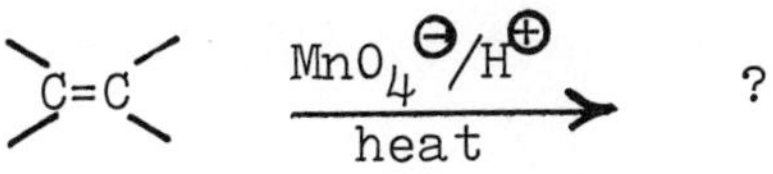

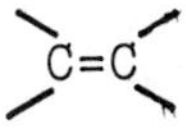

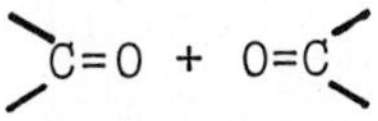

a. $(CH_3)_2C{=}CHCH_2CH_3$	a. $(CH_3)_2C{=}O + CH_3CH_2\overset{O}{\overset{\Vert}{C}}OH$
b. 1-methylcyclohexene (ring with CH_3)	b. $CH_3\overset{O}{\overset{\Vert}{C}}(CH_2)_4\overset{O}{\overset{\Vert}{C}}OH$
17. $>C{=}C<$ $\xrightarrow{O_3}$ $\xrightarrow{H_2O_2/H^{\oplus}}$? Write the structure of the products obtained when $>C{=}C<$ is: a. $(CH_3)_2C{=}CHCH_2CH_3$ b. 1-methylcyclohexene (ring with CH_3)	17. $>C{=}O + O{=}C<$ (<u>oxidative</u> <u>conditions</u>: H attached to C=O is oxidized to OH) a. $(CH_3)_2C{=}O + CH_3CH_2\overset{O}{\overset{\Vert}{C}}OH$ b. $CH_3\overset{O}{\overset{\Vert}{C}}(CH_2)_4\overset{O}{\overset{\Vert}{C}}OH$
18. $>C{=}C<$ $\xrightarrow{O_3}$ $\xrightarrow{Zn/HCl}$? Write the structure of the products obtained when $>C{=}C<$ is: a. $(CH_3)_2C{=}CHCH_2CH_3$	18. $>C{=}O + O{=}C<$ (reductive conditions) a. $(CH_3)_2C{=}O + CH_3CH_2{-}\overset{O}{\overset{\Vert}{C}}{-}H$ (continued next page)

b. 1-methylcyclohexene (ring with CH_3 on double-bond carbon)	b. $CH_3\overset{O}{\overset{\|}{C}}(CH_2)_4\overset{O}{\overset{\|}{C}}\text{-H}$
19. $>C=C<$ $\xrightarrow{CH_2N_2/h\nu}$?	19. cyclopropane ($-C—C-$ bridged by CH_2) + N_2 (stereospecific addition of singlet CH_2 (solution) and nonstereospecific addition of triplet CH_2 (gas phase)
Write the structure of the possible products in the following:	
a. $CH_3(H)C=C(H)CH_3$ (cis) $\xrightarrow[\text{solvent}]{CH_2N_2/h\nu}$?	a. cis-1,2-dimethylcyclopropane (CH_3 CH_3) (meso)
b. $CH_3(H)C=C(H)CH_3$ (cis) $\xrightarrow[\text{gas phase}]{CH_2N_2/h\nu}$?	b. cis-1,2-dimethylcyclopropane (CH_3 CH_3) (meso) + trans-1,2-dimethylcyclopropane (CH_3, CH_3) (d,l)
20. $>C=C<$ $\xrightarrow{CHCl_3/base}$?	20. $-C—C-$ bridged by CCl_2 (stereospecific addition of singlet CCl_2)

(continued next page)

Write the structure of the possible products when >C=C< is: a. CH_3(H)C=C(H)CH_3 b. [cyclohexene]	a. [cyclopropane: CH_3, Cl, Cl, CH_3] (d,l) b. [bicyclic: H, Cl, C, Cl, H] (<u>meso</u>)
21. >C=C< $\xrightarrow{CH_2I_2/Zn(Cu)}$?	21. -C—C- bridged by CH_2 (ICH_2ZnI from $CH_2I_2/Zn(Cu)$)
Write the structure of the product when >C=C< is: a. $CH_3CH{=}CH_2$ b. [cyclopentene]	a. CH_3CH—CH_2 bridged by CH_2 b. [bicyclo[3.1.0]hexane]
22. R-H $\xrightarrow[solvent]{CH_2N_2/h\nu}$?	22. $R{-}CH_3$ (insertion of singlet CH_2) (continued next page)

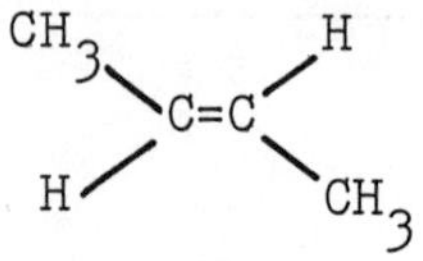

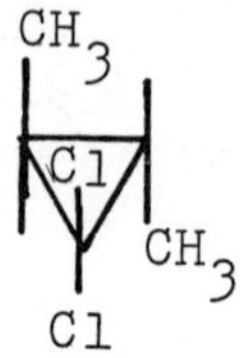

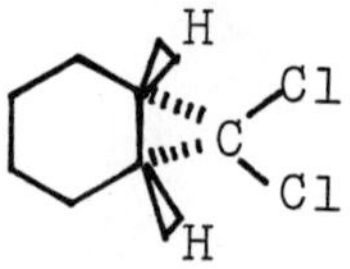

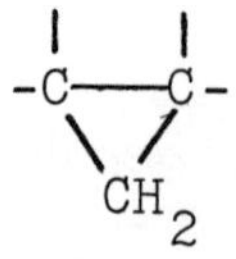

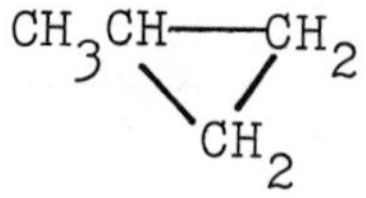

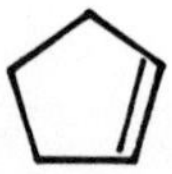

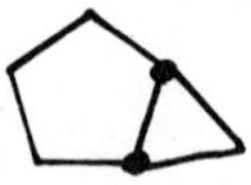

Write the structure of the products when RH is $CH_3CH_2CH_3$	$CH_3CH_2CH_2CH_3$ $(CH_3)_2CHCH_3$
23. R-H $\xrightarrow[\text{(gas phase)}]{CH_2N_2/h\nu}$? Write the structures of the products possible when RH is CH_3CH_3	23. [R• + •CH_3] → R-R + RCH_3 + CH_3CH_3 $CH_3CH_2CH_2CH_3$ $CH_3CH_2CH_3$ CH_3CH_3

B. Write structures for all letters.

1. (cyclohexane ring)–Cl $\xrightarrow{KOH/alcohol}$ A $\xrightarrow{H_2/Pt}$ B	1. A: (cyclohexene ring) B: (cyclohexane ring)
2. $CH_3CH{=}CH_2$ $\xrightarrow{HBr}$ A $\xrightarrow[ether]{Mg}$ B $\xrightarrow{D_2O}$ C	2. A: $CH_3CHBrCH_3$ B: $CH_3CH(MgBr)CH_3$ C: CH_3CHDCH_3
3. (cyclopentane ring) $\xrightarrow[h\nu]{Br_2}$ A $\xrightarrow[NH_3(\ell)]{NaNH_2}$ B $\xrightarrow{Br_2/CCl_4}$ C (indicate the stereochemistry of C)	3. A: (cyclopentane ring)–Br B: (cyclopentene ring) (continued next page)

	C: Br, Br
4. $CH_3CH{=}CH_2 \xrightarrow[\text{peroxide}]{HBr}$ A $\xrightarrow{Li}$ B $\xrightarrow{CuCl}$ C $\xrightarrow{CH_3CH_2Cl}$ D	4. A: $CH_3CH_2CH_2Br$ B: $CH_3CH_2CH_2Li$ C: $Li\left[(CH_3CH_2CH_2)_2Cu\right]$ D: $CH_3CH_2CH_2CH_2CH_3$
5. OH $\xrightarrow[\text{heat}]{H_2SO_4}$ A $\xrightarrow[\text{peroxide}]{NBS}$ B $\xrightarrow[\text{heat}]{KOH/alcohol}$ C	5. A: B: Br C:
6. $(CH_3)_2C{=}CH_2 \xrightarrow[300^\circ]{Br_2}$ A $\xrightarrow[CCl_4]{Br_2}$ B	6. A: $CH_2BrC(CH_3){=}CH_2$ B: $CH_2BrCBr(CH_3)CH_2Br$

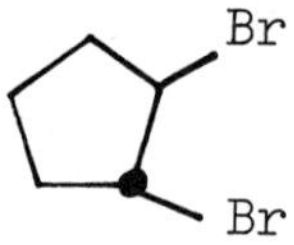

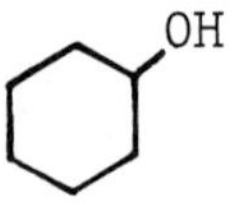

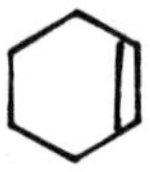

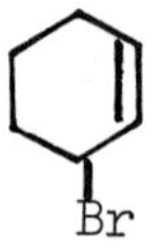

7. (methylcyclopentane) $\xrightarrow[h\nu]{Br_2}$ A $\xrightarrow[heat]{KOH/alcohol}$ B $\xrightarrow[heat]{KMnO_4/H^{\oplus}}$ C	7. A: 1-bromo-1-methylcyclopentane (CH_3, Br) (major) B: 1-methylcyclopentene (CH_3) C: $CH_3\overset{O}{\overset{\Vert}{C}}CH_2CH_2CH_2\overset{O}{\overset{\Vert}{C}}$-OH
8. $CH_3CH_2CH_3$ $\xrightarrow{Cl_2/h\nu}$ A $\xrightarrow[heat]{KOH/alcohol}$ B $\xrightarrow{C_6H_5\overset{O}{\overset{\Vert}{C}}O_2H}$ C $\xrightarrow{H_2O/OH^{\ominus}}$ D	8. A: $CH_3CHClCH_3$ + $CH_3CH_2CH_2Cl$ B: $CH_3CH{=}CH_2$ C: CH_3CH—CH_2 (epoxide, O bridging) D: $CH_3CH(OH)CH_2OH$
9. (3-bromocyclohexene, Br) $\xrightarrow{OH^{\ominus}/H_2O}$ A $\xrightarrow{MnO_4^{\ominus}\ (aq)}$ B (indicate the stereochemistry of all products)	9. A: cyclohex-2-en-1-ol (OH) (S_N2) B: cyclohexane-1,2,3-triol (OH, OH, OH) + cyclohexane-1,2,3-triol (OH, OH, OH) (meso) (d,l)

10. 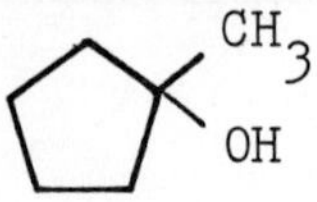 1-methylcyclopentanol (CH_3, OH) $\xrightarrow[\text{heat}]{H_2SO_4}$ A $\xrightarrow[\text{solvent}]{CH_2N_2/h\nu}$ B (indicate the stereochemistry of B)	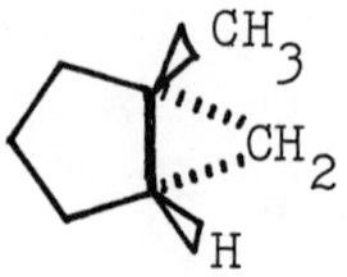 10. A: 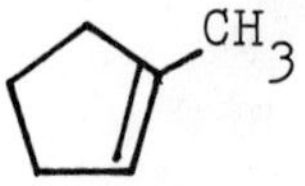 1-methylcyclopentene (CH_3) B: bicyclic product (CH_3, CH_2, H) (d,l)
11. 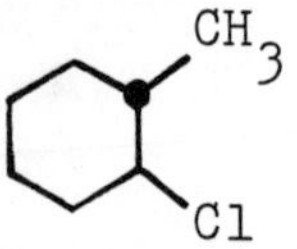 2-chloro-1-methylcyclohexane (CH_3, Cl) $\xrightarrow[NH_3(\ell)]{NaNH_2}$ A $\xrightarrow{O_3}$ $\xrightarrow[H^{\oplus}]{Zn}$ B	11. A: 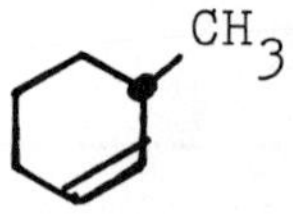 3-methylcyclohexene (CH_3) (only product by E2) B: 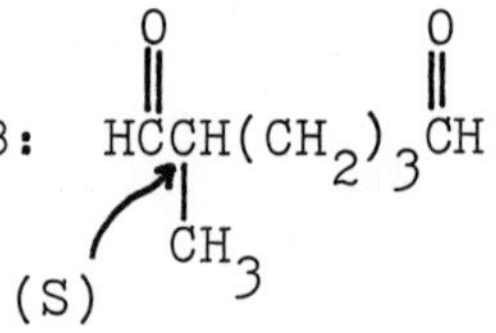 $\overset{O}{\overset{\Vert}{HC}}\underset{CH_3}{\underset{\vert}{CH}}(CH_2)_3\overset{O}{\overset{\Vert}{C}}H$ (S)
12. $CH_3CH{=}CH_2$ $\xrightarrow{Cl_2/H_2O}$ A $\xrightarrow{HI}$ B	12. A: $CH_3CH(OH)CH_2Cl$ B: CH_3CHICH_2Cl
13.  cyclohexene $\xrightarrow[\text{peroxide}]{NBS}$ A $\xrightarrow{BH_3}$ $\xrightarrow{CH_3CO_2H}$ B $\xrightarrow[DMSO]{Na^{\oplus}:CN^{\ominus}}$ C	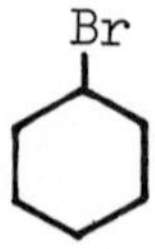 13. A: 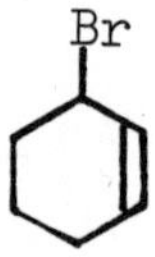 3-bromocyclohexene (Br) B: bromocyclohexane (Br) C: cyclohexanecarbonitrile (CN) 

CHAPTER 9. ALKYNES

LEARNING OBJECTIVES

When you have completed this chapter, you should be able to:

1. name alkynes (problems 1, 25);
2. compare the acidic and basic properties of alkynes and other compounds (problems 3, 4, 7, 8 and related problems 34, 35);
3. use simple chemical reactions as a means of identifying and separating alkynes and other compounds (problems 5, 6, 9, 24, 31);
4. write equations which illustrate the synthesis of alkynes (problems 10, 30 and related problems 11, 12, 29);
5. write equations which illustrate addition reactions of alkynes (problems 13, 16, 26, 27, 28 and related problem 33);
6. predict reactants or products in oxidation reactions of alkynes (problems 22, 23);
7. write equations which illustrate multistep syntheses (problems 14, 30, 31);
8. write mechanisms which explain addition reactions of alkynes (problems 15, 20, 37, 38);
9. use IR and NMR to identify or distinguish compounds (problems 40, 41, 42, 43);
10. account for the change $-\overset{|}{C}(OH)=\overset{|}{C}- \longrightarrow -C(=O)\overset{|}{C}H-$ by writing a mechanism and by bond energy calculations (problems 17, 39);
11. calculate, from heat of formation data, the amount of heat produced by combustion of hydrocarbons (problem 36).

ANSWERS TO QUESTIONS

1. (a) propyne (methylacetylene)
   (b) 1-deuteriopropyne
   (c) 1,3,3,3-tetradeuteriopropyne
   (d) 3-chloropropyne (propargyl chloride)
   (e) phenylethyne (phenylacetylene)
   (f) 1-buten-3-yne (vinylacetylene)
   (g) 2-methyl-1-buten-3-yne
   (h) 2,2-dimethyl-3-hexyne

2. $Cl_2C=C=CCl_2$

3.

acids	bases
(a) $HCl > H_3O^{\oplus}$	$H_2O > Cl^{\ominus}$
(b) $HCl > H_2O$	$OH^{\ominus} > Cl^{\ominus}$
(c) $HC{\equiv}CH > NH_3$	$NH_2^{\ominus} > HC{\equiv}C^{\ominus}$
(d) $NH_4^{\oplus} > HC{\equiv}CH$	$HC{\equiv}C^{\ominus} > NH_3$

4. Any substance possessing a nonbonding electron pair is potentially a base. Whether or not it functions as a base depends on its strength relative to the proton donor. For example NH_3 <u>is</u> a base in the reaction:

$$NH_3 + H_2O \rightleftharpoons NH_4^{\oplus} + OH^{\ominus}$$

However, NH_3 is not strong enough a base to remove a proton from ethane.

5. Bubble the mixture through $Na^{\oplus}NH_2^{\ominus}/NH_{3(\ell)}$. 1-Butyne forms the salt $CH_3CH_2C{\equiv}C^{\ominus}Na^{\oplus}$ and dissolves. 2-Butyne does not react and passes through. 1-Butyne is recovered by acidifying the mixture.

6. Both 1-butyne and 1-butene will react with (decolorize) Br_2 in CCl_4; octane will not. Only 1-butyne will react with $Na^{\oplus}NH_2^{\ominus}$ forming an acetylide and liberating ammonia.

7. $$CH_3CH_2{:}^{\ominus} + H{-}\ddot{N}H_2 \longrightarrow CH_3CH_3 + {:}\ddot{N}H_2^{\ominus}$$

$CH_3CH_2{:}^{\ominus}$ is a stronger base than ${:}\ddot{N}H_2^{\ominus}$ and ${:}NH_3$ a stronger acid than CH_3CH_3.

8. (a) methane (b) $CH_3CH_2CH_2CH_2C{\equiv}CMgBr$

Reaction: $CH_3\text{-}MgBr + H\text{-}C{\equiv}C(CH_2)_3CH_3 \longrightarrow CH_4 + CH_3(CH_2)_3C{\equiv}CMgBr$

9. (a) 1-butyne forms a heavy metal salt precipitate with $Ag(NH_3)_2NO_3$; 2-butyne does not.
   (b) 1-butyne forms a precipitate with $Ag(NH_3)_2NO_3$; 1-butene does not.
   (c) 1-butene decolorizes Br_2 in CCl_4 (brown $\rightarrow$ colorless) or cold, dilute $KMnO_4$ (purple $\rightarrow$ colorless); butane does not.
   (d) 1-butyne forms a precipitate with $Ag(NH_3)_2NO_3$; butane does not.

10. (a) $CH_3CH_3 \xrightarrow{Cl_2,\ light} CH_3CH_2Cl \xrightarrow{KOH,\ alcohol} CH_2{=}CH_2$

   (b) $CH_3CH_3 \xrightarrow{Br_2,\ light} CH_3CH_2Br \xrightarrow{KOH,\ alcohol} CH_2{=}CH_2$

   $\xrightarrow{Br_2, CCl_4} CH_2BrCH_2Br \xrightarrow{Na^{\oplus}\ NH_2^{\ominus},\ NH_{3(\ell)}} HC{\equiv}CH$

   (c) $HC{\equiv}CH$ (from b) $\xrightarrow[NH_{3(\ell)}]{Na^{\oplus}\ NH_2^{\ominus}} \xrightarrow{CH_3I} CH_3C{\equiv}CH$

   (d) $C_6H_5MgBr \xrightarrow{CH_2{=}CHCH_2Cl} C_6H_5CH_2CH{=}CH_2 \xrightarrow{Br_2, CCl_4}$

   $C_6H_5CH_2CHBrCH_2Br \xrightarrow{Na^{\oplus}\ NH_2^{\ominus}/NH_{3(\ell)}} C_6H_5CH_2C{\equiv}CH$

11. The reaction is S_N2. Cl is a better leaving group than F.

12. The substrate is a 3° halide, and crowding (steric hindrance) about the C-Br bond prevents an S_N2 reaction from occurring. An E2 reaction will occur giving $CH_3C{\equiv}CH + CH_2{=}C(CH_3)_2$.

(Text: the S_N2 reaction, S. 5.2)

13. Halogens add <u>anti</u> to alkenes and predominantly <u>anti</u> to alkynes.

(i) (a) <u>meso</u>-$CH_3CHClCHClCH_3$

(b) (<u>d</u>,<u>l</u>) $CH_3CHClCHClCH_3$

(c) (<u>E</u>)-$CH_3CH_2CCl{=}CClCH_3$

(d) (<u>E</u>)-$CH_3CH_2CH_2CCl{=}CHCl$

(ii) $CH_3C{\equiv}CCH_2CH_3 \xrightarrow{:\ddot{Cl}-\ddot{Cl}:}$ $(Cl)(CH_3)C{=}\overset{\oplus}{C}-CH_2CH_3$

$\downarrow\ :\ddot{Cl}:^{\ominus}$

$(Cl)(CH_3)C{=}C(CH_2CH_3)(Cl)$

14. (a) $CH_3C{\equiv}CH \xrightarrow{HBr/peroxide}$

(b) $CH_3C{\equiv}CH \xrightarrow{HBr}$

(c) $CH_3CH{=}CHBr$ (from a) $\xrightarrow{H_2/Pt}$

(d) $CH_3C{\equiv}CH \xrightarrow[\text{Lindlar's Cat.}]{H_2}$ (or $\xrightarrow{Na/NH_{3(\ell)}}$)

(e) $CH_3CH{=}CH_2$ (from d) $\xrightarrow{Cl_2}$

15. The mechanism is outlined below for the major products.

$CH_3C{\equiv}CH \xrightarrow{DCl}$ $CH_3CD{=}\overset{\oplus}{C}H$ (not formed)

$\rightarrow CH_3\overset{\oplus}{C}{=}CHD$ (more stable) $\xrightarrow{Cl^{\ominus}} CH_3C(Cl){=}CHD \xrightarrow{DCl}$ $CH_3\text{-}CD(Cl)\text{-}\overset{\oplus}{C}HD$ A (not formed)

$\rightarrow \left[CH_3\text{-}\overset{\oplus}{C}(\ddot{Cl}:)\text{-}CHD_2 \leftrightarrow CH_3\text{-}C(={:}\ddot{Cl}^{\oplus})\text{-}CHD_2 \right]$ B (more stable)

$\xrightarrow{Cl^{\ominus}} CH_3CCl_2CHD_2$

16. (a) $CH_3CH_2CH_2\overset{O}{\overset{\|}{C}}CH_3$

(b) cyclopentyl$\text{-}\overset{O}{\overset{\|}{C}}\text{-}CH_2\text{-}$cyclopentyl

(c) $CH_3CH_2\overset{O}{\overset{\|}{C}}CH_2CH_2CH_2CH_3$ and $CH_3CH_2CH_2\overset{O}{\overset{\|}{C}}CH_2CH_2CH_3$

(d) $CH_3\overset{O}{\overset{\|}{C}}(CH_2)_8\overset{O}{\overset{\|}{C}}CH_3$

17. two step: $(CH_3)_2C{=}C(CH_3)\text{-}O\text{-}H \xrightarrow{H^{\oplus}} CH_3\text{-}CH(CH_3)\text{-}\overset{\oplus}{C}(CH_3)\text{-}O\text{-}H \xrightarrow{:\ddot{O}H_2} CH_3\text{-}CH(CH_3)\text{-}C({=}O)CH_3$

(continued next page)

concerted: $(CH_3)_2C{=}C(CH_3){-}\ddot{O}{-}H$ (with $H^{\oplus}$ adding; arrows) $\xrightarrow{:\ddot{O}H_2}$ $CH_3{-}C(H)(CH_3){-}C({=}O)CH_3$

18. The addition of diborane, as shown in the example, is syn.

19. (a) $(CH_3)_2CHCH(CH_3)OH$ (b) $(CH_3)_2CHCH(CH_3)OH$

20. Review the mechanisms of free-radical and ionic addition of HBr to alkenes and alkynes. In free-radical addition the first species to add to the π bond is Br· (from R· + H-Br — RH + Br·). In ionic addition the first species to add is $H^{\oplus}$.

(a) $CH_3C{\equiv}CH \xrightarrow[R\cdot]{HBr} CH_3CH{=}CHBr \xrightarrow[(ionic)]{HBr} CH_3CH_2CHBr_2$

Most stable intermediates: $CH_3\dot{C}{=}CHBr$ (radical addition);

$\left[CH_3CH_2\overset{\oplus}{C}H{-}\ddot{\underset{..}{Br}}: \leftrightarrow CH_3CH_2CH{=}\overset{\oplus}{\underset{..}{Br}}:\right]$ (ionic addition)

(b) $CH_3C{\equiv}CH \xrightarrow[(ionic)]{HBr} CH_3CBr{=}CH_2 \xrightarrow[R\cdot]{HBr} CH_3CHBr{-}CH_2Br$

Most stable intermediates: $CH_3\overset{\oplus}{C}{=}CH_2$ (ionic addition);

$CH_3\dot{C}(:\ddot{Br}:){-}CH_2Br$ (radical addition)

(c) $CH_3C{\equiv}CH \xrightarrow[(ionic)]{HBr} CH_3CBr{=}CH_2 \xrightarrow[(ionic)]{HBr} CH_3CBr_2CH_3$

(continued next page)

(d) $CH_3C{\equiv}CH \xrightarrow[(R\cdot)]{HBr} CH_3CH{=}CHBr \xrightarrow[(R\cdot)]{HBr} CH_3CH_2CHBr_2$

21. Vinyl Grignard reagents are configurationally unstable.

22. (a) $CH_3C{\equiv}CCH_3$

(b) $CH_3C{\equiv}CCH_2CH_3$

(c) $(CH_3)_3CCH_2C{\equiv}CCH_3$

(d) $CH_3C{\equiv}C(CH_2)_4C{\equiv}CCH_3$

(e) cyclic $(CH_2)_{12}$ ring closed by C≡C

23. $HC{\equiv}CH$

24. (a) $CH_3(CH_2)_3CH{=}CHCH_3$ will decolorize Br_2/CCl_4 (brown to colorless); $CH_3(CH_2)_6CH_3$ does not.

(b) cyclohexene decolorizes Br_2/CCl_4 (brown to colorless); chlorocyclohexane does not.

(c) $CH_3(CH_2)_5C{\equiv}CH$ forms a precipitate with $Ag(NH_3)_2^{\oplus}$; $CH_3(CH_2)_5C{\equiv}CCH_3$ does not.

(d) $(CH_3)_2CHOH$ is oxidized by $CrO_3/H^{\oplus}$ (orange to green color change); $(CH_3)_2CHCl$ is not oxidized by $CrO_3/H^{\oplus}$.

(e) $CH_2{=}CHBr$ decolorizes Br_2/CCl_4 (brown to colorless); CH_3CH_2Br does not.

25. $CH_3CH_2CH_2CH_2C{\equiv}CH$

1-hexyne
n-butylacetylene

$CH_3CH_2CH_2C{\equiv}CCH_3$

2-hexyne
methyl-n-propylacetylene

$CH_3CH_2C{\equiv}CCH_2CH_3$

3-hexyne
diethylacetylene

$CH_3CH_2\underset{\displaystyle CH_3}{\underset{|}{C}}HC{\equiv}CH$

3-methyl-1-pentyne
<u>sec</u>-butylacetylene

$CH_3\underset{\displaystyle CH_3}{\underset{|}{C}}HCH_2C{\equiv}CH$

4-methyl-1-pentyne
isobutylacetylene

$CH_3\underset{\displaystyle CH_3}{\underset{|}{C}}HC{\equiv}CCH_3$

4-methyl-2-pentyne
methylisopropylacetylene

$(CH_3)_3CC{\equiv}CH$

3,3-dimethyl-1-butyne
<u>t</u>-butylacetylene

(Text S. 9.2)

26.

	1-butyne	2-butyne
(a)	$CH_3CH_2CH{=}CH_2$	$CH_3CH{=}CHCH_3$ (<u>trans</u>)
(b)	$CH_3CH_2\overset{\displaystyle O}{\overset{\|}{C}}CH_3$	$CH_3CH_2\overset{\displaystyle O}{\overset{\|}{C}}CH_3$
(c)	$CH_3CH_2CBr_2CHD_2$	$CH_3CBr_2CD_2CH_3$

(continued next page)

	1-butyne	2-butyne
(d)	$CH_3CH_2CCl_2CHCl_2$	$CH_3CCl_2CCl_2CH_3$
(e)	$CH_3CH_2C{\equiv}C{:}^{\ominus}\ Na^{\oplus}$	no reaction
(f)	$CH_3CH_2C{\equiv}CAg$	no reaction
(g)	$CH_3CH_2CH{=}CH_2$	$CH_3CH{=}CHCH_3$ (_cis_)
(h)	$CH_3CH_2C{\equiv}CMgCl$	no reaction
(i)	$CH_3CH_2CH_2\overset{O}{\overset{\Vert}{C}}H$	$CH_3CH_2\overset{O}{\overset{\Vert}{C}}CH_3$
(j)	$CH_3CH_2CH{=}CHSCH_3$	$CH_3CH{=}\underset{SCH_3}{\underset{\vert}{C}}CH_3$

(Ref: See Summary of Reactions and S. 9.4-9.8, Text)

27. A = acetylene

(a) $A \xrightarrow{H_2,Pd,BaSO_4} CH_2{=}CH_2$

(b) $A \xrightarrow{H_2,Pt} CH_3CH_3$

(c) $A \xrightarrow{Na^{\oplus}\ NH_2^{\ominus}} H{-}C{\equiv}C{:}^{\ominus}\ Na^{\oplus} \xrightarrow{CH_3I} HC{\equiv}CCH_3$

(d) $HC{\equiv}CCH_3$ (from c) $\xrightarrow{CH_3MgI} CH_3C{\equiv}CMgI \xrightarrow{D_2O} CH_3C{\equiv}CD$

(e) $CH_3C{\equiv}CH$ (from c) $\xrightarrow{Na^{\oplus}\ NH_2^{\ominus}} CH_3C{\equiv}C{:}^{\ominus}\ Na^{\oplus} \xrightarrow{CH_3I} CH_3C{\equiv}CCH_3$

$\xrightarrow{H_2,\ \text{Lindlar's cat.}} CH_3CH{=}CHCH_3$ (_cis_)

(continued next page)

(f) $CH_3C\equiv CCH_3$ (from e) $\xrightarrow{Na, NH_{3(\ell)}}$ $CH_3CH{=}CHCH_3$ (<u>trans</u>)

(g) $CH_3C\equiv CH$ (from c) $\xrightarrow{HCl}$ $CH_3CHCl{=}CH_2$

(h) $CH_3C\equiv C:^{\ominus}\ Na^{\oplus}$ (from e) $\xrightarrow{Cl_2}$ $CH_3C\equiv CCl$ $\xrightarrow{H_2, Pd, BaSO_4}$

$CH_3CH{=}CHCl$ (<u>cis</u>)

(i) $CH_3C\equiv CH$ (from c) $\xrightarrow{H_2,\ \text{Lindlar's catalyst}}$ $CH_3CH{=}CH_2$

$\xrightarrow{Cl_2}$ $CH_3CHClCH_2Cl$ $\xrightarrow[\text{alcohol}]{KOH}$ $CH_3CH{=}CHCl$

(j) $HC\equiv CH$ $\xrightarrow{Na^{\oplus}\ NH_2^{\ominus}}$ $Na^{\oplus}\ ^{\ominus}:C\equiv C:^{\ominus}\ Na^{\oplus}$ $\xrightarrow{D_2O}$ $DC\equiv CD$

(k) $CH_2{=}CH_2$ (from a) $\xrightarrow{H_2O/H^{\oplus}}$ CH_3CH_2OH

(l) A $\xrightarrow{H_2O, H^{\oplus}, Hg^{+2}}$ $CH_3C(=O)H$

(m) A $\xrightarrow{Na^{\oplus}\ NH_2^{\ominus}}$ $HC\equiv C:^{\ominus}\ Na^{\oplus}$ $\xrightarrow{D_2O}$ $HC\equiv CD$ $\xrightarrow{H_2O, H^{\oplus}, Hg^{+2}}$

$CH_3C(=O)D$ + $CH_2DC(=O)H$

(n) $CH_3C\equiv CH$ (from e) $\xrightarrow{H_2O, H^{\oplus}, Hg^{+2}}$ $CH_3C(=O)CH_3$

(o) $Na^{\oplus}\ ^{\ominus}:C\equiv C:^{\ominus}\ Na^{\oplus}$ (from j) $\xrightarrow{2Cl_2}$ $ClC\equiv CCl$ $\xrightarrow{Cl_2}$ $Cl_2C{=}CCl_2$

(continued next page)

(p) A $\xrightarrow{CH_3MgI}$ $HC{\equiv}CMgI$ $\xrightarrow{CH_2{=}CHCH_2Br}$ $HC{\equiv}CCH_2CH{=}CH_2$

(q) $CH_3C{\equiv}CCH_3$ (from e) $\xrightarrow{H_2O, H^{\oplus}, Hg^{+2}}$ $CH_3\overset{O}{\overset{\|}{C}}CH_2CH_3$

(r) A $\xrightarrow{HCl}$ $CH_2{=}CHCl$ $\xrightarrow{R\cdot}$ $\text{-(}CH_2\underset{Cl}{\underset{|}{CH}}\text{)}_n$

(Ref: review Summary of Reactions)

28. (i) ionic conditions: (<u>trans</u>(<u>anti</u>)-addition predominates)

(a) $CH_3CBr{=}CH_2$

(b) $CH_2{=}CBrCl$

(c) $CHBr{=}CHCF_3$

(d) $CH{=}CBr$ ring closed by $(CH_2)_8$

(ii) radical conditions:

(a) $CH_3CH{=}CHBr$

(b) $CHBr{=}CHCl$

(c) $CHBr{=}CHCF_3$

(d) $CH{=}CBr$ ring closed by $(CH_2)_8$

(Text S. 9.5, 9.6)

29. The tetrahalides required in this synthesis are themselves prepared from alkynes by the addition of halogen.

$$RC{\equiv}CR \quad + \quad 2X_2 \longrightarrow RCX_2CX_2R$$

30. (a) $CH_3CH_2Br \xrightarrow[NH_{3(\ell)}]{Na^{\oplus}\ NH_2^{\ominus}} CH_2{=}CH_2 \xrightarrow{Br_2/CCl_4} \xrightarrow[NH_{3(\ell)}]{2Na^{\oplus}\ NH_2^{\ominus}}$

(b) $CH_3C{\equiv}CCH_3 \xrightarrow[\text{Lindlar's cat.}]{H_2} CH_3CH{=}CHCH_3 \xrightarrow{O_3} \xrightarrow[H^{\oplus}]{Zn}$

(c) $HC{\equiv}CH \xrightarrow[NH_{3(\ell)}]{Na^{\oplus}\ NH_2^{\ominus}} \xrightarrow{CH_3I} CH_3C{\equiv}CH \xrightarrow[NH_{3(\ell)}]{Na^{\oplus}\ NH_2^{\ominus}} \xrightarrow{CH_3I}$

$\xrightarrow{Br_2} \xrightarrow{Cl_2}$

(d) $(CH_3)_2CHC{\equiv}CH \xrightarrow{Sia_2BH} \xrightarrow[0^{\circ}]{H_2O_2,\ OH^{\ominus}}$

(e) $(CH_3)_2CHC{\equiv}CH \xrightarrow{Hg^{2+}/H_2O/H^{\oplus}}$

(f) $CH_3CH_3 \xrightarrow[\text{light}]{Br_2} CH_3CH_2Br \longrightarrow HC{\equiv}CH$ (see a) $\longrightarrow$

$CH_3C{\equiv}CCH_3$ (see b) $\xrightarrow[OH^{\ominus}]{O_3} \xrightarrow{H_2O_2/OH^{\ominus}}$

(g) $CH_3(CH_2)_3CH{=}CH_2 \xrightarrow{Br_2/CCl_4} CH_3(CH_2)_3CHBrCH_2Br \xrightarrow[NH_{3(\ell)}]{2Na^{\oplus}\ NH_2^{\ominus}}$

(h) $CH_3C{\equiv}CH \xrightarrow[\text{Lindlar's cat.}]{H_2} CH_3CH{=}CH_2 \xrightarrow{BH_3} \xrightarrow{CH_3COOD}$

(or $CH_3C{\equiv}CH \xrightarrow{CH_3MgI} \xrightarrow{D_2O} CH_3C{\equiv}CD \xrightarrow[\text{Lindlar's cat.}]{H_2}$)

31. (a) Methylacetylene gives a precipitate with $Ag(NH_3)_2^{\oplus}$; dimethylacetylene does not.

(b) 1-Chloropropyne declorizes Br_2/CCl_4 (brown to colorless) or cold, dilute $KMnO_4$ (purple to colorless); 1-chloropropane does not react.

(c) Propyne forms a precipitate with $Ag(NH_3)_2^{\oplus}$; propene does not.

(d) Cyclodecyne will decolorize Br_2/CCl_4 or cold, dilute $KMnO_4$ solution. Cyclodecyne will form a precipitate with $Ag(NH_3)_2^{\oplus}$. Cyclodecane will not react with any of these reagents.

(e) Cyclohexene will decolorize Br_2/CCl_4 or cold, dilute $KMnO_4$ solution; cyclohexane will not.

(f) Vinyl chloride will decolorize Br_2/CCl_4 or cold, dilute $KMnO_4$ solution; ethyl chloride will not. Ethyl chloride reacts with alcoholic silver nitrate, vinyl chloride does not.

(g) Propyne will decolorize Br_2/CCl_4 or cold, dilute $KMnO_4$ solution. Propyne will form a precipitate with $Ag(NH_3)_2^{\oplus}$. Propane gives none of these reactions.

(h) 1-Butyne forms a precipitate with $Ag(NH_3)_2^{\oplus}$, 2-bromo-2-butene does not.

(i) These compounds cannot be distinguished by simple chemical means. They can be distinguished by analysis of the ozonolysis products, the uptake of hydrogen in catalytic hydrogenation etc.

(j) n-Butyl alochol will dissolve in cold, concentrated sulfuric acid, n-hexane will not.

32. A = $CH_3CHBrCHBrCH_3$ (d,l), B = $CH_3C{\equiv}CCH_3$, C = trans-$CH_3CH{=}CHCH_3$

D = $CH_3CHBrCHBrCH_3$ (meso), E = $CH_3C{\equiv}CCH_3$

33. (a) $CH_3CH{=}CHOH \rightleftharpoons CH_3CH_2CHO$ (b) $CH_3CH{=}C(OH)CH_3 \rightleftharpoons CH_3CH_2C(=O)CH_3$

(c) CH_3OH

(continued next page)

(d) $CH_2{=}CH(OH) \rightleftharpoons CH_3C(=O)H$

and $(CH_3)_2CHCH(CH_3)OH$

34. $CH_2{=}C{=}CH_2 \underset{}{\overset{Li^{\oplus}\ NH_2^{\ominus}}{\rightleftharpoons}} \left[CH_2{=}C{=}\ddot{C}H^{\ominus} \leftrightarrow \overset{\ominus}{\ddot{C}}H_2{-}C{\equiv}CH\right] \overset{NH_3}{\rightleftharpoons}$

$CH_3C{\equiv}CH \overset{NH_2^{\ominus}}{\rightleftharpoons} CH_3C{\equiv}C{:}^{\ominus} \overset{D_2O}{\rightleftharpoons} CH_3C{\equiv}CD + OD^{\ominus}$

35. $^{13}CH_3C{\equiv}CH \overset{NH_2^{\ominus}}{\rightleftharpoons} \left[{}^{13}\overset{\ominus}{:}CH_2C{\equiv}CH \leftrightarrow {}^{13}CH_2{=}C{=}\underset{\ominus}{\ddot{C}}H\right] \overset{NH_3}{\rightleftharpoons}$

$^{13}CH_2{=}C{=}CH_2 \overset{NH_2^{\ominus}}{\rightleftharpoons} \left[{}^{13}\overset{\ominus}{:}CH{=}C{=}CH_2 \leftrightarrow {}^{13}CH{\equiv}C{-}\underset{\ominus}{\ddot{C}}H_2\right] \overset{NH_3}{\rightleftharpoons}$

$^{13}CH{\equiv}C{-}CH_3 + NH_2^{\ominus}$

The similarity of this reaction to base-catalyzed conversion of an enol to a keto form is illustrated by the reactions below:

$B{:}^{\ominus} \curvearrowright H{-}\ddot{\underset{..}{O}}{-}\overset{|}{C}{=}\underset{|}{C}{-} \rightarrow \left[{:}\overset{\ominus}{\underset{..}{\ddot{O}}}{-}C{=}\underset{|}{C}{-} \leftrightarrow {:}\underset{..}{O}{=}C{-}\overset{\ominus}{\underset{|}{\ddot{C}}}{-}\right] \xrightarrow{HB} O{=}\overset{|}{C}{-}\overset{H}{\underset{|}{\overset{|}{C}}}{-}$

$B{:}^{\ominus} \curvearrowright H{-}CH_2{-}C{\equiv}CH \rightarrow \left[\overset{\ominus}{:}CH_2{-}C{\equiv}CH \leftrightarrow CH_2{=}C{=}\overset{\ominus}{\ddot{C}}H\right] \xrightarrow{HB} CH_2{=}CH{=}CH_2$

For the conversion $^{13}CH_3C{\equiv}CH \rightarrow {}^{13}CH{\equiv}CCH_3$, K = 1; for

$HO{-}\overset{|}{C}{=}\overset{|}{C}{-} \rightarrow O{=}\overset{|}{C}{-}\underset{H}{\underset{|}{\overset{|}{C}}}{-}$ K is generally quite large.

36. These data provide a basis for comparing the relative energies or stabilities of these compounds.

stabilities: $CH_3CH_3 > CH_2{=}CH_2 > HC{\equiv}CH$

$$2CH_3CH_3 + 7O_2 \longrightarrow 4CO_2 + 6H_2O$$

$$\Delta H = \left[4(-94) + 6(-58)\right] - 2(-20) = -684 \text{ kcal} = -342 \text{ kcal/mole}$$

$$CH_2{=}CH_2 + 3O_2 \longrightarrow 2CO_2 + 2H_2O \qquad \Delta H = -316 \text{ kcal/mole}$$

$$2HC{\equiv}CH + 5O_2 \longrightarrow 4CO_2 + 2H_2O \qquad \Delta H = -300 \text{ kcal/mole}$$

37. $(CH_3)_3CC{\equiv}CH \xrightarrow{HCl} CH_3\text{-}C(CH_3)_2\text{-}\overset{\oplus}{C}{=}CH_2 \longrightarrow CH_3\text{-}\overset{\oplus}{C}(CH_3)\text{-}C(CH_3){=}CH_2 \xrightarrow{:\ddot{Cl}:^{\ominus}} CH_3\text{-}C(CH_3)(Cl)\text{-}C(CH_3){=}CH_2$

(compare with the rearrangement in problem 8.21)

38. (a) $CH_3\ddot{N}H_2$ + $HC{\equiv}C{-}C(CH_3)_2{-}I$ ⟶ $CH_3{-}\overset{\oplus}{N}H_2{-}CH{=}C{=}C(CH_3)_2$ + $:\ddot{I}:^{\ominus}$

$\xrightarrow{-HI}$ $CH_3{-}\ddot{N}H{-}CH{=}C{=}C(CH_3)_2$

(b) $CH_3{-}C(CH_3)(H){-}C{\equiv}C{-}C(CH_3)_2{-}Br$ + $CH_3\ddot{N}H_2$ ⟶ $(CH_3)_2C{=}C{=}C{=}C(CH_3)_2$ + $CH_3NH_3^{\oplus}\,Br^{\ominus}$

39. $H_2C{=}CH{-}O{-}H$ ⟶ $H_3C{-}CH{=}O$ $\quad \Delta H = -13$ kcal/mole

Based on these calculations, the aldehyde is more stable.

3 C-H = 3 x 99 = 297	4 C-H = 4 x 99 = 396
1 C=C = 1 x 150 = 150	1 C=O = 1 x 175 = 175
1 C-O = 1 x 81 = 81	1 C-C = 1 x 80 = 80
1 O-H = 1 x 110 = 110	
total 638	total 651

40. (a) $C_6H_5C\equiv CH$ ir: sp C-H stretch, near 3300 cm^{-1}
nmr: acetylenic hydrogen, 1.8 - 3.1δ

These absorptions are absent in the other compound.

(b) $CH_3C\equiv CCH_2CD_3$ ir: C-D stretch, near 2100 cm^{-1}

(c) ir: C≡C stretch near 2100 cm^{-1} for $CH_3C\equiv CCH_2CH_3$.

This absorption will not appear for symmetrical $CH_3C\equiv CCH_3$

nmr: single absorption for methyl protons in $CH_3C\equiv CCH_3$

(d) same as (a)

(e) *cis*-$CH_3C\equiv CCH=CHCH_3$ ir: C-H bend 735-670 cm^{-1} for *cis* RHC=CHR

trans-$CH_3C\equiv CCH=CHCH_3$ ir: C-H bend 965 cm^{-1} for *trans* RHC=CHR

(Text S. 9.10)

41. (a) $\underset{a}{CH_3}-C\equiv C-\underset{a}{CH_3}$

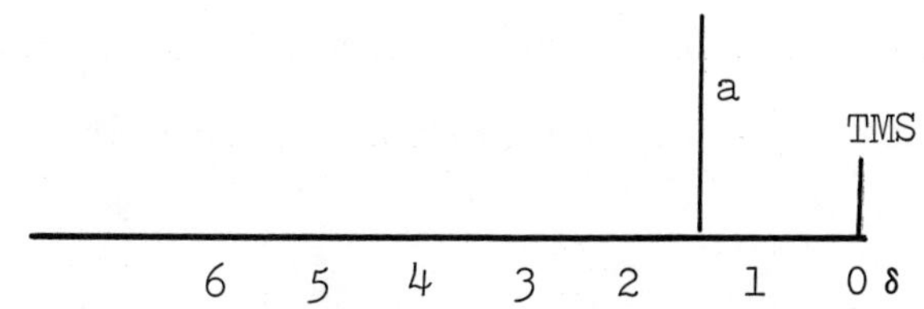

(continued next page)

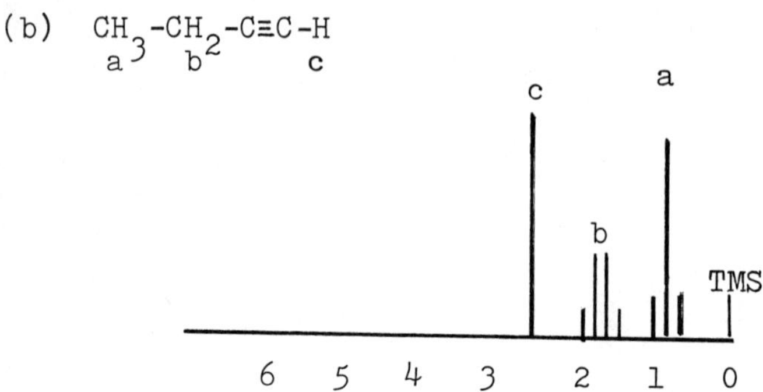

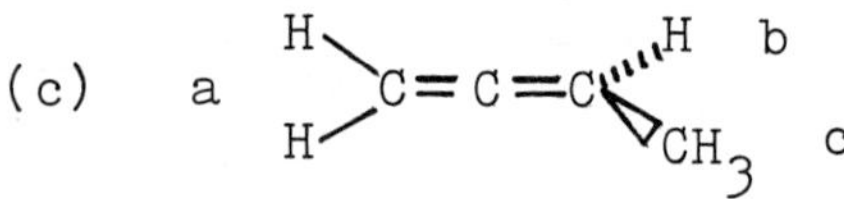

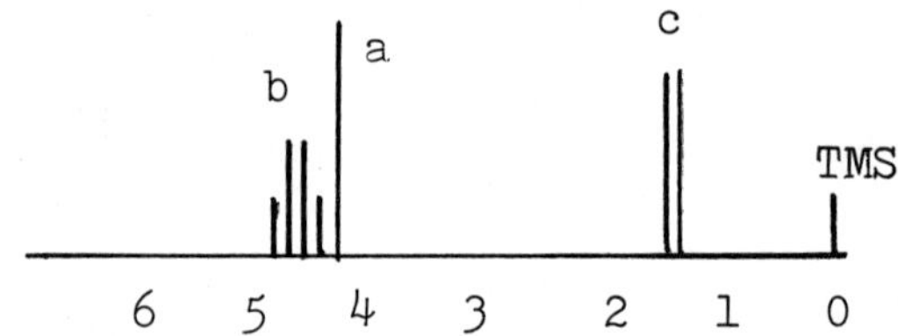

A simplified spectrum which ignores splitting due to long range coupling is shown here. Actually these signals are further split due to coupling of protons on non-adjacent carbons.

42. The infrared spectrum of 1-butyne will contain absorptions for ≡C-H stretch near 3300 cm^{-1} and C≡C stretch near 2200 cm^{-1}. 2-Butyne, a symmetric alkyne which also lacks an acetylenic hydrogen, will not have either of these absorptions in its spectrum.

43. The strong absorption near 3300 cm^{-1} in phenylacetylene (sp-C-H stretch) is missing in diphenylacetylene which lacks

(continued next page)

an acetylenic hydrogen. The absorption at 2100 cm^{-1} in phenylacetylene (C≡C stretch) is missing in symmetrical diphenylacetylene.

SUMMARY OF CHEMICAL REACTIONS

(specific examples of the general reactions which follow are provided in the Reaction Review which follows).

1. Acidity of terminal alkynes (S. 9.3)

$$RC\equiv CH \xrightarrow{Na^{\oplus}\ NH_2^{\ominus}} Na^{\oplus}\ RC\equiv C:^{\ominus} + NH_3 \text{ (sodium acetylide)}$$

$$RC\equiv CH \xrightarrow{R'MgX} RC\equiv C\text{-}MgX + R'H \text{ (acetylenic Grignard)}$$

$$RC\equiv CH \xrightarrow{Ag(NH_3)_2^{\oplus}} RC\equiv CAg \text{ (heavy metal acetylide)}$$

$$RC\equiv CH \xrightarrow{O_2/CuCl} RC\equiv C\text{-}C\equiv CR \text{ (oxidative coupling)}$$

2. Reactions of metal acetylides (S. 9.3, 9.4)

$$Na^{\oplus}\ RC\equiv C:^{\ominus} \xrightarrow{R'\text{-}L} RC\equiv CR' + Na^{\oplus}L^{\ominus} (S_N2)$$

(reactivity of R'L: $1^{o} > 2^{o} > 3^{o}$)

$$Na^{\oplus}\ RC\equiv C:^{\ominus} \xrightarrow{D_2O} RC\equiv CD + NaOD$$

$$RC\equiv CAg \xrightarrow{2CN^{\ominus}/H_2O} RC\equiv CH + Ag(CN)_2^{\ominus} + OH^{\ominus}$$

3. Electrophilic addition reactions (S. 9.5, 9.6, 9.7, 9.8)

$RC\equiv CH$ reacts with:

- HX → $RCX{=}CH_2$ $\xrightarrow{HX}$ RCX_2CH_3
- X_2 → $RCX{=}CHX$ $\xrightarrow{X_2}$ RCX_2CHX_2
- H_2O, $H^{\oplus}/Hg^{2+}$ → $RC(OH){=}CH_2$ (enol) ⇌ $RC(=O)CH_3$ (keto)
- Sia_2BH^{*} $\xrightarrow[OH^{\ominus}]{H_2O_2}$ $RCH{=}CH(OH)$ ⇌ $RCH_2C(=O)H$

  *(BH_3 can be used with internal alkynes, $RC\equiv CR$)

- $R'SH$, BF_3 → $RC(SR'){=}CH_2$ $\xrightarrow[BF_3]{R'SH}$ $RC(SR')_2CH_3$
- $RC(=O)OH$, $H^{\oplus}/Hg^{2+}$ → $RC(OC(=O)R){=}CH_2$ $\xrightarrow[H^{\oplus}/Hg^{2+}]{RC(=O)OH}$ $RC(OC(=O)R)_2CH_3$

4. Free-radical addition reactions (S. 9.8)

$$RC{\equiv}CH \xrightarrow[R\cdot]{R'SH} RCH{=}CHSR' \xrightarrow[R\cdot]{R'SH} RCH_2CH(SR')_2$$

$$RC{\equiv}CH \xrightarrow[R\cdot]{X_2} (R)(X)C{=}C(X)(H) \xrightarrow[R]{X_2} RCX_2CHX_2$$

$$RC{\equiv}CH \xrightarrow[R\cdot]{HBr} RCH{=}CHBr \xrightarrow[R\cdot]{HBr} RCH_2CHBr_2$$

5. Reduction (S. 9.9)

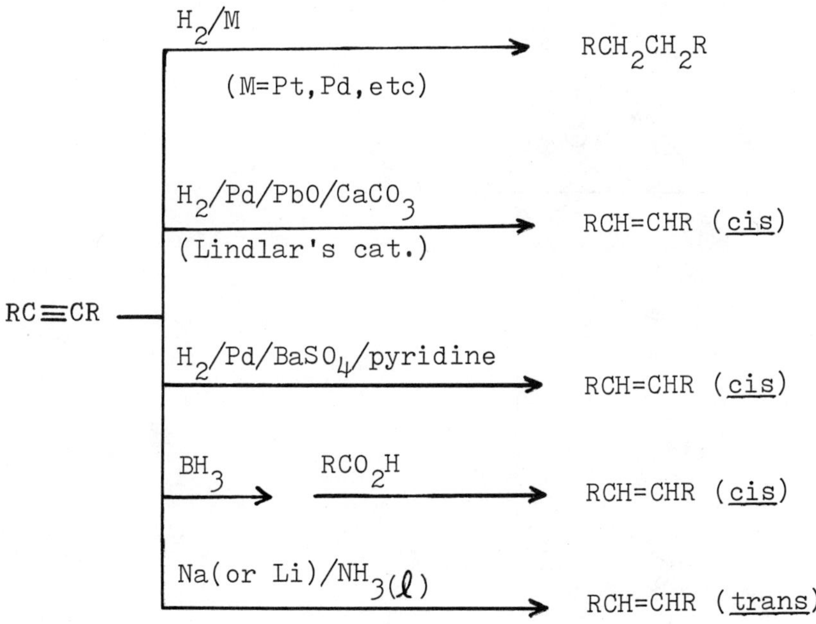

6. Carbene addition reactions (S. 9.8)

$$RC\equiv CR \xrightarrow{CH_2N_2/h\nu} \underset{\text{(cyclopropene, } CH_2\text{ bridge)}}{RC{=}CR} \xrightarrow{CH_2N_2/h\nu} \text{R-C-C-R (bicyclobutane, two } CH_2 \text{ bridges)}$$

$$RC\equiv CR \xrightarrow{CHCl_3/base} RC{=}CR \text{ bridged by } CCl_2$$

7. Oxidation reactions (S. 9.10)

$$RC\equiv CR \xrightarrow{O_3/CCl_4} \xrightarrow{H_2O} 2R\text{-}C(=O)OH \quad *$$

$$RC\equiv CR \xrightarrow{MnO_4^{\ominus}/OH^{\ominus}} \xrightarrow{H^{\oplus}} 2R\text{-}C(=O)\text{-}OH \quad *$$

$* \ (\equiv CH \longrightarrow CO_2)$

8. Alkenes via vinyl halides

$$RC\equiv CR \xrightarrow{HX} \text{(R)(H)C=C(X)(R)} \xrightarrow{Li(R')_2Cu} \text{(R)(H)C=C(R')(R)}$$

9. Synthesis of alkynes

a. Dehydrohalogenation

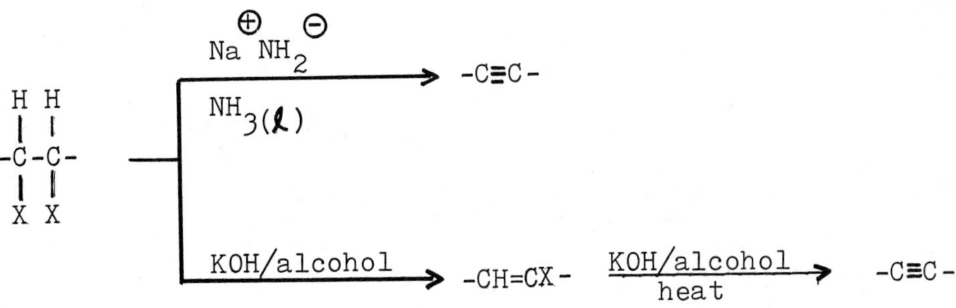

b. From sodium acetylides

$$RC\equiv C{:}^{\ominus}\ Na^{\oplus} \xrightarrow{R'-X} RC\equiv CR' + NaX \quad (S_N2)$$

(reactivity of R'X: $1^\circ > 2^\circ > 3^\circ$)

REACTION REVIEW

A. Reactions, Chapter 9

Questions	Answers
1. $RC\equiv CH \xrightarrow[NH_{3(\ell)}]{Na^{\oplus}NH_2^{\ominus}}$?	1. $Na^{\oplus}\ RC\equiv C{:}^{\ominus} + NH_3$
Write the structure of the product when the alkyne is:	
a) $CH_3CH_2C\equiv CH$	a) $Na^{\oplus}\ CH_3CH_2C\equiv C{:}^{\ominus}$
b) $CH_3C\equiv CCH_3$	b) no reaction

2. $Na^{\oplus}\ RC{\equiv}C{:}^{\ominus} \xrightarrow{R'X}$?	2. $RC{\equiv}CR'$
Write the structure of the products in the following:	
a) $Na^{\oplus}\ CH_3C{\equiv}C{:}^{\ominus} \xrightarrow{CH_3CH_2Br}$	a) $CH_3C{\equiv}CCH_2CH_3$
b) $\overset{\oplus}{Na}\ \overset{\ominus}{:}C{\equiv}\overset{\ominus}{C}{:}\ \overset{\oplus}{Na} \xrightarrow{2CH_3I}$	b) $CH_3C{\equiv}CCH_3$
c) $Na^{\oplus}\ HC{\equiv}C{:}^{\ominus} \xrightarrow{(CH_3)_3CCl}$	c) $HC{\equiv}CH + CH_2{=}C(CH_3)_2$ (RX is 3°, elimination occurs)
3. $Na^{\oplus}\ RC{\equiv}C{:}^{\ominus} \xrightarrow{D_2O}$?	3. $RC{\equiv}CD + Na^{\oplus}OD^{\ominus}$
Write the structure of the products in the following:	
a) $Na^{\oplus}\ CH_3C{\equiv}C{:}^{\ominus} \xrightarrow{D_2O}$?	a) $CH_3C{\equiv}CD$
b) $Na^{\oplus}\ C_6H_5{-}C{\equiv}C{:}^{\ominus} \xrightarrow{CH_3OD}$?	b) $C_6H_5{-}C{\equiv}CD$
4. $RC{\equiv}CH \xrightarrow{Ag(NH_3)_2^{\oplus}}$?	4. $RC{\equiv}CAg$
Write the structure of the product when the alkyne is:	
a) $CH_3C{\equiv}CH$	a) $CH_3C{\equiv}CAg$
b) $CH_3C{\equiv}CCH_3$	b) no reaction
	(continued next page)

c) What is the product of: $CH_3C\equiv CAg \xrightarrow[H_2O]{2CN^\ominus}$?	c) $CH_3C\equiv CH$
5. $RC\equiv CH \xrightarrow{HX}$? Write the structure of the product in the following: a) $CH_3C\equiv CCH_3 \xrightarrow{HBr}$ b) cyclopentyl$-C\equiv CH \xrightarrow{2HCl}$	5. $RCX=CH_2$ a) $CH_3CH=CBrCH_3$ (*trans*) b) cyclopentyl$-CCl_2CH_3$
6. $RC\equiv CH \xrightarrow{X_2}$? Write the structure of the products in the following: a) $CH_3C\equiv CCH_3 \xrightarrow[CCl_4]{Br_2}$ b) $CH_3C\equiv CH \xrightarrow{2Cl_2}$	6. $RCX=CHX$ (*trans*) a) $CH_3CBr=CBrCH_3$ (*trans*) b) $CH_3CCl_2CHCl_2$
7. $RC\equiv CH \xrightarrow{H_2O/H^\oplus/Hg^{2+}}$?	7. $RC(OH)=CH_2 \rightleftharpoons RC(=O)-CH_3$ (a ketone) (continued next page)

Write the structure of the ketone formed when $RC{\equiv}CH$ is:

a) $C_6H_5C{\equiv}CH$ (phenyl–C≡CH)

b) $CH_3C{\equiv}CCH_2CH_3$

a) $C_6H_5\overset{O}{\overset{\|}{C}}{-}CH_3$

b) $CH_3\overset{O}{\overset{\|}{C}}CH_2CH_2CH_3$ +

$CH_3CH_2\overset{O}{\overset{\|}{C}}CH_2CH_3$

---

8. $RC{\equiv}CH \xrightarrow{Sia_2BH^{*}} \xrightarrow[OH^{\ominus}]{H_2O_2}$?

*(BH_3 can be used with $RC{\equiv}CR$)

Write the structure of the ketone formed in the following:

a) $(CH_3)_2CHC{\equiv}CH \xrightarrow{Sia_2BH} \xrightarrow[OH^{\ominus}]{H_2O_2}$

b) $(CH_3)_2CHC{\equiv}CCH_3 \xrightarrow{BH_3} \xrightarrow[OH^{\ominus}]{H_2O_2}$

8. $RCH{=}\overset{OH}{\overset{|}{C}}H \rightleftharpoons RCH_2\overset{O}{\overset{\|}{C}}H$

a) $(CH_3)_2CHCH_2\overset{O}{\overset{\|}{C}}H$

b) $(CH_3)_2CH\overset{O}{\overset{\|}{C}}CH_2CH_3$ +

$(CH_3)_2CHCH_2\overset{O}{\overset{\|}{C}}CH_3$

9. $RC\equiv CR \xrightarrow[\text{catalyst}]{H_2}$? Write the structure for the products in the following: a) $CH_3C\equiv CCH_3 \xrightarrow[PbO/CaCO_3]{H_2/Pd}$? (Lindlar's) b) cyclopentyl–$C\equiv CCH_3 \xrightarrow{H_2/Pt}$?	9. $RCH{=}CHR$ (*cis*) $\xleftarrow{}$ … $RCH{=}CHR \xrightarrow[\text{cat.}]{H_2} RCH_2CH_2R$* (*cis*) *(product if *only* a metal catalyst such as Pt or Pd is used) a) $CH_3CH{=}CHCH_3$ (*cis*) b) cyclopentyl–$CH_2CH_2CH_3$
10. $RC\equiv CR \xrightarrow{BH_3} \xrightarrow{RCO_2H}$? Write the structure of the product when $RC\equiv CR$ is: a) $CH_3C\equiv CCH_3$ b) $CH_3CH_2C\equiv CCH_3$	10. $RCH{=}CHR$ (*cis*) a) $CH_3CH{=}CHCH_3$ (*cis*) b) $CH_3CH_2CH{=}CHCH_3$ (*cis*)
11. $RC\equiv CR \xrightarrow[NH_{3(\ell)}]{\text{Na (or Li)}}$?	11. $RCH{=}CHR$ (*trans*) (continued next page)

Write the structure of the product when RC≡CR is: a) $CH_3C{\equiv}CCH_3$ b) $CH_3CH_2C{\equiv}CCH_3$	a) $CH_3CH{=}CHCH_3$ (trans) b) $CH_3CH_2CH{=}CHCH_3$ (trans)
12. $RC{\equiv}CR \xrightarrow[CCl_4]{O_3} \xrightarrow{H_2O}$ Write the structure of the product when the alkyne is: a) $(CH_3)_3CC{\equiv}CCH_2CH_3$ b) $CH_3CH_2C{\equiv}CH$	12. $2RC({=}O){-}OH$ ($\equiv CH \longrightarrow CO_2$) a) $(CH_3)_3CCO_2H + CH_3CH_2CO_2H$ b) $CH_3CH_2CO_2H + CO_2$
13. $RC{\equiv}CR \xrightarrow[OH^{\ominus}]{MnO_4^{\ominus}} \xrightarrow{H^{\oplus}}$? Write the structure of the product when the alkyne is: a) $(CH_3)_2CHC{\equiv}CH$ b) $CH_3C{\equiv}CCH_3$	13. $2RC({=}O)OH$ ($\equiv CH \longrightarrow CO_2$) a) $(CH_3)_2CHCO_2H + CO_2$ b) $2CH_3CO_2H$

14. $-CHXCHX- \xrightarrow[NH_{3(\ell)}]{Na^{\oplus}NH_2^{\ominus}\ *}$?

*(or KOH/alcohol/heat)

Write the structure of the product when -CHXCHX- is:

a) $CH_3CHClCHClCH_3$

b) (cyclohexyl)–$CHBrCH_2Br$

14. $-C\equiv C-$

a) $CH_3C\equiv CCH_3$

b) (cyclohexyl)–$C\equiv C{:}^{\ominus}\ Na^{\oplus}$

$\downarrow H^{\oplus}$

(cyclohexyl)–$C\equiv CH$

15. Write the structure of the product of the free-radical reactions of $CH_3C\equiv CH$ with:

a) $CH_3SH/R\bullet$

b) $Br_2/R\bullet$

c) $HBr/R\bullet$

15. <u>anti</u>-addition in each gives:

a) $CH_3CH{=}CHSCH_3$ (cis)

b) $CH_3CBr{=}CHBr$ (trans)

c) $CH_3CH{=}CHBr$ (cis)

16. Write the structure of the product in the addition reactions of $CH_3C\equiv CH$ with:

a) CH_3SH/BF_3

b) $2CH_3\overset{O}{\overset{\|}{C}}OH/H^{\oplus}/Hg^{2+}$

a) $CH_3C(SCH_3){=}CH_2$

b) $CH_3C(O\overset{O}{\overset{\|}{C}}CH_3)_2CH_3$

B. Write structures for all letters

1. $CH_3CH{=}CHCH_3 \xrightarrow[CCl_4]{Br_2} A$ $\xrightarrow[NH_{3(\ell)}]{2Na^{\oplus}\ NH_2^{\ominus}} B \xrightarrow[\text{Lindlar's cat.}]{H_2/Pd} C$	1. A: $CH_3CHBrCHBrCH_3$ B: $CH_3C{\equiv}CCH_3$ C: $CH_3CH{=}CHCH_3$ (<u>cis</u>)
2. $CH_3C{\equiv}CH \xrightarrow[NH_{3(\ell)}]{Na^{\oplus}\ NH_2^{\ominus}} A \xrightarrow{CH_3I} B$ $\xrightarrow{H_2O/H^{\oplus}/Hg^{2+}} C$	2. A: $CH_3C{\equiv}C{:}^{\ominus}\ Na^{\oplus}$ B: $CH_3C{\equiv}CCH_3$ C: $CH_3\overset{O}{\overset{\Vert}{C}}CH_2CH_3$
3. $CH_3CH_2CH_2CH_3 \xrightarrow[h\nu]{Br_2} A$ $\xrightarrow[\text{heat}]{\text{KOH/alcohol}} B \xrightarrow{Cl_2} C$ $\xrightarrow[NH_{3(\ell)}]{2Na\ NH_2} D$	3. A: $CH_3CH_2CHBrCH_3$ B: $CH_3CH{=}CHCH_3$ C: $CH_3CHClCHClCH_3$ D: $CH_3C{\equiv}CCH_3$
4. $CH_3C{\equiv}CH \xrightarrow[\text{peroxide}]{HBr} A$ $\xrightarrow{Li(CH_3)_2Cu} B \xrightarrow{H_2O/H^{\oplus}} C$	4. A: $CH_3CH{=}CHBr$ (<u>cis</u>) B: $CH_3CH{=}CHCH_3$ (<u>cis</u>) C: $CH_3CH(OH)CH_2CH_3$

5. $CH_3C{\equiv}CH \xrightarrow{CH_3MgI} A \xrightarrow{D_2O} B$	5. A: $CH_3C{\equiv}CMgI$ B: $CH_3C{\equiv}CD$
6. $HC{\equiv}CH \xrightarrow[NH_{3(\ell)}]{2Na^{\oplus}NH_2^{\ominus}} A$ $\xrightarrow{2CH_3CH_2Br} B \xrightarrow[NH_{3(\ell)}]{Na} C$	6. A: $Na^{\oplus}\ {:}\overset{\ominus}{C}{\equiv}\overset{\ominus}{C}{:}\ Na^{\oplus}$ B: $CH_3CH_2C{\equiv}CCH_2CH_3$ C: $CH_3CH_2CH{=}CHCH_2CH_3$ (trans)
7. $CH_3CH{=}CH_2 \xrightarrow[peroxide]{NBS} A$ $\xrightarrow{HC{\equiv}CMgCl} B \xrightarrow[NH_{3(\ell)}]{Na^{\oplus}NH_2^{\ominus}} C$ $\xrightarrow{CH_3I} D$	7. A: $BrCH_2CH{=}CH_2$ B: $HC{\equiv}CCH_2CH{=}CH_2$ C: $Na^{\oplus}\ {:}\overset{\ominus}{C}{\equiv}CCH_2CH{=}CH_2$ D: $CH_3C{\equiv}CCH_2CH{=}CH_2$

CHAPTER 10, ALCOHOLS

LEARNING OBJECTIVES

When you have completed this chapter, you should be able to:

1. write names for alcohols and classify alcohols as 1°, 2° or 3° (problems 1, 2, 18 and related problem 19);
2. predict reactants or products in the synthesis of alcohols via Grignard reagents (problems 5, 7, 20);
3. predict the reactants or products in the synthesis of alcohols via aldehydes or ketones and hydride reagents (problems 8, 9);

(continued next page)

4. predict the reactants or products in the synthesis of alcohols by hydration of alkenes (hydration via $H_3O^{\oplus}$, hydroboration-oxidation, hydroboration-carbonylation-oxidation, oxymercuration-demercuration) (problems 4, 10, 11, 21 and related problem 36);
5. predict reactants or products in the oxidation of alcohols (problem 13);
6. describe simple tests which permit you to distinguish alcohols from other compounds (problems 15, 24);
7. account for the reactivity of epoxides (cyclic oxides) with Grignard reagents (problems 6, 34);
8. write mechanisms to account for the reactions of alcohols (problems 3, 32, 33, 37);
9. write equations which illustrate the multistep syntheses of alcohols and other compounds utilizing reactions from this and previous chapters (problems 21, 22, 23 and related problems 28, 29, 31);
10. account for the reactivity of alcohols in various reactions (problems 26, 27, 35);
11. predict the number of lines in the OH resonance of the NMR spectra of alcohols in DMSO (problems 16, 17 and related problem 42);
12. interpret IR and NMR spectra or identify structures on the basis of their spectra (problems 40, 41, 43, 44, 45, 46, 47).

ANSWERS TO QUESTIONS

1. (a) 2-buten-1-ol, 1°

   (b) 2-pentanol, 2°

   (c) (R)-3-phenyl-1-butanol, 1°

   (d) (2S,5R)-2,5-heptanediol, 2°

   (e) (2S,5R)-5-methyl-5-phenyl-2-heptanol, 2°

   (f) (Z)-3-penten-1-ol, 1°

2. 1° alcohol: $CH_3(CH_2)_5CH_2OH$ (or any other structure of the type RCH_2OH)

   1-heptanol

(continued next page)

2° alcohol: $CH_3(CH_2)_4CHCH_3$ with OH on the CH (or any other structure of the type R_2CHOH)

2-heptanol

3° alcohol: $CH_3(CH_2)_3C(CH_3)(OH)-CH_3$ (or any other structure of the type R_3COH)

2-methyl-2-hexanol

3. The mechanism is S_N1.

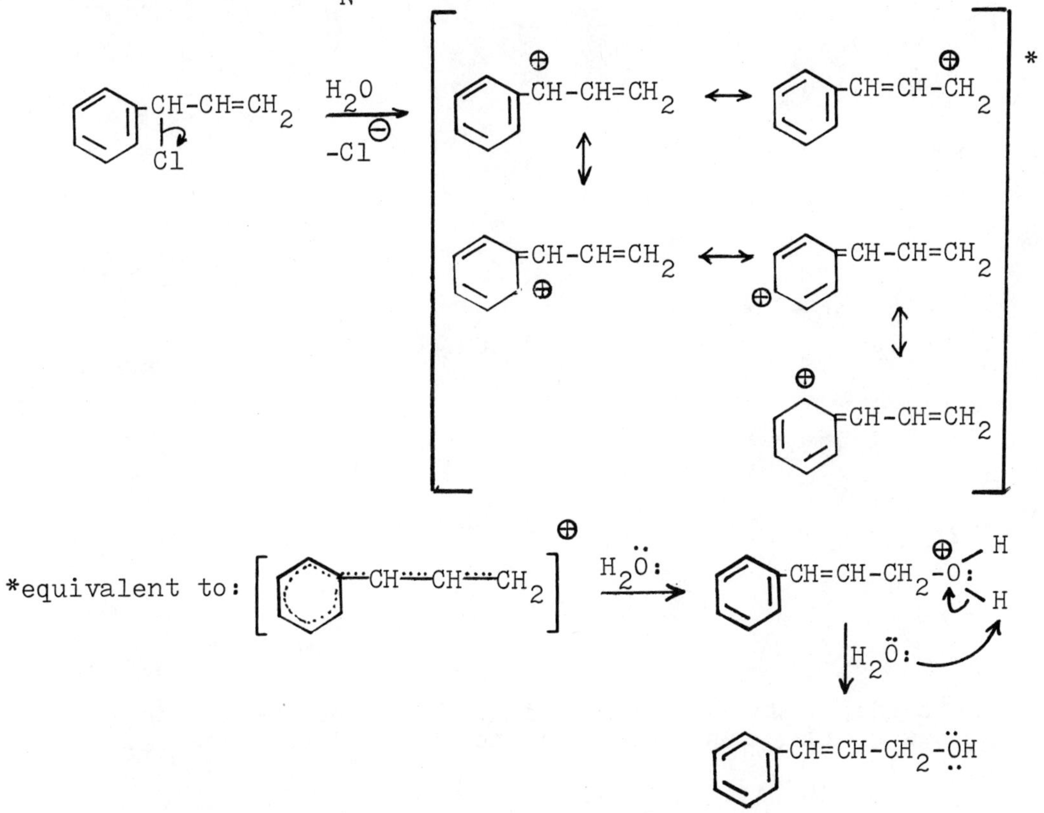

$C_6H_5-CH(Cl)-CH=CH_2 \xrightarrow[-Cl^{\ominus}]{H_2O}$ [$C_6H_5-\overset{\oplus}{C}H-CH=CH_2 \longleftrightarrow C_6H_5-CH=CH-\overset{\oplus}{C}H_2 \longleftrightarrow$ ring-delocalized forms $=CH-CH=CH_2$ with $\oplus$ at ortho, para and ortho' positions]*

*equivalent to: $[C_6H_5\cdots CH\cdots CH\cdots CH_2]^{\oplus} \xrightarrow{H_2\ddot{O}:} C_6H_5-CH=CH-CH_2-\overset{\oplus}{O}H_2 \xrightarrow{H_2\ddot{O}:} C_6H_5-CH=CH-CH_2-\ddot{O}H$

4. General reaction: $RCH{=}CH_2 \xrightarrow{BH_3} \xrightarrow[OH^{\ominus}]{H_2O_2} RCH_2CH_2OH$

(a) $D_2C(OH)CH_2CH_3$ (b) (d,l)-$CH_3CH_2CH(OH)CH_3$

(c) (d,l)-$CH_3CH_2CH(OH)CH_3$ (d) 2-methylcyclohexanol, CH_3 and OH trans (d,l)

(e) cyclohexyl-CH_2OH

5. (a) $D_2C{=}O$ (b) $CH_3C(=O)CH_3$

(c) $CH_3CH_2C(=O)CH_3$ (d) $CH_3CH_2C(=O)CH_2CH_3$

(e) CH_2—CH_2 bridged by O (ethylene oxide) (f) $CH_3C(=O)H$

6. Ethylene oxide, a three-membered ring, possesses greater angle strain, has greater energy and is, therefore, more reactive.

7. General reaction: $RMgX + \gt C{=}O \rightarrow R\text{-}\overset{|}{\underset{|}{C}}\text{-}OMgX \xrightarrow{H_2O} R\text{-}\overset{|}{\underset{|}{C}}\text{-}OH + MgXOH$

(a) $CH_3CH_2MgBr \xrightarrow{CH_3\overset{O}{\overset{\|}{C}}CH_3} \xrightarrow{H_2O}$ (or $CH_3MgI \xrightarrow{CH_3\overset{O}{\overset{\|}{C}}CH_2CH_3} \xrightarrow{H_2O}$)

(b) 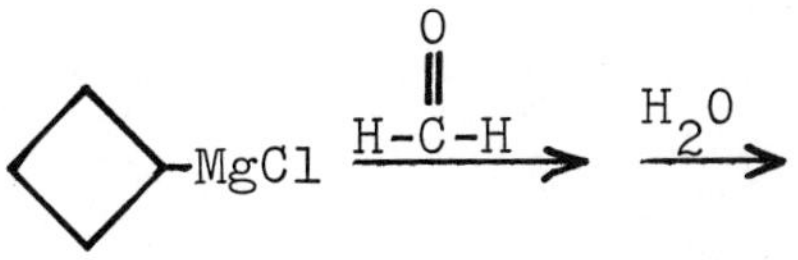 cyclobutyl-MgCl $\xrightarrow{H\text{-}\overset{O}{\overset{\|}{C}}\text{-}H} \xrightarrow{H_2O}$

(c) $CH_3MgCl \xrightarrow{\text{cyclobutanone}} \xrightarrow{H_2O}$

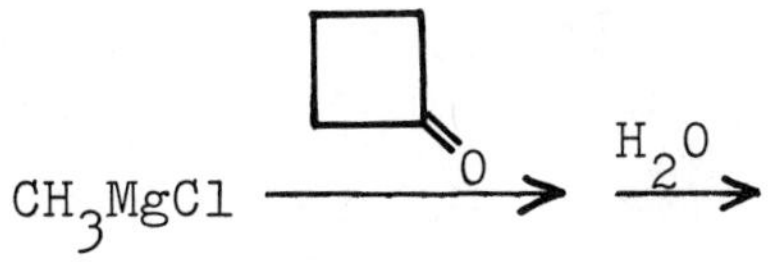

(d) 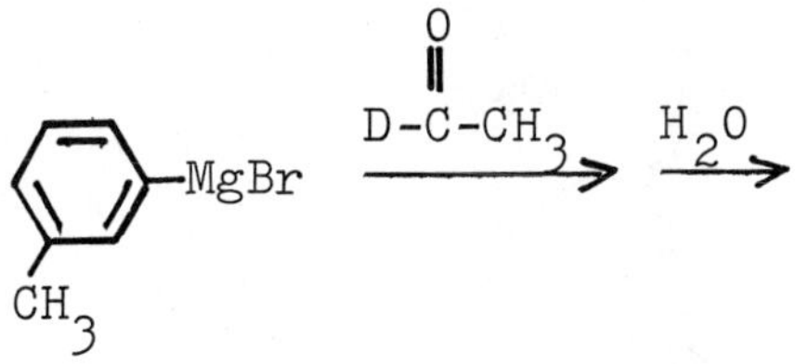 3-methylphenyl-MgBr (CH_3 on ring) $\xrightarrow{D\text{-}\overset{O}{\overset{\|}{C}}\text{-}CH_3} \xrightarrow{H_2O}$

(e) 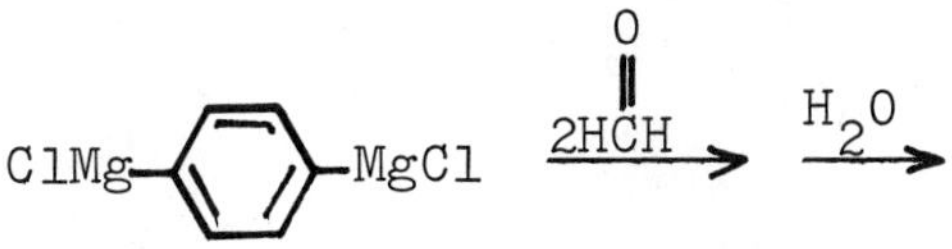 ClMg–C_6H_4–MgCl (para) $\xrightarrow{2H\overset{O}{\overset{\|}{C}}H} \xrightarrow{H_2O}$

(f) $(CH_3)_2CHMgBr \xrightarrow{(CH_3)_3CCHO} \xrightarrow{H_2O}$

8. Sodium borohydride reduces only aldehydes and ketones.

(a) N.R.

(b) $CH_3CH(OH)CH_3$

(c) 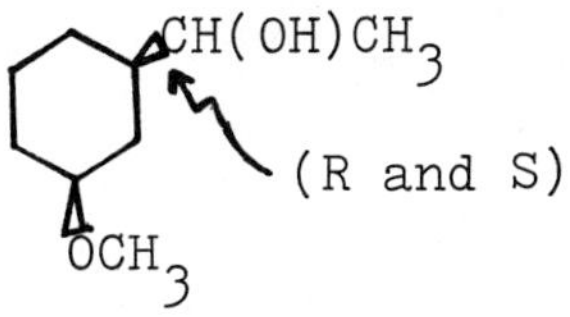 cyclohexane ring bearing $CH(OH)CH_3$ and OCH_3 (1,3-positions); $CH(OH)CH_3$ carbon: (R and S)

(continued next page)

(d) $CH_3CH(OH)-(CH_2)_4-CH(OH)CH_3$

(meso, R,R and S,S)

(e) 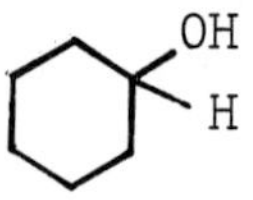 OH, H

(f) N.R.

9. (a) N.R.

(b) $CH_3CD(OD)CH_3$

(c) 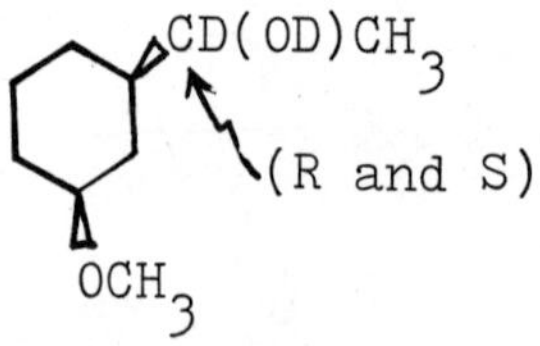 $CD(OD)CH_3$ (R and S), OCH_3

(d) $CH_3CD(OD)-(CH_2)_4-CD(OD)CH_3$

(meso, R,R and S,S)

(e) 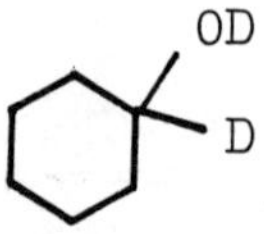 OD, D

(f) N.R.

10.

	(i) hydroboration-oxidation	(ii) oxymercuriation-deoxymercuriation
(a)	$CH_3CH_2CH_2OH$	$CH_3CH(OH)CH_3$
(b)	$CH_3CH_2CH(OH)CH_3$	$CH_3CH_2CH(OH)CH_3$
(c)	$CH_3CH_2CH(OH)CH_3$	$CH_3CH_2CH(OH)CH_3$
(d)	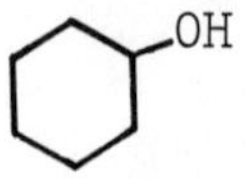 OH	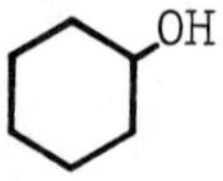 OH
(e)	(d,l) trans- 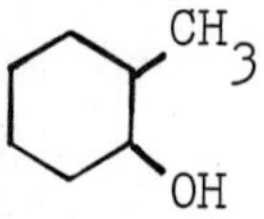 CH_3, OH	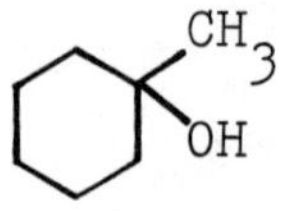 CH_3, OH

11. (a) $(CH_3)_2\underset{|}{C}CH_3$ with OCH_3 on the central carbon

(c) $CH_3\underset{|}{C}HCH_2CH_3$ with OC_2H_5 on C-2

(b) $CH_3CH_2\underset{|}{C}(CH_3)_2$ with OC_2H_5 on the quaternary carbon

12. The two chlorosulfites are diastereomers. The sulfur in each is chiral and configurationally stable.

$$(R)\text{-}CH_3CH_2CH(OH)CH_3 \xrightarrow{SOCl_2} \left[\begin{array}{l} (R)\ (R)\ CH_3CH_2CH(CH_3)\text{-O-S(O)Cl} \\ (R)\ (S)\ CH_3CH_2CH(CH_3)\text{-O-S(O)Cl} \end{array}\right] \text{diastereomers}$$

$$(R)\text{-}CH_3CH_2CHClCH_3 \xleftarrow{-SO_2} \text{chlorosulfites}$$

13. (a) $CH_3CH(OH)CH_3 + H_2CrO_4$

(b) $CH_2CH_2CH_2OH + CrO_3\cdot$ pyridine

(c) cyclohexanol + $CrO_3/H^{\oplus}$

(d) HO–(cyclohexane)–OH + $K_2Cr_2O_7/H^{\oplus}$

(e) (cyclohexenyl)–OH + $CrO_3/H^{\oplus}$

(f) $CH_3CH{=}CHCH_2OH + MnO_2$

(any of the reagents H_2CrO_4, $CrO_3/H^{\oplus}$, CrO_3 pyridine, $K_2Cr_2O_7/H^{\oplus}$ can be used in a, c, d, e)

14. $CH_3\overset{\oplus}{O}H_2\ HSO_4^{\ominus}$

15. (a) Only CH_3CH_2OH reacts with $CrO_3/H^{\oplus}$ (orange → green color change), only CH_3CH_2OH is soluble in cold, conc. H_2SO_4.

(b) Only $CH_3(CH_2)_6CH_2OH$ is soluble in H_2SO_4 ($CH_3(CH_2)_6CH_3$ forms a separate layer); only $CH_3(CH_2)_6CH_2OH$ reacts with $CrO_3/H^{\oplus}$ (orange → green color change).

(c) Only $CH_3(CH_2)_6CH{=}CH_2$ decolorizes Br_2/CCl_4.

(d) Lucas test: $(CH_3CH_2)_2CHOH$ reacts (forms separate layer) within 5 minutes; $(CH_3)_3COH$ reacts immediately.

(e) Only $(CH_3)_2CHOH$ reacts with $CrO_3/H^{\oplus}$ (orange → green color change); $CH_3C{\equiv}CH$ forms precipitate with $Ag(I)/NH_4OH$, $CH_3C{\equiv}CCH_3$ does not.

(f) Only $CH_3CH_2CH{=}CH_2$ and $CH_3CH_2C{\equiv}CH$ decolorize Br_2/CCl_4 (brown → colorless); $CH_3CH_2C{\equiv}CH$ forms a precipitate with $Ag(I)/NH_4OH$, $CH_3CH_2CH{=}CH_2$ does not.

16. (a) quartet
(b) triplet
(c) triplet
(d) doublet
(e) triplet
(f) doublet
(g) singlet

17. (a) doublet ($CH_3CH(OH)CH_3$)
(b) doublet ($CH_3CH_2CH(OH)CH_3$)
(c) singlet $(CH_3)_3COH$
(d) triplet (CH_3CH_2OH)
(e) triplet ($CH_3CH_2CH_2OH$)

18. $CH_3CH_2CH_2CH_2CH_2OH$, 1°

1-pentanol
n-pentyl alcohol

$CH_3CH_2CH(CH_3)CH_2OH$, 1°

2-methyl-1-butanol
sec-butylcarbinol

$CH_3CH(CH_3)CH_2CH_2OH$, 1°

3-methyl-1-butanol
isopentyl alcohol

$CH_3CH_2CH_2CH(OH)CH_3$, 2°

2-pentanol
methyl-n-propylcarbinol

$CH_3CH_2CH(OH)CH_2CH_3$, 2°

3-pentanol
diethylcarbinol

$(CH_3)_2C(OH)CH_2CH_3$, 3°

2-methyl-2-butanol
ethyldimethylcarbinol

$CH_3CH(OH)CH(CH_3)_2$, 2°

3-methyl-2-butanol
methylisopropylcarbinol

$(CH_3)_3CCH_2OH$, 1°

2,2-dimethyl-1-propanol
neopentyl alcohol
t-butylcarbinol

(Text: nomenclature, S. 10.3)

19. Each of the isomeric octyl alcohols contains eight carbons. Those not belonging to this group are (b) and (f) which contain nine carbons each.

(Text S. 10.3)

20. (a) $(CH_3)_2C{=}O \xrightarrow{CH_3MgI} \xrightarrow{H_2O} (CH_3)_3COH$ (2-methyl-2-propanol)

(b) $(CH_3)_2C{=}O \xrightarrow{CH_3CH_2MgI} \xrightarrow{H_2O} (CH_3)_2C(OH)CH_2CH_3$

(2-methyl-2-butanol)

(c) $CH_3C(=O){-}OCH_3 \xrightarrow{2CH_3MgCl*} \xrightarrow{H_2O} (CH_3)_3COH$ (2-methyl-2-propanol)

$*(CH_3{-}C(=O)OCH_3 \xrightarrow{CH_3MgCl} CH_3{-}C(=O){-}CH_3 \xrightarrow{CH_3MgCl} \xrightarrow{H_2O} (CH_3)_3COH)$

(d) $CH_3C(=O)OC_2H_5 \xrightarrow{2CH_3MgI} (CH_3)_3COH$ (2-methyl-2-propanol)

(e) cyclohexanone $\xrightarrow{CH_3CH_2MgBr} \xrightarrow{H_2O}$ 1-ethylcyclohexanol (ring C bearing OH and CH_2CH_3)

(1-ethylcyclohexanol)

(f) cyclohexanone $\xrightarrow{\text{cyclohexyl-MgCl}} \xrightarrow{H_2O}$ cyclohexyl–cyclohexane ring bearing OH

(1-cyclohexylcyclohexanol)

(continued next page)

(g) $CH_3\overset{O}{\overset{\|}{C}}CH_3 \xrightarrow{NaBH_4} CH_3\underset{OH}{\underset{|}{C}}HCH_3 \xrightarrow{H_2SO_4,\ heat} CH_3CH{=}CH_2 \xrightarrow{BH_3}$

$\xrightarrow[OH^{\ominus}]{H_2O_2} CH_3CH_2CH_2OH \xrightarrow{CrO_3\ pyridine} CH_3CH_2C(=O)H \xrightarrow{CH_3MgCl}$

$\xrightarrow{H_2O} CH_3CH_2\underset{OH}{\underset{|}{C}}HCH_3$ (2-butanol)

(Review Summary of Reactions and S. 10.4)

21. (a) $CH_3CH_2CH_2CH{=}CH_2 \xrightarrow{H_3O^{\oplus}}$

(b) $CH_3CH_2CH_2CH{=}CH_2 \xrightarrow{BH_3} \xrightarrow[OH^{\ominus}]{H_2O_2} CH_3CH_2CH_2CH_2CH_2OH$

(c) $CH_3CH{=}CH_2 \xrightarrow{BH_3} \xrightarrow[OH^{\ominus}]{H_2O_2} CH_3CH_2CH_2OH \xrightarrow[pyridine]{CrO_3} CH_3CH_2C(=O)H$

$\xrightarrow{CH_3CH_2MgBr} \xrightarrow{H_3O^{\oplus}} CH_3CH_2\underset{OH}{\underset{|}{C}}HCH_2CH_3$

(d) $CH_3CH_2CH_2\underset{CH_3}{\underset{|}{C}}{=}CH_2 \xrightarrow{H_3O^{\oplus}} CH_3CH_2CH_2\underset{OH}{\underset{|}{C}}(CH_3)_2$

(e) $CH_3CH_2\underset{CH_3}{\underset{|}{C}}{=}CH_2 \xrightarrow{H_3O^{\oplus}} CH_3CH_2\underset{OH}{\underset{|}{C}}(CH_3)_2$

(continued next page)

(f) $3CH_3CH{=}CH_2 \xrightarrow{BH_3} \xrightarrow[heat]{CO} \xrightarrow{H_2O_2/OH^{\ominus}} (CH_3CH_2CH_2)_3COH$

22. (a) $CH_3(CH_2)_6CH_2OH + Na \longrightarrow Na^{\oplus}\ {}^{\ominus}OCH_2(CH_2)_6CH_3 + \frac{1}{2}H_2$

(b) $(CH_3)_2CHCH_2\underset{\displaystyle OH}{\underset{|}{C}}HCH_3 + SOCl_2 \longrightarrow (CH_3)_2CHCH_2\underset{\displaystyle Cl}{\underset{|}{C}}HCH_3 \xrightarrow[ether]{Mg}$

$(CH_3)_2CHCH_2\underset{\displaystyle MgCl}{\underset{|}{C}}HCH_3 \xrightarrow{H_2O} (CH_3)_2CHCH_2CH_2CH_3$

(c) $CH_3CH_2\underset{\displaystyle CH_3}{\underset{|}{C}}H\overset{\displaystyle CH_3}{\overset{|}{C}}HCH_2OH \xrightarrow{ClSO_2OH\ *} CH_3CH_2\underset{\displaystyle CH_3}{\underset{|}{C}}H\overset{\displaystyle CH_3}{\overset{|}{C}}HCH_2OSO_2OH$

*(or HNO_3 to give the nitrate ester, see text p. 411)

(d) $3CH_3CH{=}CH_2 \xrightarrow{BH_3} \xrightarrow[heat]{CO} \xrightarrow{H_2O_2/OH^{\ominus}} (CH_3CH_2CH_2)_3COH$

(or $CH_3CH{=}CH_2 \xrightarrow{H_3O^{\oplus}} CH_3CH(OH)CH_3 \xrightarrow{CrO_3/H^{\oplus}} CH_3\overset{\displaystyle O}{\overset{\|}{C}}CH_3$

$\xrightarrow{RMgX} \xrightarrow{H_2O} CH_3\underset{\displaystyle R}{\underset{|}{C}}(OH)CH_3$)

23. (a) $CH_3CH{=}CH_2 \xrightarrow{H_2, Pt} CH_3CH_2CH_3$

(b) $CH_3CH{=}CH_2 \xrightarrow{BH_3} \xrightarrow[OH^{\ominus}]{H_2O_2} CH_3CH_2CH_2OH$

(continued next page)

(c) $CH_3CH{=}CH_2 \xrightarrow{H_3O^{\oplus}} CH_3CH(OH)CH_3$

(d) $CH_3CH_2CH_2OH$ (from b) $\xrightarrow{SOCl_2} CH_3CH_2CH_2Cl$

(e) $CH_3CH{=}CH_2 \xrightarrow{HCl} CH_3CHClCH_3$

(f) $CH_3CH_2CH_2Cl$ (from d) $\xrightarrow{Mg/ether} \xrightarrow{CH_2O} \xrightarrow{H_2O}$

$CH_3CH_2CH_2CH_2OH \xrightarrow[heat]{H_2SO_4} CH_3CH_2CH{=}CH_2$

(g) $CH_3CH{=}CH_2 \xrightarrow{DCl} \xrightarrow{OH^{\ominus},\ H_2O} CH_3CH(OH)CH_2D$

(h) $CH_3CH{=}CH_2 \xrightarrow{BD_3} \xrightarrow{H_2O_2,\ OH^{\ominus}} CH_3CHDCH_2OH$

(i) $CH_3CH_2CH_2OH$ (from b) $\xrightarrow{CrO_3\cdot pyridine} CH_3CH_2C(=O)H$

(j) $CH_3CH(OH)CH_3$ (from c) $\xrightarrow{K_2Cr_2O_7,\ H^{\oplus}} CH_3C(=O)CH_3$

(k) $CH_3CH{=}CH_2 \xrightarrow{CH_3C(=O)\text{-}OOH}$ $CH_3CH\text{—}CH_2$ (epoxide, bridged by O)

(l) $CH_3CH{=}CH_2 \xrightarrow[25^{o}]{KMnO_4} CH_3CH(OH)CH_2OH$

(or $CH_3CH\text{—}CH_2$ (epoxide, bridged by O) (from k) $\xrightarrow[H_2O]{OH^{\ominus}}$)

(continued next page)

(m) $CH_3CH_2CH{=}CH_2$ (from f) $\xrightarrow{Br_2,\ CCl_4}$ $\xrightarrow{2Na^{\oplus}\ NH_2^{\ominus}}$ $\xrightarrow{H_2O}$

$CH_3CH_2C{\equiv}CH$

(n) $CH_3CH_2C{\equiv}CH$ (from m) $\xrightarrow{Hg^{2+},\ H^{\oplus},\ H_2O}$ $CH_3CH_2\overset{O}{\overset{\|}{C}}CH_3$

(o) $CH_3CHClCH_3$ (from e) $\xrightarrow[ether]{Mg}$ $\xrightarrow{H-\overset{O}{\overset{\|}{C}}-H}$ $\xrightarrow{H_2O}$ $(CH_3)_2CHCH_2OH$

(p) $(CH_3)_2CHCH_2OH$ (from o) $\xrightarrow{SOCl_2}$ $\xrightarrow{KOH,alcohol,heat}$

$(CH_3)_2C{=}CH_2$

(q) $(CH_3)_2C{=}CH_2$ (from p) $\xrightarrow[peroxide]{HBr}$ $\xrightarrow[ether]{Mg}$ $\xrightarrow{CH_2O}$ $\xrightarrow{H_2O}$

$(CH_3)_2CHCH_2CH_2OH$

(r) $(CH_3)_2CHCH_2CH_2OH$ (from q) $\xrightarrow{PBr_3}$ $(CH_3)_2CHCH_2CH_2Br$

$\xrightarrow{KOH,alcohol,heat}$ $(CH_3)_2CHCH{=}CH_2$

(s) $(CH_3)_2CHCH{=}CH_2$ (from r) $\xrightarrow{H_3O^{\oplus}}$ $(CH_3)_2CH\underset{OH}{\underset{|}{C}}HCH_3$

(t) $(CH_3)_2CH\underset{OH}{\underset{|}{C}}HCH_3$ (from s) $\xrightarrow{SOCl_2}$ $(CH_3)_2CH\underset{Cl}{\underset{|}{C}}HCH_3$

(continued next page)

(u) $(CH_3)_2CHCH(OH)CH_3$ (from s) $\xrightarrow{PBr_3}$ $(CH_3)_2CHCH(Br)CH_3$

(v) $(CH_3)_2CHCH(Br)CH_3$ (from u) $\xrightarrow{Mg, ether}$ $\xrightarrow{D_2O}$ $(CH_3)_2CHCHDCH_3$

(other syntheses are possible in many of these questions)

24. (a) 1-Propanol reacts with sodium with the evolution of hydrogen (bubbling occurs). Propane does not react. (b) Propene decolorizes Br_2/CCl_4 solution (brown to colorless). 1-Propanol does not react. (c) 2-Propanol (2°) reacts with Lucas reagent within 5 minutes (a separate layer of insoluble 2-chloropropane forms). 1-Propanol (1°) does not react. (d) 4-Ethyl-3-hexanol (2° alcohol) is oxidized by $CrO_3/H^{\oplus}$ (orange to green color change). 2,3,4-Trimethyl-2-pentanol (3° alcohol) does not react. (e) t-Butyl alcohol (3°) reacts with Lucas reagent immediately (separate layer of insoluble t-butyl chloride forms). t-Butylcarbinol (1°) does not react. (f) Allyl alcohol decolorizes Br_2/CCl_4 solution (brown to colorless). t-Butyl alcohol does not react. Allyl alcohol is oxidized by $CrO_3/H^{\oplus}$ (orange to green), t-butyl alcohol does not react. Lucas reagent will not distinguish between these alcohols. Both react immediately with Lucas reagent since both form relatively stable carbocations. (g) Propyne decolorizes Br_2/CCl_4 solution (brown to colorless). Propanol does not react. (h) Cyclohexanol (2°) reacts with $CrO_3/H^{\oplus}$ (orange to green).

(continued next page)

1-Methylcyclohexanol (3°) does not react. (i) Propargyl alcohol forms a precipitate with $Ag(NH_3)_2^{\oplus}$. Allyl alcohol does not react.

(Text: Chemical Analysis of Alcohols, S. 10.8)

25. $CH_3CH{=}CHCHO \xrightarrow{CH_3MgCl} \xrightarrow{H_2O} CH_3CH{=}CHCH(OH)CH_3 \xrightarrow[heat]{Al_2O_3} CH_3CH{=}CHCH{=}CH_2$

piperylene
(*cis* and *trans*)

$CH_3CH{=}CHCH{=}CH_2 \xrightarrow{O_3} \xrightarrow[H^{\oplus}]{Zn} CH_3CHO + HC(O){-}C(O)H + HCHO$

26. With Lucas' reagent, benzyl alcohol gives a resonance stabilized carbocation.

$C_6H_5CH_2\ddot{O}H \xrightarrow{H^{\oplus}} C_6H_5CH_2{-}\overset{\oplus}{O}H_2 \xrightarrow{-H_2O}$

$C_6H_5\overset{\oplus}{C}H_2 \leftrightarrow \leftrightarrow \leftrightarrow$ (ortho, para, ortho positive charge resonance structures with exocyclic $=CH_2$)

equivalent to $[C_6H_5CH_2]^{\oplus}$

(continued next page)

the stability of carbocations is:

allyl, benzyl $> \underbrace{3^\circ > 2^\circ > 1^\circ}_{\text{alkyl}}$

(Text: resonance stabilization of the benzyl cation, S. 5.5)

27. The ease of acid-catalyzed dehydration is related to how easily a carbocation is formed in the process $\overset{\oplus}{R}OH_2 \rightarrow \overset{\oplus}{R} + H_2O$. This, in turn, is related to the stability of the activated complex leading to the cation (approximated by the stability of the cation: $3^\circ > 2^\circ > 1^\circ$) and to anchimeric assistance which can increase the rate of ionization of $\overset{\oplus}{R}OH_2$.

2-methyl-2-propanol leads to a 3° cation. Both 3-methyl-2-butanol and 2-butanol are 2° alcohols. The increased reactivity of the former suggests anchimeric assistance by the methyl group. The same comparison applies to the two 1° alcohols, 2-methyl-1-propanol and 1-butanol.

(Text: carbocation stability S. 5.5; anchimeric assistance p. 193).

28. (a) BH_3 then CH_3CO_2H gives $(CH_3)_2CHCH_3$ (BH_3 then $H_2O_2/OH^\ominus$ gives the product shown).

(b) HI (with or without peroxide) gives CH_3CHICH_3. Only HBr/peroxide gives anti-Markownikoff addition.

(c) Acids are not reduced by $NaBH_4$.

(d) The first step gives $CH_3CH(OH)CH_3$; the oxidation shown in the second step does not occur (a 1° alcohol <u>adjacent to a double bond</u> is oxidized by MnO_2). The final product has an extra carbon.

29. (a) A: $CH_2=CH_2$, B: CH_2ClCH_2OH , C: $CH_2(OH)CH_2OH$

(b) A: $CH_2BrCHBrCH_2OH$, B: $CH_2BrCHBrCOOH$, C: $CH_2=CHCOOH$

(c) A: cyclohexanone (ring with C=O) , B: 1-methylcyclohexanol (ring carbon bearing HO and CH_3) , C: sodium 1-methylcyclohexoxide (ring carbon bearing $Na^{\oplus}\,O^{\ominus}$ and CH_3) , D: 1-methoxy-1-methylcyclohexane (ring carbon bearing CH_3O and CH_3)

(d) A: CH_3CH_2OH , B: CH_3CH_2Cl , C: $CH_3CH_2C(OH)(CH_3)_2$

(e) A: $CH_3CH(OH)CH_3$, B: $CH_3C(=O)CH_3$, C: $(CH_3)_2C(OH)CH_2CH_2CH_3$

D: $(CH_3)_2CClCH_2CH_2CH_3$, E: $(CH_3)_2CLiCH_2CH_2CH_3$,

F: $(CH_3)_2CDCH_2CH_2CH_3$

30. A: $CH_3CH=CH_2 + Cl_2 \xrightarrow{h\nu} ClCH_2CH=CH_2 + HCl$ <u>mechanism</u>:

(1) $Cl_2 \xrightarrow{\text{light}} 2Cl\cdot$

(2) $Cl\cdot + CH_3CH=CH_2 \longrightarrow HCl + [\dot{C}H_2\text{-}CH=CH_2 \longleftrightarrow CH_2=CH\text{-}\dot{C}H_2]$

(3) $[CH_2 \cdots CH \cdots CH_2]^{\cdot} + Cl_2 \longrightarrow CH_2=CHCH_2Cl + Cl\cdot$

(4) termination: $2Cl\cdot \longrightarrow Cl_2$

B: $CH_2=CHCH_2Cl \xrightarrow{OH^{\ominus}} CH_2=CHCH_2OH + Cl^{\ominus}$ <u>mechanism</u>:

$$HO^{\ominus} + CH_2=CH\text{-}CH_2\text{-}Cl \longrightarrow \left[HO^{\delta\ominus} \cdots CH_2(\text{-}CH=CH_2) \cdots Cl^{\delta\ominus} \right] \longrightarrow CH_2CH\text{-}CH_2OH + :Cl:^{\ominus}$$

(continued next page)

C: $CH_2{=}CHCH_2OH \xrightarrow{Cl_2,\,H_2O} CH_2(Cl){-}CH(OH){-}CH_2OH$ mechanism:

$CH_2{=}CHCH_2OH \xrightarrow{Cl_2} CH_2{-}\overset{\delta\oplus}{C}HCH_2OH$ (bridged by :Cl: $\delta\oplus$) $\xrightarrow{H_2\ddot{O}} CH_2(\ddot{C}l){-}CH(\overset{\oplus}{O}H_2){-}CH_2{-}OH$ (+ :$\ddot{O}H_2$)

$\longrightarrow CH_2(\ddot{C}l){-}CH(\ddot{O}H){-}CH_2OH + H_3O^{\oplus}$

D: $CH_2ClCH(OH)CH_2OH \xrightarrow[\text{heat}]{Ca(OH)_2} CH_2{-}CH{-}CH_2OH$ (epoxide, ring O) mechanism:

$CH_2(\ddot{C}l){-}CH(\ddot{O}H){-}CH_2OH \underset{}{\overset{OH^{\ominus}}{\rightleftharpoons}} CH_2(\ddot{C}l){-}CH(\ddot{O}^{\ominus}){-}CH_2OH \longrightarrow CH_2{-}CH{-}CH_2OH$ (epoxide, ring $\ddot{O}$) $+ :\ddot{C}l:^{\ominus}$

E: $CH_2{-}CHCH_2OH$ (epoxide, ring O) $\xrightarrow{H_2O} CH_2(OH){-}CH(OH){-}CH_2OH$ mechanism:

$CH_2{-}CHCH_2OH$ (epoxide, ring :O:) $\xrightarrow{H\ddot{O}H} CH_2(\ddot{O}^{\ominus}){-}CH(\overset{\oplus}{O}H_2){-}CH_2OH \longrightarrow CH_2(\ddot{O}H){-}CH(\ddot{O}H){-}CH_2OH$

31. (a) (1) H_2SO_4 (or H_3PO_4)/heat, (2) BH_3 then $H_2O_2/OH^{\ominus}$

(b) (1) $SOCl_2$ (or PCl_3 or HCl), (2) Mg/ether, (3) CH_2O, (4) H_2O

(c) (1) PBr_3 (or HBr), (2) Mg/ether, (3) CH_3CHO, (4) H_2O, (5) $CrO_3/H^{\oplus}$ (or some other Cr(VI) reagent

(d) (1) BH_3, (2) $H_2O_2/OH^{\ominus}$, (3) CrO_3/pyridine

(e) (1) CH_3MgI, (2) H_2O, (3) $K_2Cr_2O_7/H^{\oplus}$ (or $CrO_3/H^{\oplus}$)
(product is CH_3CH_2COOH)

32.

$$CH_3\text{-}\underset{CH_3}{\overset{:\ddot{O}H}{C}}\text{—}\underset{CH_3}{\overset{:\ddot{O}H}{C}}\text{-}CH_3 \xrightarrow{H^{\oplus}} CH_3\text{-}\underset{CH_3}{\overset{:\ddot{O}H}{C}}\text{—}\underset{CH_3}{\overset{H\overset{\oplus}{O}H}{C}}\text{-}CH_3 \longrightarrow CH_2\text{-}\underset{CH_3}{\overset{:\ddot{O}\text{-}H}{C}}\text{—}\underset{CH_3}{\overset{\oplus}{C}}\text{-}CH_3$$

$$\downarrow$$

$$CH_3\text{-}\overset{:\ddot{O}}{\overset{\|}{C}}\text{-}C(CH_3)_3 + H_3O^{\oplus} \longleftarrow CH_3\text{-}\overset{\oplus:O\text{-}H}{\overset{\|}{C}}\text{-}C(CH_3)_3 \quad :\ddot{O}H_2$$

33. (a) [reaction scheme: tetrahydrofurfuryl alcohol (H, CH_2OH) with $H^{\oplus}$ → protonated alcohol ($CH_2-\ddot{O}^{\oplus}H_2$); $-H_2O$ → ring-expanded carbocation (CH_2) ≡ tetrahydropyranyl cation; $:\ddot{O}H_2$ removes H → dihydropyran + $H_3O^{\oplus}$]

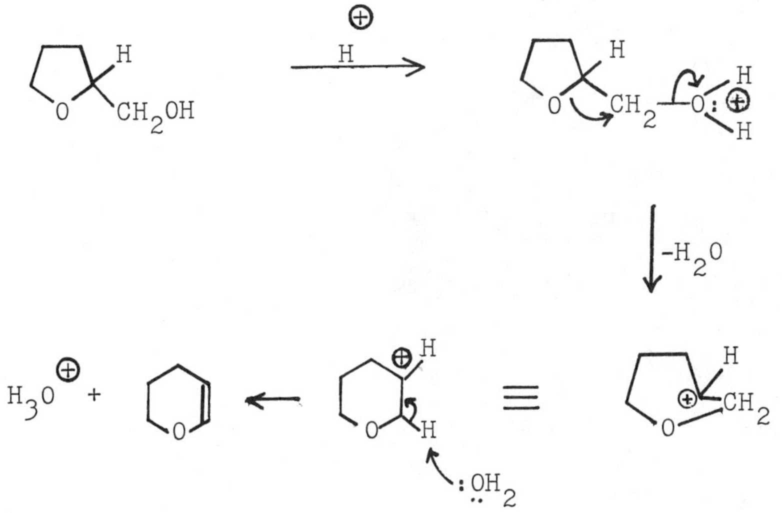

(b) [structure: 3-hydroxytetrahydropyran, OH] (d,l)

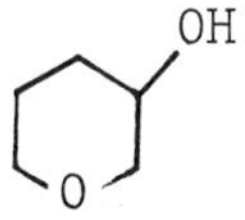

34. The reaction is S_N2. The nucleophile ($:CH_3$) attacks the less sterically hindered carbon.

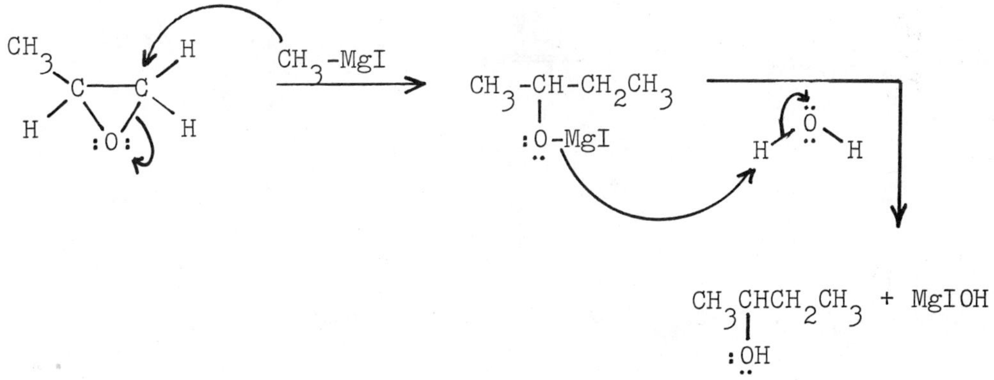

[reaction scheme: 1,2-epoxypropane (CH_3, H, H, H, :O:) + CH_3-MgI → CH_3-CH(:Ö-MgI)-CH_2CH_3; + $H-\ddot{O}-H$ → $CH_3CHCH_2CH_3$ (:ÖH) + MgIOH]

35. The 3° alcohol lacks the hydrogen (on the hydroxyl-bearing carbon) necessary for this reaction to occur. The ease of oxidation of the 1° and 2° alcohols parallels decreasing

(continued next page)

steric hindrance at the C-OH function.

(Text: S. 10.7, mechanism of oxidation)

36.

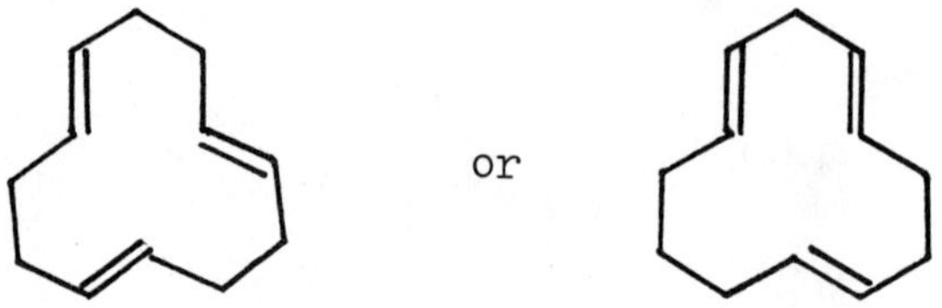

or

37. 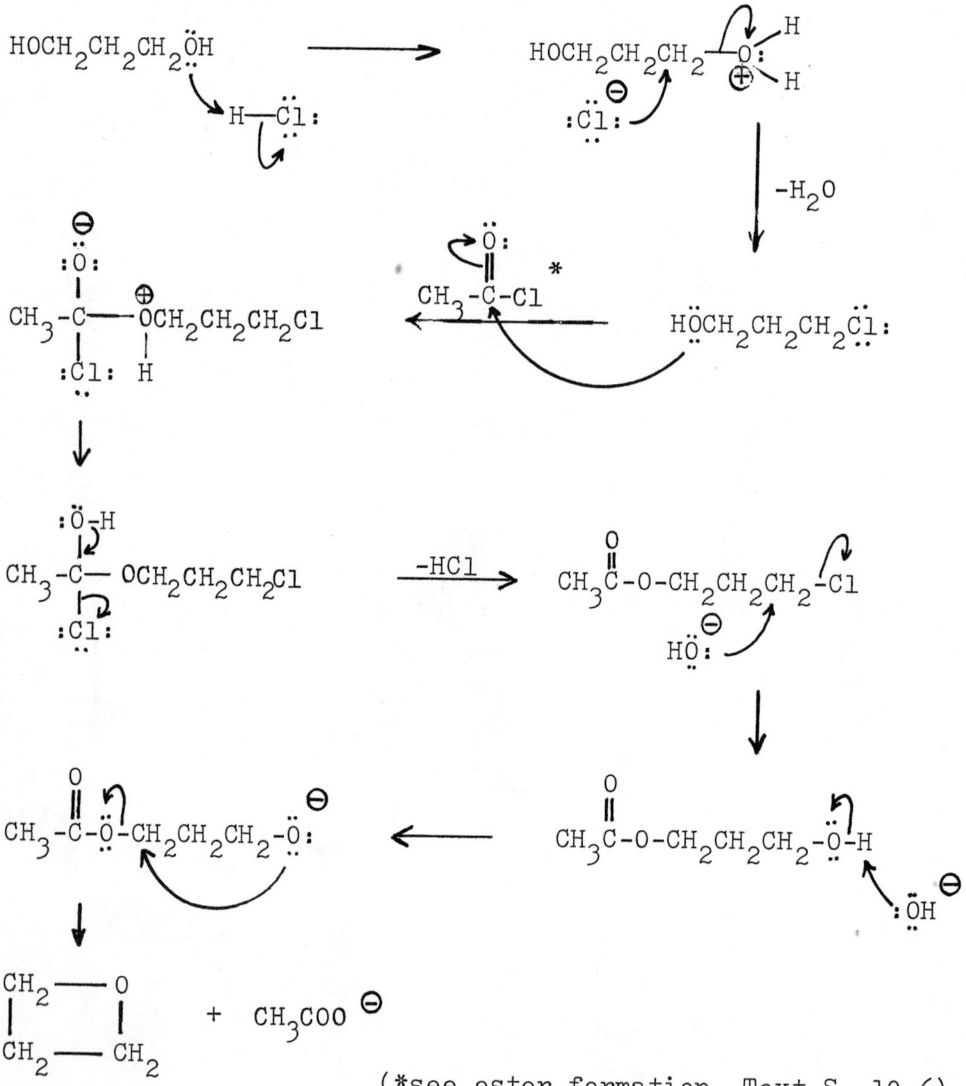 $HOCH_2CH_2CH_2\ddot{O}H \longrightarrow HOCH_2CH_2CH_2-\overset{\oplus}{O}H_2$ (with $H-Cl$, $:\ddot{Cl}:^{\ominus}$)

$\xrightarrow{-H_2O}$ $H\ddot{O}CH_2CH_2CH_2\ddot{Cl}:$

$CH_3-\overset{O}{\overset{\|}{C}}-Cl$ *

$CH_3-C(:\ddot{O}:^{\ominus})(:\ddot{Cl}:)-\overset{\oplus}{O}(H)CH_2CH_2CH_2Cl$

$\downarrow$

$CH_3-C(:\ddot{O}-H)(:\ddot{Cl}:)-OCH_2CH_2CH_2Cl \xrightarrow{-HCl} CH_3\overset{O}{\overset{\|}{C}}-O-CH_2CH_2CH_2-Cl$ (with $H\ddot{O}:^{\ominus}$)

$\downarrow$

$CH_3\overset{O}{\overset{\|}{C}}-O-CH_2CH_2CH_2-\ddot{O}-H$ (with $:\ddot{O}H^{\ominus}$)

$\longrightarrow CH_3-\overset{O}{\overset{\|}{C}}-\ddot{O}-CH_2CH_2CH_2-\ddot{O}:^{\ominus}$

$\downarrow$

CH_2-O / CH_2-CH_2 (oxetane) + $CH_3COO^{\ominus}$

(*see ester formation, Text S. 10.6)

38. Use an R in which the carbon attached to silicon is chiral. An S_N2 reaction <u>at carbon</u> will give inversion of configuration; an S_N2 reaction <u>at silicon</u> will give retention of configuration at the chiral center.

39. 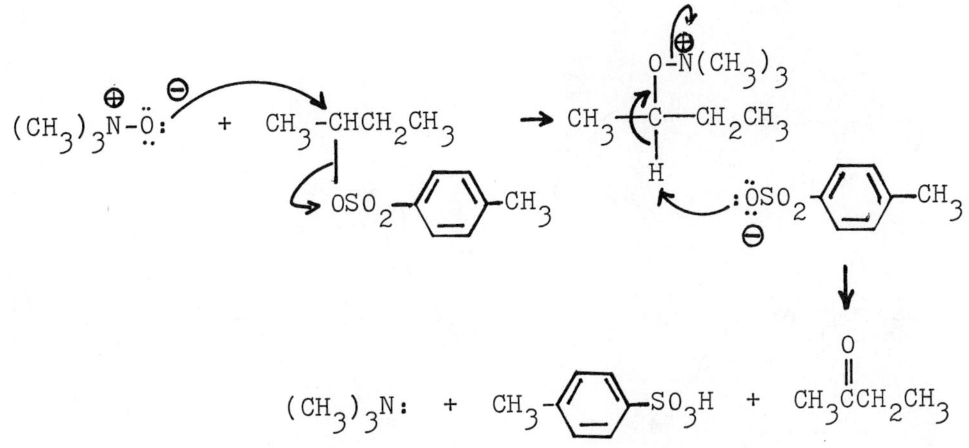 $(CH_3)_3\overset{\oplus}{N}-\ddot{O}:^{\ominus}$ + $CH_3-CH(OSO_2-C_6H_4-CH_3)CH_2CH_3$ → $CH_3-C(H)(O-\overset{\oplus}{N}(CH_3)_3)-CH_2CH_3$ + $:\ddot{O}SO_2^{\ominus}-C_6H_4-CH_3$

↓

$(CH_3)_3N:$ + $CH_3-C_6H_4-SO_3H$ + $CH_3\overset{O}{\overset{\|}{C}}CH_2CH_3$

40. The nmr spectrum of t-butyl hypochlorite will have only one absorption (a singlet) for the methyl protons. The nmr spectrum of 1-chloro-2-methyl-2-propanol will have three absorptions (all singlets) for protons of $(CH_3)_2-$, $-CH_2-$, and OH. An OH stretch near 3500 cm^{-1} will appear in the ir spectrum of 1-chloro-2-methyl-2-propanol but not in that of t-butyl hypochlorite.

<u>Bond energies (kcal/mole)</u>:

O-H = 110,	O-Cl = 52
C-H = 99,	C-Cl = 77
Cl_2 = 58,	H-Cl = 103

<u>For the reactions</u>:

(a) $(CH_3)_3COH + Cl_2 \longrightarrow (CH_3)_3COCl + HCl$

(b) $(CH_3)_3COH + Cl_2 \longrightarrow (CH_3)_2C(OH)CH_2Cl + HCl$

ΔH: (a) +13 kcal , (b) -23 kcal

41. $HC{\equiv}CH \xrightarrow{Na^{\oplus}\ NH_2^{\ominus}} HC{\equiv}\overset{\ominus}{C}{:}\ Na^{\oplus} \xrightarrow{CH_3\overset{O}{\overset{\|}{C}}CH_3} \xrightarrow{H_3O^{\oplus}} HC{\equiv}C\underset{\displaystyle OH}{\underset{|}{C}}(CH_3)_2$

A

$\xrightarrow{H_2O,\ H^{\oplus},\ Hg^{+2}} CH_3\overset{O}{\overset{\|}{C}}\underset{\displaystyle OH}{\underset{|}{C}}(CH_3)_2 \xrightarrow{NaBH_4} CH_3\underset{\displaystyle OH}{\underset{|}{C}}H\text{-}\underset{\displaystyle OH}{\underset{|}{C}}(CH_3)_2$

B

NMR spectrum of B consists of singlets for the protons of:

(a) $CH_3\text{-}\overset{O}{\overset{\|}{C}}\text{-}$, (b) -OH , (c) $-(CH_3)_2$.

42.

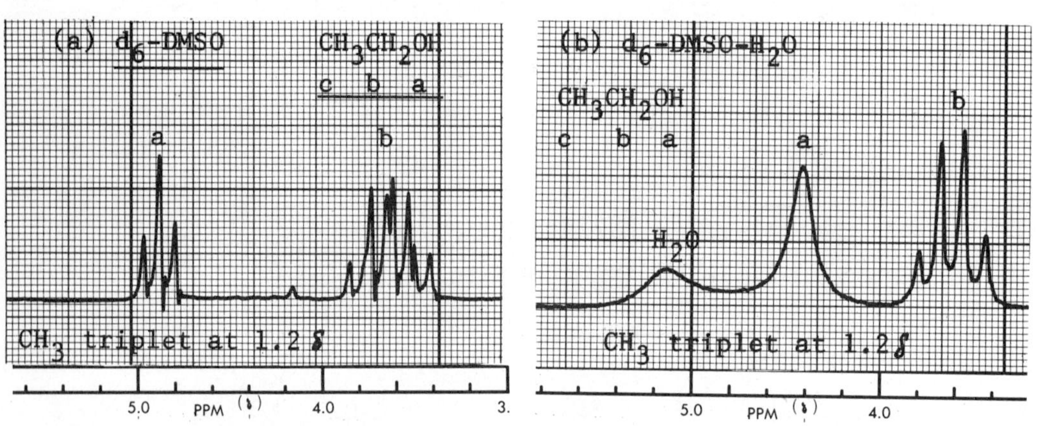

In the sample containing only ethanol and DMSO, ethanol hydrogen bonds to DMSO, and exchange of the OH protons of ethanol is prevented. Thus, the signal for the OH protons is split by the protons of $-CH_2-$, and the signal for the protons of $-CH_2-$ is split by the OH protons.

In the sample containing water, water rather than ethanol hydrogen bonds to DMSO. Rapid exchange of the OH protons occurs, and the NMR "sees" an average enviornment for these protons. Thus, the signal for the OH protons is not split, and the OH protons do not split the signal for the protons of $-CH_2-$.

43.

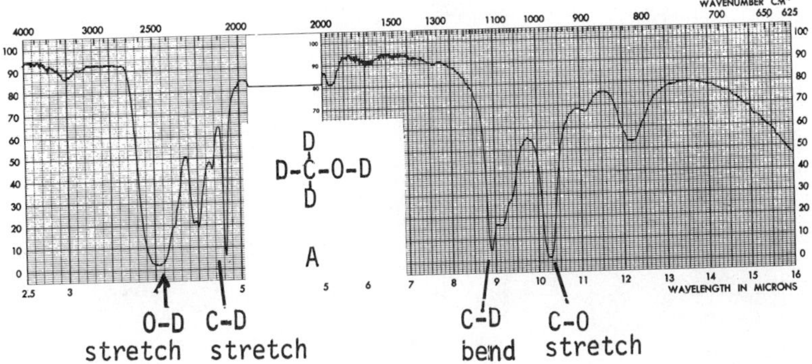
D
D-C-O-D
D
A
O-D stretch
C-D stretch
C-D bend
C-O stretch
WAVENUMBER CM⁻¹
WAVELENGTH IN MICRONS

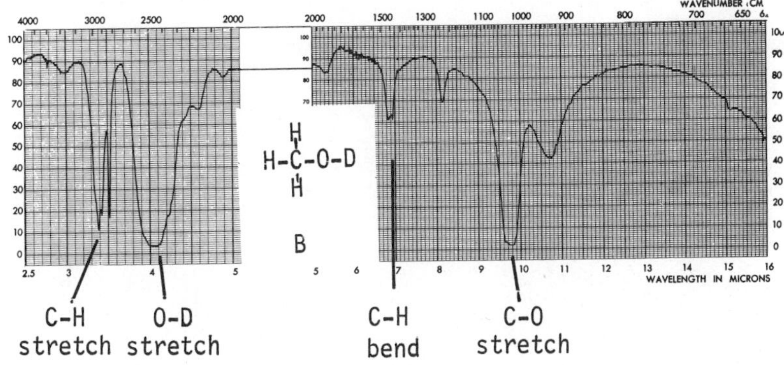
H
H-C-O-D
H
B
C-H stretch
O-D stretch
C-H bend
C-O stretch
WAVENUMBER CM⁻¹
WAVELENGTH IN MICRONS

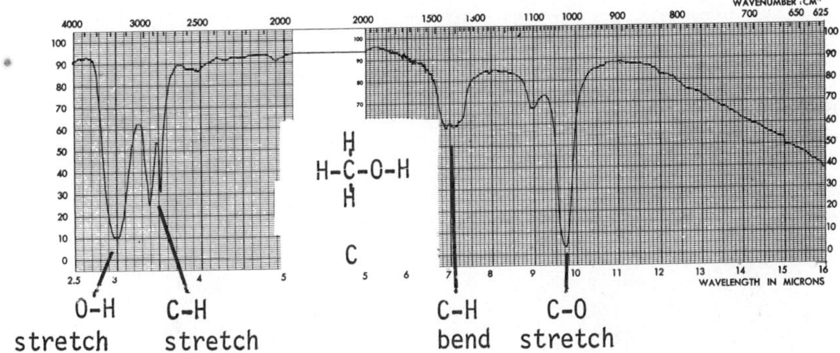
H
H-C-O-H
H
C
O-H stretch
C-H stretch
C-H bend
C-O stretch
WAVENUMBER CM⁻¹
WAVELENGTH IN MICRONS

44. (a) Isoamyl alcohol

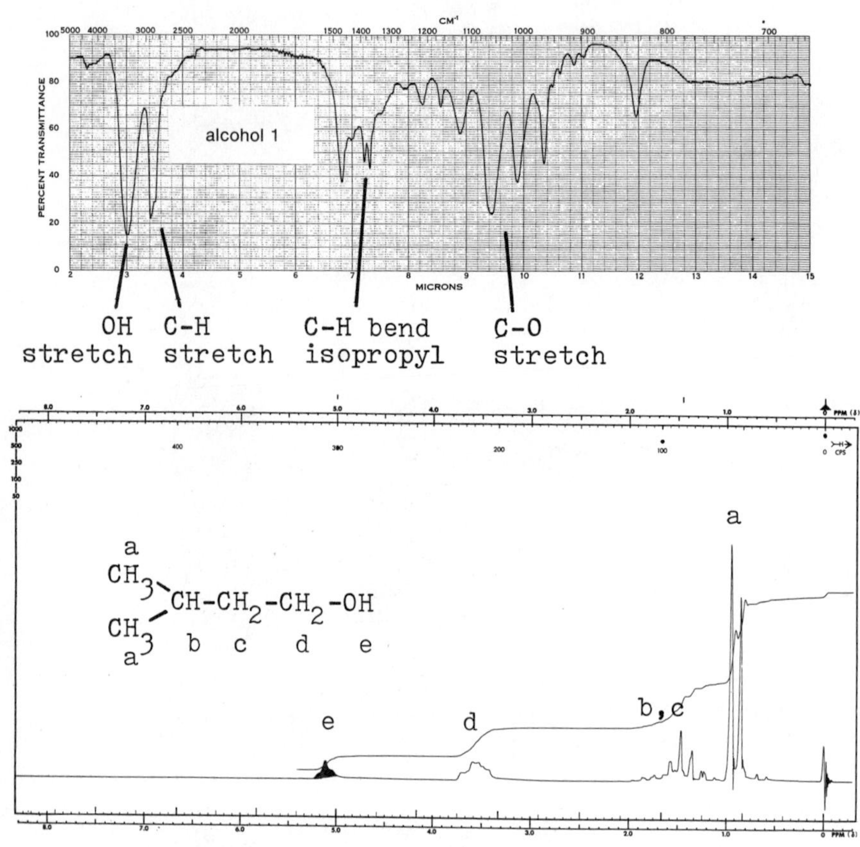

(b) 2-Methyl-1-butanol

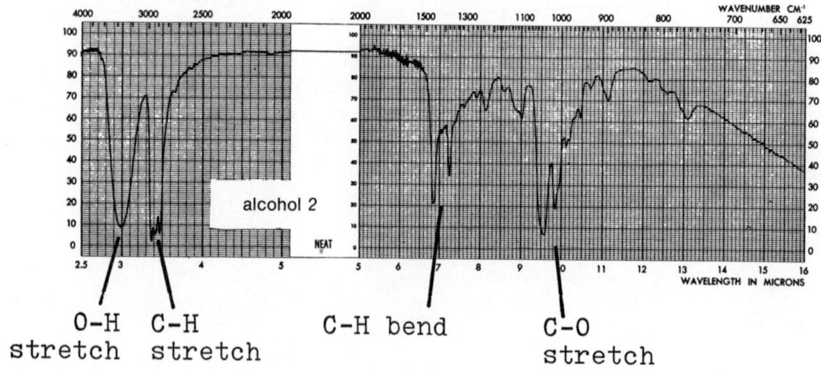

(continued next page)

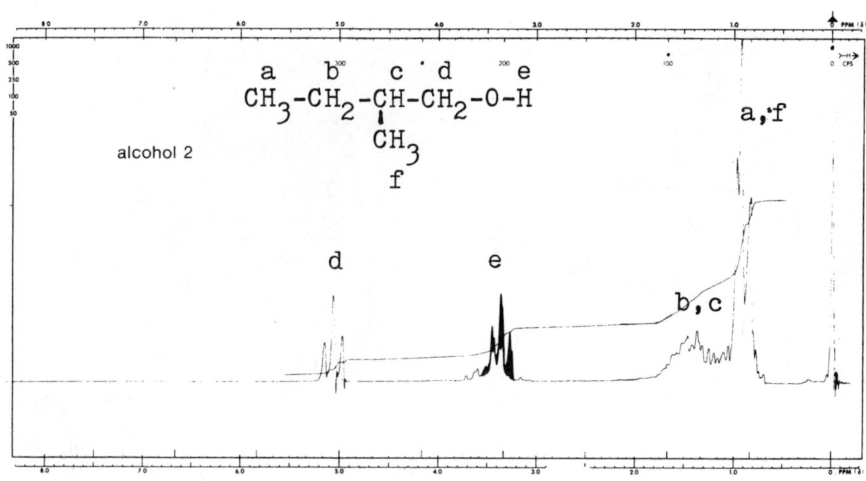

45.

$$CH_2{=}CHCH_2OH \xrightarrow{Cl_2,H_2O} \underset{\underline{A}}{\underset{Cl\quad OH\quad OH}{CH_2\text{-}CH\text{-}CH_2}} \xrightarrow{NaOH} \underset{\text{glycidol}}{\underset{\backslash O /}{CH_2\text{—}CHCH_2OH}}$$

$$CH_2{=}CHCH_2Cl \xrightarrow{Cl_2,H_2O} \underset{\underline{B}}{\underset{Cl\quad HO\quad Cl}{CH_2\text{-}CH\text{-}CH_2}} \xrightarrow{NaOH} \underset{\text{epichlorohydrin}}{\underset{\backslash O /}{CH_2\text{—}CHCH_2Cl}}$$

Methane is formed from glycidol and CH_3MgCl by the reaction: $-CH_2OH + CH_3MgCl \longrightarrow CH_4 + MgClOCH_2-$. Epichlorohydrin lacking an OH does not give this reaction.

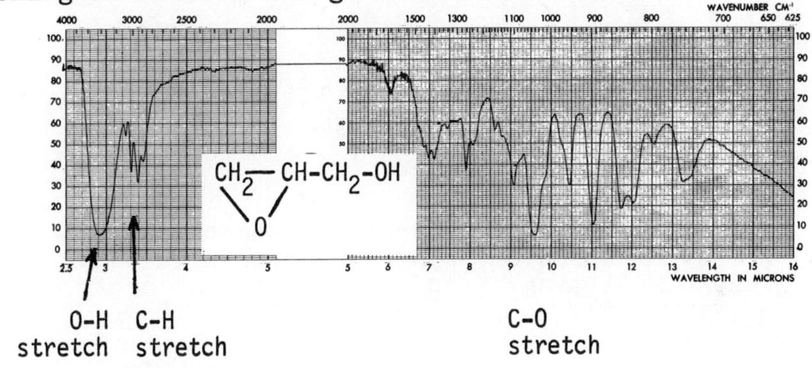

(continued next page)

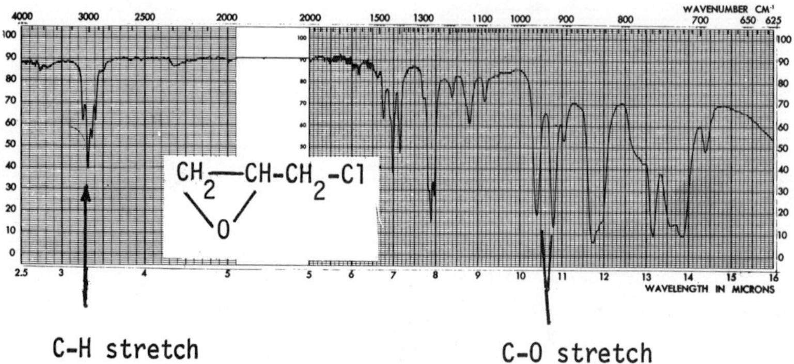

46.

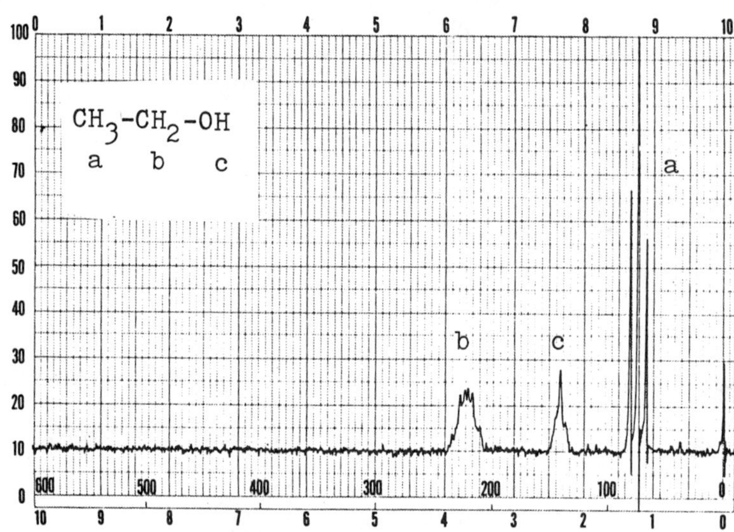

signal for b a poorly resolved multiplet (splitting by protons a and c.

signal for c a poorly resolved triplet (splitting by protons b.

see problem 42 for the spectrum of ethanol taken in DMSO (better resolution) and a spectrum with water added where splitting of and by OH does not occur.

47. (a) The species formed from methanol in acidic solution is $CH_3\overset{\oplus}{O}H_2$. The protons of the $-CH_3$ appear as a triplet (signal split by the two protons on oxygen), and the protons of the $-\overset{\oplus}{O}H_2$ appear as a quartet (signal split by the three protons on carbon).

The cation formed from 2-methyl-1-propanol is:

$$\underset{a}{(CH_3)_2}\underset{b}{CH}-\underset{c}{CH_2}-\underset{d}{\overset{\oplus}{O}H_2}$$

The six methyl protons ("a") appear as a doublet (signal split by proton "b"), proton "b" appears as a multiplet (signal split by the "a" protons and "c" protons), protons "c" appear as two overlapping triplets (signal split by the "b" proton and "d" protons), and protons "d" appear as a triplet (signal split by protons "c").

(b) At -30°, the cation from 2-methyl-1-propanol rearranges to the same oxonium ion as that formed from 2-methyl-2-propanol at -60°.

$$(CH_3)_2CHCH_2OH \xrightarrow{H^\oplus} (CH_3)_2\underset{\mid \atop H}{C}-CH_2-\overset{\oplus}{O}H_2 \xrightarrow{-H_2O} (CH_3)_2\overset{\oplus}{C}CH_3$$

2-methyl-1-propanol

$$(CH_3)_2\overset{\oplus}{C}CH_3 \xrightarrow{H_2O} \ldots$$

$$(CH_3)_2\underset{\mid \atop OH}{C}CH_3 \xrightarrow{H^\oplus} (CH_3)_2\underset{\mid \atop \oplus OH_2}{C}CH_3$$

2-methyl-2-propanol

SUMMARY OF REACTIONS

(specific examples of the general reactions are presented in the Reaction Review)

1. Synthesis of Alcohols

(a) from alkyl halides:

$$RX \xrightarrow{OH^{\ominus}/H_2O} ROH$$ (S. 10.4 and Chapter 5)

(b) from alkenes:

$RCH{=}CH_2$ +

$\xrightarrow{H_3O^{\oplus}}$ $RCH(OH)CH_3$ (S. 8.6)

$\xrightarrow{BH_3} \xrightarrow[OH^{\ominus}]{H_2O_2}$ RCH_2CH_2OH (S. 10.4)

$\xrightarrow{BH_3} \xrightarrow[heat]{CO} \xrightarrow[OH^{\ominus}]{H_2O_2}$ $(RCH_2CH_2)_3COH$ (S. 10.4)

$\xrightarrow[H_2O]{Hg(OAc)_2} \xrightarrow[H_2O]{NaBH_4}$ $RCH(OH)CH_3$ (S. 10.4)

(continued next page)

(c) from aldehydes and ketones: (S. 10.4)

$>C=O$ +
- $\xrightarrow[CH_3OH/H_2O]{NaBH_4}$ $-\overset{|}{C}HOH$
- $\xrightarrow{LiAlH_4} \xrightarrow{H_2O}$ $-\overset{|}{C}HOH$
- $\xrightarrow{B_2H_6}$ $-\overset{|}{C}HOH$
- $\xrightarrow{H_2/Pt}$ $-\overset{|}{C}HOH$
- $\xrightarrow[\text{(or RLi)}]{RMgX} \xrightarrow{H_2O}$ *

* $(H)(H)C=O \xrightarrow{RMgX} \xrightarrow{H_2O} RCH_2OH$ — 1° alcohol

$(R')(H)C=O \xrightarrow{RMgX} \xrightarrow{H_2O} RR'CHOH$ — 2° alcohol

$(R')(R'')C=O \xrightarrow{RMgX} \xrightarrow{H_2O} RR'R''COH$ — 3° alcohol

(RLi can be substituted for RMgX)

(d) from epoxides:

$-\overset{|}{C}-\overset{|}{C}-$ (bridged by O) $\xrightarrow[\text{(or RLi)}]{RMgX} \xrightarrow{H_2O} R-\overset{|}{\underset{|}{C}}-\overset{|}{\underset{|}{C}}-OH$ (S. 10.4)

(R- attacks less sterically hindered carbon of the epoxide)

(continued next page)

(e) from esters:

$$\begin{matrix} R \\ R'O \end{matrix}\!\!>\!C{=}O \xrightarrow{2R''MgX} \xrightarrow{H_2O} RR''_2COH + R'OH \quad (S.\ 10.4)$$

(3° alcohol)

2. Alcohol Reactions

(a) replacement of acidic hydrogen

$$ROH \xrightarrow{Na} Na^{\oplus}\,RO^{\ominus} + H_2 \qquad (S.\ 10.5)$$

(b) conversion of C-OH to C-L (L = leaving group) (S 10.6)

$$ROH + \begin{cases} \xrightarrow{HX} RX + H_2O \\ \xrightarrow{SOCl_2} RCl + HCl + SO_2 \\ \xrightarrow{PX_3} RX + P(OH)_3 \\ \xrightarrow[\text{(oxyacid)}]{ZOH\ (\text{or } ZCl)} ROZ + H_2O\ (\text{or } HCl) \end{cases}$$

(ester)

examples: $ROH + R'\overset{O}{\overset{\|}{C}}{-}OH \xrightarrow{H^{\oplus}} RO\overset{O}{\overset{\|}{C}}R' + H_2O$

$$ROH + CH_3{-}C_6H_4{-}SO_2{-}Cl \longrightarrow RO{-}SO_2{-}C_6H_4{-}CH_3 + HCl$$

(continued next page)

(c) oxidation (S. 10.7)

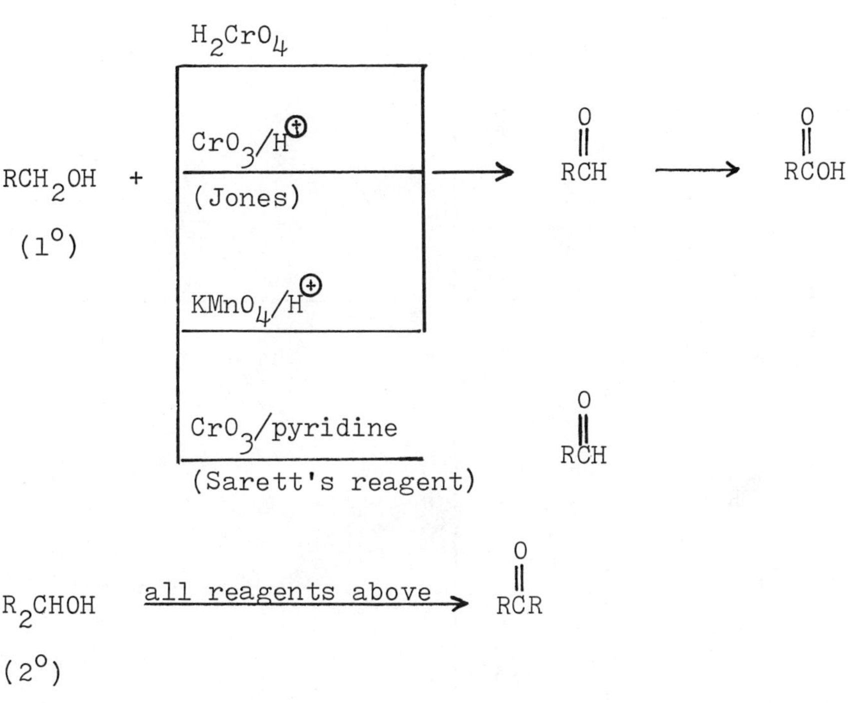

REACTION REVIEW

A. Reactions from Chapter 10.

Questions	Answers
1. $RCH{=}CH_2 \xrightarrow{BH_3} \xrightarrow[OH^{\ominus}]{H_2O_2}$?	1. RCH_2CH_2OH (continued next page)

Write the structure of the product when $RCH=CH_2$ is: a) $CH_3CH_2CH=CH_2$ b) [cyclohexene structure]	a) $CH_3CH_2CH_2CH_2OH$ b) [cyclohexanol structure, OH]
2. $RCH=CH_2 \xrightarrow{BH_3} \xrightarrow[heat]{CO} \xrightarrow[OH^{\ominus}]{H_2O_2}$? Write the structure of the product when $RCH=CH_2$ is: a) $CH_3CH=CH_2$ b) [cyclopentene structure]	2. $(RCH_2CH_2)_3COH$ a) $(CH_3CH_2CH_2)_3COH$ b) (cyclopentyl)$_3$COH
3. $RCH=CH_2 \xrightarrow[H_2O]{Hg(OAc)_2} \xrightarrow[H_2O]{NaBH_4}$? Write the structure of the product when $RCH=CH_2$ is: a) $(CH_3)_2C=CH_2$ b) [1-methylcyclohexene structure, CH_3]	3. $RCH(OH)CH_3$ a) $(CH_3)_2C(OH)CH_3$ b) [1-methylcyclohexanol structure, CH_3, OH]

4. $>C{=}O \xrightarrow[CH_3OH/H_2O]{NaBH_4}$? Write the structure of the product when $>C{=}O$ is: a) CH_3CH_2CHO b) 2,2-dimethylcyclopentanone (ring C=O, with CH_3, CH_3 on adjacent carbon)	4. H-C-OH a) $CH_3CH_2CH_2OH$ b) 2,2-dimethylcyclopentanol (ring C bearing OH and H, adjacent carbon bearing CH_3, CH_3) (d,l)
5. $>C{=}O \xrightarrow{LiAlH_4} \xrightarrow{H_2O}$? Write the structure of the product when C=O is: a) $CH_3CH_2\overset{O}{\overset{\Vert}{C}}CH_3$ b) 2-methylcyclopentanone (ring C=O, adjacent stereocenter bearing CH_3)	5. H-C-OH a) $CH_3CH_2CH(OH)CH_3$ (d,l) b) 2-methylcyclopentanol (OH, CH_3) + 2-methylcyclopentanol (OH, CH_3) (diastereomers)
6. $>C{=}O \xrightarrow{B_2H_6}$? Write the structure of the product when C=O is: a) $CH_3\overset{O}{\overset{\Vert}{C}}CH_3$	6. H-C-OH a) $CH_3CH(OH)CH_3$ (continued next page)

b) 1,4-cyclohexanedione (O=⬡=O)	b) HO–⬡–OH (*cis* and *trans*)
7. $\rangle C{=}O \xrightarrow{H_2/Pt}$? Write the structure of the product when $\rangle C{=}O$ is: a) $(CH_3)_2CH\overset{O}{\overset{\Vert}{C}}CH_3$ b) cyclohexanone	7. $H{-}\overset{\vert}{\underset{\vert}{C}}{-}OH$ a) $(CH_3)_2CHCH(OH)CH_3$ b) cyclohexanol
8. $\rangle C{=}O \xrightarrow[\text{(or RLi)}]{RMgX} \xrightarrow{H_2O}$? Write the structure of the product in the following: a) $CH_3CH_2\overset{O}{\overset{\Vert}{C}}CH_3 \xrightarrow{CH_3MgI} \xrightarrow{H_2O}$? b) cyclohexanone $\xrightarrow{n\text{-}C_4H_9Li} \xrightarrow{H_2O}$? c) $C_6H_5\overset{O}{\overset{\Vert}{C}}{-}H \xrightarrow{CH_3CH_2MgCl} \xrightarrow{H_2O}$?	8. $R{-}\overset{\vert}{\underset{\vert}{C}}{-}OH + MgXOH$ (or LiOH) a) $CH_3CH_2C(OH)(CH_3)_2$ b) 1-(n-C_4H_9)cyclohexanol (OH and n-C_4H_9 on same carbon) c) $C_6H_5CH(OH)CH_2CH_3$

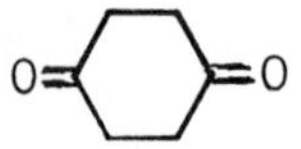

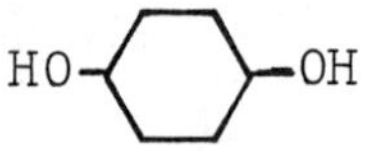

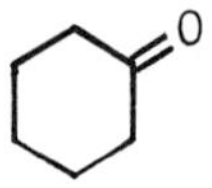

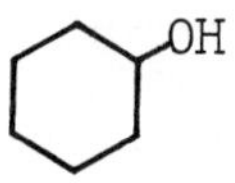

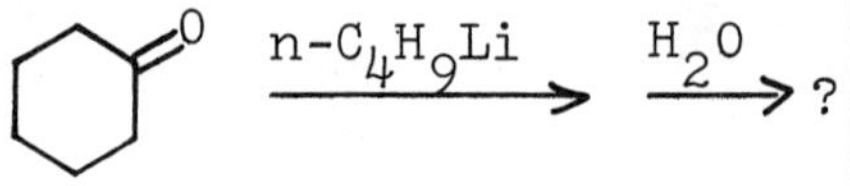

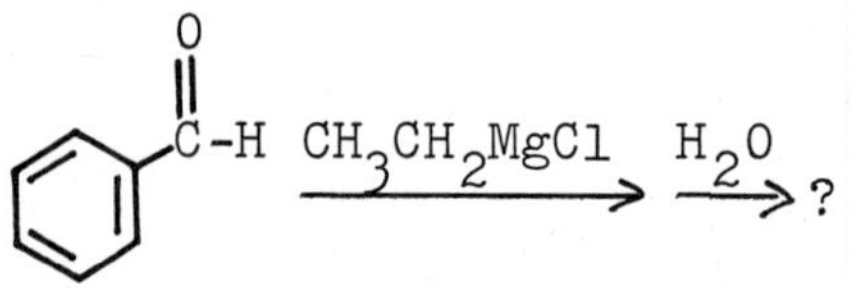

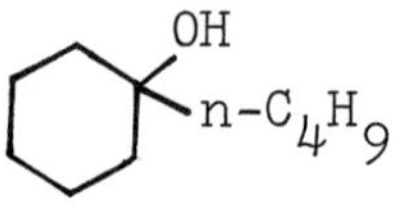

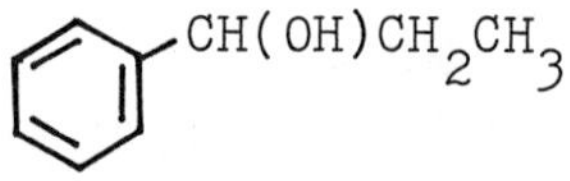

9. $-\overset{\vert}{C}-\overset{\vert}{C}-$ (epoxide, bridged by O) $\xrightarrow[\text{(or RLi)}]{RMgX}$ $\xrightarrow{H_2O}$?	9. $R-\overset{\vert}{\underset{\vert}{C}}-\overset{\vert}{\underset{\vert}{C}}-OH$ + MgXOH (or LiOH)
Write the structure of the product in the following:	
a) $CH_3CH—CH_2$ (epoxide, bridged by O) $\xrightarrow{C_2H_5MgCl}$ $\xrightarrow{H_2O}$?	a) (d,1)$CH_3CH(OH)CH_2C_2H_5$
b) (cyclopentene oxide, H, H, O) $\xrightarrow{C_6H_5Li}$ $\xrightarrow{H_2O}$?	b) (trans-2-phenylcyclopentanol: H, OH, H, phenyl) (d,1)
10. $R-\overset{O}{\overset{\Vert}{C}}-OR'$ $\xrightarrow{R''MgX}$ $\xrightarrow{H_2O}$?	10. RR''_2COH + MgXOH + R'OH
Write the structure of the product in the following:	
a) $CH_3-\overset{O}{\overset{\Vert}{C}}-OCH_3$ $\xrightarrow{C_2H_5MgCl}$ $\xrightarrow{H_2O}$?	a) $CH_3C(OH)(C_2H_5)_2$ + CH_3OH
b) $C_6H_5-\overset{O}{\overset{\Vert}{C}}-OCH_2CH_3$ $\xrightarrow{C_6H_5MgCl}$ $\xrightarrow{H_2O}$?	b) $(C_6H_5)_3COH$ + CH_3CH_2OH

11. R-OH + Na ⟶ ? Write the structure of the product when ROH is: a) CH_3CH_2OH b) $(CH_3)_3COH$	11. $\overset{\oplus}{Na}\ \overset{\ominus}{RO}$ + H_2 a) $\overset{\oplus}{Na}\ CH_3CH_2\overset{\ominus}{O}$ b) $\overset{\oplus}{Na}\ (CH_3)_3C\overset{\ominus}{O}$
12. ROH + HX ⟶ ? Write the structure of the product in the following: a) $(CH_3)_3COH \xrightarrow{HCl}$? b) 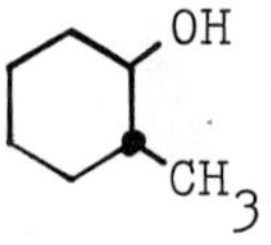 [cyclohexane with OH and CH_3 on adjacent carbons] $\xrightarrow{HBr}$?	12. RX + H_2O a) $(CH_3)_3CCl$ b) 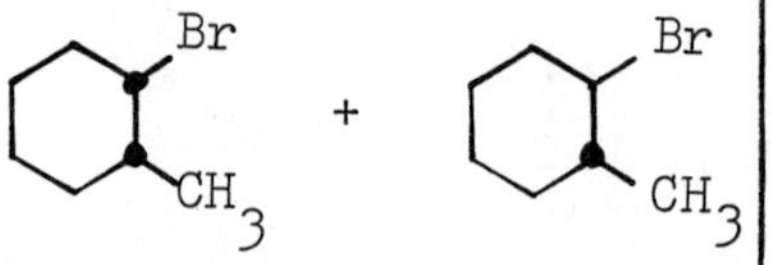 [cyclohexane with Br and CH_3] + [cyclohexane with Br and CH_3] (diastereomers)
13. ROH + $SOCl_2$ ⟶ ? Write the structure of the product when ROH is: a) $CH_3CH_2CH_2OH$ b) 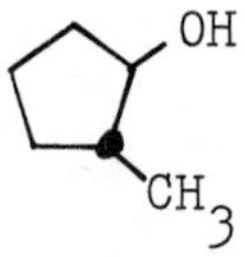 [cyclopentane with OH and CH_3 on adjacent carbons]	13. RCl + HCl + SO_2 a) $CH_3CH_2CH_2Cl$ b) 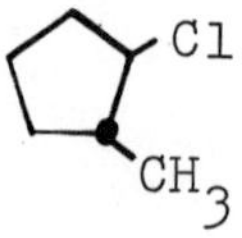 [cyclopentane with Cl and CH_3 on adjacent carbons] (retention of configuration)

14. ROH + PX_3 ⟶ ? Write the structure of the product in the following: a) cyclohexanol (OH) $\xrightarrow{PI_3}$? b) (R)-$CH_3CH_2C(CH_3)(OH)CH_2OCH_3$ $\xrightarrow{PBr_3}$?	14. RX + $P(OH)_3$ a) iodocyclohexane (I) b) (R)and(S)-$CH_3CH_2C(CH_3)(Br)CH_2OCH_3$
15. $R\overset{*}{O}H$ + Z-OH ⟶ ? oxyacid or ROH + ZCl ⟶ ? Write the structure of the ester formed in each of the following: a) $CH_3CH_2OH + CH_3\overset{O}{\overset{\Vert}{C}}OH \xrightarrow{H^{\oplus}}$? b) cyclohexanol (OH) + CH_3-C_6H_4-SO_2-Cl ⟶ ?	15. $R\overset{*}{O}Z + H_2O$ ester or ROZ + HCl a) $CH_3CH_2O\overset{O}{\overset{\Vert}{C}}CH_3$ b) cyclohexyl-O-SO_2-C_6H_4-CH_3 (or cyclohexyl-OTs)

16. $-\overset{\vert}{\underset{H}{C}}-OH$ $\xrightarrow{\text{Cr(VI) species}}$? Write the structure of the product in the following: a) cyclohexanol (OH) $\xrightarrow{H_2CrO_4}$? b) $CH_3CH_2CH_2OH$ $\xrightarrow{CrO_3/H^{\oplus}}$? c) $CH_3CH_2CH_2OH$ $\xrightarrow[\text{pyridine}]{CrO_3}$? d) $CH_3CH(OH)CH_3$ $\xrightarrow{CrO_3/H^{\oplus}}$?	16. $\rangle C=O$ (H attached to $\rangle C=O$ is oxidized to OH) a) cyclohexanone (=O) b) $[CH_3CH_2\overset{O}{\overset{\Vert}{C}}H] \rightarrow CH_3CH_2\overset{O}{\overset{\Vert}{C}}OH$ c) $CH_3CH_2\overset{O}{\overset{\Vert}{C}}H$ (no further oxidation) d) $CH_3\overset{O}{\overset{\Vert}{C}}CH_3$
17. $-\overset{\vert}{\underset{H}{C}}-OH$ $\xrightarrow{MnO_4^{\ominus}/H^{\oplus}}$? Write the structure of the product when $-\overset{\vert}{\underset{H}{C}}-OH$ is: a) cyclohexyl–CH_2OH b) $CH_3CH(OH)CH_2CH_3$	17. $\rangle C=O$ (H attached to C=O is oxidized to OH) a) [cyclohexyl–$\overset{O}{\overset{\Vert}{C}}H$] $\rightarrow$ cyclohexyl–$\overset{O}{\overset{\Vert}{C}}OH$ b) $CH_3\overset{O}{\overset{\Vert}{C}}CH_2CH_3$

18. $\diagdown C=C-C(H)(OH)-$ $\xrightarrow{MnO_2}$? Write the structure of the product when the alcohol is: a) $CH_2=CH-CH(OH)CH_3$ b) 2-cyclohexen-1-ol (ring with OH)	18. $\diagdown C=C-C(=O)-$ (no oxidation of C=C) a) $CH_2=CHC(=O)CH_3$ b) 2-cyclohexen-1-one (ring with =O)

B. Write structures for all letters

Questions	Answers
1. $CH_3CH(OH)CH_3 \xrightarrow[heat]{H_2SO_4}$ A $\xrightarrow{BH_3} \xrightarrow[OH^{\ominus}]{H_2O_2}$ B $\xrightarrow{SOCl_2}$ C	1. A: $CH_3CH=CH_2$ B: $CH_3CH_2CH_2OH$ C: $CH_3CH_2CH_2Cl$
2. cyclohexanone (ring with =O) $\xrightarrow{LiAlH_4} \xrightarrow{H_2O}$ A $\xrightarrow{HBr}$ B $\xrightarrow[ether]{Mg} \xrightarrow{D_2O}$ C	2. A: cyclohexanol (ring with OH) B: bromocyclohexane (ring with Br) C: deuteriocyclohexane (ring with D)

3. $CH_3CH(OH)CH_3 \xrightarrow{CrO_3/H^{\oplus}} A \xrightarrow{CH_3MgI} \xrightarrow{H_2O} B$	3. A: $CH_3\overset{O}{\overset{\Vert}{C}}CH_3$ B: $(CH_3)_3COH$
4. 2-bromo-1-methylcyclohexane (ring with Br and CH_3) $\xrightarrow[\text{alcohol}]{KOH} A \xrightarrow{H_3O^{\oplus}} B \xrightarrow{PBr_3} C$	4. A: 1-methylcyclohexene (ring with CH_3) B: cyclohexane ring with OH and CH_3 on same carbon C: cyclohexane ring with Br and CH_3 on same carbon
5. $CH_3CH{=}CH_2 \xrightarrow[\text{peroxide}]{HBr} A \xrightarrow{OH^{\ominus}/H_2O} B \xrightarrow[\text{pyridine}]{CrO_3} C$	5. A: $CH_3CH_2CH_2Br$ B: $CH_3CH_2CH_2OH$ C: $CH_3CH_2\overset{O}{\overset{\Vert}{C}}H$
6. cyclohexane $\xrightarrow[h\nu]{Br_2} A \xrightarrow[\text{ether}]{Mg} B \xrightarrow{\text{ethylene oxide (O)}} \xrightarrow{H_2O} C \xrightarrow[\text{heat}]{KMnO_4/H^{\oplus}} D$	6. A: cyclohexane ring–Br B: cyclohexane ring–MgBr C: cyclohexane ring–CH_2CH_2OH D: cyclohexane ring–$CH_2\overset{O}{\overset{\Vert}{C}}OH$

7. $CH_3CH_2\overset{O}{\overset{\|\|}{C}}CH_3 \xrightarrow[CH_3OH/H_2O]{NaBH_4}$ A $\xrightarrow[heat]{H_2SO_4}$ B $\xrightarrow{O_3}$ $\xrightarrow[HCl]{Zn}$ C	7. A: $CH_3CH_2CH(OH)CH_3$ B: $CH_3CH{=}CHCH_3$ (<u>trans</u> > <u>cis</u>) C: $2CH_3\overset{O}{\overset{\|\|}{C}}H$
8.  cyclopentene $\xrightarrow[H_2O]{Hg(OAc)_2}$ $\xrightarrow[H_2O]{NaBH_4}$ A $\xrightarrow[heat]{MnO_4^-/H^{\oplus}}$ B $\xrightarrow{CH_3CH_2Li}$ $\xrightarrow{H_2O}$ C	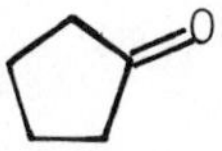 8. A: 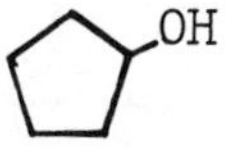 cyclopentanol (ring–OH) B: cyclopentanone (ring=O) C: 1-ethylcyclopentanol (ring with OH and CH_2CH_3) 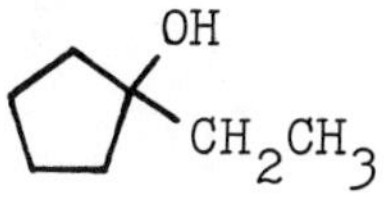
9. $CH_3\overset{O}{\overset{\|\|}{C}}OH \xrightarrow[H^{\oplus}]{CH_3OH}$ A $\xrightarrow{2CH_3MgBr}$ $\xrightarrow{H_2O}$ B $\xrightarrow{Na}$ C	9. A: $CH_3\overset{O}{\overset{\|\|}{C}}OCH_3$ B: $(CH_3)_3COH$ C: $Na^{\oplus}\ (CH_3)_3CO^{\ominus}$
10.  cyclohexene $\xrightarrow{NBS}$ A $\xrightarrow{Li(CH_3)_2Cu}$ B $\xrightarrow{H_2/Pt}$ C $\xrightarrow{Br_2/h\nu}$ D $\xrightarrow{H_2O}$ E	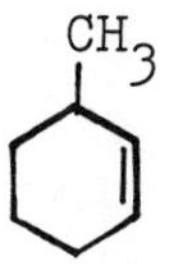 10. A: 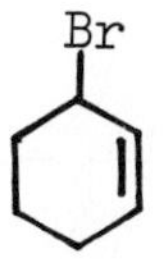 3-bromocyclohexene (Br on ring) B: 3-methylcyclohexene (CH_3 on ring) (continued next page)

	C: methylcyclohexane (CH_3 on cyclohexane ring) D: 1-bromo-1-methylcyclohexane (Br, CH_3 on same ring carbon) E: 1-methylcyclohexanol (HO, CH_3 on same ring carbon)
11. cyclohexyl bromide (Br on cyclohexane) $\xrightarrow[NH_{3(\ell)}]{Na^{\oplus}\ NH_2^{\ominus}}$ A $\xrightarrow{MnO_4^{\ominus}(aq)}$ B	11. A: cyclohexene B: cyclohexane-1,2-diol (OH, OH) (<u>meso</u>)
12. $CH_3C{\equiv}CH \xrightarrow[NH_{3(\ell)}]{Na^{\oplus}\ NH_2^{\ominus}}$ A $\xrightarrow{CH_3I}$ B $\xrightarrow{H_2O/H^{\oplus}/Hg^{2+}}$ C $\xrightarrow{CH_3Li}\ \xrightarrow{H_2O}$ D	12. A: $Na^{\oplus}\ CH_3C{\equiv}C{:}^{\ominus}$ B: $CH_3C{\equiv}CCH_3$ C: $CH_3\overset{\overset{O}{\parallel}}{C}CH_2CH_3$ D: $(CH_3)_2\underset{\underset{OH}{\mid}}{C}CH_2CH_3$

CHAPTER 11 - ETHERS, EPOXIDES AND DIOLS

LEARNING OBJECTIVES

When you have completed this chapter, you should be able to:

1. name or draw structures for ethers and epoxides (problems 1, 12);
2. write equations which illustrate the multistep syntheses of ethers and other compounds (problems 2, 8, 15 and related problems 7, 17, 18);
3. predict the products resulting from cleavage of ethers by HX (problem 3);
4. predict the structures of epoxides formed in epoxidation of alkenes (problem 4);
5. predict the products of reactions of epoxides with nucleophiles (problems 5, 14);
6. predict the products of oxidation reactions of diols and related compounds (problems 10, 11, 16);
7. suggest chemical tests which distinguish between ethers and other compounds (problems 13, 25);
8. account for the reactions of ethers and epoxides (problems 20, 21, 22);
9. identify or distinguish compounds based on their IR and NMR spectra (problems 26, 27 and related problem 28).

ANSWERS TO QUESTIONS

1. (a) methyl n-propyl ether
(1-methoxypropane)

(b) vinyl ether

(c) allyl ethynyl ether

(d) 1,2-epoxy-2-methylpropane
(isobutylene oxide)

(e) <u>cis</u>-1,2-cyclohexanediol

(f) (1R,2S,3S)-3-methylcyclohexene oxide

2. (a) $CH_2{=}CH_2 \xrightarrow{H_2O/H^{\oplus}} CH_3CH_2OH \xrightarrow{Na} CH_3CH_2\overset{\ominus}{O}\overset{\oplus}{Na}$

$CH_3CH_2OH \xrightarrow{PBr_3 \text{ (or HBr)}} CH_3CH_2Br$

$CH_3CH_2ONa + CH_3CH_2Br \rightarrow CH_3CH_2OCH_2CH_3$

or $2CH_3CH_2OH$ (as above) $\xrightarrow{H_2SO_4/\text{heat}} CH_3CH_2OCH_2CH_3$

(b) $CH_3CH{=}CH_2 \xrightarrow{H_2O/H^{\oplus}} CH_3CH(OH)CH_3$

$CH_3CH(OH)CH_3 + CH_3(CH_2)_3CH{=}CH_2 \xrightarrow{Hg(CF_3CO_2)_2} \xrightarrow{NaBH_4} CH_3(CH_2)_3CHCH_3$ (with $OCH(CH_3)_2$ on CH)

(c) $CH_2{=}CH_2 \xrightarrow{O_3} \xrightarrow{Zn/CH_3COOH} 2CH_2O \xrightarrow{LiAlH_4} \xrightarrow{H_3O^{\oplus}} 2CH_3OH$

$CH_3OH \xrightarrow{Na} CH_3\overset{\ominus}{O}\ \overset{\oplus}{Na}$

$CH_3OH \xrightarrow{PBr_3} CH_3Br$

$CH_3ONa + CH_3Br \rightarrow CH_3OCH_3$

(d) cyclohexene $\xrightarrow{H^{\oplus}/H_2O}$ cyclohexanol (OH) $\xrightarrow{Na}$ cyclohexyl $\overset{\ominus}{O}\overset{\oplus}{Na}$

cyclopentene $\xrightarrow{HCl}$ cyclopentyl Cl

cyclohexyl ONa + cyclopentyl Cl → cyclohexyl–O–cyclopentyl

(continued next page)

(e) $CH_3\text{-}\underset{\displaystyle CH_3}{\underset{|}{C}}\text{=}CH_2 \xrightarrow{H_2O/H^{\oplus}} (CH_3)_3COH \xrightarrow{Na} (CH_3)_3CO^{\ominus}\ Na^{\oplus}$

$CH_2\text{=}CH_2 \xrightarrow{HBr} CH_3CH_2Br \xrightarrow[\text{(from above)}]{(CH_3)_3CO^{\ominus}\ Na^{\oplus}} (CH_3)_3COCH_2CH_3$

not $CH_3CH_2O^{\ominus}Na^{\oplus}$ + $(CH_3)_3CBr$ which gives

$CH_3CH_2OH + (CH_3)_2C\text{=}CH_2$ (E2)

3. (a) Peroxides with HI:

$R\text{-}O\text{-}R' + H\text{-}I \longrightarrow RI + R'OH$ (no excess HI)

a) $CH_3CH_2OH + CH_3CH_2I$

b) $CH_3CH_2CH_2CH_2\underset{\displaystyle CH_3}{\underset{|}{C}}HOH + CH_3\underset{\displaystyle CH_3}{\underset{|}{C}}HI$ and $CH_3CH_2CH_2CH_2\underset{\displaystyle CH_3}{\underset{|}{C}}HI + CH_3\underset{\displaystyle CH_3}{\underset{|}{C}}HOH$

c) CH_3OH and CH_3I

d) cyclohexanol (OH) + iodocyclopentane (I) + iodocyclohexane (I) + cyclopentanol (OH)

e) $(CH_3)_3CI + CH_3CH_2OH$ (S_N1)

(b) Reaction with excess HI converts the alcohol to an alkyl iodide.

4. Recall that epoxidation of a carbon-carbon double bond will occur at both faces of the π bond as shown below.

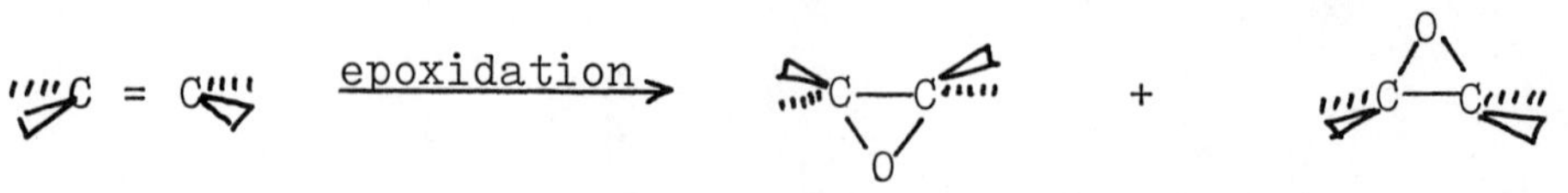

C = C —epoxidation→ C—C (O below) + C—C (O above)

(a) (R and S) 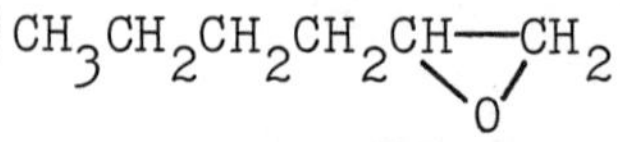 $CH_3CH_2CH_2CH_2CH—CH_2$ (epoxide, O)

(b) 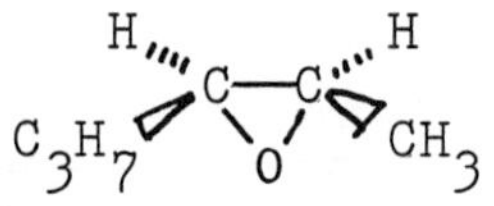 H, H, C_3H_7, CH_3 epoxide ; H, CH_3, C_3H_7, H epoxide

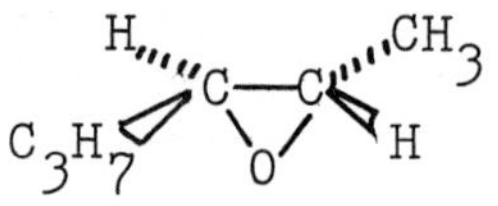

(d,l) (d,l)

(from cis-2-hexene) (from trans-2-hexene)

(c) 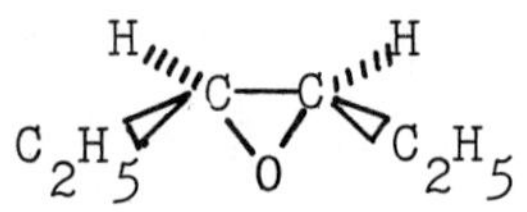 H, H, C_2H_5, C_2H_5 epoxide ; H, C_2H_5, C_2H_5, H epoxide

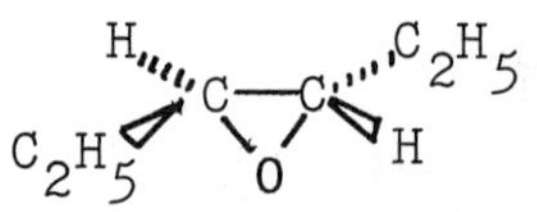

meso (d,l)

(from cis-3-hexene) (from trans-3-hexene)

(d) 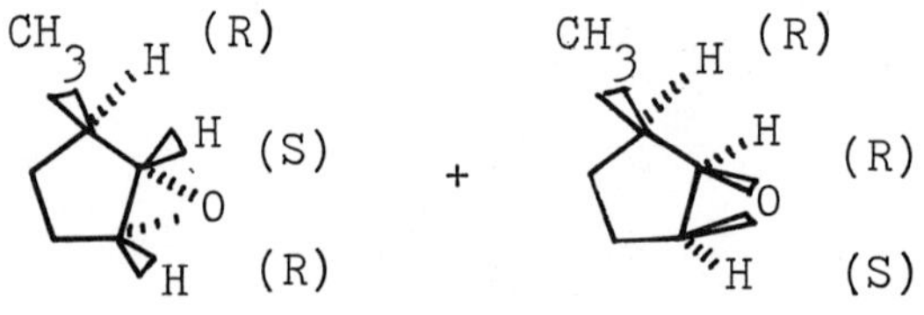 CH_3 H (R), H (S), O, H (R) + CH_3 H (R), H (R), O, H (S)

diastereomers

(e) 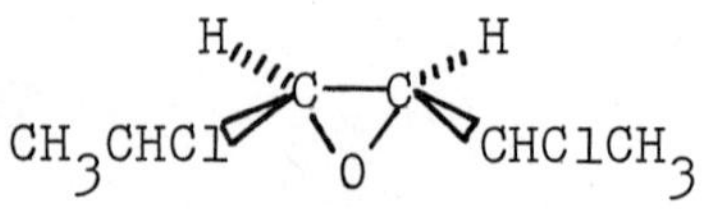 H, H, CH_3CHCl—C—C—$CHClCH_3$ (epoxide, O) and (2R,3S,4R, 5R)

(2R,3R,4S, 5R)

5. acid: $-\overset{|}{C}-\overset{|}{C}-$ (epoxide, O bridging) $\xrightarrow{H^{\oplus}}$ $-\overset{|}{C}-\overset{|}{C}-$ (protonated epoxide, $\overset{+}{\ddot{O}}-H$) $\xrightarrow{:Nu-H}$ $-\underset{:\ddot{O}H}{\overset{|}{C}}-\overset{\overset{\oplus}{Nu-H}}{C}-$ $\xrightarrow{-H^{\oplus}}$ $-\underset{:\ddot{O}H}{\overset{|}{C}}-\overset{Nu:}{C}-$

absence of acid: $-\overset{|}{C}-\overset{|}{C}-$ (epoxide, $\ddot{O}$ bridging) $\xrightarrow{:Nu^{\ominus}}$ $-\underset{:\ddot{O}:^{\ominus}}{\overset{|}{C}}-\overset{Nu}{C}-$ $\xrightarrow{H_2O}$ $-\underset{:\ddot{O}H}{\overset{|}{C}}-\overset{Nu}{C}-$

(a) CH_2OHCH_2OH

(b) CH_2OHCH_2OH

(c) CH_2OHCH_2Cl

(d) $CH_3SCH_2CH_2O^{\ominus}\ Na^{\oplus}$

(e) $CH_3CH_2CH_2CH_2CH_2OH$

(f) $CH_3CH_2CH_2OH$

(g) $NCSCH_2CH_2O^{\ominus}\ K^{\oplus}$

(h) $HOCH_2CH_2OSO_2OH$

(i) $CH_3C{\equiv}CCH_2CH_2O^{\ominus}\ Na^{\oplus}$

6. S_N2 attack of $:P(C_6H_5)_3$ gives the $-P(C_6H_5)_3$ and -O groups <u>trans</u>.

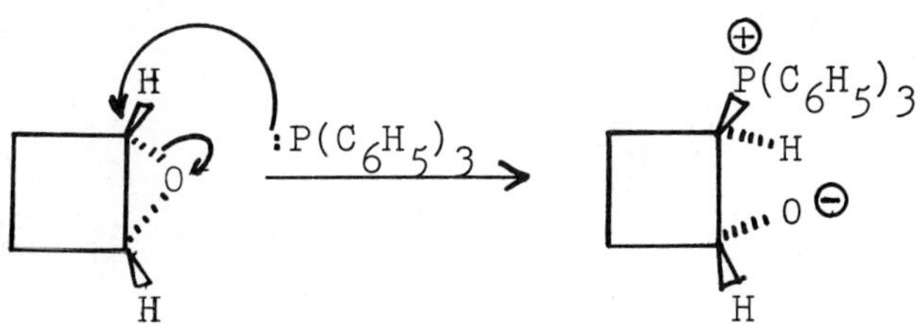

Rotation of these groups to permit <u>cis</u>-elimination is not possible.

7. (a) A: $CH_3CH_2CH(Cl)CH_3$ B: $CH_3CH=CHCH_3$ (<u>trans</u>) C: $CH_3CH=CHCH_3$ (<u>cis</u>)

D: H, O, CH_3 / CH_3 C—C H

E: H, O, H / CH_3 C—C CH_3

(b) F: $CH_3CHBrCH_3$ G: $CH_3CH=CH_2$

H: $CH_3CH_2CH_2Br$ I: $CH_3CH_2CH_2OCH_3$

(c) J: OH K: O

L: I

8. (a) $CH_2—CH_2$ (O) $\xrightarrow{CH_3S^{\ominus}\ Na^{\oplus}}$ $CH_3SCH_2CH_2O^{\ominus}\ Na^{\oplus}$ $\xrightarrow{\text{(O)}}$

$CH_3SCH_2CH_2OCH_2CH_2O^{\ominus}\ Na^{\oplus}$ $\xrightarrow{H_3O^{\oplus}}$ $CH_3SCH_2CH_2OCH_2CH_2OH$

(b) $CH_2—CH_2$ (O) $\xrightarrow{CH_3O^{\ominus}\ Na^{\oplus}}$ $CH_3OCH_2CH_2O^{\ominus}\ Na^{\oplus}$ $\xrightarrow{\text{(O)}}$

$CH_3OCH_2CH_2OCH_2CH_2O^{\ominus}\ Na^{\oplus}$ $\xrightarrow{\text{(O)}}$ $CH_3OCH_2CH_2OCH_2CH_2OCH_2CH_2O^{\ominus}\ Na^{\oplus}$

$\xrightarrow{\text{(O)}}$ $CH_3OCH_2CH_2OCH_2CH_2OCH_2CH_2OCH_2CH_2O^{\ominus}\ Na^{\oplus}$ $\xrightarrow{H_3O^{\oplus}}$

$CH_3O(CH_2CH_2O)_3CH_2CH_2OH$

(continued next page)

(c) Oxetane (CH_2—CH_2 / CH_2—O ring) $\xrightarrow{CH_3(CH_2)_3O^{\ominus}\ Na^{\oplus}}$ $CH_3(CH_2)_3OCH_2CH_2CH_2O^{\ominus}\ Na^{\oplus}$

$\xrightarrow{\text{ethylene oxide}}$ $CH_3(CH_2)_3OCH_2CH_2CH_2OCH_2CH_2O^{\ominus}\ Na^{\oplus}$ $\xrightarrow{H_3O^{\oplus}}$

$CH_3(CH_2)_3OCH_2CH_2CH_2OCH_2CH_2OH$

9. (a) $2CH_3\overset{O}{\overset{\|}{C}}H$

(b) $2CH_3\overset{O}{\overset{\|}{C}}H$

(c) $(CH_3)_2C{=}O + H\overset{O}{\overset{\|}{C}}H$

(d) $(CH_3)_2C{=}O + CH_3\overset{O}{\overset{\|}{C}}H$

(e) $2(CH_3)_2C{=}O$

(f) $H\overset{O}{\overset{\|}{C}}(CH_2)_3\overset{O}{\overset{\|}{C}}H$

10. (a) $2CH_2O + 2H\overset{O}{\overset{\|}{C}}OH$

(b) $3H\overset{O}{\overset{\|}{C}}OH + CH_3\overset{O}{\overset{\|}{C}}H$

(c) $2H\overset{O}{\overset{\|}{C}}OH + H\overset{O}{\overset{\|}{C}}CH_2CH_2OH$

11. (a) CH_3OCH_3

(b) $CH_3CH_2OCH_2CH_3$

(c) $CH_3OCH_2CH_3$

(d) $(CH_3)_2CHCH_2O\underset{CH_3}{\underset{|}{C}}HCH_2CH_3$

(e) $H_2C{=}CHCH_2OCH_2CH{=}CH_2$

(continued next page)

(f) $H_2C{=}CHOCH{=}CH_2$

(g) Cl, H, CH_3 / C—CH_2—O—CH_2—C / CH_3, H, Cl

(h)

(i) C_2H_5, H / C—C (O) / CH_3, H (or enantiomer)

(j) C_2H_5, H / C—C (O) / H, CH_3 (or enantiomer)

(Text S. 11.1)

12. (a) Ethanol reacts with sodium (bubbling due to H_2 release); ethyl ether does not. (Text S. 10.5)

(b) Ethyl iodide reacts with $AgNO_3$ giving a precipitate of AgI, ethyl alcohol does not. (Text S. 5.7)

(c) Cyclohexene rapidly decolorizes Br_2/CCl_4 solution, cyclohexene oxide does not. (Text S. 8.6)

(d) Vinyl ether rapidly decolorizes Br_2/CCl_4 solution, ethyl ether does not. (Text S. 8.6)

(e) Ethanol and isopropyl alcohol both react with sodium (evolution of hydrogen), isopropyl ether does not. Ethanol and isopropyl alcohol can be distinguished by Lucas reagent (isopropyl alcohol reacts within 5 minutes, ethyl alcohol does not react). (Text S. 10.5, 10.9)

(continued next page)

(f) Ethynyl ether forms a precipitate with $Ag(NH_3)_2NO_3$, vinyl ether does not. (Text S. 9.3)

(g) Ethylene oxide reacts (via the diol) with HNO_3/HIO_4 followed by $Ag^{\oplus}_{(aq)}$ giving a precipitate of $AgIO_3$ (Malaprade reaction), cyclohexane does not. (Text S. 11.10)

(h) 1,2-Octanediol reacts with HNO_3/HIO_4 followed by $Ag^{\oplus}_{(aq)}$ giving a precipitate of $AgIO_3$, 1,6-octanediol does not. (Text S. 11.10)

(i) Each chloride reacts with alcoholic $AgNO_3$ giving a precipitate of AgCl. t-Butyl chloride reacts faster. n-Propyl ether does not react. (See electrophilic catalysis, S. 5.7)

13. (a) $CH_3CH_2CH_2OMgCl$

(b) CH_4 + CH_2ClCH_2OMgCl*

(*this can react further by an intramolecular S_N2 to give CH_2—CH_2 (ring closed through O)).

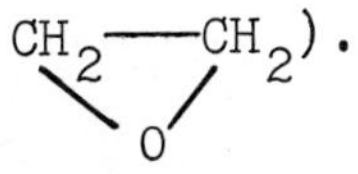

(c) CH_4 + $ClCH_2CH_2CH_2OMgCl$*

(*this can react further by an intramolecular S_N2 to give CH_2—O / CH_2—CH_2 (four-membered ring)).

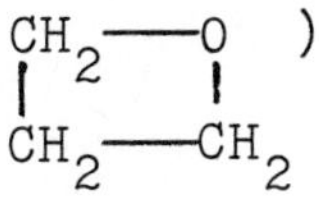

(d) 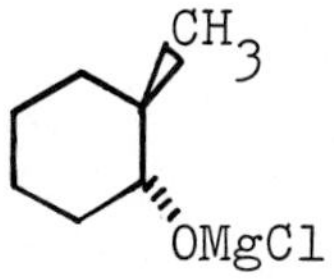 [cyclohexane ring bearing CH_3 (wedge) and OMgCl (hashed) on adjacent carbons] (d,l)

(e) 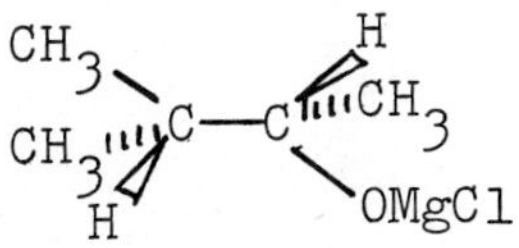 CH_3, CH_3, H on C—C bearing H, CH_3, OMgCl (d,l)

(continued next page)

(f) no reaction

(g) $CH_4 + CH_2(OH)CH_2OMgCl$

(h) $H \cdots C(OMgCl)(CH_3)CH_2CH_3$ (S_N2 attack at the less sterically hindered carbon)

14. (a) $CH_3CH_2OCH_2CH_3 + NaI$

(b) $H_2C{=}C(CH_3)_2 + CH_3CH_2OH + NaI$ (E2)

(c) $(CH_3)_3COCH_2CH_2O^{\ominus}\ K^{\oplus}$

(d) $CH_3CH_2CH_2\text{-}\overset{\oplus}{O}(H)\text{-}CH_2CH_2CH_3\ \ HSO_4^{\ominus}$

(e) $CH_3CH{=}CH_2$ (from $CH_3CH_2CH_2\text{-}\overset{\oplus}{O}(H)\text{-}CH_2\text{-}CH(H)CH_3 \longrightarrow$) with $^{\ominus}OSO_3H$ removing H

(f) $(CH_3)_2CHI + CH_3OH$ (from S_N1)

$(CH_3)_2CHOH + CH_3I$ (from S_N2); $S_N1 > S_N2$

15. (a) $H\text{-}C(H){=}O$

(b) $CH_3C(H){=}O + H\text{-}C(H){=}O$

(Text S. 11.10)

(c) $2HCH(=O) + HCOH(=O)$, i.e. $2\,HCHO + HCOOH$

(d) $O{=}C(H)CH_2CH_2CH_2C(H){=}O$

16. (a) A: CH_2—O, CH_2—CH_2 (four-membered ring: oxetane)

(b) B: cyclohexane-1,2-diol (OH, OH)
(d,l)

(c) CH_2=$CHCH_2CH_2CH$=CH_2

(d) D: CH_3, H, C—C, CH_3, H, O (epoxide)

E: CH_3, H, C—C, OCH_3, H, CH_3, ONa
(d,l)

F: CH_3, H, C—C, OCH_3, H, CH_3, C_2H_5O
(d,l)

17. (a) $CH_3\overset{O}{\overset{\|}{C}}CH_2CH_3$

(b) $(CH_3)_2CH\overset{O}{\overset{\|}{C}}H$

(c) CH_2OHCH_2OH

(d) $(CH_3)_2C$—$C(CH_3)_2$ with OH, OH

(e) $CH_3CH_2OCH_2CH_3$

(f) cyclopentene

(g) C_6H_5CH—CH_2 (bridged by O)

(h) CH_2=CH_2

(i) $(CH_3)_3CO^{\ominus}\ Na^{\oplus}$

(j) Cl, H, CH_3, HO (cyclopentane ring)

(k) H, CH_3, C—C, CH_3, H, O (epoxide)

(continued next page)

(l) 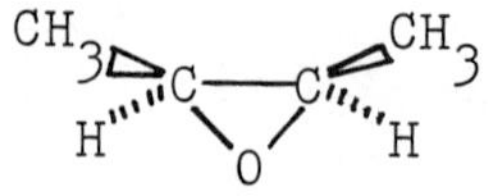 [structure: trans-2,3-dimethyloxirane; CH_3 and H on each ring carbon, O bridging]

18. (a) 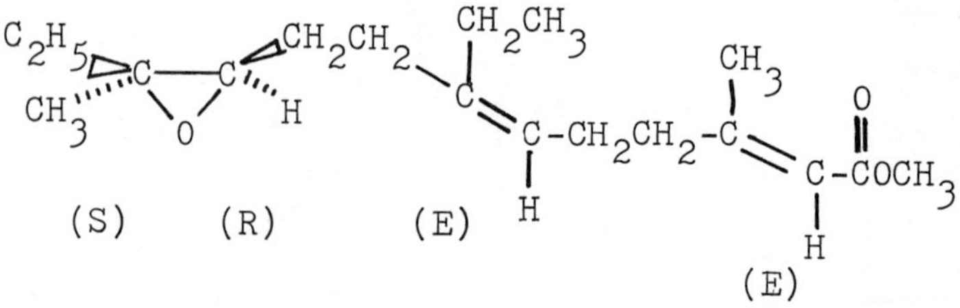 [structure: C_2H_5, CH_3 epoxide carbon — epoxide carbon with H and CH_2CH_2-C(CH_2CH_3)=C(H)-CH_2CH_2-C(CH_3)=C(H)-$COCH_3$ (C=O)]

(S) (R) (E) (E)

(b) 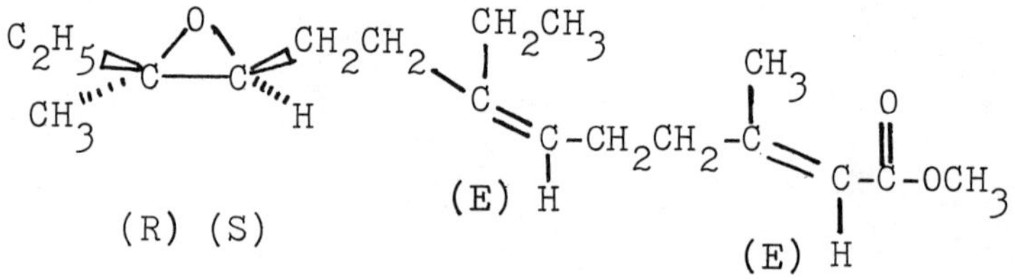 [structure: same chain, epoxide O drawn above]

(R) (S) (E) (E)

(c) [structure: C_2H_5, CH_3 epoxide carbon — epoxide carbon with H and CH_2CH_2-C(CH_2CH_3)=C-H, -CH_2CH_2-C(CH_3)=C(H)-C(=O)-OCH_3]

(S) (R) (Z) (E)

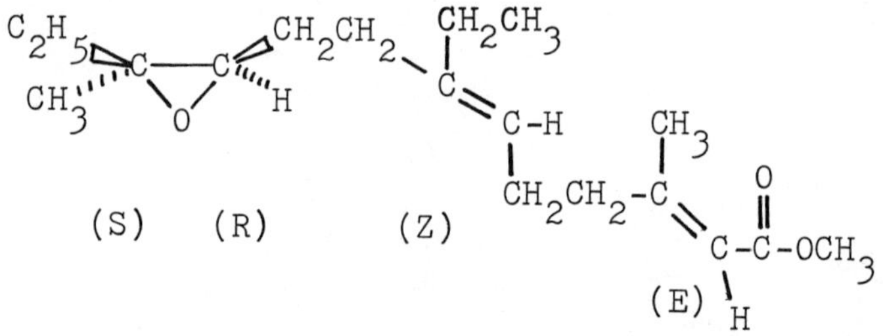

also:

(d) (S)(R) (Z) (Z)
(e) (S)(R) (E) (Z)
(f) (R)(S) (E) (Z)
(g) (R)(S) (Z) (E)
(h) (R)(S) (Z) (Z)
(i) (S)(S) (E) (E)
(j) (S)(S) (Z) (Z)
(k) (S)(S) (Z) (E)
(l) (S)(S) (E) (Z)
(m) (R)(R) (E) (E)
(n) (R)(R) (Z) (Z)
(o) (R)(R) (Z) (E)
(p) (R)(R) (E) (Z)

(Text S. 8.3)

19. Grignard reagents react with 2-chloroethers to give an activated complex which is resonance stabilized by the nonbonding electron pair on oxygen (see the related example of S_N2 reactivity of allyl halides, p. 164, text)

20. In neutral solution, the relative amounts of products reflect the attack of $Cl^{\ominus}$ primarily at the less sterically hindered carbon. In acid solution, however, the intermediate undergoing nucleophilic attack has considerable positive charge residing on the more highly substituted carbon. Thus nucleophilic attack occurs to a greater extent at C-2.

$CH_3\overset{\delta\oplus}{C}H—CH_2$, $\overset{\delta\oplus}{O}$—H

21. (a) H⊕; R–Ö–H; ROH; H⊕; -ROH; $H_2\ddot{O}$:; :OH_2; + $H_3O^{\oplus}$

(continued next page)

(b) The reaction of (R)-2-butanol with dihydropyran produces the diastereomers (R,R) and (R,S)-$CH_3CH_2\overset{*}{C}H(CH_3)$-O-(tetrahydropyran-2-yl, * C bearing H)

In addition unreacted (R)-2-butanol is present.

22. The use of cyclopentene would not allow one to clearly distinguish between (1) a one-step addition of oxygen in which both carbon-oxygen bonds are formed simultaneously and (2) the two-step addition:

H O⊖ H ⊕ → H O H

rotation prohibited

Either process gives the same stereochemical result. <u>cis</u>-2-Butene can be used to distinguish these pathways since a two-step process would lead to <u>cis</u>- and <u>trans</u>-oxides.

CH_3(H)C=C(H)CH_3 —RCOOH→ CH_3(H)C—C⊕(H)CH_3 with :O:⊖ —rotation→ CH_3(H)C—C⊕(CH_3)H with :O:⊖

↓ CH_3(H)C—C(H)CH_3 bridged by O; ↓ CH_3(H)C—C(CH_3)H bridged by O

(only the <u>cis</u>-product is actually obtained from <u>cis</u>-2-butene -- evidence that a two-step process does not occur)

23. Ring strain (with C-O bonds <u>trans</u>) makes A impossible. The greater flexibility in larger rings permits <u>trans</u> C-O bonds in B. Models are helpful here.

24. (Note: Consider the kinds of compounds containing a single oxygen considered up to this point -- alcohols, ethers, and epoxides. Of these, an epoxide can not be derived from the formula $C_5H_{12}O$.)

The compound which reacts with sodium is an alcohol, $C_5H_{11}OH + Na \longrightarrow C_5H_{11}O^{\ominus} Na^{\oplus} + \frac{1}{2}H_2$. The alcohol reacts with concentrated HCl giving a water-insoluble alkyl halide. The second compound is an ether. Acid-catalyzed cleavage of the ether gives the alcohol and a water-insoluble alkyl halide, $ROR + HCl \longrightarrow ROH + RCl$.

Rapid reaction of both compounds with hydrochloric acid suggests a 3° alcohol ($(CH_3)_2C(OH)CH_2CH_3$) and an ether containing a 3° group ($(CH_3)_3COCH_3$).

25. (A)

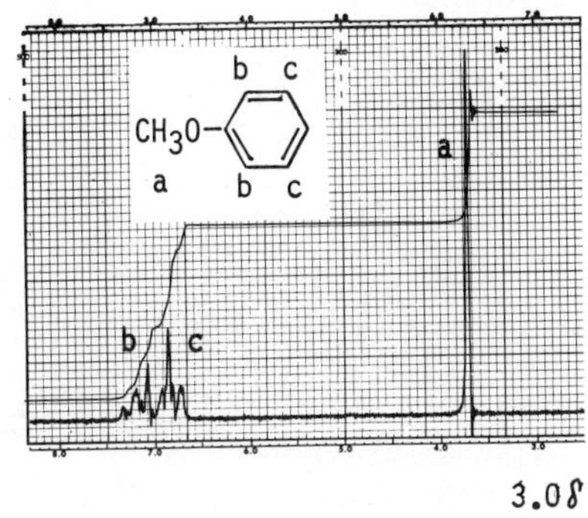

3.0δ

(B)

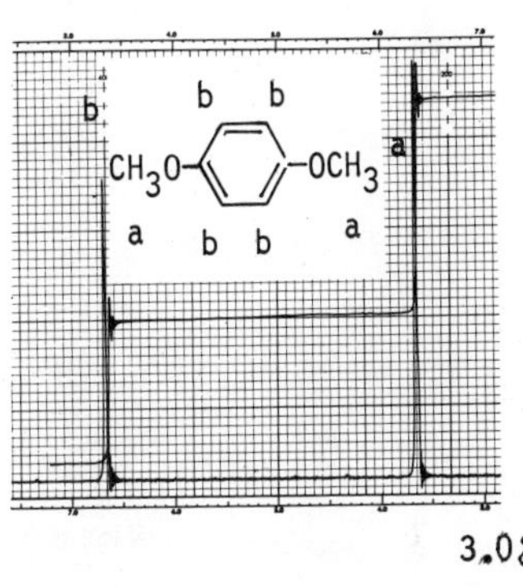

3.0δ

(C)

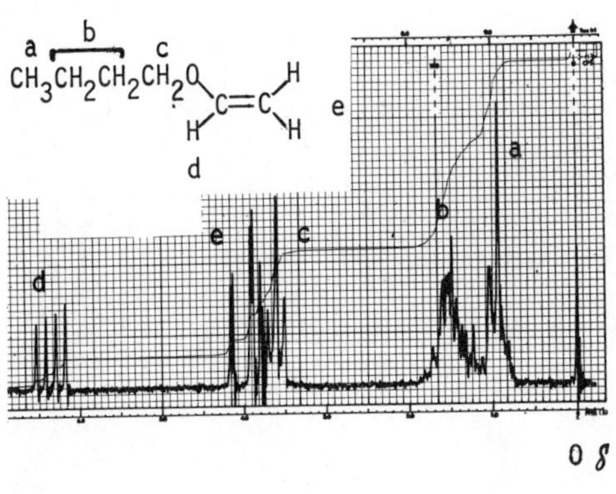

0 δ

(D)

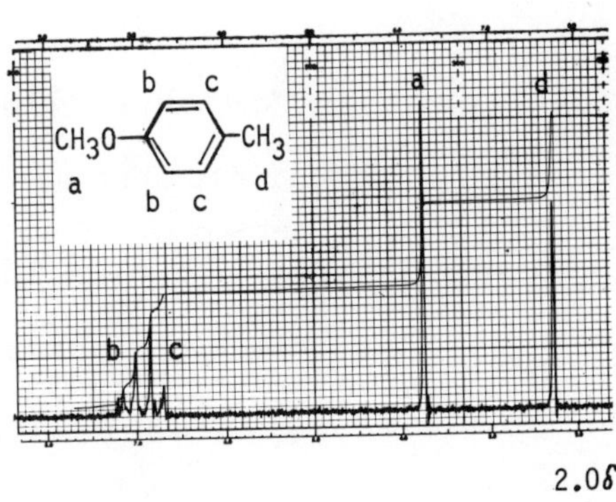

2.0δ

26. (a) ir: OH stretch (3400-3600 cm^{-1}) in ir spectrum of CH_3CH_2OH is absent in ir spectrum of $(C_2H_5)_2O$.

nmr: $(CH_3CH_2)_2O$: triplet 1δ (CH_3-); quartet 3.5δ (-CH_2-)
C_2H_5OH: triplet ~1δ (CH_3-); quartet ~3.5δ (-CH_2-); singlet 2-4δ (-OH)

(b) ir: C_2H_5OH shows only non-hydrogen-bonded OH stretch in the ir spectrum of a dilute solution (3550-3650 cm^{-1}); the ir spectrum of a dilute solution of $CH_2(OH)CH_2OH$ shows both non-hydrogen-bonded and intramolecular-hydrogen-bonded OH stretches (3550-3650 and 3200 cm^{-1}).

nmr: C_2H_5OH: triplet ~1δ (CH_3-); quartet ~3.5δ (-CH_2-) and singlet 2-4δ (-OH).
$CH_2(OH)CH_2OH$: singlet ~3.5δ (-CH_2CH_2-); singlet 2-4δ (-OH).

(c) ir: OH stretch (3400-3600 cm^{-1}) in the ir spectrum of $(HOCH_2CH_2)_2O$ is absent in the spectrum of the ether.

nmr: $(HOCH_2CH_2)_2O$: singlet ~ 4.5δ (-CH_2CH_2-); singlet 2-4δ (-OH)
$(CH_3OCH_2CH_2)_2O$: singlet ~3δ (CH_3-); broad singlet ~3.5δ (-CH_2CH_2-).

(d) ir: C-O stretch (~1100 cm^{-1}) in spectrum of $[CH_3(CH_2)_5]_2O$ absent in spectrum of $CH_3(CH_2)_4CH_3$

nmr: triplet ~3.5δ (-CH_2-O) in spectrum of $[CH_3(CH_2)_5]_2O$.
No downfield absorption in spectrum of $CH_3(CH_2)_4CH_3$.

(continued next page)

(e) ir: C=C stretch (~1650 cm^{-1}) in spectrum of $CH_3(CH_2)_3CH{=}CH_2$ is absent in spectrum of $CH_3(CH_2)_3CH—CH_2$ (epoxide, with O bridging the two carbons).

nmr: vinyl protons (5-6δ) in spectrum of $CH_3(CH_2)_3CH{=}CH_2$ are absent in the spectrum of $CH_3(CH_2)_3CH—CH_2$ (epoxide, with O bridging the two carbons).

(f) ir: C=C stretch (~1650 cm^{-1}) in the spectrum of $CH_3CH{=}CH_2$ absent in $CH_3CH—CH_2$ (epoxide, with O bridging the two carbons).

nmr: vinyl protons (5-6δ) in the spectrum of $CH_3CH{=}CH_2$ are absent in the spectrum of $CH_3CH—CH_2$ (epoxide, with O bridging the two carbons).

(g) ir: OH stretch in the spectrum of $CH_3CH(OH)CH_2Cl$ is absent in the spectrum of $CH_3CH—CH_2$ (epoxide, with O bridging the two carbons).

nmr: signal for OH proton in $CH_3CH(OH)CH_2Cl$ will decrease in intensity in the presence of D_2O.

27. Intermolecular hydrogen bonding is absent in both isomers in dilute solution where molecules are relatively far apart. However, intramolecular hydrogen bonding occurs in cis-1,2-cyclopentanediol and is not affected by dilution.

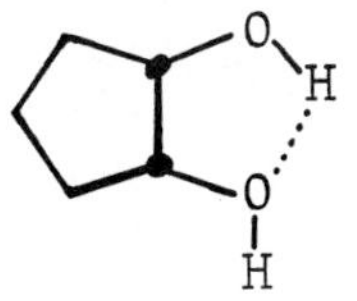

SUMMARY OF CHEMICAL REACTIONS

1. Synthesis of Ethers (S. 11.3)

(a) $Na^{\oplus} RO^{\ominus} + R'X \longrightarrow ROR' + Na^{\oplus} X^{\ominus}$

$Na^{\oplus}$ $C_6H_5O^{\ominus}$ + RX ⟶ C_6H_5–O-R + $Na^{\oplus} X^{\ominus}$

(Williamson synthesis)

These are S_N2; elimination reactions compete.

(b) $>C=C< \xrightarrow[ROH]{Hg^{2+}\ CF_3CO_2^{\ominus}}$ -C(OR)-C(HgOC(=O)CF$_3$)- $\xrightarrow{NaBH_4}$ -C(OR)-C(H)-

(alkoxymercuriation - demercuriation)

(c) $2ROH \xrightarrow{H^{\oplus}} R\text{-}O\text{-}R + H_2O$ (dehydration of alcohols)

Preparation of symmetric ethers.

2. Reactions of Ethers (S. 11.4)

(a) $R\text{-}O\text{-}R' + HX \longrightarrow ROH + R'X$ (acid-catalyzed cleavage)

Reaction may be S_N1 or S_N2 depending on structure of ether.

(b) H-C-O-C- $\xrightarrow{O_2}$ H-O-O-C-O-C- (hydroperoxide formation)

3. Synthesis of Epoxides (S. 11.5)

(a) $RCH{=}CHR \xrightarrow{R\text{-}C(=O)\text{-}O\text{-}O\text{-}H}$ epoxide (R and H cis on C—C ring with O) (epoxidation)

(b) $RCH{=}CH_2 \xrightarrow[H_2O]{Cl_2} R\text{-}CH(OH)\text{-}CH_2Cl \xrightarrow{OH^{\ominus}} RCH(:\ddot{O}:^{\ominus})\text{-}CH_2Cl \longrightarrow RCH\text{—}CH_2\ (\text{epoxide, O bridge}) + Cl^{\ominus}$

(dehydrohalogenation of a halohydrin)

4. Reactions of Epoxides (S. 11.6)

(a) General

$RCH\text{—}CH_2\ (\text{epoxide}) \xrightarrow[S_N2]{Nu:^{\ominus}\ (\text{or } Nu:)} \xrightarrow{H_2O} RCH(OH)CH_2Nu$

(attack of $Nu:^{\ominus}$ at the less sterically hindered position

common nucleophiles: $OH^{\ominus}$, $OR^{\ominus}$, RMgX, RLi, $LiAlH_4$)

$RCH\text{—}CH_2\ (\text{epoxide}) \xrightarrow{H^{\oplus}} RCH\text{—}\overset{*}{C}H_2\ (\text{protonated epoxide, } O^{\oplus}\text{—H}) \xrightarrow{Nu:\ (\text{or } Nu:^{\ominus})} RCH(Nu)CH_2OH$

*(involves $RCH\text{—}CH_2\ (\text{protonated epoxide, } O^{\oplus}\text{—H}) \longrightarrow \overset{\oplus}{R}CH\text{-}CH_2\ddot{O}H$)

(common nucleophiles: H_2O, ROH, $X^{\ominus}$)

(continued next page)

(b) Reduction

$$RCH\text{—}CH_2 \text{ (epoxide, O bridge)} \xrightarrow{LiAlH_4} \xrightarrow{H_2O} RCH(OH)CH_3$$

$$RCH\text{—}CH_2 \text{ (epoxide, O bridge)} \xrightarrow{B_2H_6} \xrightarrow{H_2O} RCH_2CH_2OH$$

(c) R,H-C—C-R,H (epoxide, O bridge) $\xrightarrow{(C_6H_5)_3P}$ R,H-C=C-H,R (deoxygenation)

Useful in converting E to Z isomers.

5. Synthesis of vic-Diols (S. 11.8)

$$>C=C< \xrightarrow{KMnO_4} \quad >C=C< \xrightarrow{OsO_4} \quad \text{-C(OH)—C(OH)-}$$

(syn-addition of hydroxyl groups)

$$\text{—C—C— (epoxide, O bridge)} \xrightarrow{OH^{\ominus}/H_2O} \text{-C(OH)—C(OH)-}$$

(anti-addition of hydroxyl groups)

6. <u>Oxidation of vic-Diols</u> (S. 11.9)

$$\text{-C(OH)-C(OH)-} \xrightarrow{KMnO_4} 2\ \rangle C{=}O\ ^{*}$$

* $-CH_2OH \rightarrow CO_2$

$-CRHOH \rightarrow RC(=O)-OH$

$-CR_2OH \rightarrow RC(=O)R$

$$\text{-C(OH)-C(OH)-} \xrightarrow{HIO_4 \text{ or } Pb(OAc)_4} 2\ \rangle C{=}O\ ^{\#}$$

\# $-CH_2OH \rightarrow H_2C{=}O$

$-CRHOH \rightarrow RHC{=}O$

$-CR_2OH \rightarrow R_2C{=}O$

$$\text{-C(HO)-C(OH)-} \xrightarrow{HNO_3,\, HIO_4} \xrightarrow{AgNO_3} 2\ \rangle C{=}O + AgIO_3 \downarrow$$

(test for <u>vic</u>-diol)

REACTION REVIEW

A. Reactions from Chapter 11

Questions	Answers
1. $Na^{\oplus}\ RO^{\ominus} + R'\text{-}X \longrightarrow$?	1. $ROR' + Na^{\oplus}\ X^{\ominus}$ (reaction S_N2)
Write the structure of the product in the following:	
a) $Na^{\oplus}\ CH_3O^{\ominus} + CH_3CH_2CH_2Cl \longrightarrow$?	a) $CH_3CH_2CH_2OCH_3$
b) $Na^{\oplus}\ CH_3O^{\ominus} + (CH_3)_3C\text{-}Cl \longrightarrow$?	b) $CH_2{=}C(CH_3)_2$ (substrate 3°, E2 reaction occurs)
c) $Na^{\oplus}$ 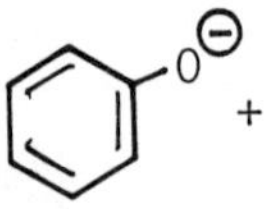 $C_6H_5O^{\ominus}$ (phenoxide) $+ CH_3I \longrightarrow$?	c) 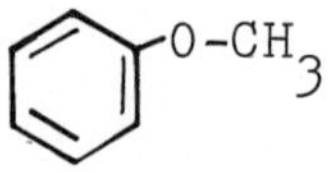 $C_6H_5\text{-}O\text{-}CH_3$
2. $RCH{=}CH_2 \xrightarrow[Hg(CF_3CO_2)_2]{ROH}$?	2. $RCHCH_3$ with OR on the CH ($RCH(OR)CH_3$)
Write the structure of the product in the following:	
a) $CH_3CH{=}CH_2 \xrightarrow[Hg(CF_3CO_2)_2]{CH_3CH_2OH}$?	a) CH_3CHCH_3 with OCH_2CH_3 on the CH ($CH_3CH(OCH_2CH_3)CH_3$)
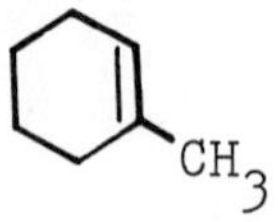 b) 1-methylcyclohexene (CH_3 on the ring double bond) $\xrightarrow[Hg(CF_3CO_2)_2]{(CH_3)_3COH}$?	b) 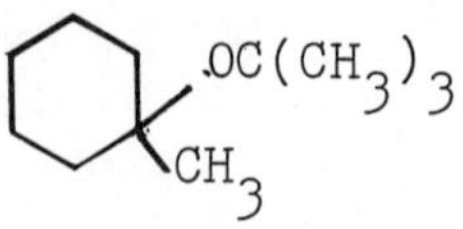 cyclohexane with $OC(CH_3)_3$ and CH_3 on the same carbon

3. $2ROH \xrightarrow[\text{(or } Al_2O_3)]{H_2SO_4}$? Write the structure of the product when ROH is: a) CH_3CH_2OH b) 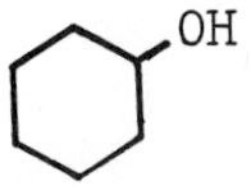 cyclohexanol (cyclohexane ring with OH)	3. $ROR + H_2O$ (symmetric ethers only) a) $CH_3CH_2OCH_2CH_3$ b) 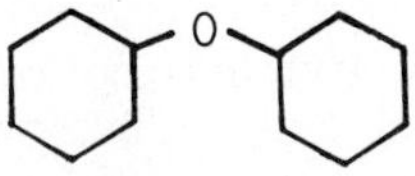 dicyclohexyl ether (two cyclohexane rings joined by O)
4. $ROR' + HX \longrightarrow$? (HX is HI or HBr) Write the structure of the product in the following: a) $(CH_3)_2CHOCH_3 \xrightarrow{HI}$? b) $(CH_3)_3COCH(CH_3)_2 \xrightarrow{HBr}$? c) $CH_3CH_2OCH_3 \xrightarrow[\text{excess}]{HI}$?	4. $RX + R'OH$ a) S_N2 $(CH_3)_2CHOH + CH_3I$ b) S_N1 $(CH_3)_3CBr + (CH_3)_2CHOH$ c) $CH_3CH_2I + CH_3I$
5. 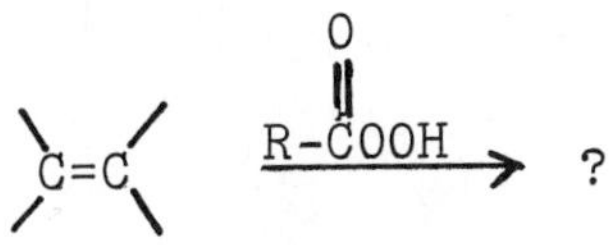 $>C=C< \xrightarrow{R-\overset{O}{\overset{\Vert}{C}}OOH}$? Write the structure of the product when 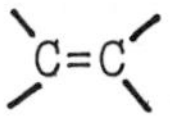 $>C=C<$ is: a) $CH_3CH=CH_2$	5. 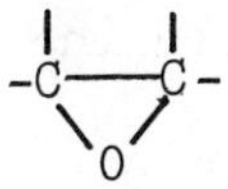 epoxide: $-\overset{\vert}{C}-\overset{\vert}{C}-$ with O bridging the two carbons a) CH_3CH-CH_2 with O bridging the two CH carbons (propylene oxide) 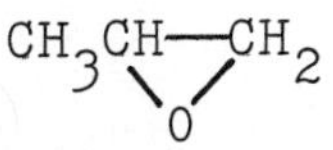

(continued next page)

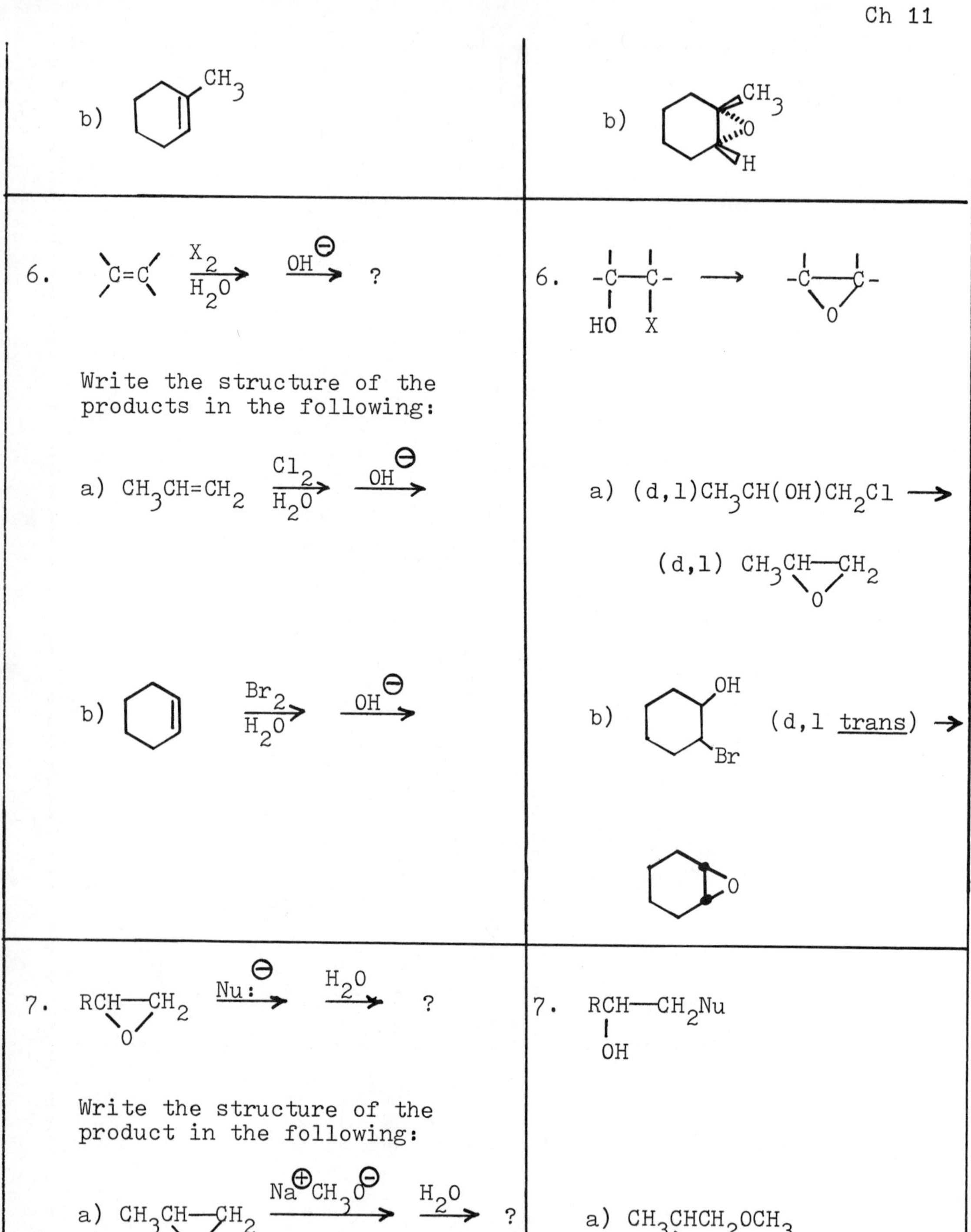

b) 1-methylcyclohexene (CH_3 on the ring double bond)

b) epoxide of 1-methylcyclohexene (CH_3, O, H)

6. $\rangle C{=}C\langle \xrightarrow[H_2O]{X_2} \xrightarrow{OH^{\ominus}}$?

Write the structure of the products in the following:

a) $CH_3CH{=}CH_2 \xrightarrow[H_2O]{Cl_2} \xrightarrow{OH^{\ominus}}$

b) cyclohexene $\xrightarrow[H_2O]{Br_2} \xrightarrow{OH^{\ominus}}$

6. $-\overset{|}{\underset{HO}{C}}-\overset{|}{\underset{X}{C}}- \longrightarrow$ epoxide $-C-C-$ bridged by O

a) (d,l)$CH_3CH(OH)CH_2Cl$ ⟶

(d,l) $CH_3CH{-}CH_2$ (bridged by O)

b) 2-bromocyclohexanol, OH and Br (d,l *trans*) ⟶

cyclohexene oxide (epoxide bridged by O)

7. $RCH{-}CH_2$ (bridged by O) $\xrightarrow{Nu:^{\ominus}} \xrightarrow{H_2O}$?

Write the structure of the product in the following:

a) $CH_3CH{-}CH_2$ (bridged by O) $\xrightarrow{Na^{\oplus}CH_3O^{\ominus}} \xrightarrow{H_2O}$?

7. $RCH{-}CH_2Nu$, with OH on the RCH carbon

a) $CH_3CHCH_2OCH_3$, with OH on the CH carbon

(continued next page)

b) 1-methyl-1,2-epoxycyclohexane (CH$_3$, O, H) $\xrightarrow{CH_3CH_2MgBr}$ $\xrightarrow{H_2O}$?	b) cyclohexane with CH$_3$, OH, H, CH$_2$CH$_3$ (d,l)
c) $CH_3CH\text{—}CH_2$ (epoxide, O) $\xrightarrow{LiAlH_4}$ $\xrightarrow{H_2O}$?	c) CH_3CHCH_3 with OH
8. R-CH—CH$_2$ (epoxide, O) $\xrightarrow{H^{\oplus}}$ $\xrightarrow{Nu:^{\ominus}}$?	8. a) RCHCH$_2$OH with Nu
Write the structure of the product in the following:	
a) $CH_3\text{—}CH\text{—}CH_2$ (epoxide, O) $\xrightarrow{HBr}$?	a) CH_3CHCH_2OH with Br
b) 1,2-dimethyl-1,2-epoxycyclohexane (CH$_3$, O, CH$_3$) $\xrightarrow{CH_3OH/H^{\oplus}}$?	b) cyclohexane with CH$_3$, OH, CH$_3$, OCH$_3$ (d,l)
9. -C—C- (epoxide, O) $\xrightarrow{(C_6H_5)_3P}$?	9. >C=C<
Write the structure of the product when the epoxide is:	
CH$_3$(H)C—C(H)CH$_3$ (epoxide, O) $\xrightarrow{(C_6H_5)_3P}$?	H(CH$_3$)C=C(CH$_3$)H (E)

10. $>C=C<$ $\xrightarrow[\text{(or } OsO_4)]{MnO_4^{\ominus} \text{ (aq)}}$?	10. $-\overset{\mid}{\underset{HO}{C}}-\overset{\mid}{\underset{OH}{C}}-$ (syn-addition of OH groups)
Write the structure of the product when $>C=C<$ is:	
a) $CH_3(H)C=C(H)CH_3$ (cis)	a) $CH_3-\underset{HO}{CH}-\underset{OH}{CH}-CH_3$ (meso)
b) 1-methylcyclohexene	b) cis-1-methylcyclohexane-1,2-diol (OH, CH_3, H, OH) (d,l)
11. $>C=C<$ $\xrightarrow[\text{(reaction 5)}]{RCOOH (O \text{ over C})}$ $\xrightarrow{OH^{\ominus}/H_2O}$?	11. $-\underset{HO}{C}-\overset{OH}{C}-$ (anti-orientation of OH groups)
Write the structure of the product when $>C=C<$ is:	
a) $CH_3(H)C=C(H)CH_3$ (cis)	a) $CH_3-\underset{HO}{CH}-\underset{OH}{CH}CH_3$ (d,l)
b) cyclohexene	b) trans-cyclohexane-1,2-diol (OH, OH) (d,l)

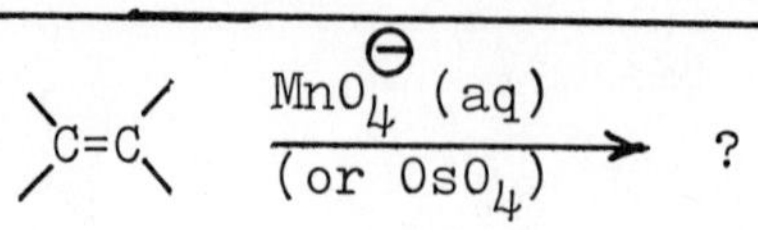

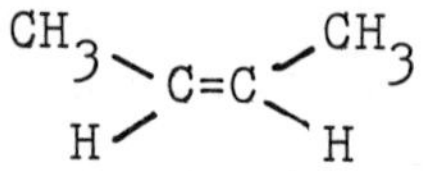

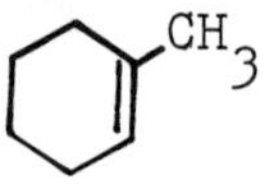

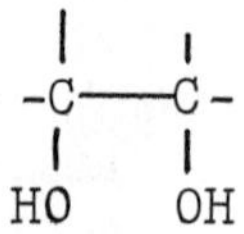

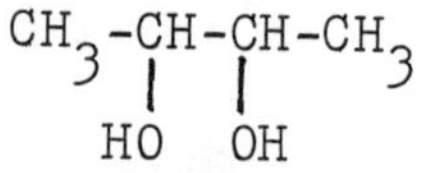

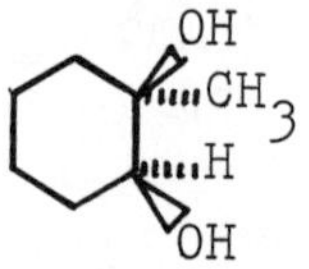

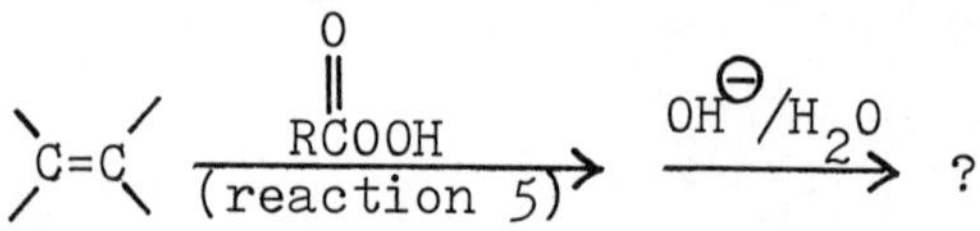

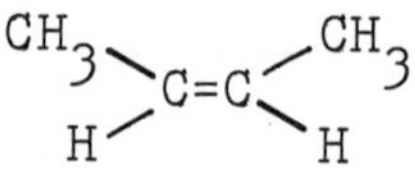

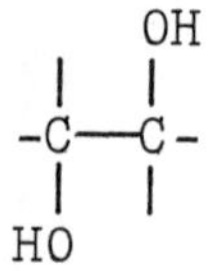

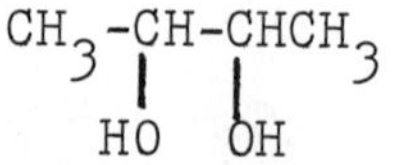

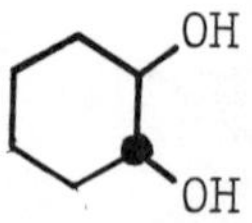

12. $-\overset{\vert}{\underset{HO}{C}}-\overset{\vert}{\underset{OH}{C}}-$ $\xrightarrow[\text{heat}]{KMnO_4}$? Write the structure of the product when the diol is: a) $CH_3\underset{OH}{\underset{\vert}{C}}HCH_2OH$ b) 1-methylcyclohexane-1,2-diol (ring with OH on C-1 and C-2, CH_3 on C-1)	12. 2 $\rangle C{=}O$ (H attached to C=O is oxidized to OH) a) $CH_3\overset{O}{\overset{\Vert}{C}}{-}OH + CO_2$ b) $CH_3\overset{O}{\overset{\Vert}{C}}(CH_2)_4\overset{O}{\overset{\Vert}{C}}OH$
13. $-\overset{\vert}{\underset{HO}{C}}-\overset{\vert}{\underset{OH}{C}}-$ $\xrightarrow[\text{or } Pb(OAc)_4]{HIO_4}$? Write the structure of the products when the diol is: a) $CH_3\underset{OH}{\underset{\vert}{C}}HCH_2OH$ b) 1-methylcyclohexane-1,2-diol (ring with OH on C-1 and C-2, CH_3 on C-1)	13. 2 $\rangle C{=}O$ a) $CH_3\overset{O}{\overset{\Vert}{C}}H + H\overset{O}{\overset{\Vert}{C}}H$ b) $CH_3\overset{O}{\overset{\Vert}{C}}(CH_2)_4\overset{O}{\overset{\Vert}{C}}H$
14. $-\overset{\vert}{\underset{HO}{C}}-\overset{\vert}{\underset{OH}{C}}-$ $\xrightarrow[HNO_3]{HIO_4}$ $\xrightarrow{AgNO_3}$?	14. 2 $\rangle C{=}O$ + $AgIO_3\downarrow$ (test for _vic_-diols)

CHAPTER 12. ORGANIC SYNTHESIS

ANSWERS TO QUESTIONS

1. Compounds which do not form stable Grignard reagents are:

   b) Grignard reagent undergoes α-elimination:

   $BrMg-CH_2-Br \longrightarrow :CH_2 + MgBr_2$

   d) A stable Grignard reagent is formed but not the one derived directly from the alkyl halide given. As each molecule of Grignard reagent derived from $HC{\equiv}C(CH_2)_3Br$ forms, it abstracts a proton from another molecule of $HC{\equiv}C(CH_2)_3Br$ giving $BrMgC{\equiv}C(CH_2)_3Br$. $HC{\equiv}C(CH_2)_3MgBr + HC{\equiv}C(CH_2)_3Br \longrightarrow HC{\equiv}C(CH_2)_2CH_3 + BrMgC{\equiv}C(CH_2)_3Br$.

   g) As each molecule of Grignard reagent forms, it abstracts a proton.

   intermolecular:

   OH, MgI-cyclohexane + I, H-O-cyclohexane ⟶ OH, H-cyclohexane + I, IMgO-cyclohexane

   intramolecular:

   cyclohexane with O–H, MgI, H ⟶ cyclohexane with OMgI, H, H

   h) The Grignard reagent undergoes β-elimination.

   $ClMg-CH_2-CH_2-Br \longrightarrow CH_2{=}CH_2 + MgClBr$

2. 

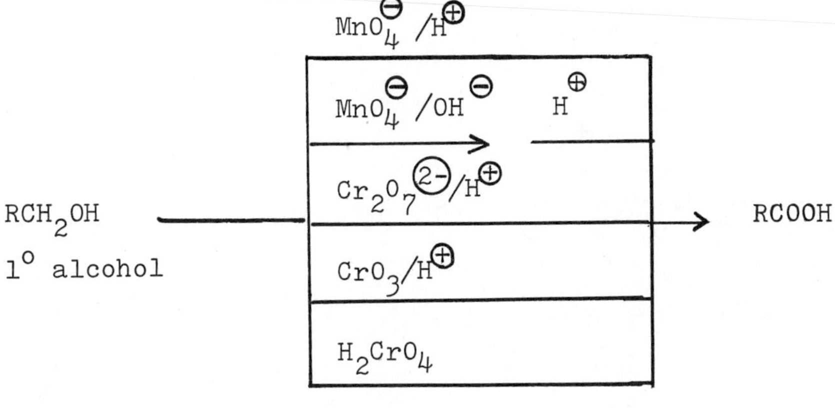

$RCHO \xrightarrow{\text{each of the above}} RCOOH$

aldehyde

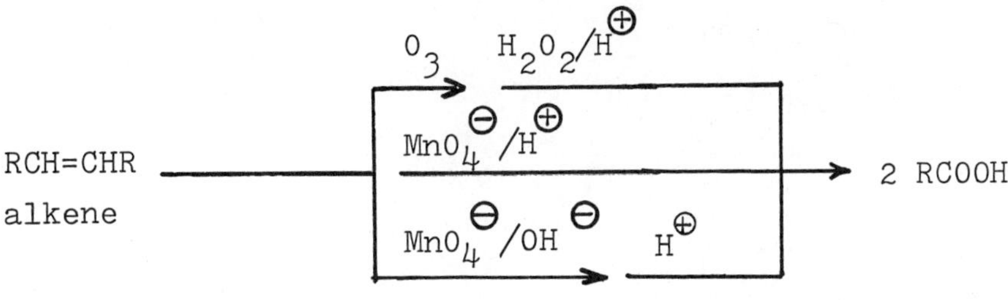

$RC{\equiv}CR \xrightarrow{O_3} \xrightarrow{H_2O} RCOOH$

alkyne

3. For most of the compounds in this question, more than one synthesis is given (indicated by (1), (2), etc). In some cases, syntheses are possible other than those shown.

a) <u>CH_3CH_3</u> (1) $CH_2{=}CH_2$ + H_2/cat. (or $CH_2{=}CH_2$ + BH_3 then RCOOH/heat); (2) $HC{\equiv}CH$ + $2H_2$/cat.; (3) CH_3CH_2MgX + H_2O

b) <u>$CH_3CH_2CH_2CH_3$</u> (1) $CH_3CH_2CH{=}CH_2$ (or $CH_3CH{=}CHCH_3$) + H_2/cat. (or +BH_3 then RCOOH/heat); (2) $CH_3CH_2C{\equiv}CH$ (or $CH_3C{\equiv}CCH_3$) + $2H_2$/cat.; (3) $CH_3CH_2CH_2CH_2MgX$ (or $CH_3CH_2CH(MgX)CH_3$) + H_2O; (4) CH_3CH_2X + $Li(CH_3CH_2)_2Cu$ (or $CH_3CH_2CH_2Li$ + $Li(CH_3)_2Cu$ etc).

(continued next page)

(c) $\underline{CH_3C{\equiv}CCH_2CH_3}$ (1) $CH_3CHBrCHBrCH_2CH_3$ + $Na^{\oplus}NH_2^{\ominus}/NH_{3(\ell)}$; (2) $CH_3CH_2C{\equiv}CNa$ + CH_3I (or $CH_3C{\equiv}CNa$ + CH_3CH_2Br)

(d) $\underline{CH_3\overset{O}{\overset{\|}{C}}CH_3}$ (1) $CH_3CH(OH)CH_3$ + $CrO_3/H^{\oplus}$ (or any other Cr(VI) reagent or $MnO_4^{\ominus}/H^{\oplus}$); (2) $(CH_3)_2C{=}C(CH_3)_2$ + O_3 then $Zn/H^{\oplus}$ (or $H_2O_2/H^{\oplus}$); (3) $CH_3C{\equiv}CH$ + $H^{\oplus}/H_2O/Hg^{2+}$; (4) $(CH_3)_2CHCl$ + $DMSO/OH^{\ominus}$

(e) $\underline{CH_3CH_2CH_2CHO}$ (1) $CH_3CH_2CH_2CH_2OH$ + CrO_3/pyridine; (2) $CH_3CH_2CH_2CH{=}CHCH_2CH_2CH_3$ + O_3 then $Zn/H^{\oplus}$; (3) $CH_3CH_2C{\equiv}CH$ + BH_3 then $H_2O_2/OH^{\ominus}$; (4) $CH_3CH_2CH_2CH_2Cl$ + $DMSO/NaHCO_3$

(f) $\underline{CH_3\overset{O}{\overset{\|}{C}}CH_2CH_3}$ (1) $CH_3CH(OH)CH_2CH_3$ + $CrO_3/H^{\oplus}$ (or any other Cr(VI) reagent or $MnO_4^{\ominus}/H^{\oplus}$); (2) $CH_3C{\equiv}CCH_3$ + BH_3 then $H_2O_2/OH^{\ominus}$; (3) $CH_3CH_2(CH_3)C{=}C(CH_3)CH_2CH_3$ + O_3 then $H_2O_2/H^{\oplus}$; (4) $CH_3CH(Cl)CH_2CH_3$ + $DMSO/NaHCO_3$

(g) $\underline{trans\text{-}CH_3CH{=}CHCH_3}$ (1) $CH_3C{\equiv}CCH_3$ + Na/liq NH_3; (2) $\underline{trans\text{-}}$$CH_3CH{=}CHCl$ + $Li(CH_3)_2Cu$

(h) $\underline{CH_3CH_2CH_2Cl}$ (1) $CH_3CH_2CH_2OH$ + $SOCl_2$ (or PCl_3 or HCl); (2) $CH_3CH{=}CH_2$ + BH_3 then $Cl_2/OH^{\ominus}$

(i) $\underline{CH_3CH_2OCH_2CH_3}$ (1) $CH_3CH_2O^{\ominus}Na$ + CH_3CH_2Br; (2) CH_3CH_2OH + H_2SO_4/heat; (3) $CH_2{=}CH_2$ + $CH_3CH_2OH/Hg(OAc)_2$ then $NaBH_4$

(j) $\underline{CH_3CH(OH)CH_2OH}$ (1) $CH_3CH{=}CH_2$ + $MnO_4^{\ominus}/OH^{\ominus}/25^{\circ}$; (2) $CH_3CH{-}CH_2$ (epoxide, with O bridging the two carbons) + $H_3O^{\oplus}$ (or $OH^{\ominus}/H_2O$)

(continued next page)

(k) $\underline{CH_3CO_2H}$ (1) $CH_3CH_2OH + K_2Cr_2O_7/H^{\oplus}$ (or any other Cr(VI) reagent, or $MnO_4^{\ominus}/H^{\oplus}$); (2) CH_3CHO + reagents shown in 1 ; (3) $CH_3CH{=}CHCH_3 + O_3$ then $H_2O_2/H^{\oplus}$; (4) $CH_3CH{=}CHCH_3$ + $MnO_4^{\ominus}/OH^{\ominus}$ then $H_3O^{\oplus}$; (5) $CH_3C{\equiv}CCH_3 + O_3$ then H_2O

(l) $\underline{HC(=O)(CH_2)_8C(=O)H}$ (1) $HOCH_2(CH_2)_8CH_2OH + CrO_3$/pyridine; (2) $BrCH_2(CH_2)_8CH_2Br$ + DMSO/$NaHCO_3$; (3) $HC{\equiv}C(CH_2)_6C{\equiv}CH$ + Sia_2BH then $H_2O_2/OH^{\ominus}$; (4) $CH_2{=}CH(CH_2)_8CH{=}CH_2 + O_3$ then $Zn/H^{\oplus}$;

(5) [cyclodecene] $\xrightarrow{O_3}$ $\xrightarrow{Zn/H^{\oplus}}$

(m) $\underline{CH_3CH_2CHICH_3}$ (1) $CH_3CH_2CH(OH)CH_3 + PI_3$ (or HI); (2) $CH_3CH_2CHClCH_3 + Na^{\oplus}I^{\ominus}$/acetone; (3) $CH_3CH{=}CHCH_3 + BH_3$ then $I_2/OH^{\ominus}$; (4) $CH_3CH{=}CHCH_3$ + HI

(n) $\underline{CH_3CH_2CHBrCH_3}$ (1) $CH_3CH_2CH(OH)CH_3 + PBr_3$ (or HBr); (2) $CH_3CH_2CHClCH_3 + Na^{\oplus}Br^{\ominus}$/acetone; (3) $CH_3CH{=}CHCH_3 + BH_3$ then $Br_2/OH^{\ominus}$; (4) $CH_3CH{=}CHCH_3$ (or $CH_3CH_2CH{=}CH_2$) + HBr; (5) $CH_3CH_2CH_2CH_3 + Br_2/h\nu$

(o) [cyclohexanone] (1) [cyclohexanol] + H_2CrO_4 (or any other Cr (VI) reagent or $MnO_4^{\ominus}/H^{\oplus}$); (2) [chlorocyclohexane] + DMSO/$NaHCO_3$; (3) [methylenecyclohexane] + O_3 then $Zn/H^{\oplus}$ (or $H_2O_2/H^{\oplus}$)

(continued next page)

(p) (1-methylcyclohexanol: HO, CH_3 on cyclohexane ring) (1) cyclohexanone + CH_3MgI then H_2O ; (2) 1-chloro-1-methylcyclohexane + H_2O (S_N1);

(3) 1-methylcyclohexene + $H_2O/H^{\oplus}$, (4) 1-methylcyclohexene + $Hg(OAc)_2/H_2O$ then $NaBH_4$

(q) cyclohexene (1) chlorocyclohexane + KOH/alcohol/heat (or $Na^{\oplus}NH_2^{\ominus}/NH_{3(\ell)}$);

(2) cyclohexanol + H_2SO_4/heat

(r) $CH_3CH_2CH{=}CH_2$ (1) $CH_3CH_2CH_2CH_2Cl$ + KOH/alcohol/heat (or $Na^{\oplus}NH_2^{\ominus}/NH_{3(\ell)}$); (2) $CH_3CH_2CH_2CH_2OH + H_2SO_4$/heat; (3) $CH_3CH_2C{\equiv}CH + H_2$/Lindlar's catalyst (or BH_3 then CH_3CH_2COOH); (4) $CH_3CH_2CH{-}CH_2$ (epoxide, O bridging) + $(C_6H_5)_3P$/heat; (5) $CH_2{=}CHCl$ + $Li(CH_3CH_2)_2Cu$

(s) $CH_3OCH_2CH_3$ (1) $CH_3O^{\ominus}Na^{\oplus} + CH_3CH_2Br$ (or $CH_3CH_2O^{\ominus}Na^{\oplus}$ + CH_3I); (2) $CH_2{=}CH_2 + CH_3OH/Hg(OAc)_2$ then $NaBH_4$

(t) 2-methylcyclohexanone (1) 2-methylcyclohexanol + $CrO_3/H^{\oplus}$ (or any other Cr(VI) reagent or $MnO_4^{\ominus}/H^{\oplus}$); (2) 1-chloro-2-methylcyclohexane + DMSO/$NaHCO_3$;

(continued next page)

(3) [2-methyl-1-methylenecyclohexane: ring with =CH_2 and CH_3] 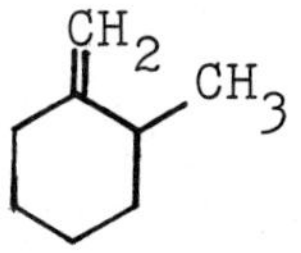 + O_3 then $Zn/H^{\oplus}$

(u) 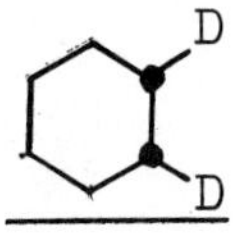 [cis-1,2-dideuteriocyclohexane: D, D] (1) [cyclohexene] + D_2/cat. (or BD_3 then CH_3COOD); (2) 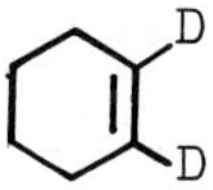 [1,2-dideuteriocyclohexene: D, D] + H_2/cat.

(v) 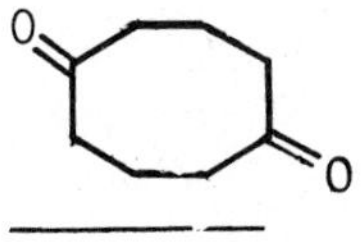 [cyclooctane-1,5-dione: O, O] (1) [cyclooctane-1,5-diol: HO, OH] + $K_2Cr_2O_7/H^{\oplus}$ (or any other Cr(VI) reagent, or $MnO_4^{\ominus}/H^{\oplus}$); (2) 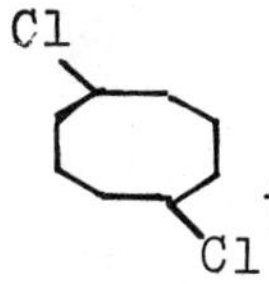 [1,5-dichlorocyclooctane: Cl, Cl] + DMSO/$NaHCO_3$;

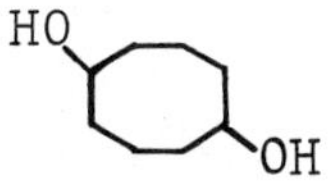

(3) 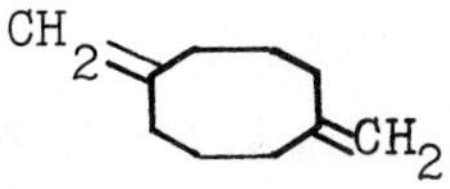 [1,5-dimethylenecyclooctane: CH_2=, =CH_2] + O_3 then $Zn/H^{\oplus}$

(w) $CH_3-\overset{OH}{\underset{D}{C}}-CH_3$ (1) $CH_3\overset{O}{\overset{\|}{C}}-D$ + CH_3MgI then H_2O ; (2) $CH_3\overset{O}{\overset{\|}{C}}CH_3$ + $LiAlD_4$ then H_2O (or $NaBD_4/H_2O$, or B_2D_6)

4. (a) $CH_2=CH_2$ (or CH_3CH_2OH)
   (b) $CH_3C{\equiv}CCH_3$
   (c) CH_3MgI
   (d) $CH_2=CHCH_3$
   (e) $HOCH_2(CH_2)_6CH_2OH$
   (f) <u>trans</u>-$CH_3CH=CHCH_3$
   (g) <u>cis</u>-$CH_3CH=CHCH_2CH_3$
   (h) <u>trans</u>-$CH_3CH=CHCH_3$
   (i) $CH_3CH_2C{\equiv}CCH_2CH_3$
   (j) $CH_3C{\equiv}CH$
   (k) $CH_3C{\equiv}CCH_3$ (or $HC{\equiv}CCH_2CH_3$)
   (l) $HOCH_2CH_2Cl$
   (m) CH_3CH_2Br
   (n) 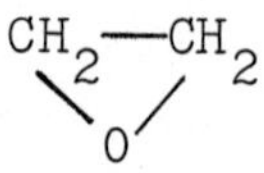 CH_2—CH_2 (epoxide, bridged by O) (or $ClCH_2CH_2Cl$)
   (o) 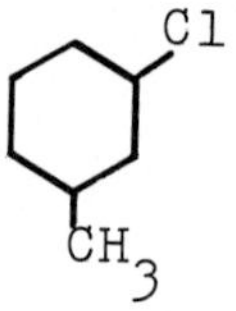 cyclohexane ring bearing Cl and CH_3 (1-chloro-3-methylcyclohexane)

5. (a) last step: hydroxyl protons are acidic enough to be exchanged, but the methyl protons are not. The reactions shown give $(CH_3)_3COD$.

   (b) first step: product is $CH_3CH_2CH_2OH$. The first step can be accomplished using $CH_3CH=CH_2 + H_2O/H^{\oplus}$.

   (c) second step: allylic bromination occurs with NBS. Product is $BrCH_2CH=CH_2$. The sequence of reactions shown gives $CH_3CH_2CH=CH_2$.

   (d) second step: the Grignard reagent initially formed ($CH_3CH_2CH(MgBr)CH_2Br$ or $CH_3CH_2CHBrCH_2MgBr$) undergoes elimination to give $CH_3CH_2CH=CH_2 + MgBr_2$.

(continued next page)

(e) first step: the product is $CH_3CH(OH)CH_2CH_3$. In the last two-step sequence, the ether should be formed from an alkoxide, $RO^{\ominus}$ (from ROH + Na → $RO^{\ominus}Na^{\oplus}$ + $\frac{1}{2}H_2$) and CH_3Cl.

(f) first step: the product is the diol, [1-(hydroxymethyl)cyclohexanol: cyclohexane ring bearing CH_2OH and OH on the same carbon]

$KMnO_4/H^{\oplus}$/heat should be used to produce [cyclohexanone].

In step 3, cyclohexanol does not react appreciably with NaOH. This step is best accomplished using Na. The reaction of $C_6H_{11}O^{\ominus}Na^{\oplus}$ + $CH_2{=}CHCH_2Br$ (last step) gives $C_6H_{11}OCH_2CH{=}CH_2$ (an ether).

6. In some parts, syntheses are possible other than those shown.

(a) $CH_3CH_2\overset{O}{\overset{\|}{C}}CH_3 \xrightarrow{CH_3CH_2MgBr} \xrightarrow{H_2O}$

(b) $CH_3C{\equiv}C{:}^{\ominus}\ Na^{\oplus} \xrightarrow{\text{(ethylene oxide)}} \xrightarrow{H_2O} \xrightarrow{SOCl_2} \xrightarrow{\text{Mg/dry ether}}$

$CH_3C{\equiv}CCH_2CH_2MgCl \xrightarrow{\text{(cyclobutanone)}} \xrightarrow{H_2O}$

(c) $CH_3CH_2C{\equiv}C{:}^{\ominus}\ Na^{\oplus} \xrightarrow{CH_3I} \xrightarrow{Na/NH_3(\ell)}$ trans-$CH_3CH_2CH{=}CHCH_3$

$\xrightarrow{CH_3\overset{O}{\overset{\|}{C}}OOH}$ (epoxide) $\xrightarrow{CH_3O^{\ominus}\ Na^{\oplus}}$ $CH_3CH_2\underset{ONa}{\underset{|}{C}}HCH(OCH_3)CH_3 \xrightarrow{CH_3I}$

(d) $CH_3CH_2C{\equiv}C{:}^{\ominus}\ Na^{\oplus} \xrightarrow{CH_3I} \xrightarrow[\text{Lindlar's cat.}]{H_2}$ cis-$CH_3CH_2CH{=}CHCH_3$

then as in (c)

(e) $CH_3CH_2\overset{O}{\overset{\|}{C}}CH_3 \xrightarrow{CH_3CH_2\underset{CH_3}{\underset{|}{C}}HMgBr} \xrightarrow{H_2O}$

(continued next page)

(f) $HOCH_2CH_2OCH_2CH_2OH \xrightarrow{Na} \xrightarrow{\text{ethylene oxide}} \xrightarrow{H_3O^{\oplus}}$

(g) $CH_3CH_2CH_2CHO \xrightarrow{CH_3CH_2CH_2MgCl} \xrightarrow{H_2O} CH_3CH_2CH_2CH(OH)CH_2CH_2CH_3$

$\xrightarrow{CrO_3/H^{\oplus}} \xrightarrow{CH_3(CH_2)_2CH_2MgCl} \xrightarrow{H_2O}$

(h) $CH_3CH{=}CH_2 \xrightarrow{BH_3} \xrightarrow[\text{heat}]{CO} \xrightarrow[OH^{\ominus}]{H_2O_2} (CH_3CH_2CH_2)_3COH$

(i) $CH_2{=}CHCH_2MgCl \xrightarrow{\text{ethylene oxide}} \xrightarrow{H_2O} CH_2{=}CHCH_2CH_2CH_2OH \xrightarrow[\text{peroxide}]{HBr}$

$\xrightarrow{OH^{\ominus}/H_2O} HOCH_2(CH_2)_3CH_2OH \xrightarrow{CrO_3/H^{\oplus}} \xrightarrow{2CH_3MgCl} \xrightarrow{H_2O}$

(j) $CH_3\overset{O}{\overset{\|}{C}}CH_3 \xrightarrow{CH_2{=}CHCH_2MgBr} \xrightarrow{H_2O} CH_3\underset{OH}{\overset{CH_3}{C}}{-}CH_2CH{=}CH_2 \xrightarrow{H_2O/H^{\oplus}}$

(k) cyclobutane $\xrightarrow{Br_2/h\nu} \xrightarrow{\text{Mg/dry ether}}$ cyclobutyl–MgBr $\xrightarrow{CH_2O} \xrightarrow{H_2O}$ cyclobutyl–CH_2OH

(l) $CH_3CH_2CH_2MgCl \xrightarrow{\text{ethylene oxide}} \xrightarrow{H_2O} \xrightarrow{CrO_3/\text{pyridine}} CH_3CH_2CH_2CH_2CHO$

$\xrightarrow{HC{\equiv}CMgCl} \xrightarrow{H_2O}$

7. In some parts, syntheses other than those shown are possible.

(a) $CH_3CHO \xrightarrow{CH_3CH_2CH_2MgCl} \xrightarrow{H_2O}$ (or $CH_3CH_2CHO \xrightarrow{CH_3CH_2MgCl}$

$\xrightarrow{H_2O}$)

(continued next page)

(b) $CH_3CH{=}CH_2 \xrightarrow{CH_3CH(OH)CH_3/Hg(OCOCF_3)_2} \xrightarrow{NaBH_4}$ (less desirable because of competing elimination are (1) $CH_3CH(OH)CH_3 \xrightarrow{H_2SO_4/heat}$ and (2) $CH_3CH(Cl)CH_3 \xrightarrow{(CH_3)_2CHO^{\ominus}Na^{\oplus}}$).

(c) $CH_3CH_2CHO \xrightarrow{CH_3MgI} \xrightarrow{H_2O}$ (or $CH_3CHO \xrightarrow{CH_3CH_2MgBr} \xrightarrow{H_2O}$)

$CH_3CH_2CH(OH)CH_3 \xrightarrow{H_2SO_4/heat} CH_3CH{=}CHCH_3 \xrightarrow[Hg(CF_3CO_2)_2]{CH_3CH_2CH(OH)CH_3}$

(d) $CH_3C{\equiv}CH \xrightarrow[NH_{3(\ell)}]{Na^{\oplus}NH_2^{\ominus}} \xrightarrow{CH_3I} CH_3C{\equiv}CCH_3 \xrightarrow{H_2/\text{Lindlar's cat.}}$

cis-$CH_3CH{=}CHCH_3 \xrightarrow[25^{o}]{MnO_4^{\ominus}/OH^{\ominus}}$ (or $CH_3C{\equiv}CCH_3$ as in first part of (d) $\xrightarrow{Na/NH_{3(\ell)}}$ trans-$CH_3CH{=}CHCH_3 \xrightarrow{CH_3CO_3H}$ $\xrightarrow{OH^{\ominus}/H_2O}$)

(e) $CH_3C{\equiv}CCH_3$ (as in d) $\xrightarrow{H_2/\text{Lindlar's cat.}}$ cis-$CH_3CH{=}CHCH_3$ $\xrightarrow{CH_3CO_2H} \xrightarrow{OH^{\ominus}/H_2O}$ (or $CH_3C{\equiv}CCH_3 \xrightarrow{Na/NH_{3(\ell)}}$ trans-$CH_3CH{=}CHCH_3 \xrightarrow{MnO_4^{\ominus}/OH^{\ominus}}$)

(f) $CH_3CH_2CH_2MgCl \xrightarrow{CH_2O} \xrightarrow{H_2O}$ (or $CH_3CH_2MgCl \xrightarrow{\triangledown_{O}} \xrightarrow{H_2O}$)

$CH_3CH_2CH_2CH_2OH \xrightarrow{PCl_3} \xrightarrow{CH_3CH_2CH_2CH_2O^{\ominus}Na^{\oplus}}$

(continued next page)

(or from $CH_3CH_2CH_2CH_2OH$ $\xrightarrow{H_2SO_4}$)

(g) CH_3CHO $\xrightarrow{CH_3CH_2MgCl}$ $\xrightarrow{H_2O}$ (or CH_3CH_2CHO $\xrightarrow{CH_3MgI}$ $\xrightarrow{H_2O}$)

(or $CH_3C{\equiv}CCH_3$ from d $\xrightarrow{BH_3}$ $\xrightarrow{H_2O_2/OH^{\ominus}}$) (or $CH_3C{\equiv}CCH_3$

$\xrightarrow{H_2O/Hg^{2+}/H^{\oplus}}$) $CH_3\overset{O}{\overset{\|}{C}}CH_2CH_3$ $\xrightarrow{LiAlH_4}$ $\xrightarrow{H_2O}$

(h) cis-$CH_3CH{=}CHCH_3$ as in e $\xrightarrow{CH_3CO_3H}$

(i) trans-$CH_3CH{=}CHCH_3$ as in d $\xrightarrow{CH_3CO_3H}$

(j) (ethylene oxide) $\xrightarrow{CH_3CH_2MgCl}$ $\xrightarrow{H_2O}$ (or CH_2O $\xrightarrow{CH_3CH_2CH_2MgCl}$ $\xrightarrow{H_2O}$)

(k) $OHCCH_2CHO$ $\xrightarrow{2CH_3MgCl}$ $\xrightarrow{H_2O}$

(l) CH_3CH_2CHO $\xrightarrow{CH_3CH_2MgCl}$ $\xrightarrow{H_2O}$

(m) (1) (ethylene oxide) $\xrightarrow{CH_3CH_2MgCl}$ $\xrightarrow{H_2O}$ $CH_3CH_2CH_2CH_2OH$ (or CH_2O

$\xrightarrow{CH_3CH_2CH_2MgCl}$ $\xrightarrow{H_2O}$) $CH_3CH_2CH_2CH_2OH$ $\xrightarrow{CrO_3/pyridine}$

$\xrightarrow{CH_3CH_2CH_2MgCl}$ $\xrightarrow{H_2O}$ (2) $CH_3CH_2O\overset{O}{\overset{\|}{C}}H$ $\xrightarrow{2CH_3CH_2CH_2MgBr}$ $\xrightarrow{H_2O}$

(continued next page)

(n) $CH_3CH{=}CH_2$ $\xrightarrow{BH_3}$ $\xrightarrow[heat]{CO}$ $\xrightarrow{H_2O_2/OH^{\ominus}}$ (or $CH_3CH_2CH_2CHO$

from m $\xrightarrow{CH_3CH_2CH_2MgBr}$ $\xrightarrow{H_2O}$ $CH_3CH_2CH_2CH(OH)CH_2CH_2CH_3$

$\xrightarrow{K_2Cr_2O_7/H^{\oplus}}$ $\xrightarrow{CH_3CH_2CH_2MgBr}$ $\xrightarrow{H_2O}$)

(o) (ethylene oxide) $\xrightarrow{CH_3CH_2CH_2MgCl}$ $\xrightarrow{H_2O}$ $\xrightarrow{CrO_3/pyridine}$ $CH_3CH_2CH_2CH_2CHO$

$\xrightarrow{CH_3CH_2CH_2CH_2MgCl^{*}}$ $\xrightarrow{H_2O}$

*from $CH_3CH_2CH_2CH_2OH$ (as in m) $\xrightarrow{SOCl_2}$ $\xrightarrow{Mg/ether}$

(p) $CH_3CH_2CH_2CH_2OH$ (as in m) $\xrightarrow{H_2SO_4/heat}$ (or $\xrightarrow{PBr_3}$

$\xrightarrow[heat]{KOH/alcohol}$) $CH_3CH_2CH{=}CH_2$ $\xrightarrow{BH_3}$ $\xrightarrow[heat]{CO}$ $\xrightarrow{H_2O_2/OH^{\ominus}}$

8. In some parts, syntheses other than those shown are possible.

(a) CH_3CH_2MgCl $\xrightarrow{CH_2O}$ $\xrightarrow{H_2O}$ (or (ethylene oxide) $\xrightarrow{CH_3MgCl}$ $\xrightarrow{H_2O}$)

(b) CH_3CHO $\xrightarrow{CH_3MgI}$ $\xrightarrow{H_2O}$ (or $CH_3CH_2CH_2OH$ as in a $\xrightarrow{H_2SO_4/heat}$

$\xrightarrow{H^{\oplus}/H_2O}$)

(c) CH_3CHO $\xrightarrow{CH_3CH_2MgBr}$ $\xrightarrow{H_2O}$

(continued next page)

(d) $CH_3CH_2CH_2OH$ (as in a) $\xrightarrow{SOCl_2}$ $CH_3CH_2CH_2Cl$ $\xrightarrow{Mg/ether}$

$\xrightarrow{CH_3CHO}$ $\xrightarrow{H_2O}$

(e) $CH_3CH_2CH_2OH$ (as in a) $\xrightarrow{CrO_3/pyridine}$ CH_3CH_2CHO

$\xrightarrow{CH_3CH_2MgCl}$ $\xrightarrow{H_2O}$

(f) $CH_2{=}CH_2$ $\xrightarrow{BH_3}$ $\xrightarrow[heat]{CO}$ $\xrightarrow{H_2O_2/OH^{\ominus}}$

(g) $CH_3CH_2CH_2Cl$ (as in d) $\xrightarrow{KOH/alcohol/heat}$ $CH_3CH{=}CH_2$ $\xrightarrow{BH_3}$

$\xrightarrow[heat]{CO}$ $\xrightarrow{H_2O_2/OH^{\ominus}}$

(h) ethylene oxide (ring with O) $\xrightarrow{CH_3CH_2MgCl}$ $\xrightarrow{H_2O}$ $CH_3CH_2CH_2CH_2OH$ $\xrightarrow{SOCl_2}$

$\xrightarrow{Mg/ether}$ $\xrightarrow{CH_2O}$ $\xrightarrow{H_2O}$ $CH_3CH_2CH_2CH_2CH_2OH$ $\xrightarrow{CrO_3/pyridine}$

$CH_3CH_2CH_2CH_2CHO$ $\xrightarrow{CH_3CH_2CH_2CH_2MgCl}$ $\xrightarrow{H_2O}$

$CH_3CH_2CH_2CH_2CH(OH)CH_2CH_2CH_2CH_3$ $\xrightarrow{CrO_3/H^{\oplus}}$ $\xrightarrow{CH_3MgI}$ $\xrightarrow{H_2O}$

(i) $CH_3CH(OH)CH_2CH_2CH_3$ (from d) $\xrightarrow{K_2Cr_2O_7/H^{\oplus}}$ $\xrightarrow{CH_3CH_2MgCl}$ $\xrightarrow{H_2O}$

(continued next page)

(j) $CH_3CH(OH)CH_2CH_3$ (from c) $\xrightarrow[\text{heat}]{H_2SO_4}$ $CH_3CH{=}CHCH_3$ $\xrightarrow{Br_2}$

$\xrightarrow[NH_{3(\ell)}]{Na^{\oplus}NH_2^{\ominus}}$ $CH_3C{\equiv}CCH_3$ $\xrightarrow[\text{Lindlar's cat.}]{H_2}$ cis-$CH_3CH{=}CHCH_3$

$\xrightarrow{MnO_4^{\ominus}/OH^{\ominus}/25^{o}}$

(k) $CH_3C{\equiv}CCH_3$ (from j) $\xrightarrow[NH_{3(\ell)}]{Na}$ trans-$CH_3CH{=}CHCH_3$

$\xrightarrow{MnO_4^{\ominus}/OH^{\ominus}/25^{o}}$

9. In some parts, syntheses are possible other than those shown.

(a) CH_3CH_2OH $\xrightarrow{Na}$ $\xrightarrow{\text{(ethylene oxide)}}$ $\xrightarrow{H_2O}$ $CH_3CH_2OCH_2CH_2OH$ $\xrightarrow{SOCl_2}$

(b) $CH_2{=}CHCH_2Br$ $\xrightarrow[\text{ether}]{Mg}$ $\xrightarrow{CH_2{=}CHCH_2Br}$ $\xrightarrow{2Br_2}$

(c) $CH_3\overset{O}{\overset{\|}{C}}CH_3$ $\xrightarrow{CH_3MgI}$ $\xrightarrow{H_2O}$ $\xrightarrow{PBr_3}$

(d) $CH_2{=}CH_2$ $\xrightarrow{BH_3}$ $\xrightarrow[\text{heat}]{CO}$ $\xrightarrow{H_2O_2/OH^{\ominus}}$ $(CH_3CH_2)_3COH$ $\xrightarrow{PBr_3}$

(e) CH_3CH_2CHO $\xrightarrow{CH_3CH_2CH_2MgCl}$ $\xrightarrow{H_2O}$ $\xrightarrow{K_2Cr_2O_7/H^{\oplus}}$

$CH_3CH_2\overset{O}{\overset{\|}{C}}CH_2CH_2CH_3$ $\xrightarrow{CH_3CH_2MgBr}$ $\xrightarrow{H_2O}$ $\xrightarrow{SOCl_2}$

(f) (E)-BrCH=CHBr $\xrightarrow{Li(CH_3CH_2)_2Cu}$

(continued next page)

(g) $\underset{O}{\bigtriangledown}$ $\xrightarrow{CH_3MgCl}$ $\xrightarrow{H_2O}$ (or $CH_3CH{=}CH_2$ $\xrightarrow{BH_3}$ $\xrightarrow{H_2O_2/OH^{\ominus}}$)

(h) $CH_3CH{=}CH_2$ $\xrightarrow{BH_3}$ $\xrightarrow[heat]{CO}$ $\xrightarrow{H_2O_2/OH^{\ominus}}$

(i) $BrCH_2CH{=}CH_2$ $\xrightarrow[ether]{Mg}$ $\xrightarrow{CH_3CH\text{—}CH_2 \text{ (epoxide, O)}}$ $\xrightarrow{H_2O}$

$CH_3CH(OH)CH_2CH_2CH{=}CH_2$ $\xrightarrow{BH_3}$ $\xrightarrow{H_2O_2/OH^{\ominus}}$

(j) CH_3CH_2OH $\xrightarrow{CrO_3 \text{ pyridine}}$ (or CH_3CH_2Cl $\xrightarrow[NaHCO_3]{DMSO}$)

(k) $CH_3CH(OH)CH_3$ $\xrightarrow{CrO_3/H^{\oplus}}$ (or $CH_3CHClCH_3$ $\xrightarrow[NaHCO_3]{DMSO}$)

(l) CH_3CH_2CHO $\xrightarrow{CH_3MgCl}$ $\xrightarrow{H_2O}$ $CH_3CH_2CHOHCH_3$ $\xrightarrow{K_2Cr_2O_7/H^{\oplus}}$

(or (1) $CH_3C{\equiv}C\!:^{\ominus}Na^{\oplus}$ $\xrightarrow{CH_3I}$ $CH_3C{\equiv}CCH_3$ $\xrightarrow{H^{\oplus}/H_2O/Hg^{2+}}$

or (2) $CH_3C{\equiv}CCH_3$ $\xrightarrow[\text{Lindlar's cat.}]{H_2}$ $CH_3CH{=}CHCH_3$ $\xrightarrow{H^{\oplus}/H_2O}$

$\xrightarrow{CrO_3/H^{\oplus}}$

(m) $HOCH_2CH_2OH$ $\xrightarrow{CrO_3/pyridine}$

(n) $CH_2{=}CHCH_2MgBr$ $\xrightarrow{\underset{O}{\bigtriangledown}}$ $\xrightarrow{H_2O}$ $HOCH_2CH_2CH_2CH{=}CH_2$ $\xrightarrow{H^{\oplus}/H_2O}$

$HOCH_2CH_2CH_2\underset{OH}{CH}CH_3$ $\xrightarrow{CrO_3/H^{\oplus}}$

(continued next page)

(o) $CH_3CH{=}CH_2 \xrightarrow{CH_3CH(OH)CH_3/Hg(OCOCF_3)_2} \xrightarrow{NaBH_4}$

(or $CH_3CH(OH)CH_3 \xrightarrow[heat]{H_2SO_4}$)

(p) $CH_3C{\equiv}CMgBr \xrightarrow{CH_2O} \xrightarrow{H_2O}$

10. (a) $CH_2{=}CH_2 \xrightarrow{KMnO_4/OH^{\ominus}/25^{o}}$

(or $CH_2{=}CH_2 \xrightarrow{Cl_2/OH^{\ominus}} CH_2ClCH_2OH \xrightarrow{OH^{\ominus}/H_2O}$)

(b) $CH_2{=}CH{-}CH_2{-}CH{=}CH_2 \xrightarrow{BH_3} \xrightarrow{H_2O_2/OH^{\ominus}}$

(c) $CH_2{=}CHCH_3 \xrightarrow{Br_2/h\nu^{*}} CH_2{=}CHCH_2Br \xrightarrow[dry\ ether]{Mg} \xrightarrow{CH_2CH{=}CH_2Br}$

$CH_2{=}CHCH_2CH_2CH{=}CH_2 \xrightarrow{2H^{\oplus}/H_2O}$

*(or NBS/peroxide)

(d) $CH_3CH_2CH{=}CH_2 \xrightarrow{H^{\oplus}/H_2O}$

(e) $CH_3CH_2CH{=}CH_2 \xrightarrow{Br_2} \xrightarrow[NH_3(\ell)]{NaNH_2} CH_3CH_2C{\equiv}CH \xrightarrow{Na}$

$\xrightarrow{CH_3\overset{O}{\overset{\|}{C}}CH_3{}^{*}} \xrightarrow{H_2O}$

*(from $CH_3CH{=}CH_2 \xrightarrow{H^{\oplus}/H_2O} CH_3CH(OH)CH_3 \xrightarrow{K_2Cr_2O_7/H^{\oplus}}$)

11. (a) $CH_3CH_2Br \xrightarrow[\text{dry ether}]{Mg} \xrightarrow{D_2O}$

(b) $CH_2{=}CH_2 \xrightarrow{D_2/Pt}$

(c) $HC{\equiv}CH \xrightarrow{2D_2/Pt}$

(d) $CH_3OH \xrightarrow{Na} CH_3O^{\ominus}Na^{\oplus} \xrightarrow{D_2O}$

(e) $H{-}\overset{O}{\overset{\|}{C}}{-}H \xrightarrow{LiAlD_4} \xrightarrow{H_2O}$

(f) $CH_2{=}CH_2 \xrightarrow{DCl}$

(g) CH_2DCH_2Cl (from f) $\xrightarrow[\text{ether}]{Mg} \xrightarrow{CH_2O} \xrightarrow{H_2O} CH_2DCH_2CH_2OH$

$\xrightarrow{PCl_3} \xrightarrow[\text{ether}]{Mg} \xrightarrow{D_2O}$

(h) $CH_3\overset{O}{\overset{\|}{C}}CH_3 \xrightarrow{LiAlD_4} \xrightarrow{H_2O}$

(i) $CH_3C{\equiv}CH \xrightarrow{Na} \xrightarrow{D_2O} CH_3C{\equiv}CD \xrightarrow[\text{Lindlar's cat.}]{H_2}$

(j) $CH_2{=}CH_2 \xrightarrow{DCl} \xrightarrow[\text{ether}]{Mg} \xrightarrow{\triangledown_O} \xrightarrow{H_2O} DCH_2CH_2CH_2CH_2OH$

$\xrightarrow{HBr} \xrightarrow[\text{ether}]{Mg} \xrightarrow{D_2O}$

(k) $HC{\equiv}CCH_3 \xrightarrow{HBr} CH_2{=}\underset{Br}{\underset{|}{C}}CH_3 \xrightarrow[\text{ether}]{Mg} \xrightarrow{D_2O}$

(continued next page)

(l) $CH_3C{\equiv}C{:}^{\ominus}\,Na^{\oplus}$ $\xrightarrow{CH_3I}$ $CH_3C{\equiv}CCH_3$ $\xrightarrow[\text{Lindlar's cat.}]{D_2}$

<u>cis</u>-$CH_3CD{=}CDCH_3$ $\xrightarrow{MnO_4^{\ominus}/OH^{\ominus}/25^{o}}$

(m) <u>cis</u>-$CH_3CD{=}CDCH_3$ (from l) $\xrightarrow{CH_3CO_3H}$ $\xrightarrow{OH^{\ominus}/H_2O}$

(n) $CH_2{=}CH_2$ $\xrightarrow{BH_3}$ $\xrightarrow[\text{heat}]{CO}$ $\xrightarrow{H_2O_2/OH^{\ominus}}$ $(CH_3CH_2)_3COH$ $\xrightarrow{HBr}$

$\xrightarrow[\text{ether}]{Mg}$ $\xrightarrow{D_2O}$

(o) CH_2DOH (from e) $\xrightarrow{CrO_3/\text{pyridine}}$ $\underset{D}{\overset{H}{>}}C{=}O$ $\xrightarrow{CH_3CH_2MgCl}$

CH_3CH_2CHDOH $\xrightarrow{CrO_3/\text{pyridine}}$ CH_3CH_2CDO

(C-H bonds are broken more easily then C-D bonds)

(or $CH_3CH_2C\overset{\diagup O}{\underset{\diagdown H}{}}$ $\xrightarrow{LiAlD_4}$ $\xrightarrow{H_2O}$ CH_3CH_2CHDOH $\xrightarrow{CrO_3/\text{pyridine}}$

(p) CH_3CH_2MgCl $\xrightarrow{\text{(ethylene oxide)}}$ $\xrightarrow{H_3O^{\oplus}}$ (alcohol) $\xrightarrow{CrO_3/\text{pyridine}}$

$CH_3CH_2CH_2CHO$ $\xrightarrow{LiAlD_4}$ $\xrightarrow{H_2O}$ $CH_3CH_2CH_2CHDOH$ $\xrightarrow{CrO_3/\text{pyridine}}$

(q) $CH_3CH_2CH_2CDO$ (from p) $\xrightarrow{LiAlD_4}$ $\xrightarrow{H_2O}$

(r) $HC{\equiv}CH$ $\xrightarrow{Na^{\oplus}NH_2^{\ominus}}$ $Na^{\oplus}\,{:}^{\ominus}C{\equiv}CH$ $\xrightarrow{CH_3CH_2Br}$ $CH_3CH_2C{\equiv}CH$ $\xrightarrow{Na^{\oplus}NH_2^{\ominus}}$

$\xrightarrow{D_2O}$ $CH_3CH_2C{\equiv}CD$ $\xrightarrow[\text{Lindlar's cat.}]{D_2}$ $CH_3CH_2CD{=}CD_2$ $\xrightarrow{BD_3}$ $\xrightarrow{H_2O_2/OH^{\ominus}}$

12. (a) $CH_2{=}CH{-}CH_2MgCl \xrightarrow{\text{ethylene oxide (O)}} \xrightarrow{H_2O} CH_2{=}CH{-}CH_2CH_2CH_2OH \xrightarrow{BH_3}$

$\xrightarrow{H_2O_2/OH^{\ominus}} HOCH_2(CH_2)_3CH_2OH \xrightarrow{CrO_3/pyridine}$

(b) $CH_3CH_2CH_2CHO \xrightarrow{CH_3CH_2CH_2MgCl} \xrightarrow{H_2O}$ (alcohol) $\xrightarrow{K_2Cr_2O_7/H^{\oplus}}$

(c) $CH_3CH_2CH_2CH_2MgCl \xrightarrow{CH_3CH_2CHO} \xrightarrow{H_2O} CH_3CH_2CH(OH)CH_2CH_2CH_2CH_3$

$\xrightarrow{CrO_3/H^{\oplus}}$

(d) $CH_2{=}CHCH_2Cl \xrightarrow{Li(CH_2CH_2CH_2CH_3)_2Cu} CH_2{=}CHCH_2CH_2CH_2CH_2CH_3$

$\xrightarrow{H^{\oplus}/H_2O}$ (alcohol) $\xrightarrow{CrO_3/H^{\oplus}}$

(e) $CH_2{=}CH(CH_2)_4CH_3$ (from d) $\xrightarrow{BH_3} \xrightarrow{H_2O_2/OH^{\ominus}}$ (alcohol)

$\xrightarrow{CrO_3/pyridine}$

(f) $(CH_3)_2CHCH_2MgCl \xrightarrow{CH_3CHO} \xrightarrow{H_2O} (CH_3)_2CHCH_2CH(OH)CH_3$

$\xrightarrow{K_2Cr_2O_7/H^{\oplus}}$

(g) $(CH_3)_2CHCH_2MgCl \xrightarrow{CH_3CH_2CHO} \xrightarrow{H_2O} \xrightarrow{K_2Cr_2O_7/H^{\oplus}}$

(h) $(CH_3)_2CHCH_2MgCl \xrightarrow{\text{ethylene oxide (O)}} \xrightarrow{H_2O} \xrightarrow{CrO_3/pyridine}$

(continued next page)

(i) $(CH_3)_2CHCH_2CH_2CH_2OH$ (from h) $\xrightarrow{SOCl_2}$ $\xrightarrow[\text{ether}]{Mg}$ $\xrightarrow{HCOCH_2CH_3 \;(\text{C=O})}$

$((CH_3)_2CHCH_2CH_2CH_2)_2CH_2OH$ $\xrightarrow{CrO_3/H^{\oplus}}$

(j) cyclobutanone $\xrightarrow{CH_3MgBr}$ $\xrightarrow{H_2O}$ 1-methylcyclobutanol (CH_3, OH) $\xrightarrow[\text{heat}]{H_2SO_4}$ 1-methylcyclobutene (CH_3) $\xrightarrow{BH_3}$

$\xrightarrow{H_2O_2/OH^{\ominus}}$ $\xrightarrow{CrO_3/H^{\oplus}}$

(k) $CH_3CH\text{—}CH_2$ (epoxide, O) $\xrightarrow{CH_2=CH\text{-}CH_2MgBr}$ $CH_3CHCH_3CH_3CH=CH_2$ (OH) $\xrightarrow{CrO_3/H^{\oplus}}$

(l) $(CH_3)_3CLi$ $\xrightarrow{CH_2O}$ $\xrightarrow{H_2O}$ $(CH_3)_3CCH_2OH$ $\xrightarrow{CrO_3/\text{pyridine}}$

$(CH_3)_3CCHO$ $\xrightarrow{(CH_3)_3CLi}$ $\xrightarrow{H_2O}$

CHAPTER 13. DIENES AND TERPENES

LEARNING OBJECTIVES

When you have completed this chapter, you should be able to:

1. compare properties of bonds in alkanes, alkenes, alkynes and related compounds (problem 1, 2, 3, 4, 5);
2. account for the products formed in addition reactions of dienes (problems 6, 7, 31 and related problem 33);
3. predict the reactants or products in Diels-Alder reactions (problems 9, 10, 11, 12, 23 and related problems 25, 26, 35);
4. account for products formed in Diels-Alder reactions (problems 28, 29);
5. write equations using reactions from this and previous chapters which illustrate the multistep syntheses of compounds (problems 13, 24);
6. draw the isomers of and name dienes (problem 16);
7. assign absolute configurations to chiral centers in terpenes or predict the number of stereoisomers (problems 14, 15);
8. identify the conjugated systems in structures (problem 17);
9. write the products expected in the addition and oxidation reactions of dienes and simple alkenes (problems 18, 20, 21, 22);
10. write the structures for the intermediate cations formed in addition reactions of dienes and simple alkenes (problem 19 and related problem 21).

ANSWERS TO QUESTIONS

1. Hybridizations of carbons in each carbon-carbon single bond are shown.

CH_3-CH_3 ($2sp^3$-$2sp^3$) 1.53Å

CH_2=CH-CH_3 ($2sp^2$-$2sp^3$) 1.50Å

HC≡C-CH_3 (2sp-$2sp^3$) 1.46Å

HC≡C-CH=CH_2 (2sp-$2sp^2$) 1.43Å

HC≡C-C≡CH (2sp-2sp) 1.37Å

2. The central bond in 1,3-butadiene has about 12% double bond character (see p.506, text). Thus the π bond order is 12% of 1 or 0.12.

3. Bromine addition is a reaction characteristic of alkenes. Since addition occurs at the C9-C10 bond of phenanthrene, this double bond is most like that of a simple alkene. This reaction gives a product containing two aromatic rings, and stability of the product favors addition at the C_9-C_{10} position.

4. Steric replusion between the large *t*-butyl groups prevents coplanarity of the double bonds. Thus, overlap is significantly reduced and each double bond exhibits properties of an isolated double bond.

5. 1,3-Cyclohexadecadiene is large enough to accommodate an *s-trans*-diene structure; 1,3-cyclohexadiene is not large enough to permit this conformation. Models are helpful.

6. Protonation at C2 or C3 leads to the cations below. Neither is resonance stabilized; therefore, each is less stable than the resonance stabilized cations formed by protonation at C1 and C4.

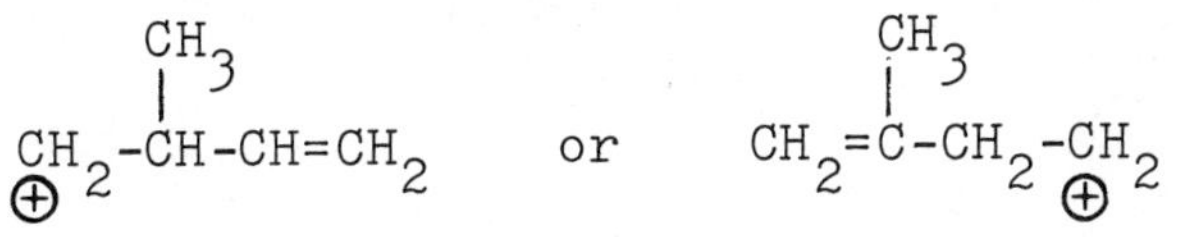

$$\overset{\oplus}{C}H_2\text{-}CH(CH_3)\text{-}CH{=}CH_2 \quad \text{or} \quad CH_2{=}C(CH_3)\text{-}CH_2\text{-}\overset{\oplus}{C}H_2$$

7. 3-Chlorocyclopentene can be formed by 1,2- and 1,4- addition.

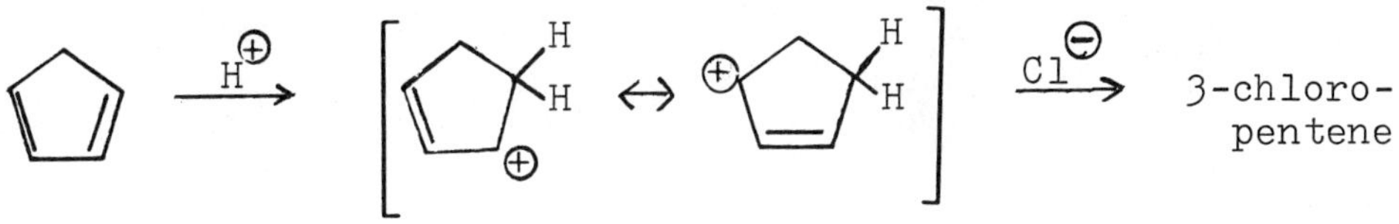

cyclopentadiene $\xrightarrow{H^{\oplus}}$ [allylic cyclopentenyl cation ↔ resonance form] $\xrightarrow{Cl^{\ominus}}$ 3-chloro-pentene

1,2- and 1,4- addition can be distinguished using DCl. The products are:

(3-chloro-4-deuteriocyclopentene) + (3-chloro-5-deuteriocyclopentene)

8. (a) 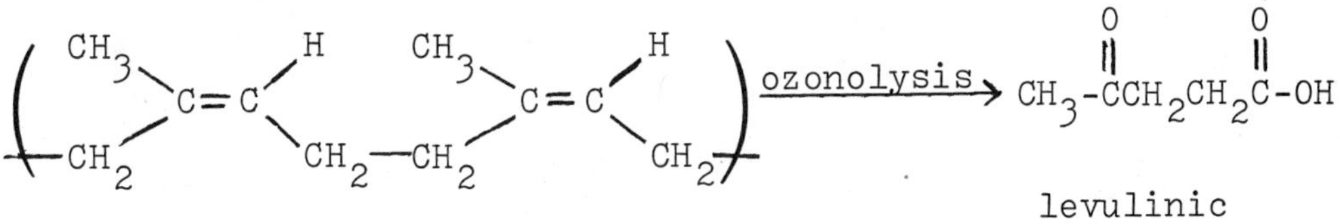 $\left(\text{—}CH_2\text{—}C(CH_3){=}CH\text{—}CH_2\text{—}CH_2\text{—}C(CH_3){=}CH\text{—}CH_2\text{—} \right) \xrightarrow{\text{ozonolysis}} CH_3\text{-}\overset{O}{\overset{\|}{C}}CH_2CH_2\overset{O}{\overset{\|}{C}}\text{-}OH$

natural rubber segment — levulinic acid

(continued next page)

(b) $CH_3\overset{O}{\overset{\|}{C}}CH_2CH_2COOH \xrightarrow{NaBH_4} CH_3\overset{OH}{\overset{|}{C}}HCH_2CH_2COOH$

(c) $CH_3\text{-}\overset{O}{\overset{\|}{C}}CH_2CH_2\overset{O}{\overset{\|}{C}}\text{-}OH \xrightarrow[\text{(excess)}]{CH_3MgCl} CH_4 + CH_3\overset{O}{\overset{\|}{C}}CH_2CH_2\overset{O}{\overset{\|}{C}}\text{-}OMgCl$

1 mole of CH_4 formed per mole of levulinic acid

9. Recall that the product in a Diels-Alder reaction can be conveniently constructed by the simple bond changes shown below.

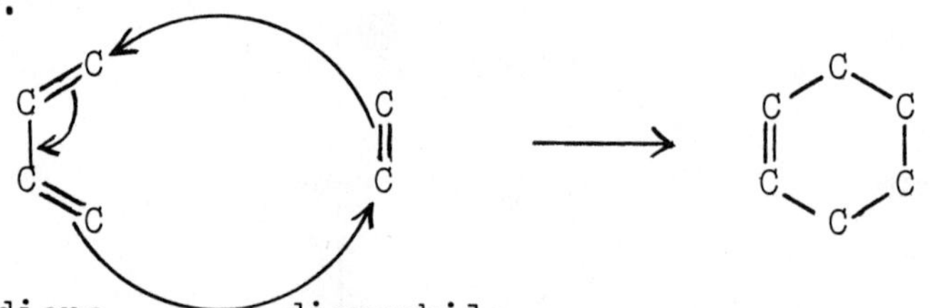

diene dienophile

diene	dienophile	product
(a) $CH_2\text{=}CH\text{-}CH\text{=}CH_2$	$CH_2\text{=}CHCl$	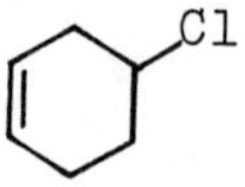 Cl
(b) $CH_2\text{=}\overset{CH_3}{\overset{\mid}{C}}\text{-}CH\text{=}CH_2$		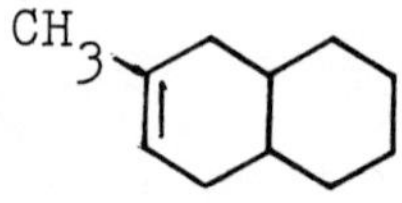 CH_3
(c) 	$CH_2\text{=}CHCCl_3$	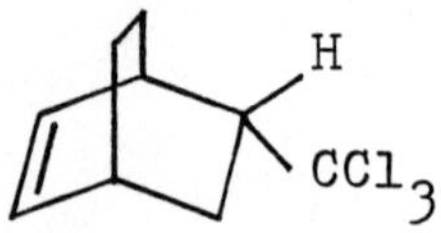 H, CCl_3

(continued next page)

	diene	dienophile	product
(d)		Cl, Cl	Cl, Cl; Cl, Cl
(e)	Cl Cl, Cl, Cl, Cl, Cl	$(CH_3)_2C{=}C(CH_3)_2$	Cl Cl, Cl, CH_3, Cl, CH_3, CH_3, Cl, Cl, CH_3
(f)		$CH_2{=}CH\text{-}CHO$	H, CHO

10. (a) Cl Cl, Cl, H, Cl, CH_3, Cl, Cl

(c) CCl, Cl, Cl

(b) Cl Cl, Cl, Cl, CH_3, CH_3, Cl, Cl

(d) CO_2CH_3, CO_2CH_3

11. (a) 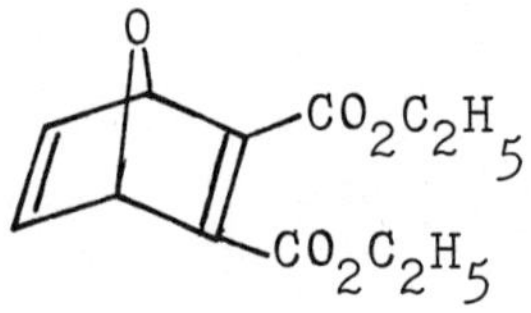 O, $CO_2C_2H_5$, $CO_2C_2H_5$

(c) 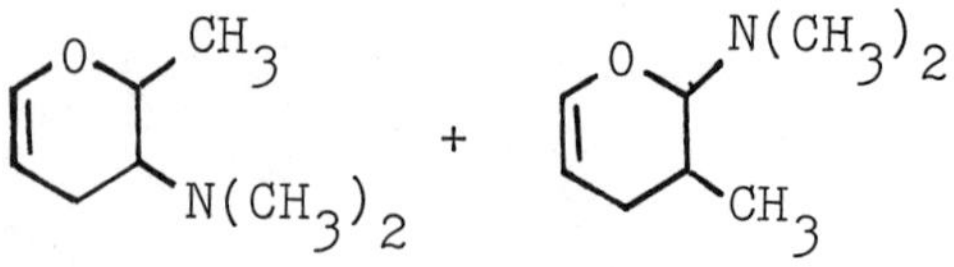 O, CH_3, $N(CH_3)_2$ + O, $N(CH_3)_2$, CH_3

(cis)

(b) 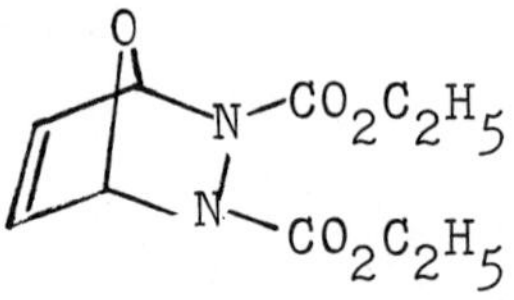 O, $N-CO_2C_2H_5$, $N-CO_2C_2H_5$

12. (a) 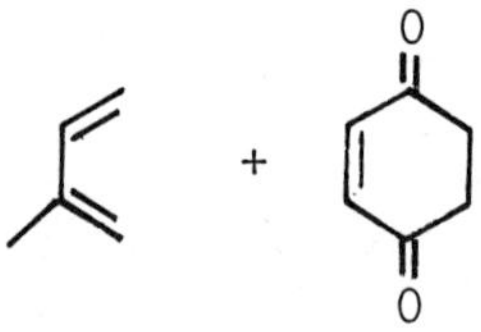 + O, O

(b) 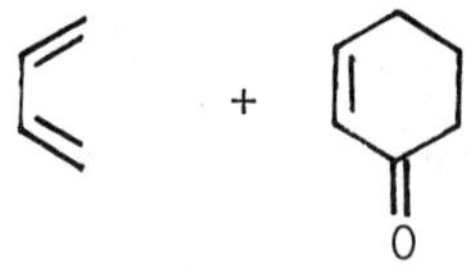 + O

(c) 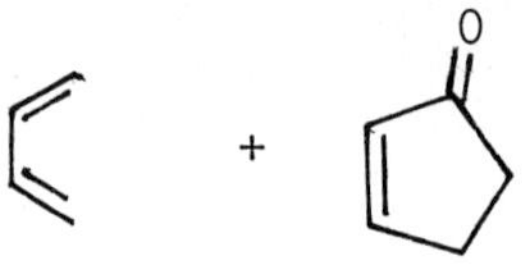 + O

(d) 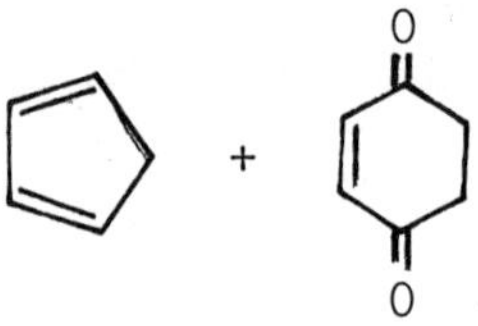 + O, O

(e) 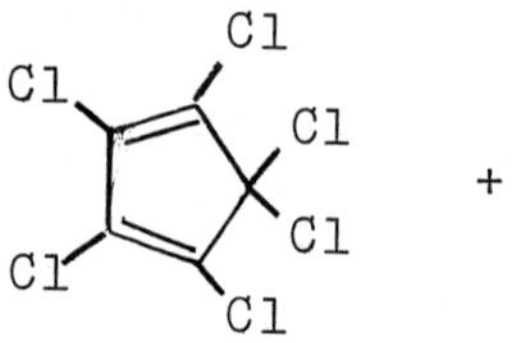 Cl, Cl, Cl, Cl, Cl, Cl + HC≡CH

13. (a) 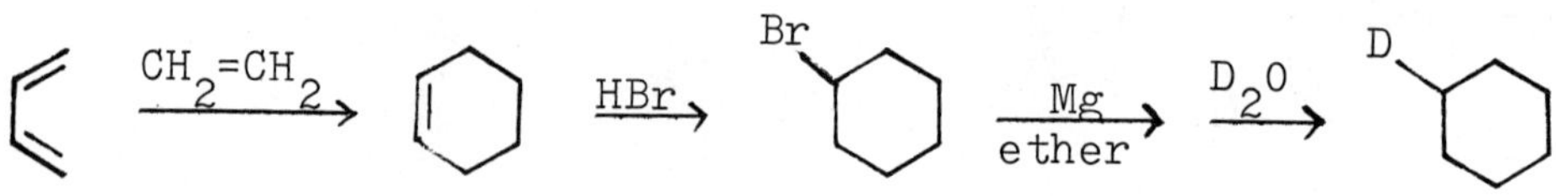 $\xrightarrow{CH_2=CH_2}$ $\xrightarrow{HBr}$ Br $\xrightarrow[\text{ether}]{Mg}$ $\xrightarrow{D_2O}$ D

(or from $\xrightarrow{BH_3}$ $\xrightarrow{CH_3COOD}$)

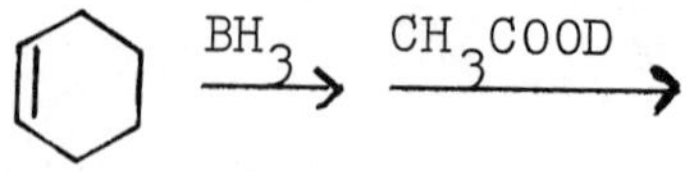

(continued next page)

(b) CH_3-diene $\xrightarrow{CH_2=CH_2}$ CH_3-cyclohexene $\xrightarrow{H_2/Pt}$ CH_3-cyclohexane

(or begin with diene $\xrightarrow{CH_2=CHCH_3}$)

(c) CH_3-cyclohexene (from b) $\xrightarrow{HCl}$ CH_3, Cl-cyclohexane

(d) CH_3-cyclohexene (from b) $\xrightarrow[\text{peroxide}]{HBr}$ CH_3, Br-cyclohexane

(e) CH_3, Br-cyclohexane (from d) $\xrightarrow{H_2O}$ CH_3, OH-cyclohexane $\xrightarrow{CrO_3/H^{\oplus}}$ CH_3-cyclohexanone (O)

(or CH_3-cyclohexene (from b) $\xrightarrow{BH_3}$ $\xrightarrow[OH^{\ominus}]{H_2O_2}$ CH_3, OH-cyclohexane $\xrightarrow{CrO_3/\ H^{\oplus}}$)

(f) cyclohexene (from a) $\xrightarrow{CH_3CO_3H}$ cyclohexene oxide (O)

(or cyclohexene $\xrightarrow{Cl_2/H_2O}$ Cl, OH-cyclohexane $\xrightarrow{Ca(OH)_2}$)

(g) cyclohexene (from a) $\xrightarrow{CH_3OH/Hg(OCCF_3)_2}$ (C=O) $\xrightarrow{NaBH_4}$ OCH_3-cyclohexane

(continued next page)

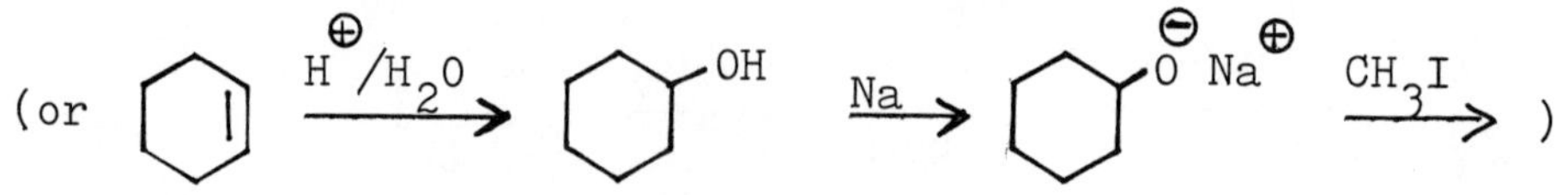

(or cyclohexene $\xrightarrow{H^{\oplus}/H_2O}$ cyclohexanol (OH) $\xrightarrow{Na}$ cyclohexyl $O^{\ominus} Na^{\oplus}$ $\xrightarrow{CH_3I}$)

14. (a) vitamin E has 3 chiral centers and 8 stereoisomers are possible.

(b) vitamin K has 2 chiral centers and can exist in <u>cis</u> and <u>trans</u> forms; thus, 8 stereoisomers are possible.

15. (a) Ecdysone is a steroid. (b) Chiral centers are starred below.

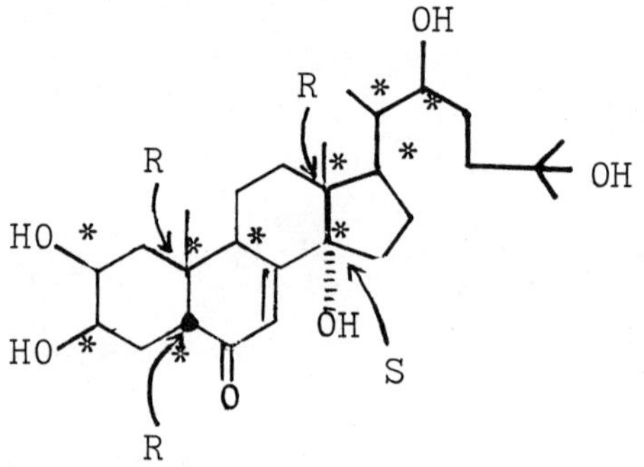

OH, R, R, HO, HO, OH, O, OH, S, R

(c) Catalytic hydrogenation (of C=C and C=O) creates two new chiral centers and four diastereomeric products.

(d) These compounds are stereoisomers.

16. (a) 1,2-pentadiene CH_3CH_2–CH=C=CH_2 (sp^3 carbons: CH_3CH_2; sp^2: terminal C=C carbons; sp: central C)

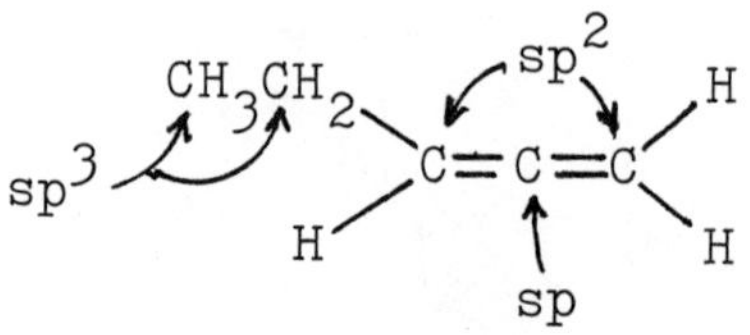

(b) 1,3-pentadiene CH_3–CH=CH–CH=CH_2 (sp^3: CH_3; sp^2: C=C carbons) and CH_3–CH=CH–CH=CH_2

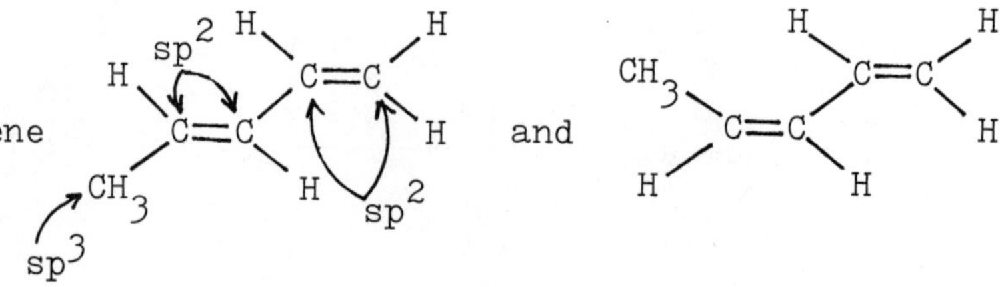

(continued next page)

(c) 1,4-pentadiene

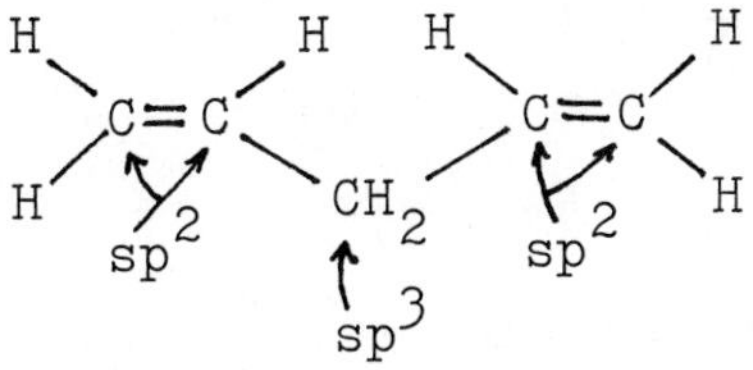

$H_2C{=}CH{-}CH_2{-}CH{=}CH_2$ (C=C carbons: sp^2; CH_2: sp^3)

(d) 2,3-pentadiene

$CH_3{-}CH{=}C{=}CH{-}CH_3$ (terminal C=C carbons: sp^2; central C: sp; CH_3: sp^3)

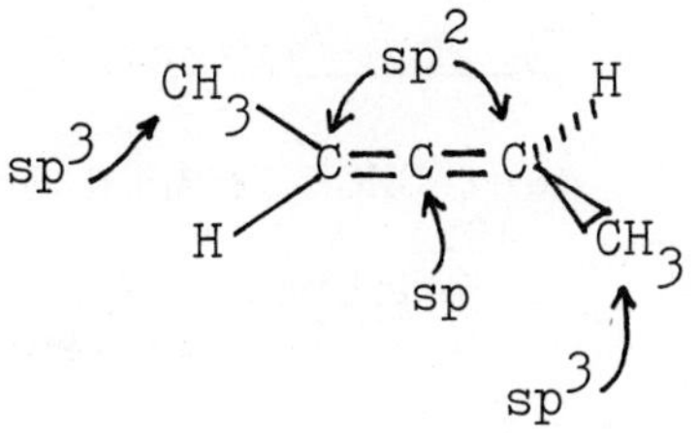

(R and S)

(e) 3-methyl-1,2-butadiene

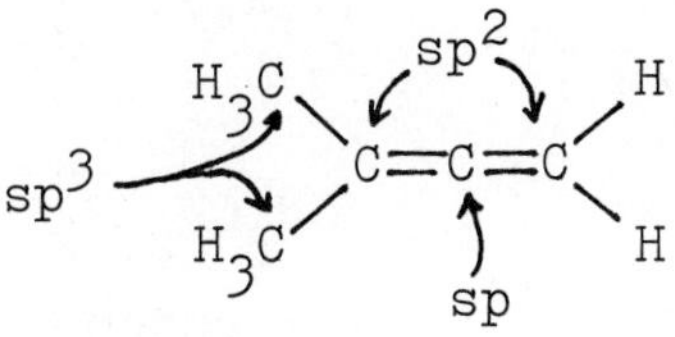

$(H_3C)_2C{=}C{=}CH_2$ (CH_3: sp^3; terminal C: sp^2; central C: sp)

(f) 2-methyl-1,3-butadiene

$H_2C{=}C(CH_3){-}CH{=}CH_2$ (C=C carbons: sp^2; CH_3: sp^3)

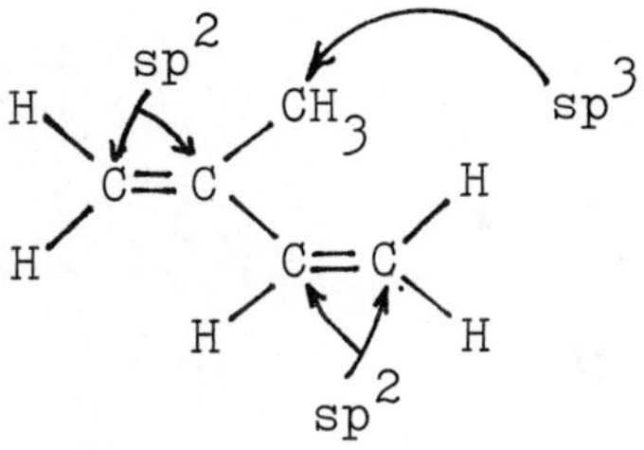

17. Conjugated systems are enclosed.

(a) 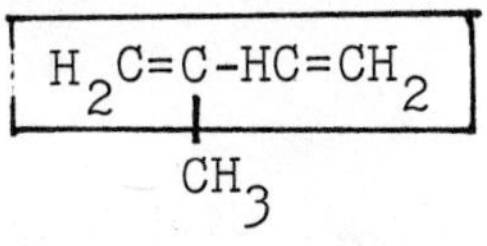 [$H_2C{=}C(CH_3){-}HC{=}CH_2$]

(b) $H_2C{=}CH{-}CH_2$ [$HC{=}CH{-}HC{=}CH$] CH_3

(c) CH_3 [$\ddot{O}{-}HC{=}C$] $(CH_3)_2$

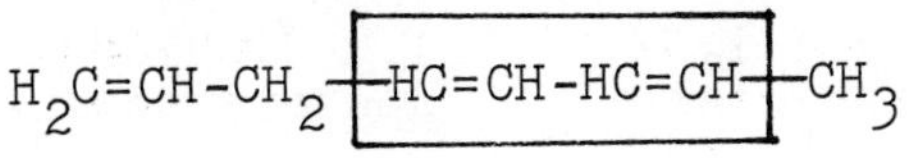

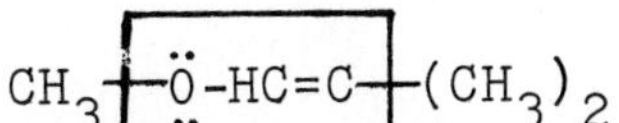

(d) 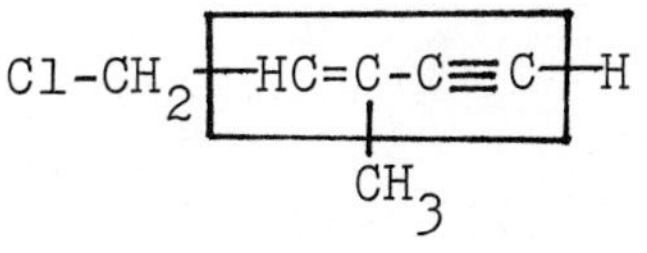 $Cl{-}CH_2$ [$HC{=}C(CH_3){-}C{\equiv}C$] H

(e) 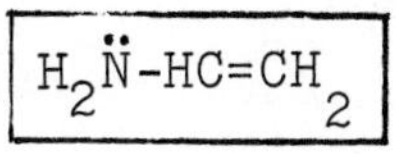 [$H_2\ddot{N}{-}HC{=}CH_2$]

(f) 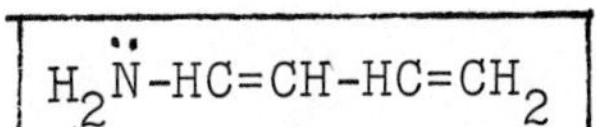 [$H_2\ddot{N}{-}HC{=}CH{-}HC{=}CH_2$]

(continued next page)

(g) $H_3\overset{\oplus}{N}$—[$HC=CH-HC=CH_2$]

(Text S. 13.1)

18. Only the major products expected are given.

	1,4-pentadiene	1,3-butadiene
(a) Br_2/CCl_4 (1 mole, 40°)	$CH_2=CHCH_2CHBrCH_2Br$ 4,5-dibromo-1-pentene	$CH_2BrCH=CHCH_2Br$ (trans > cis) 1,4-dibromo-2-butene
(b) Br_2/CCl_4 (excess)	$CH_2BrCHBrCH_2CHBrCH_2Br$ 1,2,4,5-tetrabromo-pentane	$CH_2BrCHBrCHBrCH_2Br$ 1,2,3,4-tetrabromo-butane
(c) H_2/Pt (1 mole)	$CH_3CH_2CH_2CH=CH_2$ 1-pentene	$CH_3CH_2CH=CH_2$ 1-butene
(d) D_2/Pt (1 mole)	$CH_2DCHDCH_2CH=CH_2$ 4,5-dideuterio-1-pentene	$CH_2DCHDCH=CH_2$ 3,4-dideuterio-1-butene
(e) H_2/Pt (excess)	$CH_3CH_2CH_2CH_2CH_3$ pentane	$CH_3CH_2CH_2CH_3$ butane
(f) DCl (1 mole, $+40^o$)	$CH_2DCHClCH_2CH=CH_2$ 4-chloro-5-deuterio-1-pentene	$CH_2DCH=CHCH_2Cl$ (cis and trans) 1-chloro-4-deuterio-2-butene
(g) DCl (excess)	$CH_2DCHClCH_2CHClCH_2D$ 2,4-dichloro-1,5-dideuteriopentane	$CH_2DCHClCHDCH_2Cl$ 1,3-dichloro-2,4-di-deuteriobutane

(continued next page)

(h)	excess CH_3OH/ $Hg(F_3CCO_2^{\ominus})_2$, then $BH_4^{\ominus}$	$CH_3\underset{OCH_3}{\underset{\vert}{C}}HCH_2\underset{OCH_3}{\underset{\vert}{C}}HCH_3$ 2,4-dimethoxypentane	$CH_3CH(OCH_3)CH_2CH_2OCH_3$ 1,3-dimethoxybutane
(i)	O_3 (excess) (reductive work up)	$H\overset{O}{\overset{\|}{C}}H + H\overset{O}{\overset{\|}{C}}CH_2\overset{O}{\overset{\|}{C}}H$	$H\overset{O}{\overset{\|}{C}}H + H\overset{O}{\overset{\|}{C}}-\overset{O}{\overset{\|}{C}}H$
(j)	O_3 (excess) (oxidative work up)	$CO_2 + HO\overset{O}{\overset{\|}{C}}CH_2\overset{O}{\overset{\|}{C}}OH$	$CO_2 + HO\overset{O}{\overset{\|}{C}}-\overset{O}{\overset{\|}{C}}OH$
(k)	BH_3 (excess) then $CH_3CH_2CO_2D$	$DCH_2CH_2CH_2CH_2CH_2D$ 1,5-dideuterio-pentane	$DCH_2CH_2CH_2CH_2D$ 1,4-dideuterio-butane

19. The two cations being compared in this problem are the cation formed from bromine and an <u>isolated</u> double bond (S. 8.6) and the cation from bromine and a <u>conjugated</u> double bond. The first produces a cyclic bromonium ion and the second a resonance-stabilized cation.

(a) $CH_3CH_2CH_2CH_2\overset{\delta^+}{C}H$——$CH_2$, with the CH and CH_2 bridged by $\underset{\delta^+}{Br}$ (dotted bond to CH, solid bond to CH_2)

(b) $\left[\underset{Br}{\underset{\vert}{C}}H_2-\overset{\oplus}{C}H-CH{=}CH-CH_2CH_3 \longleftrightarrow \underset{Br}{\underset{\vert}{C}}H_2-CH{=}CH-\overset{\oplus}{C}H-CH_2CH_3\right]$

(continued next page)

(c) 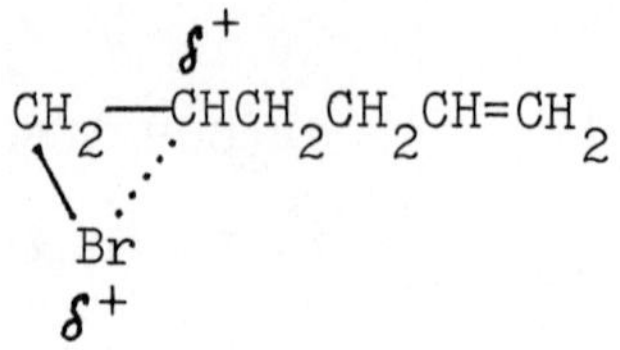 $\overset{\delta+}{CH_2 - CHCH_2CH_2CH=CH_2}$ bridged by $\overset{\delta+}{Br}$ (solid bond to CH_2, dotted bond to CH)

(d) 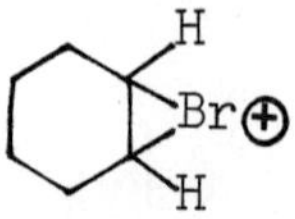 cyclohexane ring fused to bromonium ion: H, $Br^{\oplus}$, H

(e) 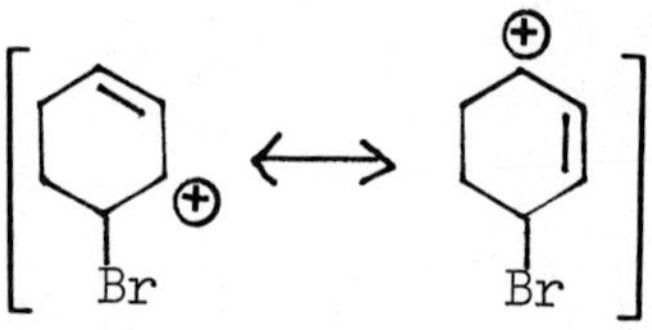 $\left[\text{3-bromocyclohexenyl cation, } \oplus \text{ adjacent to C-Br} \longleftrightarrow \text{4-bromocyclohexenyl cation, } \oplus \text{ at allylic position} \right]$ (Br)

(Text S. 8.6)

20. (a) $CH_3CH_2CH_2\overset{O}{\overset{\|}{C}}H + H\overset{O}{\overset{\|}{C}}H$

(b) $CH_3CH_2\overset{O}{\overset{\|}{C}}H + CH_3\overset{O}{\overset{\|}{C}}H$

(c) $CH_3CH_2CH_2\overset{O}{\overset{\|}{C}}OH + CO_2$

(d) $H\overset{O}{\overset{\|}{C}}H + H\overset{O}{\overset{\|}{C}}-\overset{O}{\overset{\|}{C}}H$

(e) $H\overset{O}{\overset{\|}{C}}H + H\overset{O}{\overset{\|}{C}}-\overset{O}{\overset{\|}{C}}CH_3$

(f) $H\overset{O}{\overset{\|}{C}}CH_2CH_2\overset{O}{\overset{\|}{C}}-\overset{O}{\overset{\|}{C}}H + CH_3\overset{O}{\overset{\|}{C}}CH_3 + H\overset{O}{\overset{\|}{C}}H$

(g) $H\overset{O}{\overset{\|}{C}}H + H\overset{O}{\overset{\|}{C}}\overset{O}{\overset{\|}{C}}CH_3 + H\overset{O}{\overset{\|}{C}}CH_2CH_2\overset{O}{\overset{\|}{C}}CH_3$

(h) 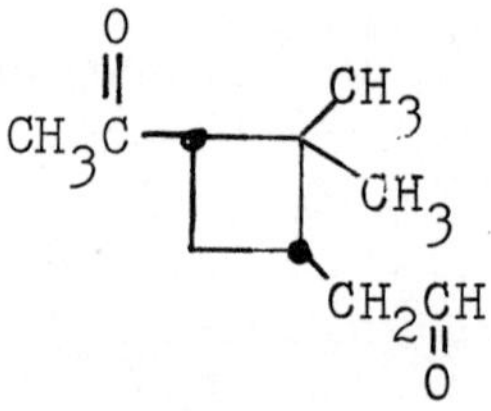 cyclobutane with $CH_3\overset{O}{\overset{\|}{C}}-$, gem-dimethyl ($CH_3$, CH_3), and $CH_2\underset{O}{\underset{\|}{C}}H$ substituents

(i) $CH_3\overset{O}{\overset{\|}{C}}CH_3 + H\overset{O}{\overset{\|}{C}}CH_2CH_2\overset{O}{\overset{\|}{C}}CH_3 + H\overset{O}{\overset{\|}{C}}CH_2OH$

(j) $CH_3\overset{O}{\overset{\|}{C}}CH_3 + H\overset{O}{\overset{\|}{C}}CH_2CH_2\overset{O}{\overset{\|}{C}}CH_3 + H\overset{O}{\overset{\|}{C}}CH_2OH$

21. (a) $CH_3C(CH_3)(Cl)-CH=CH_2$ and $CH_3C(CH_3)=CH-CH_2Cl$

(b) $CH_2BrCBr(CH_3)-CH=CH_2$ and $CH_2BrC(CH_3)=CHCH_2Br$

In (a) and (b) products are derived from the carbocation described by the resonance structures below.

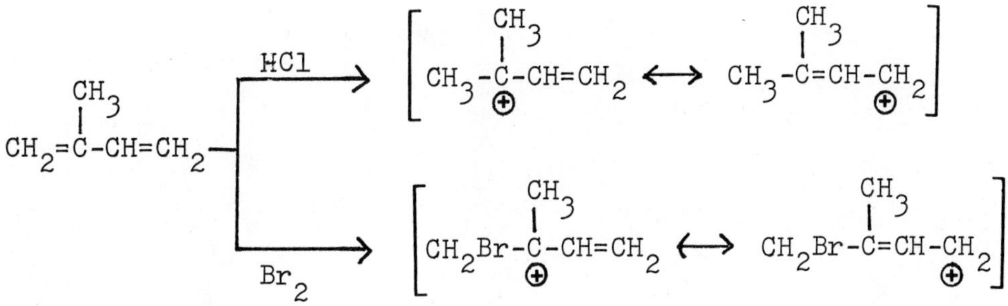

$$CH_2=C(CH_3)-CH=CH_2 \xrightarrow{HCl} \left[CH_3-\overset{\oplus}{C}(CH_3)-CH=CH_2 \longleftrightarrow CH_3-C(CH_3)=CH-\overset{\oplus}{C}H_2 \right]$$

$$CH_2=C(CH_3)-CH=CH_2 \xrightarrow{Br_2} \left[CH_2Br-\overset{\oplus}{C}(CH_3)-CH=CH_2 \longleftrightarrow CH_2Br-C(CH_3)=CH-\overset{\oplus}{C}H_2 \right]$$

These cations are more stable than those derived from addition of $H^{\oplus}$ or $Br^{\oplus}$ to other carbons of the carbon-carbon double bonds.

22. (a) $CH_3CH_2CHBrCH_2CCl_3$

(b) $BrCH_2CH=CHCH_2CCl_3$

(c) $CH_3CHBrCH=CHCH_2CCl_3$

(d) $CH_2=CH-CH_2CHBrCH_2CCl_3$

(e) $Cl_3CCH_2C(CH_3)=CHCH_2Br$

23. (a) $CH_2=CH-CH=CH_2$ + [p-benzoquinone: ring with O at top and bottom]

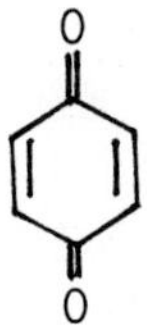

(b) $CH_2=CH-CH=CH_2$ + $CH_2=CH-CH=CH_2$

(continued next page)

(c) $CH_2{=}CH{-}CH{=}CH_2$ + [2-methyl-1,4-benzoquinone structure]

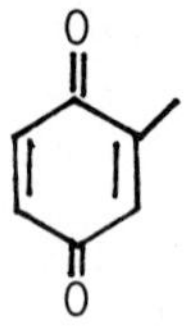

(d) 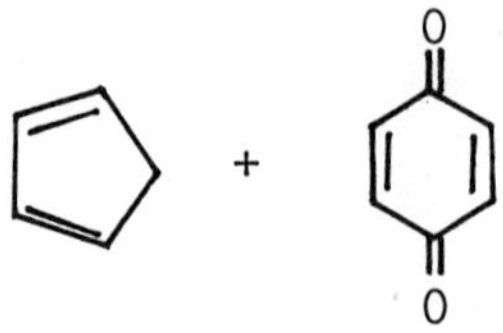 [cyclopentene structure] + [1,4-benzoquinone structure]

(e)  [cyclopentadiene structure] + $C_2H_5CH{=}CH{-}CH{=}O$ (<u>cis</u>)

(f) $CH_2{=}CH{-}CH{=}O$ + $CH_2{=}CH{-}OCH_3$

(g) 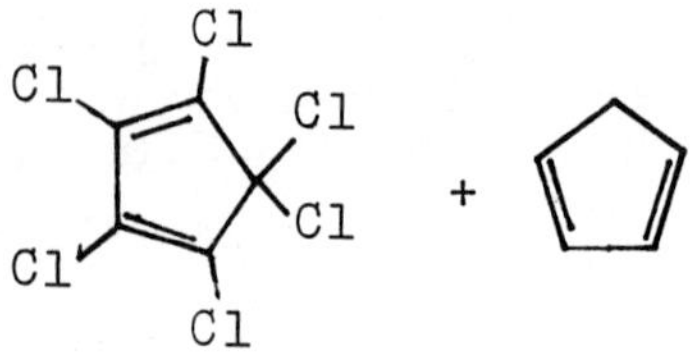 [hexachlorocyclopentadiene structure: Cl, Cl, Cl, Cl, Cl, Cl] + [cyclopentene structure]

(h) $CH_3\overset{O}{\overset{\|}{C}}OCH{=}CH{-}CH{=}CHO\overset{O}{\overset{\|}{C}}CH_3$ + $CH_2{=}CH\overset{O}{\overset{\|}{C}}OCH_3$

(i) $CH_3\overset{O}{\overset{\|}{C}}OCH{=}CH{-}CH{=}CHO\overset{O}{\overset{\|}{C}}CH_3$ + [vinylene carbonate structure: O, O, =O]

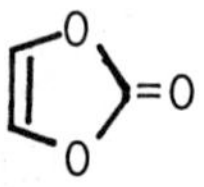

(j) $CH_2{=}\underset{CH_3}{\underset{|}{C}}{-}\underset{CH_3}{\underset{|}{C}}{=}CH_2$ + $(CH_3)_2\overset{Cl}{\overset{|}{C}}{-}N{=}O$

(k) $CH_2{=}\underset{OCH_3}{\underset{|}{C}}{-}CH{=}CH_2$ + [maleic anhydride structure: O, O, O]

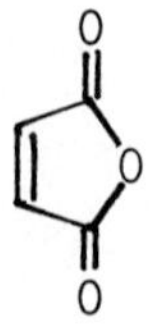

(continued next page)

(l) [benzyne structure] + [furan structure]

(benzyne)

(m) $CH_2=C(CH_3)-CH=CHCH_3$ + $H_2C=O$

(n) [2,3,4-triphenylcyclopentadienone: C_6H_5, C_6H_5, C_6H_5 substituents, =O] + $C_6H_5C\equiv CC_6H_5$

(o) [1,3-cyclohexadiene structure] + NCCH=CHCN

(*cis*)

24. (a) $CH_2=CH_2$ $\xrightarrow{CH_2=CH-CH=CH_2}$ [cyclohexene] $\xrightarrow{D_2/Pt}$ [cis-1,2-dideuteriocyclohexane, D, D]

(b) [cyclohexene] (as in a) $\xrightarrow{HCl}$ [chlorocyclohexane, Cl]

(c) [chlorocyclohexane, Cl] (as in b) $\xrightarrow[ether]{Mg}$ $\xrightarrow{H_2C=O}$ $\xrightarrow{H_2O}$ [cyclohexylmethanol, CH_2OH]

(continued next page)

(d) 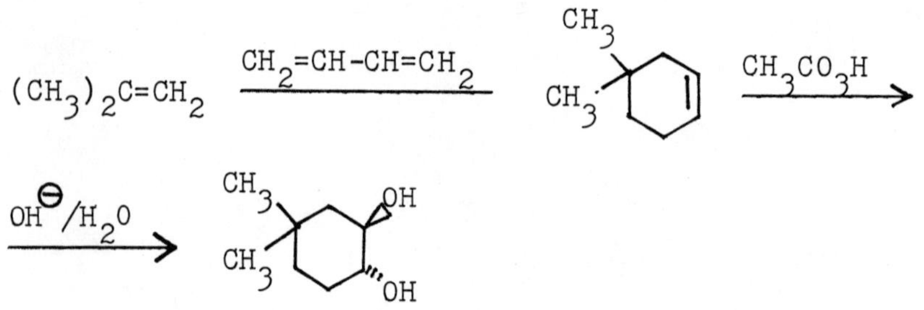 $(CH_3)_2C=CH_2$ $\xrightarrow{CH_2=CH-CH=CH_2}$ (4,4-dimethylcyclohexene, CH_3, CH_3) $\xrightarrow{CH_3CO_3H}$

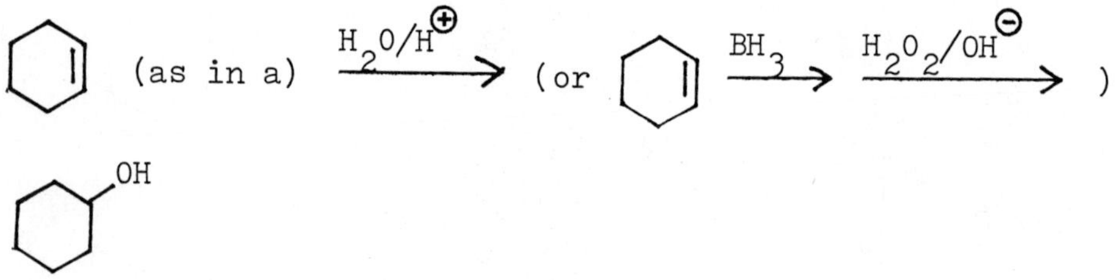

$\xrightarrow{OH^{\ominus}/H_2O}$ (4,4-dimethyl-trans-1,2-cyclohexanediol: CH_3, CH_3, OH, OH)

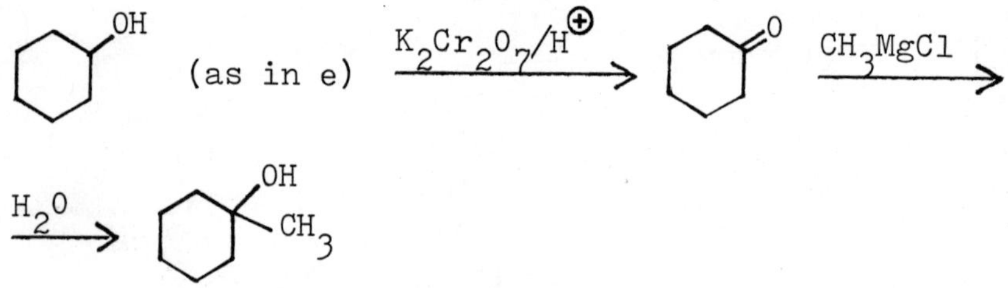

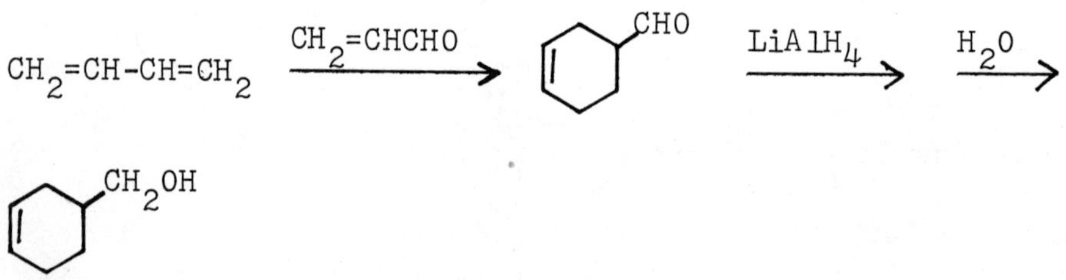

(e) (cyclohexene) (as in a) $\xrightarrow{H_2O/H^{\oplus}}$ (or (cyclohexene) $\xrightarrow{BH_3}$ $\xrightarrow{H_2O_2/OH^{\ominus}}$)

(cyclohexanol, OH)

(f) (cyclohexanol, OH) (as in e) $\xrightarrow{K_2Cr_2O_7/H^{\oplus}}$ (cyclohexanone, O) $\xrightarrow{CH_3MgCl}$

$\xrightarrow{H_2O}$ (1-methylcyclohexanol, OH, CH_3)

(g) $CH_2=CH-CH=CH_2$ $\xrightarrow{CH_2=CHCHO}$ (cyclohexene-CHO) $\xrightarrow{LiAlH_4}$ $\xrightarrow{H_2O}$

(cyclohexene-CH_2OH)

(h) (cyclohexene-CHO) (from g) $\xrightarrow{CH_3MgI}$ $\xrightarrow{H_2O}$ (cyclohexene-$CH(OH)CH_3$)

(or $CH_2=CH-CH=CH_2$ $\xrightarrow{CH_2=CH-\overset{O}{\overset{\|}{C}}CH_3}$ (cyclohexene-$\overset{O}{\overset{\|}{C}}CH_3$) $\xrightarrow[H_2O,\ CH_3OH]{NaBH_4}$)

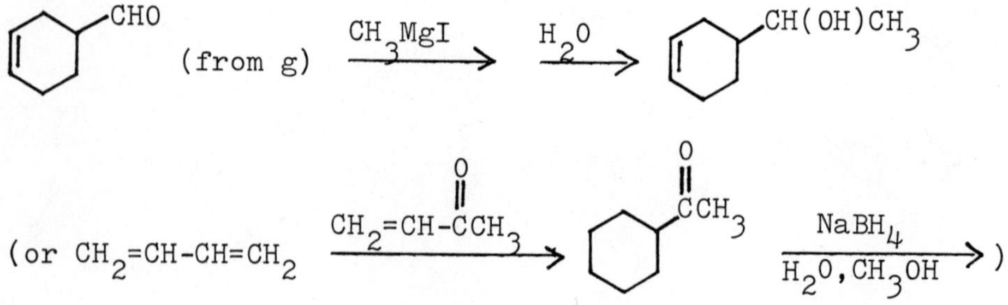

(continued next page)

(i) 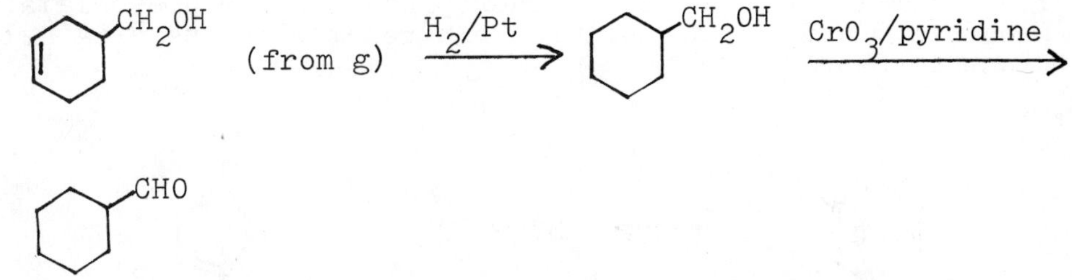 [cyclohexene ring]–CH_2OH (from g) $\xrightarrow{H_2/Pt}$ [cyclohexane ring]–CH_2OH $\xrightarrow{CrO_3/pyridine}$

[cyclohexane ring]–CHO

(j) CH_2=CH-CH=CH_2 $\xrightarrow{CH_2=CHCH_3}$ [cyclohexene ring]–CH_3 $\xrightarrow{H_2/Ni}$ [cyclohexane ring]–CH_3

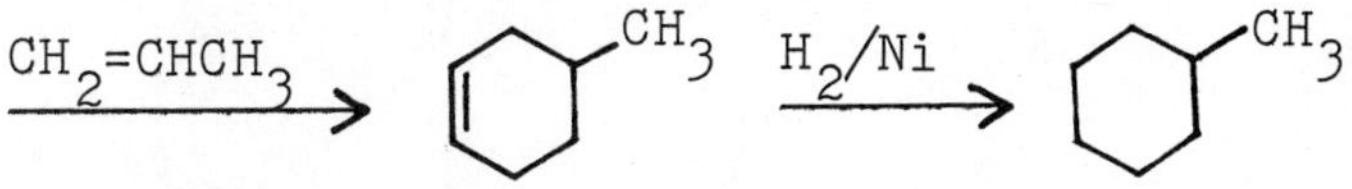

25. (e) 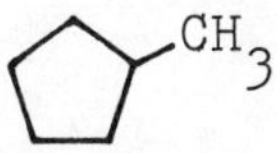 [cyclopentane ring]–CH_3 is not synthesized by Diels-Alder. The reaction between a diene and dienophile gives a six-membered ring, not a five-membered ring.

26. It is more difficult for cis-piperylene to adopt the necessary planar geometry because of greater steric hindrance in the s-cis conformation.

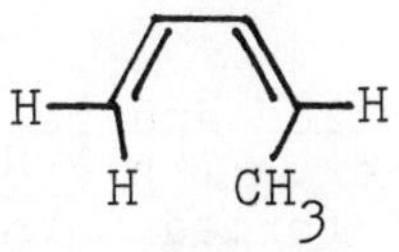

H, H, H, CH_3 — cis

H, CH_3, H, H — trans

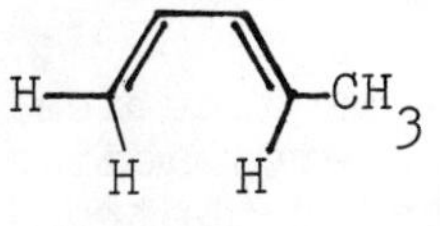

27. The diene can bond to either face of the dienophile giving:

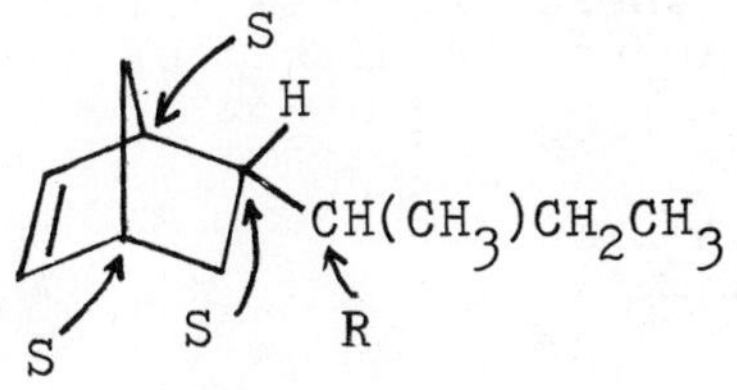

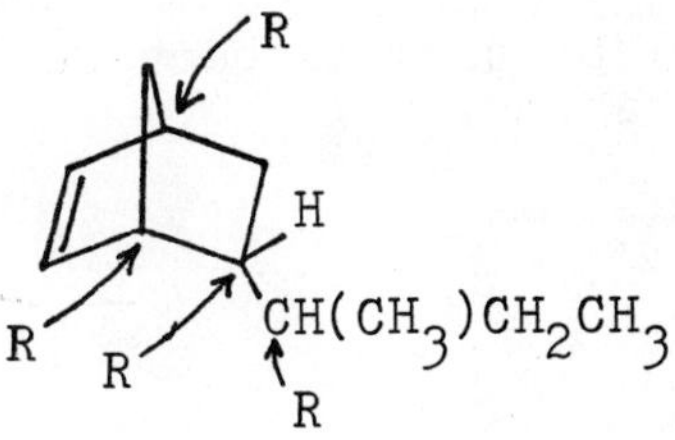

S, H, $CH(CH_3)CH_2CH_3$, S, S, R

R, H, $CH(CH_3)CH_2CH_3$, R, R, R

28. The initially formed Diels-Alder product A can dissociate to products B and C. B, in turn, can react with the original diene as shown below.

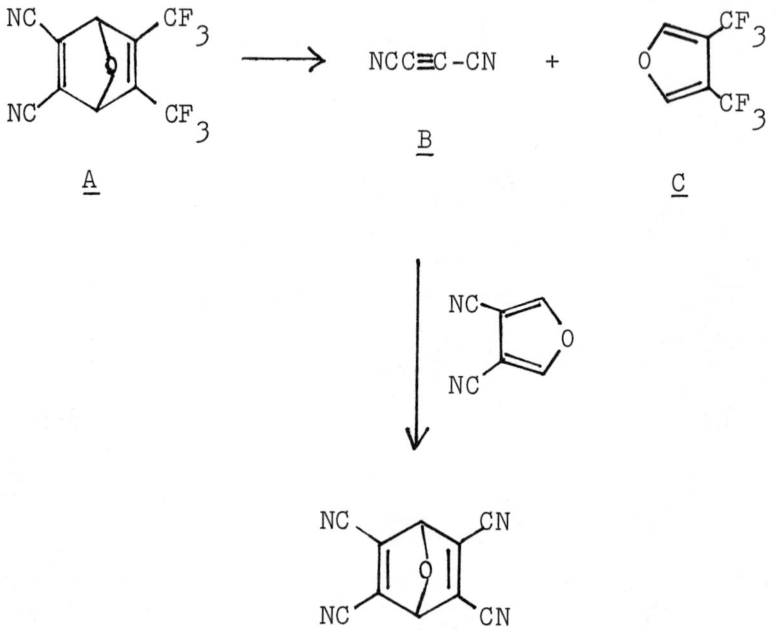

NC, CF$_3$, NC, CF$_3$ (A) ⟶ NCC≡C-CN (B) + O, CF$_3$, CF$_3$ (C)

↓ NC, NC, O

NC, CN, NC, CN

29. (a) The endo product is kinetically controlled and is formed through a lower energy activated complex. The exo product is thermodynamically controlled and is formed through a higher energy activated complex.

(b) See if the endo product formed at lower temperatures rearranges to the exo product at higher temperature.

30. (a) Formation of a conjugated diene requires initial removal of one of the "a" protons. These are less acidic than the "b" protons because of the inductive effect of the bromine atoms.

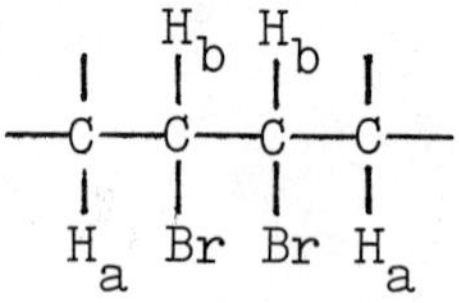

H_b H_b
—C—C—C—C—
H_a Br Br H_a

(continued next page)

(b) Angle strain does not permit formation of [cyclohexyne structure] from 1,2-dibromocyclohexane.

2,3-Dichloro-2,3-dimethylbutane lacks hydrogens on the carbons holding the chlorine atoms.

31.

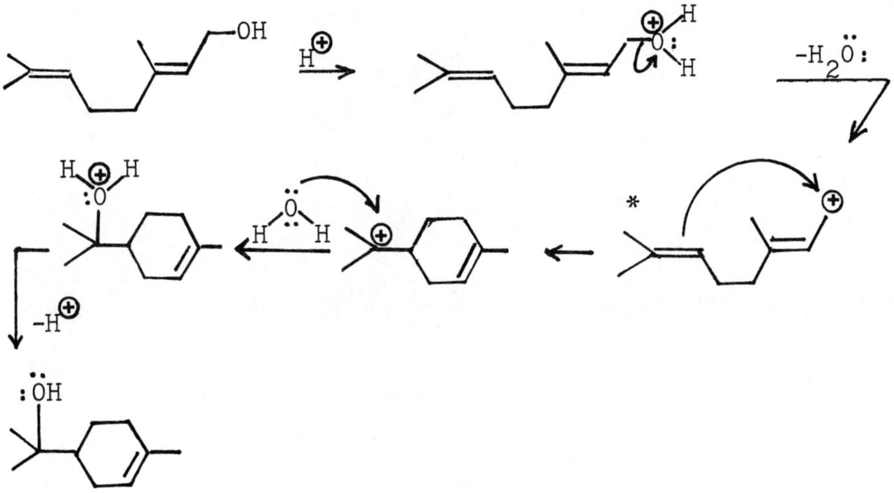

$\xrightarrow{H^{\oplus}}$ $\xrightarrow{-H_2\ddot{O}:}$ $\longleftarrow$ $\xleftarrow{H_2\ddot{O}:}$ $\xrightarrow{-H^{\oplus}}$

:ÖH

* resonance stabilized cation

32.  dihydro-myrcene

$\xrightarrow{KMnO_4}$ $CH_3\overset{O}{\overset{\|}{C}}CH_3$ + $(CH_3)C(=O)CH_2CH_2CHO$ + CH_3-COOH

$\xrightarrow{H_2/cat.}$ $(CH_3)_2CH(CH_2)_3CH(CH_3)CH_2CH_3$ ($C_{10}H_{22}$)

(Text: $KMnO_4$ oxidation S. 8.9)

33. Bromine adds to 1,3,5-hexatriene giving the resonance-stabilized carbocation shown below.

$$H_2C{=}CH{-}CH{=}CH{-}CH{=}CH_2 \xrightarrow{Br_2} \left[H_2C{=}CH{-}CH{=}CH{-}\overset{\oplus}{C}H{-}CH_2{-}Br \leftrightarrow H_2C{=}CH{-}\overset{\oplus}{C}H{-}CH{=}CH{-}CH_2Br \leftrightarrow H_2\overset{\oplus}{C}{-}CH{=}CH{-}CH{=}CHCH_2Br \right]$$

$\xrightarrow{Br^{\ominus}}$

$$H_2C{=}CH{-}CH{=}CH{-}\underset{Br}{CH}{-}\underset{Br}{CH_2} \quad + \quad \underset{Br}{H_2C}CH{=}CH{-}CH{=}CH{-}\underset{Br}{CH_2}$$

These products are more stable because they have conjugated double bonds.

(Text: see stability of 1,3-butadiene S. 13.2)

34. (a) $(CH_3)_2CHCH_2\underset{OH}{CH}CH_2\overset{CH_2}{\overset{\|}{C}}CH{=}CH_2$ 2-methyl-6-methylidene-7-octen-4-ol

(b) yes

(c) this compound and geraniol are isomers of the formula $C_{10}H_{18}O$.

(d) catalytic hydrogenation gives $(CH_3)_2CHCH_2\overset{*}{\underset{OH}{C}}HCH_2\underset{*}{\overset{CH_3}{C}}HCH_2CH_3$

containing 2 chiral centers. Thus, from (d,l)-2-methyl-6-methylidene-7-octen-4-ol, diastereomers are formed, i.e.,

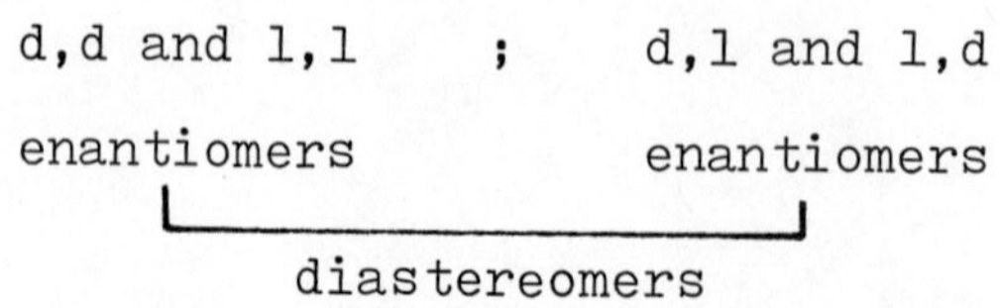

35.

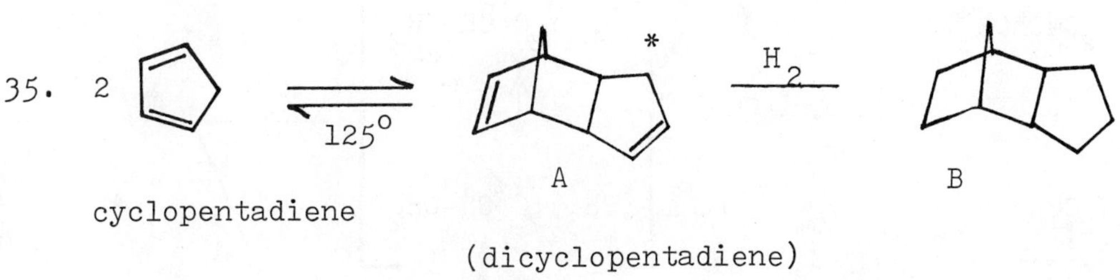

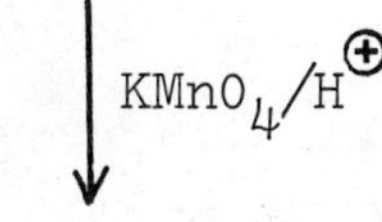

HOOC HOOC COOH COOH ≡ COOH CH_2COOH COOH COOH

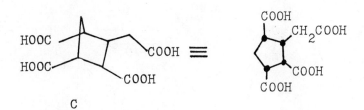

C

* In forming A, the diene can react at either face of the carbon-carbon double bond of the dienophile giving the enantiomers shown below.

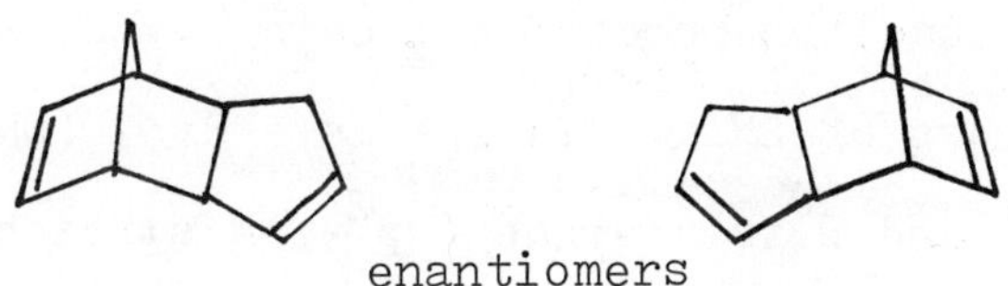

36. 2.4 kcal/mole. Calculated from $\Delta G = -RT\ln K$

$$\Delta G = -(1.987 \times 10^{-3})(298)\ln 50 = -2.3$$

37. Both 1-chloro-3-methyl-2-butene and 3-chloro-3-methyl-1-butene ionize in aqueous solution to give the same resonance-stabilized carbocation. Treatment of either A or B with acid gives the same carbocation and leads to the same mixture of A or B.

(continued next page)

$(CH_3)_2C{=}CHCH_2Cl$ or $(CH_3)_2C(Cl)CH{=}CH_2$ $\xrightarrow{\text{ionization}}$ $\left[(CH_3)_2C{=}CH{-}\overset{\oplus}{C}H_2 \leftrightarrow (CH_3)_2\overset{\oplus}{C}{-}CH{=}CH_2\right]^*$ $\xrightarrow{H_2O}$

B $(CH_3)_2C{=}CH{-}CH_2{-}OH$ $\underset{H^\oplus}{\overset{-H^\oplus}{\rightleftharpoons}}$ $(CH_3)_2C{=}CH{-}CH_2{-}\overset{\oplus}{O}H_2$

+

A $(CH_3)_2C(OH){-}CH{=}CH_2$ $\underset{H^\oplus}{\overset{-H^\oplus}{\rightleftharpoons}}$ $(CH_3)_2C(\overset{\oplus}{O}H_2){-}CH{=}CH_2$

* equivalent to $(CH_3)_2\overset{\delta^+}{C}\cdots CH \cdots \overset{\delta^+}{C}H_2$

38. A, $C_6H_{14}O_2$, is a reduction product of acetone. B, C_6H_{10}, is formed from A by loss of $2H_2O$. Thus A is a diol (OH stretch ~3400 cm^{-1} in ir) and B is a diene (sp^2-C-H stretch ~3100 cm^{-1}, C=C stretch ~1650 cm^{-1}, C-H bend ~890 cm^{-1} for $R_2C{=}CH_2$).

A reacts with $NaNH_2$ by $-O{-}H + NH_2^\ominus \rightarrow -O^\ominus + NH_3$. Reactions of B with $KMnO_4$ and Br_2 confirm that it is an alkene. Ozonolysis products allow construction of B.

$H{-}CHO + CH_3{-}CO{-}CO{-}CH_3 + HCHO \xleftarrow{\text{ozonolysis}} CH_2{=}C(CH_3){-}C(CH_3){=}CH_2$

(continued next page

Other reactions are:

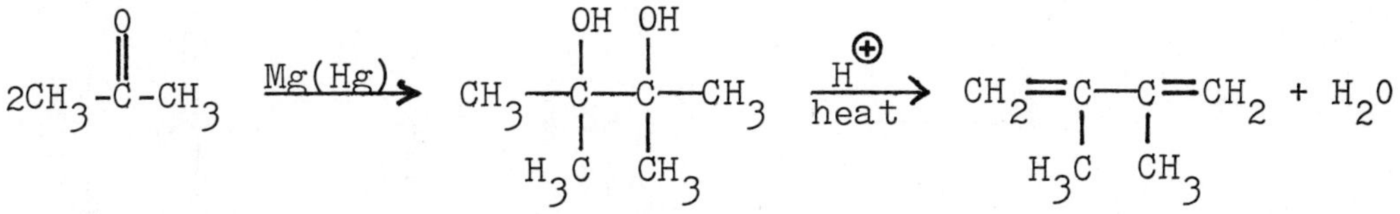

$$2CH_3\text{-}\overset{\displaystyle O}{\overset{\|}{C}}\text{-}CH_3 \xrightarrow{Mg(Hg)} \underset{A}{CH_3\text{-}\underset{H_3C}{\overset{OH}{C}}\text{-}\underset{CH_3}{\overset{OH}{C}}\text{-}CH_3} \xrightarrow[\text{heat}]{H^{\oplus}} \underset{B}{CH_2{=}\underset{H_3C}{C}\text{-}\underset{CH_3}{C}{=}CH_2} + H_2O$$

39. 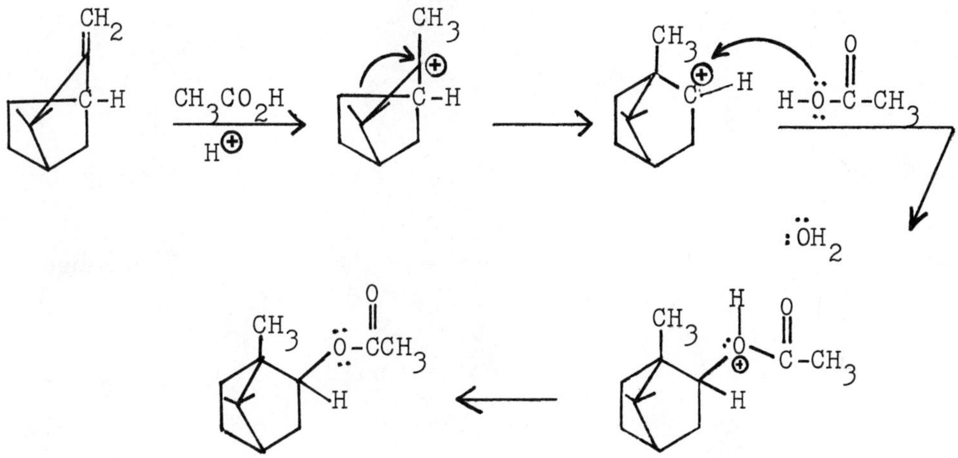 Methylene-bicyclic alkene (=CH_2, C-H) $\xrightarrow[H^{\oplus}]{CH_3CO_2H}$ tertiary carbocation (CH_3, ⊕, C-H) ⟶ rearranged secondary carbocation (CH_3, C⊕-H) attacked by H-Ö-C(=O)-CH_3 ⟶ oxonium ion (CH_3, H-O⊕-C(=O)-CH_3, H) with :$\ddot{O}H_2$ ⟶ acetate ester (CH_3, $\ddot{O}$-C(=O)CH_3, H)

40. The nmr spectrum of $CH_3\text{-}C{\lessgtr}^{CH_2}_{\overset{\oplus}{C}H_2}$ would consist of 3 signals (CH_3-, $=CH_2$, $-\overset{\oplus}{C}H_2$). The nmr data are consistent with the structure:

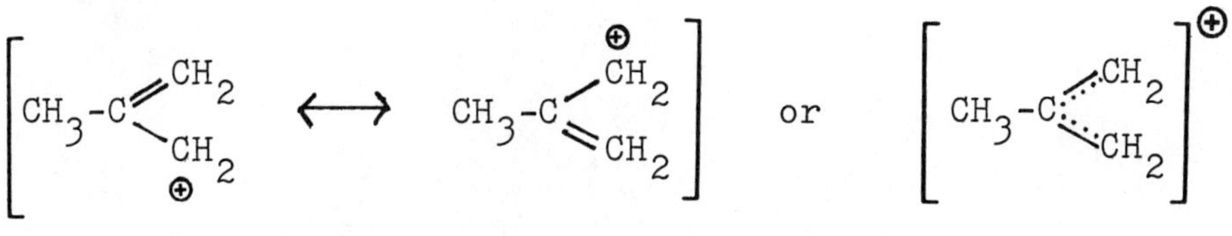

$$\left[CH_3\text{-}C{\lessgtr}^{CH_2}_{\overset{\oplus}{C}H_2} \longleftrightarrow CH_3\text{-}C{\lessgtr}^{\overset{\oplus}{C}H_2}_{CH_2}\right] \text{ or } \left[CH_3\text{-}C\overset{\cdots CH_2}{\underset{\cdots CH_2}{}}\right]^{\oplus}$$

3H, 3.85 δ 4H, 8.95 δ

41. $H-C\equiv C-H + B: \longrightarrow HB + H-C\equiv C:^{\ominus}$

$$HC\equiv C:^{\ominus} + CH_3\overset{O}{\overset{\|}{C}}CH_3 \longrightarrow \underset{A}{CH_3-\overset{CH_3}{\overset{|}{\underset{OH}{\underset{|}{C}}}}-C\equiv CH} \xrightarrow[cat.]{H_2} \underset{B}{CH_3-\overset{CH_3}{\overset{|}{\underset{OH}{\underset{|}{C}}}}-CH=CH_2}$$

$$B \xrightarrow{H^{\oplus}} \underset{C\ (isoprene)}{CH_2=\overset{CH_3}{\overset{|}{C}}-CH=CH_2} \xrightarrow{ozonolysis} CH_3\overset{O}{\overset{\|}{C}}-\overset{O}{\overset{\|}{C}}-H + H-\overset{O}{\overset{\|}{C}}-H$$

<u>ir</u>: C can be distinguished from A and B by the <u>absence</u> of an OH stretch (3200-3550 cm^{-1}) in the ir spectrum. A will show both $C\equiv C$ (2255-2100 cm^{-1}) and $\equiv C-H$ (3300 cm^{-1}) stretches. The ir spectrum of B lacks these absorptions but shows C=C ($\sim$1650 cm^{-1}) and =C-H (above 3000 cm^{-1}) stretches.

<u>nmr</u>: OH: singlets 2-4 δ (1H) in the spectra of A and B; no OH in spectrum of C.

methyl: singlet near 1 δ (6H) in spectra of A and B; singlet near 1 δ (3H) in spectrum of C.

vinyl H: absorptions near 5 δ (5H) in spectrum of C, absorptions near 5 δ (3H) for B, no vinyl hydrogens in spectrum of A.

chemical: A reacts with $Ag(NH_3)_2^{\oplus}$ forming a precipitate of $(CH_3)_2C(OH)C\equiv CAg$. B and C do not react. B (OH group) reacts with sodium with evolution of hydrogen. C does not react.

42. 1,3-butadiene
<u>anion-radical</u>

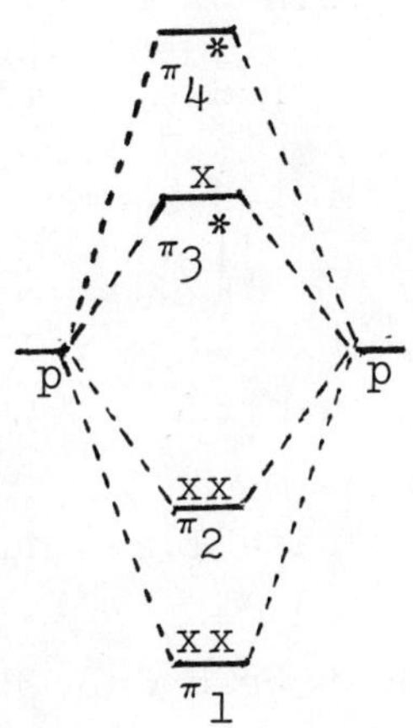

The anion should be less stable since the additional electron is in an antibonding orbital.

43.

SUMMARY OF REACTIONS

1. Electrophilic Addition (S. 13.4)

Dienes react with the same reagents that were presented for simple alkenes in Chapter 8. With one mole of reagent both 1,2- and 1,4- addition products are formed. Generally at room temperature and above, the 1,4- addition product predominates.

$$-\overset{|}{C}=\overset{|}{C}-\overset{|}{C}=\overset{|}{C}- \xrightarrow[\text{(1 mole)}]{\text{E-Nu}} -\underset{E}{\overset{|}{C}}-\overset{|}{C}=\overset{|}{C}-\underset{Nu}{\overset{|}{C}}- \; + \; -\underset{E}{\overset{|}{C}}-\underset{Nu}{\overset{|}{C}}-\overset{|}{C}=\overset{|}{C}-$$

(E-Nu = HX, $H^{\oplus}/H_2O$, $ROH/Hg(F_3CO_2)_2$ then $BH_4^{\ominus}$, X_2 etc.)

The example most commonly used to illustrate 1,2- and 1,4- addition is:

$$CH_2{=}CH{-}CH{=}CH_2 \xrightarrow[CCl_4]{Br_2} BrCH_2CH{=}CHCH_2Br + BrCH_2CHBrCH{=}CH_2$$

at 40°	80%	20%
at -80°	20%	80%

2. Free-Radical Addition of Bromine (S. 13.4)

$$-\overset{|}{C}=\overset{|}{C}-\overset{|}{C}=\overset{|}{C}- \xrightarrow[\text{heat}]{Br_2} -\underset{Br}{\overset{|}{C}}-\overset{|}{C}=\overset{|}{C}-\underset{Br}{\overset{|}{C}}- \; + \; -\underset{Br}{\overset{|}{C}}-\underset{Br}{\overset{|}{C}}-\overset{|}{C}=\overset{|}{C}-$$

3. Polymerization of Dienes (S. 13.5)

$$-\overset{|}{C}=\overset{|}{C}-\overset{|}{C}=\overset{|}{C}- \xrightarrow{R\cdot} \left(-\underset{|}{\overset{|}{C}}-\overset{|}{C}=\overset{|}{C}-\underset{|}{\overset{|}{C}}-\right)_n$$

4. Diels-Alder Reaction (S. 13.6)

diene + dienophile → (cyclohexene)

diene dienophile

(the general types of Diels-Alder reactants are summarized in the Reaction Review)

REACTION REVIEW

Reactions from Chapter 13.

Questions	Answers
1. $-\overset{\vert}{C}=\overset{\vert}{C}-\overset{\vert}{C}=\overset{\vert}{C}-$ $\xrightarrow[\text{(1 mole)}]{X_2}$?	1. $-\overset{\vert}{\underset{X}{\underset{\vert}{C}}}-\overset{\vert}{C}=\overset{\vert}{C}-\overset{\vert}{\underset{X}{\underset{\vert}{C}}}-$ (major +40°) and $-\overset{\vert}{\underset{X}{\underset{\vert}{C}}}-\overset{\vert}{\underset{X}{\underset{\vert}{C}}}-\overset{\vert}{C}=\overset{\vert}{C}-$ (major -80°) (d,l)
Write the major product when Br_2 (1 mole) reacts at room temperature with:	
(a) $CH_2=CH-CH=CH_2$	(a) $CH_2BrCH=CHCH_2Br$
(b) [cyclopentadiene]	(b) [3,5-dibromocyclopentene] (meso and d,l)
2. $-\overset{\vert}{C}=\overset{\vert}{C}-\overset{\vert}{C}=\overset{\vert}{C}-$ $\xrightarrow[40^{o}]{\text{HX(1 mole)}}$?	2. $-\overset{\vert}{\underset{H}{\underset{\vert}{C}}}-\overset{\vert}{C}=\overset{\vert}{C}-\overset{\vert}{\underset{X}{\underset{\vert}{C}}}-$ (major) and $-\overset{\vert}{\underset{H}{\underset{\vert}{C}}}-\overset{\vert}{\underset{X}{\underset{\vert}{C}}}-\overset{\vert}{C}=\overset{\vert}{C}-$ (d,l)
	(continued next page)

Write the major product in the following at room temperature: (a) cyclohexa-1,3-diene $\xrightarrow{HBr}$? (b) $CH_3CH{=}CH{-}CH{=}CH_2 \xrightarrow{HCl}$?	(a) 3-bromocyclohexene (d,l) (b) $CH_3CHClCH{=}CHCH_3$ (d,l)
3. $-\overset{\vert}{C}{=}\overset{\vert}{C}{-}\overset{\vert}{C}{=}\overset{\vert}{C}-$ $\xrightarrow[Hg(F_3CO_2)_2]{ROH(1\ mole)}$ $\xrightarrow{BH_4^{\ominus}}$? Write the major product in the following reactions at room temperature: (a) $CH_2{=}CH{-}CH{=}CH_2 \xrightarrow[Hg(F_3CO_2)_2]{C_2H_5OH} \xrightarrow{BH_4^{\ominus}}$? (b) cyclopentadiene $\xrightarrow[Hg(F_3CO_2)_2]{CH_3OH}$ $\xrightarrow{BH_4^{\ominus}}$?	3. $-\underset{H}{\overset{\vert}{C}}{-}\overset{\vert}{C}{=}\overset{\vert}{C}{-}\underset{OR}{\overset{\vert}{C}}-$ (major) and $-\underset{H}{\overset{\vert}{C}}{-}\underset{OR}{\overset{\vert}{C}}{-}\overset{\vert}{C}{=}\overset{\vert}{C}-$ (d,l) (a) $CH_3CH{=}CHCH_2OC_2H_5$ (b) 3-methoxycyclopentene (OCH_3) (d,l)
4. $-\overset{\vert}{C}{=}\overset{\vert}{C}{-}\overset{\vert}{C}{=}\overset{\vert}{C}-$ $\xrightarrow[(2\ moles)]{X_2}$?	4. $-\underset{X}{\overset{\vert}{C}}{-}\underset{X}{\overset{\vert}{C}}{-}\underset{X}{\overset{\vert}{C}}{-}\underset{X}{\overset{\vert}{C}}-$ (meso and d,l) (continued next page)

Write the structure of the <u>major</u> product of Br_2 with:

CH_2=CH-CH=CH_2

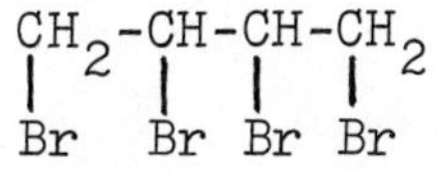

```
CH2-CH-CH-CH2
|    |  |  |
Br   Br Br Br
```

(meso and d,l)

5. 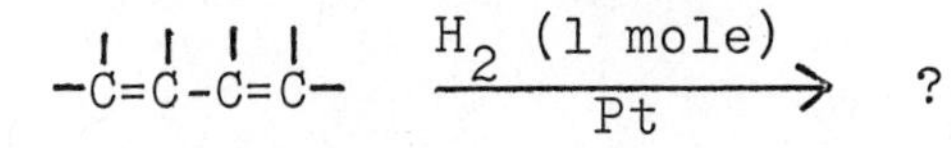 -C=C-C=C- $\xrightarrow[\text{Pt}]{H_2 \text{ (1 mole)}}$?

5. 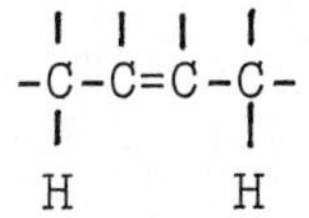
```
 |  |  |  |
-C-C=C-C-
 |        |
 H        H
```

and

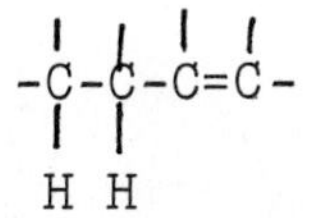

```
 |  |  |  |
-C-C-C=C-
 |  |
 H  H
```

Write the structure of the products in the following

(a) 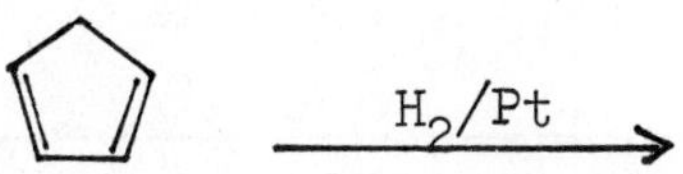 cyclopentadiene $\xrightarrow{H_2/Pt}$

(b) CH_2=CHCH=CH_2 $\xrightarrow{D_2/Pt}$

(a)  cyclopentene

(b) CH_2DCHDCH=CH_2 and
(d,l)

CH_2DCH=CHCH_2D

6. 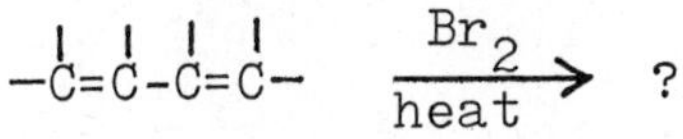 -C=C-C=C- $\xrightarrow[\text{heat}]{Br_2}$?

6. 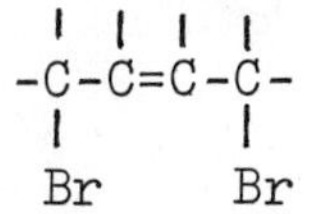
```
 |  |  |  |
-C-C=C-C-     (major)
 |        |
 Br       Br
```

and

```
 |  |  |  |
-C--C--C=C-
 |  |
 Br Br
```

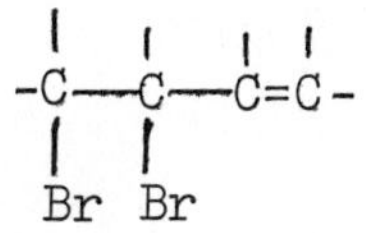

7. 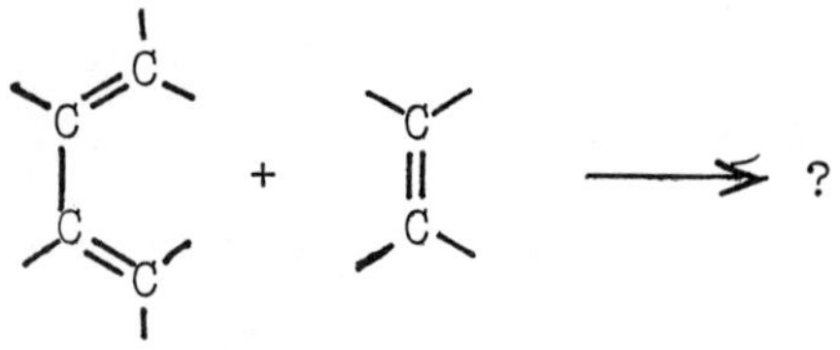 [diene: C=C–C=C] + [C=C] ⟶ ?

Write the products in the following:

(a) 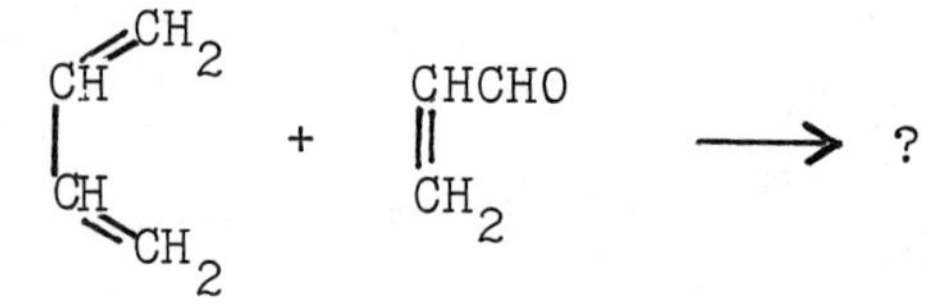 $CH_2=CH–CH=CH_2$ + $CHCHO=CH_2$ ⟶ ?

(b) $CH_2=CHCH=CHCH_3$ + [p-benzoquinone] ⟶ ?

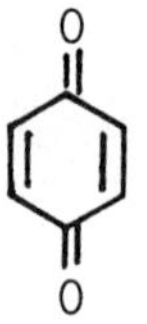

7.  [cyclohexene product]

(a) 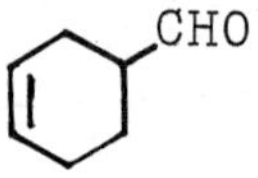 [cyclohex-3-ene-1-carbaldehyde, CHO]

(b) 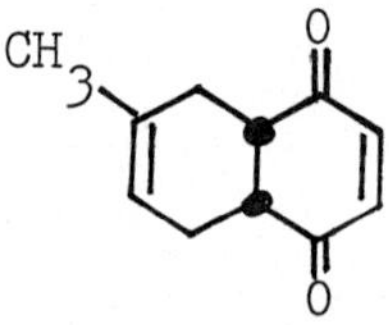 [CH_3-substituted tetrahydronaphthoquinone, O, O]

8. 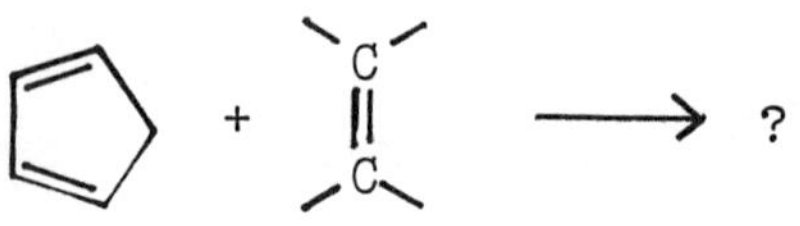 [cyclopentadiene] + [C=C] ⟶ ?

Write the products in the following:

(a) 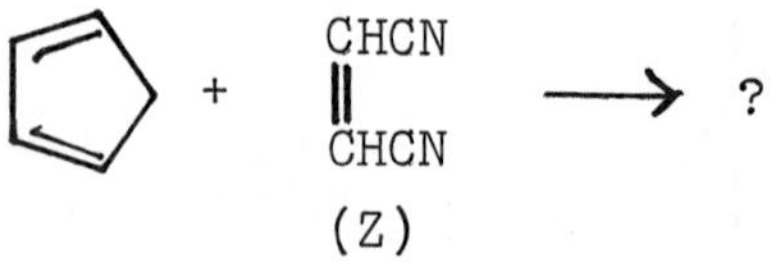 [cyclopentadiene] + CHCN ‖ CHCN (Z) ⟶ ?

8.  [norbornene product]

(a) 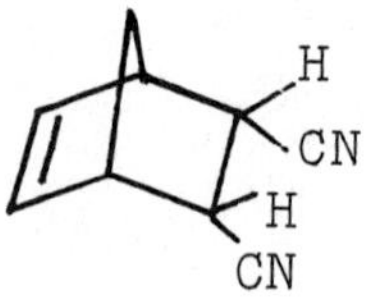 [norbornene with H, CN, H, CN]

(endo major)

(continued next page)

(b) cyclopentadiene + CHCHO ‖ CHCHO (E) ⟶ ?	(b) CHO, H, H, CHO
9. cyclohexadiene + C=C ⟶ ?	9.
Write the products in the following:	
(a) CH_3, CH_3 (dimethylcyclohexadiene) + $CHCO_2CH$ ‖ CH_2 ⟶ ?	(a) CH_3, CH_3, H, CO_2CH_3, H, H
(b) cyclohexadiene + O, O (quinone) ⟶ ?	(b) H, O, H, O
10. C=C–C=C + C≡C ⟶ ? (or other diene type)	10.

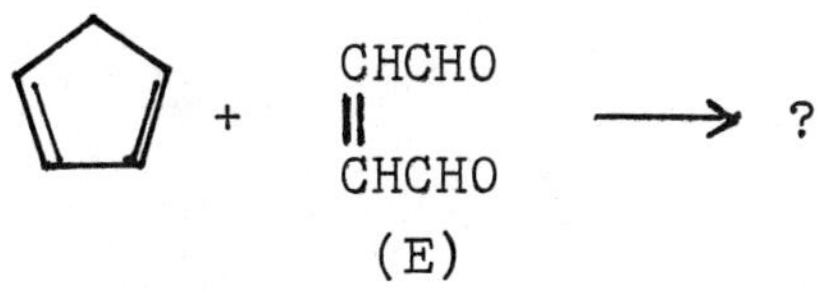

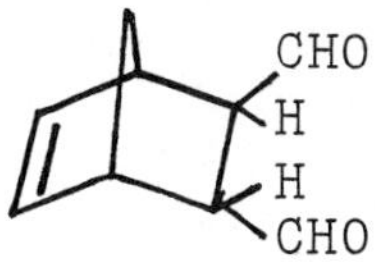

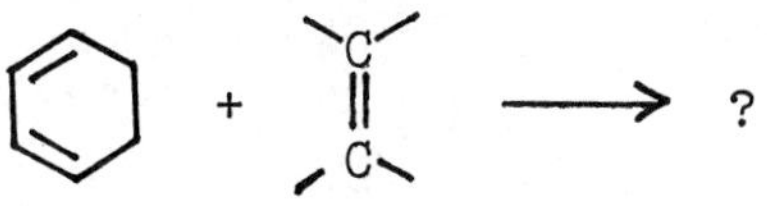

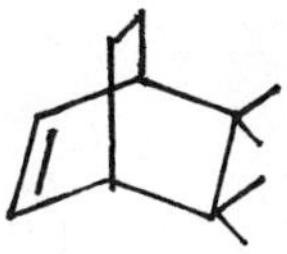

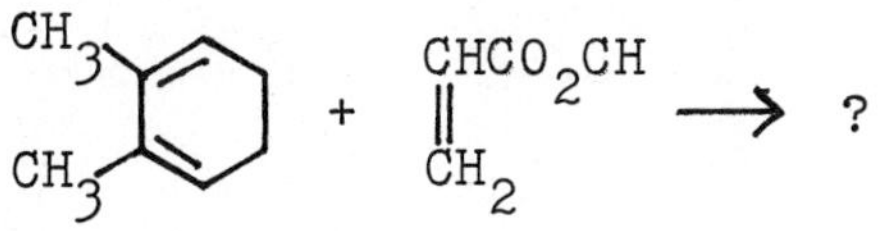

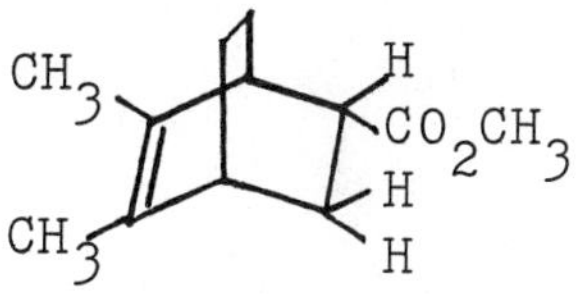

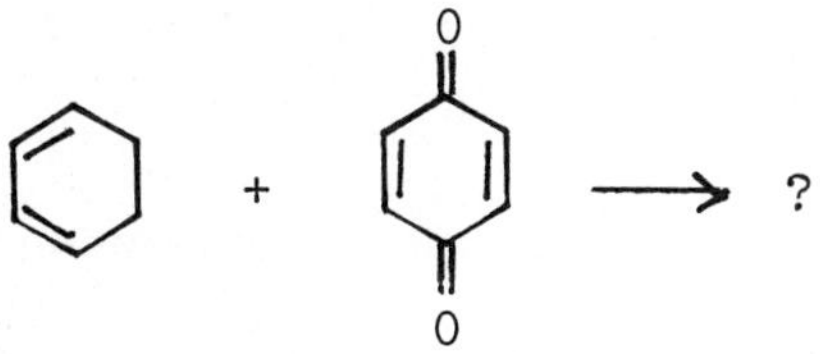

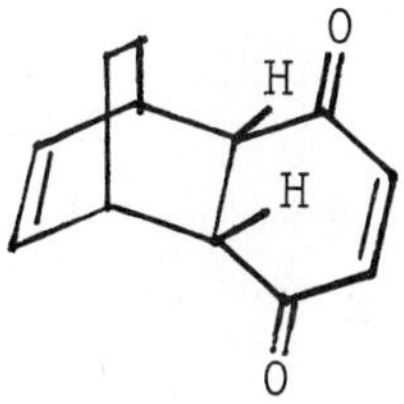

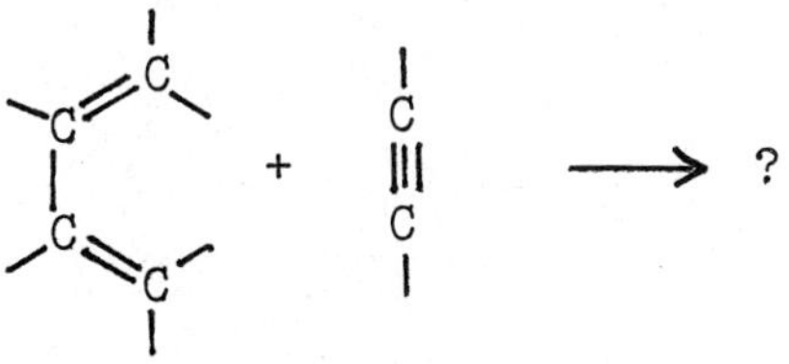

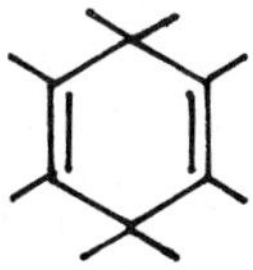

(continued next page)

Write the products in the following:

(a) $HC(=CH_2)-HC=CH_2$ + $CCHO \equiv CH$ → ?

(b) cyclopentadiene + $CH \equiv CCH_3$ → ?

(a) 1,4-cyclohexadiene with CHO

(b) bicyclic diene with CH_3

CHAPTER 14, ELECTROCYCLIC AND CYCLOADDITION REACTIONS

LEARNING OBJECTIVES

When you have completed this chapter, you should be able to:

1. predict the types of reagents which react in pericyclic reactions (problem 1);
2. write the distribution of electrons in the molecular orbitals of 1,3-butadiene (problem 2);
3. define "conrotatory" and "disrotatory" (problem 3);
4. classify reactions as suprafacial or antrafacial and predict products in these reactions (problems 6, 8);
5. predict the products of Diels-Alder reactions (problems 10, 15 and related problems 7, 12, 13 and 20);
6. predict whether cycloaddition reactions are accelerated by heat or light (problems 4, 11, 19 and related problems 5, 14);
7. predict products formed in thermal or photochemical ring closures (problems 14, 16, 17);
8. suggest mechanisms to explain cyclization and related reactions (problems 20, 21 22, 23).

ANSWERS TO QUESTIONS

1. Pericyclic reactions are concerted and involve cyclic activated complexes. The reactions of ethene with HN=NH (text, p. 301) and $C_6H_5CO_3H$ (text, p.328) are pericyclic. Other reactions involve more than one step.

2. (a) Electrons in ground state: π_1, 2; π_2, 2; π_3^* and π_4^*, 0.

   (b) Electrons in first excited state: π_1, 2; π_2, 1; π_3^*, 1; π_4^*, 0.

(Text: See in Chapter 13, Figs. 13.2 and 13.4)

3. Conrotatory corresponds to clockwise or counterclockwise motion of both groups and disrotatory motion corresponds to clockwise motion by one group and counterclockwise motion by the other.

4. The stereochemistry of the product (hydrogens <u>cis</u>) requires a disrotatory cyclization of the 1,3-butadiene moiety.

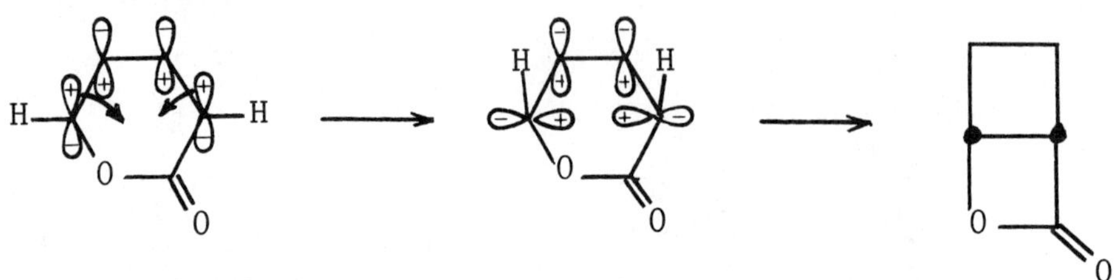

The proper array of orbitals for disrotatory cyclization is provided in the 1st excited state, thus the compound is prepared by illuminating the starting material.

(Text: also see S. 13.2)

5. Upon illumination, π_4^* becomes the HOMO of 1,3,5-hexatriene and conrotatory motion is required for cyclization.

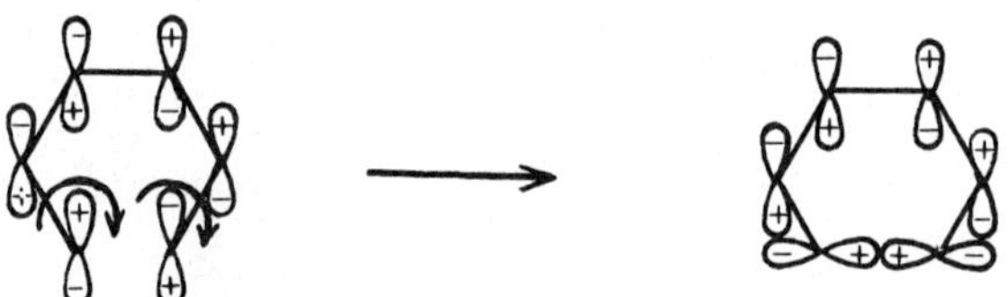

symmetry allowed

In a thermally-induced reaction the HOMO is π_3, and a disrotatory motion is required for cyclization.

symmetry allowed

(continued next page)

(2E,4Z,6E)-Octatriene could be used to show the pathways followed in these reactions. It will produce cis-5,6-dimethyl-1,3-cyclohexadiene by disrotatory closure but trans-5,6-dimethyl-1,3-cyclohexadiene by conrotatory closure.

For example:

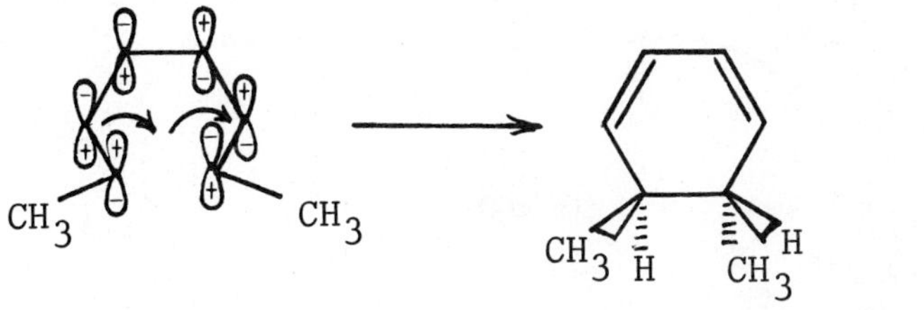

conrotatory closure

trans-5,6-dimethyl-1,3-cyclohexadiene

6. The suprafacial processes, in which bonds are formed or broken on the same face of a molecule, are (a), (b) and (c). Reaction (d), an E2 elimination reaction is antarafacial.

7. Vinylacetylene would not function readily as a diene in the Diels-Alder reaction. Because of the sp carbon, the system is too long to permit development of overlap at C-1 and C-4 of the enyne.

8. (a) 4,

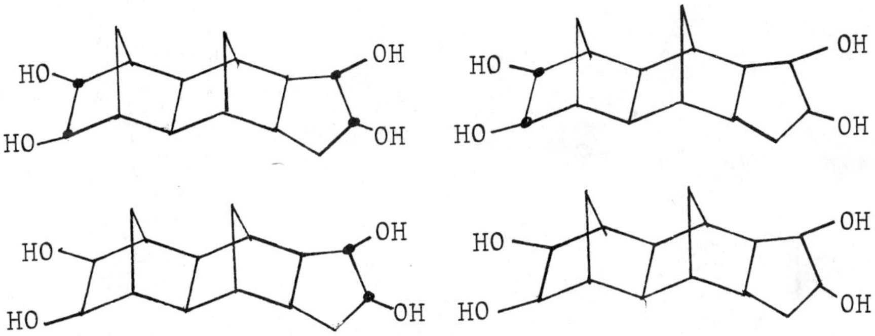

(b) No. No double bonds are present and a retro-Diels-Alder is not possible.

9. No. The hydrogenated product possesses a plane of symmetry and is optically inactive.

10. Products can be predicted on the basis of the general reaction:

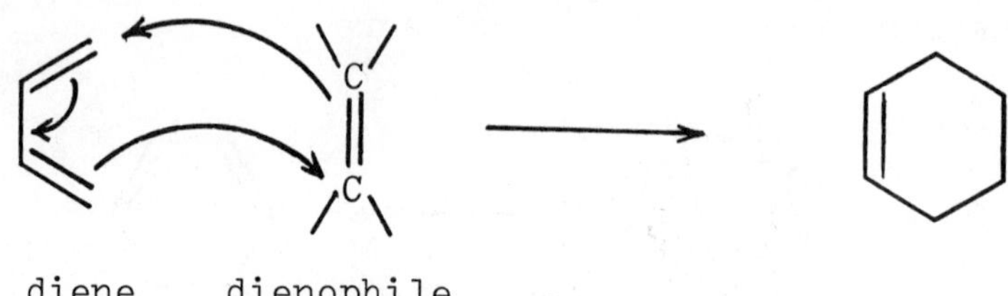

- the diene reacts in an <u>s cis</u> conformation
- addition to the dienophile is a suprafacial process
- addition of the diene to dienophile having an unsaturated group occurs in an <u>endo</u> fashion

(a)

(b) 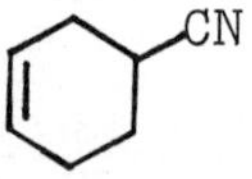 CN

(c) 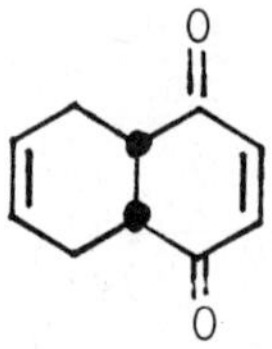 O O

(d) 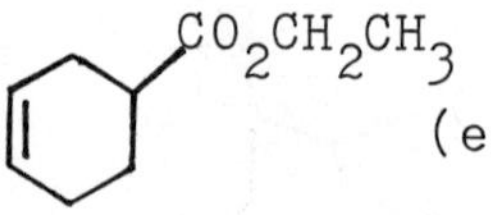 $CO_2CH_2CH_3$

(e) 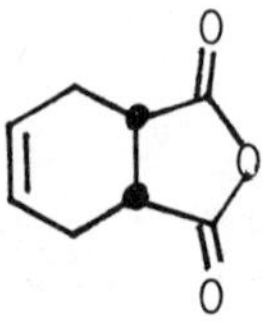 O O O

(f) 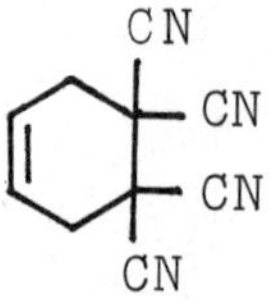 CN CN CN CN

(g) 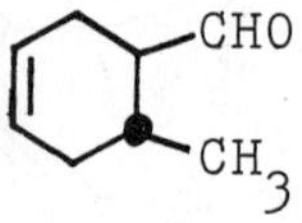 CHO CH_3

(Text S. 14.3)

11. Yes, reaction A has the lower energy of activation. Thermal [4 + 2] cycloaddition reactions are symmetry allowed (HOMO of the diene and LUMO of the dienophile or LUMO of the diene and HOMO of the dienophile).

(Text S. 14.3)

12. In cyclopentadiene, the diene structure is maintained in the required s-*cis* geometry. (2E,4E)-2,4-hexadiene can exist in s-*cis* and s-*trans* conformations.

(Text: Diels-Alder reaction S. 14.3)

13. These reactions illustrate the stereospecificity of the Diels-Alder reaction. The addition of the diene to the dienophile is a suprafacial process with respect to the dienophile.

(Text: Diels-Alder reaction S. 14.3)

14. The reverse of a conrotatory cyclization is required.

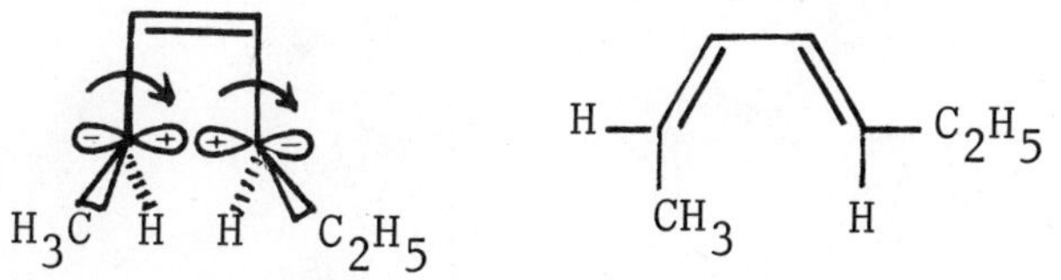

H_3C H H C_2H_5

H C_2H_5 CH_3 H (2Z,4E)

and conrotatory opening in the opposite direction gives the 2E,4Z-diene.

(Text: Review cyclization of butadiene, S. 14.2)

15. B: H Br O O + HB + :$Br^{\ominus}$

A

diene + dienophile → B

diene + dienophile → C

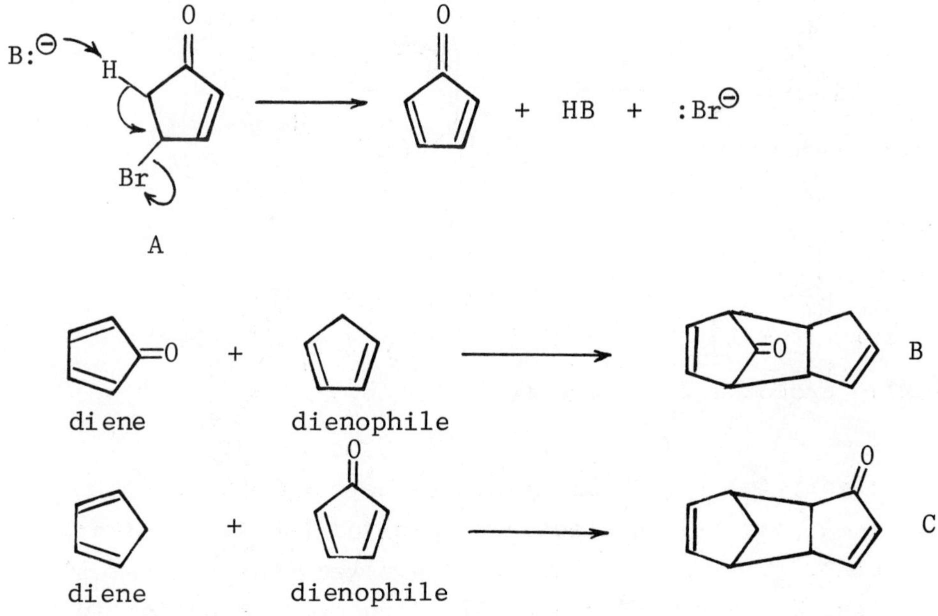

16. These are 4n systems, and they undergo thermal cyclization in a conrotatory fashion.

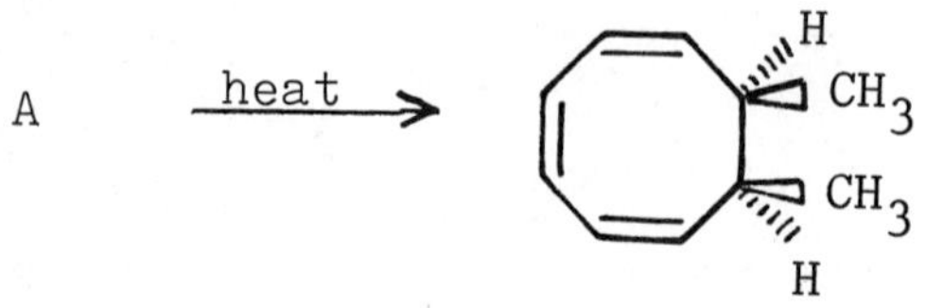

A $\xrightarrow{\text{heat}}$ H, CH_3, CH_3, H

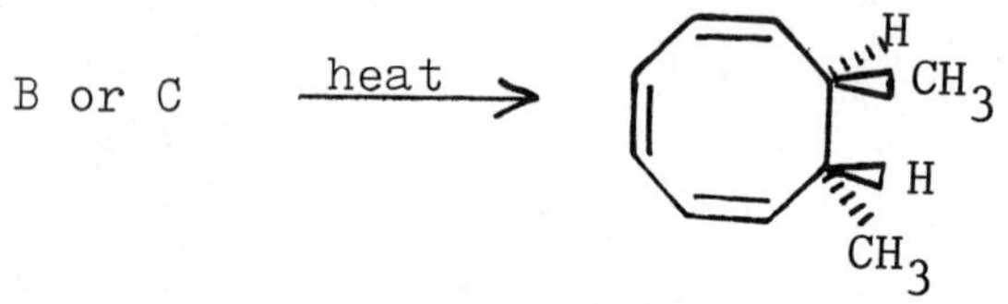

B or C $\xrightarrow{\text{heat}}$ H, CH_3, H, CH_3

(Text S. 14.2)

17. Photochemical cyclizations require disrotatory motion for 4n systems.

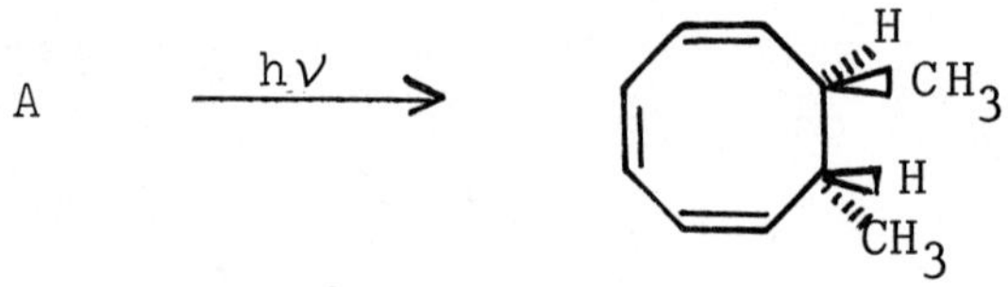

A $\xrightarrow{h\nu}$ H, CH_3, H, CH_3

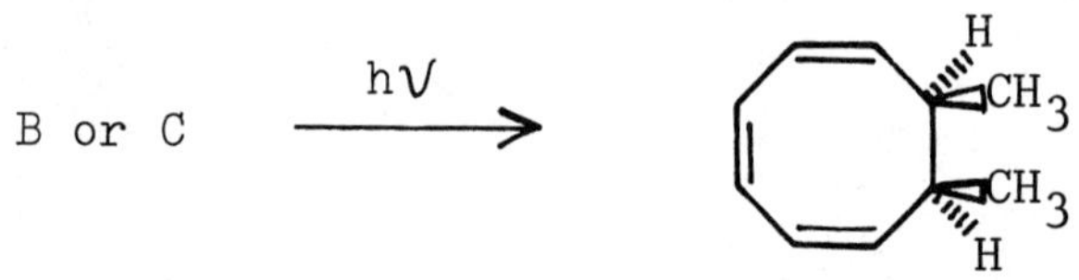

B or C $\xrightarrow{h\nu}$ H, CH_3, CH_3, H

(Text S. 14.2)

18. All are [2 + 2] cycloadditions

(Text: cycloaddition reactions S. 14.3)

19. (a) light, total of 4 electrons (4n system)
    (b) heat, total of 6 electrons involved (4n + 2 system)
    (c) heat, total of 10 electrons involved (4n + 2 system)
    (d) light, reverse of the cyclization of a 4n system in a disrotatory reaction.

(Text: electrocyclic reactions S. 14.2)

20.

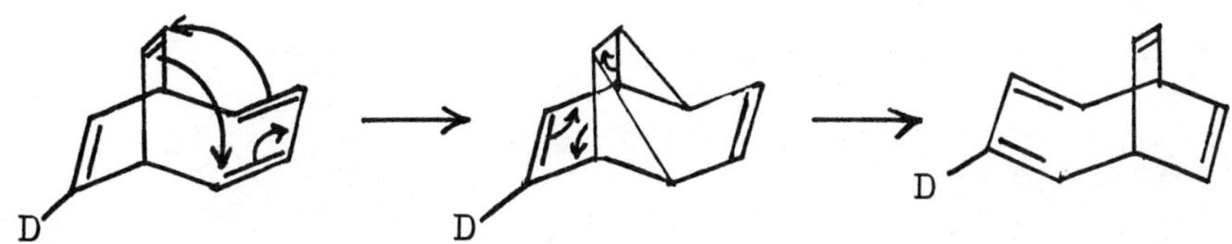

21.

f b
b: : f
f b

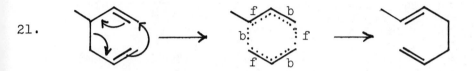

b = bond being <u>broken</u>

f = bond being <u>formed</u>

If charge-separated intermediates are involved, increasing the solvent polarity should significantly increase the rate of reaction.

22. A: CF_2—$\dot{C}Cl_2$
$\quad$ |
$\quad$ CF_2—$\dot{C}Cl_2$

B: CF_2—$\dot{C}Cl_2$
$\quad$ |
$\quad$ CH_2—$\dot{C}HC_6H_5$

23. The conversion of 7-dehydrocholesterol to pre-vitamin D_3 is an electrocyclic reaction. The conversion of pre-vitamin D_3 to vitamin D_3 can be pictured as a concerted shift involving transfer of H from one carbon to another.

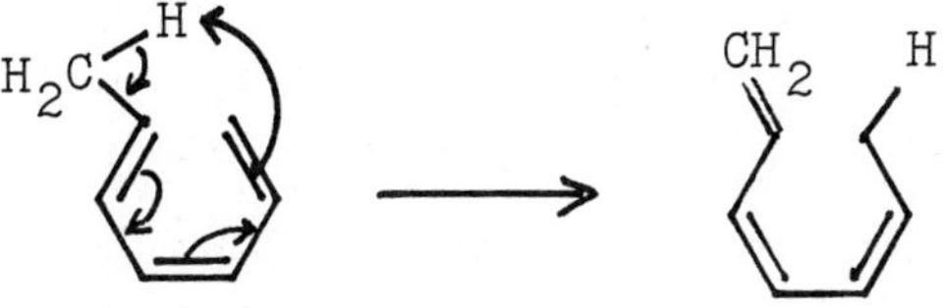

CHAPTER 15, AROMATICITY

LEARNING OBJECTIVES

When you have completed this chapter, you should be able to:

1. draw resonance structures for compounds and ions to account for aromatic properties, stability, reaction products, etc. (problems 1, 6, 16, 26, 27, 37, 38, 41 and related problem 43);
2. predict or explain the properties of benzene and other compounds based on their structures (problems 2, 3, 11 and related problems 4, 5, 31);
3. predict or account for the properties of compounds based on their aromatic, antiaromatic and nonaromatic electronic systems (problems 14, 17, 19, 29, 32 and related problem 28);
4. predict whether compounds or ions are aromatic, antiaromatic or nonaromatic based on Hückel's rule (problems 12, 13, 16, 20, 24, 25 and related problems 7, 8);
5. determine the hybridization of atoms in aromatic systems (problems 18, 20);
6. account for the NMR signals in aromatic, antiaromatic and nonaromatic systems (problems 21, 22, 39, 44, 45);
7. compare the geometry of benzene with that of the [10]-annulene (problem 9);
8. predict the linear species to be used as a model in determining aromaticity (problems 10, 34);
9. write mechanisms to explain the formation of cyclic polyenes (problems 15, 33);
10. calculate the delocalization energy of cyclooctatetrene by use of a graphic representation (problem 35);
11. account for enantiomeric forms of hexahelicene (problem 40):

ANSWERS TO QUESTIONS

1. (a)  and

(b) 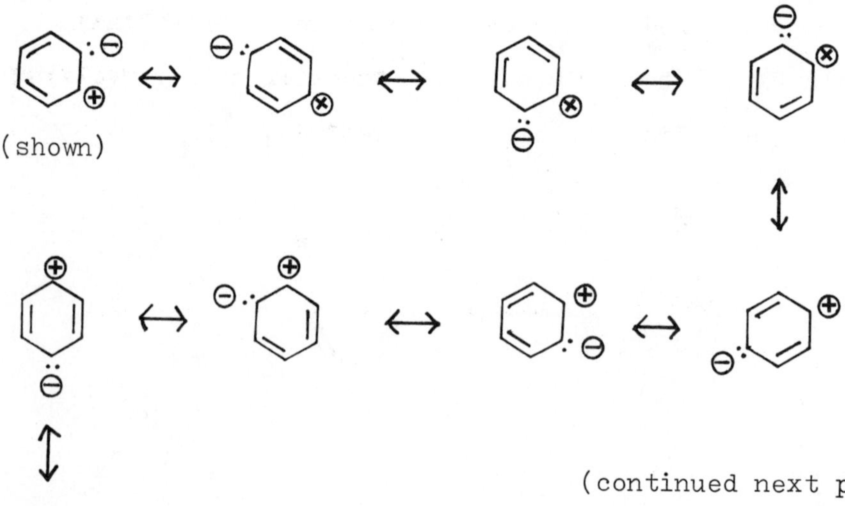 ↔ ↔ ↔

(shown)

↕

↔ ↔ ↔

↕

(continued next page)

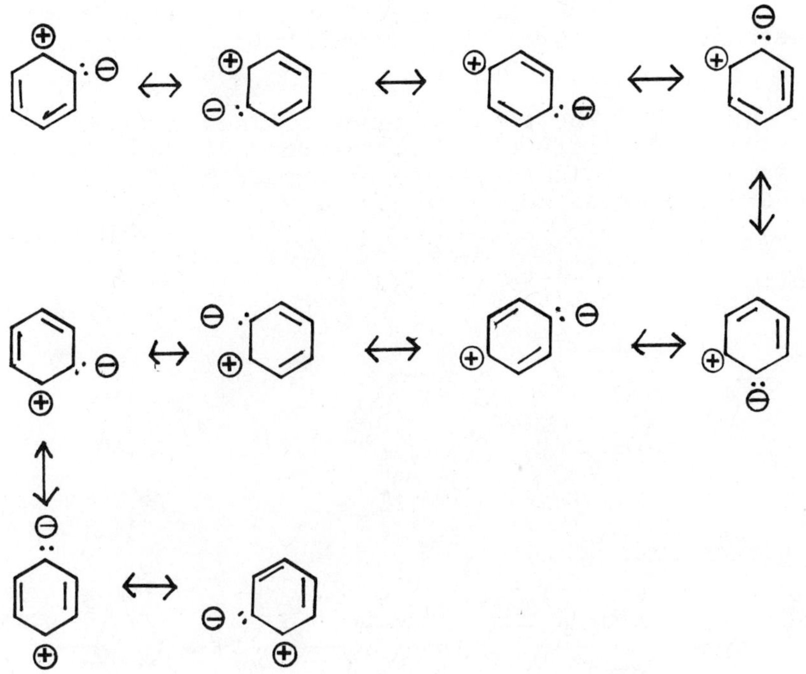

2. A molecule such as benzene which has a point of symmetry cannot have a dipole moment. The individual bond moments cancel each other.

3. (a) 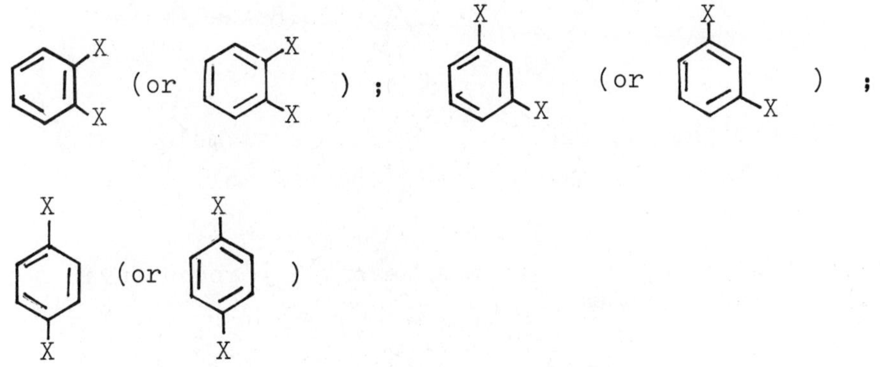 (or) ; (or) ;

(or)

(b) the same as in (a): 1,2-, 1,3-, and 1,4- disubstituted.

4. The bond distance between any two atoms in A (e.g., 1 and 2) is different from the bond distance between the same two atoms in B. Thus, atoms have been "moved", and this is not allowed in drawing resonance structures of a common hybrid.

5. Same answer as in 4. The bond distance between any two atoms in A (e.g., 1 and 2) is different from the bond distance between the same two atoms in B. Atoms have been moved, and this is not allowed in drawing resonance structures.

6. Atomic orbital representation:

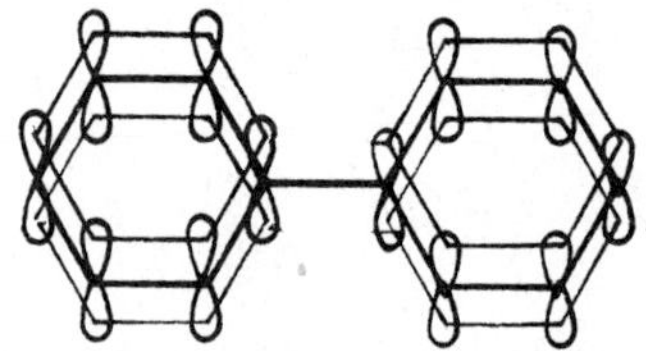

Resonance structures:

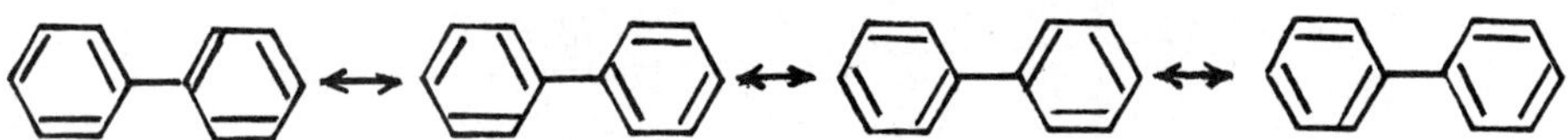

Estimated heat of hydrogenation:

For $\xrightarrow{H_2}$ $\Delta H = 28.8$ kcal/mole

For 6 C=C in biphenyl: 28.8 x 6 = 172.8 kcal/mole ;
172.8 kcal/mole - 71 kcal/mole = 101.8 kcal mole.

7. The "extra" electron pair is in an orbital perpendicular to the orbitals of the π system.

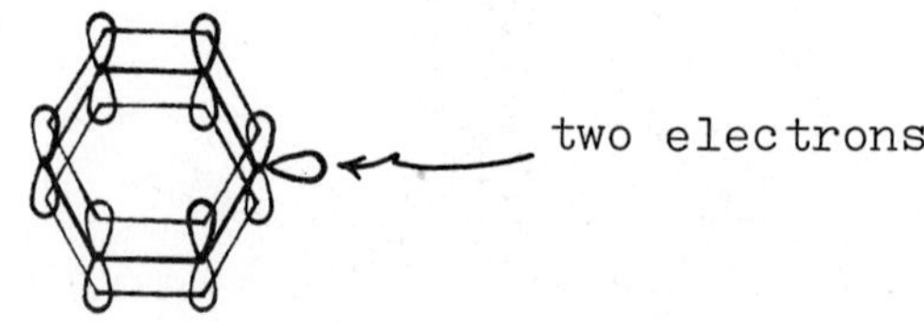

8. (a) Benzyne is aromatic because the π system contains 6 electrons. Electrons of the "third" bond are in orbitals orthogonal to those of the π system.

   (b) The reactivity of benzyne is due to these additional electrons (see Figure 22-4).

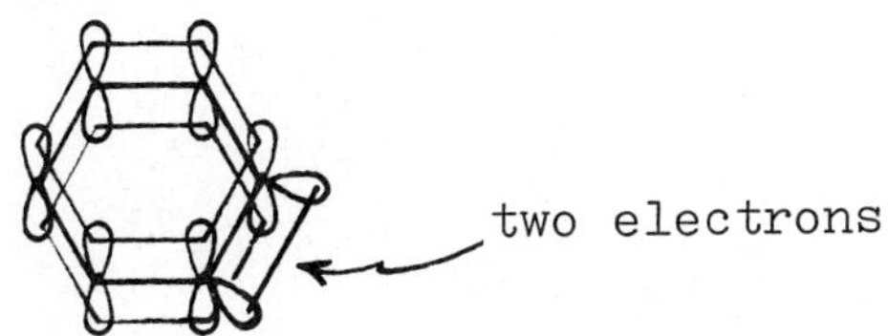

   (c) Aromaticity refers to the ground state stability while reactivity refers to the energy of activation for a process. The statements are not inconsistent.

9. An all cis-[10]-annulene is highly strained because of undesirable 144° bond angles.

10. 1,3,5,7-octatetraene

11. Three diphenylcyclobutadienes suggest a rectangular cyclobutadiene ring with isolated or distinct single and double bonds. The isomers are:

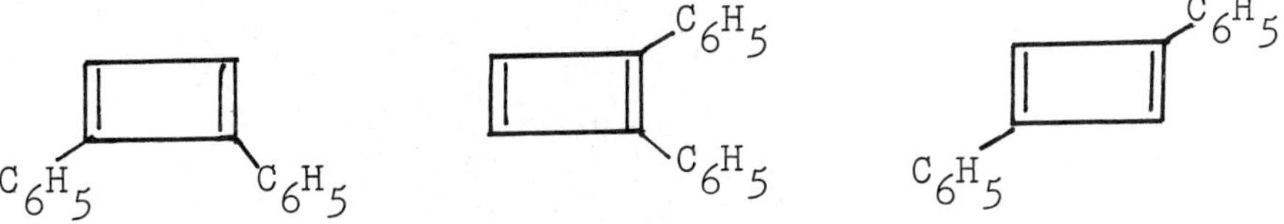

12. (a) Cyclooctatetraene contains $8\pi\ e^-$ (a 4n system), and delocalization gives an antiaromatic system.

   (b) Planar with all C-C bond distances equal.

   (c) No. Different bond lengths suggest distinct single (1.54Å) and double (1.34Å) bonds; thus, little or no delocalization occurs.

13. Cyclopentadienyl cation (A) is a 4n, antiaromatic system containing $4\pi\ e^-$ and is less stable than the cyclopentadienyl anion (B), a 4n + 2 aromatic system containing 6π electrons.

(continued next page)

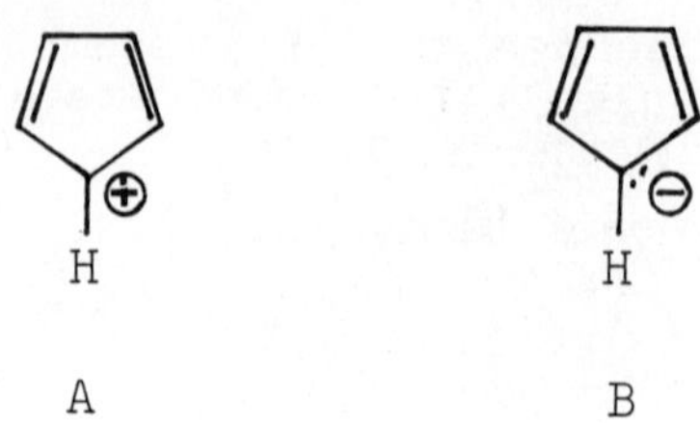

(Text: S. 15.3)

14. No. A planar cycloheptatrienide anion containing a π network of 8 electrons (a 4n system) would be antiaromatic.

15. (a) A is ($\xrightarrow{H_2/Pt}$)

(b) A Diels-Alder reaction. + CH≡CH ⟶

16. A ↔ B ↔ C ↔ ↔

E ↔ F ↔ G ↔ H ↔

I ↔ etc.

(continued next page)

All resonance structures are equivalent to 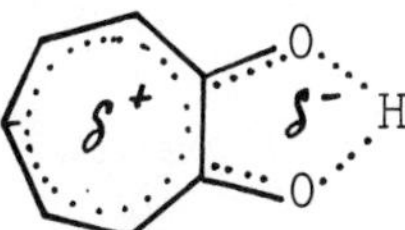

Resonance structures such as C → H show tropolone to be an aromatic compound containing 6 π electrons.

17. The rings are planar. Each has 10 electrons in the π system. The oxygen in the second compound is in a plane perpendicular to the plane of the rings.

18. Note: See p. 588 for structures.

(a) Each is an aromatic system containing 6 π electrons. Each heteroatom is sp^2 hybridized. In furan and pyrrole the heteroatom furnishes an electron pair to the π system. Some examples are shown below.

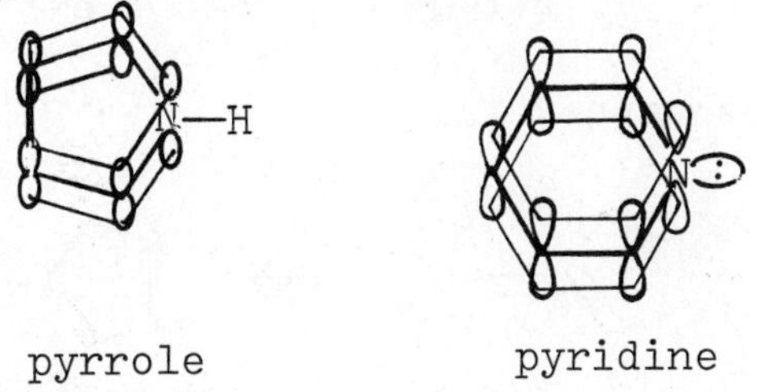

pyrrole   pyridine

(b) The hybridization of N in the pyridinium ion is sp^2.

19. The resultant cation is aromatic if the proton attaches to the C=0 oxygen (the more basic of the two oxygens).

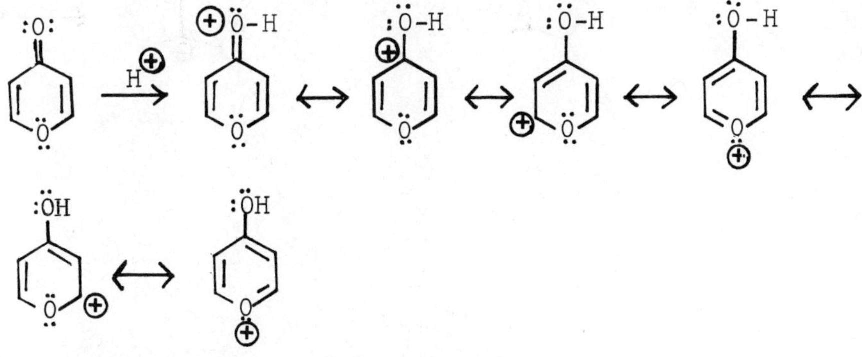

(Text: See tropone example, p. 583)

20. Delphinidin chloride has an aromatic system containing 10 π electrons. Oxygen is sp^2 hybridized.

21. At low temperature, the NMR spectrum of the [18]-annulene shows two absorption peaks, one for the inner protons and one for the outer protons (see section 15.7, text). A single line in the NMR spectrum at room temperature means the protons are equivalent at this temperature. This is thought to be because of rapid inside-outside rotation which the NMR spectrometer sees as the same average position for each proton.

22. Yes, NMR signals for outer protons of antiaromatic systems appear upfield of the protons of aromatic systems.

23. Check <u>your</u> definitions be referring to the following text sections. Hückel's rule, p. 593; annulene, p. 574; diradical, p. 578, 578; conjugated, p. 535; ring current, p. 594; resonance energy, p. 571; vertical resonance energy, p. 573; aromatic, p. 593; antiaromatic, p. 593.

24. Aromatic compounds are: a(6 π e^-), b(18 π e^-), d(14 π e^-), e(14 π e^- on the periphery), f(14 π e^-), h(6 π e^-)*, k(6 π e^-)**

*aromaticity due to resonance structures:

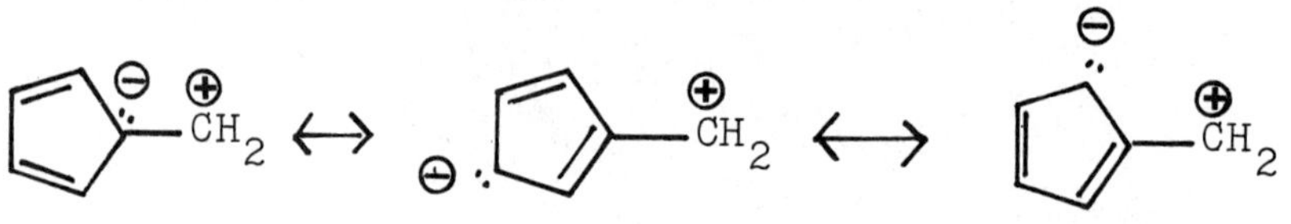

CH_2 ⟷ CH_2 ⟷ CH_2

**aromaticity due to resonance structures:

⟷ ⟷ ⟷ etc.

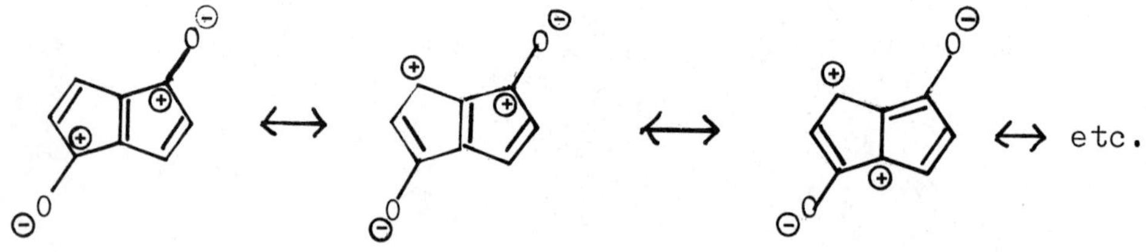

25. Compounds are aromatic if resonance contributors like B or C make both rings aromatic.

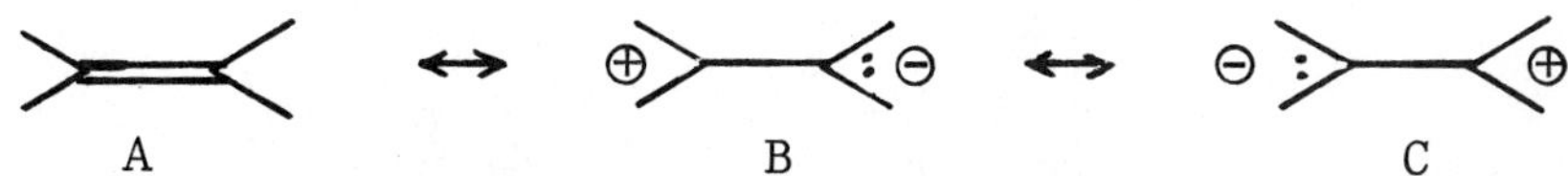

Aromatic compounds are:

(b) each ring has 6 π electrons

(e) three-membered ring has 2 π electrons; five-membered ring has 6 π electrons

Dipole moment studies could be used to demonstrate polarizations in these molecules.

26. Nitrogen and oxygen (period 2 elements) cannot accommodate more than 8 electrons; however, sulfur can <u>via</u> d orbitals. Thus, additional resonance structures can be drawn for thiophene (structures 6 and 7 below).

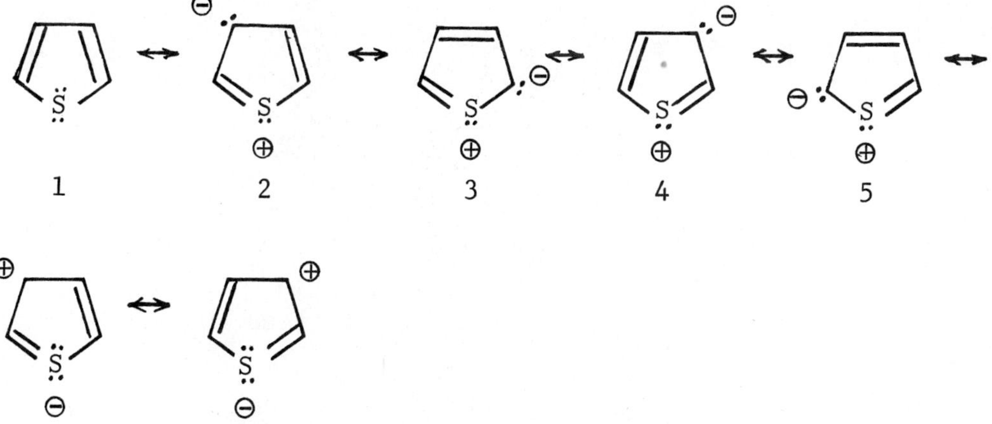

27. (a,b) A reacts with hydride to form a resonance-stabilized anion (an aromatic system containing 6 π electrons).

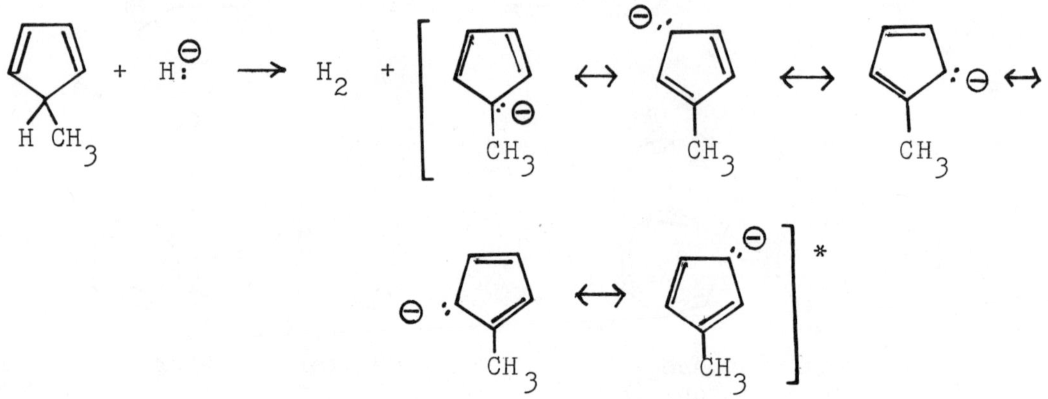

H CH3 + H:⊖ → H2 + [CH3 ↔ CH3 ↔ CH3 ↔ CH3 ↔ CH3]*

*equivalent to: [CH3]⊖

[CH3]⊖ —CH3I→ 1 + 2 + 3

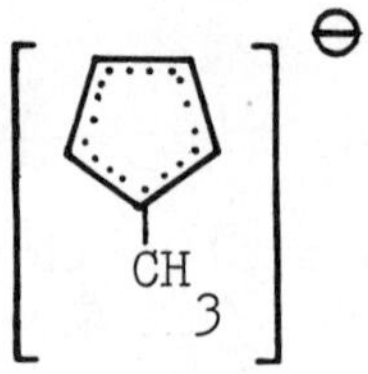

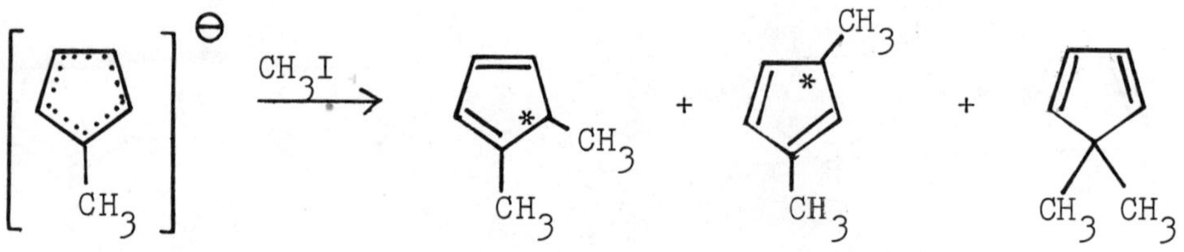

(c) Compounds 1 and 2 contain chiral centers (starred) and are capable of optical activity.

(d) Reaction of B with hydride ion gives the same resonance-stabilized anion; therefore, the same dimethylcyclopentadienes are formed.

28. No. Cycloheptatrienyl cation is aromatic and highly reactive. The two statements are not mutually exclusive since aromaticity is a ground-state (thermodynamic) argument and reactivity is a transition state (kinetic) argument.

The product is [structure: cycloheptatriene ring bearing H and OH on the same carbon]

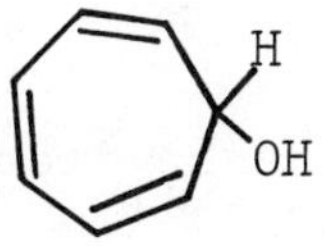

(Text: Review reactivity vs. stability, p. 181)

29. The anion of anthrone forms more easily since it is an aromatic system containing 14 π electrons as indicated by resonance contributor B.

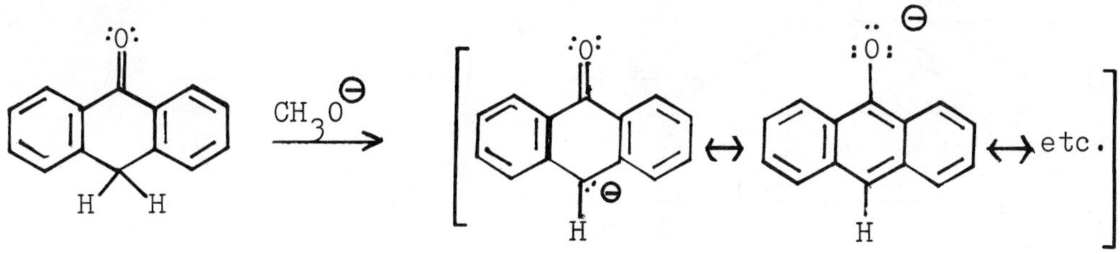

[structure: anthrone $\xrightarrow{CH_3O^{\ominus}}$ [A ↔ B ↔ etc.]]

A B

30. No. Although additional stability is associated with a system because it is aromatic, other criteria define an aromatic system, i.e., a planar, cyclic system containing $4n + 2\,\pi$ electrons with a p orbital on each ring atom permitting continuous overlap.

31. Compare the actual number of isomers formed in a reaction using benzene (such as halogenation) with the number which can be derived theoretically from the Ladenburg structure. For example, three nonresolvable dibromobenzenes are possible. The Ladenburg structure predicts three dibromo compounds one of which is resolvable.

32. A is [structure: pyrrole, N–H] The nonbonding electron pair on N is included in the π system giving 6 π electrons. The cation formed in acid solution is antiaromatic and, consequently, unstable.

33. Dimerization can occur by the Diels-Alder mechanism shown below.

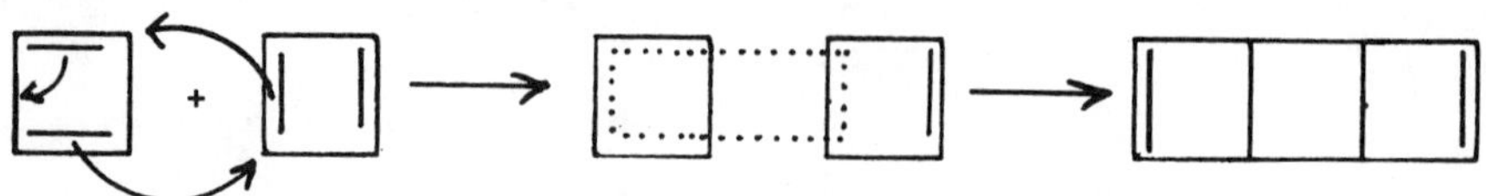

34. Recall that the resonance energy of benzene is determined by comparing 3 times the heat of hydrogenation of cyclohexene with the heat of hydrogenation of benzene (text, p. 571). If ethylene is an aromatic system, there is not a nonaromatic standard for comparison for the purpose of determining the resonance energy.

35. Results are most easily obtained by plotting the needed data on graph paper and estimating the value of β .

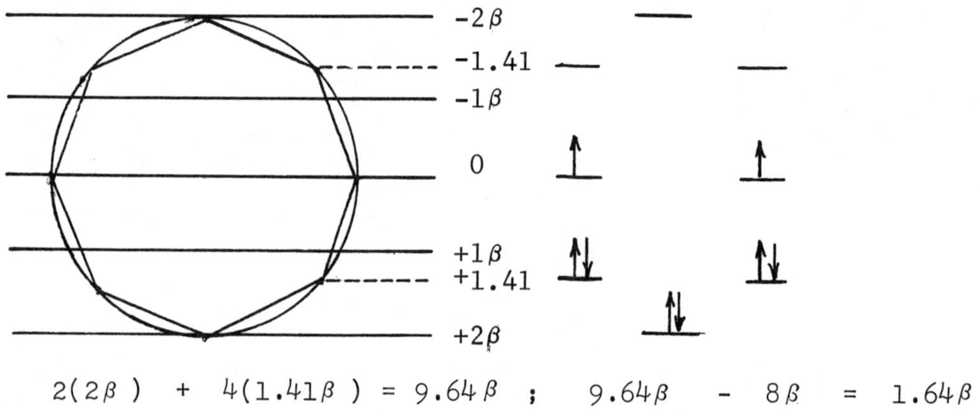

$2(2\beta) + 4(1.41\beta) = 9.64\beta$; $9.64\beta - 8\beta = 1.64\beta$

36. A planar [10]-annulene is destabilized by the repulsion of hydrogens inside the ring.

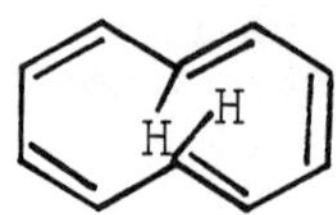

37. The major resonance contributors are:

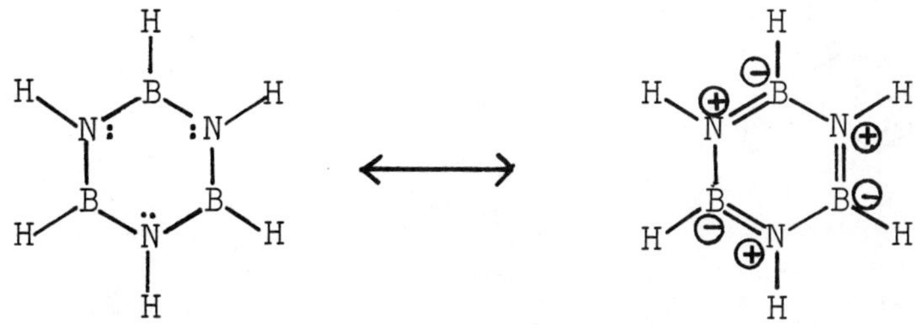

Borazole has a zero dipole moment.

38. Resonance contributors B → D provide an aromatic system and account for the unusually large dipole moment and for salt formation.

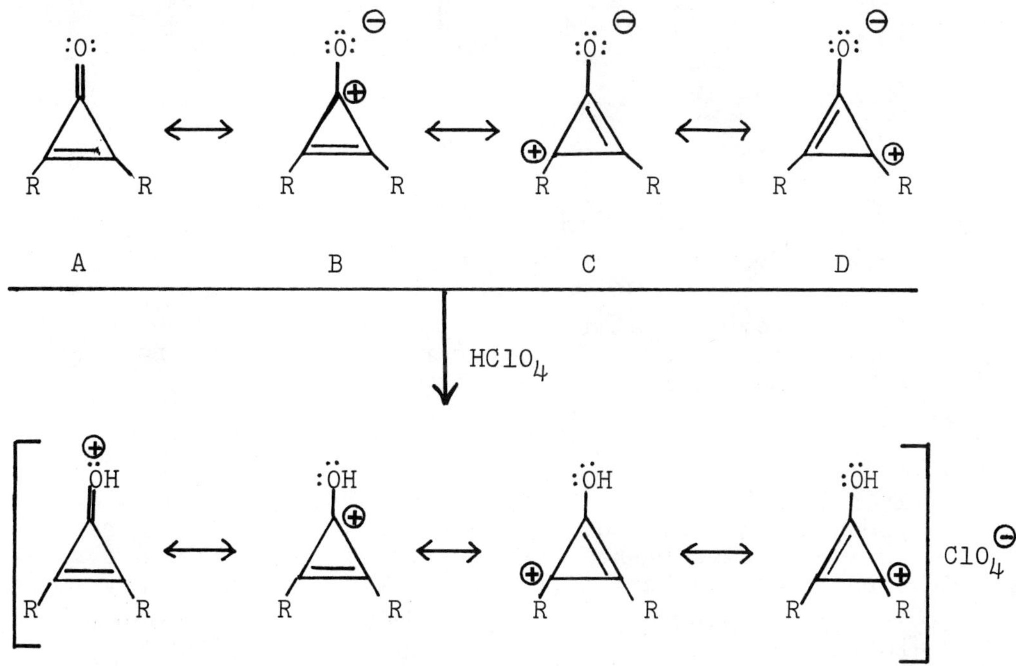

39. The isomers of pyrimidine are A and B below. Since A has four equivalent hydrogens (thus one nmr signal), it is pyrazine and B is pyridazine.

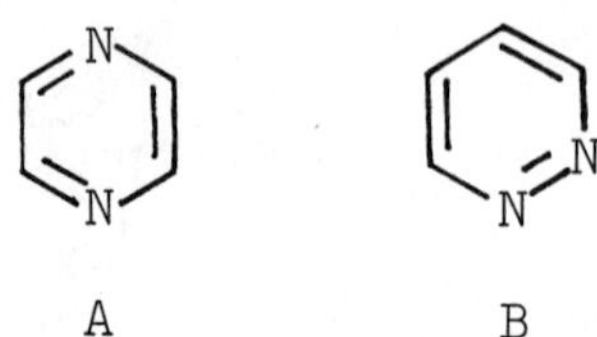

40. Hexahelicene is not planar but exists as a helix; thus the enantiomeric forms are right-handed and left-handed helices.

41. The Kekule resonance forms are:

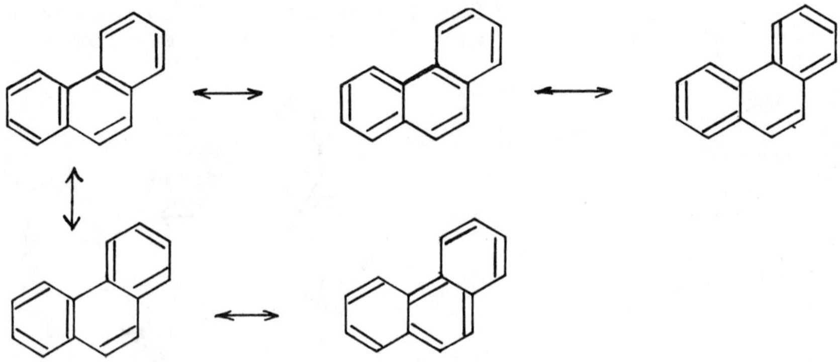

A double bond appears at the 9,10-position more often than at other positions. Thus, the 9,10 carbon-carbon bond has more double-bond character and is shorter.

42. (a) Yes.

(b) This is a 4n + 2 cyclization. The stereochemistry of the product requires a disrotatory reaction, thus the reaction is accelerated by heat (a photochemically induced cyclization involving 4n + 2 electrons follows conrotatory motion).

(c) The dihydronaphthalene is favored in the equilibrium because of crowding of the inside hydrogens in the [10]-annulene (see problem 36).

(Text S. 14.2)

43. (a) Rectangular cyclobutadiene need not be a diradical since there are no degenerate orbitals; i.e.,

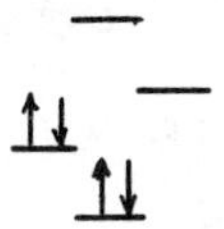

rather than

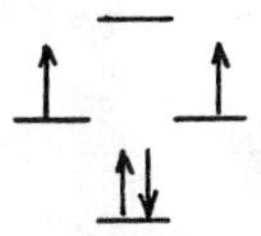

(b) The two resonance contributors of rectangular, planar cyclobutadiene are of unequal energy. The one on the right (long double, short single bonds) may be of too high energy to be significant.

44. + $SbCl_5$ → [ ] $SbCl_6^{\ominus}$

or $SbCl_6^{\ominus}$

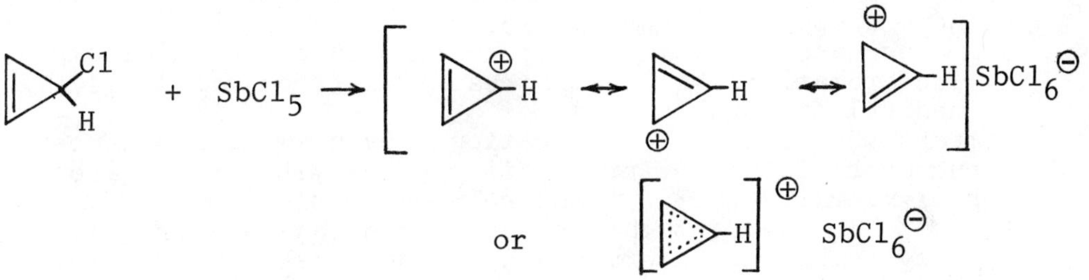

All hydrogens in this aromatic cation are equivalent and give a single line in the nmr spectrum.

45. A single line in the nmr spectrum requires that hydrogens be chemical shift equivalent. Loss of two chloride ions produces an aromatic cation (2 π electrons) in which all protons are equivalent.

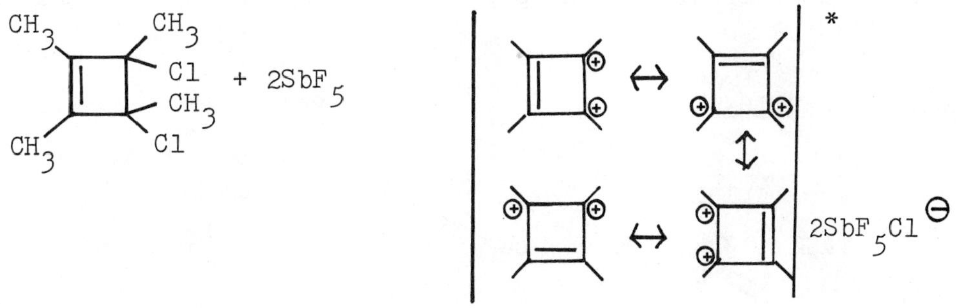

CH_3 CH_3 Cl CH_3 CH_3 Cl + $2SbF_5$

* $2SbF_5Cl^{\ominus}$

*equivalent to CH_3 CH_3 CH_3 CH_3 $2^{\oplus}$ $2SbF_5Cl^{\ominus}$

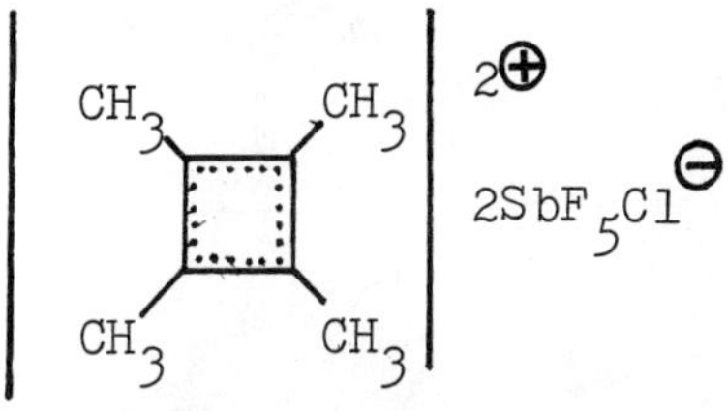

CHAPTER 16. ELECTROPHILIC AROMATIC SUBSTITUTION

LEARNING OBJECTIVES

When you have completed this chapter, you should be able to:

1. identify ortho, meta and para positions in monosubstituted benzene (problem 1 and related problem 27);
2. name or draw structures for benzene derivatives (problems 2, 3, 4, 5, 6, 7, 34, 35);
3. compare the mechanism of electrophilic aromatic substitution with that of the S_N1 reaction (problem 8);
4. write equations which illustrate the mechanisms for electrophilic aromatic substitution reactions (problems 16, 18, 19, 45, 46, 47, 54, 56, 57 and related problems 9, 10, 11, 12, 14, 40, 42);
5. write equations which illustrate the synthesis of substituted benzene compounds (problems 13, 38);
6. draw resonance structures for cations involved in electrophilic aromatic substitution reactions (problems 15, 17, 24, 55 and related problem 39);
7. compare the stabilities of cations (by drawing resonance structures) formed by protonation of alkylbenzenes (arenes) (problems 21, 22, 23, 44 and related problem 20);
8. account for the fact that groups are ortho-para or meta directors by comparing intermediate cation stabilities (problems 24, 25, 43 and related problems 26, 32, 49);
9. predict products in electrophilic aromatic substitution reactions (problems 28, 37);
10. account for the reactivity of benzylic hydrogens in free-radical halogenation (problem 29);
11. predict products in the oxidation of alkylbenzenes (arenes) (problem 30);
12. explain the NMR signals in the spectra of benzene derivatives (problems 33, 58);
13. account for the reactivity of compounds in electrophilic aromatic substitution (problems 41, 48, 50 and related problem 51).

ANSWERS TO QUESTIONS

1. 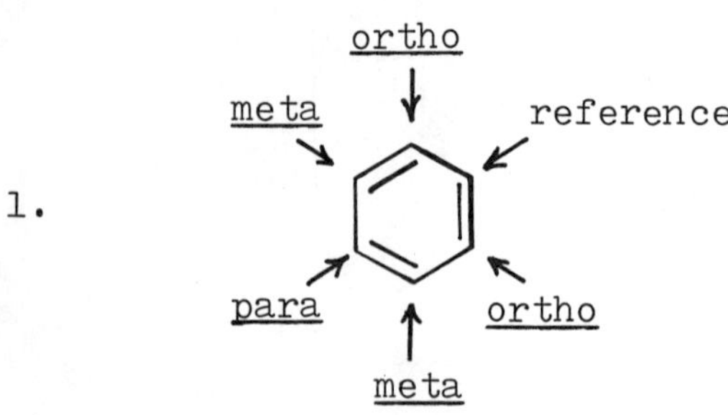

2. (a) m-dibromobenzene
   (b) o-bromonitrobenzene
   (c) p-chloroaniline
   (d) p-aminoaniline

3. (a) 1,2-bis-(p-chlorophenyl)ethane
   (b) 1-(p-bromophenyl)-2-butyne
   (c) 1-(o-chlorophenyl)ethanol
   (d) 1-(p-chlorophenyl)ethanol

4. (a) 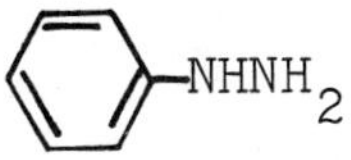 C_6H_5—NHNH$_2$ (b) 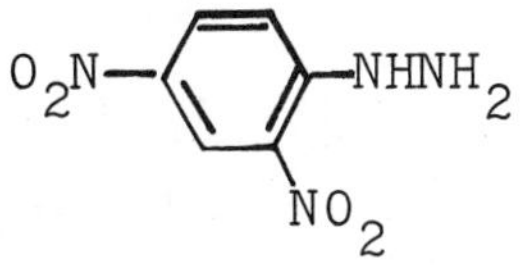 O_2N—C_6H_3(NO_2)—NHNH$_2$

5. CHO, OH

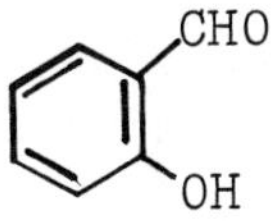

6. (a) 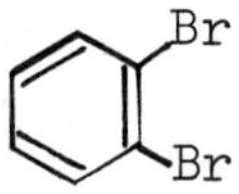 Br, Br (d) 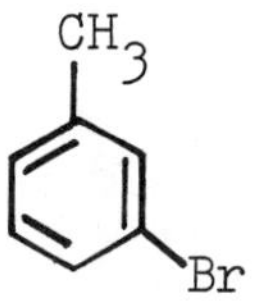 CH_3, Br

   (b) 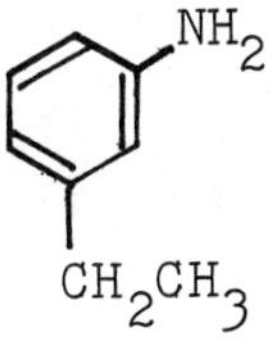 NH_2, CH_2CH_3 (e) 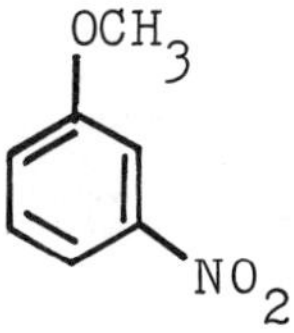 OCH_3, NO_2

   (c) 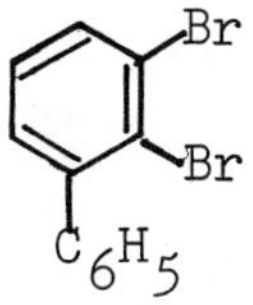 Br, Br, C_6H_5 (f) 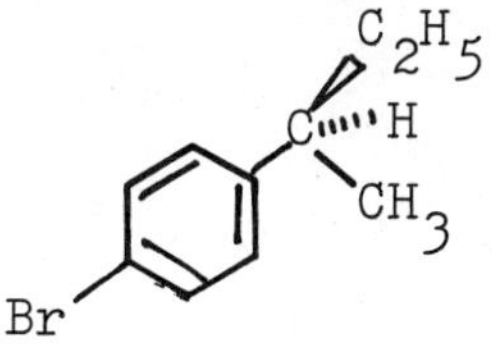 Br, C_2H_5, C, H, CH_3

(continued next page)

(g) 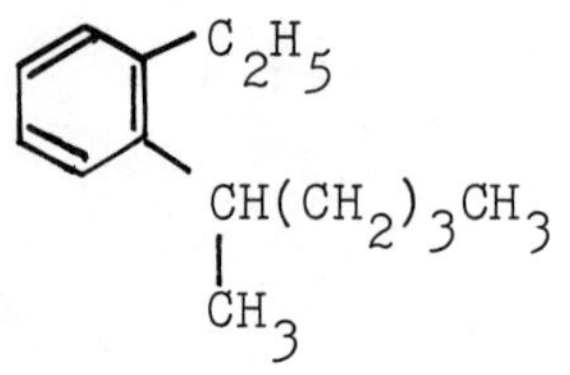 [structure: benzene ring with C_2H_5 and $CH(CH_2)_3CH_3$ / CH_3 substituents]

C_2H_5

$\underset{\displaystyle CH_3}{CH}(CH_2)_3CH_3$

7. (a) Should be named as a toluene (o-bromotoluene),
   (b) Should be named as a toluene (m-nitrotoluene),
   (c) Numbered incorrectly. Should be 1,2,4-trinitrotoluene,
   (d) Should be named as an aniline (m-chloroaniline),
   (e) The prefix "di" omitted. Should be 1,2-dichlorobenzene.

8. (a) The initial reaction step in electrophilic aromatic substitution and in S_N1 reactions produces a cation. This is the rate-determining step in each.

first step

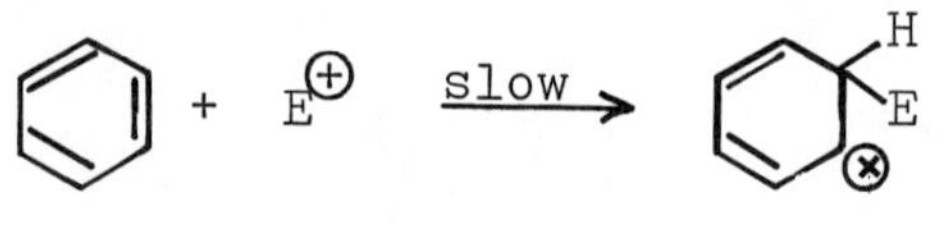

benzene + $E^{\oplus}$ $\xrightarrow{\text{slow}}$ (cyclohexadienyl cation bearing H and E, ⊕) (electrophilic aromatic substitution)

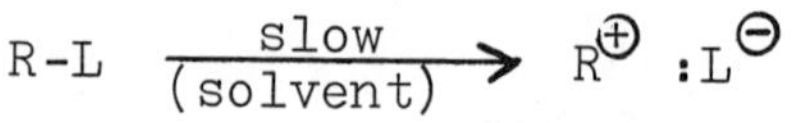

$R\text{-}L \xrightarrow[\text{(solvent)}]{\text{slow}} R^{\oplus} \; :L^{\ominus}$ (S_N1)

(b) The second step in each is different. In electrophilic aromatic substitution the cation loses a proton, and in an S_N1 reaction the cation reacts with a nucleophile.

second step

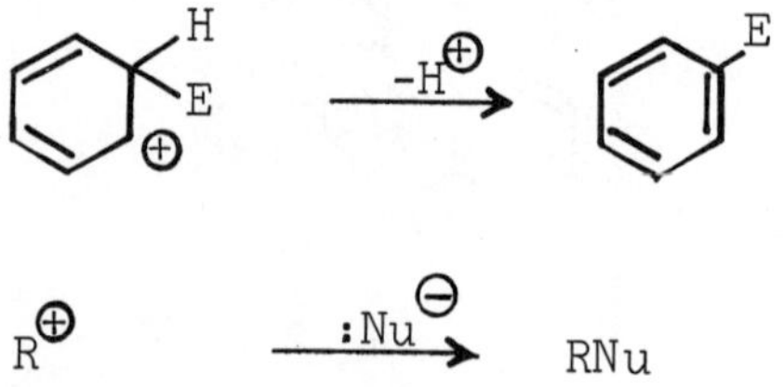

(cyclohexadienyl cation bearing H and E, ⊕) $\xrightarrow{-H^{\oplus}}$ (benzene ring bearing E)

$R^{\oplus} \xrightarrow{:Nu^{\ominus}} RNu$

(b)

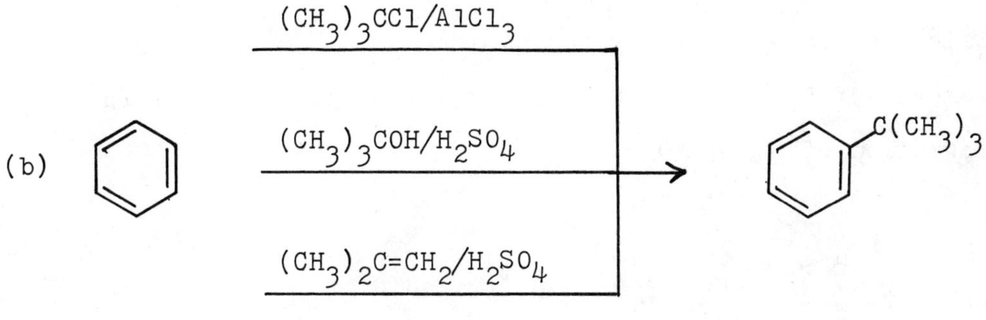

(c) Benzene $\xrightarrow{\substack{CH_3CH(Cl)CH_2CH_3/BF_3 \\ CH_3CH{=}CHCH_3/H_3PO_4 \\ CH_3CH(OH)CH_2CH_3/H_2SO_4}}$ C_6H_5–$C(CH_3)(H)$–CH_2CH_3

(d) Benzene $\xrightarrow{\substack{CH_3CH_2Br/AlCl_3 \\ CH_2{=}CH_2/H_3PO_4 \\ CH_3CH_2OH/H_2SO_4}}$ $C_6H_5CH_2CH_3$

9. (a) water is converted to hydronium ion, $H_3O^{\oplus}$

(b) removal of H_2O as $H_3O^{\oplus}$ will shift the equilibrium toward formation of $H_2O + NO_2^{\oplus}$.

10. $2H\ddot{O}NO_2 \rightleftharpoons H_2\ddot{O}^{\oplus}\text{-}NO_2 \rightleftharpoons NO_2^{\oplus} + H_2O$

$+ NO_3^{\ominus}$

11. Benzene reacts with the electrophile produced when chlorine and aluminum bromide react. Bromine is present only in the form of the nucleophile $Br^{\ominus}$.

$:\ddot{C}l\text{-}\ddot{C}l: + AlBr_3 \longrightarrow :\ddot{C}l^{\delta\oplus}\cdots\cdots\ddot{C}l\cdots\cdots AlBr_3^{\delta\ominus}$

electrophile

12. (a) $H^{\oplus}\ AlCl_3X^{\ominus} \longrightarrow HCl + AlCl_2X$

(b) No

13. (a) benzene $\xrightarrow{CH_3Cl/AlCl_3 \text{ or } CH_3OH/H_2SO_4}$ toluene ($C_6H_5CH_3$)

(continued next page)

14. Fluoride ion is a much poorer nucleophile than bromide ion. Bromide ion tends to add to initially formed isopropyl cation much faster than does fluoride ion.

$$CH_3CH{=}CH_2 \xrightarrow{HBr} CH_3\overset{\oplus}{C}HCH_3\ Br^{\ominus} \xrightarrow{fast} CH_3CH(Br)CH_3$$

$$CH_3CH{=}CH_2 \xrightarrow{HF} CH_3\overset{\oplus}{C}HCH_3\ F^{\ominus} \xrightarrow{very\ slow} CH_3CH(F)CH_3$$

Since there is less addition by $F^{\ominus}$, alkylation of benzene by the isopropyl cation occurs with higher yield.

15. $CH_3\overset{\oplus}{C}{=}\ddot{O} \longleftrightarrow CH_3C{\equiv}\overset{\oplus}{O}:$

16. $R{-}\overset{O}{\overset{\|}{C}}{-}\ddot{O}{-}H \xrightarrow{H{-}F} R{-}\overset{O}{\overset{\|}{C}}{-}\overset{\oplus}{O}H_2 \xrightarrow{-H_2O} \left[R{-}\overset{\oplus}{C}{=}\ddot{O} \longleftrightarrow R{-}C{\equiv}\overset{\oplus}{O}:\right]$

$\left[R{-}\overset{\oplus}{C}{=}\ddot{O} \longleftrightarrow R{-}C{\equiv}\overset{\oplus}{O}:\right]$ + benzene ⇌ arenium ion ($C(H)(COR)$, ⊕) + $:\ddot{O}H_2$

$\xrightarrow{-\ H_3O^{\oplus}}$ $C_6H_5{-}\overset{O}{\overset{\|}{C}}{-}R$

17. 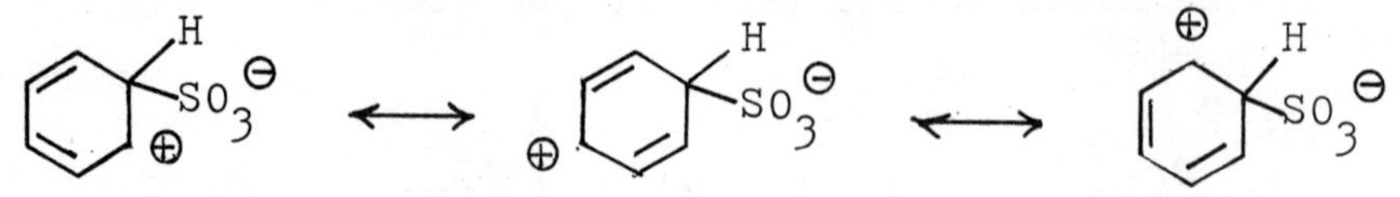 [resonance structures: cyclohexadienyl cation bearing H and $SO_3^{\ominus}$, positive charge at ortho, para, and ortho positions] $\longleftrightarrow$ $\longleftrightarrow$

18. (a) $H_2O + C_6H_5SO_3H \rightleftharpoons H_3O^{\oplus} + C_6H_5SO_3^{\ominus}$

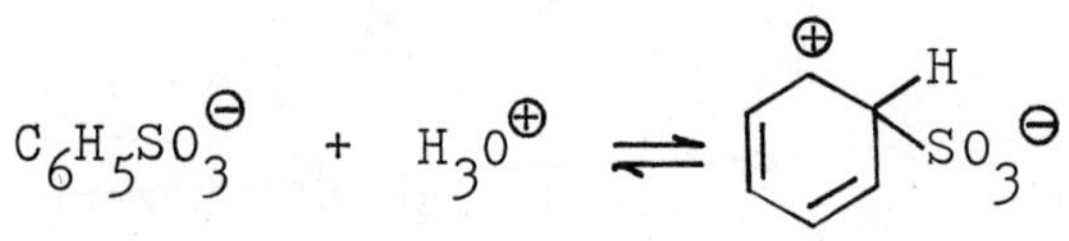

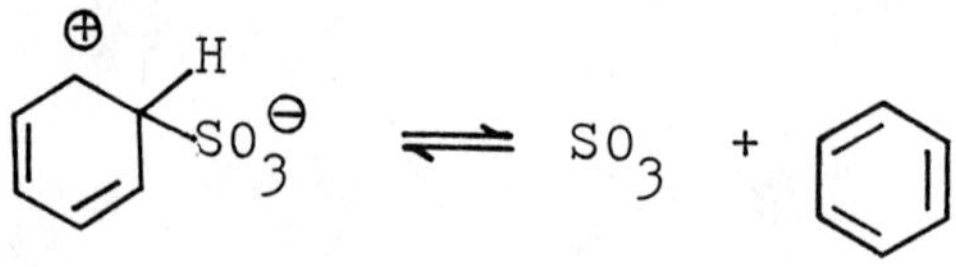

$C_6H_5SO_3^{\ominus} + H_3O^{\oplus} \rightleftharpoons$ [cyclohexadienyl cation with H and $SO_3^{\ominus}$]

[cyclohexadienyl cation with H and $SO_3^{\ominus}$] $\rightleftharpoons SO_3 +$ [benzene]

(b) $H^{\oplus}$

(c) 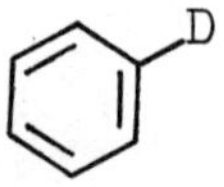 [benzene ring with D]

19. (a) $Br\text{-}Br + FeBr_3 \longrightarrow \overset{\delta\oplus}{Br}\cdots\cdots Br\cdots\cdots \overset{\delta\ominus}{FeBr_3}$

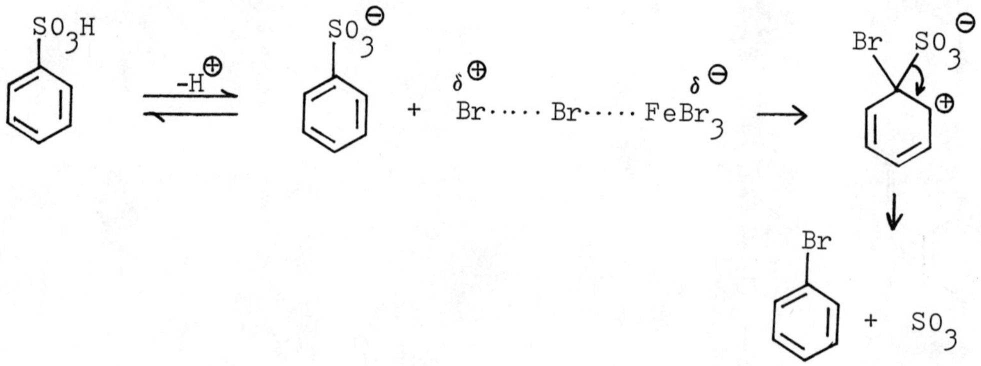

[benzene with SO_3H] $\underset{}{\overset{-H^{\oplus}}{\rightleftharpoons}}$ [benzene with $SO_3^{\ominus}$] $+ \overset{\delta\oplus}{Br}\cdots\cdots Br\cdots\cdots \overset{\delta\ominus}{FeBr_3} \longrightarrow$ [cyclohexadienyl cation with Br and $SO_3^{\ominus}$] $\downarrow$ [bromobenzene] $+ SO_3$

(continued next page)

(b) Yes. However, unlike $-SO_3^{\ominus}$, most groups attached to the benzene ring are poor leaving groups ($-SO_3^{\ominus}$ is the only group that leaves as a neutral molecule), and this reaction is not common.

20.

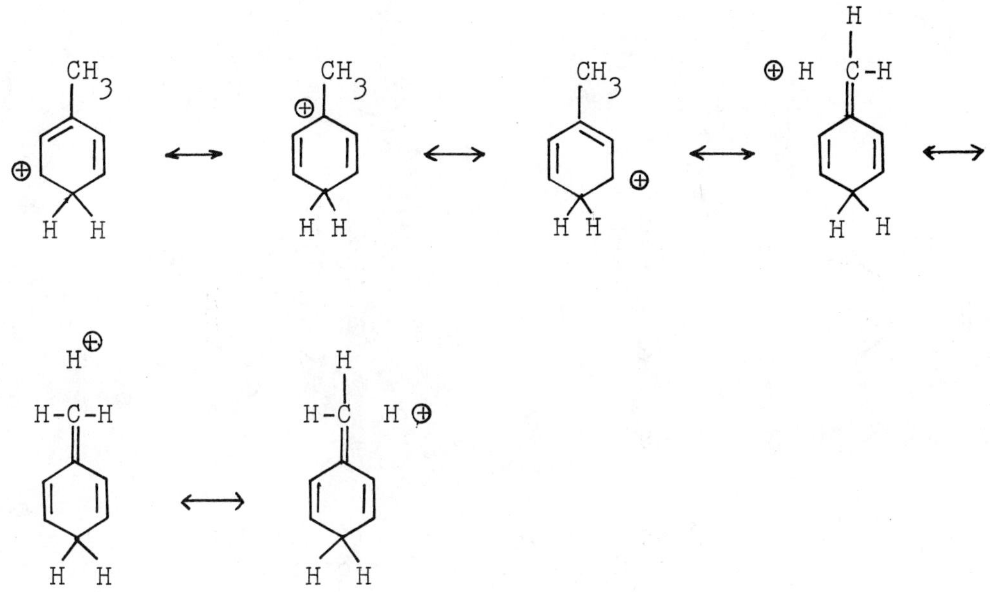

(Text: hyperconjugation, p. 186)

21.

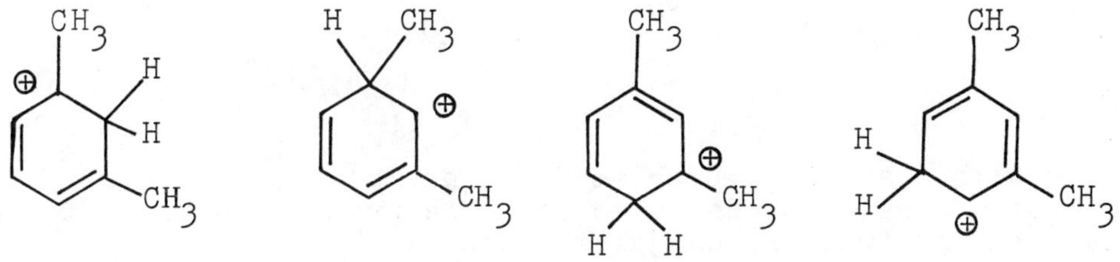

A B C D

A and C are most stable since in each the positive charge may reside on both carbons holding the methyl groups, e.g., for C:

(continued next page)

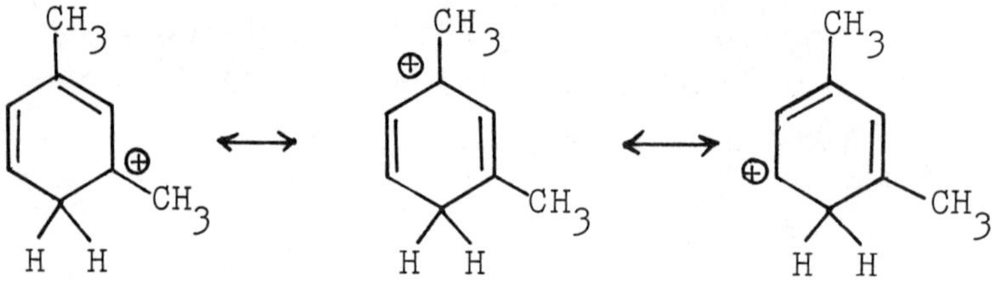

22.

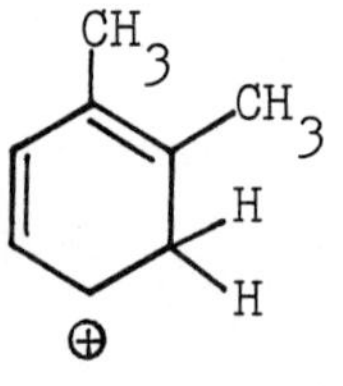

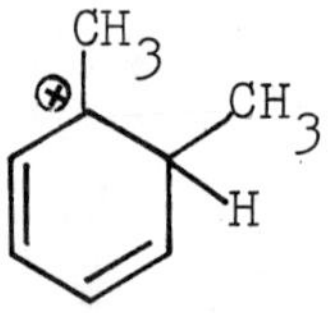

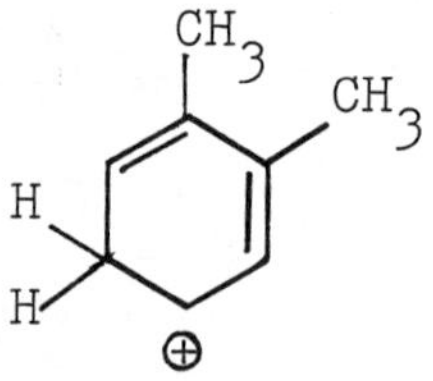

A B C

None of the protonated o-xylenes can utilize both methyl groups to delocalize the positive charge. For example for A:

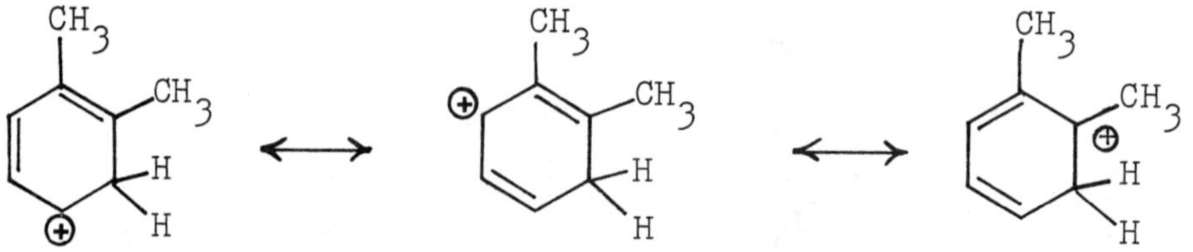

Compare with A and C in problem 21. Thus m-xylene is a stronger base (has a greater tendency to accept a proton) because it forms a lower-energy (weaker) conjugate acid.

23. (a) Only one monoprotonated 1,2,3,4,5,6-hexamethylbenzene is possible.

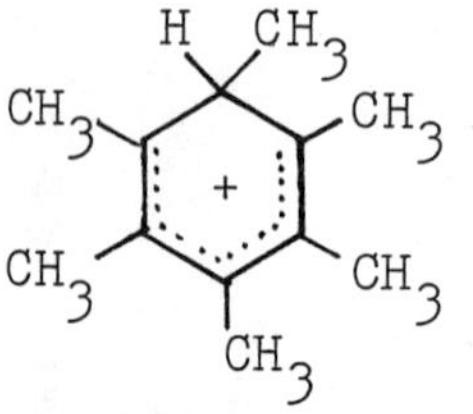

(continued next page)

(b) Diprotonation produces a higher energy cation having two units of positive charge.

24. (a) 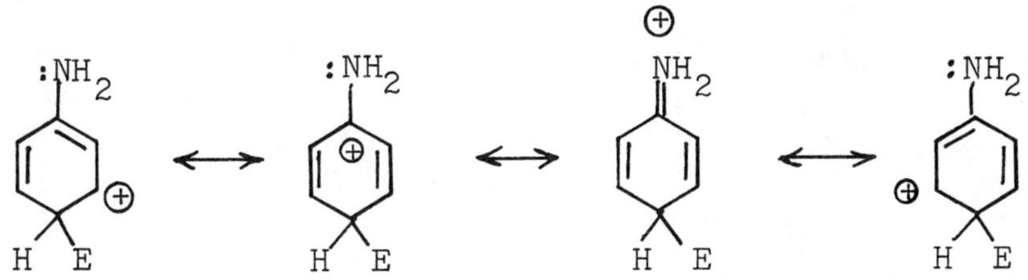 [resonance structures: $:NH_2$ ring cation with H, E; ↔ ; ↔ $\overset{\oplus}{N}H_2$; ↔]

H E ↔ H E ↔ H E ↔ H E

(b) No, since the positive charge does not appear on the carbon bonded to the NH_2 group, only three can be drawn.

(c) There is greater delocalization of charge in the intermediate in which bromine attacks <u>ortho</u> to the methoxy group. The resonance structures are:

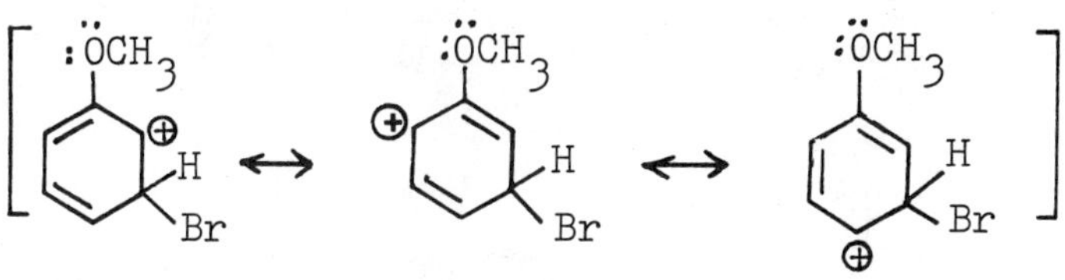

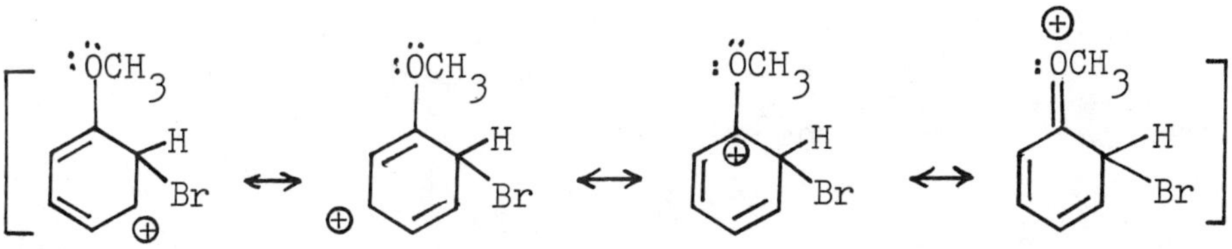

[$:\ddot{O}CH_3$ / H / Br ↔ $:\ddot{O}CH_3$ / H / Br ↔ $:\ddot{O}CH_3$ / H / Br]

[$:\ddot{O}CH_3$ / H / Br ↔ $:\ddot{O}CH_3$ / H / Br ↔ $:\ddot{O}CH_3$ / H / Br ↔ $\overset{\oplus}{:OCH_3}$ / H / Br]

25. Both -OH and $-O^{\ominus}$ groups are o,p-directors. A comparison of the cations which result from the attack of an electrophile p- and m- to the -OH group is shown on next page.

(continued next page)

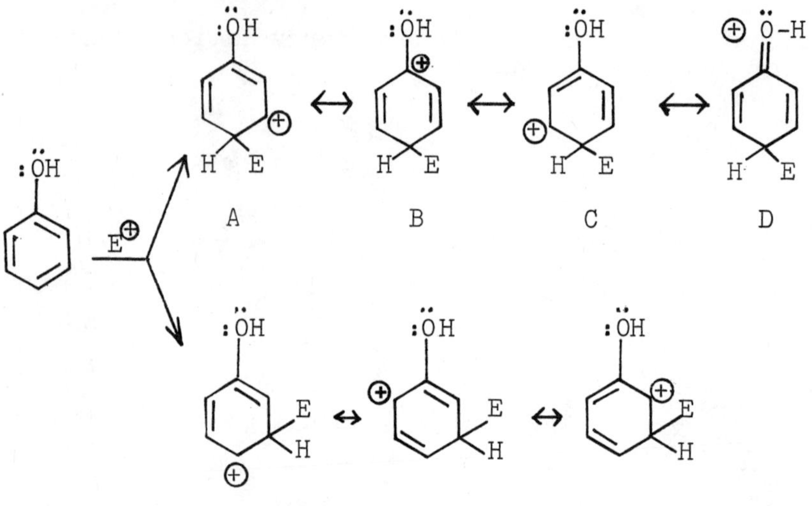

A B C D

E F G

The oxygen of the -OH group can assist in delocalizing the positive charge by donating a nonbonding electron pair to the positively charged ring only when the electrophile attaches para (or ortho) to the -OH.

A ⟶ D equivalent to

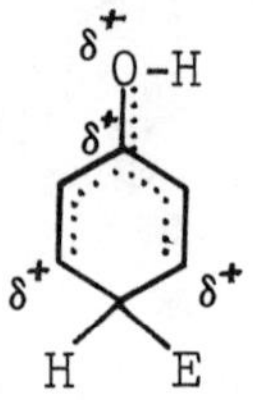

E ⟶ G equivalent to

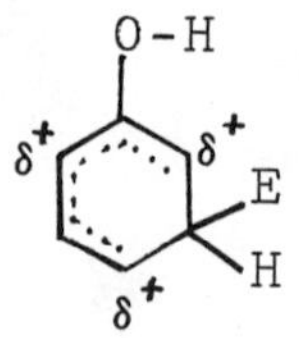

The comparison with the $-O^{\ominus}$ group is:

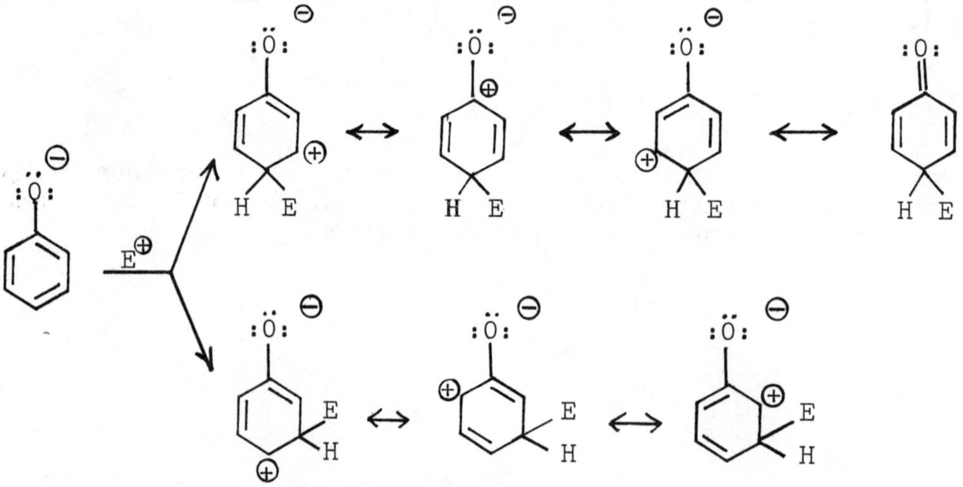

(continued next page)

The $-O^{\ominus}$ group, like the -OH group, is an ortho, para-director since the oxygen can participate in resonance stabilization if the electrophile attacks ortho or para to the $-O^{\ominus}$.

26. The bulk of the product (due to o,p-substitution) comes from aniline present in low concentration in the equilibrium mixture:

$$C_6H_5-\overset{\oplus}{N}H_3\overset{\ominus}{Cl} \rightleftharpoons C_6H_5-\ddot{N}H_2 + HCl$$

27. There are two positions ortho and two meta but only one position para to the group of a monosubstituted benzene. Thus out of 5 positions total, the statistical percent for meta is:

$$2/5 \times 100 = 40\% \quad \text{etc.}$$

28. (a) p-Bromochlorobenzene (Br and Cl para on benzene ring) + ortho

(b) CH_3 para to CCH_2CH_3 (C=O) on benzene ring + ortho

(c) $C(CH_3)_3$ para to I on benzene ring

(d) $C_6H_5-CH_2-$ ring with Br and $-OCH_3$ (Br ortho to OCH_3)

(e) SO_3H and NO_2 meta on benzene ring

29. (a) In the ethyl group of phenylethane there are three 1° hydrogens and two 2° hydrogens. The statistical amounts are:

$3/5 \times 100 = 60\%$ 2-chloro-1-phenylethane and
$2/5 \times 100 = 40\%$ 1-chloro-1-phenylethane

(b) The 2° hydrogens are also benzylic and substitution of a benzylic hydrogen gives a resonance-stabilized free radical.

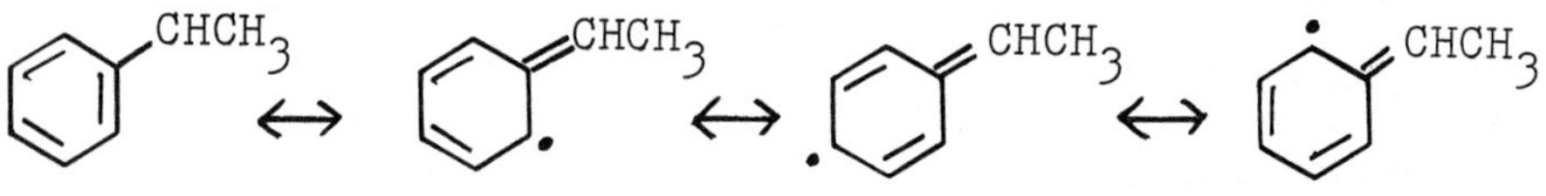

$C_6H_5\dot{C}HCH_3 \leftrightarrow$ (ring radical resonance forms) CHCH$_3$ $\leftrightarrow$ CHCH$_3$ $\leftrightarrow$ CHCH$_3$

(c) $Cl_2 \xrightarrow{h\nu} 2Cl\cdot$

$Cl\cdot + C_6H_5CH_2CH_3 \longrightarrow HCl + C_6H_5CH_2\dot{C}H_2$

$C_6H_5CH_2\dot{C}H_2 + Cl_2 \longrightarrow C_6H_5CH_2CH_2Cl + Cl\cdot$

etc.

30. (a),(b),(c) o-chlorobenzoic acid (Cl, CO_2H)

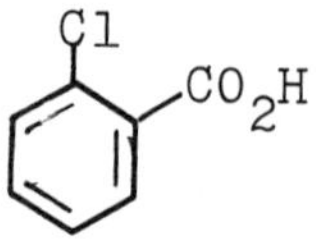

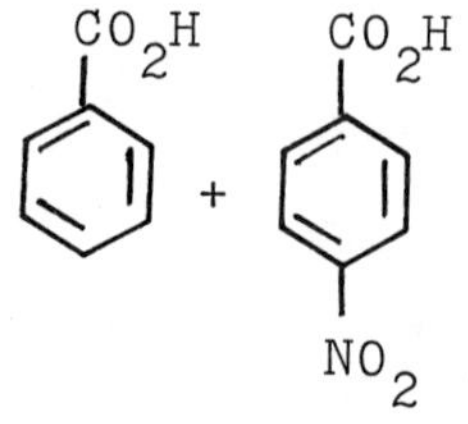

(d) phthalic acid (CO_2H, CO_2H)

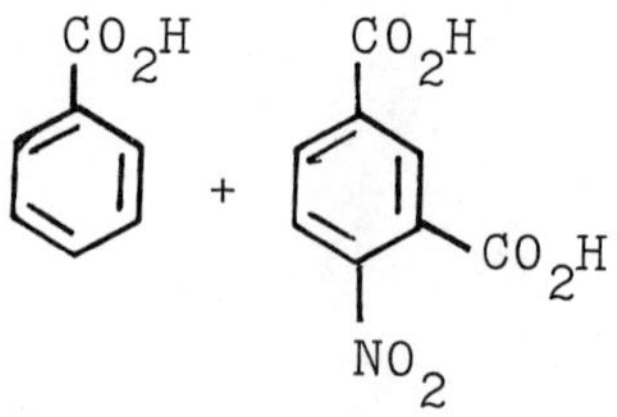

(e) benzoic acid (CO_2H)

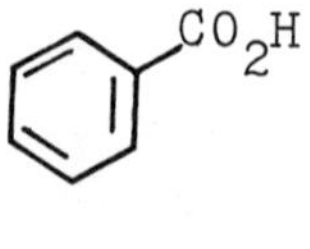

(f) $HO_2C-C_6H_4-CO_2H$ + $O_2N-C_6H_4-CO_2H$

(g) $C_6H_5CO_2H$ + p-$O_2NC_6H_4CO_2H$ (CO_2H, NO_2)

(h) $C_6H_5CO_2H$ + (CO_2H, CO_2H, NO_2)

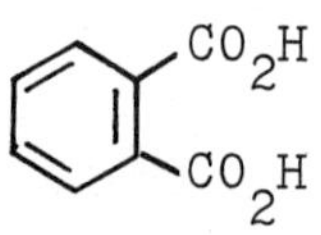

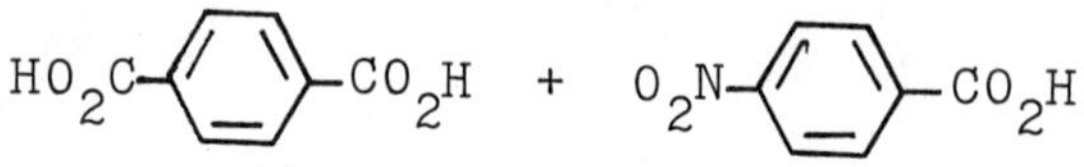

31. Thiophene is converted to a water-soluble sulfonic acid derivative and removed by washing.

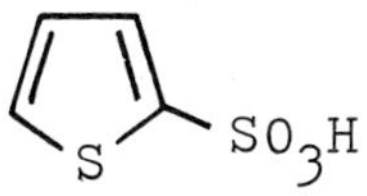

SO_3H

water soluble

Fuming sulfuric acid will sulfonate benzene as well as thiophene, and benzene sulfonic acid is also water soluble.

32. Reaction of an electrophile at C-2 of pyrrole gives a cation more stable (three resonance forms) than the one formed by attack at C-3 (two resonance forms).

Reaction of an electrophile at C-3 of indole gives an intermediate for which two resonance forms can be drawn with the aromaticity of the other ring maintained. Only one such structure can be drawn if $E^{\oplus}$ attacks at C-2.

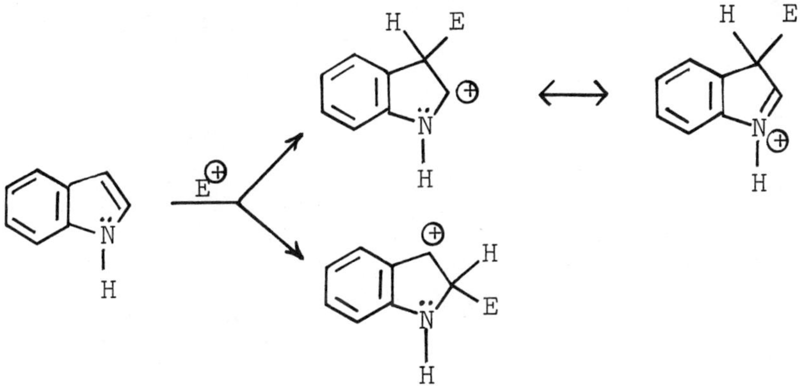

33. The signal for the methyl protons, "a", is split into a doublet by the one adjacent proton, "c". The signal for proton "c" is split into a heptet by the six adjacent methyl protons (examples of the n + 1 rule).

34. (a) 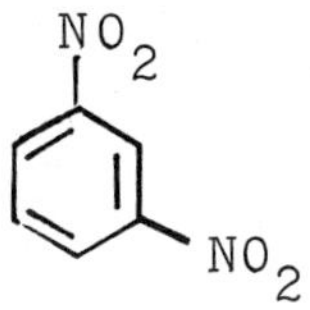 NO_2, NO_2 (b) 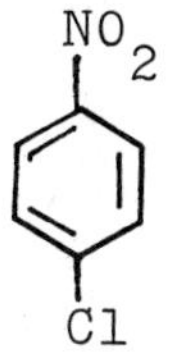 NO_2, Cl (c) same as (b)

(continued next page)

(d) [structure: 3,4-dimethylnitrobenzene; labels CH_3, CH_3, NO_2]

(e) [structure: 1,3,5-trideuteriobenzene; labels D, D, D]

(f) [structure: 2-naphthol; label OH]

(g) [structure: 1-bromonaphthalene; label Br]

(h) [structure: 2-aminopyridine; labels N, NH_2]

(i) [structure: nicotinic acid; labels CO_2H, N]

(j) [structure: pyridine with CH_2OH, $HOCH_2$, OH, N, CH_3]

(k) [structure: 9,10-dimethylanthracene; labels CH_3, CH_3]

(l) [structure: 2,4,6-trinitrotoluene; labels CH_3, O_2N, NO_2, NO_2]

(m) [structure: 2-acetylthiophene; labels O, S, $\overset{\|}{C}CH_3$]

(Text S. 16.2)

35. (a) [structure: 1,2,4-tribromobenzene; labels Br, Br, Br]

(b) [structure: 1-bromo-3,5-dichlorobenzene; labels Cl, Cl, Br] or [structure: 2-bromo-1,3-dichlorobenzene; labels Cl, Br, Cl]

(c) [structure: p-xylene; labels CH_3, CH_3]

36. The student assumed that the smoke was hydrogen bromide produced in the bromination of the aromatic compound.

37. (a) NR

(b) 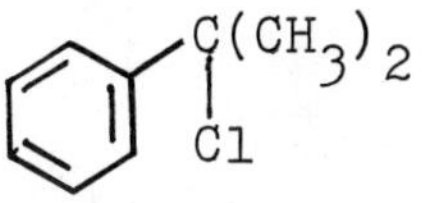 $C_6H_5C(Cl)(CH_3)_2$

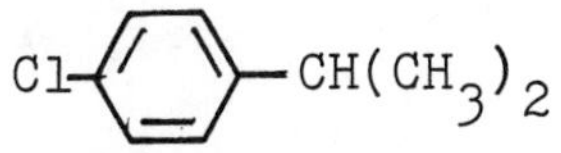

(c) Cl–C_6H_4–$CH(CH_3)_2$ (para)
(and ortho)

(d) 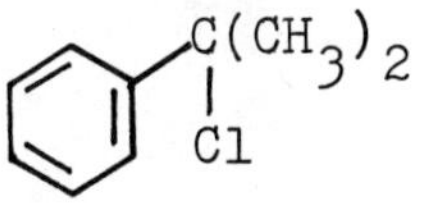 $C_6H_5C(Cl)(CH_3)_2$

(e) NR

(f) NR

(g) NR (fuming H_2SO_4 required)

(h) 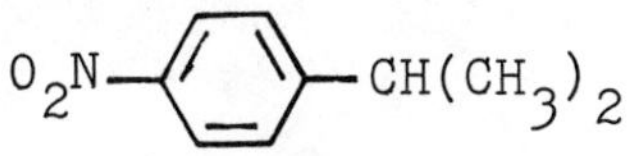 O_2N–C_6H_4–$CH(CH_3)_2$ (para)
(and ortho)

(i) 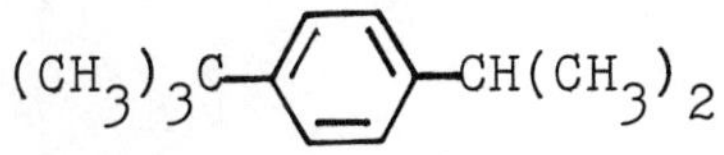 $(CH_3)_3C$–C_6H_4–$CH(CH_3)_2$ (para)

(j) 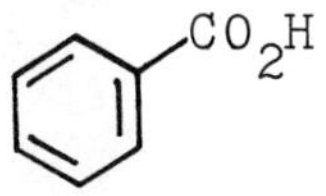 $C_6H_5CO_2H$

(k) NR

(l) NR

(m) 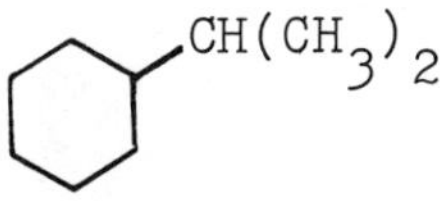 $C_6H_{11}CH(CH_3)_2$ (cyclohexyl)

(n) NR

(o) 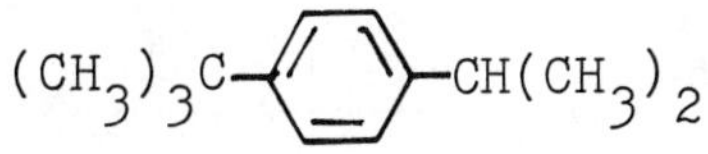 $(CH_3)_3C$–C_6H_4–$CH(CH_3)_2$ (para)

(p) NR

(q) NR

(r) 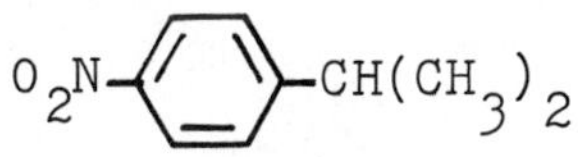 O_2N–C_6H_4–$CH(CH_3)_2$ (para)
(and ortho)

(s)  $(CH_3)_2CH$–C_6H_4–$CH(CH_3)_2$ (para)

38. benzene, C_6H_6

(a) $C_6H_6 \xrightarrow{C_2H_5Cl/AlCl_3}$

(b) $C_6H_6 \xrightarrow{CH_3Cl/AlCl_3} \xrightarrow{NO_2^{\oplus}\ BF_4^{\ominus}}$

(c) [o-nitrotoluene: benzene ring with NO_2 and CH_3 on adjacent carbons] (from b) $\xrightarrow{Br_2/FeBr_3}$

(d) $C_6H_6 \xrightarrow{D_2SO_4}$

(e) $C_6H_6 \xrightarrow{CH_3Cl/AlCl_3} \xrightarrow{H_2SO_4 \cdot SO_3}$

(f) $CH_3-C_6H_4-SO_3H$ (from e) $\xrightarrow[heat]{D_2O}$

(g) $C_6H_5CH_3$ (from b) $\xrightarrow[heat]{KMnO_4/OH^{\ominus}} \xrightarrow{H^{\oplus}}$

(h) $C_6H_5CH_3$ (from b) $\xrightarrow{Cl_2/FeCl_3}$ $Cl-C_6H_4-CH_3$

$\xrightarrow{KMnO_4/OH^{\ominus}} \xrightarrow{H^{\oplus}}$

(i) $C_6H_5CH_3$ (from b) $\xrightarrow{Cl_2/light}$

(continued next page)

(j) $C_6H_5CH_2Cl$ (from i) $\xrightarrow{Na^{\oplus}I^{\ominus}/acetone}$

(k) $C_6H_5CH_2Cl$ (from i) $\xrightarrow{Mg/ether}$ $\xrightarrow{D_2O}$

(l) $C_6H_5CH_3$ (from b) $\xrightarrow{CH_3Cl/AlCl_3}$ $\xrightarrow{KMnO_4/OH^{\ominus}}$ $\xrightarrow{H^{\oplus}}$

(m) $C_6H_5CH_2Cl$ (from i) $\xrightarrow{C_6H_6/AlCl_3}$ $C_6H_5CH_2C_6H_5$ $\xrightarrow{Cl_2/FeCl_3}$

(n) C_6H_6 $\xrightarrow{CH_3CH_2Cl/AlCl_3}$ $\xrightarrow{Cl_2/h\nu}$ $C_6H_5CH(Cl)CH_3$ $\xrightarrow{C_6H_6/AlCl_3}$

(o) [phthalic acid: benzene ring with two adjacent COOH groups] (from l) $\xrightarrow[-H_2O]{heat}$ [phthalic anhydride] $\xrightarrow[AlCl_3]{C_6H_6}$

[2-benzoylbenzoic acid: C(=O) linking phenyl and a ring bearing COOH] $\xrightarrow[HCl]{Zn(Hg)}$ [2-benzylbenzoic acid: CH_2 linking phenyl and a ring bearing COOH] $\xrightarrow{HF}$

[anthrone (C=O)] $\xrightarrow[HCl]{Zn(Hg)}$ [9,10-dihydroanthracene] $\xrightarrow[heat]{Pt}$ [anthracene]

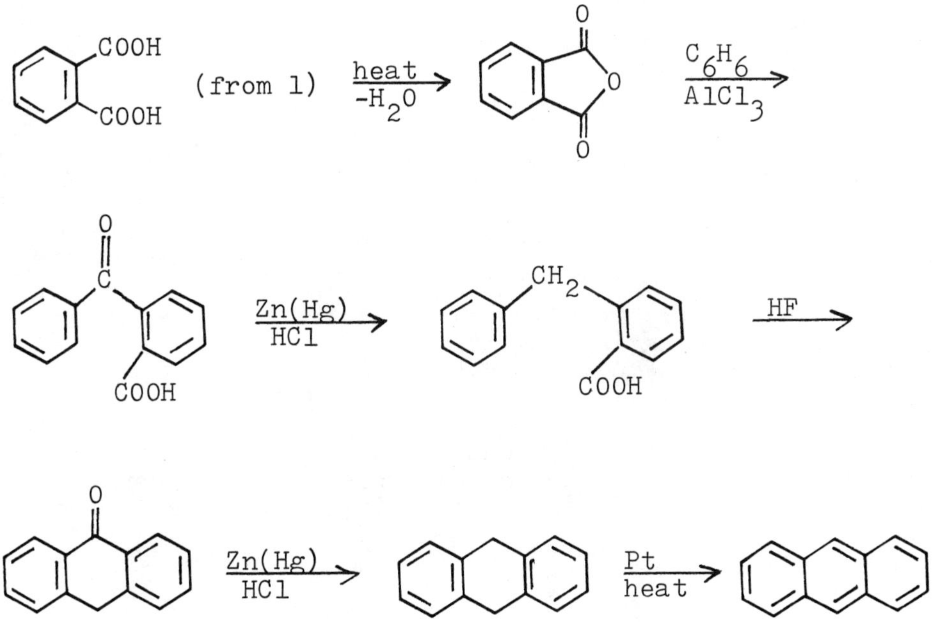

(continued next page)

(p) $C_6H_5CH_3$ (from b) + [succinic anhydride] $\xrightarrow[CS_2]{AlCl_3}$ [p-$CH_3C_6H_4C(=O)(CH_2)_2COOH$]

$\xrightarrow[HCl]{Zn(Hg)}$ $\xrightarrow{HF}$

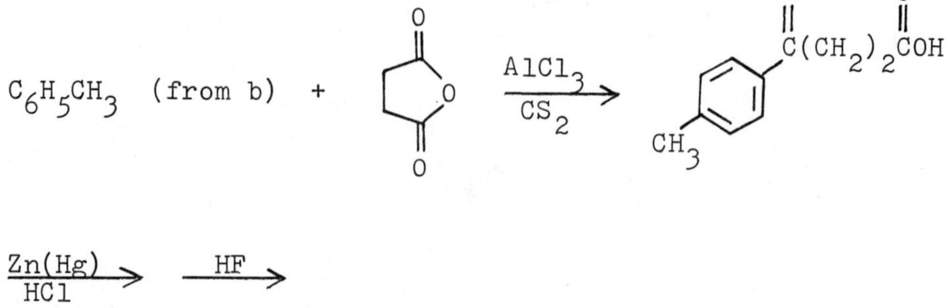

(q) [7-methyl-1-tetralone] (from p) $\xrightarrow{LiAlH_4}$ $\xrightarrow{H_2O}$ $\xrightarrow[heat]{H_2SO_4}$

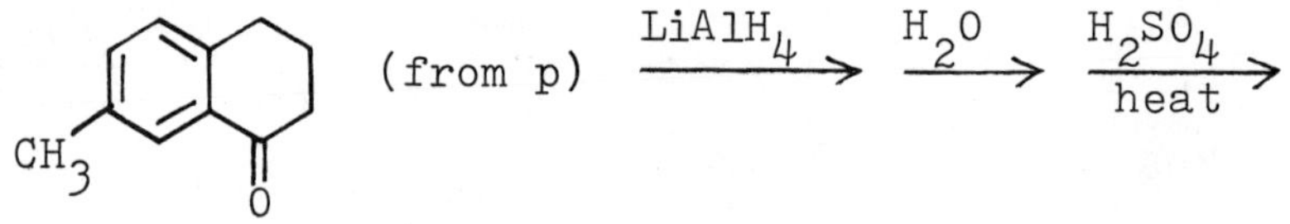

(r) [6-methyl-1,2-dihydronaphthalene] (from q) $\xrightarrow{O_3}$ $\xrightarrow[OH^{\ominus}]{H_2O_2}$ $\xrightarrow{H_3O^{\oplus}}$

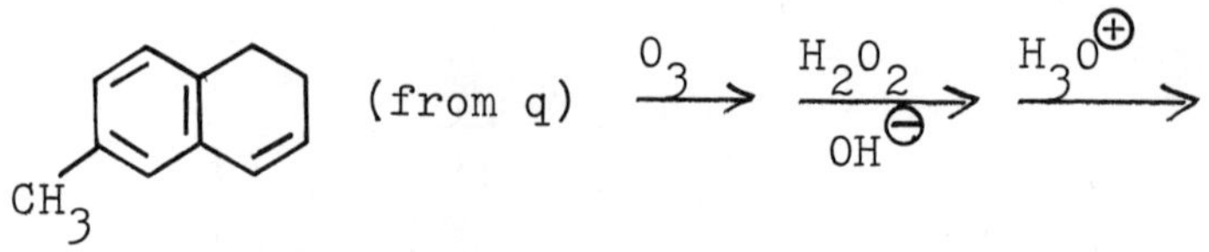

39. $NO_2^{\oplus}$, N is sp; $NO_3^{\ominus}$, N is sp^2

40. Although aromatic, the tropylium cation has a positive charge, and electrophilic aromatic substitution would require the addition of one cation to another.

41. The carbonyl group (-C=O) adjacent to the nitrogen decreases the electron density on nitrogen (and, therefore, its tendency to donate electrons to the benzene ring) via inductive and resonance effects.

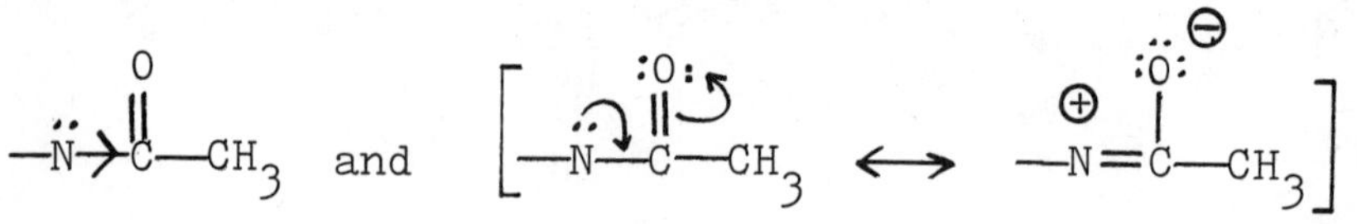

$-\ddot{N}\rightarrow C(=O)-CH_3$ and $[-\ddot{N}-C(=\ddot{O}:)-CH_3 \longleftrightarrow -\overset{\oplus}{N}=C(-\ddot{O}:^{\ominus})-CH_3]$

42. Chlorine is more electronegative than iodine, and the polarization is $\underset{\delta\oplus}{\text{I}}\text{—}\underset{\delta\ominus}{\text{Cl}}$.

Thus the Lewis acid catalyst (such as $FeCl_3$) reacts with the more negative end of the molecule

$$:\underset{\delta\oplus}{\ddot{\underset{..}{\text{I}}}}\text{—}\underset{\delta\ominus}{\ddot{\underset{..}{\text{Cl}}}}: \; + \; FeCl_3 \rightleftharpoons \underset{\delta\oplus}{\text{I}}\cdots\cdots\text{Cl}\cdots\cdots\underset{\delta\ominus}{\text{FeCl}_3}$$

43. (a) The nitroso group has a nonbonding electron pair on nitrogen ($-\ddot{N}=\ddot{O}:$) which can stabilize an intermediate resulting from attack of an electrophile ortho or para to -NO.

(b) The phenyl group may also act as an electron-pair donor to stabilize an intermediate resulting from attack of an electrophile ortho or para to the other benzene ring.

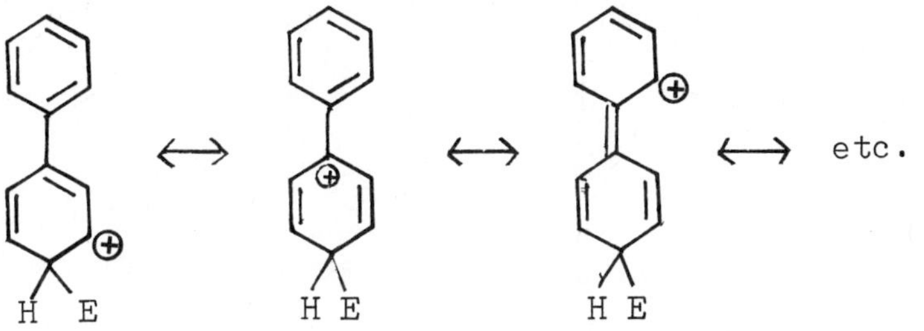

44. The reaction between 1,2,3,4-tetramethylbenzene and hydrogen fluoride leads to three cations (all stabilized by two methyl groups) while the reaction with 1,2,4,5-tetramethylbenzene and hydrogen fluoride gives only two cations (each stabilized by two methyl groups). The cations from 1,2,3,4-tetramethylbenzene are shown below.

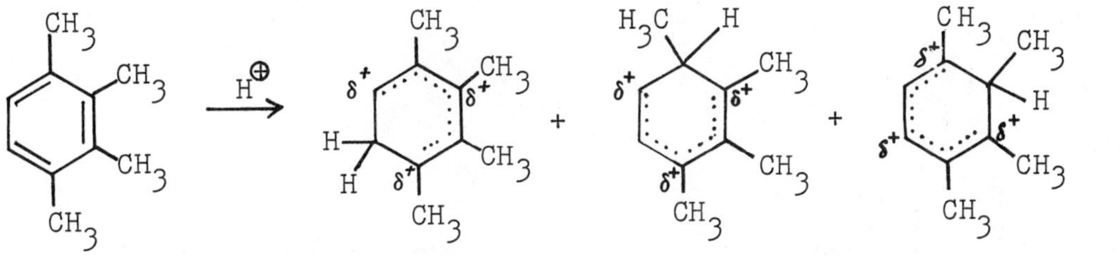

45. $Cl_3CCHO \underset{}{\overset{H^{\oplus}}{\rightleftharpoons}} [Cl_3CCH{=}\overset{\oplus}{O}H \longleftrightarrow Cl_3C\overset{\oplus}{C}H(OH)] \xrightarrow{C_6H_5Cl}$ $Cl_3CCH(OH)$–(H)$C_6H_4^{+}$–Cl $\xrightarrow{-H^{\oplus}}$ $Cl_3CCH(OH)$–C_6H_4–Cl $\xrightarrow{H^{\oplus}}$ Cl_3C–CH($\overset{\oplus}{O}H_2$)–C_6H_4–Cl $\xrightarrow{-H_2O}$ Cl_3C–$\overset{\oplus}{C}H$–C_6H_4–Cl $\xrightarrow{C_6H_5Cl}$ Cl_3C–CH(C_6H_4–Cl)–(H)$C_6H_4^{+}$–Cl $\xrightarrow{-H^{\oplus}}$ $Cl_3CCH(C_6H_4Cl)_2$

46. (a)
$(CH_3)_3C\ddot{O}H \overset{H^{\oplus}}{\rightleftharpoons} (CH_3)_3C$–$\overset{\oplus}{O}H_2 \overset{-H_2O}{\rightleftharpoons} (CH_3)_3C^{\oplus} \xrightarrow{CH_3C_6H_5}$ $(CH_3)_3C$–(H)$C_6H_4^{+}$–$CH_3 \overset{-H^{\oplus}}{\rightleftharpoons} (CH_3)_3C$–$C_6H_4$–$CH_3$

(continued next page)

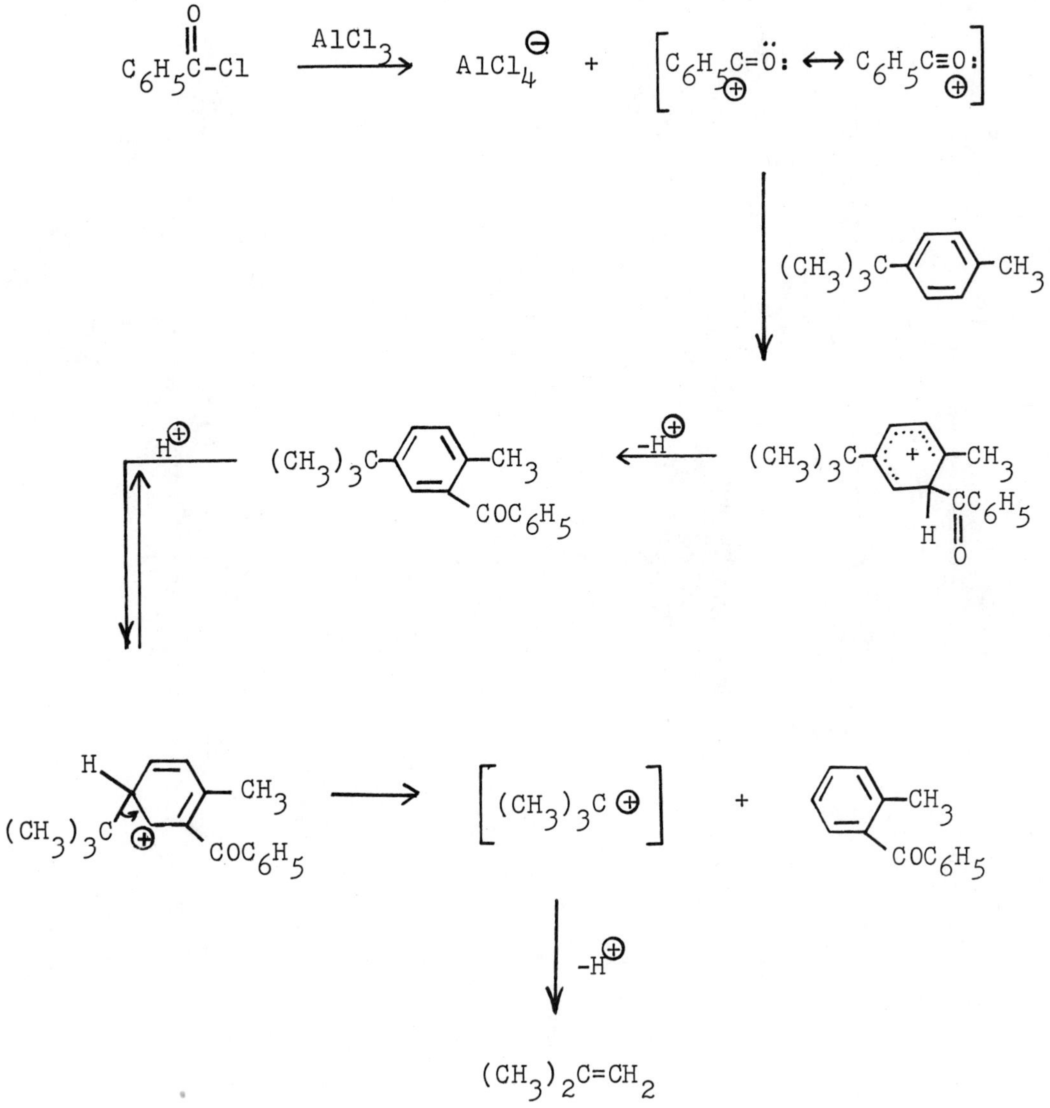

$C_6H_5\overset{O}{\overset{\|}{C}}\text{-Cl} \xrightarrow{AlCl_3} AlCl_4^{\ominus} + \left[C_6H_5\overset{\oplus}{C}=\ddot{O}: \longleftrightarrow C_6H_5C\equiv\overset{\oplus}{O}:\right]$

$\downarrow$ $(CH_3)_3C-C_6H_4-CH_3$

$(CH_3)_3C-C_6H_3(CH_3)(COC_6H_5) \xleftarrow{-H^{\oplus}} (CH_3)_3C-[C_6H_3^{+}](CH_3)(H)(\overset{O}{\overset{\|}{C}}C_6H_5)$

$\downarrow\uparrow$ $H^{\oplus}$

$(CH_3)_3C(H)-[C_6H_3^{\oplus}](CH_3)(COC_6H_5) \longrightarrow \left[(CH_3)_3C^{\oplus}\right] + C_6H_4(CH_3)(COC_6H_5)$

$\downarrow -H^{\oplus}$

$(CH_3)_2C=CH_2$

47. The electrophile may be formed by the process:

$$H_2C{=}O \xrightarrow{H\text{-}Cl} \left[H_2C{=}\overset{\oplus}{O}H \longleftrightarrow H_2\overset{\oplus}{C}{-}OH \right] \xrightarrow{:Cl:^{\ominus}} H_2C(OH)Cl$$

$$H_2C(OH)Cl \xrightarrow{ZnCl_2} H_2C(Cl){-}\overset{\oplus}{O}H{-}\overset{\ominus}{Z}nCl_2 \longrightarrow \left[H_2C{=}\overset{\oplus}{Cl}: \longleftrightarrow H_2\overset{\oplus}{C}{-}Cl: \right]^{*}$$

$$\xrightarrow{C_6H_6} C_6H_6^{+}(H)(CH_2Cl) \xrightarrow{-H^{\oplus}} C_6H_5CH_2Cl$$

*or possibly $\overset{\delta\ominus}{Z}nCl_2 \cdots OH \cdots \overset{\delta\oplus}{C}H_2{-}Cl$

48. OCH$_3$, CH$_3$ (A); OCH$_3$ (B); CH$_3$, OCH$_3$, CH$_3$ (C)

A B C

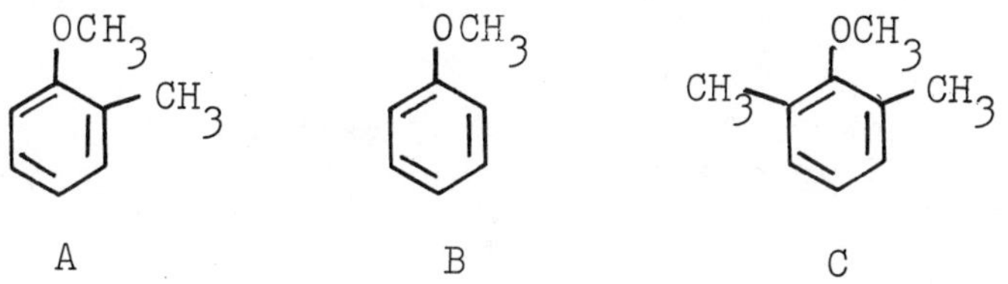

"A" with two electron-releasing groups can stabilize more effectively than "B" the cation intermediate formed in bromination. Although "C" possesses three electron-releasing groups, steric hindrance prevents effective resonance stabilization of the intermediate cation by the $-OCH_3$ group which requires that oxygen attain sp^2 hydridization (see below).

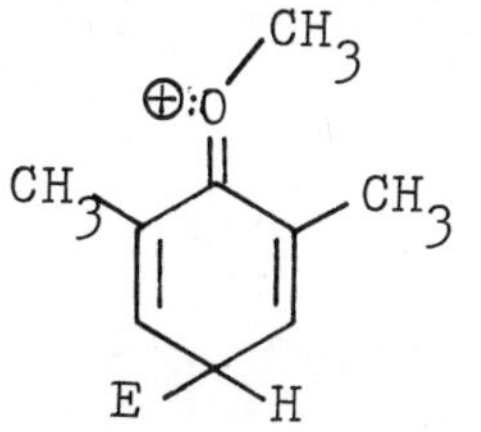

$\oplus$:O–CH$_3$, CH$_3$, CH$_3$, E, H

Overall, C is less reactive than A or B.

49. Both alkyl groups are o-, p- directors; however, substitution occurs ortho to the methyl rather than to the t-butyl for steric reasons (thus > 90% 1-methyl-2-chloro-4-t-butylbenzene and no 1-methyl-3-chloro-4-t-butylbenzene).

p-Chlorotoluene and t-butyl chloride may be formed by the process:

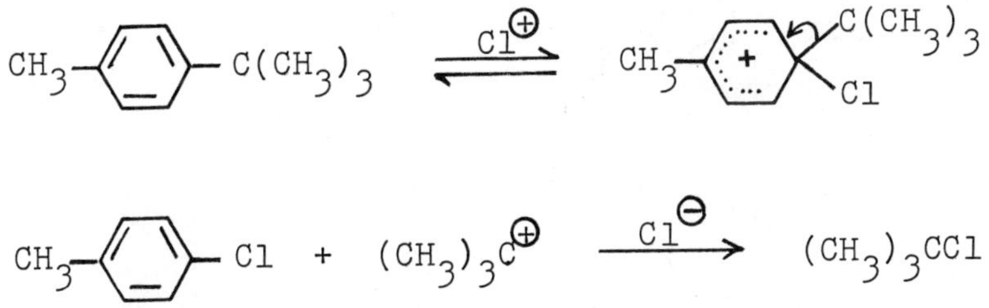

$CH_3-C_6H_4-C(CH_3)_3 \xrightleftharpoons{Cl^{\oplus}} CH_3-C_6H_4^{+}(C(CH_3)_3)(Cl)$

$CH_3-C_6H_4-Cl + (CH_3)_3C^{\oplus} \xrightarrow{Cl^{\ominus}} (CH_3)_3CCl$

50. These data are consistent with σ-complex formation as the rate-determining step since toluene and ethylbenzene possess alkyl (electron-releasing) groups which may stabilize the developing positive charge in the transition state (and the cations subsequently formed). The data suggest that ethyl is a better electron-releasing group than methyl.

51. If a methyl cation is formed in the product-determining step of both reactions (and the halogens have no effect on the substitution), product distributions should be the same.

52. With hydrogens at the 9,10 positions of anthracene, a 9,10-dibromo adduct could undergo elimination of HBr to form 9-bromoanthracene.

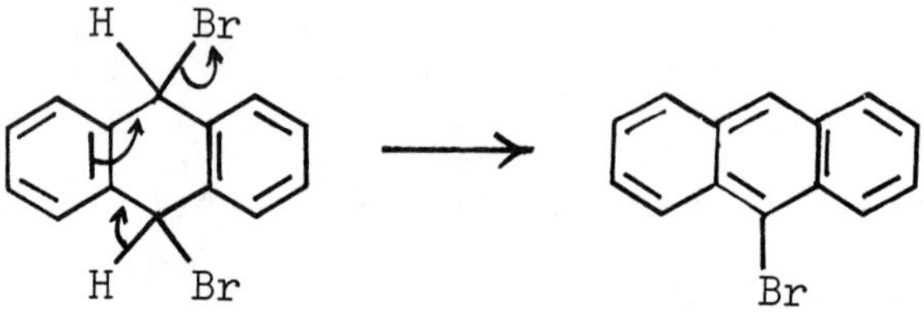

9,10-Dimethylanthracene cannot undergo elimination. The two products formed are:

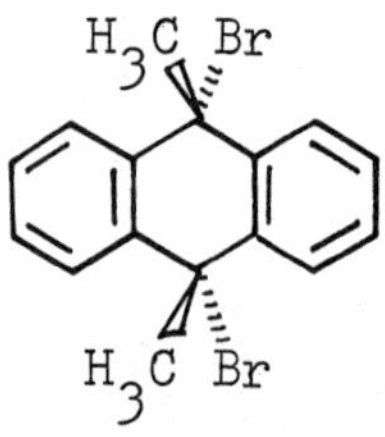

cis-methyls

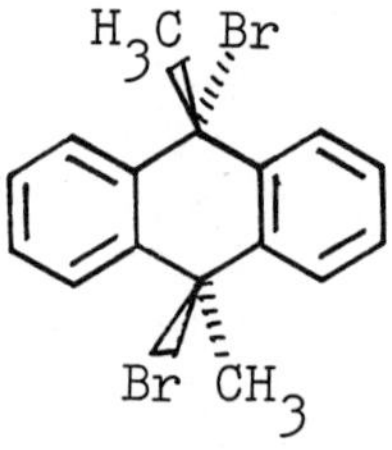

trans-methyls

53. indene $\xrightarrow{Br_2/CCl_4}$

indene

$\downarrow H_2/Pt$

indane $\xrightarrow{H_2/Pt}$

indane

$\downarrow KMnO_4$

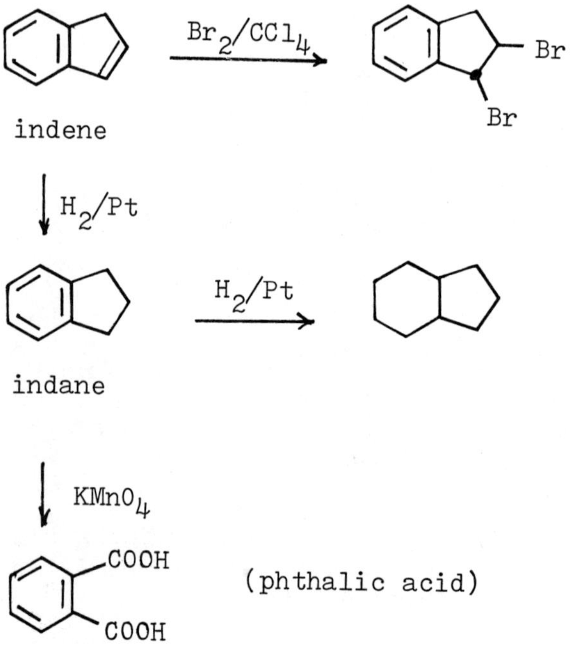

(phthalic acid)

54. (a) $\left[C_6H_5\text{-}\overset{\oplus}{N}{\equiv}N{:} \longleftrightarrow C_6H_5\text{-}\ddot{N}{=}\overset{\oplus}{N}{:} \right]$ $\xrightarrow{\text{pyrrole (N-H)}}$ (2-H, 2-$N{=}NC_6H_5$ pyrrolium cation, N-H) $\xrightarrow{-H^{\oplus}}$ 2-($N{=}NC_6H_5$)pyrrole (N-H)

(b) $CH_3\text{-}\overset{O}{\overset{\|}{C}}\text{-}O\text{-}\overset{O}{\overset{\|}{C}}\text{-}CH_3 \underset{}{\overset{SnCl_4}{\rightleftharpoons}} \left[CH_3\overset{\oplus}{C}{=}\ddot{O}{:} \leftrightarrow CH_3C{\equiv}\overset{\oplus}{O}{:} \right] + SnCl_4OCOCH_3^{\ominus}$

$\xrightarrow{\text{pyrrole (N-H)}}$ (2-H, 2-($\overset{O}{\overset{\|}{C}}\text{-}CH_3$) pyrrolium cation, N-H) $\xrightarrow{-H^{\oplus}}$ 2-($\overset{O}{\overset{\|}{C}}CH_3$)pyrrole (N-H)

(continued next page)

(c) $Cl\text{-}\ddot{S}\text{-}Cl + AlCl_3 \rightleftharpoons AlCl_4^{\ominus} + \overset{\oplus}{S}\text{-}Cl$ → (benzene) → [arenium ion with H and SCl] $\xrightarrow{-H^{\oplus}}$ C_6H_5SCl $\xrightarrow{AlCl_3}$ $C_6H_5S^{\oplus}$ → (benzene) → [arenium ion with H and S-Ph] $\xrightarrow{-H^{\oplus}}$ $C_6H_5\text{-}S\text{-}C_6H_5$ $\xrightarrow{\text{repeat}}$ thianthrene (S, S)

(d) $F\text{-}\overset{O}{\overset{\|}{C}}\text{-}H \underset{}{\overset{BF_3}{\rightleftharpoons}} BF_4^{\ominus} + \left[H\text{-}\overset{\oplus}{C}\text{=}\ddot{O} \longleftrightarrow H\text{-}C\equiv\overset{\oplus}{O}\right]$ → (benzene) → [arenium ion with H and C(=O)H] $\xrightarrow{-H^{\oplus}}$ $C_6H_5\text{-}\overset{O}{\overset{\|}{C}}\text{-}H$

(continued next page)

(e)

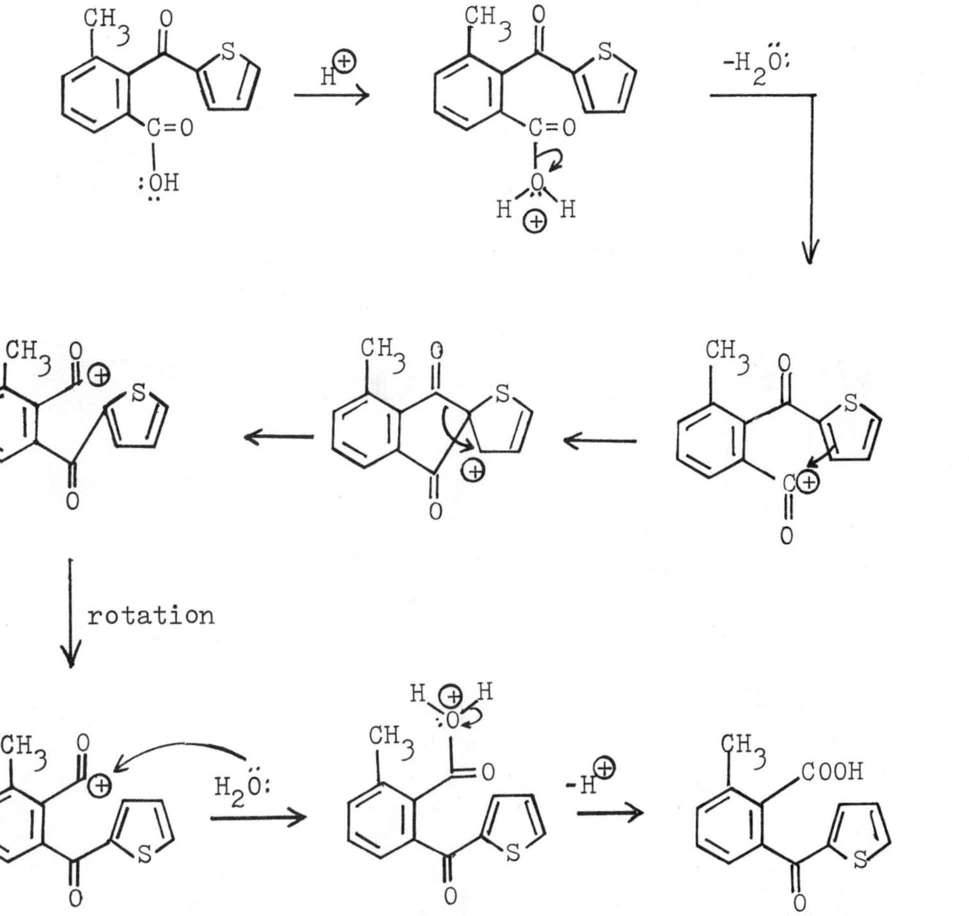

55. At short reaction times, the product distribution reflects the relative stabilities of the initially formed cations,

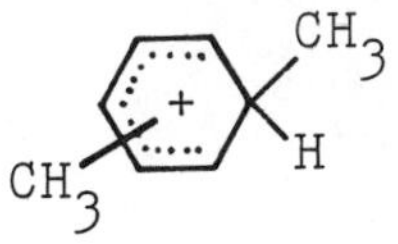

. At long reaction times, the initially formed dimethylbenzenes may rearrange (by reprotonation, etc.) to thermodynamically more stable compounds.

56. 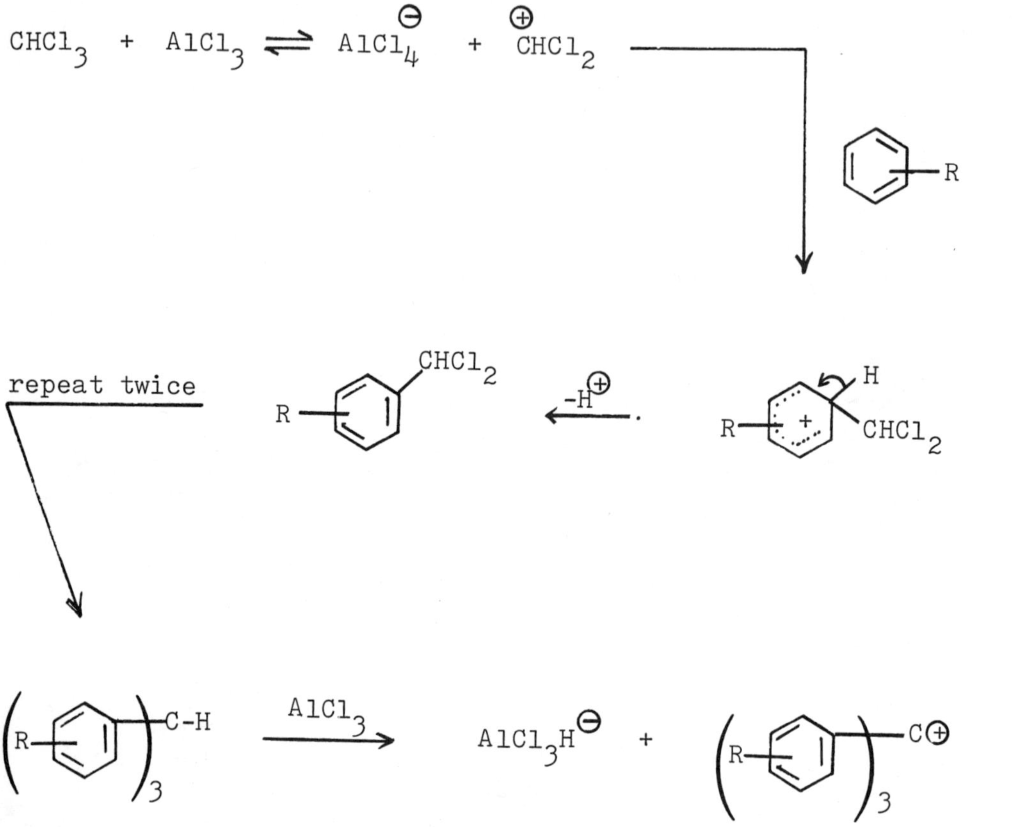 $CHCl_3 + AlCl_3 \rightleftharpoons AlCl_4^{\ominus} + \overset{\oplus}{C}HCl_2$ → (benzene–R) ↓

repeat twice

R–C$_6$H$_4$–CHCl$_2$ $\xleftarrow{-H^{\oplus}}$ R–C$_6$H$_5^{+}$(H)(CHCl$_2$)

$(R\text{–}C_6H_4)_3C\text{–}H \xrightarrow{AlCl_3} AlCl_3H^{\ominus} + (R\text{–}C_6H_4)_3C^{\oplus}$

57. Ether A is thought to form a complex (between the oxygen of the ether and N_2O_5) which furnishes $NO_2^{\oplus}$ to the ring.

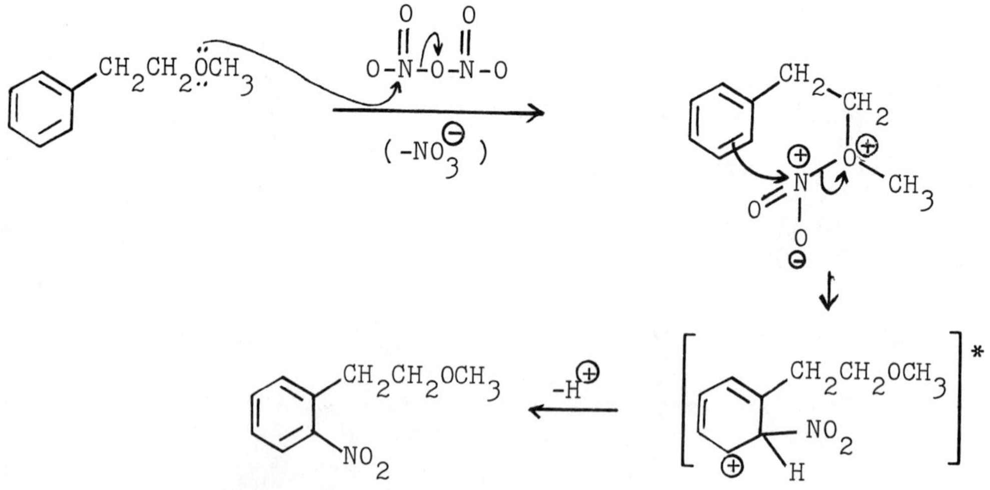

$C_6H_5CH_2CH_2\ddot{O}CH_3$ + O$_2$N–O–NO$_2$ $\xrightarrow{(-NO_3^{\ominus})}$ [cyclic complex: CH$_2$–CH$_2$–$\overset{\oplus}{O}$(CH$_3$)–$\overset{\oplus}{N}$(=O)–O$^{\ominus}$] ↓

o-NO$_2$–C$_6$H$_4$–CH$_2$CH$_2$OCH$_3$ $\xleftarrow{-H^{\oplus}}$ [C$_6$H$_5^{\oplus}$(H)(NO$_2$)–CH$_2$CH$_2$OCH$_3$]*

*resonance stabilized

58. $CH_3CH_2F \xrightarrow{BF_3} BF_4^{\ominus} + CH_3\overset{\oplus}{C}H_2$

$$\text{1,3,5-}(CH_3)_3C_6H_3 \xrightarrow{CH_3\overset{\oplus}{C}H_2} \text{[ring with } CH_3,\ CH_2CH_3,\ H,\ CH_3,\ H_3C;\ \delta^+,\ \delta^+,\ \delta^+] \xrightarrow{\text{heat}} \text{ring with } CH_3\ (a),\ CH_2CH_3\ (b),\ CH_3\ (a),\ CH_3\ (d)$$

nmr spectrum: 6H (a), singlet near 2δ, 3H (d), singlet near 2δ (slightly downfield from a), 3H (b), triplet near 1δ, 2H (c), quartet near 1.5δ

59. (a) A

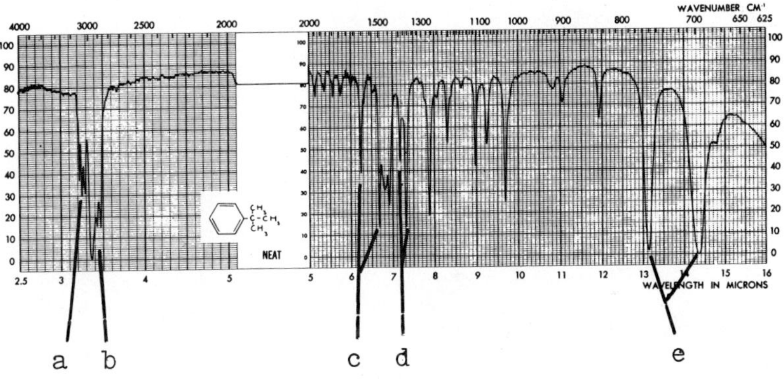

a: =C-H stretch (benzene)

b: C-H stretch (alkyl)

c: C=C stretch (benzene)

d: C-H bending (unsymmetric doublet for t-butyl)

e: =C-H out-of-plane bending (monosubstituted benzene)

NMR: 1.3δ, t-butyl; 7.2δ aromatic

(continued next page)

(b) B

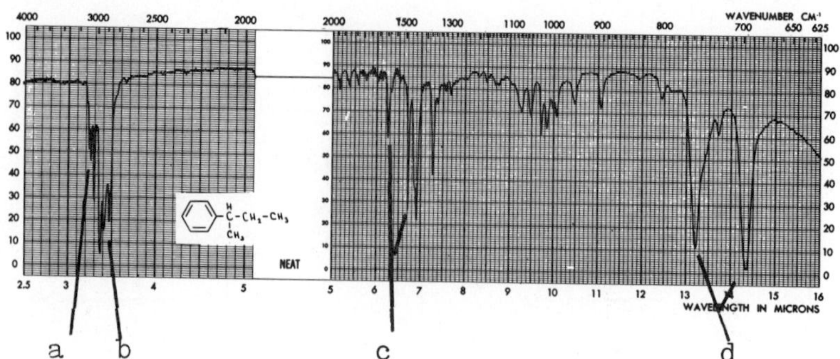

IR a: =C-H stretch (benzene)

b: C-H stretch (alkyl)

c: C=C stretch (benzene)

d: =C-H out-of-plane bending (monosubstituted benzene)

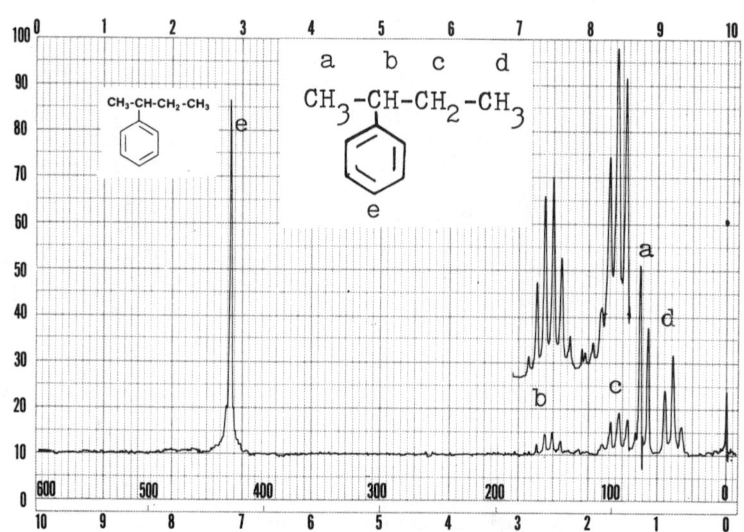

NMR a: doublet c: multiplet

b: multiplet d: triplet

(c) C

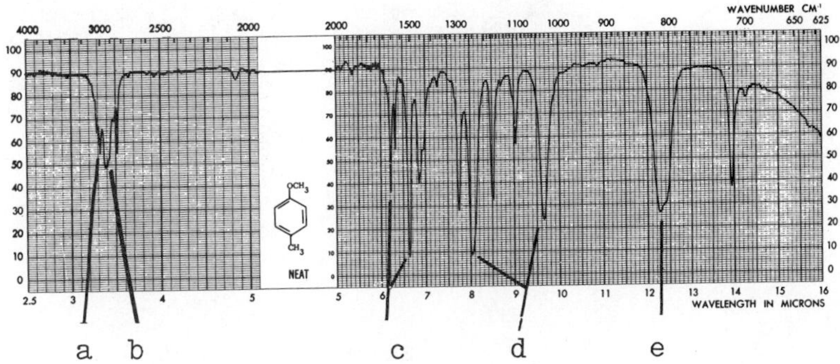

IR a: =C-H stretch (benzene)

b: C-H stretch (alkyl)

c: C=C stretch (benzene)

d: C-O stretch

e: =C-H out-of-plane bending (p-disubstituted benzene)

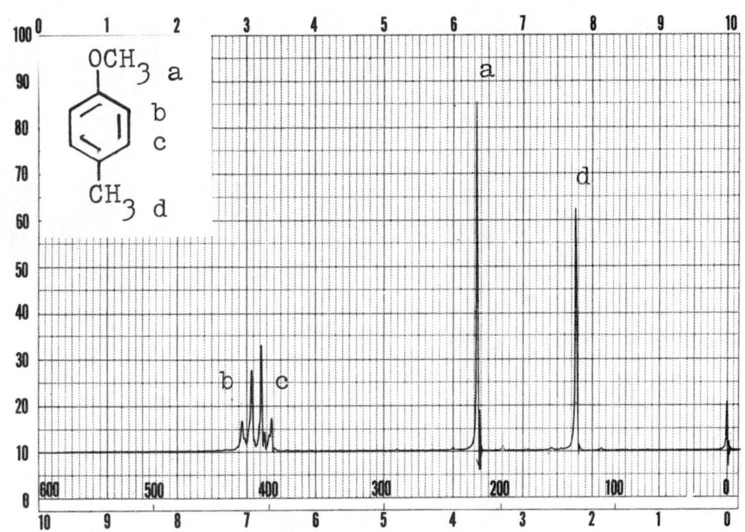

The pattern near 7 δ for the aromatic protons is characteristic of p-disubstituted benzenes where two different groups are attached.

(d) D

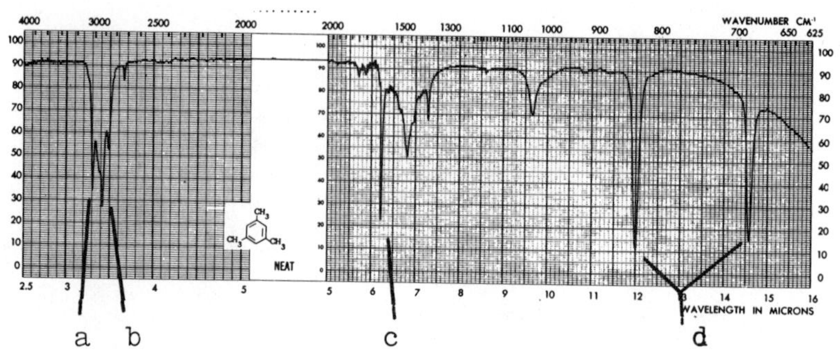

IR a: =C-H stretch (benzene)

b: C-H stretch (alkyl)

c: C=C bending (benzene)

d: =C-H out-of-plane bending

NMR 2.3 δ methyl groups; 6.8 δ aromatic

(e) E

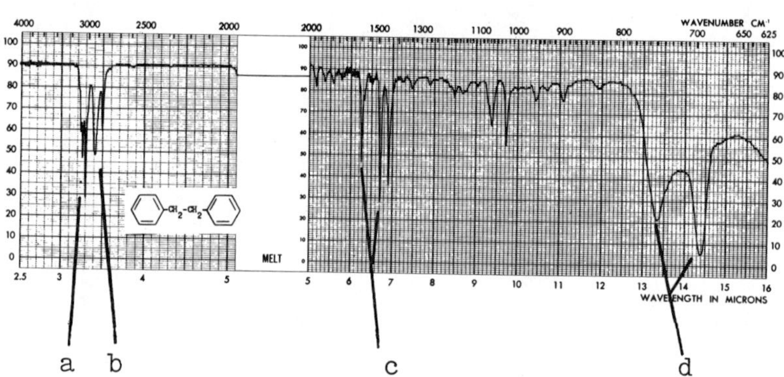

IR a: =C-H stretch (benzene)

b: C-H stretch (alkyl)

c: C=C stretch (benzene)

d: =C-H out-of-plane bending (monosubstituted benzene)

NMR 2.9 δ methylene groups; 7.2 δ aromatic

(f) F

IR

a: =C-H stretch (benzene)

b: C-H stretch (alkyl)

c: C=C stretch (benzene)

d: C-O stretch

e: =C-H out-of-plane bending (o-disubstituted benzene)

NMR 5.9 methylene; 6.8 aromatic

(g) G

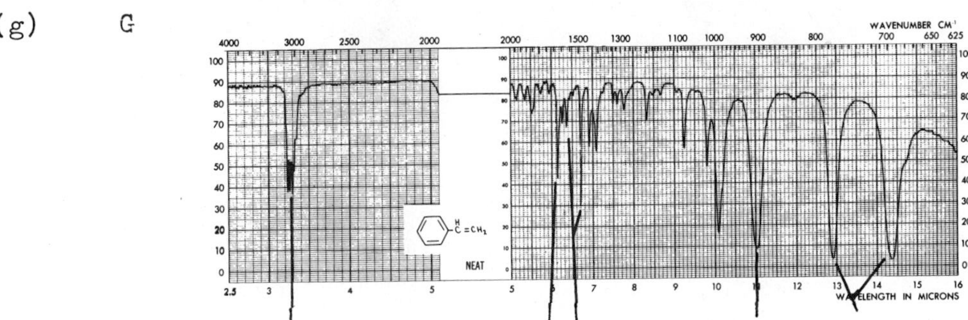

a: =C-H stretch (alkene and benzene)

b: C=C stretch (alkene)

c: C=C stretch (benzene)

d: =C-H out-of-plane bending (alkene, $RHC=CH_2$)

e: =C-H out-of-plane bending (benzene, monosubstituted)

(g) continued

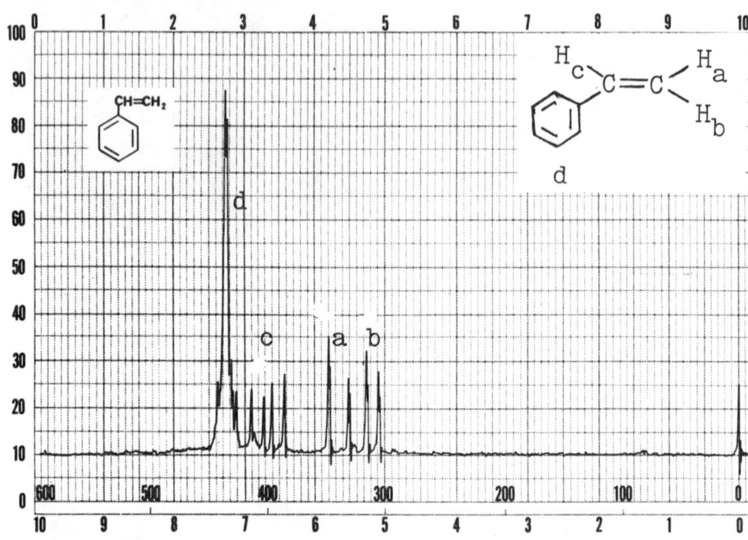

(h) H

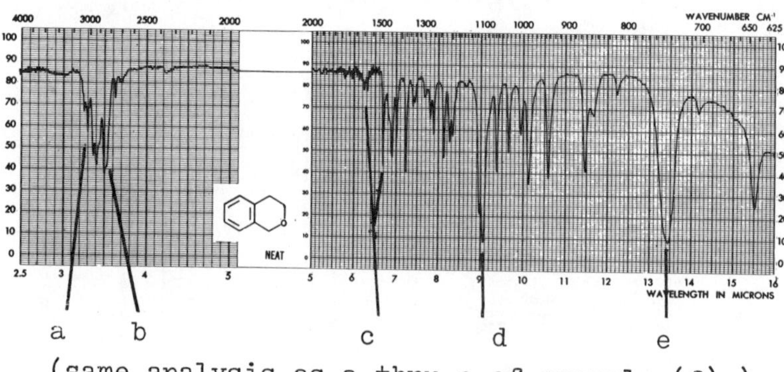

(same analysis as a thru e of example (f))

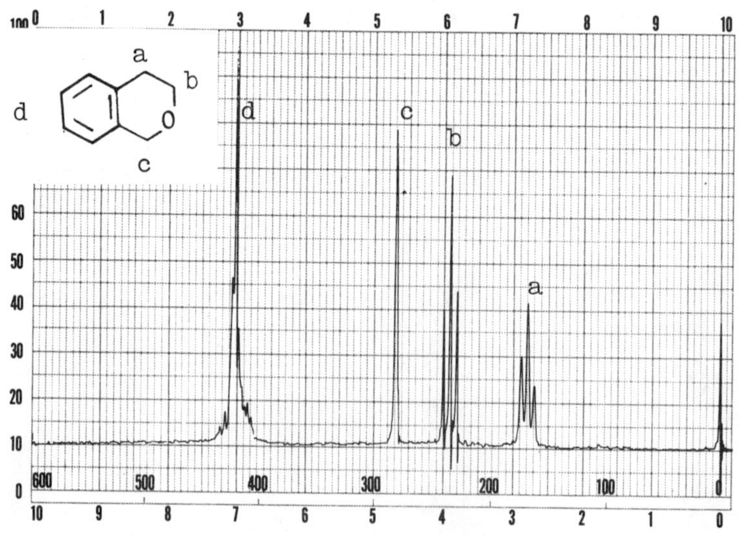

SUMMARY OF REACTIONS

Benzene →

- HNO_3/H_2SO_4 or $NO_2^{\oplus}\ BF_4^{\ominus}$ or $CH_3\overset{O}{\overset{\|}{C}}-ONO_2$ → nitrobenzene (NO_2) (nitration p. 609)
- X_2/Lewis acid; X usually Cl or Br; Lewis acid usually FeX_3 or $AlCl_3$ → halobenzene (X) (halogenation p. 611)
- $-\overset{|}{\underset{|}{C}}-\overset{|}{\underset{|}{C}}-X/AlCl_3$ or $-\overset{|}{\underset{|}{C}}-\overset{|}{\underset{|}{C}}-OH/H_2SO_4$ or $>C=C</H_2SO_4$ → $-\overset{|}{\underset{|}{C}}-\overset{|}{\underset{|}{C}}-$ substituted benzene (alkylation p. 612 , rearrangement may occur)
- $R-\overset{O}{\overset{\|}{C}}-Cl/AlCl_3$ → $\overset{O}{\overset{\|}{C}}-R$ substituted benzene (acylation p. 614)
- fuming H_2SO_4 → SO_3H substituted benzene (sulfonation p. 618)
- $ClSO_3H$ → SO_2Cl substituted benzene (chlorosulfonation p. 621)
- D_2SO_4 → D substituted benzene (proton-deuterium exchange p. 621)

(continued next page)

$$\text{C}_6\text{H}_5\text{-C(=O)-} \xrightarrow{Zn(Hg)/HCl} \text{C}_6\text{H}_5\text{-CH}_2\text{-}$$

(or R-C(=O)-)

(Clemmenson reduction p. 617)

$$\text{C}_6\text{H}_5\text{-SO}_3\text{H} \xrightarrow{H_2O\ (steam)} \text{C}_6\text{H}_6$$

(desulfonation p. 620)

$$\text{C}_6\text{H}_5\text{-C(H)-} \xrightarrow{X_2/light} \text{C}_6\text{H}_5\text{-C(X)-}$$

(side chain halogenation p. 635)

$$\text{C}_6\text{H}_5\text{-C(H)-} \xrightarrow{KMnO_4/OH^{\ominus}} \xrightarrow{H_3O^{\oplus}} \text{C}_6\text{H}_5\text{-CO}_2\text{H}$$

(side chain oxidation p. 636)

The Influence of Groups Already on the Ring on Electrophilic Substitution

1. Groups of the type shown below are <u>ortho-para</u> directors and, with the exception of the halogens, are activators.

   a) -R (alkyl): $-CH_3$, $-CH_2CH_3$, $-CH(CH_3)_2$, etc.

   b) $-\ddot{Z}$ (the atom <u>attached to the ring</u> has one or more nonbonding electron pairs):

   $-\ddot{O}H$, $-\ddot{O}R$, $-\ddot{N}H_2$, $-\ddot{N}HR$, $-\ddot{N}R_2$, $-\ddot{N}HCCH_3$ (with C=O), $-\ddot{X}:$ (halogens).

(continued next page)

2. Groups of the type $-\overset{\oplus \text{ (or } \delta^{\oplus})}{Z}$ (the atom attached to the ring has no nonbonding electron pairs but does possess a ⊕ or $\delta^{\oplus}$) are <u>meta</u> directors and are deactivating. These include:

$-\overset{\oplus}{N}(=O)O^{\ominus}$, $-\overset{\oplus}{N}R_3$, $-\overset{\delta+}{S}(=\overset{\delta-}{O})_2-OH$, $-\overset{\delta+}{C}\equiv\overset{\delta-}{N}$, $-\overset{\delta+}{C}(=\overset{\delta-}{O})-H(R)$, etc.

(see Table 16-2)

REACTION REVIEW

A. Reactions of Benzene and Derivatives

Questions	Answers
1. $C_6H_6 \xrightarrow{HNO_3/H_2SO_4\ *} ?$ *(or $CH_3\overset{O}{\overset{\Vert}{C}}ONO_2/H^{\oplus}$ or $N_2O_5/H^{\oplus}$)	1. $C_6H_5NO_2$
Write the structure of the major mononitration product of:	
a) $C_6H_5-NO_2$	a) m-$C_6H_4(NO_2)_2$
b) $C_6H_5-\ddot{O}-CH_3$	b) $O_2N-C_6H_4-OCH_3$ (para) (and <u>ortho</u>)
	(continued next page)

c) $CH_3O-C_6H_4-NO_2$	c) $CH_3O-C_6H_3(NO_2)-NO_2$ (2,4-dinitro) (CH_3O- stronger activator)
2. $C_6H_6 \xrightarrow[AlX_3 \text{ or } FeX_3 \text{ or Fe etc.}]{X_2 \ (X=Cl \text{ or } Br)}$? Write the major monohalogenation product in the following: a) $C_6H_5-SO_3H \xrightarrow[FeBr_3]{Br_2}$? b) $CH_3\overset{O}{\overset{\parallel}{C}}-C_6H_4-CH_3 \xrightarrow[AlCl_3]{Cl_2}$?	2. C_6H_5X a) $C_6H_4(Br)-SO_3H$ (meta) b) $CH_3\overset{O}{\overset{\parallel}{C}}-C_6H_3(Cl)-CH_3$ (Cl ortho to CH_3) (CH_3- stronger activator)
3. $C_6H_6 \xrightarrow[HNO_3]{I_2}$? Write the structure of the major product in the reaction: $C_6H_6 \xrightarrow[FeCl_3]{Cl_2} \xrightarrow[HNO_3]{I_2}$?	3. C_6H_5I $I-C_6H_4-Cl$ (and ortho)

Question	Answer
4. Benzene $\xrightarrow[AlCl_3]{R-X}$?	4. Benzene–R
Write the monoalkylation products in the following:	
a) Benzene–CH_3 $\xrightarrow[AlCl_3]{CH_3CH_2Cl}$?	a) CH_3CH_2–Benzene–CH_3 (and ortho)
b) Benzene–Cl $\xrightarrow[AlCl_3]{(CH_3)_2CHCH_2Cl}$?	b) $(CH_3)_3C$–Benzene–Cl (rearrangement)
c) Benzene–NO_2 $\xrightarrow[AlCl_3]{CH_3CH_2Br}$?	c) no reaction, $-NO_2$ strongly deactivating
5. (i) Benzene $\xrightarrow[H^{\oplus}]{C=C}$?	5. (i) Benzene–C–C–H
(ii) Benzene $\xrightarrow[H^{\oplus}]{ROH}$?	(ii) Benzene–R + H_2O
Write the products for:	
a) Benzene $\xrightarrow[H^{\oplus}]{(CH_3)_2C=CH_2}$?	a) Benzene–$C(CH_3)_3$
b) Benzene $\xrightarrow[H^{\oplus}]{(CH_3)_3COH}$?	b) Benzene–$C(CH_3)_3$

6. $C_6H_6 \xrightarrow[AlCl_3]{R-\overset{O}{\overset{\parallel}{C}}-Cl}$? Write the structure of the major product of $CH_3\overset{O}{\overset{\parallel}{C}}Cl/AlCl_3$ with: a) $C_6H_5-\ddot{O}CH_3$ b) $CH_3-C_6H_4-OCH_3$ (para)	6. $C_6H_5-\overset{O}{\overset{\parallel}{C}}-R$ + HCl a) $CH_3\overset{O}{\overset{\parallel}{C}}-C_6H_4-OCH_3$ (para) b) $CH_3-C_6H_3(OCH_3)(\overset{O}{\overset{\parallel}{C}}-CH_3)$ (acetyl ortho to OCH_3)
7. (i) $C_6H_6 \xrightarrow{R\overset{O}{\overset{\parallel}{C}}OH/HF}$? (ii) $C_6H_6 \xrightarrow{R\overset{O}{\overset{\parallel}{C}}-O-\overset{O}{\overset{\parallel}{C}}R/AlCl_3}$? Write the structure of the major product in the following: a) $Cl-C_6H_5 \xrightarrow[AlCl_3]{CH_3\overset{O}{\overset{\parallel}{C}}-O-\overset{O}{\overset{\parallel}{C}}CH_3}$?	7. (i) $C_6H_5-\overset{O}{\overset{\parallel}{C}}-R$ (ii) $C_6H_5-\overset{O}{\overset{\parallel}{C}}-R$ a) $Cl-C_6H_4-\overset{O}{\overset{\parallel}{C}}CH_3$ (para) (and *ortho*) (continued next page)

b) $C_6H_5-(CH_2)_3\overset{O}{\overset{\|}{C}}OH \xrightarrow{HF}$?	b) [structure: 1-tetralone]
8. $C_6H_6 \xrightarrow{H_2SO_4 \cdot SO_3}$? Write the structure of the major monosulfonation product of: a) $CH_3CH_2-C_6H_5$ b) $O_2N-C_6H_4-CH_3$	8. $C_6H_5-SO_3H$ a) $CH_3CH_2-C_6H_4-SO_3H$ b) $O_2N-C_6H_3(SO_3H)-CH_3$
9. $C_6H_5-SO_3H \xrightarrow{H_2O/heat}$? Write the structure of the major product in the following: a) $CH_3-C_6H_4-SO_3H \xrightarrow[heat]{H_2O}$? b) $O_2N-C_6H_5 \xrightarrow{H_2SO_4 . SO_3}$ $\xrightarrow[heat]{D_2O}$?	9. C_6H_6 + SO_3 a) $CH_3-C_6H_5$ b) $O_2N-C_6H_4-D$

10. C_6H_5–C(H)< $\xrightarrow{X_2/h\nu}$?

10. C_6H_5–C(X)< + HX

Write the structure of the major product in the following:

a) $C_6H_5CH_2CH_3 \xrightarrow[h\nu]{Cl_2}$?

b) CH_3–C_6H_4–$C(CH_3)_3 \xrightarrow[h\nu]{Br_2}$?

a) C_6H_5–$CH(Cl)CH_3$

b) $BrCH_2$–C_6H_4–$C(CH_3)_3$

11. C_6H_5–C(H)< $\xrightarrow{MnO_4^{\ominus}/OH^{\ominus}}$ $\xrightarrow{H_3O^{\oplus}}$?

11. C_6H_5–C(=O)–OH

(an acid)

Write the structure of the major product when C_6H_5–C(H)< is:

a) $C_6H_5CH_2CH_2CH_3$

b) CH_3–C_6H_4–CH_2CH_3

c) C_6H_5–$C(CH_3)_3$

a) C_6H_5–CO_2H

b) HO_2C–C_6H_4–CO_2H

c) no reaction

12. C_6H_5-C(=O)-H(R) $\xrightarrow[HCl]{Zn(Hg)}$? (or R-C(=O)-H(R)) Write the structure of the major product when C_6H_5C(=O)H(R) is: a) CH_3-C_6H_4-C(=O)CH_2CH_3 b) C_6H_4(C(=O)CH_3)(C(=O)CH_3) [CH_3-C(=O)- and -C(=O)CH_3]	12. C_6H_5-CH_2-H(R) (or $RCH_2H(R)$) a) CH_3-C_6H_4-$CH_2CH_2CH_3$ b) C_6H_4(CH_2CH_3)(CH_3CH_2)

B. Write a structure (major product) for each letter.

1. C_6H_6 $\xrightarrow[AlCl_3]{CH_3Cl}$ A $\xrightarrow[H_2SO_4]{HNO_3}$ B $\xrightarrow[FeCl_3]{Cl_2}$ C	1. A: C_6H_5-CH_3 B: O_2N-C_6H_4-CH_3 C: O_2N-C_6H_3(Cl)-CH_3

2. $C_6H_5CH_2CH_3 \xrightarrow[FeBr_3]{Br_2}$ A $\xrightarrow[h\nu]{Cl_2}$ B	2. A: $Br-C_6H_4-CH_2CH_3$ (para) B: $Br-C_6H_4-CHClCH_3$ (para)
3. $C_6H_6 \xrightarrow[H^{\oplus}]{CH_3CH(OH)CH_3}$ A $\xrightarrow[HNO_3]{I_2}$ B	3. A: $(CH_3)_2CH-C_6H_5$ B: $(CH_3)_2CH-C_6H_4-I$ (para)
4. $C_6H_5CH_2CH_3 \xrightarrow{H_2SO_4 \cdot SO_3}$ A $\xrightarrow[heat]{D_2O}$ B $\xrightarrow{KMnO_4/OH^{\ominus}} \xrightarrow{H_3O^{\oplus}}$ C	4. A: $HO_3S-C_6H_4-CH_2CH_3$ (para) B: $D-C_6H_4-CH_2CH_3$ (para) C: $D-C_6H_4-CO_2H$ (para)

CHAPTER 17

SYNTHESIS AND REACTIONS OF ALDEHYDES AND KETONES

LEARNING OBJECTIVES

When you have completed this chapter, you should be able to:

1. write names for aldehydes and ketones (problems 2, 20);
2. write equations using reactions from this chapter and previous chapters which illustrate the synthesis of compounds (problems 4, 24, 25, 27, 37, 38);
3. predict the products from reactions of aldehydes, ketones and from reactions of compounds formed from aldehydes and ketones (problem 22);
4. predict products from the reaction of Grignard reagents with nitriles (problem 7 and related problem 8);
5. suggest substrate-oxidant combinations for the synthesis of aldehydes and ketones (problem 21);
6. suggest substrate-Grignard reagent combinations for the synthesis of alcohols (problem 23);
7. account for (write mechanisms for) nucleophilic addition reactions of aldehydes and ketones (problems 11, 12, 34, 35, 40, 42, 43, 47, 48, 49, 51, 57, 58 and related problems 17, 19, 31, 33, 36, 53);
8. account for (write mechanisms for) acid-catalyzed hydrolysis of aldimines and ketimines (problem 9 and related problem 18);
9. account for the hydrolysis of hemiacetals and acetals (problems 13, 14 and related problems 10, 12, 15, 50, 56);
10. summarize the reactions of organoboranes (problem 26);
11. distinguish aldehydes, ketones, alcohols and other compounds by chemical tests (problem 29);
12. predict the stereochemistry of products formed in reactions of aldehydes and ketones with Grignard reagents and lithium aluminum hydride (problem 32);
13. account for (write mechanisms for) addition-elimination reactions (problems 44, 45, 47 and related problem 46).

ANSWERS TO QUESTIONS

1. Hydrogen bonding occurs between the carbonyl oxygen of acetone and water. Hydrogen bonding is not possible with propene.

2. (a) acetaldehyde
   (b) propenal
   (c) 3-phenylpropenal
   (d) 2,2,2-trichloroethanal (chloral)
   (e) dimethyl ketone (propanone)
   (f) 2-butanone
   (g) 4-methyl-3-penten-2-one

3. $LiAlH_4 + 3CH_3CH_2OH \longrightarrow LiAlH(OCH_2CH_3)_3 + 3H_2$

4. (a) $C_6H_5CH_3$ (toluene) $\xrightarrow[H_2SO_4]{CrO_3/Ac_2O}$ $\xrightarrow{H_2O}$ C_6H_5CHO

(b) $C_6H_5CH_3$ $\xrightarrow[ZnCl_2]{HCN/HCl}$ $\xrightarrow{H_2O}$ $CH_3-C_6H_4-CHO$

(c) $CH_3-C_6H_4-CHO$ (from b) $\xrightarrow[H_2O]{NaBH_4}$ $CH_3-C_6H_4-CH_2OH$

(d) $CH_3-C_6H_4-CHO$ (from b) $\xrightarrow[ether]{C_6H_5MgBr}$ $\xrightarrow{H_3O^{\oplus}}$

$CH_3-C_6H_4-CH(OH)-C_6H_5$

(e) $CH_3-C_6H_4-CHO$ (from b) $\xrightarrow{Zn \cdot Hg/HCl}$ $CH_3-C_6H_4-CH_3$

(f) $CH_3-C_6H_4-CH_3$ (from e) $\xrightarrow[H_2SO_4]{CrO_3/Ac_2O}$ $\xrightarrow{H_2O}$ $OHC-C_6H_4-CHO$

(g) $C_6H_5CH_3$ $\xrightarrow{Br_2/h\nu}$ $C_6H_5CH_2Br$

5. (a) $C_6H_5\overset{O}{\overset{\|}{C}}CH_3$

(b) 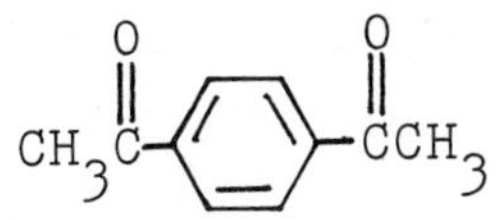 $CH_3\overset{O}{\overset{\|}{C}}-C_6H_4-\overset{O}{\overset{\|}{C}}CH_3$ (para)

(c) $CH_3CH_2NH_2$

(d) CH_3CHO (after hydrolysis on work up)

(e) 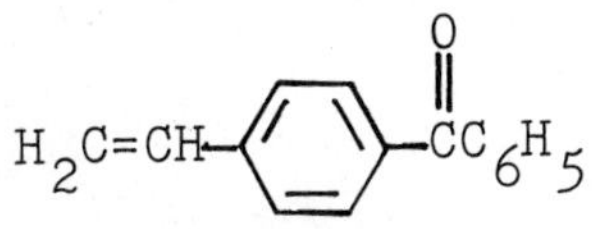 $H_2C=CH-C_6H_4-\overset{O}{\overset{\|}{C}}C_6H_5$ (para)

(f) $(CH_3)_2\underset{OH}{\underset{|}{C}}C_6H_5$

6. Ethylmagnesium iodide (source of the strong base $CH_3CH_2{:}^{\ominus}$) removes a proton from CH_3CH_2CN (α to CN). The anion formed is resonance stabilized.

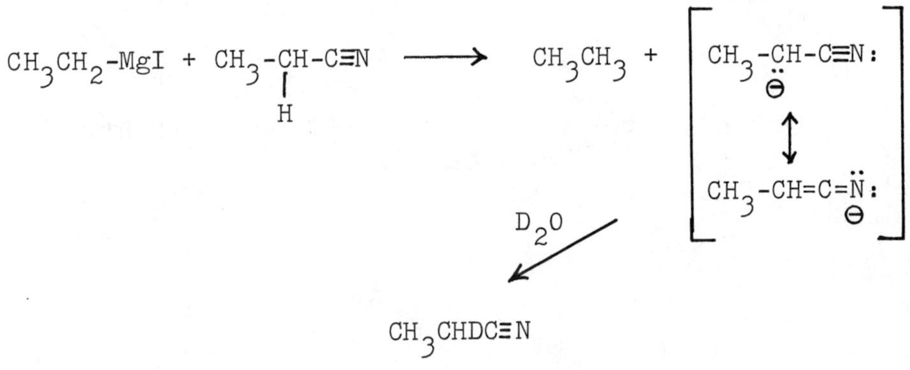

$$CH_3CH_2\text{-}MgI + CH_3\text{-}\underset{H}{\underset{|}{C}}H\text{-}C{\equiv}N \longrightarrow CH_3CH_3 + \left[CH_3\text{-}\overset{..}{\underset{\ominus}{C}}H\text{-}C{\equiv}N{:} \longleftrightarrow CH_3\text{-}CH{=}C{=}\overset{..}{\underset{\ominus}{N}}{:} \right]$$

$$\xrightarrow{D_2O} CH_3CHDC{\equiv}N$$

Propane does not react with ethylmagnesium iodide because it cannot form a resonance-stabilized anion.

7. No. An achiral intermediate carbocation is formed which leads to a racemic modification.

(continued next page)

$$(+)\text{-}C(CH_3)(OH) \xrightarrow{H^{\oplus}} (+)\text{-}C(CH_3)(OH_2^{\oplus}) \xrightarrow{-H_2O} C_6H_5\text{-}\overset{\oplus}{C}(CH_3)(CH_2OH)$$

achiral

$$\downarrow$$

$$C_6H_5\text{-}CH(CH_3)\text{-}CHO$$

(d,l)

8. $CH_3\text{-}C(OH)(C_2H_5)\text{-}C(OH)(C_2H_5)\text{-}CH_3$ (<u>meso</u> and <u>d,l</u>)

9. $R\text{-}C(=\ddot{N}\text{-}R)\text{-}H(R') \xrightarrow{H^{\oplus}} R\text{-}C(=\overset{\oplus}{N}H\text{-}R)\text{-}H(R') \xrightarrow{H_2\ddot{O}:} R\text{-}C(\ddot{N}H\text{-}R)(\overset{\oplus}{O}H_2)\text{-}H(R')$

aldimine (H)
or ketimine (R')

$$\downarrow$$

$$R\text{-}C(\overset{\oplus}{N}H_2\text{-}R)(:\ddot{O}\text{-}H)\text{-}H(R') + :\ddot{O}H_2$$

$$\longrightarrow H_3O^{\oplus} + :NH_2R + R\text{-}C(=\ddot{O}:)\text{-}H(R')$$

10. The reaction would have to occur by S_N2. Since the substrate is 3° and sterically hindered, an S_N2 reaction cannot occur. In addition, $OCH_3^{\ominus}$ is a poor leaving group.

11. $HO(CH_2)_4C(=O)-H \underset{}{\overset{H^{\oplus}}{\rightleftharpoons}} HO(CH_2)_4C(=\overset{\oplus}{O}H)-H$ * $\rightarrow$ [cyclic oxonium: H, OH on C; $\oplus$O-H in ring] $\xrightarrow{-H^{\oplus}}$ [cyclic hemiacetal: H, OH on C; :O: in ring]

* one resonance form of $\left[\overset{\oplus}{O}H{=}C{-}H \leftrightarrow :\ddot{O}H{-}\overset{\oplus}{C}{-}H \right]$

12. Formation of the hemiacetal creates a new chiral center at C-1. Closing the ring at both faces of the carbon of the C=O gives diastereomers which equilibrate via the free aldehyde, i.e.,

[ring O–C(H)(OH)] $\rightleftharpoons$ [OH / CHO open chain] $\rightleftharpoons$ [ring O–C(OH)(H)]

See Chapter 26A for detail.

13. Acid-catalyzed cleavage of 2,2-dimethoxypropane is accelerated by the formation of a resonance-stabilized cation.

$CH_3-C(OCH_3)_2-CH_3 \xrightarrow{H^{\oplus}} CH_3-C(\overset{\oplus}{O}(H)CH_3)(OCH_3)-CH_3 \xrightarrow{-CH_3OH} \left[CH_3-\overset{\oplus}{C}(-\ddot{O}-CH_3)-CH_3 \leftrightarrow CH_3-C(=\overset{\oplus}{O}-CH_3)-CH_3 \right]$

Cleavage of diethyl ether and 1,2-dimethoxypropane does not give resonance-stabilized cations.

14.

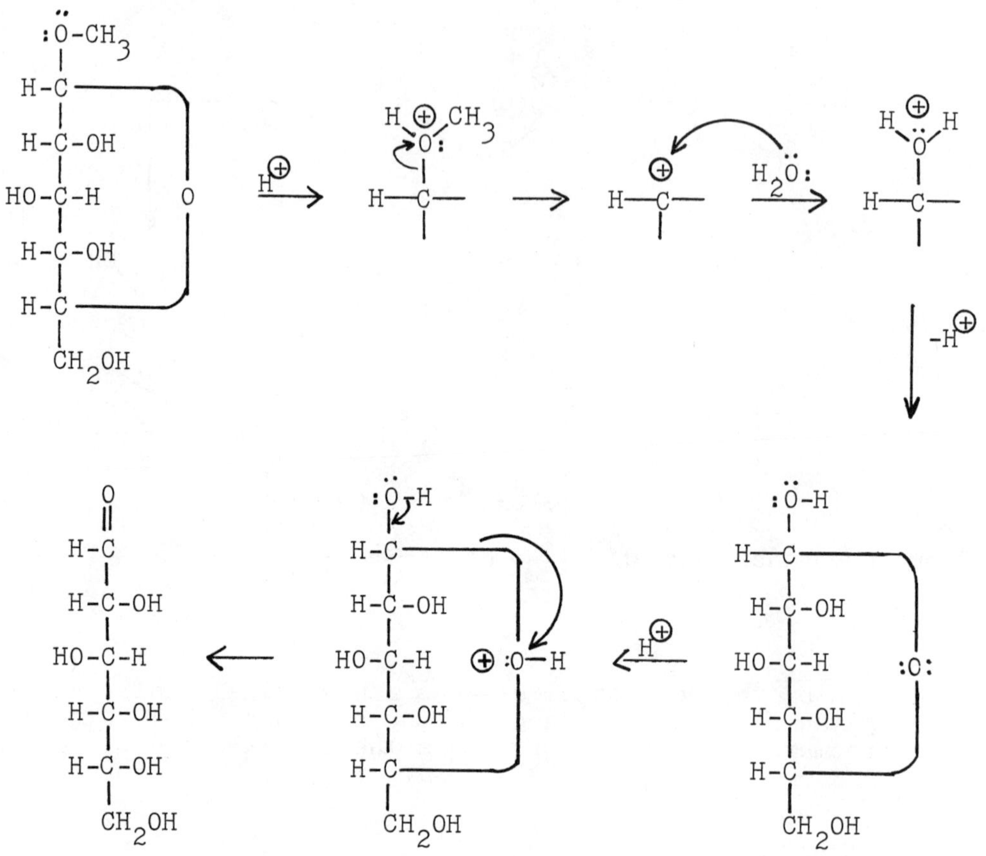

```
:Ö-CH3
 |
H-C ──────┐
 |        |
H-C-OH    |
 |        |
HO-C-H    O      --H⊕-->   H ⊕ CH3          -->   ⊕         --H2Ö:-->   H ⊕ H
 |        |                  O:                   H─C─                     O
H-C-OH    |                  |                      |                      |
 |        |                H─C─                                          H─C─
H-C ──────┘                  |                                             |
 |
CH2OH                                                                    | -H⊕
                                                                         v

  O                   :Ö-H                      :Ö-H
  ||                   |                         |
H-C                  H-C ───┐                  H─C ───┐
  |                    |    |                    |    |
H-C-OH               H-C-OH |                  H-C-OH |
  |                    |    |                    |    |
HO-C-H     <--      HO-C-H  ⊕ :O─H   <--H⊕--   HO-C-H  :O:
  |                    |    |                    |    |
H-C-OH               H-C-OH |                  H-C-OH |
  |                    |    |                    |    |
H-C-OH               H-C ───┘                  H-C ───┘
  |                    |                         |
CH2OH                CH2OH                     CH2OH
```

15. B. This mechanism involves formation of a resonance-stabilized cation similar to the one formed in acetal formation (see p. 473).

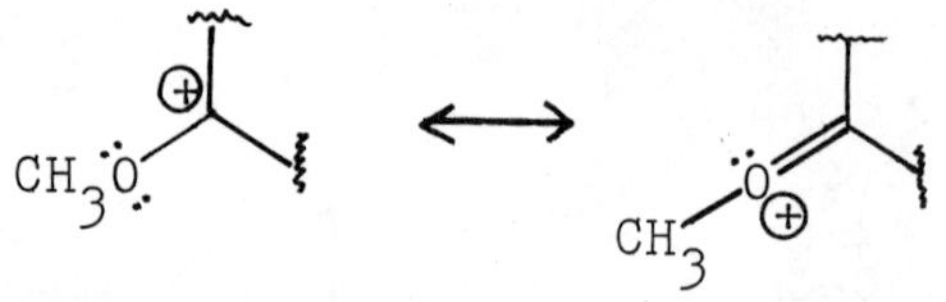

```
     ⊕                            |
 CH3Ö─C<          <-->      CH3─Ö⊕=C<
```

16. (a) about 120°

(b) about 109°

(c) In forming the bisulfite adduct, the two isopropyl groups are forced much closer together. Thus, there is greater steric hindrance in the bisulfite adduct.

17. (a) $CH_3-\underset{CH_3}{\overset{OH}{C}}-SO_3^{\ominus}\ Na^{\oplus}$ $\qquad$ $CH_3-\underset{CH_3}{\overset{OH}{C}}-CN$

(sodium bisulfite addition compound) $\qquad$ (cyanohydrin)

(b) $CH_3-\underset{CH_3}{\overset{:\ddot{O}-H}{C}}-SO_3^{\ominus}\ Na^{\oplus} \underset{}{\overset{CN^{\ominus}}{\rightleftharpoons}} HCN + CH_3-\underset{CH_3}{\overset{:\ddot{O}:^{\ominus}}{C}}-SO_3^{\ominus}\ Na^{\oplus}$

$\rightleftharpoons$

$Na^{\oplus}\ SO_3^{2-} + CH_3-\overset{:O:}{\overset{\|}{C}}-CH_3$

$\rightleftharpoons$ $:CN^{\ominus}$

$CN^{\ominus} + CH_3-\underset{CN}{\overset{OH}{C}}-CH_3 \xleftarrow{HCN} CH_3-\underset{CN}{\overset{:\ddot{O}:^{\ominus}}{C}}-CH_3$

$CN^{\ominus}$ is a better nucleophile than SO_3^{2-}

18. One possible mechanism is the concerted process shown below.

$H_2\ddot{O}:$ $\quad R-\underset{R}{\overset{H}{C}}-N=C\begin{smallmatrix}R'\\R'\end{smallmatrix} \quad H-\overset{\oplus}{O}\begin{smallmatrix}H\\H\end{smallmatrix} \longrightarrow \begin{smallmatrix}R\\R\end{smallmatrix}C=N-\underset{R'}{\overset{R'}{C}}-H$

19. (a)

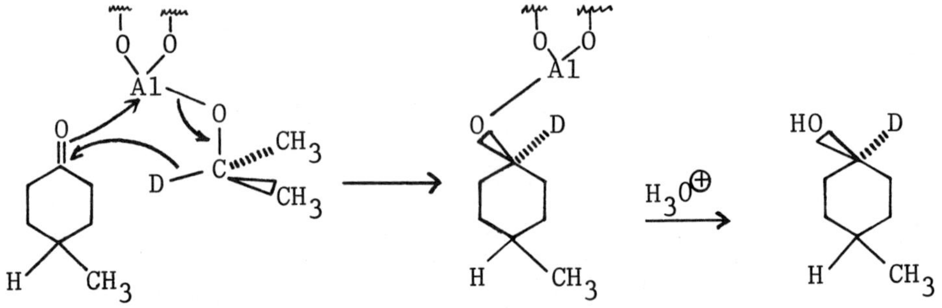

$$\text{4-methylcyclohexanone} + (\text{Al-O-CD}(CH_3)_2) \longrightarrow \text{Al-O-C(D)-4-methylcyclohexane} \xrightarrow{H_3O^{\oplus}} \text{HO-C(D)-4-methylcyclohexane}$$

(b)

$$R_2CH(OH) + Al(OCH(CH_3)_2)_3 \rightleftarrows (R_2CH{-}O)_3Al$$

$$R_2CH{-}O{-}Al + CH_3COCH_3 \longrightarrow RCOR + (CH_3)_2CH{-}O{-}Al$$

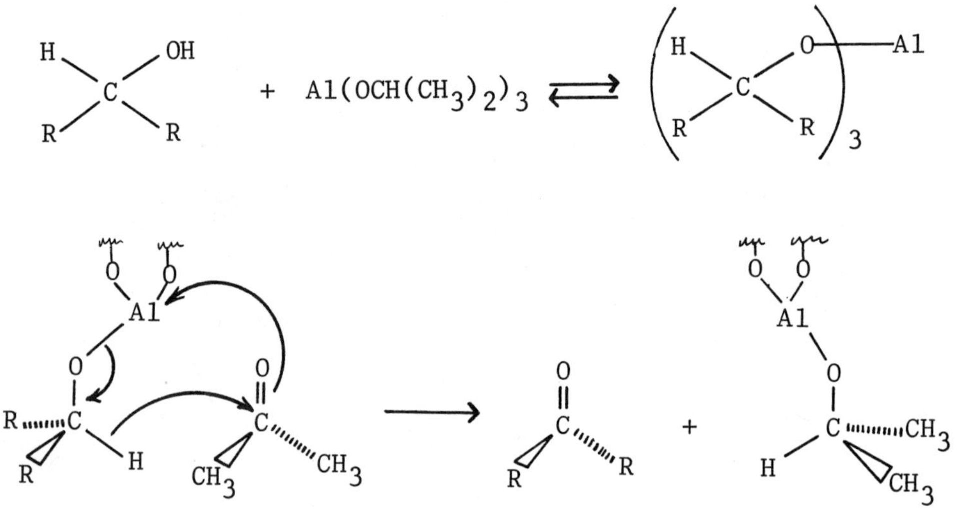

In the MPV reduction, the reaction is driven to completion by using an excess of reductant (isopropyl alcohol). Use of an excess of acetone assures that the Oppenauer oxidation goes to completion.

20. (a) ethanal
(acetaldehyde)

(b) methanal
(formaldehyde)

(c) phenylmethanal
(benzaldehyde)

(d) dicyclohexyl ketone

(e) methylcyclohexyl ketone

(continued next page)

(f) cis-4-chlorocyclohexanecarboxaldehyde (cis-4-chlorocyclohexylmethanal)

(g) methanimine (methyleneimine)

(h) 2-cyclohexenone

(i) benzyl ethyl ketone (1-phenyl-2-butanone)

(j) 1-chloro-3-penten-2-one

(k) trichloroethanal (trichloroacetaldehyde) (chloral)

(l) 1,3-dioxolane (1,3-dioxacyclopentane)

(m) acetone phenylhydrazone

(n) 4-methyl-3-penten-2-one (mesityl oxide)

(o) 3-phenylpropenal (cinnamaldehyde)

(Text S. 17.2)

21. (a) $CH_3CH(OH)CH_3 + CrO_3/H^{\oplus}$

(b) $CH_3CH(OH)CH_2CH_3 + CrO_3/H^{\oplus}$

(c) $CH_2{=}CHCH_2OH + MnO_2$

(d) $CH_3CH(OH)(CH_2)_4CH(OH)CH_3 + K_2Cr_2O_7/H^{\oplus}$

(e) $HOCH_2(CH_2)_4CH_2OH + CrO_3$/pyridine

(f) 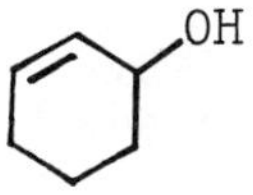 2-cyclohexen-1-ol (ring structure with OH) + MnO_2

(g) 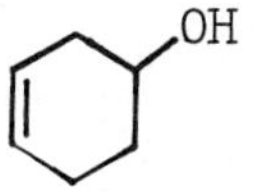 3-cyclohexen-1-ol (ring structure with OH) + $CrO_3/H^{\oplus}$

(h) 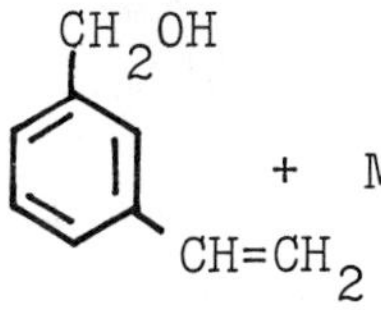 benzene ring bearing CH_2OH and $CH{=}CH_2$ (meta) + MnO_2

(Text: S. 17.3 and summary Figure 17-1)

22. (a) $(CH_3)_2C(OH)-C(OH)(CH_3)_2$

(after work up with H_2O)

(b) 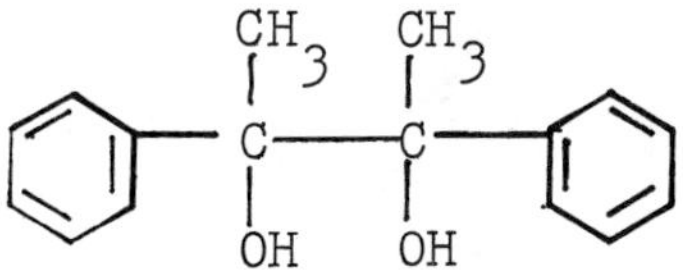 $C_6H_5-C(CH_3)(OH)-C(CH_3)(OH)-C_6H_5$

(after work up with H_2O)

(continued next page)

(c) NR

(d) 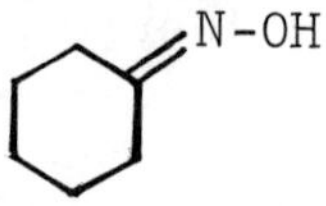 [cyclohexanone oxime: cyclohexane ring with =N-OH]

(e) NR

(f) NR

(g) $CH_2(OH)_2$ (simple addition) or $H(OCH_2)_nOH$

(h) CH_3CH_2CHO

(i) $CH_2{=}CHCH_2OH$ (after work up with H_2O)

(j)  [cyclopentane-1,3-dione bis-hydrazone: $H_2NN{=}$ and $={NNH_2}$ on cyclopentane ring]

(k) $CH_3\overset{O}{\overset{\|}{C}}CH_3$

(l) $CH_3\overset{O}{\overset{\|}{C}}(CH_2)_3CH_3$ and $CH_3CH_2\overset{O}{\overset{\|}{C}}(CH_2)_2CH_3$

(m) 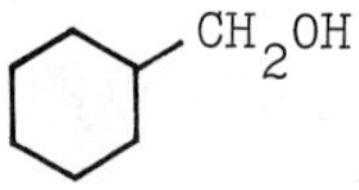 [cyclohexane ring with CH_2OH]

(n) NR

(o) $CH_3CH_2CH_2\underset{CH_2CH_3}{\underset{|}{C}}{=}NNHCONH_2$

(p) NR

(q) $CH_2{=}CH\underset{OH}{\underset{|}{C}}HCH_3$ (after work up with H_2O)

(r) $H\overset{O}{\overset{\|}{C}}(CH_2)_7\overset{O}{\overset{\|}{C}}H$

(s) $CH_3CH(OCH_2CH_3)_2$

(t) $CH_3\underset{OH}{\underset{|}{C}}HOCH_2CH_3$ (unstable)

(continued next page)

(u) $(CH_3)_3CCH_2OH + (CH_3)_3CCOOH$
(Cannizzaro)

(v) $CH_2(OCH_3)_2$

(w) 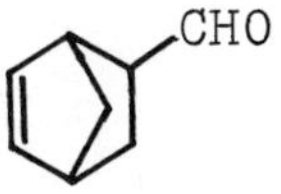 [bicyclic structure with CHO substituent]

(x) $CH_3CH_2CH_2OH$
(after work up with H_2O)

(y) CH_3CHO
(after work up with H_2O)

(z) 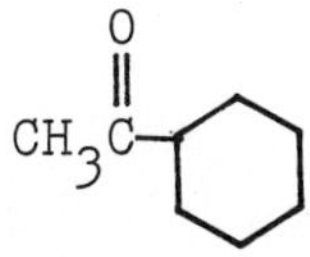 $CH_3\overset{O}{\overset{\|}{C}}$–(cyclohexyl)

(aa) $CH_3CH_2NH_2$

(bb) CH_3CHO

(cc) 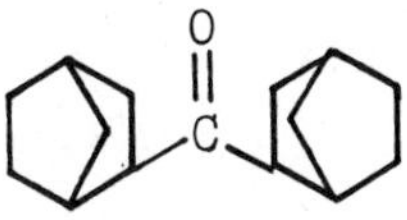 (bicyclic)–$\overset{O}{\overset{\|}{C}}$–(bicyclic)

(dd) $CH_3\underset{CH_3}{\underset{|}{C}}HCH_2\underset{CH_3}{\underset{|}{C}}HCH_2\overset{O}{\overset{\|}{C}}CH_3$
(after work up with H_2O)

(ee) $(C_6H_5)_3CCH_2OH + HCO_2H$

(ff) $(CH_3)_2C=C(CH_3)_2$

(gg) NR

(hh) $CH_2O + CH_3\underset{HO}{\underset{|}{C}}H\text{-}\underset{OH}{\underset{|}{C}}HCH_3$
(meso)

(ii) $CH_2O + CH_3\underset{HO}{\underset{|}{C}}H\text{-}\underset{OH}{\underset{|}{C}}HCH_3$
(d,l)

(jj) $H(OCH_2CH_2)_nOH$

(kk) CH_2O

23. (i) (a) $CH_2O + CH_3MgI$

(b) $CH_3CHO + CH_3MgI$

(c) $C_6H_5CHO + C_6H_5MgCl$

(d) $CH_3\overset{O}{\overset{\|}{C}}CH_3 + CH_3MgCl$

(e) cyclohexanone (ring with C=O) + CH_3MgI

(f) CH_2O + cyclohexyl–MgBr

(g) $CH_2{=}CH{-}CHO + CH_3MgI$

(or $CH_3CHO + CH_2{=}CHMgCl$)

(ii) (a) <u>Oxidized by MnO_2:</u> c, product $C_6H_5\overset{O}{\overset{\|}{C}}C_6H_5$ and

g, product $CH_2{=}CH\overset{O}{\overset{\|}{C}}CH_3$

(b) <u>Oxidized by Sarett's reagent (CrO_3·pyridine):</u>

all but t-butyl alcohol and 1-methylcyclohexanol (both 3°). Products:

(a) CH_3CHO

(b) $CH_3\overset{O}{\overset{\|}{C}}CH_3$

(c) $(C_6H_5)_2C{=}O$

(f) cyclohexyl–CHO

(g) $CH_2{=}CH{-}\overset{O}{\overset{\|}{C}}CH_3$

(Text: review Chapter 10)

24. (a) cyclohexanone $\xrightarrow{LiAlH_4}$ $\xrightarrow{H_2O}$ cyclohexanol (OH)

(b) cyclohexanol (OH) (from a) $\xrightarrow[heat]{H_2SO_4}$ cyclohexene

(c) cyclohexene (from b) $\xrightarrow{D_2/Pt}$ cis-1,2-dideuteriocyclohexane (D, D)

(d) cyclohexene (from b) $\xrightarrow{HCl}$ chlorocyclohexane (Cl)

(e) chlorocyclohexane (Cl) (from d) $\xrightarrow[ether]{Mg}$ cyclohexylmagnesium chloride (MgCl) $\xrightarrow{D_2O}$ deuteriocyclohexane (D)

(f) cyclohexanone $\xrightarrow{CH_3MgI}$ $\xrightarrow{H_3O^{\oplus}}$ 1-methylcyclohexanol (OH, CH_3)

(g) cyclohexanone $\xrightarrow{Mg \cdot Hg}$ 1,1'-bicyclohexyl-1,1'-diol (HO, OH)

(h) cyclohexylmagnesium chloride (MgCl) (from e) $\xrightarrow{\text{cyclohexanone}}$ $\xrightarrow{H_3O^{\oplus}}$ 1-cyclohexylcyclohexanol (OH)

(i) cyclohexanone $\xrightarrow{LiAlD_4}$ $\xrightarrow{H_3O^{\oplus}}$ 1-deuteriocyclohexanol (OH, D)

(continued next page)

(j) (from b) $\xrightarrow[\text{heat}]{\text{Pt}}$

(k) $\xrightarrow{NH_2NH_2}$ $=N-NH_2$

(l) 2 $\xrightarrow{NH_2NH_2}$ $=N-N=$

(m) $\xrightarrow{CH_3OH/H^{\oplus}}$ OCH_3, OCH_3

(n) (from c) $\xrightarrow{NBS}$ Br $\xrightarrow[\text{heat}]{\text{KOH/alcohol}}$

(o) (from c) $\xrightarrow{H_2/Pt}$

(p) (from c) $\xrightarrow{O_3}$ $\xrightarrow[OH^{\ominus}]{H_2O_2}$ $HO_2C(CH_2)_4CO_2H$

(q) OH (from a) $\xrightarrow{C_6H_5SO_2Cl}$ $O-SO_2-C_6H_5$

(r) OH $\xrightarrow{\text{conc. } H_2SO_4}$ $O-SO_2O$

(continued next page)

(s) [cyclohexyl–Cl] (from d) $\xrightarrow[\text{DMF}]{Na^{\oplus}\ CN^{\ominus}}$ [cyclohexyl–CN]

(t) [cyclohexyl–CN] (from t) $\xrightarrow{LiAlH(OEt)_3}$ $\xrightarrow{H_2O}$ [cyclohexyl–CHO]

25. (a) CH_3CHO $\xrightarrow{CH_3MgCl}$ $\xrightarrow{H_2O}$

(b) CH_2—CH_2 (bridged by O) $\xrightarrow{CH_3MgCl}$ $\xrightarrow{H_2O}$

(c) HC≡CH $\xrightarrow[NH_{3(\ell)}]{Na^{\oplus}\ NH_2^{\ominus}}$ $\xrightarrow{CH_3I}$ (or $CH_3\underset{OH}{CH}CH_3$ (from a)

$\xrightarrow[\text{heat}]{H_2SO_4}$ $CH_3CH{=}CH_2$ $\xrightarrow{Br_2}$ $\xrightarrow[NH_{3(\ell)}]{2Na^{\oplus}\ NH_2^{\ominus}}$)

(d) $CH_3CH(OH)CH_3$ (from a) $\xrightarrow{CrO_3/H^{\oplus}}$

(e) $CH_3\overset{O}{\overset{\|}{C}}CH_3$ (from d) $\xrightarrow{2CH_3OH/H^{\oplus}}$

(f) $2CH_3\overset{O}{\overset{\|}{C}}CH_3$ (from d) $\xrightarrow{Mg \cdot Hg}$

(continued next page)

(g) $CH_3CH(OH)CH_3$ (from a) $\xrightarrow[\text{heat}]{H_2SO_4}$ $CH_3CH{=}CH_2$ $\xrightarrow{NBS}$

$BrCH_2CH{=}CH_2$ $\xrightarrow{OH^{\ominus}/H_2O}$ (OR $CH_2{=}CHMgCl$ $\xrightarrow{CH_2O}$ $\xrightarrow{H_2O}$)

(h) $CH_2{=}CHCH_2OH$ (from g) $\xrightarrow{MnO_2}$

(i) same as (g)

(j) $CH_3CH_2CH_2OH$ (from b) $\xrightarrow{SOCl_2}$ $CH_3CH_2CH_2Cl$ $\xrightarrow[\text{ether}]{Mg}$

$\xrightarrow{CH_2O}$ $\xrightarrow{H_2O}$ $CH_3CH_2CH_2CH_2OH$ $\xrightarrow{CrO_3/\text{pyridine}}$

(k) $CH_2{=}CHCHO$ (from h) $\xrightarrow{HBr}$

(l) CH_2O $\xrightarrow{HOCH_2CH_2OH/H^{\oplus}}$

(m) CH_3CHO $\xrightarrow{HOCH_2CH_2OH/H^{\oplus}}$

(n) $CH_2{=}CHCH_2OH$ (from g) $\xrightarrow{H^{\oplus}/H_2O}$ $CH_3CH(OH)CH_2OH$ $\xrightarrow[H^{\oplus}]{CH_2O}$

(o) $(CH_3)_2CO$ $\xrightarrow{HOCH_2CH_2OH/H^{\oplus}}$

26. $RCH{=}CH_2 \xrightarrow{BH_3} (RCH_2CH_2)_3B$

- $\xrightarrow{H_2O_2/OH^{\ominus}}$ RCH_2CH_2OH (S. 10.4)
- $\xrightarrow{CO} \xrightarrow{H_2O_2/OH^{\ominus}}$ $(RCH_2CH_2)_3COH$ (S. 10.4)
- $\xrightarrow{CO/H_2O} \xrightarrow{H_2O_2/OH^{\ominus}}$ $(RCH_2CH_2)_2C{=}O$ (S. 17.3)

$RC{\equiv}CH(R) \xrightarrow{BH_3} (RCH{=}C(H)-)_3B \xrightarrow{H_2O_2/OH^{\ominus}} RCH_2C(=O)H(R)$ (S. 9.8)

$R_3B \xrightarrow{-\overset{|}{C}=\overset{|}{C}-\overset{|}{C}=O} R-\overset{|}{\underset{|}{C}}-\overset{|}{C}=C(-)OBR_2 \xrightarrow{H_2O} R-\overset{|}{\underset{|}{C}}-\overset{|}{C}=C(-)OH \rightleftharpoons R-\overset{|}{\underset{|}{C}}-\overset{|}{\underset{H}{C}}-\underset{O}{\overset{\|}{C}}-$

(S. 17.9)

27. (a) benzene $\xrightarrow[AlCl_3]{CH_3Cl}$ toluene (CH_3) $\xrightarrow[FeCl_3]{Cl_2}$ $\xrightarrow[FeCl_3]{Cl_2}$ 2,4-dichlorotoluene (CH_3, Cl, Cl)

$\xrightarrow[H_2SO_4]{CrO_3/(CH_3CO)_2O}$ $\xrightarrow{H_2O}$ 2,4-dichlorobenzaldehyde (CHO, Cl, Cl) $\xrightarrow{LiAlH_4}$ $\xrightarrow{H_3O^{\oplus}}$

2,4-dichlorobenzyl alcohol (CH_2OH, Cl, Cl)

(b) and (c) benzaldehyde (CHO) or benzoic acid (CO_2H) $\xrightarrow{LiAlH_4}$ $\xrightarrow{H_2O}$ $C_6H_5CH_2OH$ $\xrightarrow[H^{\oplus}]{\text{dihydropyran}^*}$

$C_6H_5CH_2$-O-(tetrahydropyranyl) $\xrightarrow[FeCl_3]{2Cl_2}$ 2,4-dichlorophenyl-CH_2-O-(tetrahydropyranyl) (Cl, Cl) $\xrightarrow{H_3O^{\oplus}}$ 2,4-dichlorobenzyl alcohol (CH_2OH, Cl, Cl)

*direct halogenation of $C_6H_5CH_2OH$ with X_2/cat. cannot be used because oxidation to the aldehyde occurs (see problem 52).

28. 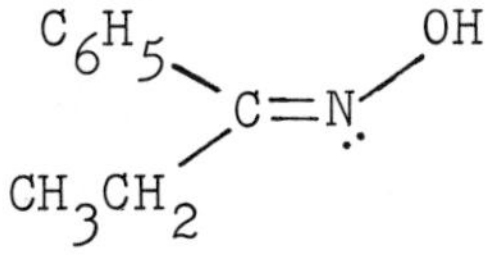 and 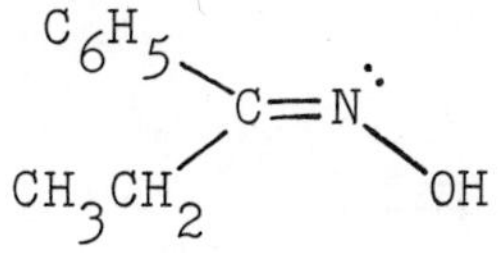

-OH group can be <u>cis</u> (<u>syn</u>) to CH_3CH_2 or to C_6H_5

29. Only isopropyl bromide reacts with aqueous $AgNO_3$ to give a ppt. of AgBr. Both methyl ethyl ketone and butanal form crystalline phenylhydrazones and bisulfite addition products but only butanal is oxidized (to butyric acid) by dilute $KMnO_4$ (purple → colorless) or $CrO_3/H^{\oplus}$ (orange → green). Hexane gives none of these reactions.

30. Puberulic acid forms an aromatic cation upon protonation (resonance structures B ⟶ D).

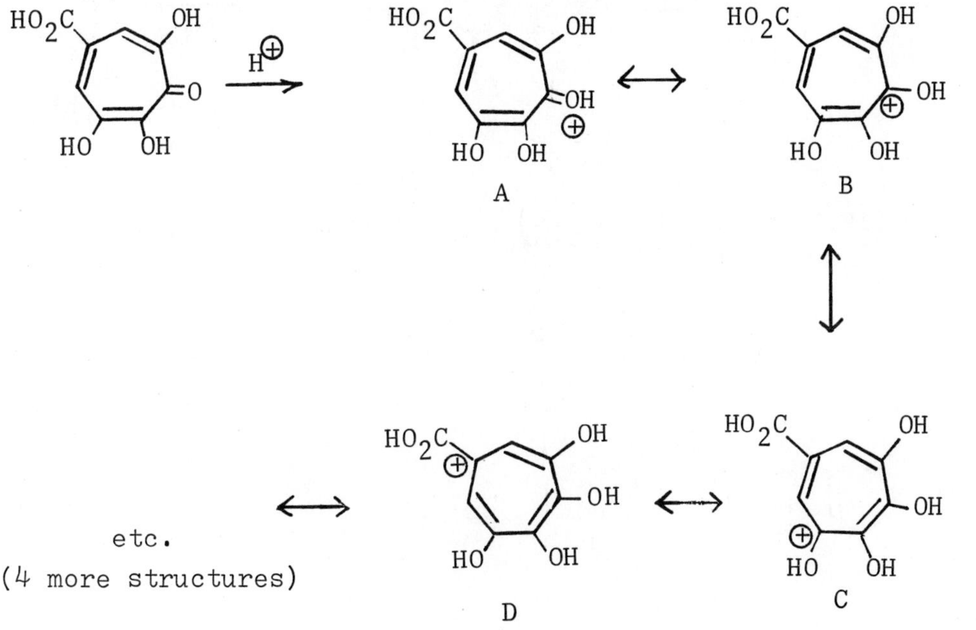

31. The reaction which occurs is:

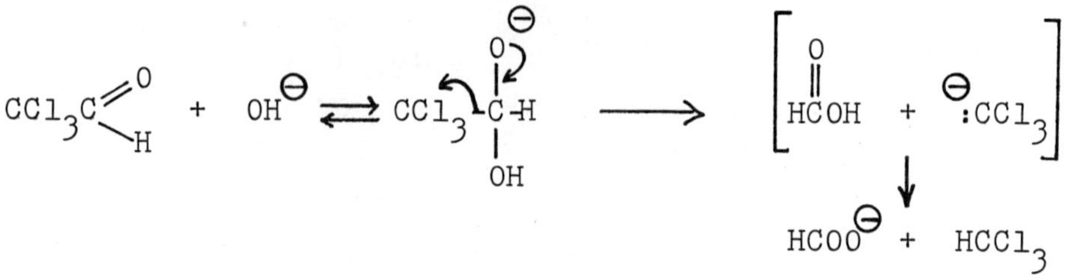

$$CCl_3C(=O)H + OH^{\ominus} \rightleftarrows CCl_3-C(O^{\ominus})(H)(OH) \longrightarrow \left[HC(=O)OH + {}^{\ominus}\!:CCl_3 \right] \downarrow HCOO^{\ominus} + HCCl_3$$

${}^{\ominus}\!:CCl_3$ is a good leaving group because of the stabilizing effect of the electronegative chlorines.

32. (a) (2R,4R)- and (2S,4R)-C_6H_5-CH(CH_3)-CH_2-CH(OH)-CH_3

(b) 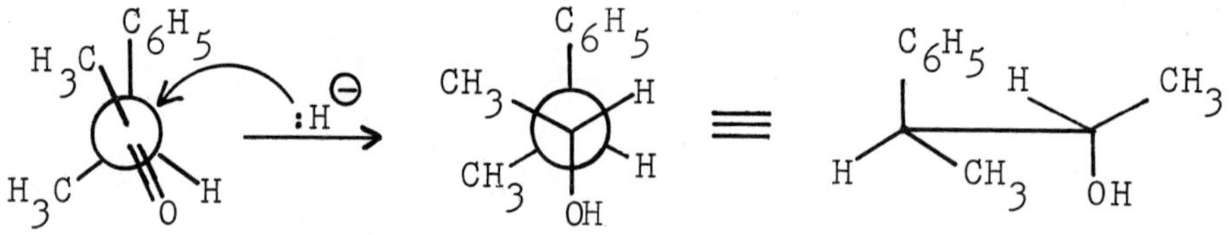 H₃C, H₃C, C₆H₅ Newman projection with C=O and H; :H⊖ attack → Newman projection (C_6H_5, CH_3, CH_3, H, H, OH) ≡ zigzag structure with C_6H_5, H, CH_3 (S) and H, CH_3, OH (R)

(S) (R)

(c) (1R,3R) and (1S,3R)-C_6H_5-CH(CH_3)-CH_2-CHD(OH)

(d) 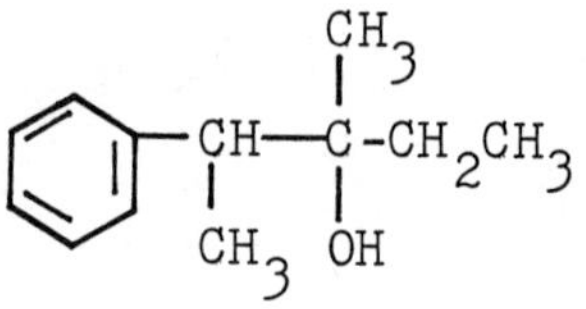 C_6H_5-CH(CH_3)-C(CH_3)(OH)-CH_2CH_3

(S) (R)

(continued next page)

(e) CH_3 ... OH, H H, (R) (S)

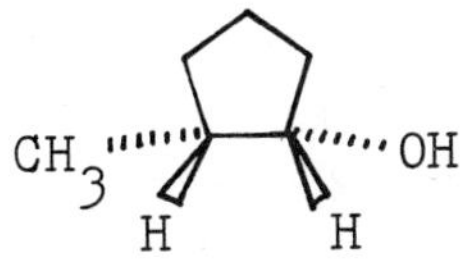

(f) CH_3 ... OH, H D, (R) (S)

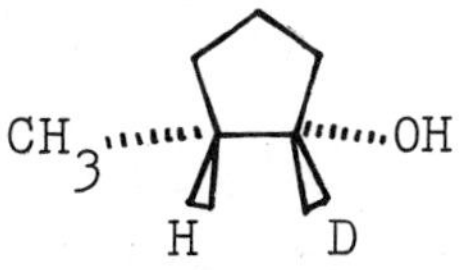

(Text: S. 17.6)

33. Excess benzaldehyde remained at the end of the reaction with phenylmagnesium bromide. On washing the reaction mixture with bisulfite solution, the bisulfite addition product of benzaldehyde precipitated.

34. (a)

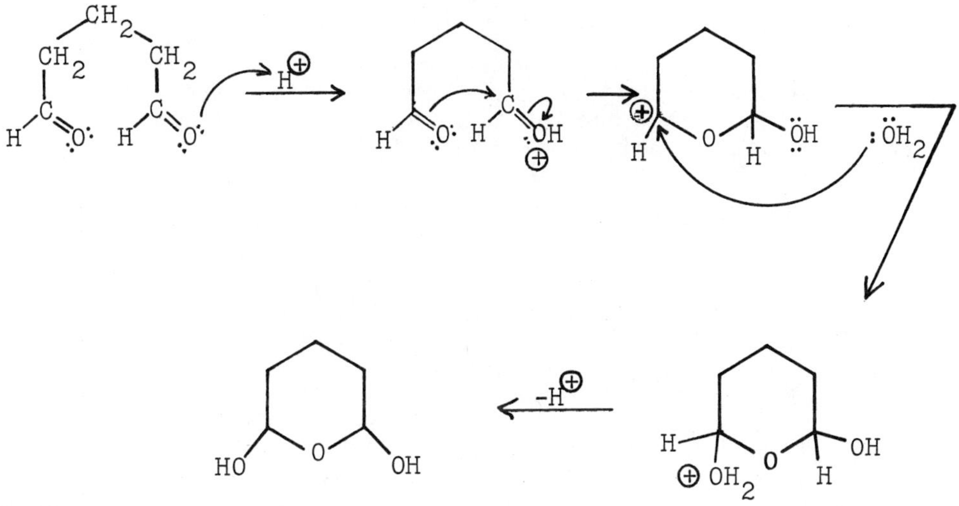

$$\text{OHC-CH}_2\text{CH}_2\text{CH}_2\text{-CHO} \xrightarrow{H^{\oplus}} \ldots \rightarrow \ldots \xrightarrow{:\ddot{O}H_2} \ldots \xrightarrow{-H^{\oplus}} \text{2,6-dihydroxytetrahydropyran}$$

(b) There are two chiral centers produced. The cis-isomer is meso. The trans-isomer is d,l.

35. (a) yes

(b) carbonyl carbon

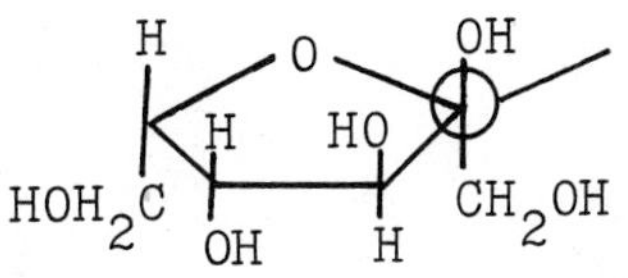

(continued next page)

(c) [structure: diacetone glucose-type bis-isopropylidene furanose with CH_2OH] (the same reaction as:

$$CH_3\overset{O}{\overset{\|}{C}}CH_3 \xrightarrow[H^{\oplus}]{2ROH} (CH_3)_2C(OR)_2$$)

36. The electron deficiency of carbon is actually increased in the protonated species as shown by resonance form B.

$$\delta^{\oplus}\,C{=}\ddot{O}\,\delta^{\ominus} \xrightarrow{H^{\oplus}} \left[\, C{=}\ddot{O}^{\oplus}{-}H \;(A) \longleftrightarrow {}^{\oplus}C{-}\ddot{O}{-}H \;(B) \,\right]$$

The preference for nucleophilic attack at carbon also can be explained by comparing the two products which result from nucleophilic attack at carbon and oxygen.

$$Nu{:}^{\ominus} + C{=}\ddot{O}^{\oplus}{-}H \longrightarrow Nu{-}C{-}\ddot{O}{-}H \text{ (neutral)}$$

$$C{=}\ddot{O}^{\oplus}{-}H + {:}Nu^{\ominus} \not\longrightarrow {}^{\ominus}{:}C{-}O^{\oplus}(H){-}Nu$$

(charge separation, higher energy)

37. $CH_3(CH_2)_7CH_2OH \xrightarrow{SOCl_2} CH_3(CH_2)_7CH_2Cl \xrightarrow[ether]{Mg}$ [ethylene oxide] $\longrightarrow$

$\xrightarrow{H_2O} CH_3(CH_2)_7CH_2CH_2CH_2OH \xrightarrow{CrO_3/pyridine}$

38. $CH_4 \xrightarrow{Cl_2/h\nu} CH_3Cl \xrightarrow[AlCl_3]{C_6H_6}$ $C_6H_5CH_3$ $\xrightarrow[H_2SO_4]{CrO_3/(CH_3CO)_2O}$

$\xrightarrow{H_3O^{\oplus}}$ C_6H_5-CHO $\xrightarrow{HCN}$ C_6H_5-CH(OH)CN

39. The 2° cation is destabilized to a greater extent by the C=O group since positive charges appear on adjacent carbons. The 1° cation is more stable because the positive charge is further from the -CHO group, and it is the cation actually formed in this reaction (anti-Markownikoff addition is observed).

CH_3-$\overset{\oplus}{C}$H-$\overset{\delta\oplus}{C}$(=$O^{\delta\ominus}$)-H

40. cyclohexanone $\xrightarrow{Li-C\equiv CH}$ 1-(OLi)-1-(C≡CH)cyclohexane $\xrightarrow{H_2O}$ 1-(OH)-1-(C≡CH)cyclohexane + $H^{\oplus}$ → cation: $\oplus$ on ring carbon, -C(OH)=CH_2; loss of H to :$\ddot{O}H_2$ → cyclohexenyl-C(OH)=CH_2 ⇌ cyclohexenyl-C(=O)-CH_3

41. Dehydration of a 1,1-diol involves the change

$$>C(OH)_2 \xrightarrow{-H_2O} >C=O$$

The ideal bond angles for the diol and carbonyl compound are ~109° and ~120° respectively. The greatest angle strain is introduced in converting 1,1-cyclopropanediol to cyclopropanone. Thus, it has the least tendency to undergo dehydration.

(continued next page)

1,1-Cyclobutanediol and 1,1-cyclopentanediol which form cycloalkanones having more favorable ring bond angles (90° and 108° respectively) follow in reactivity in that order.

42. Unfavorable dipole-dipole repulsions of carbonyl groups in the tricarbonyl compound are relieved in the hydrate.

Hydration at the middle carbonyl relieves both carbonyl to carbonyl repulsions. Hydration at either of the other carbonyl groups gives a product with one repulsion remaining.

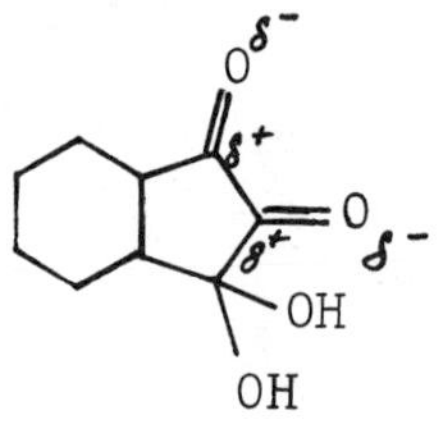

O δ-, δ+, O δ-, δ+, OH, OH

43. (a)

H-C-OH, HO-C-H, H-C-OH, H-C-OH, CH_2OH = R

$H^\oplus$ ⇌ $R-C(=\overset{\oplus}{O}H)-H$ + $H-\ddot{S}-CH_2CH_3$ ⇌ $R-C(OH)(H)-\overset{\oplus}{S}(H)CH_2CH_3$ ⇌ $R-C(H)(\overset{\oplus}{O}H_2)-S-CH_2CH_3$ ⇌ ($-H_2O$) $R-\overset{\oplus}{C}H-S-CH_2CH_3$ * → ($CH_3CH_2\ddot{S}H$) $R-CH(\overset{\oplus}{S}(H)CH_2CH_3)(S-CH_2CH_3)$ ⇌ ($-H^\oplus$) $\underset{R}{CH}(\ddot{S}CH_2CH_3)_2$

* resonance stabilized

$[-\overset{\oplus}{C}-\ddot{S}- \leftrightarrow -C=\overset{\oplus}{S}-]$

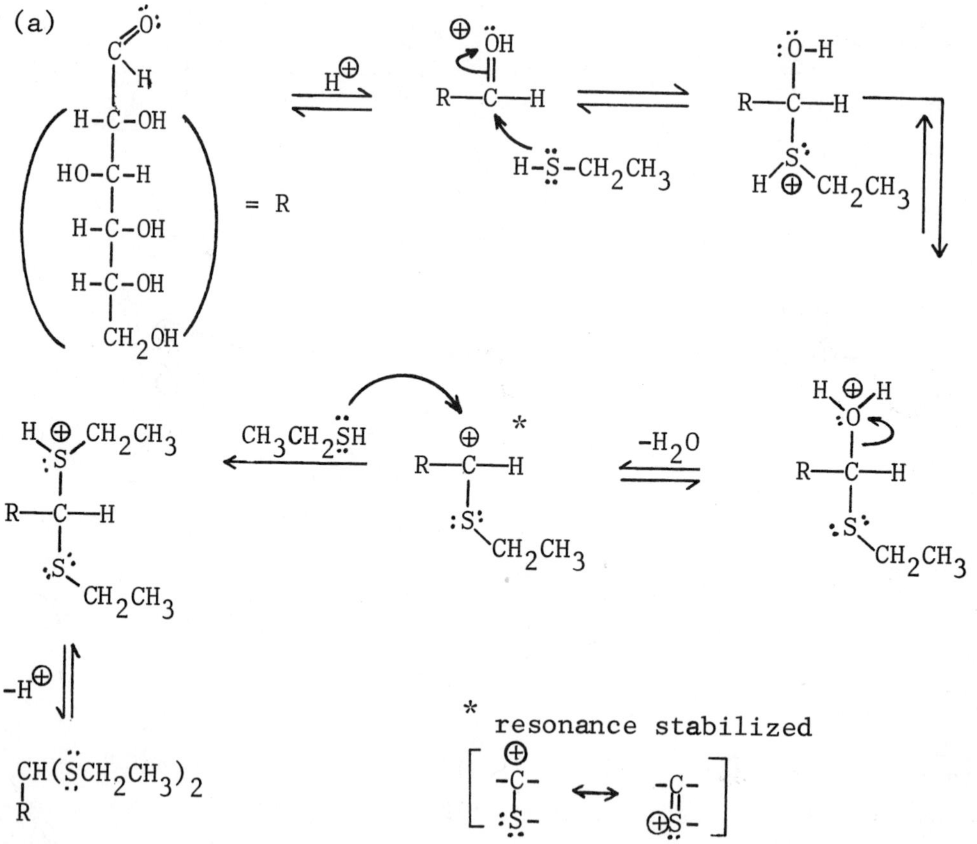

(continued next page)

(b)

$$HOC(=O)H\text{-}CHOH\text{-}CH_2OPO_3H_2 \quad \underset{}{\overset{H\ddot{S}\text{-enzyme}}{\rightleftharpoons}} \quad H\text{-}\underset{CH_2OPO_3H_2}{\underset{|}{\underset{CHOH}{\underset{|}{\overset{O^{\ominus}}{\overset{|}{C}}}}}}\text{-}\overset{H}{\overset{|}{S^{\oplus}}}\text{-enzyme} \longrightarrow H\text{-}\underset{CH_2OPO_3H_2}{\underset{|}{\underset{CHOH}{\underset{|}{\overset{OH}{\overset{|}{C}}}}}}\text{-}\ddot{S}\text{-enzyme}$$

44. $CH_3(CH_2)_4C{\equiv}C\text{—}MgBr^{*} + CH_3\text{-}\overset{O}{\overset{\|}{C}}\text{-}\ddot{O}\text{-}\overset{O}{\overset{\|}{C}}\text{-}CH_3 \longrightarrow CH_3\text{—}\underset{CH_3(CH_2)_4C{\equiv}C}{\underset{|}{\overset{\ddot{O}MgBr}{\overset{|}{C}}}}\text{—}\ddot{O}\text{—}\overset{O}{\overset{\|}{C}}\text{-}CH_3$

$$\downarrow$$

$$CH_3\overset{O}{\overset{\|}{C}}\text{—}\ddot{O}:^{\ominus} + CH_3(CH_2)_4C{\equiv}C\text{-}\overset{:O:}{\overset{\|}{C}}\text{-}CH_3$$

*source of the nucleophile $CH_3(CH_2)_4C{\equiv}C:^{\ominus}$

(Text: see addition-elimination, p. 427)

45. $C_6H_5\text{-}\overset{:O:}{\overset{\|}{C}}\text{-H} \xrightarrow{:NH_3} C_6H_5\text{-}\overset{:\ddot{O}:^{\ominus}}{\overset{|}{\underset{\oplus NH_3}{\underset{|}{C}}}}\text{-H} \longrightarrow C_6H_5\text{-}\overset{:\ddot{O}H}{\overset{|}{\underset{\ddot{N}H\text{-}H}{\underset{|}{C}}}}\text{-H} \quad (H_3N:) \xrightarrow{-H_2O}$

$$C_6H_5\text{-}\underset{CN}{\underset{|}{CH}}\text{-}\ddot{N}H_2 + CN^{\ominus} \xleftarrow{HCN} C_6H_5\text{-}\underset{CN}{\underset{|}{CH}}\text{-}\ddot{N}H^{\ominus} \xleftarrow{:N{\equiv}C:^{\ominus}} C_6H_5\text{-}CH{=}\ddot{N}H$$

46.

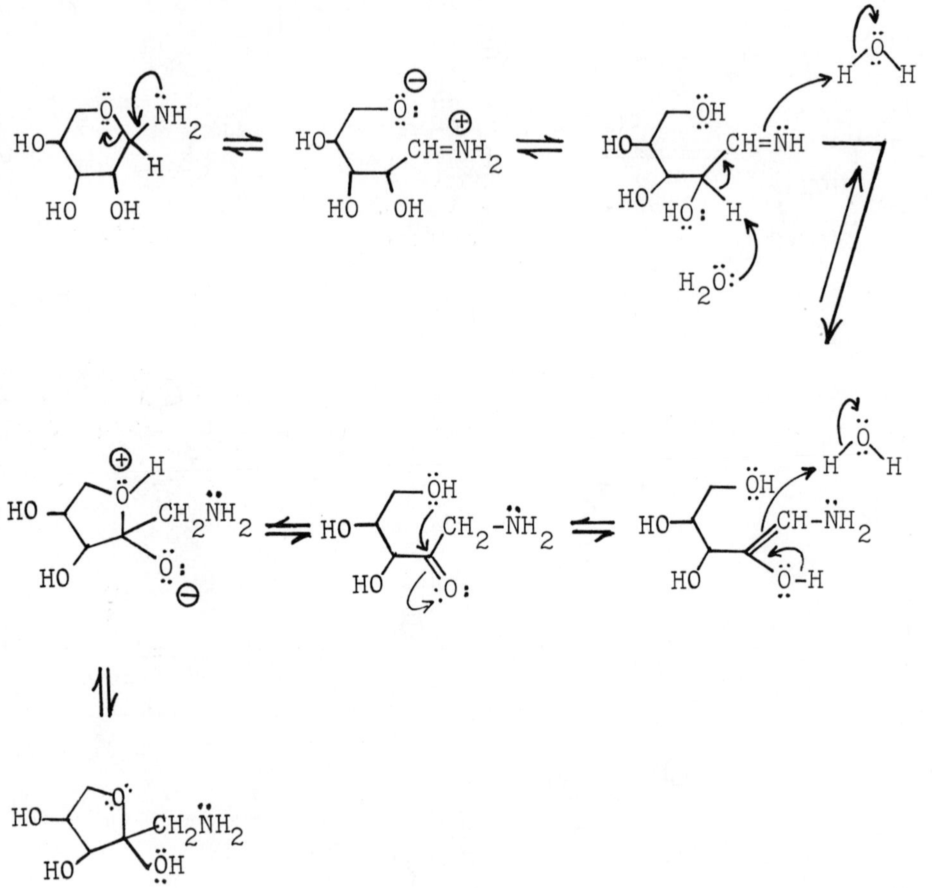

HO, HO, OH, NH$_2$, H ⇌ HO, HO, OH, $CH=\ddot{N}H_2$ ⇌ HO, HO, HO:, H, $\ddot{O}H$, CH=N̈H, $H_2\ddot{O}$:, H–Ö–H

HO, HO, $CH_2\ddot{N}H_2$, H ⇌ HO, HO, ÖH, CH_2-N̈H$_2$, :O: ⇌ HO, HO, ÖH, CH-N̈H$_2$, Ö-H, H–Ö–H

⇅

HO, HO, $CH_2\ddot{N}H_2$, ÖH

47.

$$CH_2=CH-\overset{:O:}{\overset{\|}{C}}-H \xrightarrow{H-Cl} \overset{\oplus}{C}H_2-CH_2-\overset{O}{\overset{\|}{C}}-H \xrightarrow{:Cl^{\ominus}} ClCH_2-CH_2-\overset{:O:}{\overset{\|}{C}}-H \quad (A)$$

↓ H-Cl

$$Cl-CH_2CH_2-\overset{\oplus\ddot{O}H}{\overset{\|}{C}}-H \xleftarrow{C_2H_5\ddot{O}H} \;$$

$$Cl-CH_2CH_2-\underset{C_2H_5\overset{\oplus}{O}H}{\overset{:\ddot{O}H}{C}}-H \longleftarrow$$

$$Cl-CH_2CH_2-\underset{C_2H_5\ddot{O}}{\overset{H\overset{\oplus}{O}H}{C}}-H \longleftarrow$$

↓ $-H_2O$

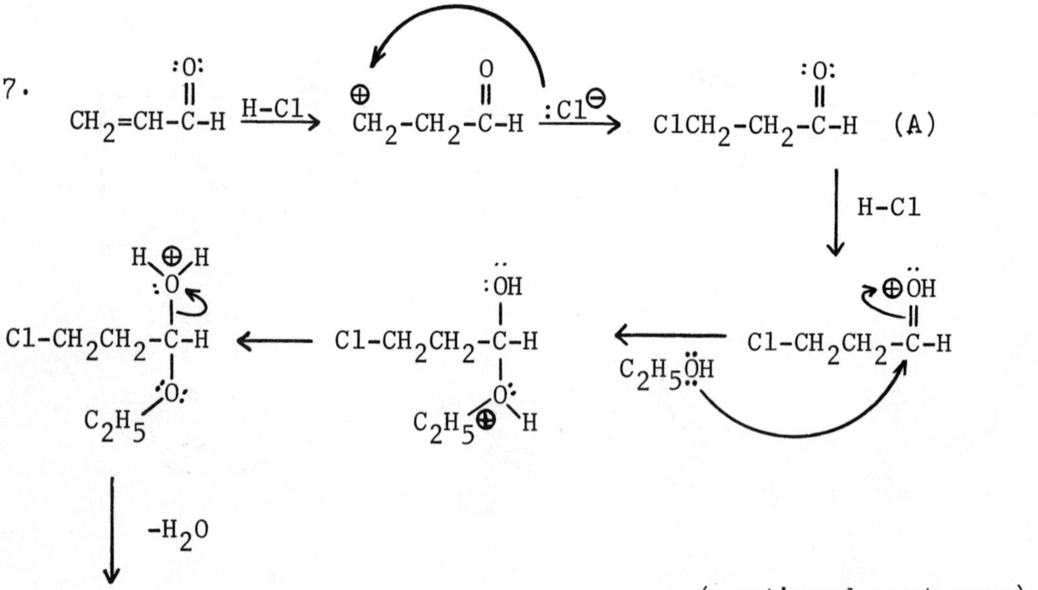

(continued next page)

$$Cl\text{-}CH_2CH_2\text{-}\overset{\oplus *}{C}H(OC_2H_5) \xrightarrow{C_2H_5\ddot{O}H} Cl\text{-}CH_2CH_2\text{-}CH(OC_2H_5)(\overset{\oplus}{O}(H)C_2H_5) \xrightarrow{-H^{\oplus}} Cl\text{-}CH_2\text{-}CH(H)CH(OC_2H_5)_2 \;(B) \xleftarrow{}\; :\ddot{O}H^{\ominus}$$

$$\downarrow$$

$$CH_2(OH)\text{-}CH(OH)\text{-}CH(OC_2H_5)_2 \;(D) \xleftarrow{\text{dil } KMnO_4} CH_2{=}CHCH(OC_2H_5)_2 \;(C)$$

$$\xrightarrow{H^{\oplus}}$$

$$CH_2(OH)\text{-}CH(OH)\text{-}CH(OC_2H_5)(\ddot{O}^{\oplus}(H)C_2H_5) \xrightarrow{-C_2H_5OH} CH_2(OH)\text{-}CH(OH)\text{-}\overset{\oplus}{C}H\text{-}\ddot{O}C_2H_5{}^{*} \xrightarrow{H_2\ddot{O}:} CH_2(OH)\text{-}CH(OH)\text{-}CH(\ddot{O}C_2H_5)(\overset{\oplus}{O}H_2)$$

$$\rightleftharpoons$$

$$CH_2(OH)\text{-}CH(OH)\text{-}C(H){=}\ddot{O}: \xleftarrow[-H^{\oplus}]{-C_2H_5OH} CH_2(OH)\text{-}CH(OH)\text{-}CH(:\ddot{O}\text{-}H)\text{-}\overset{\oplus}{O}(H)C_2H_5$$

*resonance stabilized cation $\left[-\overset{\oplus}{C}H\text{-}\ddot{O}\text{-} \longleftrightarrow -CH{=}\overset{\oplus}{\ddot{O}}\text{-}\right]$

48. $R\text{-}C(=O)\text{-}R \xrightarrow{PCl_4^{\oplus}} R\text{-}C(=\overset{\oplus}{O}\text{-}PCl_4)\text{-}R$ (+ $Cl\text{-}PCl_5^{\ominus}$) $\longrightarrow R\text{-}C(O\text{-}PCl_4)(Cl)\text{-}R$

$\xrightarrow{-Cl^{\ominus}} R\text{-}C(\overset{\oplus}{O}\text{-}PCl_3)(Cl)\text{-}R$ *

$\xrightarrow{Cl:^{\ominus}} POCl_3 + R\text{-}CCl_2\text{-}R$

* or $R\text{-}C(\overset{\oplus}{O}PCl_3)(Cl)\text{-}R \xrightarrow{-POCl_3} \left[R\text{-}\overset{\oplus}{C}(:\ddot{Cl}:)\text{-}R \longleftrightarrow R\text{-}C(=\overset{\oplus}{Cl}:)\text{-}R \right] \xrightarrow{:\ddot{Cl}:^{\ominus}} R\text{-}CCl_2\text{-}R$

49. (a) $(CH_3)_2C=\ddot{O}: \xrightarrow{H^{\oplus}} (CH_3)_2C=\overset{\oplus}{O}H$

HOCH$_2$ / O / OCH$_3$ / H / H / H / H / HO: / OH

$\longrightarrow$ H–$\overset{\oplus}{O}$: / OH / $CH_3\text{-}C\text{-}CH_3$ / :OH

(continued next page)

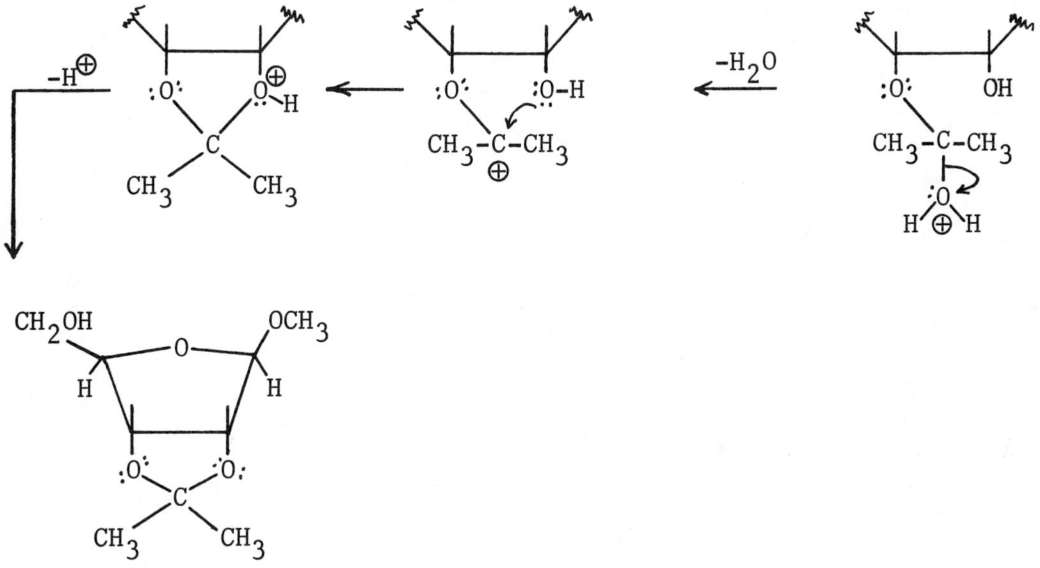

(b) The two hydroxy groups which reacted are <u>cis</u>.

50.

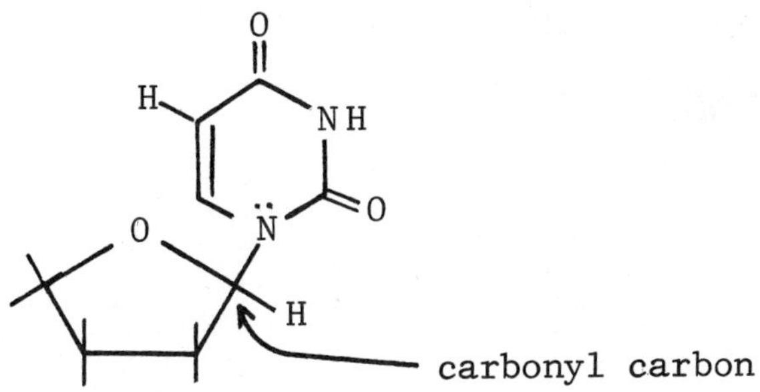

carbonyl carbon

The carbonyl group can be liberated by acid hydrolysis.

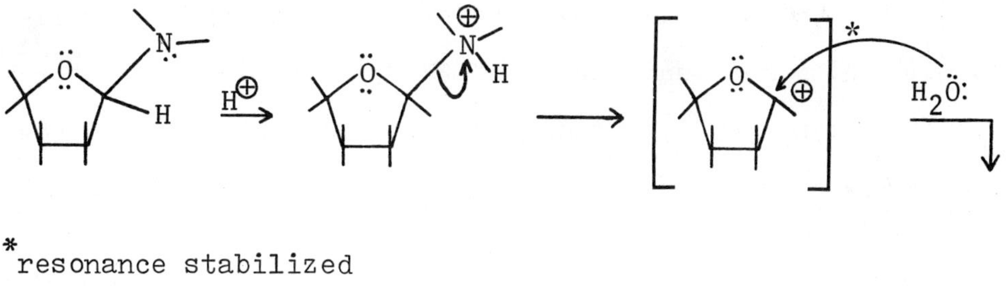

*resonance stabilized

(continued next page)

51. The 94% exists as a cyclic hemiacetal

52.

H_2O + $:\ddot{Br}:^{\ominus}$ + R-C=O(OR')

53. The change is from >C=O (120° bond angles preferred) to >CHOH (109° bond angles preferred). This occurs faster for cyclobutanone which has bond angles of 90° (greater deviation from 120° than either cyclopentanone or cyclohexanone). Also see problem 41.

54. Yes. The first product formed in any reaction is the kinetically controlled one. In an irreversible process the product formed is kinetically controlled (it may also be the thermodynamically controlled product).

55. Species 3 has greater energy and is more reactive because of two negative charges. It has a greater tendency to transfer a hydride in forming $RCOO^{\ominus}$ than does 2 in forming RCOOH.

56. 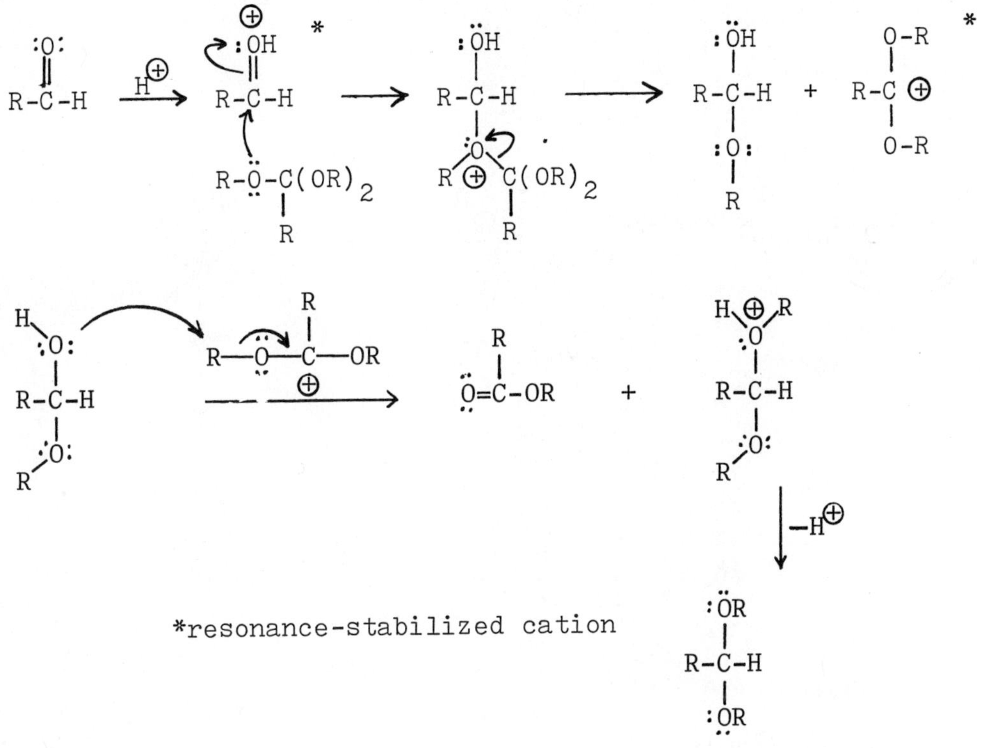 R-C(=O)-H $\xrightarrow{H^{\oplus}}$ R-C(=$\overset{\oplus}{O}$H)-H* + R-O-C(OR)$_2$(R) ⟶ R-C(OH)(H)-$\overset{\oplus}{O}$(R)-C(OR)$_2$(R) ⟶ R-C(OH)(H)-O-R + R-C$^{\oplus}$(O-R)(O-R)*

R-C(OH)(H)(OR) + R-O-C$^{\oplus}$(R)-OR ⟶ O=C(R)-OR + R-C(H)($\overset{\oplus}{O}$HR)(OR) $\xrightarrow{-H^{\oplus}}$ R-C(H)(OR)(OR)

*resonance-stabilized cation

57. 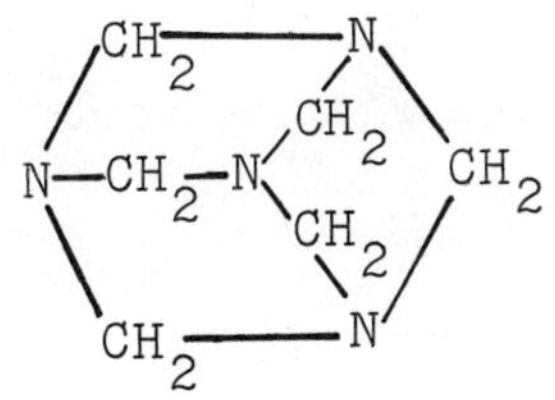 N—CH$_2$—N (hexamethylenetetramine cage structure: four N atoms linked by six CH$_2$ groups) OR (adamantane-like cage with N at four bridgehead positions)

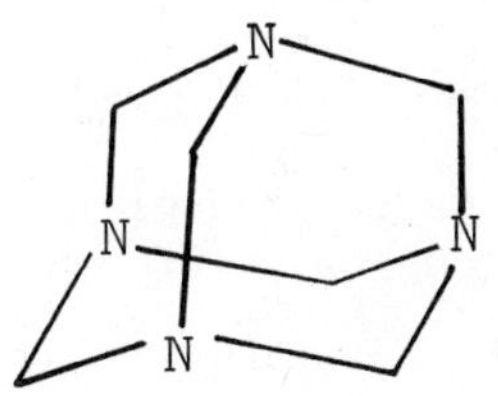

58. $CH_2{=}CH{-}\underset{H}{C}{=}\ddot{O}$ $\xrightarrow{H^{\oplus}}$ $CH_2{=}CH{-}\underset{H}{C}{=}\overset{\oplus}{O}H$ $\xrightarrow{:NH_2{-}NH_2}$

$NH_2{-}\overset{\oplus}{N}H_2{-}CH_2{-}CH{=}\underset{H}{C}{-}\ddot{O}H$ (+ $H^{\oplus}$) $\longrightarrow$ $NH_2{-}\overset{\oplus}{N}H_2{-}CH_2{-}CH_2CH{=}\overset{\oplus}{O}H$

$\rightleftharpoons$ $\ddot{N}H_2NHCH_2CH_2CH{=}\overset{\oplus}{O}H$ $\longrightarrow$ (ring: OH, H, $\overset{\oplus}{N}{-}H$, H, $\ddot{N}$–H) $\rightleftharpoons$ (ring: $\overset{\oplus}{O}H_2$, H, $\ddot{N}$–H, $\ddot{N}$–H; $:\ddot{O}H_2$)

$\downarrow$

(ring: N:, $\ddot{N}$–H)

SUMMARY OF REACTIONS

1. Synthesis of Aldehydes and Ketones

   a. review methods from previous chapters, see Figure 17.1

   b. oxidation of aryl methyl groups (p. 674)

$$C_6H_5CH_3 \xrightarrow[H_2SO_4]{CrO_3/(CH_3CO)_2O} C_6H_5CH(OAc)_2 \xrightarrow{H_3O^{\oplus}} C_6H_5CHO$$

(continued next page)

c. reduction of acid chlorides (p. 675)

$$\text{R-}\overset{\overset{\text{O}}{\|}}{\text{C}}\text{-Cl} \xrightarrow{\text{LiAlH(t-BuO)}_3} \text{R-}\overset{\overset{\text{O}}{\|}}{\text{C}}\text{-H}$$

d. reduction of nitriles (p. 675)

$$\text{R-C}\equiv\text{N} \xrightarrow{\text{LiAlH}_4\text{(1 equiv.)}^*} \xrightarrow{\text{H}_2\text{O}} \text{R-}\overset{\overset{\text{O}}{\|}}{\text{C}}\text{-H}$$

*or $\text{LiAlH(OC}_2\text{H}_5)_3$

e. addition of Grignard reagents to nitriles (p. 678)

$$\text{R-C}\equiv\text{N} \xrightarrow{\text{R'MgX}} \xrightarrow{\text{H}_2\text{O}} \text{R-}\overset{\overset{\text{O}}{\|}}{\text{C}}\text{-R'}$$

f. carbonylation-oxidation of trialkylboranes (p. 680)

$$\text{RCH=CH}_2 \xrightarrow{\text{BH}_3} \xrightarrow[\text{heat}]{\text{CO/H}_2\text{O}} \xrightarrow{\text{H}_2\text{O}_2\text{/OH}^{\ominus}} (\text{RCH}_2\text{CH}_2)_2\text{C=O}$$

g. rearrangement of <u>vic</u>-diols (pinacol rearrangement) (p. 678)

$$-\underset{\text{OH}}{\overset{|}{\underset{|}{\text{C}}}}-\underset{\text{OH}}{\overset{|}{\underset{|}{\text{C}}}}- \xrightarrow{\text{H}^{\oplus}} -\overset{|}{\underset{|}{\text{C}}}-\underset{\text{O}}{\underset{\|}{\text{C}}}-$$

2. Nucleophilic Addition Reactions

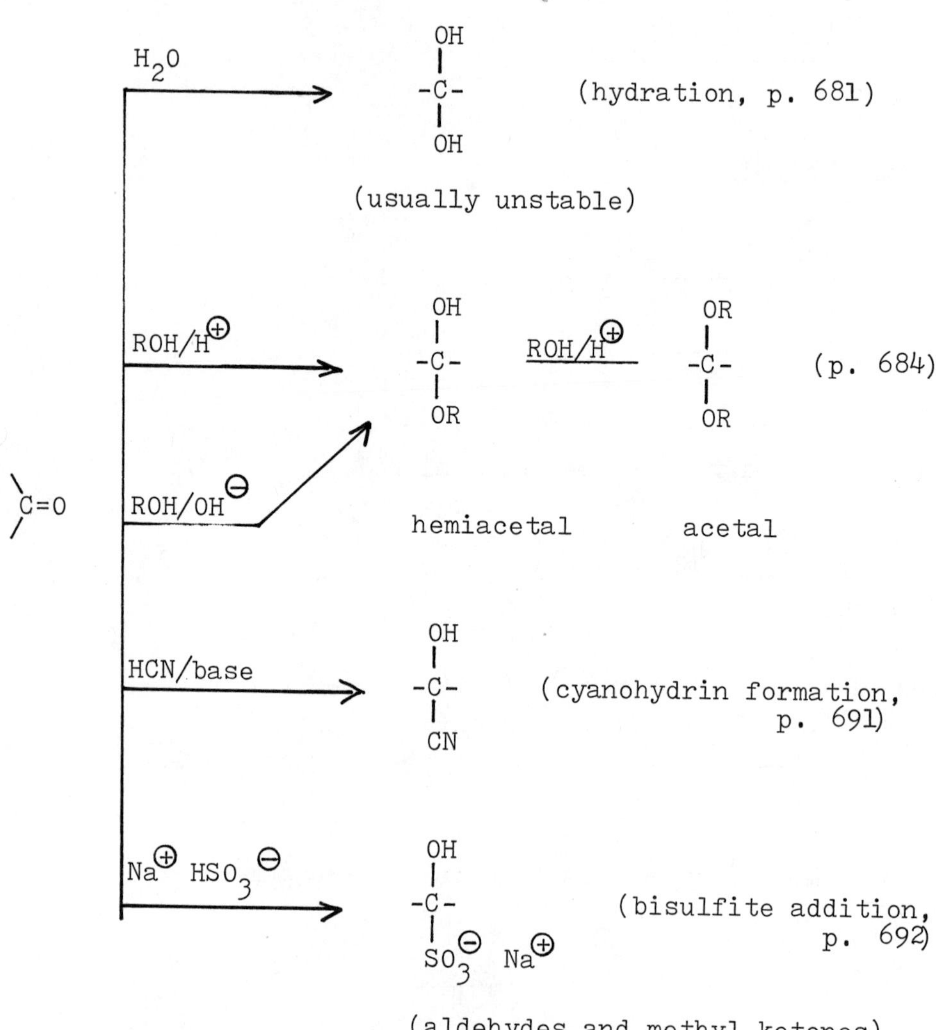

3. Addition-Elimination Reactions (p. 694)

General Reaction:

$$>C{=}O \xrightarrow{NH_2\text{-}G} G\text{-}HN\text{-}\overset{|}{\underset{|}{C}}\text{-}OH \rightleftharpoons\ >C{=}NG + H_2O$$

(continued next page)

$>C=O$ —

- NH_3 → $>C=NH$ (imine)
- NH_2OH (hydroxylamine) → $>C=N-OH$ (an oxime)
- $NH_2NHC_6H_5$ (phenylhydrazine) → $>C=N-NHC_6H_5$ (a phenylhydrazone)
- $NH_2NH\overset{O}{\overset{\|}{C}}NH_2$ (semicarbazide) → $>C=N-NH\overset{O}{\overset{\|}{C}}NH_2$ (a semicarbazone)

4. <u>Reduction of Carbonyl Compounds</u>

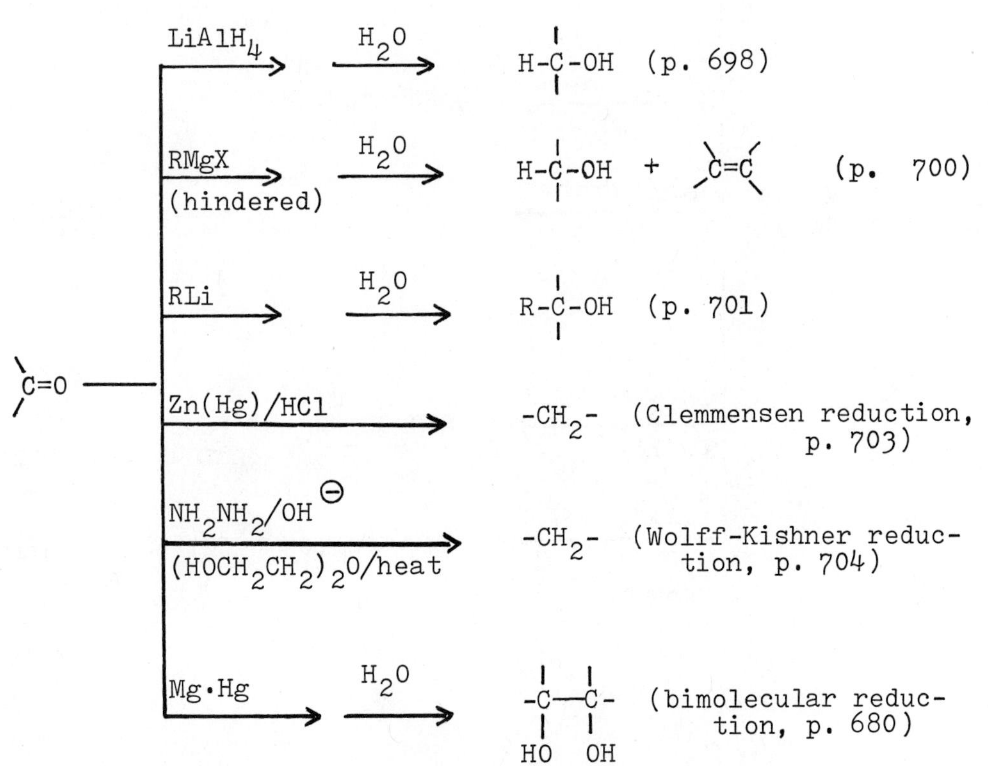

5. <u>Disproportionation</u> (Cannizzaro Reaction)

$$C_6H_5\text{-}\overset{O}{\overset{\|}{C}}\text{-}H \xrightarrow{NaOH/H_2O/heat} C_6H_5CH_2OH + C_6H_5COO^{\ominus}\ Na^{\oplus}$$

(no hydrogens on α carbon) (p. 795)

6. <u>Benzylic Acid Rearrangement</u>

$$C_6H_5\text{-}\overset{O}{\overset{\|}{C}}\text{-}\overset{O}{\overset{\|}{C}}\text{-}C_6H_5 \xrightarrow{NaOH/H_2O} (C_6H_5)_2\overset{OH}{\overset{|}{C}}\text{-}CO_2^{\ominus}\ Na^{\oplus}$$ (p. 706)

7. <u>Conjugate Addition</u> (p. 707)

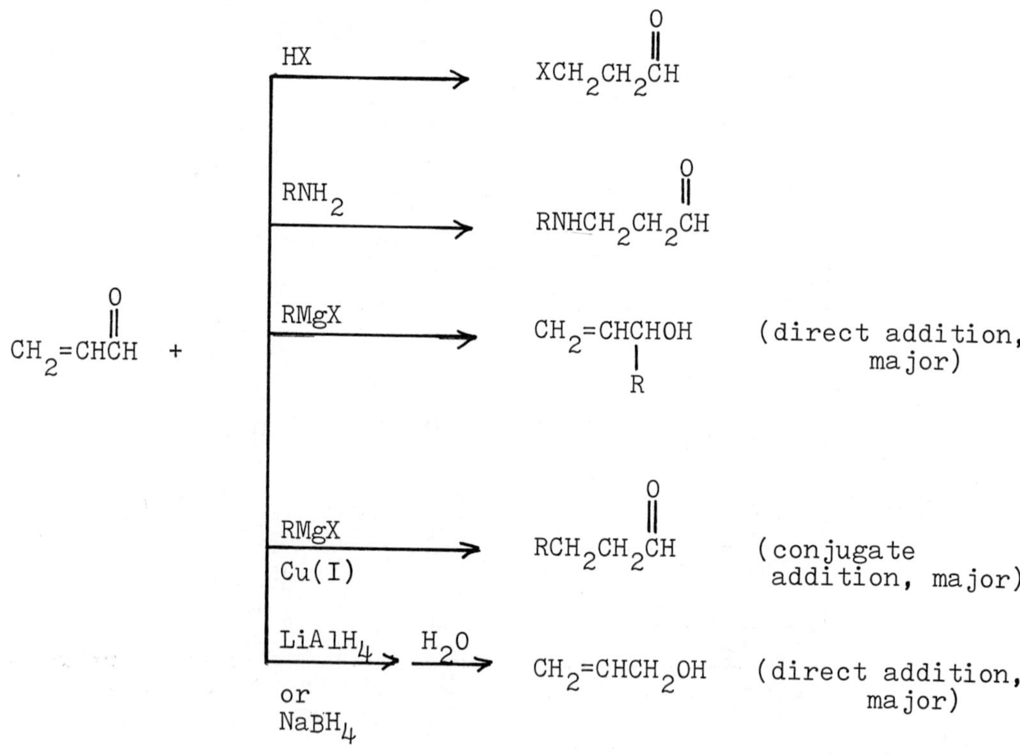

REACTION REVIEW

A. Reactions from Chapter 17

Questions	Answers
1. $RC{\equiv}N \xrightarrow[\text{(1 equiv.)}]{LiAlH_4\,*} \xrightarrow{H_2O}$? *or $LiAlH(OC_2H_5)_3$ Write the structure of the product when $RC{\equiv}N$ is: a) $CH_3-C_6H_4-C{\equiv}N$ (para) b) $CH_3CH_2C{\equiv}N$	1. $R-\overset{O}{\overset{\Vert}{C}}-H$ a) $CH_3-C_6H_4-\overset{O}{\overset{\Vert}{C}}H$ (para) b) $CH_3CH_2\overset{O}{\overset{\Vert}{C}}H$
2. $RC{\equiv}N \xrightarrow{R'MgX} \xrightarrow{H_2O}$? Write the structure of the products in the following: a) $C_6H_5C{\equiv}N \xrightarrow{CH_3MgCl} \xrightarrow{H_2O}$? b) $CH_3CH_2C{\equiv}N \xrightarrow{C_6H_5MgBr} \xrightarrow{H_2O}$?	2. $R-\overset{O}{\overset{\Vert}{C}}-R'$ a) $C_6H_5\overset{O}{\overset{\Vert}{C}}CH_3$ b) $CH_3CH_2\overset{O}{\overset{\Vert}{C}}C_6H_5$
3. $C_6H_5-CH_3 \xrightarrow{CrO_3/(CH_3CO)_2O} \xrightarrow{H_3O^{\oplus}}$?	3. $C_6H_5-\overset{O}{\overset{\Vert}{C}}-H$ (continued next page)

Write the structure of the product when $C_6H_5-CH_3$ is:	
$CH_3-C_6H_4-CH_3$	$HC(=O)-C_6H_4-C(=O)H$
4. $RC(=O)Cl \xrightarrow{LiAlH(t\text{-}BuO)_3}$?	4. $RC(=O)H$
Write the structure of the product when $RC(=O)Cl$ is:	
a) $CH_3CH_2C(=O)\text{-}Cl$	a) $CH_3CH_2C(=O)\text{-}H$
b) $Cl-C_6H_4-C(=O)\text{-}Cl$	b) $Cl-C_6H_4-C(=O)\text{-}H$
5. $RCH{=}CH_2 \xrightarrow{BH_3} \xrightarrow[heat]{CO/H_2O} \xrightarrow[OH^{\ominus}]{H_2O_2}$?	5. $(RCH_2CH_2)_2C{=}O$
Write the structure of the product when $RCH{=}CH_2$ is:	
a) $CH_3CH{=}CH_2$	a) $CH_3CH_2CH_2C(=O)CH_2CH_2CH_3$
	(continued next page)

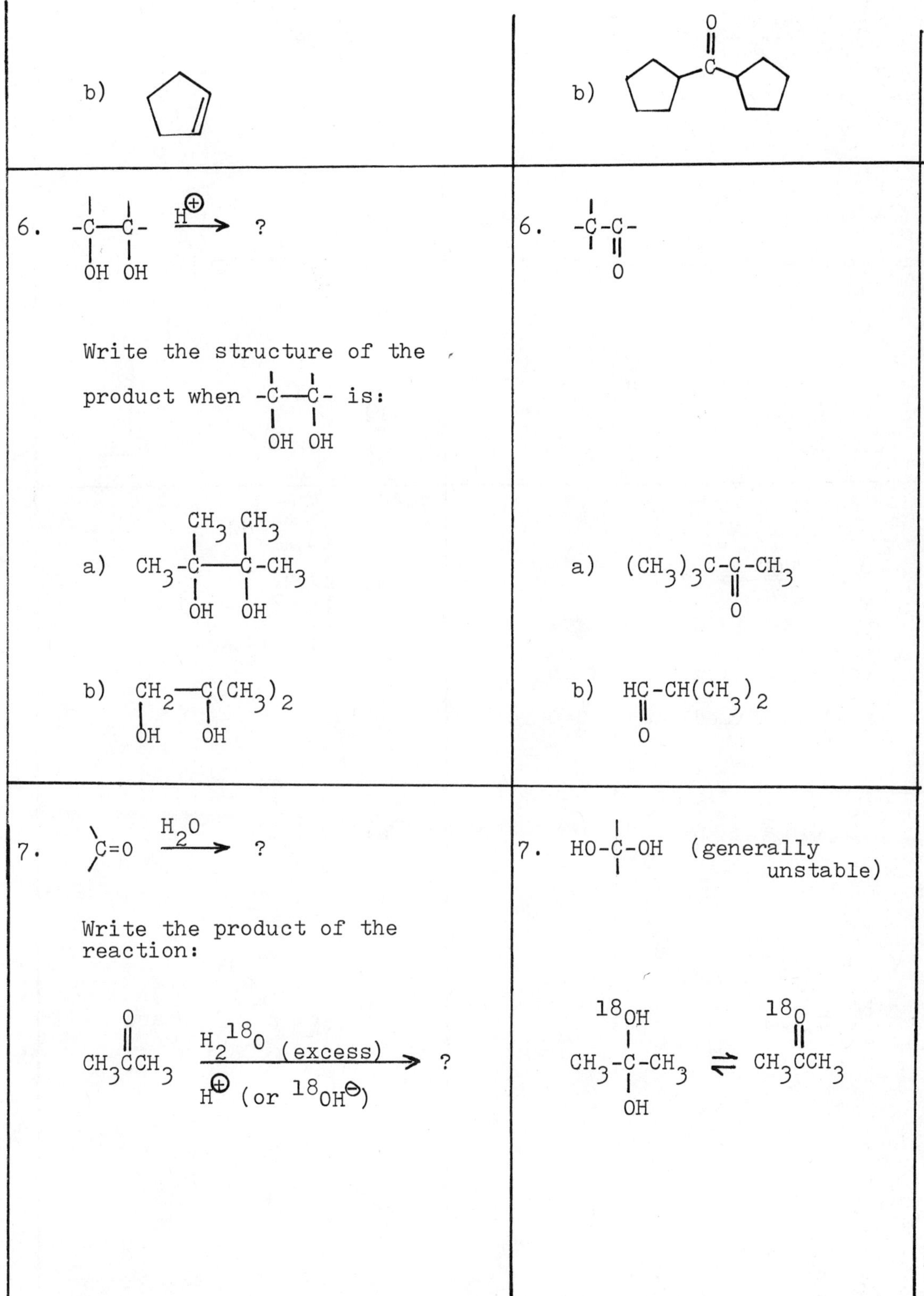

b)
b)
6. $-\overset{|}{\underset{|}{C}}-\overset{|}{\underset{|}{C}}-$ with OH OH, $\xrightarrow{H^{\oplus}}$?
6. $-\overset{|}{\underset{|}{C}}-\underset{\|}{C}-$ O
Write the structure of the product when $-\overset{|}{\underset{|}{C}}-\overset{|}{\underset{|}{C}}-$ (OH OH) is:
a) $CH_3-C(CH_3)(OH)-C(CH_3)(OH)-CH_3$
a) $(CH_3)_3C-C(=O)-CH_3$
b) $CH_2(OH)-C(OH)(CH_3)_2$
b) $HC(=O)-CH(CH_3)_2$
7. $\rangle C=O \xrightarrow{H_2O}$?
7. $HO-\overset{|}{\underset{|}{C}}-OH$ (generally unstable)
Write the product of the reaction:
$CH_3C(=O)CH_3 \xrightarrow[H^{\oplus}\ (\text{or}\ ^{18}OH^{\ominus})]{H_2{}^{18}O\ (\text{excess})}$?
$CH_3-C(^{18}OH)(OH)-CH_3 \rightleftharpoons CH_3C(=^{18}O)CH_3$

8. $\rangle C{=}O \xrightarrow[H^{\oplus}]{2ROH}$?	8. $RO{-}\overset{\vert}{\underset{\vert}{C}}{-}OR$
Write the structure of the acetal formed in the following:	
a) $C_6H_5\overset{O}{\overset{\Vert}{C}}H \xrightarrow[H^{\oplus}]{2CH_3OH}$?	a) $C_6H_5CH(OCH_3)_2$
b) $CH_3CH_2\overset{O}{\overset{\Vert}{C}}CH_3 \xrightarrow[H^{\oplus}]{HOCH_2CH_2OH}$?	b) 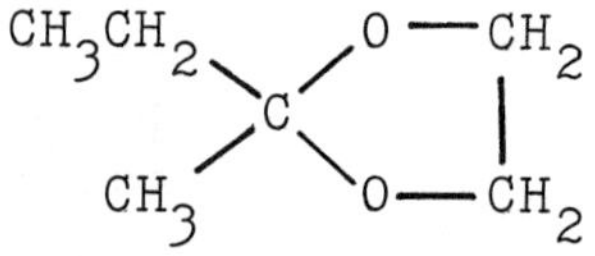 CH_3CH_2 and CH_3 on a C that is bonded to two ring O atoms; O—CH_2—CH_2—O (ring)
9. $\rangle C{=}O \xrightarrow[base]{HCN}$?	9. $N{\equiv}C{-}\overset{\vert}{\underset{\vert}{C}}{-}OH$
Write the structure of the product formed when $\rangle C{=}O$ is:	
a) $C_6H_5\overset{O}{\overset{\Vert}{C}}CH_3$	a) $C_6H_5\overset{OH}{\overset{\vert}{C}}HCN$ (d,l)
b) 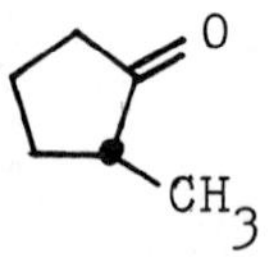 2-methylcyclopentanone (C=O on the ring, CH_3 on the adjacent carbon)	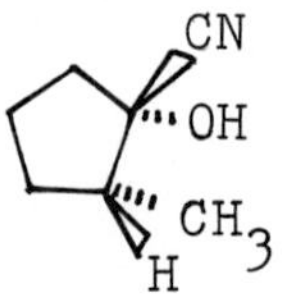 b) 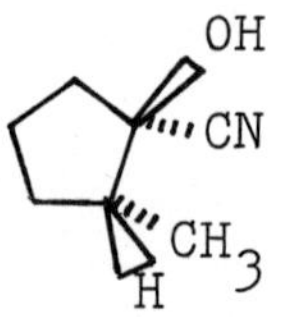 cyclopentane ring bearing OH, CN and, on the adjacent carbon, CH_3, H; and the same with CN and OH interchanged (diastereomers)

10. $>C=O \xrightarrow{Na^{\oplus}\ HSO_3^{\ominus}}$?	10. $Na^{\oplus}\ {}^{\ominus}OSO_2-\overset{\vert}{\underset{\vert}{C}}-OH$ (aldehydes and methyl ketones react)
Write the structure of the product when $>C=O$ is:	
a) $C_6H_5\overset{O}{\overset{\Vert}{C}}H$	a) $C_6H_5\overset{OH}{\overset{\vert}{\underset{\underset{H}{\vert}}{C}}}-SO_3^{\ominus}\ Na^{\oplus}$
b) $CH_3CH_2\overset{O}{\overset{\Vert}{C}}CH_3$	b) $CH_3CH_2\overset{OH}{\overset{\vert}{\underset{\underset{CH_3}{\vert}}{C}}}-SO_3^{\ominus}\ Na^{\oplus}$
11. $>C=O \xrightarrow{NH_2R\ *}$?	11. $RN=C<\ + H_2O$
*derivatives of NH_3	
Write the structure of the products in the following:	
a) $C_6H_5\overset{O}{\overset{\Vert}{C}}H \xrightarrow{NH_2OH}$?	a) $C_6H_5\underset{\underset{H}{\vert}}{C}=NOH$
b) $CH_3\overset{O}{\overset{\Vert}{C}}CH_3 \xrightarrow{NH_2NH_2}$?	b) $(CH_3)_2C=NNH_2$
c) cyclohexanone (cyclohexane ring with =O) $\xrightarrow{C_6H_5NHNH_2}$?	c) cyclohexane ring $=NNHC_6H_5$
	(continued next page)

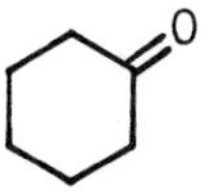

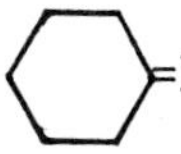

d) $CH_3CH_2\overset{O}{\overset{\|\|}{C}}H \xrightarrow{NH_2\overset{O}{\overset{\|\|}{C}}NHNH_2}$?	d) $CH_3CH_2C(H)=NNH\overset{O}{\overset{\|\|}{C}}NH_2$
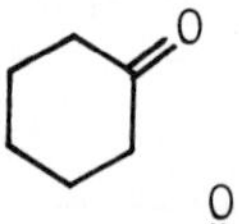 12. $>C=0 \xrightarrow{NH_2NH_2/KOH}$? (or $\xrightarrow{Zn(Hg)/HCl}$, Clemmensen) Write the structure of the product when $>C=0$ is: a) cyclohexanone b) $C_6H_5\overset{O}{\overset{\|\|}{C}}-CH_3$ 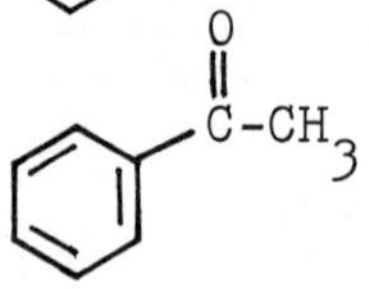	12. $-CH_2-$ Wolff-Kishner reduction a)  cyclohexane b) $C_6H_5CH_2CH_3$ 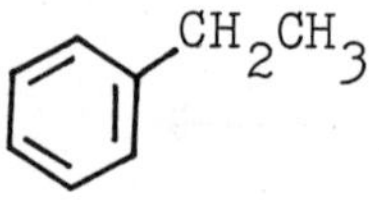
13. $R_2C=0 \xrightarrow{Mg(Hg)} \xrightarrow{H_2O}$? Write the structure of the product when $R_2C=0$ is: a) $CH_3\overset{O}{\overset{\|\|}{C}}CH_3$ b) 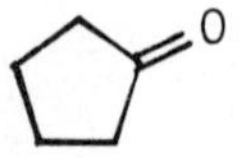 cyclopentanone	13. $R_2C(OH)—C(OH)-R_2$ a) $(CH_3)_2C(OH)—C(OH)(CH_3)_2$ b) 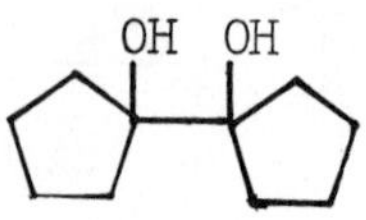 1,1'-bi(cyclopentyl)-1,1'-diol (two cyclopentane rings joined, each bearing OH)

14. $R_3C-\overset{O}{\overset{\Vert}{C}}-H \xrightarrow[H_2O]{NaOH}$? (no hydrogens on α carbon) Write the structure of the product when $R_3C\overset{O}{\overset{\Vert}{C}}H$ is:	14. R_3CCH_2OH and $R_3CCO_2^{\ominus}\ Na^{\oplus}$ (Cannizzaro)
a) $(CH_3)_3C\overset{O}{\overset{\Vert}{C}}H$	a) $(CH_3)_3CCH_2OH$ and $(CH_3)_3CCO_2^{\ominus}\ Na^{\oplus}$
b) $C_6H_5\overset{O}{\overset{\Vert}{C}}H$	b) $C_6H_5CH_2OH$ and $C_6H_5CO_2^{\ominus}\ Na^{\oplus}$
15. see text to review benzilic acid rearrangement	
B. Write the structure for each letter.	
1. $CH_3CH_2CH_2OH \xrightarrow{HBr} A \xrightarrow[ether]{Mg} B \xrightarrow{CH_3CH=O} \xrightarrow{H_2O} C \xrightarrow{CrO_3/H^{\oplus}} D$	1. A: $CH_3CH_2CH_2Br$ B: $CH_3CH_2CH_2MgBr$ C: $CH_3CH_2CH_2\underset{OH}{\underset{\vert}{C}}HCH_3$ D: $CH_3CH_2CH_2\underset{O}{\underset{\Vert}{C}}CH_3$

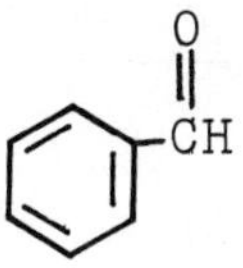

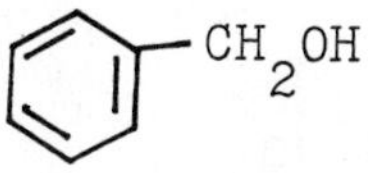

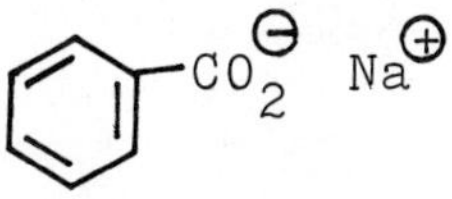

2. Toluene ($C_6H_5CH_3$) $\xrightarrow[\text{heat}]{Br_2}$ A $\xrightarrow[\text{DMF}]{Na^{\oplus}\ CN^{\ominus}}$ B $\xrightarrow{CH_3MgI}$ $\xrightarrow{H_2O}$ C	2. A: $C_6H_5\text{-}CH_2Br$ B: $C_6H_5CH_2C{\equiv}N$ C: $C_6H_5CH_2C(=O)CH_3$
3. $CH_3CH{=}CH_2$ $\xrightarrow{BH_3}$ $\xrightarrow[OH^{\ominus}]{H_2O_2}$ A $\xrightarrow[\text{pyridine}]{CrO_3}$ B $\xrightarrow{C_6H_5NHNH_2}$ C	3. A: $CH_3CH_2CH_2OH$ B: $CH_3CH_2CH(=O)$ C: $CH_3CH_2C(H){=}NNHC_6H_5$
4. Benzene $\xrightarrow[AlCl_3]{CH_3Cl}$ A $\xrightarrow[FeCl_3]{Cl_2}$ B $\xrightarrow[(CH_3CO)_2O]{CrO_3}$ C	4. A: toluene (benzene ring with CH_3) B: p-chlorotoluene (ring with CH_3 and Cl para) C: p-chlorobenzaldehyde (ring with $C(=O)H$ and Cl para)
5. $CH_3C(=O)Cl$ $\xrightarrow{LiAlH(t\text{-}BuO)_3}$ A $\xrightarrow[H^{\oplus}]{2C_2H_5OH}$ B	5. A: $CH_3C(=O)H$ B: $CH_3CH(OC_2H_5)_2$

6. $C_6H_6 \xrightarrow[AlCl_3]{CH_3CH_2\overset{O}{\overset{\Vert}{C}}Cl}$ A $\xrightarrow[OH^{\ominus},\ heat]{NH_2NH_2}$ B $\xrightarrow[h\nu]{Br_2}$ C $\xrightarrow[heat]{KOH/alcohol}$ D	6. A: $C_6H_5\overset{O}{\overset{\Vert}{C}}CH_2CH_3$ B: $C_6H_5CH_2CH_2CH_3$ C: $C_6H_5\underset{Br}{\underset{\vert}{C}}HCH_2CH_3$ D: $C_6H_5CH{=}CHCH_3$ (<u>trans</u> > <u>cis</u>)
7. $(CH_3)_2C{=}C(CH_3)_2$ $\xrightarrow{C_6H_5\overset{O}{\overset{\Vert}{C}}OOH}$ $\xrightarrow{OH^{\ominus}/H_2O}$ A $\xrightarrow{H^{\oplus}}$ B $\xrightarrow{NH_2OH}$ C	7. A: $(CH_3)_2\underset{OH}{\underset{\vert}{C}}-\underset{OH}{\underset{\vert}{C}}(CH_3)_2$ B: $(CH_3)_3C-\underset{O}{\underset{\Vert}{C}}CH_3$ C: $(CH_3)_3C-\underset{NOH}{\underset{\Vert}{C}}CH_3$
8. $CH_2{=}CHCH_3 \xrightarrow{NBS}$ A $\xrightarrow{Li(CH_3)_2Cu}$ B $\xrightarrow{H_3O^{\oplus}}$ C $\xrightarrow[H^{\oplus}]{K_2Cr_2O_7}$ D $\xrightarrow{Mg\cdot Hg}$ E	8. A: $CH_2{=}CHCH_2Br$ B: $CH_2{=}CHCH_2CH_3$ C: $CH_3\underset{OH}{\underset{\vert}{C}}HCH_2CH_3$ D: $CH_3\underset{O}{\underset{\Vert}{C}}CH_2CH_3$

(continued next page)

$$\text{E:}\quad CH_3CH_2\overset{\overset{\Large CH_3}{|}}{\underset{\underset{\Large OH}{|}}{C}}-\overset{\overset{\Large CH_3}{|}}{\underset{\underset{\Large OH}{|}}{C}}-CH_2CH_3$$

CHAPTER 18
ALDEHYDES AND KETONES - THEIR CARBANIONS AND SPECTRA

LEARNING OBJECTIVES

When you have completed this chapter, you should be able to:

1. write equations which illustrate the mechanisms of hydrogen-deuterium exchange in aldehydes and ketones (problems 1, 2, 3, 4, 5, 6, 30);
2. write equations which illustrate the mechanism of halogenation of ketones (problems 7, 8, 16 and related problems 10, 52);
3. draw resonance structures for enolate ions and draw enol forms (problems 9, 11, 12, 13 and related problems 25, 26, 27);
4. identify products from the reactions of methyl ketones and related compounds with halogens in alkaline media (problem 15 and related problem 37);
5. write equations which illustrate the mechanisms of aldol and related condensation reactions (problems 17, 18, 19, 21, 24, 39, 40, 41);
6. predict products formed in aldol condensation reactions (problems 19, 31, 34 and related problem 48);
7. write equations which include an aldol condensation as one step for the synthesis of compounds (problems 23, 45);
8. write equations for the multistep synthesis of compounds using reactions (other than condensations) covered in this and previous chapters (problems 33, 35, 36, 44);
9. account for products formed in reactions which involve enol forms or enolate ions (problems 20, 22, 28, 29, 32, 38, 42, 43, 46, 53);
10. write mechanisms which involve various addition reactions of aldehydes and ketones (problems 49, 50, 51);
11. suggest spectroscopic methods for distinguishing compounds (problems 55, 58, 59, 60);
12. account for the IR and NMR spectra of compounds (problems 56, 57);
13. predict conformations or account for properties of cyclic ketones based on optical rotatory dispersion data (problems 62, 63, 64).

ANSWERS TO QUESTIONS

FORMATION OF ENOLS AND ENOLATE IONS

1. enol formation (acid solution):

$$-\overset{|}{\underset{H}{C}}-\overset{:O:}{\overset{\|}{C}}- \xrightarrow{H^{\oplus}} \left[-\overset{|}{\underset{H}{C}}-\overset{\oplus:O-H}{\overset{\|}{C}}- \leftrightarrow -\overset{|}{\underset{H}{C}}-\overset{:\ddot{O}-H}{\overset{|}{\underset{\oplus}{C}}}- \right]^{*} \rightleftharpoons -\overset{|}{C}=\overset{:\ddot{O}-H}{\overset{|}{C}}-$$

a b enol

*for brevity only resonance contributor "a" is often shown in problems in which this ion is formed, and enol formation is represented as:

$$-\overset{|}{\underset{H}{C}}-\overset{\oplus:O-H}{\overset{\|}{C}}- \;\; (H_2\ddot{O}:) \rightleftharpoons -\overset{|}{C}=\overset{:\ddot{O}-H}{\overset{|}{C}}- + H_3O^{\oplus}$$

2. enolate ion formation (basic solution):

$$-\overset{|}{\underset{H}{C}}-\overset{:O:}{\overset{\|}{C}}- \;\; (B:^{\ominus}) \rightleftharpoons \left[-\overset{|}{\underset{\ddot{\ominus}}{C}}-\overset{\ddot{O}:}{\overset{\|}{C}}- \leftrightarrow -\overset{|}{C}=\overset{:\ddot{O}:^{\ominus}}{\overset{|}{C}}- \right]^{\#} + HB$$

d e

for brevity only resonance contributor "d" or "e" is often used to represent the enolate ion in problems in which this ion is formed.

1. $C_6H_5(CH_3)CD-\overset{\overset{:O:}{\|}}{C}-CD_2H \xrightarrow{D^{\oplus}}$ (protonated intermediate: $\oplus\ddot{O}-D$, $C-CD_2$ with H removed by $D_2\ddot{O}:$) $\longrightarrow$ $:\ddot{O}-D$, $C=CD_2$ attacking $D-\overset{\oplus}{O}D_2$ $\downarrow$ $\oplus \ddot{O}-D$, $C-CD_3$ (D removed by $:OD_2$) $\xrightarrow{-D_3O^{\oplus}}$ $\overset{:O:}{\|}$ $C-CD_3$

2. $CH_3-\overset{\overset{:O:}{\|}}{C}-CH_2D \xrightarrow{H^{\oplus}}$ $\oplus :O-H$, CH_3-C-CH_2 (D removed by $:\ddot{O}H_2$) $\longrightarrow$ $:\ddot{O}-H$, $CH_3-C=CH_2$ attacking $H-\overset{\oplus}{O}H_2$ $\downarrow$ $\oplus :O-H$, CH_3-C-CH_3 (H removed by $:\ddot{O}H_2$) $\longrightarrow$ $CH_3-\overset{\overset{:O:}{\|}}{C}-CH_3$

3. $\overset{\overset{:O:}{\|}}{C}-CH_2D \xrightarrow{OD^{\ominus}}$ HOD + $\left[\overset{\overset{:O:}{\|}}{C}-\ddot{C}HD^{\ominus} \longleftrightarrow \overset{:\ddot{O}:^{\ominus}}{|}C=CHD \right]$ $\xrightarrow{D_2O}$ $\overset{\overset{:O:}{\|}}{C}-CHD_2$ + $OD^{\ominus}$

Note:

Although the anion formed in step 1 is a hybrid of both resonance structures, either can be used as a convenient way of showing "where the electrons go" in step 2.

(continued next page)

$$\sim C(=O)-\overset{\ominus}{C}HD + D_2O \longrightarrow \sim C(=O)-CHD_2 + :\ddot{O}D^{\ominus}$$ OR

$$\sim C(-O^{\ominus})=CHD + D_2O \longrightarrow \sim C(=O)-CHD_2 + :\ddot{O}D^{\ominus}$$

4. $CH_3-C(=O)-CH_2D + OH^{\ominus} \longrightarrow HOD + [CH_3-C(=O)-\overset{\ominus}{C}H_2 \leftrightarrow CH_3-C(-O^{\ominus})=CH_2]$

$\xrightarrow{H_2O}$ $CH_3-C(=O)-CH_3 + OH^{\ominus}$

(see mechanism note in question 3)

5. (a) $CH_3-C(=O)-CH(CH_3)C_6H_5 \xrightarrow{OD^{\ominus}} HOD + [CH_3-C(=O)-\overset{\ominus}{C}(CH_3)-C_6H_5 \leftrightarrow CH_3-C(-O^{\ominus})=C(CH_3)-C_6H_5 \leftrightarrow$ (negative charge delocalized to the ortho, para and other ortho ring positions, $CH_3-C(=O)-C(CH_3)=C_6H_5^{\ominus}$) $]$

(b) $[{}^{\ominus}:CH_2-C(=O)\sim \;\leftrightarrow\; CH_2=C(-O^{\ominus})\sim]$

(c) The anion formed by removal of a proton from C-3. Greater delocalization of charge.

(d) The more stable (lower energy) anion should form more easily. It is formed through a lower energy activated complex.

(continued next page)

(e) The hydrogen at C-3. The hydrogen most easily removed by a base ($OD^{\ominus}$) is the most acidic.

6. This reaction requires formation of the anion:

$$D-\overset{\overset{H_5C_6}{|}}{\underset{\underset{\ominus\,:CH_2}{|}}{C}}-\overset{\overset{O}{\|}}{C}-CD_3$$

Since there is no stabilization for this anion, it does not form and exchange does not occur at this position.

7. (a) $$Br-\overset{\overset{H_5C_6}{|}}{\underset{\underset{CH_3}{|}}{C}}-\overset{\overset{:O:}{\|}}{C}-CH_3 \xrightarrow{OH^{\ominus}} H_2O + \left[\sim\overset{\overset{:O:}{\|}}{C}-\ddot{C}H_2^{\ominus} \longleftrightarrow \sim\overset{\overset{:\ddot{O}:^{\ominus}}{|}}{C}=CH_2 \right]$$

$$\downarrow Br-Br$$

$$\sim\overset{\overset{:O:}{\|}}{C}-CH_2-Br + Br^{\ominus}$$

Last step can be shown as:

$$\sim\overset{\overset{:O:}{\|}}{C}-\ddot{C}H_2^{\ominus} \xrightarrow{Br-Br} \sim\overset{\overset{:O:}{\|}}{C}-CH_2-Br + Br^{\ominus}$$ OR

$$\sim\overset{\overset{:\ddot{O}:^{\ominus}}{|}}{C}=CH_2 \xrightarrow{Br-Br} \sim\overset{\overset{:O:}{\|}}{C}-CH_2-Br + Br^{\ominus}$$

(b) $$Br-\overset{\overset{H_5C_6}{|}}{\underset{\underset{CH_3}{|}}{C}}-\overset{\overset{:O:}{\|}}{C}-CH_3 \xrightarrow{H_3O^{\oplus}} \sim\overset{\overset{\oplus:O-H}{\|}}{C}-\underset{\underset{H\;\leftarrow\;:\ddot{O}H_2}{|}}{C}H_2 \rightleftharpoons \sim\overset{\overset{:\ddot{O}-H}{|}}{C}=CH_2 \rightarrow Br-Br$$

$\downarrow$

(continued next page)

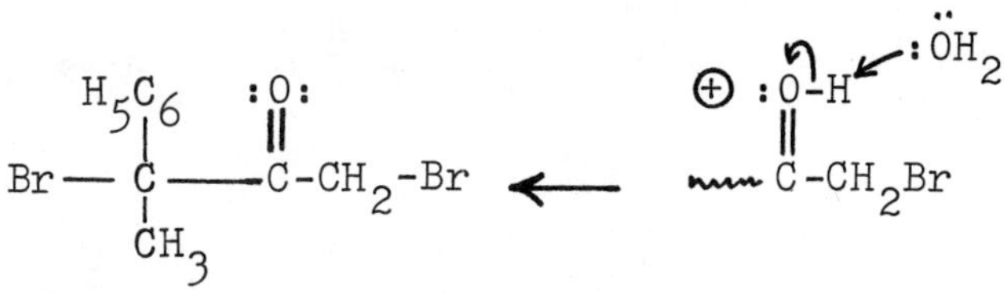

$$Br-\underset{CH_3}{\overset{H_5C_6}{C}}-\overset{:O:}{\overset{\|}{C}}-CH_2-Br \longleftarrow \overset{\oplus\ :O-H \leftarrow :\ddot{O}H_2}{\overset{\|}{C}}-CH_2Br$$

(also see problem 16)

8. The reaction shown requires formation of the anion:

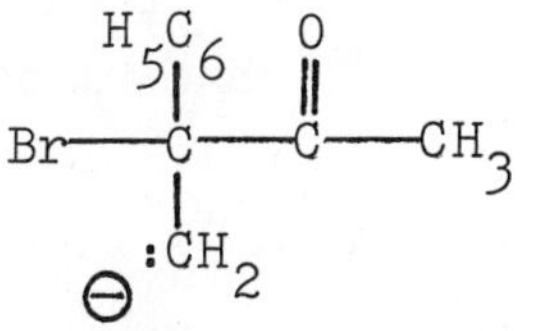

$$Br-\underset{\ominus :CH_2}{\overset{H_5C_6}{C}}-\overset{O}{\overset{\|}{C}}-CH_3$$

Since no stabilization for this anion is possible, it does not form, and bromination does not occur at this position (also see problem 6).

9. This structure shows the delocalization of charge over oxygen and carbon, i.e., the resonance hybrid of:

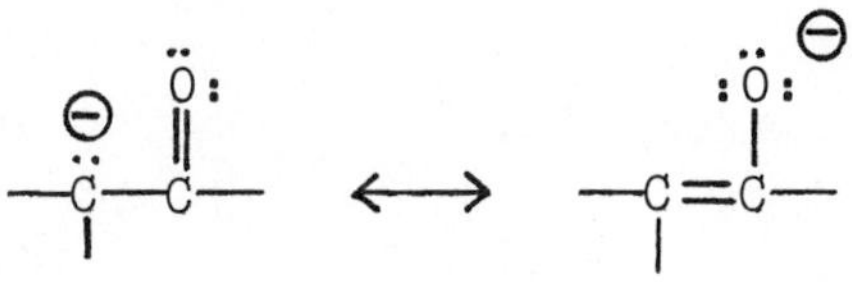

$$-\overset{\ominus}{\ddot{C}}-\overset{\ddot{O}:}{\overset{\|}{C}}- \longleftrightarrow -C=\overset{:\ddot{O}:^{\ominus}}{\overset{|}{C}}-$$

10. The rate-determining step of each is enol formation. Halogenation occurs after the rate determining step.

$$C_6H_5\overset{O}{\overset{\|}{C}}CH_3 \xrightarrow{H^{\oplus}} C_6H_5\overset{\oplus OH}{\overset{\|}{C}}-CH_3 \xrightarrow{slow} C_6H_5\overset{OH}{\overset{|}{C}}=CH_2 \xrightarrow[fast]{X_2} C_6H_5\overset{O}{\overset{\|}{C}}CH_2X$$

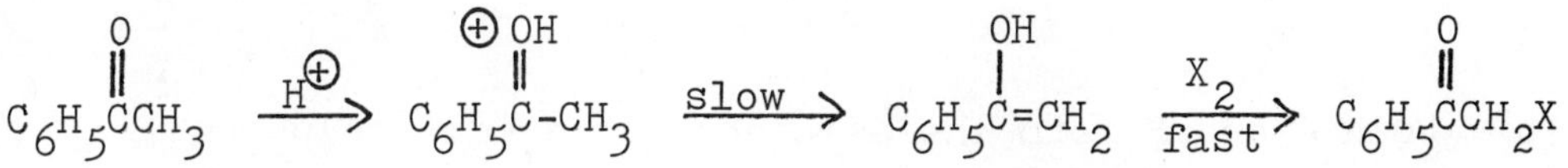

11. The most acidic hydrogen is one of those attached to the carbon between the two carbonyl groups of 2,4-pentanedione. Removal of one of these protons gives an anion which is resonance stabilized by both carbonyl groups.

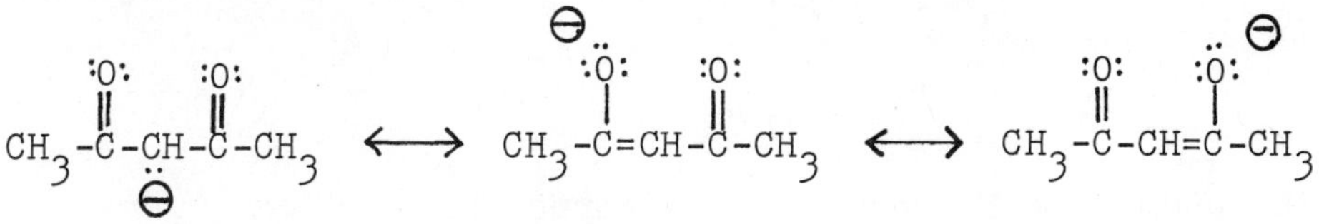

$$CH_3-\overset{:O:}{\overset{\|}{C}}-\underset{\ominus}{\ddot{C}H}-\overset{:O:}{\overset{\|}{C}}-CH_3 \longleftrightarrow CH_3-\overset{\ominus :\ddot{O}:}{\overset{|}{C}}=CH-\overset{:O:}{\overset{\|}{C}}-CH_3 \longleftrightarrow CH_3-\overset{:O:}{\overset{\|}{C}}-CH=\overset{:\ddot{O}:^{\ominus}}{\overset{|}{C}}-CH_3$$

12. 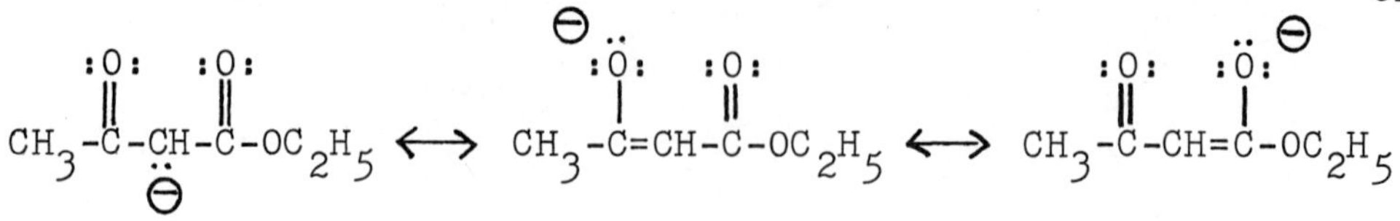 $CH_3-\overset{\overset{:O:}{\|}}{C}-\underset{\ominus}{\ddot{C}}H-\overset{\overset{:O:}{\|}}{C}-OC_2H_5 \longleftrightarrow CH_3-\overset{\overset{\ominus\,:\ddot{O}:}{|}}{C}=CH-\overset{\overset{:O:}{\|}}{C}-OC_2H_5 \longleftrightarrow CH_3-\overset{\overset{:O:}{\|}}{C}-CH=\overset{\overset{:\ddot{O}:\,\ominus}{|}}{C}-OC_2H_5$

13. (a) 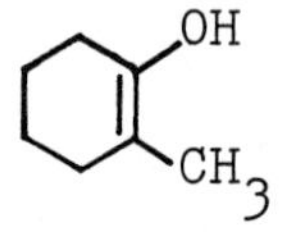 [structure: 2-methylcyclohex-1-en-1-ol, OH and CH_3 on the ring double bond]

more highly substituted double bond

(b) 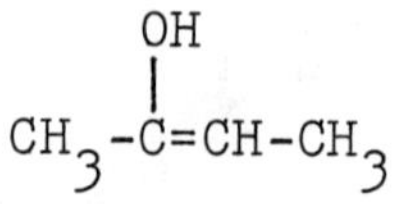 $CH_3-\overset{\overset{OH}{|}}{C}=CH-CH_3$

more highly substituted double bond

(c) 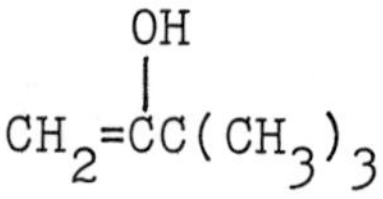 $CH_2=\overset{\overset{OH}{|}}{C}C(CH_3)_3$

only possible product

(d) 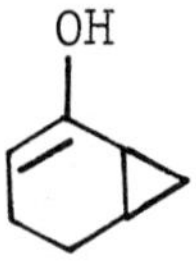 [structure: bicyclic enol with OH]

double bond at other position prohibited (Bredt's rule)

14. (a) The enol (phenol) is aromatic.

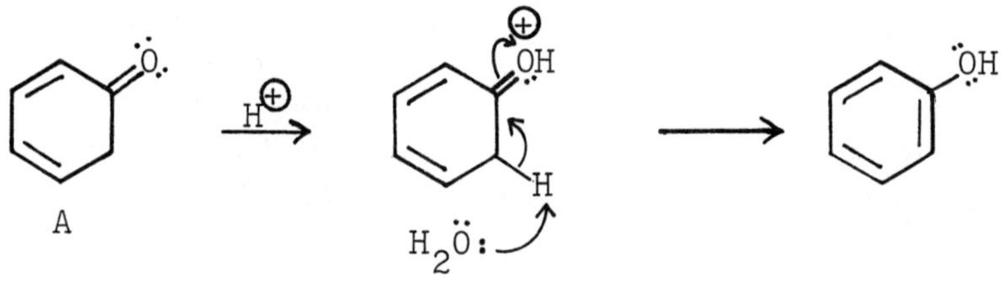

A $\xrightarrow{H^{\oplus}}$ [protonated ketone, $\overset{\oplus}{O}H$; $H_2\ddot{O}:$ removes H] $\longrightarrow$ [phenol, $\ddot{O}H$]

(b) Both A and B form the same enol in acidic solution.

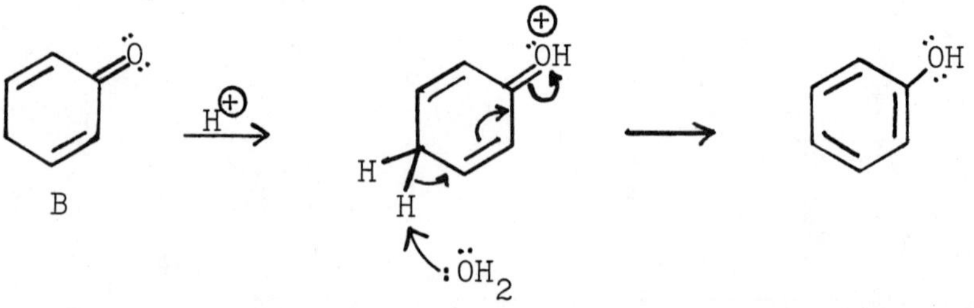

B $\xrightarrow{H^{\oplus}}$ [protonated ketone, $\overset{\oplus}{O}H$; $:\ddot{O}H_2$ removes H] $\longrightarrow$ [phenol, $\ddot{O}H$]

(c) The enol form is phenol. The structure is shown above.

15. (a) $C_6H_5CO_2H$

(b) $HO_2C(CH_2)_5CO_2H$

(c) $(CH_3)_2C{=}CHCO_2H$

(d) m-$CH_3C_6H_4CO_2H$

(e) $CH_3CH_2CO_2H$

(f) no acid formed

16. $Br{-}C_6H_4{-}C(=O){-}CH_3$ $\xrightarrow{CH_3CO_2H}$ $\sim C(=\overset{\oplus}{O}{-}H){-}CH_2{-}H$ (deprotonated by $^{\ominus}:O_2CCH_3$) $\longrightarrow$ $\sim C(\ddot{O}{-}H){=}CH_2$ + $Br{-}Br$ $\longrightarrow$ $\sim C(=\overset{\oplus}{O}{-}H){-}CH_2{-}Br$ + $:\ddot{Br}:^{\ominus}$ $\longrightarrow$ HBr + $Br{-}C_6H_4{-}C(=O){-}CH_2{-}Br$

(also see problem 7)

17. CH_3CH_2CHO $\underset{}{\overset{:\ddot{O}H^{\ominus}}{\rightleftharpoons}}$ H_2O + $CH_3\ddot{C}^{\ominus}HCHO$ * + $CH_3CH_2C(=\ddot{O}:)H$ $\longrightarrow$ $CH_3CH_2CH(:\ddot{O}:^{\ominus})CH(CH_3)CHO$ + $H{-}O{-}H$ $\longrightarrow$ $OH^{\ominus}$ + $CH_3CH_2CH(:\ddot{O}H)CH(CH_3)CHO$

*this resonance form of $CH_3\ddot{C}^{\ominus}HC(=\ddot{O})-H \longleftrightarrow CH_3CH{=}C(-\ddot{O}:^{\ominus})-H$ is used for convenience in showing bond changes.

18. 2,2-Dimethylpropanal, $(CH_3)_3CCHO$, lacks a proton adjacent to the carbonyl group and cannot form the anion necessary for condensation.

19. (a)

$$\underset{\displaystyle}{CH_3CH_2}\overset{OH}{\overset{|}{C}}H\underset{\underset{CH_3}{|}}{C}HCHO \quad \text{and} \quad CH_3\overset{OH}{\overset{|}{C}}H\underset{\underset{CH_3}{|}}{C}HCHO$$

(b)

$$CH_3CH_2CHO \xrightarrow{B:^{\ominus}} HB + \left[CH_3\underset{\ominus}{\ddot{C}}H\text{-}\overset{:O:}{\overset{\|}{C}}H \longleftrightarrow CH_3CH\text{=}\overset{:\ddot{O}:^{\ominus}}{\overset{|}{C}}H \right]$$

$$CH_3CH_2\overset{:O:}{\overset{\|}{C}}H \xrightarrow{CH_3\ddot{C}^{\ominus}HCHO\ *} CH_3CH_2\overset{:\ddot{O}:^{\ominus}}{\overset{|}{C}}H\underset{\underset{CH_3}{|}}{C}HCHO \xrightarrow[(-OH^{\ominus})]{H\text{-}\ddot{O}\text{-}H} CH_3CH_2\overset{:\ddot{O}H}{\overset{|}{C}}H\underset{\underset{CH_3}{|}}{C}HCHO$$

and

$$CH_3\overset{:O:}{\overset{\|}{C}}H \xrightarrow{CH_3\ddot{C}^{\ominus}HCHO\ *} CH_3\overset{:\ddot{O}:^{\ominus}}{\overset{|}{C}}H\underset{\underset{CH_3}{|}}{C}HCHO \xrightarrow[(-OH^{\ominus})]{H\text{-}\ddot{O}\text{-}H} CH_3\overset{:\ddot{O}H}{\overset{|}{C}}H\underset{\underset{CH_3}{|}}{C}HCHO$$

*this resonance form of anion used for convenience

20. This reaction is the reverse of an aldol condensation (retroaldol). The leaving group is a resonance-stabilized enolate ion.

$$CH_3\text{-}\overset{:O:}{\overset{\|}{C}}\text{-}CH_2\text{-}\overset{:\ddot{O}\text{-}H}{\overset{|}{C}}\text{-}CH_3 \ (\text{with } \underset{|}{C}\text{-}CH_3) \xrightarrow{:\ddot{O}H^{\ominus}} CH_3\text{-}\overset{:O:}{\overset{\|}{C}}\text{-}CH_3 + \left[CH_3\text{-}\overset{:O:}{\overset{\|}{C}}\text{-}\ddot{C}H_2^{\ominus} \longleftrightarrow CH_3\text{-}\overset{:\ddot{O}:^{\ominus}}{\overset{|}{C}}\text{=}CH_2 \right]$$

(enolate $\xrightarrow{H_2O}$ $CH_3\text{-}\overset{O}{\overset{\|}{C}}\text{-}CH_3$)

21. The equations below illustrate the general mechanism of the base-catalyzed aldol condensations in this problem.

(1) $-\overset{\overset{\displaystyle :O:}{\|}}{C}-\overset{|}{\underset{\underset{\displaystyle H}{|}}{C}}- \;\xrightarrow{\;H\ddot{O}:^{\ominus}\;}\; \left[-\overset{\overset{\displaystyle :O:}{\|}}{C}-\overset{|}{\underset{\ddot{\ominus}}{C}}- \longleftrightarrow -\overset{\overset{\displaystyle :\ddot{O}:^{\ominus}}{|}}{C}=\overset{|}{C}- \right] + H_2O$

(2) $-\overset{\overset{\displaystyle O:}{\|}}{C}-\overset{|}{\underset{\ddot{\ominus}}{C}}-{}^{*} \;+\; -\overset{\overset{\displaystyle :O:}{\|}}{C}-\overset{|}{C}H- \longrightarrow -\overset{\overset{\displaystyle O}{\|}}{C}-\overset{|}{\underset{|}{C}}-\overset{\overset{\displaystyle :\ddot{O}:^{\ominus}}{|}}{C}-\overset{|}{C}H- \;\xrightarrow{\;H-\ddot{O}-H\;}\; -\overset{\overset{\displaystyle O}{\|}}{C}-\overset{|}{\underset{|}{C}}-\overset{\overset{\displaystyle :\ddot{O}H}{|}}{C}-\overset{|}{C}H- \;+\; :\ddot{O}H^{\ominus}$

*this resonance form will be used for convenience in illustrating bond changes in the condensation step.

(a) $(CH_3)_2\underset{\underset{\displaystyle H}{|}}{C}-\overset{\overset{\displaystyle O}{\|}}{C}-H \xrightarrow{H\ddot{O}:^{\ominus}}$

$(CH_3)_2\underset{\ddot{\ominus}}{C}-\overset{\overset{\displaystyle O}{\|}}{C}-H \;+\; H-\overset{\overset{\displaystyle :O:}{\|}}{C}CH(CH_3)_2 \longrightarrow (CH_3)_2\underset{\underset{\displaystyle CHO}{|}}{C}\overset{\overset{\displaystyle :\ddot{O}:^{\ominus}}{|}}{C}HCH(CH_3)_2$

$(CH_3)_2\underset{\underset{\displaystyle CHO}{|}}{C}\overset{\overset{\displaystyle :\ddot{O}:^{\ominus}}{|}}{C}HCH(CH_3)_2 \xrightarrow{H-\ddot{O}-H} H\ddot{O}:^{\ominus} + (CH_3)_2\underset{\underset{\displaystyle CHO}{|}}{C}\overset{\overset{\displaystyle :\ddot{O}H}{|}}{C}HCH(CH_3)_2$

(continued next page)

(b) $C_6H_5\overset{O}{\overset{\|}{C}}CH_3 \xrightarrow{B:^{\ominus}} C_6H_5\overset{O}{\overset{\|}{C}}\overset{\ominus}{\ddot{C}}H_2 \xrightarrow{C_6H_5\overset{O}{\overset{\|}{C}}CH_3} C_6H_5\overset{O}{\overset{\|}{C}}CH_2\overset{:\ddot{O}:^{\ominus}}{\overset{|}{C}}(CH_3)\text{-}C_6H_5$

$\downarrow H_2O$

$C_6H_5\overset{O}{\overset{\|}{C}}CH{=}\underset{CH_3}{\underset{|}{C}}C_6H_5 \xleftarrow{-H_2O} \left[C_6H_5\overset{O}{\overset{\|}{C}}CH_2\overset{:\ddot{O}H}{\overset{|}{C}}(CH_3)\text{-}C_6H_5 \right]$

(c) cyclopentanone $\xrightarrow{B:^{\ominus}}$ enolate (carbanion α to C=O) $+$ cyclopentanone $\longrightarrow$ 2-(1-oxido-cyclopentyl)cyclopentanone ($:\ddot{O}:^{\ominus}$) $\xrightarrow{H_2O}$ 2-(1-hydroxycyclopentyl)cyclopentanone ($:\ddot{O}H$)

(d) (R)-2-methylbutanal racemizes in the presence of base:

CH_3, H, C_6H_5, CHO on C $\underset{H_2O}{\overset{B:^{\ominus}}{\rightleftharpoons}}$ CH_3, C_2H_5 on $\overset{\delta^-}{C}\text{····}\underset{H}{\underset{|}{C}}\text{····}\overset{\delta^-}{O}$ $\underset{B:^{\ominus}}{\overset{H_2O}{\rightleftharpoons}}$ CH_3, H, OHC, C_2H_5 on C

(R) achiral (S)

(continued next page)

Thus the condensation involves (R)- and (S)-2-methylbutanal.

$$2CH_3CH_2\underset{\underset{CH_3}{|}}{CH}CHO \xrightarrow{B:^{\ominus}} \xrightarrow{H_2O} CH_3CH_2\underset{\underset{CH_3}{|}}{CH}-\overset{\overset{OH}{|}}{CH}-\underset{\underset{CHO}{|}}{C}(CH_3)CH_2CH_3$$

(R and S)

R and S (arrow to the CH bearing CH_3); R and S (arrow to the $C(CH_3)$ bearing CHO); new chiral center, R and S configurations (arrow to the CH bearing OH)

possible stereoisomers are:

R R R S S S	enantiomers	R S R S R S	enantiomers
R R S S S R	enantiomers	S R R R S S	enantiomers

Only (b) does not give a β-hydroxycarbonyl compound because the enolate leading to the dehydration is resonance stabilized.

22. (a,b) Carbanion formation is slow. As soon as the anion is formed, it may add to the carbonyl group of a second acetaldehyde molecule or may abstract a deuteron (giving CH_2DCHO). In concentrated solution where the carbanions and acetaldehyde molecules are close together, addition to the carbonyl group predominates. In dilute solution abstraction of a deuteron is more competitive. Both results are consistent with carbanion formation being the rate-determining step.

(c) The reaction between an anion formed from acetone and another molecule of acetone is slower than the reaction between an anion of acetaldehyde and another acetaldehyde molecule for steric reasons. Thus the anion formed from acetone has a greater tendency to react with D_2O.

23. All condensations are shown in acid solution since dehydration of the aldol product occurs spontaneously (a necessary condition in condensing less reactive ketones).

(a) $C_6H_5CHO + CH_3\overset{O}{\overset{\|}{C}}CH_3 \xrightarrow[-H_2O]{H_3O^{\oplus}} C_6H_5CH{=}CH\overset{O}{\overset{\|}{C}}CH_3 \xrightarrow{H_2/Pt} C_6H_5CH_2CH_2CH(OH)CH_3$

(the intermediate α, β-unsaturated carbonyl compound is not shown in subsequent examples)

(b) $C_6H_5CHO + CH_3\overset{O}{\overset{\|}{C}}H \xrightarrow[-H_2O]{H_3O^{\oplus}} \xrightarrow{H_2/Pt} C_6H_5CH_2CH_2CH_2OH$

(c) $C_6H_5CHO + C_6H_5\overset{O}{\overset{\|}{C}}CH_3 \xrightarrow[-H_2O]{H_3O^{\oplus}} \xrightarrow{H_2/Pt} C_6H_5CH_2CH_2\underset{OH}{\underset{|}{C}}HC_6H_5$

(d) $CH_3CHO \xrightarrow[-H_2O]{H_3O^{\oplus}} \xrightarrow{H_2/Pt} CH_3CH_2CH_2CH_2OH$

(e) $CH_3CH_2CH_2CHO \xrightarrow[-H_2O]{H_3O^{\oplus}} \xrightarrow{H_2/Pt} CH_3CH_2CH_2CH_2\underset{CH_2CH_3}{\underset{|}{C}}HCH_2OH$

(f) $CH_3\overset{O}{\overset{\|}{C}}CH_3 \xrightarrow[-H_2O]{H_3O^{\oplus}} \xrightarrow{H_2/Pt} CH_3\underset{OH}{\underset{|}{C}}HCH_2\underset{CH_3}{\underset{|}{C}}HCH_3$

(g) $C_6H_5CH_2\overset{O}{\overset{\|}{C}}CH_2C_6H_5 + CH_2O \xrightarrow[-H_2O]{H_3O^{\oplus}} \xrightarrow{H_2/Pt} C_6H_5\overset{CH_3}{\overset{|}{C}}H\underset{OH}{\underset{|}{C}}HCH_2C_6H_5$

24. 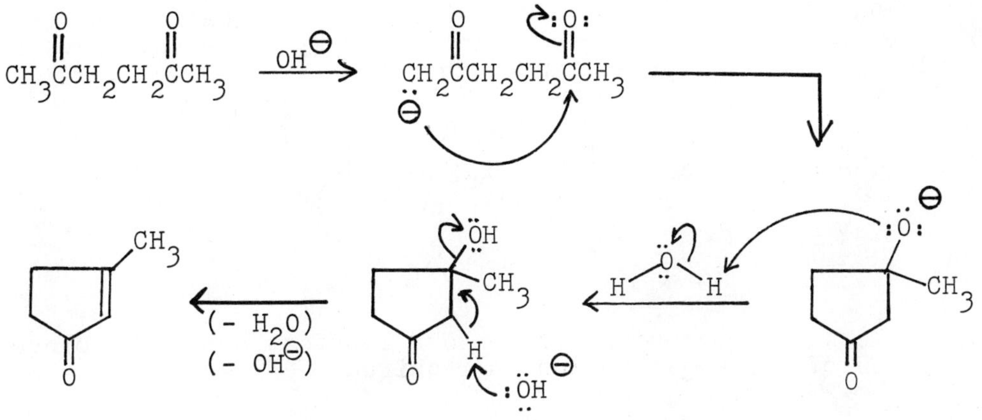 $CH_3\overset{O}{\overset{\|}{C}}CH_2CH_2\overset{O}{\overset{\|}{C}}CH_3 \xrightarrow{OH^{\ominus}} \overset{\ominus}{C}H_2\overset{O}{\overset{\|}{C}}CH_2CH_2\overset{:O:}{\overset{\|}{C}}CH_3 \longrightarrow$

(- H_2O)
(- $OH^{\ominus}$)

25. (a) the enol is not aromatic since crowding within molecule prevents it from being planar (see p. 576).

(b) the enol is antiaromatic and has greater ring strain.

(c) the enol has a double bond at the bridgehead (violates Bredt's rule).

(d) the enol is antiaromatic.

26. (a) 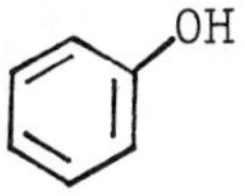 (b) 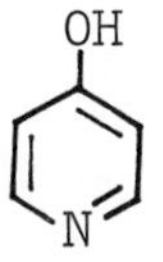 (c) 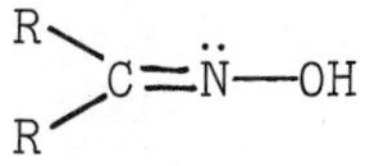 $R_2C{=}\ddot{N}{-}OH$

(d) 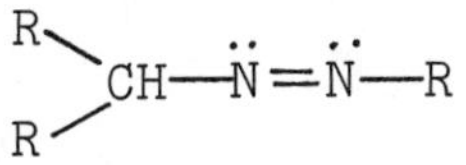 $R_2CH{-}\ddot{N}{=}\ddot{N}{-}R$ (e) 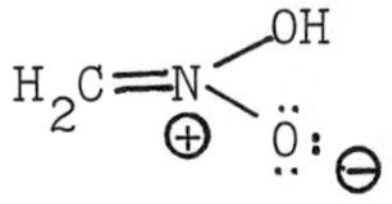 $H_2C{=}\overset{\oplus}{N}(OH){-}\ddot{O}:^{\ominus}$ (f) 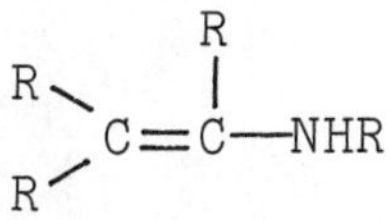 $R_2C{=}C(R){-}NHR$

27. In water solution, A is stabilized by hydrogen bonding with water. In toluene the enol form (B) is stabilized by intramolecular hydrogen bonding.

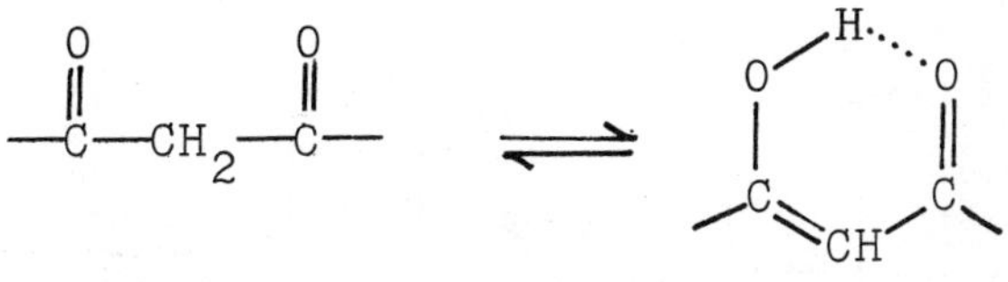

$-\overset{O}{\overset{\|}{C}}-CH_2-\overset{O}{\overset{\|}{C}}- \rightleftharpoons$

A B

28. After the first iodination step, subsequent iodinations occur on the same carbon since anion A (stabilized by the electron-withdrawing iodine) is formed faster than B.

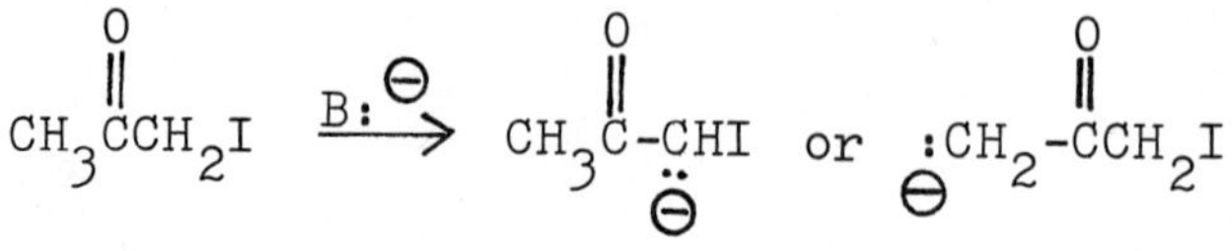

$$CH_3\overset{O}{\overset{\|}{C}}CH_2I \xrightarrow{B:^{\ominus}} CH_3\overset{O}{\overset{\|}{C}}-\underset{\ominus}{\ddot{C}}HI \text{ (A)} \quad \text{or} \quad {}^{\ominus}:CH_2-\overset{O}{\overset{\|}{C}}CH_2I \text{ (B)}$$

After cleavage, iodination of acetate ion does not occur since formation of a dinegative ion is required.

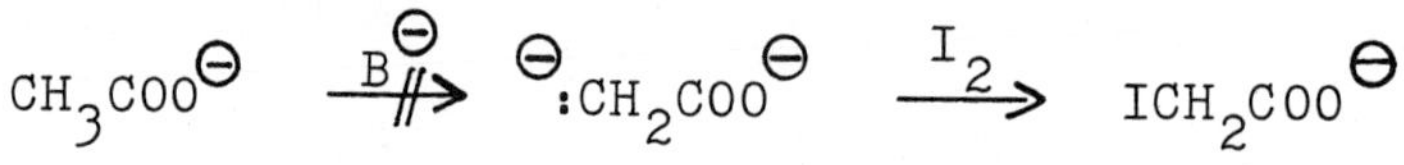

$$CH_3COO^{\ominus} \overset{B^{\ominus}}{\not\longrightarrow} {}^{\ominus}:CH_2COO^{\ominus} \xrightarrow{I_2} ICH_2COO^{\ominus}$$

not formed

29. (a) $CH_2OH-C{=}O$ $\xrightarrow[H_2O]{OH^{\ominus}}$ $\left[{}^{\ominus}:CHOH-C{=}\ddot{O}: \longleftrightarrow CHOH{=}C-\ddot{O}:^{\ominus} \right]$ $\xrightarrow{H_2O}$ $HC(-\ddot{O}-H){=}C-\ddot{O}H$ (H–O–H) $\longrightarrow$ $HC{=}\ddot{O}$ / $CHOH$

(b)

CHO
HO-C-H
HO-C-H
H-C-OH
H-C-OH
CH_2OH

(c) glucose $\underset{H_2O}{\overset{OH^{\ominus}/H_2O}{\rightleftharpoons}}$ $\left[HC{=}\ddot{O} / {}^{\ominus}:COH \longleftrightarrow HC-\ddot{O}:^{\ominus} / {=}COH \right]$ $\underset{OH^{\ominus}/H_2O}{\overset{H_2O}{\rightleftharpoons}}$ mannose

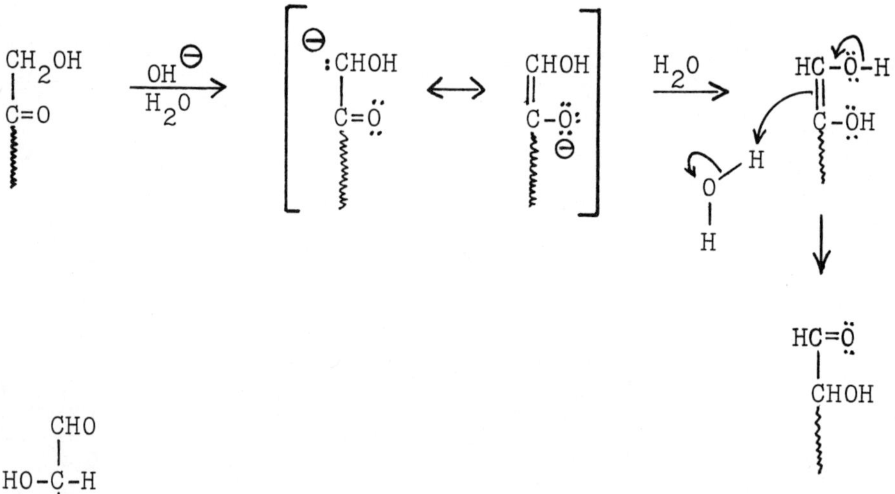

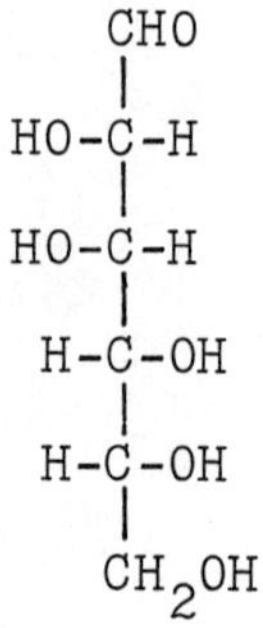

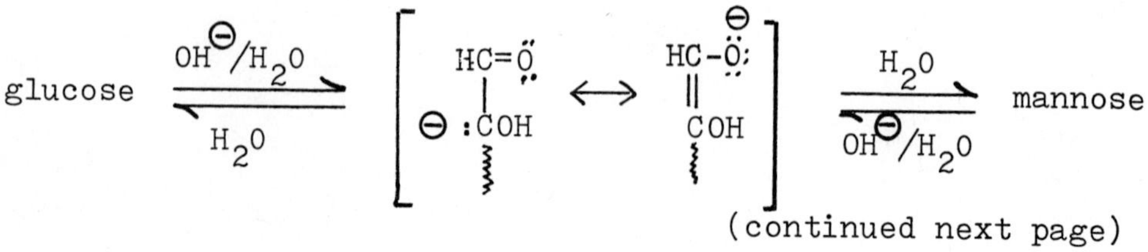

(continued next page)

(d) glucose and mannose differ only in the configuration at the chiral center adjacent to the -CHO group. They are diastereomers.

30. $CH_3COCH_3 \xrightarrow{D-Cl} CH_2(H)-C(=\overset{\oplus}{O}D)CH_3$ (H removed by $D_2\ddot{O}$:) ⇌ $CH_2=C(-\ddot{O}-D)-CH_3$ → attacks D–O(D)–D → (-OD⊖) → $CH_2D-C(=\overset{\oplus}{O}D)CH_3$

repeat → $CD_3C(=\overset{\oplus}{O}D)CD_3$ + $D_2\ddot{O}$:* → $CD_3-C(\ddot{O}D)(\overset{\oplus}{O}^*D_2)-CD_3$

⇅

$CD_3-C(\overset{\oplus}{O}D_2)(\ddot{O}^*-D)-CD_3 \xrightarrow{(-D_2O)} CD_3C(=\overset{\oplus}{O}^*-D)CH_3$ (D removed by :$\ddot{O}D_2$) $\xrightarrow{(-D_3O^{\oplus})} CD_3-C(=O)-CD_3$

O^* is ^{18}O

There is no required order of exchange processes. In some molecules, ^{18}O exchange may precede D incorporation.

31. (a) $CH_3CH_2CH(OH)CH(CH_3)CHO$

(b) $CH_3CH_2CH_2CH(OH)CH(CH_2CH_3)CHO$

(c) $C_6H_5CH_2CH_2CH(OH)CH(CH_2C_6H_5)CHO$

(d) $CD_3CH_2CH(OH)CH(CD_3)CHO$

(continued next page)

(e) $(CH_3)_2CHCH_2CH(OH)CH(CH(CH_3)_2)CHO$

(i) $C_6H_5CH{=}C(CH_3)CHO$

(f) (±) $CH_3CH_2CH(CH_3)CH(OH)-C(CH_3)(CH_2CH_3)-CHO$

(j) $C_6H_5CH{=}CHC(=O)CH_3$

(g) cyclohexyl–CH(OH)–C(CHO) (cyclohexane ring)

(k) 2-naphthyl–$CH{=}C(CH_3)CHO$

(h) $C_6H_5CH{=}CHCHO$

(spontaneous dehydration of $C_6H_5CH(OH)CH_2CHO$)

32. (a) (R)-2-methylbutanal racemizes in the presence of base. See problem 21d for detail.

(b) (i) $CH_3CH_2CH(CH_3)CH(OH)-C(CH_3)(CH_2CH_3)-CHO$

Each of the three stereocenters (marked by arrows): R and S

Isomers:

R R R
S S S

S R R
R S S

S S R
R R S

R S R
S R S

(ii) same as (i) since racemization of (R)-2-methylbutanal produces (R,S)-2-methylbutanal.

(continued next page)

(iii) (R)-3-methylpentanal retains its configuration in base since the chiral center is not involved in forming the π system of the enolate ion.

$CH_3CH_2CHCH_2CHO$ (CH_3 on C-3; chiral center (R)) $\xrightarrow{B:^{\ominus}}$ $CH_3CH_2CHC\cdots C\cdots O$ (H on each enolate carbon; δ^- on carbon and δ^- on O; CH_3 on chiral center (R))

condensation gives: $CH_3CH_2CH(CH_3)CH_2CH(OH)C(CHO)(H)CH(CH_3)CH_2CH_3$

chiral centers: (R), (R and S), (R and S), (R)

Stereoisomers are: RRRR, RSRR, RRSR and RSSR

(iv) $CH_3CH_2CH(CH_3)CH_2CH(OH)C(CHO)(H)CH(CH_3)CH_2CH_3$

chiral centers: (R and S), (R and S), (R and S), (R and S)

Isomers: RRRR, SSSS, SRRR, RSSS, SSRR, RRSS, etc.
(16 isomers possible)

33. (a) cyclohexanone $\xrightarrow{D_2O/Na^{\oplus}\ OD^{\ominus}}$ 2,2,6,6-tetradeuteriocyclohexanone (D, D on each α carbon)

(continued next page)

(b) cyclohexanone $\xrightarrow{CH_3MgI}$ $\xrightarrow{H_2O}$ 1-methylcyclohexanol (OH, CH_3)

(c) cyclohexanone $\xrightarrow{LiAlD_4}$ $\xrightarrow{H_2O}$ cyclohexanol-1-d (OH, D)

(d) cyclohexanone $\xrightarrow{Br_2/CH_3COOH}$ 2-bromocyclohexanone (Br) $\xrightarrow[\text{heat}]{\text{KOH, alcohol}}$ 2-cyclohexenone

(e) cyclohexanone $\xrightarrow{HOCH_2CH_2OH/H^{\oplus}}$ ethylene ketal (O–CH₂CH₂–O)

(f) 2-cyclohexenone (from d) $\xrightarrow{CH_3MgI}$ $\xrightarrow{H_3O^{\oplus}}$ 1-methyl-2-cyclohexenol (OH, CH_3)

(g) 2-cyclohexenone (from d) $\xrightarrow[Cu(I)]{CH_3MgI}$ $\xrightarrow{H_2O}$ 3-methylcyclohexanone (CH_3) (see conjugate addition, p. 707)

(h) cyclohexanone $\xrightarrow[\text{ether}]{Na^{\oplus}\ NH_2^{\ominus}}$ $\xrightarrow{CH_3I}$ 2-methylcyclohexanone (CH_3)

(i) 2-methylcyclohexanone (CH_3) $\xrightarrow{Br_2/H_3O^{\oplus}}$ 2-bromo-2-methylcyclohexanone (Br, CH_3) $\xrightarrow[NH_3\ (l)]{Na^{\oplus}NH_2^{\ominus}}$ 2-methyl-2-cyclohexenone (CH_3)

(continued next page)

(j) 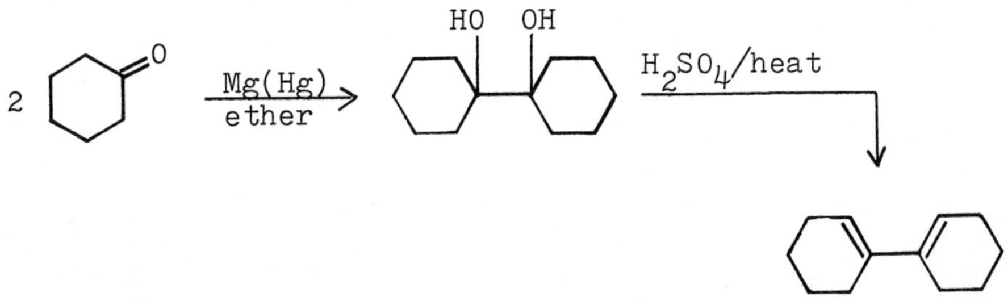 2 cyclohexanone $\xrightarrow[\text{ether}]{Mg(Hg)}$ HO OH (pinacol) $\xrightarrow{H_2SO_4/\text{heat}}$ bi(cyclohexenyl)

(k) 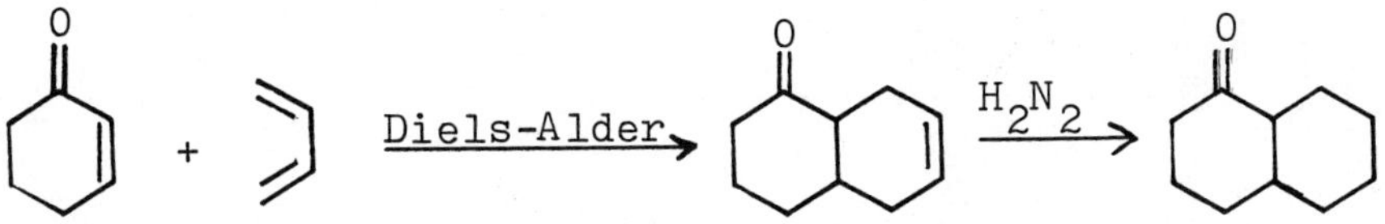 cyclohexenone + butadiene $\xrightarrow{\text{Diels-Alder}}$ $\xrightarrow{H_2N_2}$

34. (a) NR

(b) 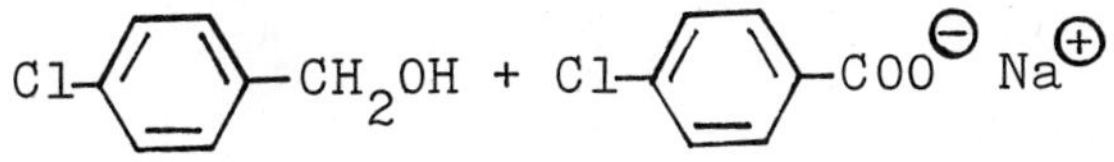 Cl–C₆H₄–CH_2OH + Cl–C₆H₄–$COO^{\ominus}$ $Na^{\oplus}$ (Cannizzaro)

(c) 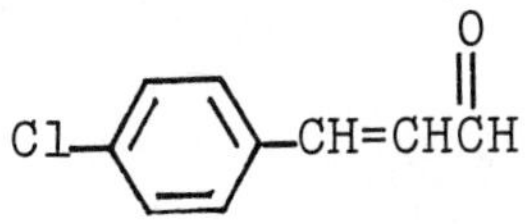 Cl–C₆H₄–CH=CHCH (C=O)

(d) 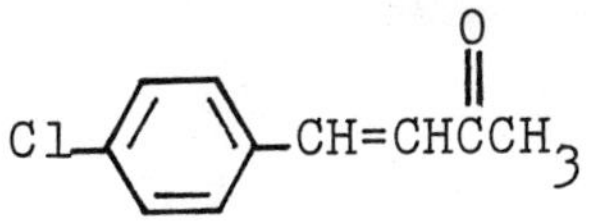 Cl–C₆H₄–CH=CHCCH_3 (C=O)

(e) 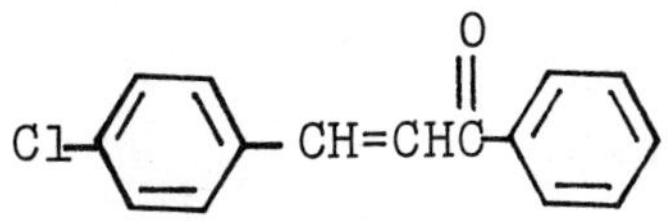 Cl–C₆H₄–CH=CHC(=O)–C₆H₅

(f) 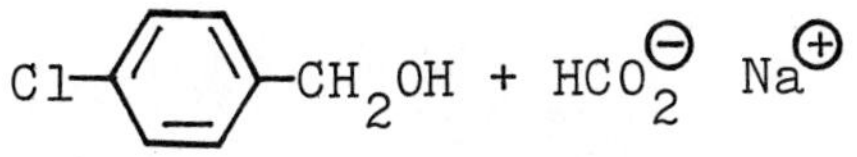 Cl–C₆H₄–CH_2OH + $HCO_2^{\ominus}$ $Na^{\oplus}$

(g) 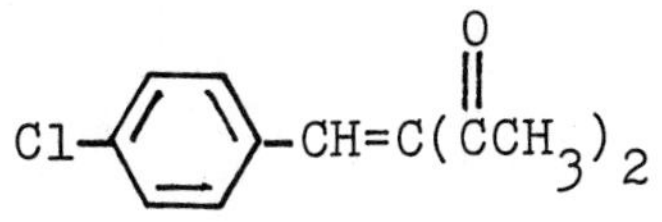 Cl–C₆H₄–CH=C(CCH_3)$_2$ (C=O)

(h) NR

35. (a) $C_6H_5\overset{O}{\overset{\|}{C}}CH_3 \xrightarrow{LiAlH_4} \xrightarrow{H_3O^{\oplus}} C_6H_5\overset{OH}{\overset{|}{C}}HCH_3$

(b) same as (a)

(c) $C_6H_5\overset{O}{\overset{\|}{C}}CH_3 \xrightarrow[H_2O]{NH_2NH_2/OH^{\ominus}}$ (or $\xrightarrow{Zn(Hg)/HCl}$)

(d) $C_6H_5\overset{O}{\overset{\|}{C}}CH_3 \xrightarrow{CH_3MgI} \xrightarrow{H_3O^{\oplus}} C_6H_5\underset{OH}{\underset{|}{C}}(CH_3)_2$

(e) $C_6H_5\overset{O}{\overset{\|}{C}}CH_3 \xrightarrow[H_2O]{Br_2/OH^{\ominus}} C_6H_5\overset{O}{\overset{\|}{C}}\text{-}O^{\ominus} + CHBr_3 \xrightarrow{H_3O^{\oplus}} C_6H_5CO_2H$

(f) $C_6H_5\overset{O}{\overset{\|}{C}}CH_3 \xrightarrow{Na^{\oplus}\ NH_2^{\ominus}} \xrightarrow{CH_3I} C_6H_5\overset{O}{\overset{\|}{C}}CH_2CH_3$

(g) $C_6H_5\overset{O}{\overset{\|}{C}}CH_3 \xrightarrow{Mg\cdot Hg} \xrightarrow{H_2O} C_6H_5\underset{OH}{\underset{|}{\overset{CH_3}{\overset{|}{C}}}}\text{—}\underset{OH}{\underset{|}{\overset{CH_3}{\overset{|}{C}}}}\text{-}C_6H_5$

(h) product of (g) $\xrightarrow{H^{\oplus}} (C_6H_5)_2\overset{CH_3}{\overset{|}{C}}\text{—}\overset{O}{\overset{\|}{C}}\text{-}CH_3$ (pinacol rearrangement)

(i) product of (h) $\xrightarrow[OH^{\ominus}]{Br_2} \xrightarrow{H_3O^{\oplus}} (C_6H_5)_2\overset{CH_3}{\overset{|}{C}}\text{-}COOH$

36. (a) $CH_3CH_2\overset{O}{\overset{\|}{C}}H \xrightarrow{CH_3CH_2MgCl} \xrightarrow{H_2O} CH_3CH_2\overset{OH}{\overset{|}{C}}HCH_2CH_3 \xrightarrow{CrO_3/H^{\oplus}}$

$CH_3CH_3\overset{O}{\overset{\|}{C}}CH_2CH_3 \xrightarrow{CH_3CH_2MgCl} \xrightarrow{H_2O} (CH_3CH_2)_3COH$

(or $CH_2{=}CH_2 \xrightarrow{BH_3} \xrightarrow[heat]{CO} \xrightarrow[OH^{\ominus}]{H_2O_2}$)

(b) $(CH_3)_3CCl \xrightarrow{Mg/ether} (CH_3)_3CMgCl \xrightarrow{H_2C=O} \xrightarrow{H_3O^{\oplus}}$

$(CH_3)_3CCH_2OH \xrightarrow{K_2Cr_2O_7/H^{\oplus}} (CH_3)_3CCO_2H$

(c) $(CH_3)_3CCH_2OH$ (from b) $\xrightarrow{CrO_3 \cdot pyridine} (CH_3)_3CCHO$

(d) $C_6H_5\text{-}\underset{H}{\underset{|}{C}}{=}O \xrightarrow{CH_3\overset{O}{\overset{\|}{C}}CH_3/OH^{\ominus}} C_6H_5\text{-}CH{=}CH\overset{O}{\overset{\|}{C}}CH_3$

(e) $CH_3\overset{O}{\overset{\|}{C}}CH_3 \xrightarrow{H^{\oplus}/H_2O} CH_3\underset{CH_3}{\underset{|}{C}}{=}CH\overset{O}{\overset{\|}{C}}CH_3 \xrightarrow[1\ atm]{H_2/Pt} CH_3\underset{CH_3}{\underset{|}{C}}HCH_2\overset{O}{\overset{\|}{C}}CH_3$

(f) benzene $\xrightarrow{CH_3Cl/AlCl_3}$ $C_6H_5CH_3$ $\xrightarrow[H^{\oplus}]{CrO_3/(CH_3CO)_2O}$ C_6H_5CHO

$\xrightarrow[H^{\oplus}]{HOCH_2CH_2OH}$ $C_6H_5\text{-}CH(OCH_2CH_2O)$ (cyclic acetal)

(continued next page)

(g) benzene $\xrightarrow{H_2/Pt}$ cyclohexane $\xrightarrow[\text{light}]{Br_2}$ bromocyclohexane $\xrightarrow[\text{heat}]{\text{KOH/alcohol}}$ cyclohexene

$\xrightarrow{O_3}$ $\xrightarrow{H_2/Pt}$ $OHCCH_2CH_2CH_2CH_2CHO$

(h) toluene (from f) $\xrightarrow{Cl_2/Fe}$ $Cl-C_6H_4-CH_3$ $\xrightarrow[H^{\oplus}]{CrO_3/(CH_3CO)_2O}$

$Cl-C_6H_4-CHO$

(i) $Cl-C_6H_4-CHO$ (from f) $\xrightarrow[OH^{\ominus}]{CH_3\overset{O}{\overset{\|}{C}}CH_3}$ $Cl-C_6H_4-CH{=}CH\overset{O}{\overset{\|}{C}}CH_3$

(j) $CH_2{=}CHCl$ $\xrightarrow{Mg/THF}$ $CH_2{=}CHMgCl$ $\xrightarrow{CH_3\overset{O}{\overset{\|}{C}}CH_3}$ $\xrightarrow{H_3O^{\oplus}}$ $CH_2{=}CH\underset{OH}{\underset{|}{C}}(CH_3)_2$

(k) chlorocyclobutane $\xrightarrow{Mg/ether}$ cyclobutyl-MgCl $\xrightarrow{H_2C=O}$ $\xrightarrow{H_2O}$ cyclobutyl-CH_2OH

$\xrightarrow{CrO_3\cdot\text{pyridine}}$ cyclobutyl-CHO $\xrightarrow{\text{cyclobutyl-MgCl}}$ $\xrightarrow{H_2O}$ (cyclobutyl)$_2$CH-OH

$\xrightarrow{CrO_3/\text{pyridine}}$ (cyclobutyl)$_2$C=O (OR $H\overset{O}{\overset{\|}{C}}OEt$ $\xrightarrow{2\ \text{cyclobutyl-MgCl}}$

$\xrightarrow{H_2O}$ (cyclobutyl)$_2$CH-OH $\xrightarrow[H^{\oplus}]{CrO_3}$)

37. Ethanol (the solvent for each) is oxidized by $I_2/OH^\ominus$ to acetaldehyde which reacts with additional $I_2/OH^\ominus$ to give iodoform.

38. Cyclooctanone $\xrightarrow{Na^\oplus\ :H^\ominus}$ enolate (carbanion α to C=O) $\xrightarrow{C_2H_5OCOC_2H_5\ (C=O)}$ tetrahedral intermediate: ring-CH–C($:\ddot{O}:^\ominus$)(OC_2H_5)(OC_2H_5)

$\downarrow$ $(-\ C_2H_5O^-)$

2-carbethoxycyclooctanone: ring(C=O)–CH–C(=O)–OC_2H_5

39. $CH_3C(CH_3)=CHCH(=O)$ $\xrightarrow{OH^\ominus}$ $[\ ^\ominus:CH_2–C(CH_3)=CH–CH=:O:\ \longleftrightarrow\ CH_2=C(CH_3)–CH=CH–:\ddot{O}:^\ominus\ ]$

$\downarrow$ $(CH_3)_2C=CHCH(=O)$

$(CH_3)_2C=CHCH=CHC(CH_3)=CHCH(=O)$ $\xleftarrow[(-\ OH^\ominus)]{(-H_2O)}$ $(CH_3)_2C=CH–CH(OH)–CH(H)–C(CH_3)=CHCH(=O)$ ← $:\ddot{O}H^\ominus$

40. $N\equiv CCH_2C\equiv N$ $\xrightarrow{base}$ $[\ :N\equiv C–\overset{\ominus}{\ddot{C}}H–C\equiv N:\ \longleftrightarrow\ :N\equiv C–CH=C=\ddot{N}:^\ominus\ \updownarrow\ ^\ominus:\ddot{N}=C=CH–C\equiv N\ ]$

(continued next page)

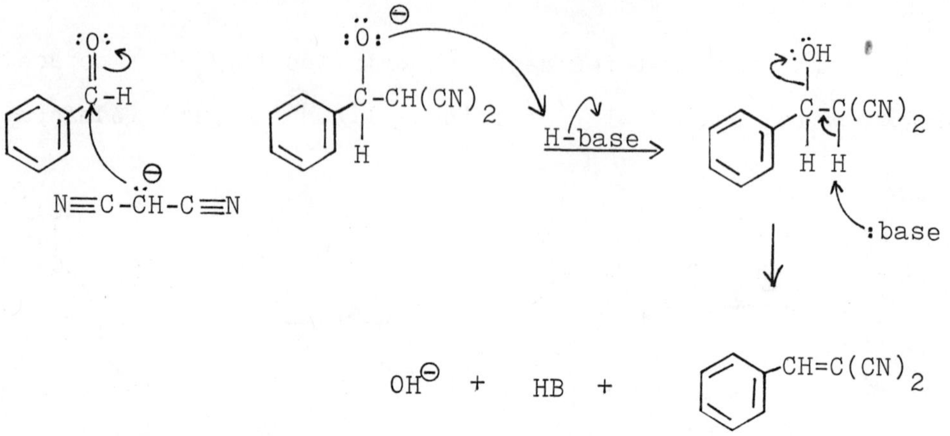

C_6H_5-CH=O + $N{\equiv}C\text{-}\ddot{C}H^{\ominus}\text{-}C{\equiv}N$ → C_6H_5-CH($O^{\ominus}$)-CH(CN)$_2$ $\xrightarrow{\text{H-base}}$ C_6H_5-CH(OH)-CH-C(CN)$_2$ (H, :base)

↓

$OH^{\ominus}$ + HB + C_6H_5-CH=C(CN)$_2$

41. $H_2C{=}\overset{\oplus}{N}{=}\ddot{N}\!:^{\ominus}$ ⟷ $H_2\overset{\ominus}{\ddot{C}}\text{-}\overset{\oplus}{N}{\equiv}N\!:$

cyclohexanone + $\overset{\ominus}{\ddot{C}}H_2\text{—}\overset{\oplus}{N}{\equiv}N$ → 1-($^{\ominus}$:Ö:)-1-($CH_2\text{-}\overset{\oplus}{N}{\equiv}N$)cyclohexane

1-($^{\ominus}$:Ö:)-1-($CH_2\text{-}\overset{\oplus}{N}{\equiv}N$)cyclohexane → spiro epoxide (:Ö—CH_2) + N_2

1-($^{\ominus}$:Ö:)-1-($CH_2\text{-}\overset{\oplus}{N}{\equiv}N$)cyclohexane → cycloheptanone + N_2

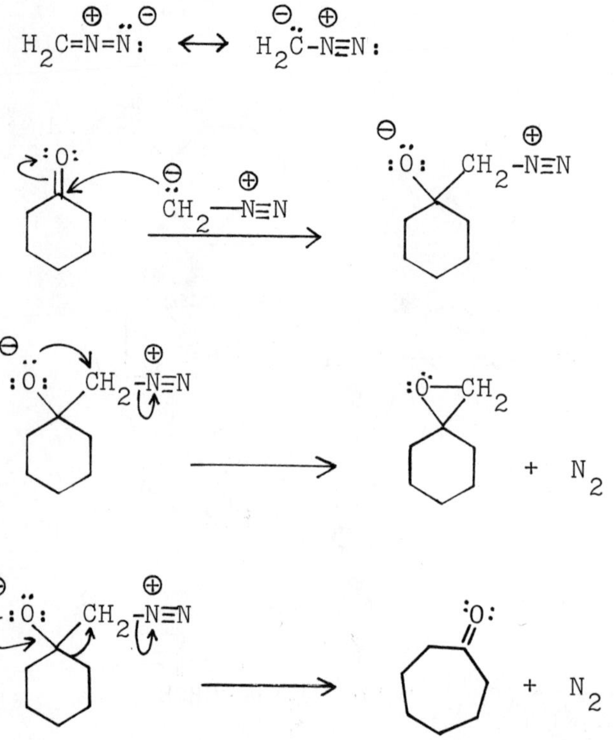

42. 1-Cl-1-(C(=O)-CH_3)cyclohexane $\xrightarrow{OH^{\ominus}}$ 1-Cl-1-(C(=O)-$\ddot{C}H_2^{\ominus}$)cyclohexane → spiro cyclopropanone (CH_2, C=Ö:) + $:\ddot{O}H^{\ominus}$ ↓

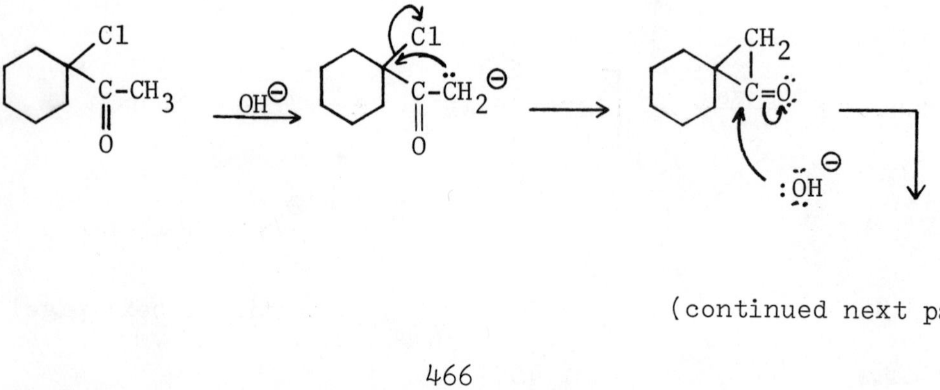

(continued next page)

$$\text{1-methylcyclohexanecarboxylic acid} \xleftarrow{H^{\oplus}} \text{(cyclohexane)}(\ddot{C}H_2^{\ominus})COOH \longleftarrow \text{(cyclohexane)}(CH_2)C(OH)\ddot{O}^{\ominus}$$

(Text: Favorskii rearrangement, p. 734)

43. (a) $CH_3\overset{O}{\overset{\|}{C}}CH_2\overset{O}{\overset{\|}{C}}OC_2H_5 \xrightarrow{C_2H_5O^{\ominus}} CH_3\overset{O}{\overset{\|}{C}}\overset{\ominus}{\ddot{C}}H\overset{O}{\overset{\|}{C}}OC_2H_5{}^{*} \xrightarrow{CH_3-I} CH_3\overset{O}{\overset{\|}{C}}\underset{CH_3}{\underset{|}{C}}H\overset{O}{\overset{\|}{C}}OC_2H_5$

*charge delocalized by both C=O groups.

$$CH_3\overset{:O:}{\overset{\|}{C}}-\overset{\ominus}{\ddot{C}}H-\overset{:O:}{\overset{\|}{C}}OCH_3 \longleftrightarrow CH_3\overset{:O:}{\overset{\|}{C}}-CH=\overset{:\ddot{O}:^{\ominus}}{\overset{|}{C}}OCH_3 \longleftrightarrow CH_3\overset{:\ddot{O}:^{\ominus}}{\overset{|}{C}}=CH-\overset{:O:}{\overset{\|}{C}}OCH_3$$

(Text: see acetoacetic ester synthesis, p. 847)

(b) $CH_3\overset{O}{\overset{\|}{C}}OC_2H_5 \xrightarrow{C_2H_5O^{\ominus}} \overset{\ominus}{\ddot{C}}H_2-\overset{O}{\overset{\|}{C}}OC_2H_5{}^{\ddagger} \longrightarrow CH_3-\overset{:\ddot{O}:^{\ominus}}{\overset{|}{C}}(OC_2H_5)-CH_2\overset{O}{\overset{\|}{C}}OC_2H_5$

(enolate attacks $CH_3\overset{:\ddot{O}}{\overset{\|}{C}}OC_2H_5$)

$$\xrightarrow{-OC_2H_5^{\ominus}} CH_3\overset{O}{\overset{\|}{C}}CH_2\overset{O}{\overset{\|}{C}}OC_2H_5$$

‡resonance stabilized

(continued next page)

(c) The $-\overset{O}{\overset{\|}{C}}OC_2H_5$ group can stabilize the charge via resonance.

$$-\overset{|}{\underset{|}{\overset{\ominus}{\ddot{C}}}}-\overset{:O:}{\overset{\|}{C}}OC_2H_5 \longleftrightarrow -\overset{|}{\underset{|}{C}}=\overset{:\ddot{O}:^{\ominus}}{\overset{|}{C}}OC_2H_5$$

44. $C_6H_5CHO \xrightarrow[CH_3COONa]{(CH_3\overset{O}{\overset{\|}{C}})_2O} C_6H_5CH=CHCOOH$ (Perkin)

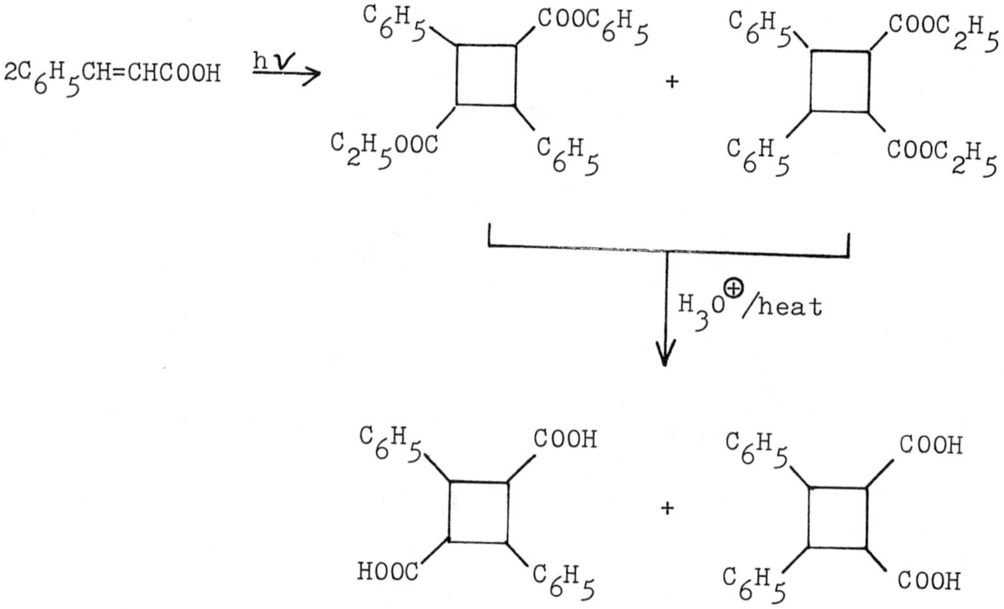

$2C_6H_5CH=CHCOOH \xrightarrow{h\nu}$ [cyclobutane: C_6H_5, $COOC_6H_5$, C_6H_5, C_2H_5OOC] + [cyclobutane: C_6H_5, $COOC_2H_5$, $COOC_2H_5$, C_6H_5]

$\xrightarrow{H_3O^{\oplus}/\text{heat}}$

[cyclobutane: C_6H_5, $COOH$, C_6H_5, $HOOC$] + [cyclobutane: C_6H_5, $COOH$, $COOH$, C_6H_5]

45. (a) $CH_3\overset{O}{\overset{\|}{C}}CH_3 \xrightarrow{H_3O^{\oplus}} CH_3\underset{CH_3}{\underset{|}{C}}=CH\overset{O}{\overset{\|}{C}}CH_3 \xrightarrow{NaBH_4/H_2O} CH_3\underset{CH_3}{\underset{|}{C}}=CH\underset{OH}{\underset{|}{C}}HCH_3$

$\xrightarrow{D_2/Pt} CH_3\underset{CH_3}{\underset{|}{C}}DCHD\underset{OH}{\underset{|}{C}}HCH_3$

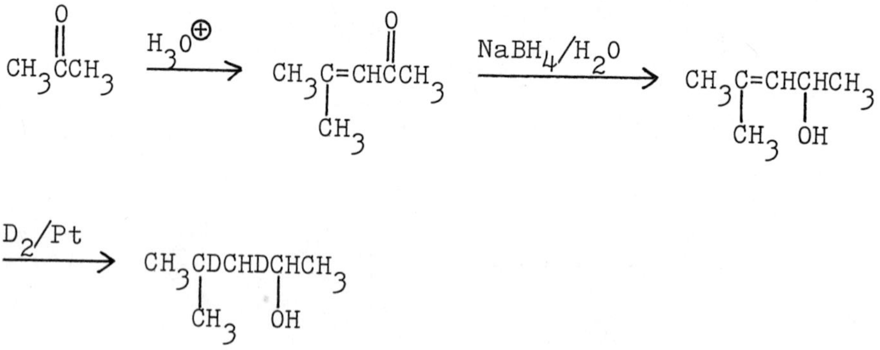

(continued next page)

(b) $CH_3\overset{CH_3}{\underset{|}{C}}=CH\overset{O}{\overset{||}{C}}CH_3$ (from a) $\xrightarrow{NaBD_4/H_2O}$ $CH_3C(CH_3)=CHCD(OH)CH_3$

$\xrightarrow{H_2/Pt}$ $CH_3CH(CH_3)CH_2CD(OH)CH_3$

(c) $CH_3\overset{O}{\overset{||}{C}}CH_2CH_2\overset{O}{\overset{||}{C}}CH_3$ $\xrightarrow{H_3O^{\oplus}}$ [3-hydroxy-3-methylcyclopentanone (HO, CH_3, =O)] $\xrightarrow{-H_2O}$ 3-methylcyclopent-2-enone (CH_3, =O)

$\xrightarrow[H_2O]{NaBH_4}$ 3-methylcyclopent-2-enol (CH_3, H, OH)

(d) $CH_3-C_6H_4-\overset{O}{\overset{||}{C}}H$ + $CH_3\overset{O}{\overset{||}{C}}CH_3$ $\xrightarrow[H_2O]{OH^{\ominus}}$ $[CH_3-C_6H_4-CH(OH)CH_2\overset{O}{\overset{||}{C}}CH_3]$

$\xrightarrow{-H_2O}$ $CH_3-C_6H_4-CH=CH-\overset{O}{\overset{||}{C}}CH_3$ $\xrightarrow{NaBH_4/H_2O}$

$CH_3-C_6H_4-CH=CHCH(OH)CH_3$

(e) $CH_3\overset{O}{\overset{||}{C}}-H$ $\xrightarrow{OH^{\ominus}/H_2O}$ $CH_3CH(OH)CH_2\overset{O}{\overset{||}{C}}-H$ $\xrightarrow{LiAlH_4}$ $\xrightarrow{H_2O}$ $CH_3CH(OH)CH_2CH_2OH$

(continued next page)

(f) $CH_3CH_2\overset{O}{\overset{\|}{C}}\text{-}H \xrightarrow{OH^{\ominus}/H_2O} CH_3CH_2\underset{OH}{\underset{|}{C}H}\overset{CH_3}{\overset{|}{C}H}CHO \xrightarrow{H_2/Pt} CH_3CH_2\underset{OH}{\underset{|}{C}H}\overset{CH_3}{\overset{|}{C}H}CH_2OH$

(g) $(CH_3)_3CCHO + CH_3CHO \xrightarrow{H_3O^{\oplus}} (CH_3)_3CCH{=}CHCHO \xrightarrow{H_2/Pt}$

$(CH_3)_3CCH_2CH_2CH_2OH$

(h) $CH_3\overset{O}{\overset{\|}{C}}\text{-}H \xrightarrow{H_3O^{\oplus}} CH_3CH{=}CH\overset{O}{\overset{\|}{C}}\text{-}H$

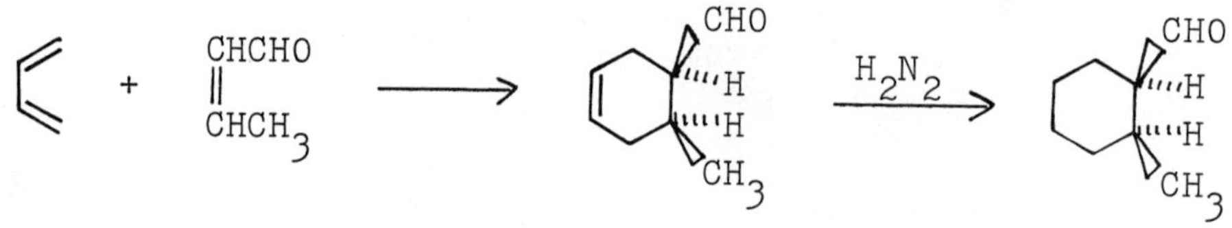

46.

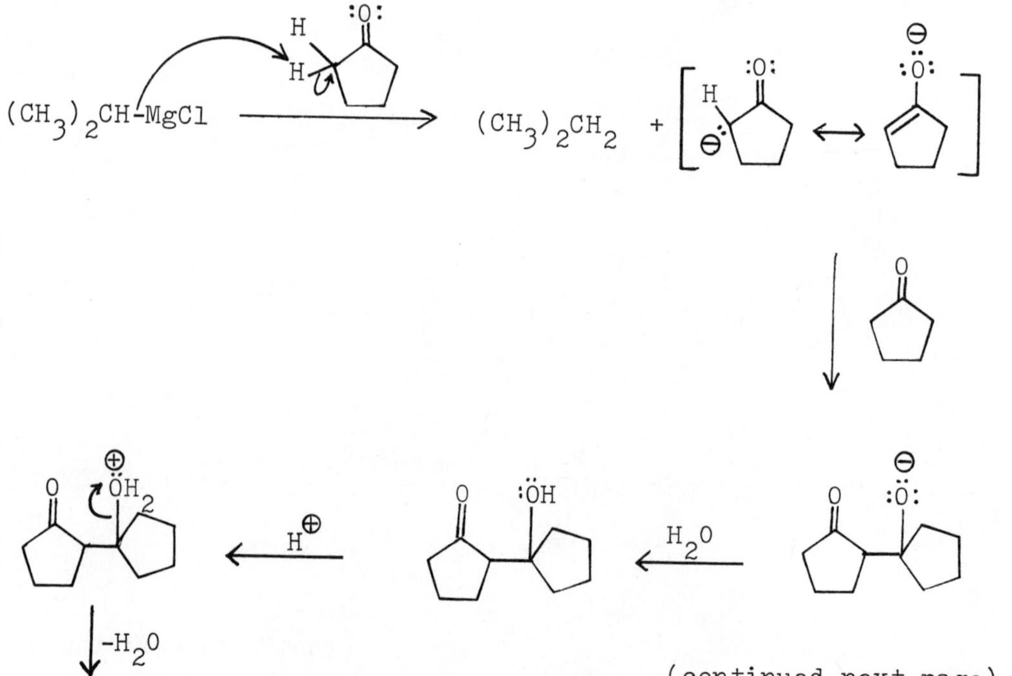

(continued next page)

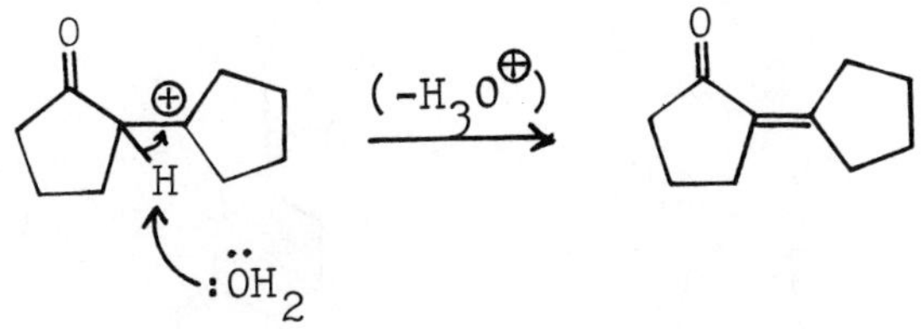

(−$H_3O^{\oplus}$)

:$\ddot{O}H_2$

47. 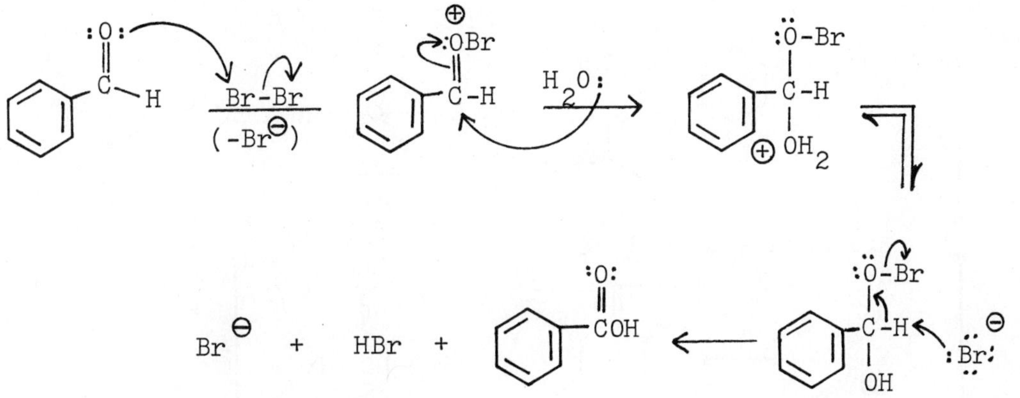 Br–Br (−$Br^{\ominus}$) H_2O:

:O–Br C–H OH $\oplus OH_2$

$Br^{\ominus}$ + HBr + COOH ← :Br:$^{\ominus}$

48. $CH_3\overset{O}{\overset{\|}{C}}CH_3$

49. (a) 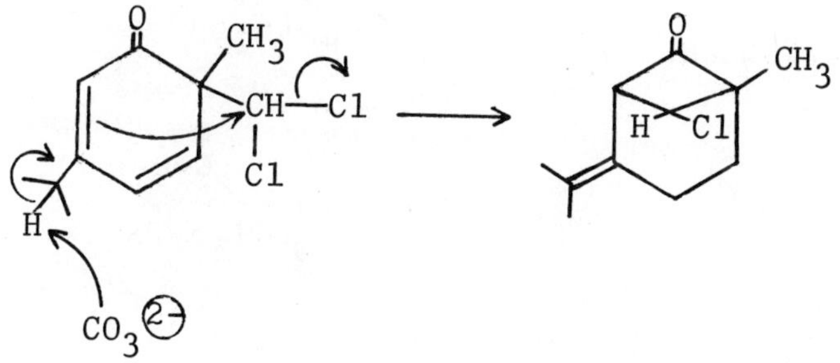 CH_3 CH—Cl Cl H CO_3^{2-} → CH_3 H Cl

(b) 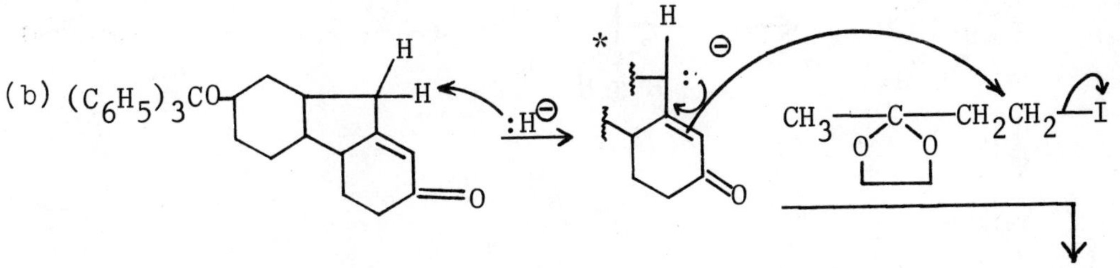 $(C_6H_5)_3CO$ H H :$H^{\ominus}$ → * H CH_3—C—CH_2CH_2—I O O

*resonance stabilized

(continued next page)

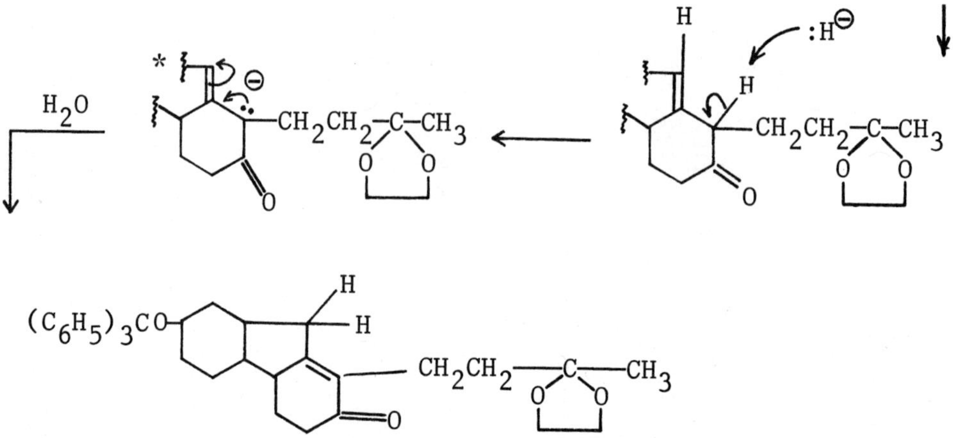

H_2O

$-CH_2CH_2-C-CH_3$

$-CH_2CH_2-C-CH_3$

$:H^{\ominus}$

$(C_6H_5)_3CO-$

$CH_2CH_2-C-CH_3$

*resonance stabilized

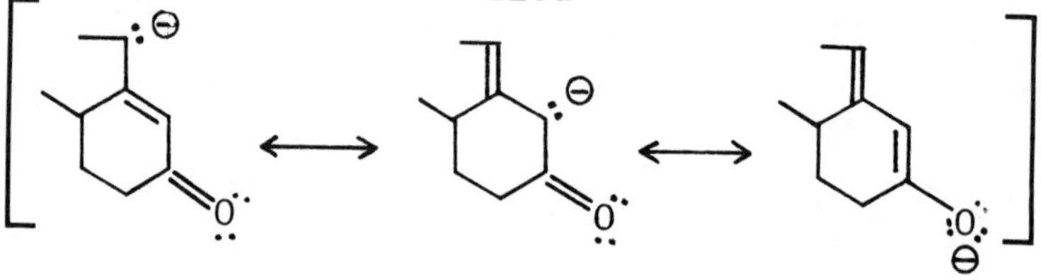

(c)

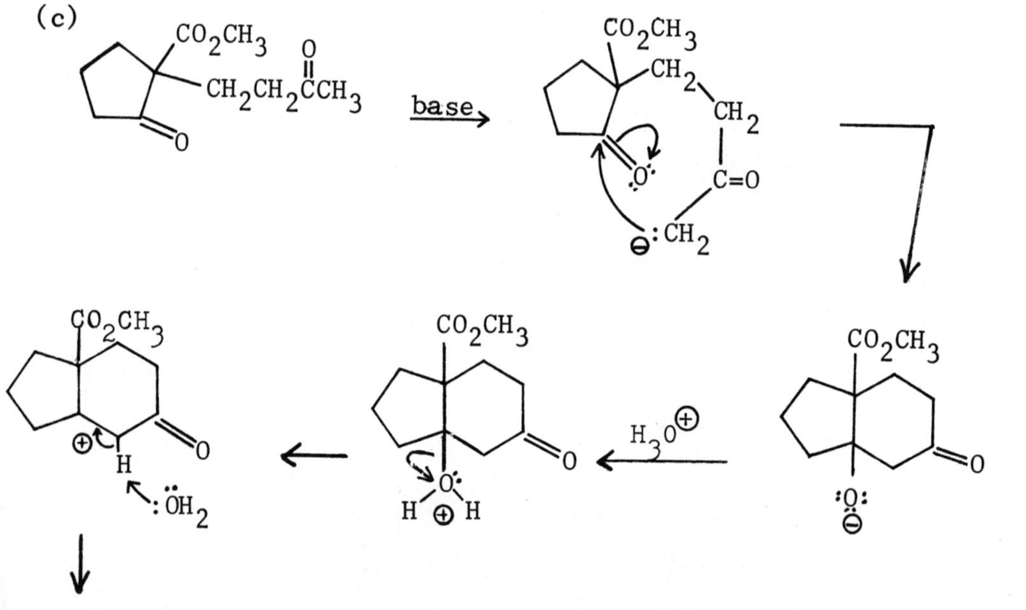

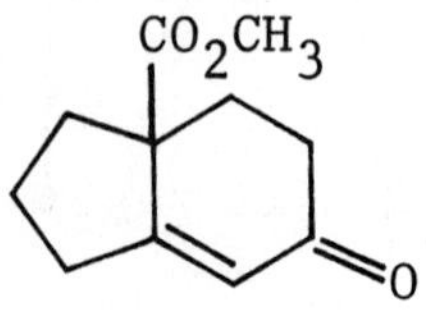

CO_2CH_3

$CH_2CH_2\overset{O}{\overset{\|}{C}}CH_3$

base

CO_2CH_3 CH_2 CH_2 $C=O$ $\ominus:CH_2$

CO_2CH_3

$H_3O^{\oplus}$

CO_2CH_3

CO_2CH_3

$:\ddot{O}H_2$

CO_2CH_3

(continued next page)

(d) $\mathrm{C_6H_{10}}[\mathrm{C(=O)CH_3}][(\mathrm{CH_2})_2\mathrm{CHO}]$ $\xrightarrow{OH^{\ominus}}$ $\mathrm{C(=O)\ddot{C}H_2^{\ominus}}$ ... $(\mathrm{CH_2})_2\text{-CH=}\ddot{O}$: $\longrightarrow$ spiro ketone $-\ddot{O}:^{\ominus}$

$\downarrow H_2O$

$:\ddot{O}H^{\ominus}$, H, $\ddot{O}H$

$\xleftarrow{(-2H_2O)}$

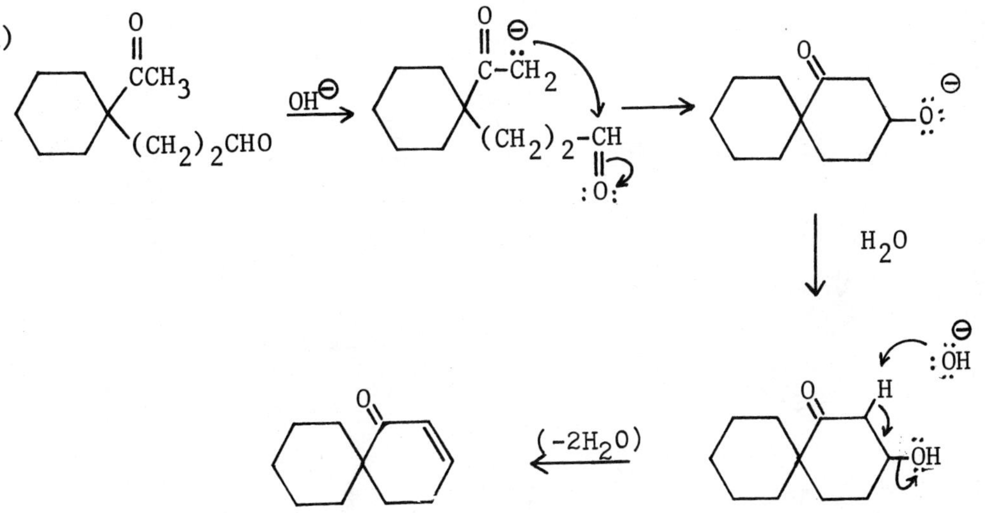

(e) $(\mathrm{CH_2})_2\mathrm{CCl}$ (O) $\xrightarrow{AlCl_3}$ $AlCl_4^{\ominus}$ + $\oplus$C=O

$\oplus$ ... =O $\equiv$ $\oplus$ ... =O

$\xleftarrow{-H^{\oplus}}$ ... =O

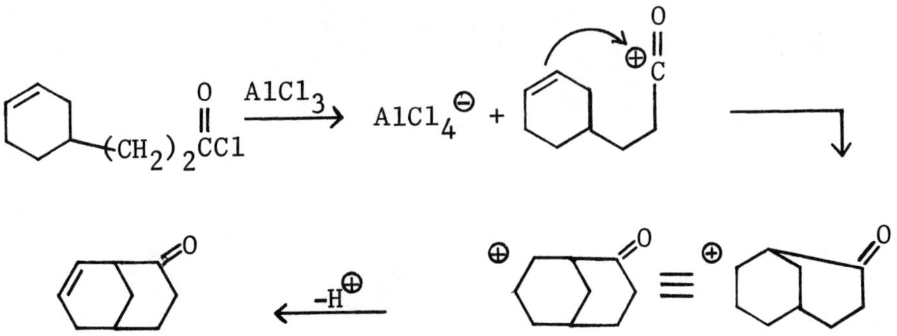

(f) O, H, H $\xrightarrow{base}$ O, H, $\ominus$ $C_6H_5CH{=}CH{-}C({=}O){-}CO_2^{\ominus}$ H–O–H

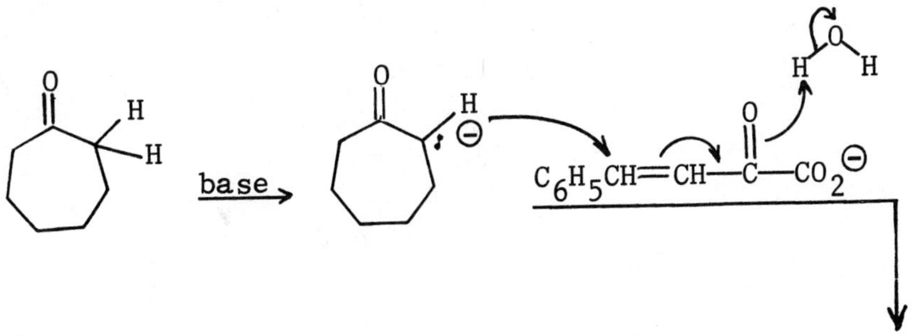

(continued next page)

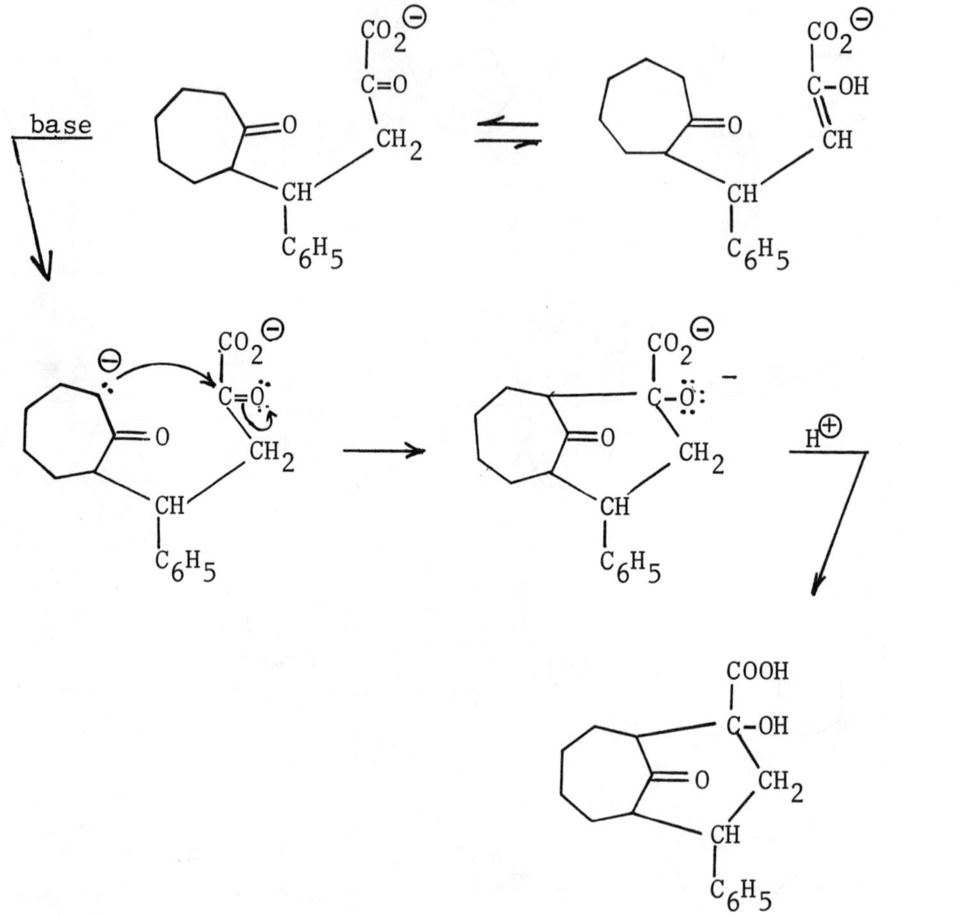

base

$CO_2^{\ominus}$ C=O CH_2 CH C_6H_5 ⇌ $CO_2^{\ominus}$ C–OH CH CH C_6H_5

→ $H^{\oplus}$

COOH C–OH CH_2 CH C_6H_5

(g) $C_6H_5C—CH{=}CHOC_2H_5$ + $C_6H_5—MgBr$ ⟶ $C_6H_5C{=}CH—CH(C_6H_5)—OC_2H_5$ (O—MgBr) ⟶ $(-C_2H_5O^{\ominus})$

$C_6H_5C—CH{=}CH—C_6H_5$ + $C_6H_5—MgBr$ ⟶ $C_6H_5C{=}CH—CH(C_6H_5)—C_6H_5$ ($O^{\ominus}$) ⟶ H_2O

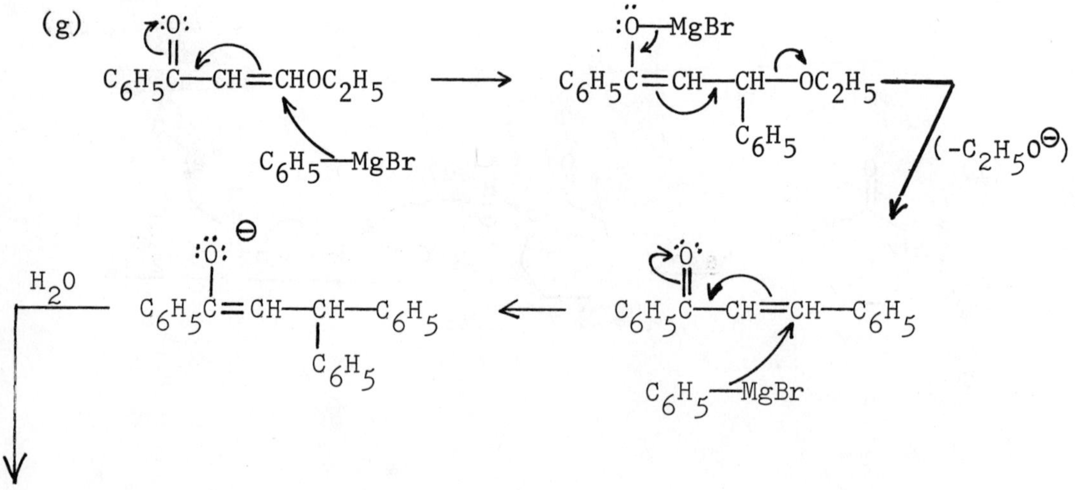

(continued next page)

$$C_6H_5\overset{:\ddot{O}H}{\overset{|}{C}}{=}CH{-}CH(C_6H_5)_2 \rightleftarrows C_6H_5\overset{:O:}{\overset{\|}{C}}{-}CH_2CH(C_6H_5)_2$$

(h)

O, O–C–CH$_3$ (=:O:), CH$_3$, CH$_3$, CH$_3$, Br, H–Ö–H → CH$_3$, CH$_3$, Br, CH$_3$, O, O–C–CH$_3$ (:Ö:$^\ominus$), Ö$^\oplus$ H H ⇌ :Ö–H, O, O–C–CH$_3$, O–H, CH$_3$, CH$_3$, Br, CH$_3$

→ CH$_3$COOH + Br$^\ominus$ + H, :Ö:, Ö:, CH$_3$, CH$_3$, CH$_3$; H$_2$Ö:

→ :Ö:$^\ominus$, Ö:, CH$_3$, CH$_3$, CH$_3$ $\xrightarrow{H_2O}$:ÖH, O, CH$_3$, CH$_3$, CH$_3$

(i)

:Ö: + :NH$_2$CH$_2$C$_6$H$_5$ → Ö:$^\ominus$, $\overset{\oplus}{N}$H$_2$CH$_2$C$_6$H$_5$ ⇌ Ö–H, N̈–CH$_2$C$_6$H$_5$, H $\xrightarrow{-H_2O}$ N̈CH$_2$C$_6$H$_5$

(j)

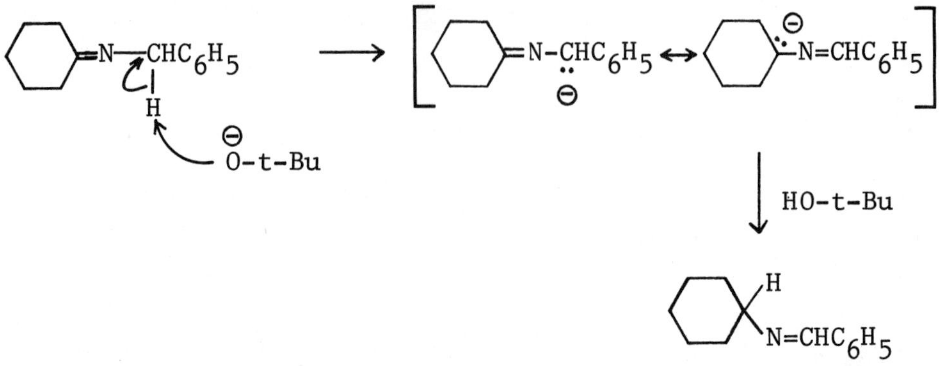

$\longrightarrow$ N–CHC$_6$H$_5$ ↔ N=CHC$_6$H$_5$

O-t-Bu

HO-t-Bu

N=CHC$_6$H$_5$

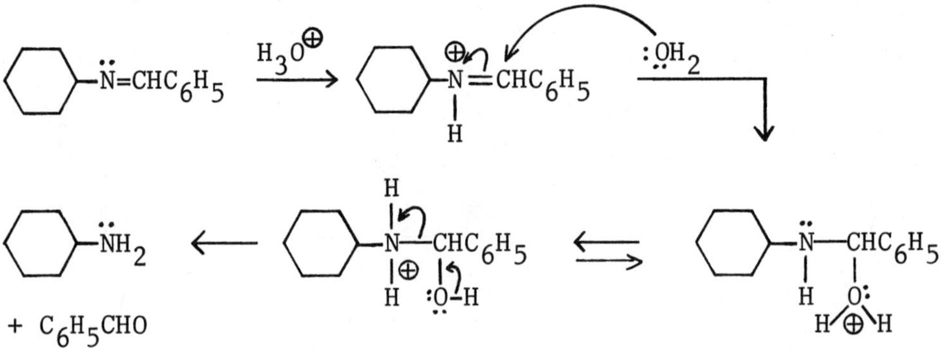

(k)

N=CHC$_6$H$_5$ $\xrightarrow{H_3O^{\oplus}}$ N=CHC$_6$H$_5$

:OH$_2$

NH$_2$ + C$_6$H$_5$CHO ← N–CHC$_6$H$_5$ ⇌ N–CHC$_6$H$_5$

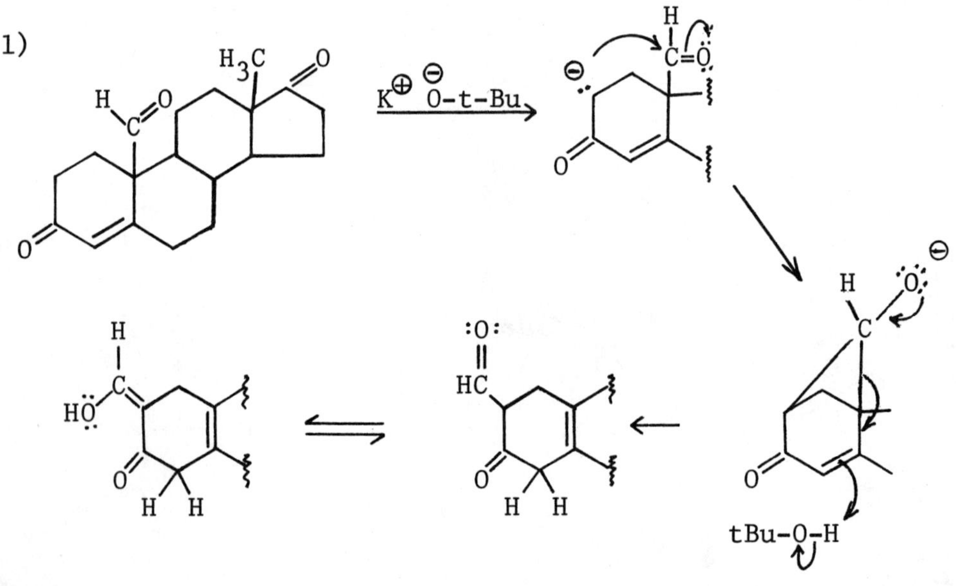

(l)

H$_3$C

$\xrightarrow{K^{\oplus}\ \overset{\ominus}{O}\text{-t-Bu}}$

tBu-O-H

50. $CH_2{=}C{=}\ddot{O}$ + $H_2\ddot{O}$ → $\left[\ \overset{\ominus}{C}H_2{-}C(\overset{\oplus}{O}H_2){=}O: \longleftrightarrow CH_2{=}C(\overset{\oplus}{O}H_2){-}\ddot{O}:^{\ominus}\ \right]$ $\xrightarrow{H_2O}$ $CH_3{-}C({:}\ddot{O}H){=}\ddot{O}$

$CH_3{-}C(=O){-}CH_3$ $\xrightarrow{H^{\oplus}}$ $CH_2{-}C(=\overset{\oplus}{O}{-}H){-}CH_3$ (H on CH_2 removed by $H_2\ddot{O}:$) $\xrightarrow{-H_3O^{\oplus}}$ $CH_2{=}C(\ddot{O}{-}H){-}CH_3$ $\xrightarrow{O{=}C{=}CH_2}$

$CH_3{-}C(=CH_2){-}\overset{\oplus}{O}(H){-}C(\ddot{O}:^{\ominus}){=}CH_2$ ⟶ $CH_3{-}C(=CH_2){-}\ddot{O}{-}C(\ddot{O}{-}H){=}CH_2$ ⇌ $CH_3{-}C(=CH_2){-}\ddot{O}{-}C(=O){-}CH_3$

(Text S. 18.2)

51. 2-chloro-2-methylcyclohexanone $\xrightarrow{NH_2NH_2}$ hydrazone ($C{=}N{-}NH_2$, Cl, CH_3) $\overset{B^{\ominus}}{\rightleftarrows}$ ($N{=}N{-}H$, $:B^{\ominus}$, H, Cl, CH_3)

↓

HB + $Cl^{\ominus}$ + N_2 + 1-methylcyclohexene (H, CH_3)

hydrazone

52. The methyl protons are acidic because proton removal leads to a resonance-stabilized anion.

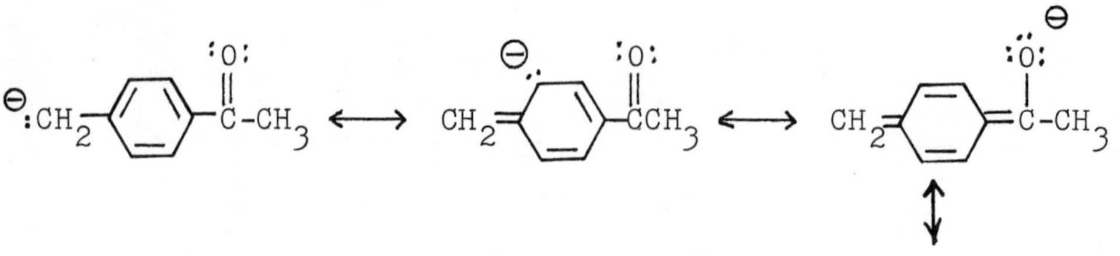

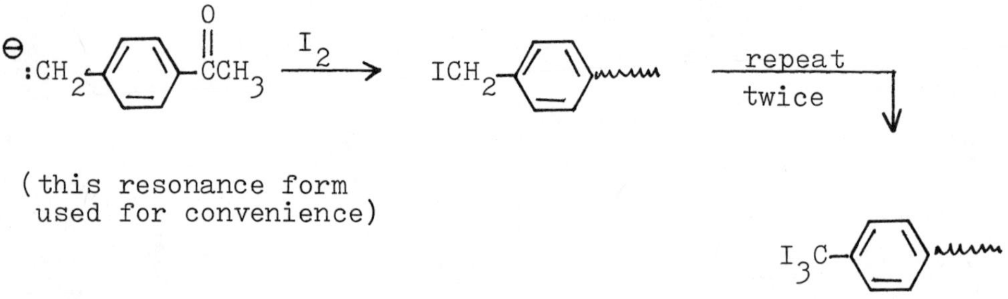

etc.

(this resonance form used for convenience)

repeat twice

Replacement of -I by $OH^{\ominus}$ gives:

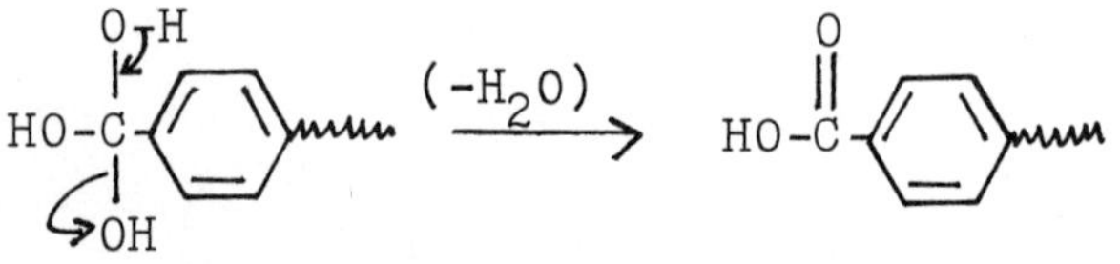

$(-H_2O)$

(as salt in basic solution)

(unstable)

53.

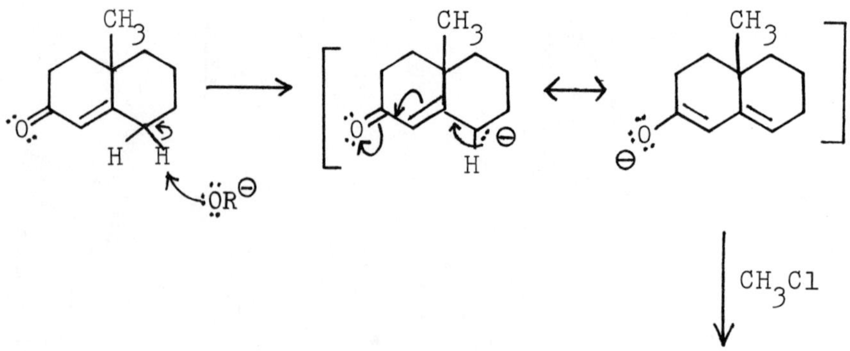

CH_3Cl

(continued next page)

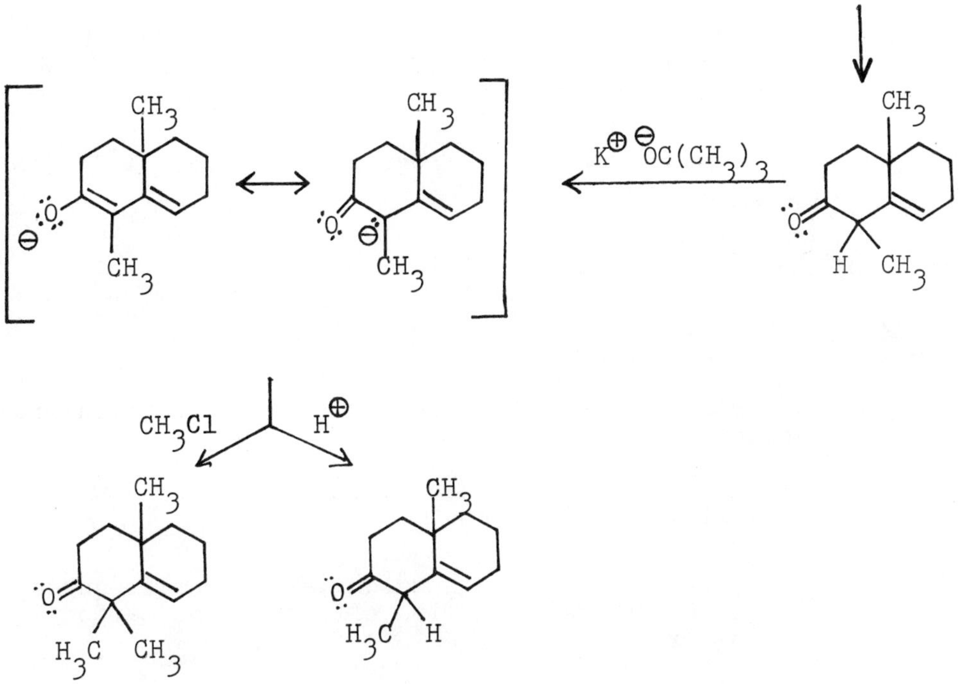

CH_3 CH_3 $K^{\oplus}\ {}^{\ominus}OC(CH_3)_3$ CH_3 CH_3 CH_3 H CH_3

CH_3Cl $H^{\oplus}$

CH_3 H_3C CH_3 CH_3 H_3C H

54. A carbonyl group associated with a solvent by hydrogen bonding has less double bond character (>C=O······H—Z) than a free carbonyl group.

55. (a) ir: C=O stretch ~ 1715 cm^{-1} for 2-methylcyclohexanone; absent for methylcyclopentane

nmr: absorption near 2 δ for hydrogens α to C=O of 2-methylcyclohexanone. All absorptions for methylcyclopentane further upfield.

uv: absorption max. near 280 nm for 2-methylcyclohexanone

(b) ir: C=O stretch ~ 1675 cm^{-1} in ir spectrum of 2-methylcyclohexenone (conjugated C=O); ~ 1715 cm^{-1} in 3-cyclohexenone

nmr: doublet ($-CH_2-$ of $-C(O)-CH_2-CH=CH-$) 3-4 δ in spectrum of 3-cyclohexenone; absent in spectrum of 2-cyclohexenone

(continued next page)

uv: absorption max. ~300 nm for 2-cyclohexenone, ~280 for 3-cyclohexenone

(c) ir: C=O stretch at higher frequency (~1745 cm^{-1}) for

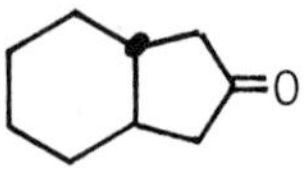

=O (C=O in five-membered ring)

nmr: not readily distinguished by nmr

(d) ir: C=O stretch at higher frequency in 1,1,1-trichloroacetone

nmr: singlet in 1,1,1-trichloroacetone further downfield than singlet in acetone

(e) ir: C=O stretch of 2,3-butanedione at higher frequency (~1720) due to inductive effect of one carbonyl on the other

nmr: one singlet for 2,3-butadione; two singlets for 2,4-pentanedione

(f) ir: C=O stretch at higher frequency in the cis-isomer (dihedral angle in Cl-C-C=O near 0°)

nmr: not readily distinguished by nmr

(g) not easily distinguished by ir and nmr

ord: O H OH gives an ord curve

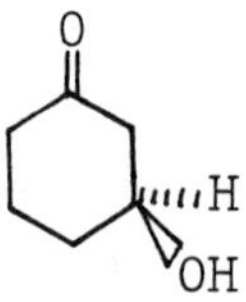

(h) C=O stretch at higher frequency in 1,2-dione (compare with e)

(i) ir: not easily distinguished

nmr: CH_3 a doublet in CH_3CH=CHCHO; CH_3's a singlet in $(CH_3)_2C$=CHCHO

uv: absorption max. 5 nm higher for $(CH_3)_2C$=CHCHO

(j) ir: C=O stretch only in cyclohexanone

(continued next page)

nmr: OH 2-4 δ for 1-hydroxycyclopentanecarbonitrile (cyclopentane ring bearing OH and CN)

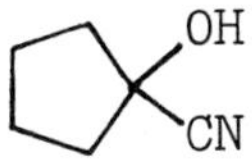

uv: max. ~ 280 for cyclohexanone and ~ 340 for 1-hydroxycyclopentanecarbonitrile (cyclopentane ring bearing OH and CN)

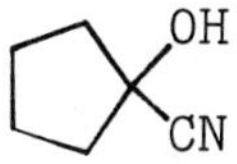

(k) ir: C=O stretch a higher frequency in five-membered ring

nmr: singlet (CH_2 of $(CH_3)_2$-CH_2-C(=O)-) ~ 2 δ

56. The methyl hydrogens of acetone are exchanged with deuterium.

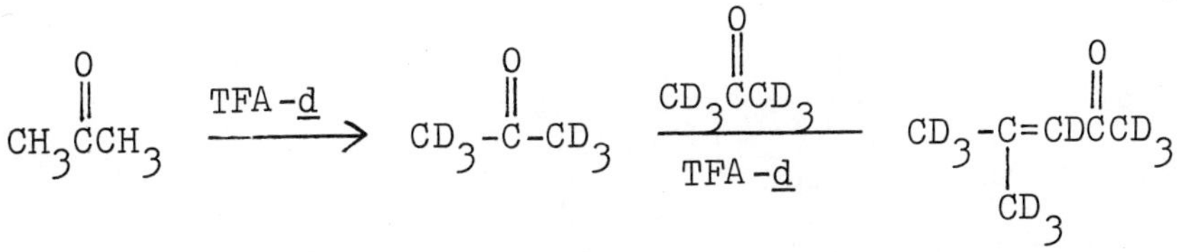

$$CH_3C(=O)CH_3 \xrightarrow{\text{TFA-}\underline{d}} CD_3\text{-}C(=O)\text{-}CD_3 \xrightarrow[\text{TFA-}\underline{d}]{CD_3C(=O)CD_3} CD_3\text{-}C(CD_3)\text{=}CDC(=O)CD_3$$

F_3CC(=O)OH formed in the hydrogen-deuterium exchange gives rise to the signal at 10 δ. The exchange occurs by the following mechanism.

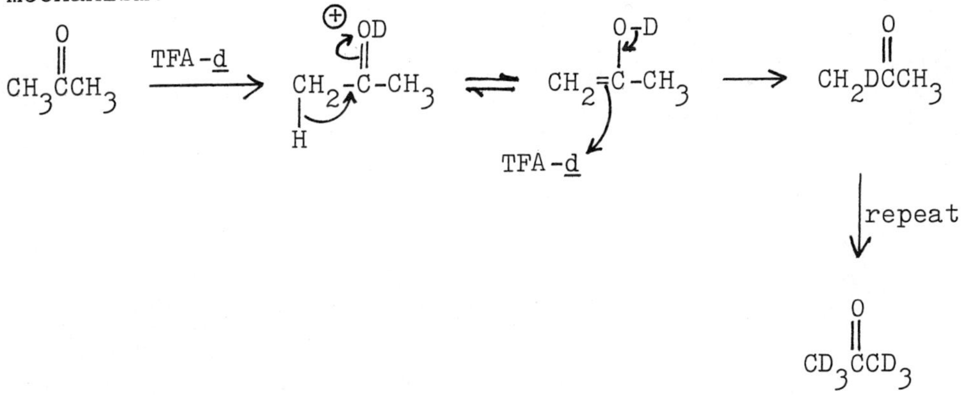

$$CH_3C(=O)CH_3 \xrightarrow{\text{TFA-}\underline{d}} H\text{-}CH_2\text{-}C(=\overset{\oplus}{O}D)\text{-}CH_3 \rightleftharpoons CH_2\text{=}C(OD)\text{-}CH_3 \;(+\ \text{TFA-}\underline{d}) \longrightarrow CH_2DC(=O)CH_3$$

$$\downarrow \text{repeat}$$

$$CD_3C(=O)CD_3$$

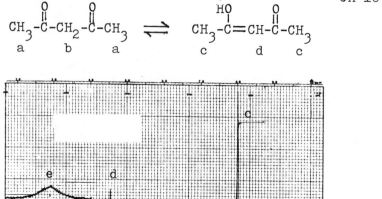

57.

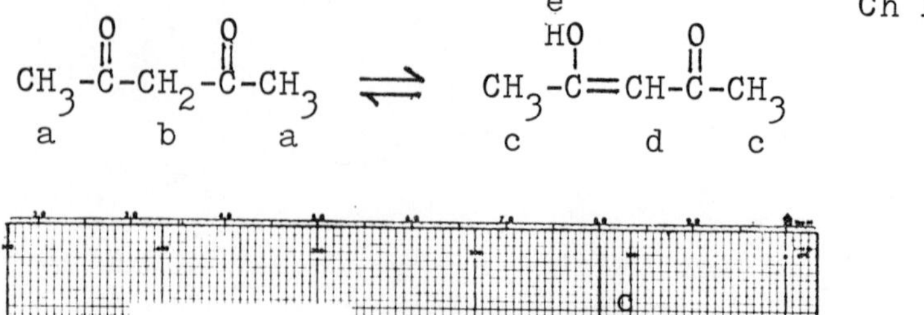

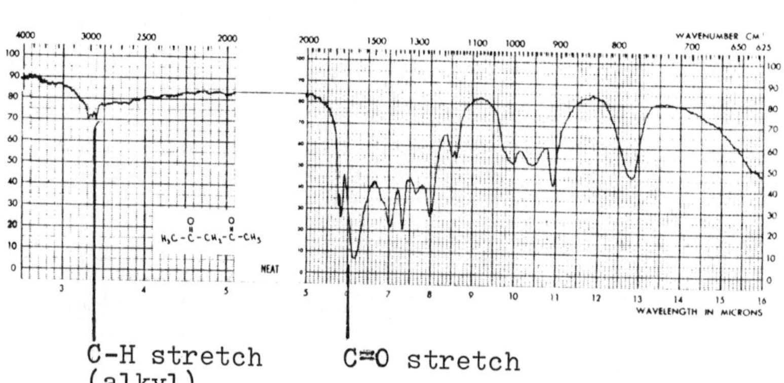

(OH and C=C are not detectable in the IR. This suggests that very little enol form is present in the sample from which the IR spectrum was obtained.)

58.

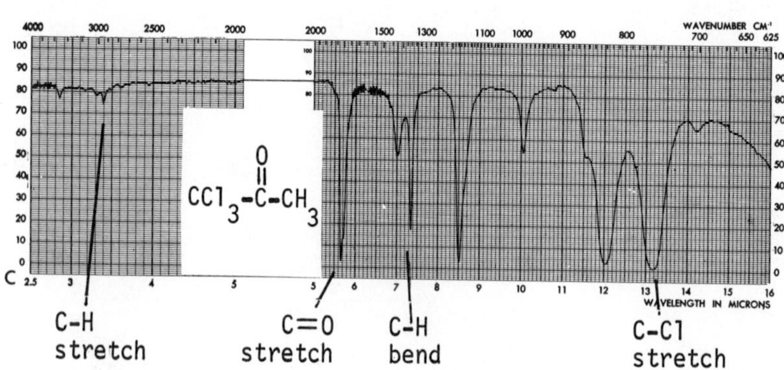

(continued next page)

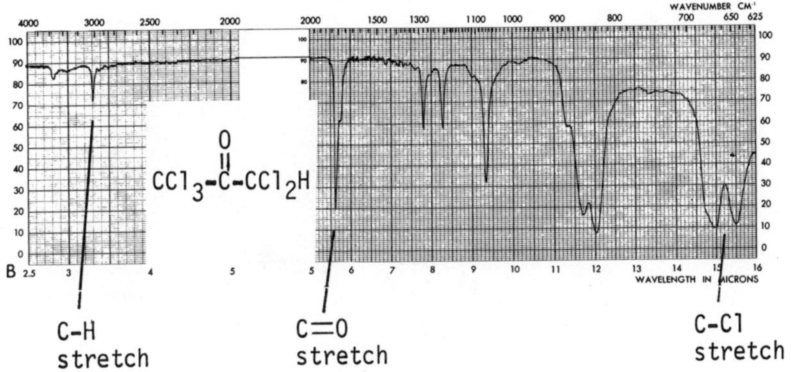

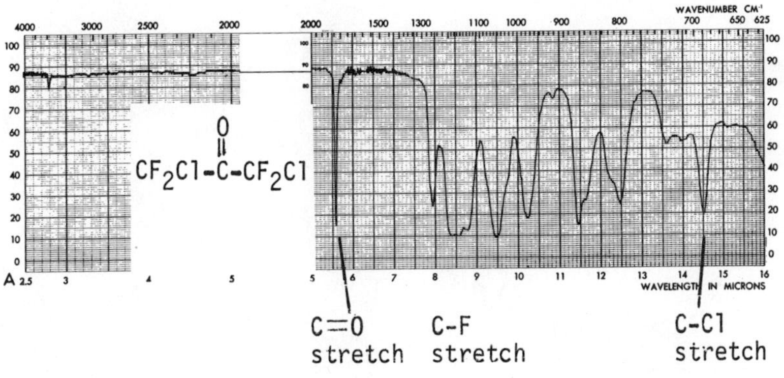

59.

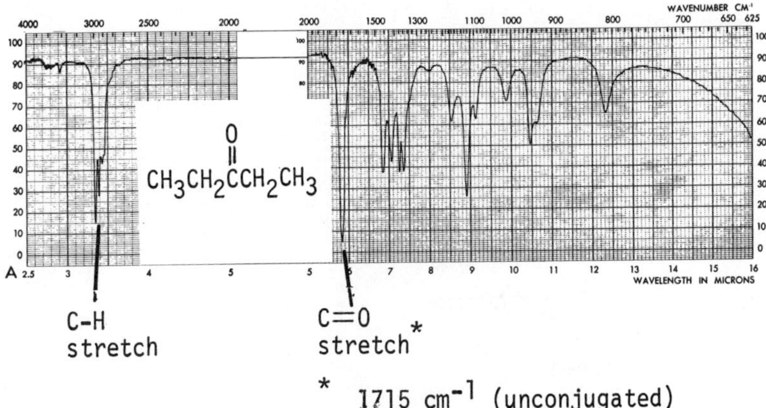

* 1715 cm^{-1} (unconjugated)

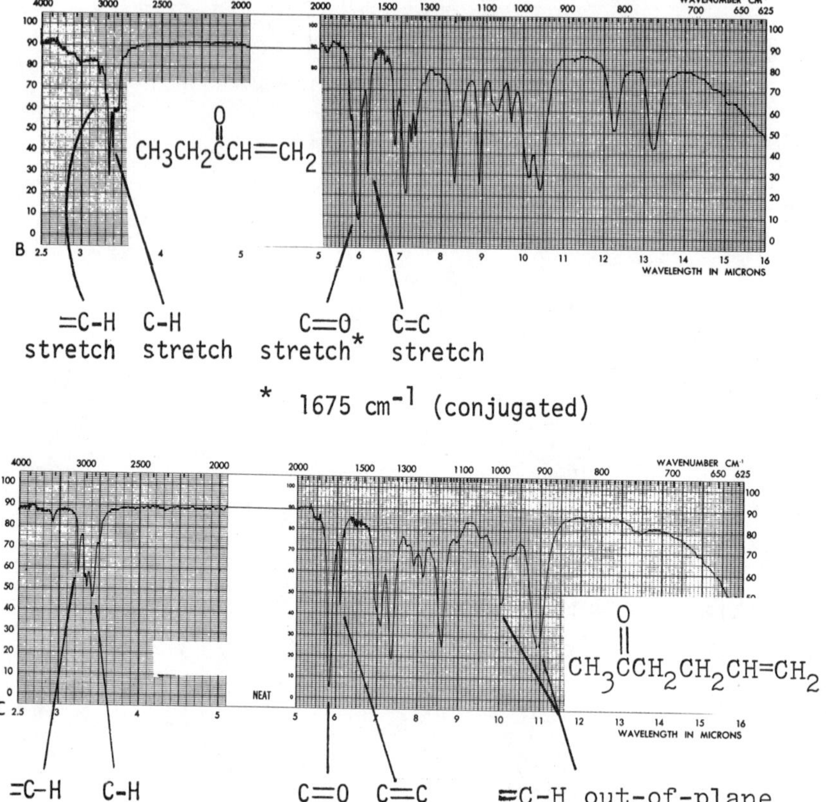

60.

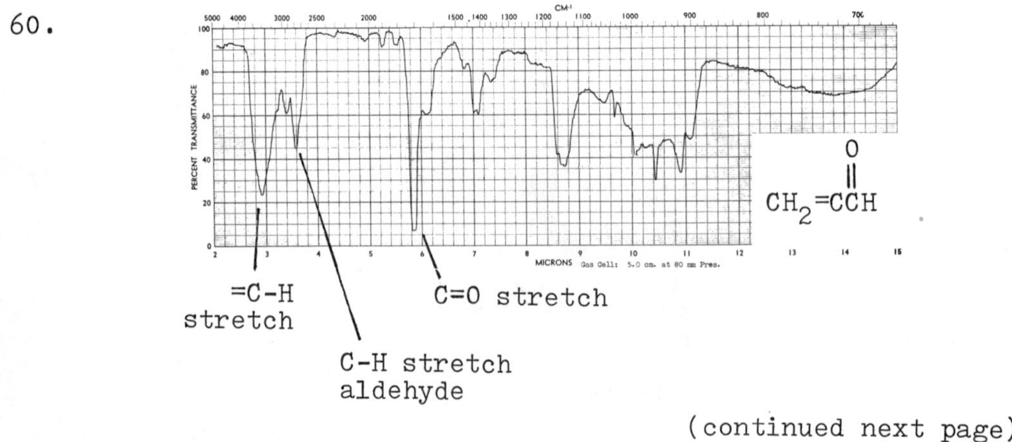

(continued next page)

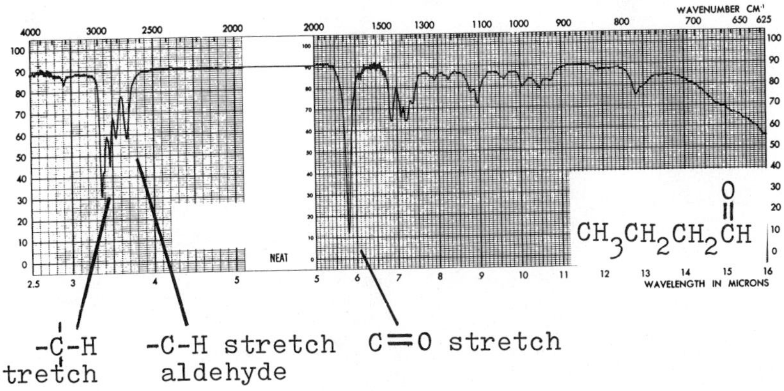

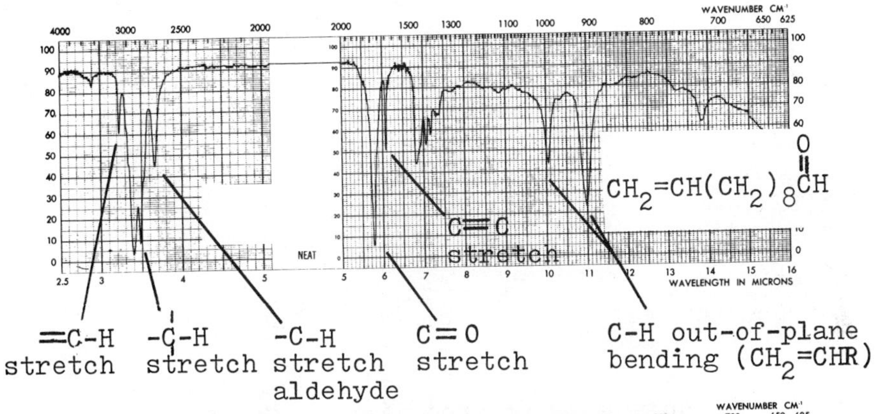

$CH_3CH_2CH_2\overset{O}{\overset{\|}{C}}H$

$-\overset{|}{\underset{|}{C}}-H$ stretch

$-C-H$ stretch aldehyde

C═O stretch

$CH_2=CH(CH_2)_8\overset{O}{\overset{\|}{C}}H$

═C-H stretch

$-\overset{|}{\underset{|}{C}}-H$ stretch

$-C-H$ stretch aldehyde

C═C stretch

C═O stretch

C-H out-of-plane bending ($CH_2=CHR$)

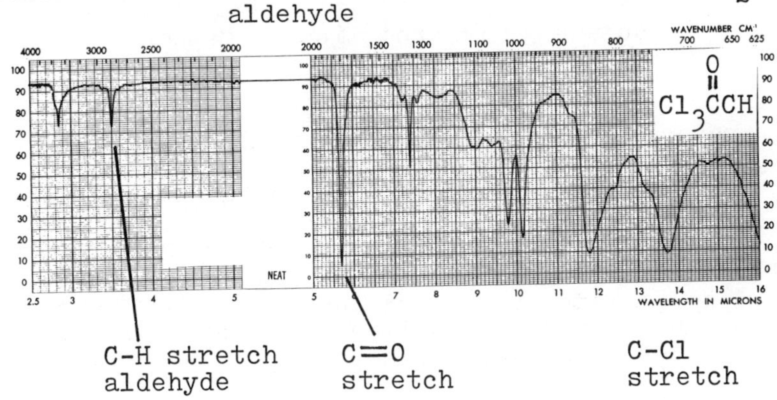

$Cl_3C\overset{O}{\overset{\|}{C}}H$

C-H stretch aldehyde

C═O stretch

C-Cl stretch

61. Use the diagram illustrated in Figure 18-6 and data in Table 18-6.

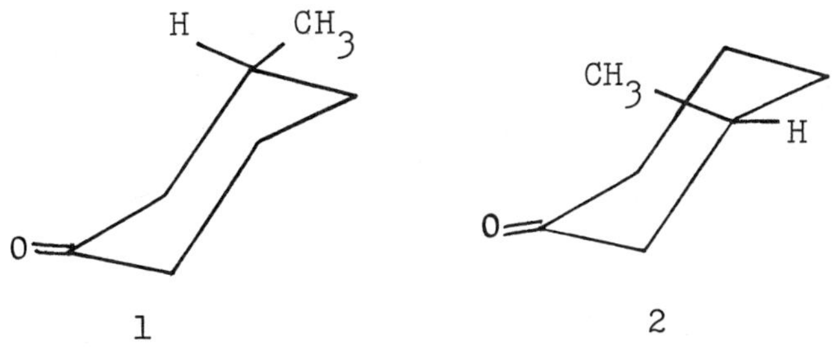

(continued next page)

1: CH_3 is L3 and e; contribution is +

2: CH_3 is R3 and a: contribution is -

Thus compound 1 is (+)-3-methylcyclohexanone.

62.

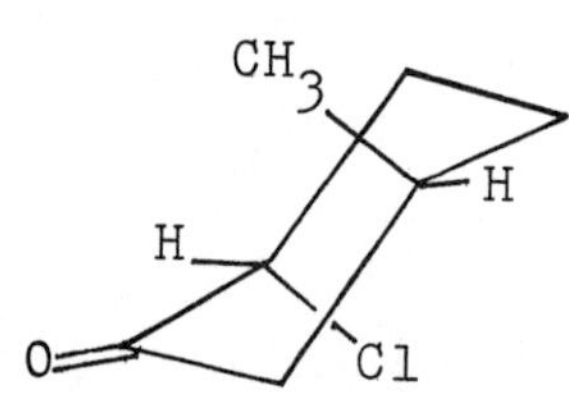

Cl: C-6(L2), e
contribution -

CH_3: C-3(R3), a
contribution -

(-) rotation in octane agrees with this structure

Cl: C-2(R2), e
contribution 0

CH_3: C-5(L3), e
contribution +

(+) rotation in methanol agrees with this structure

The conformation on the left has fewer dipole interactions and is apparently more stable in the nonpolar solvent octane. In methanol dipole interactions are less important.

63. The D line rotation may reflect the tailing from a very intense curve at shorter wavelength. That short wavelength transition may have a sign opposite to that of the nearby (weaker) cotton effect. Thus, the sign of the D line rotation need not reflect the sign of the cotton effect closest to it.

64. The 2-ketosystem (B) does not form very much methyl acetal at equilibrium because of unfavorable diaxial repulsions. Such repulsions are absent in A and so it is converted largely to acetal which (not having a carbonyl group) lacks the ord transition.

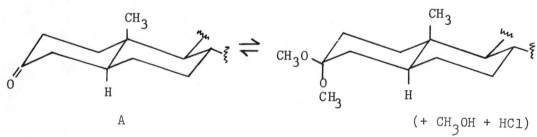

A

(+ CH_3OH + HCl)

(continued next page)

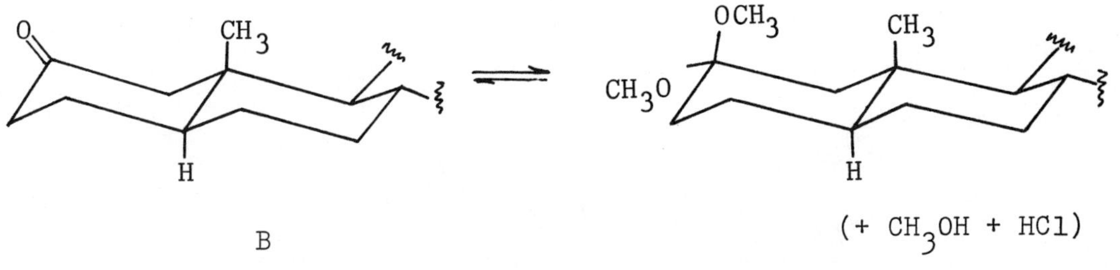

$$\text{B} \rightleftharpoons (+\ CH_3OH + HCl)$$

SUMMARY OF REACTIONS

1. Hydrogen-deuterium exchange (S. 18.2)

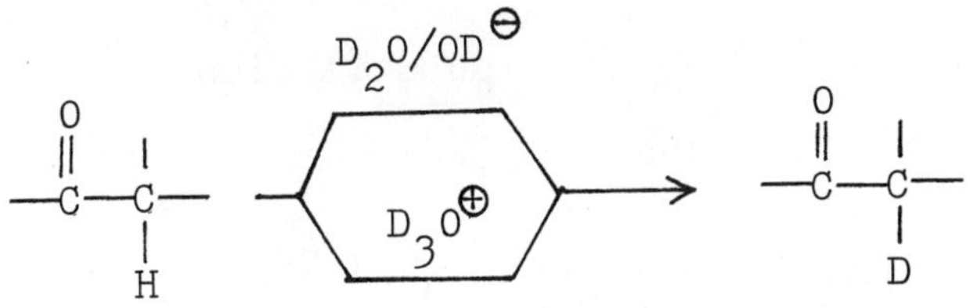

$$-\overset{O}{\overset{\|}{C}}-\underset{H}{\underset{|}{\overset{|}{C}}}- \xrightarrow[D_3O^{\oplus}]{D_2O/OD^{\ominus}} -\overset{O}{\overset{\|}{C}}-\underset{D}{\underset{|}{\overset{|}{C}}}-$$

2. Synthesis of α-halo carbonyl compounds (S. 8.5)

$$-\overset{O}{\overset{\|}{C}}-\underset{H}{\underset{|}{\overset{|}{C}}}- \xrightarrow{X_2/H_3O^{\oplus}} -\overset{O}{\overset{\|}{C}}-\underset{X}{\underset{|}{\overset{|}{C}}}-$$

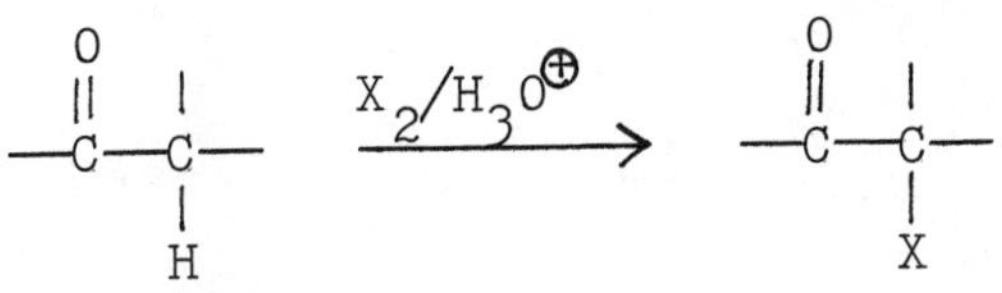

3. Haloform reaction (S. 8.3)

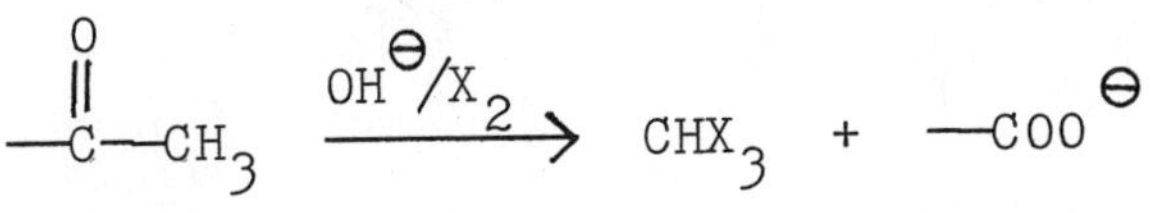

$$-\overset{O}{\overset{\|}{C}}-CH_3 \xrightarrow{OH^{\ominus}/X_2} CHX_3 + -COO^{\ominus}$$

4. Reactions of α-halo carbonyl compounds (S. 8.5)

(continued next page)

$$\underset{X}{-\overset{O}{\overset{\|}{C}}-\overset{|}{\underset{|}{C}}-} \xrightarrow{AlH_3/THF} \underset{H\;\;\;X}{-\overset{OH}{\overset{|}{C}}-\overset{|}{C}-}$$

$$\xrightarrow{Zn/HOAc} -\overset{O}{\overset{\|}{C}}-\underset{H}{\underset{|}{\overset{|}{C}}}-$$

$$-\overset{O}{\overset{\|}{C}}-\underset{X}{\underset{|}{\overset{|}{C}}}-\overset{H}{\overset{|}{\underset{|}{C}}}- \xrightarrow{Nu:^{\ominus}} -\overset{O}{\overset{\|}{C}}-\overset{|}{C}=\overset{|}{C}- \quad \text{(elimination)}$$

$$\xrightarrow{Nu:^{\ominus}} -\overset{O}{\overset{\|}{C}}-\underset{Nu}{\underset{|}{\overset{|}{C}}}-\overset{H}{\overset{|}{\underset{|}{C}}}- \quad \text{(substitution)}$$

5. Alkylation of carbonyl compounds (S. 8.6)

$$-\overset{O}{\overset{\|}{C}}-\underset{H}{\underset{|}{\overset{|}{C}}}- \xrightarrow{:B^{\ominus}} \xrightarrow{R-X} -\overset{O}{\overset{\|}{C}}-\underset{R}{\underset{|}{\overset{|}{C}}}-$$

6. Aldol condensation (S. 18.7)

$$-\overset{O}{\overset{\|}{C}}-\underset{H}{\underset{|}{\overset{H}{\overset{|}{C}}}}- \xrightarrow{OH^{\ominus}/H_2O} -\underset{H}{\underset{|}{\overset{H}{\overset{|}{C}}}}-\overset{OH}{\overset{|}{\underset{|}{C}}}-\underset{H}{\underset{|}{\overset{|}{C}}}-\overset{O}{\overset{\|}{C}}- \xrightarrow[H_2O]{OH^{\ominus}/heat} -\underset{H}{\underset{|}{\overset{H}{\overset{|}{C}}}}-\underset{|}{C}=\underset{|}{C}-\overset{O}{\overset{\|}{C}}-$$

$$\xrightarrow{H_3O^{\oplus}} -\underset{H}{\underset{|}{\overset{H}{\overset{|}{C}}}}-\overset{OH}{\overset{|}{\underset{|}{C}}}-\underset{H}{\underset{|}{\overset{|}{C}}}-\overset{O}{\overset{\|}{C}}- \xrightarrow{-H_2O}$$

(usually cannot be isolated)

(trans > cis)

7. <u>Benzoin condensation</u> (S. 18.8)

$$2\ C_6H_5CHO \xrightarrow{CN^{\ominus}/H_2O/C_2H_5OH} C_6H_5CH(OH)-C(=O)-C_6H_5$$

REACTION REVIEW

Questions	Answers
1. $-C(=O)-\underset{H}{C}- \xrightarrow[OD^{\ominus}\ (\text{or } D_3O^{\oplus})]{D_2O}$?	1. $-C(=O)-\underset{D}{C}-$
Write the product which represents complete D-H exchange:	
a) $CH_3CH_2\overset{O}{\overset{\Vert}{C}}CH_2CH_3$	a) $CH_3CD_2\overset{O}{\overset{\Vert}{C}}CD_2CH_3$
b) cyclohexanone	b) 2,2,6,6-tetradeuteriocyclohexanone (D, D on each carbon adjacent to C=O)
2. $-C(=O)-\underset{H}{C}- \xrightarrow[H_3O^{\oplus}]{X_2}$?	2. $-C(=O)-\underset{X}{C}-$
Write the structure of the monochlorination product when $-C(=O)-\underset{H}{C}-$ is:	(continued next page)

a) cyclopentanone (cyclopentane ring =O)	a) 2-chlorocyclopentanone (ring =O, Cl on adjacent carbon) (d,l)
b) $CH_3CH_2\overset{O}{\overset{\|\|}{C}}CH_2CH_3$	b) $CH_3\underset{Cl}{\underset{\|}{C}}H\overset{O}{\overset{\|\|}{C}}CH_2CH_3$ (d,l)
3. $-\overset{O}{\overset{\|\|}{C}}-CH_3 \xrightarrow[OH^{\ominus}]{X_2}$?	3. $-\overset{O}{\overset{\|\|}{C}}-O^{\ominus}$ + CHX_3 (haloform reaction)
Write the structure of the products in the following:	
a) $CH_3CH_2\overset{O}{\overset{\|\|}{C}}CH_3 \xrightarrow{Cl_2/OH^{\ominus}}$	a) $CH_3CH_2CO_2^{\ominus}$ + $CHCl_3$
b) $C_6H_5\overset{O}{\overset{\|\|}{C}}CH_3 \xrightarrow{I_2/OH^{\ominus}}$	b) $C_6H_5CO_2^{\ominus}$ + CHI_3
c) $CH_3CH_2\overset{O}{\overset{\|\|}{C}}CH_2CH_3 \xrightarrow{Cl_2/OH^{\ominus}}$	c) no reaction
4. $-\overset{O}{\overset{\|\|}{C}}-\underset{X}{\underset{\|}{C}}- \xrightarrow{a)\ AlH_3}$?	4. a) $-\overset{OH}{\overset{\|}{C}}H-\underset{X}{\underset{\|}{C}}-$
$\xrightarrow{b)\ Zn/HOAc}$?	b) $-\overset{O}{\overset{\|\|}{C}}-\underset{H}{\underset{\|}{C}}-$

(continued next page)

Write the structure of the product in the following:	
c) $CH_3CH(Cl)C(=O)CH_3 \xrightarrow{AlH_3}$?	c) $CH_3CH(Cl)CH(OH)CH_3$
d) 2-chlorocyclohexanone $\xrightarrow{Zn/HOAc}$?	d) cyclohexanone
5. $-C(=O)-C(X)-C(H)-$ $\xrightarrow{Nu:^{\ominus}}$?	5. $-C(=O)-C(Nu)-C(H)-$ and $-C(=O)-C=C-$
Write the structure of the major product in the following:	
a) $CH_3CH_2C(=O)CH(Cl)CH_3 \xrightarrow[NH_{3(\ell)}]{Na^{\oplus}\ NH_2^{\ominus}}$?	a) $CH_3CH_2C(=O)CH=CH_2$
b) $CH_3C(=O)CH_2Br \xrightarrow{Na^{\oplus}\ CH_3C(=O)-O^{\ominus}}$	b) $CH_3C(=O)CH_2OC(=O)CH_3$
6. $-C(=O)-C(H)-$ $\xrightarrow{B:^{\ominus}}$ $\xrightarrow{RX}$?	6. $-C(=O)-C(R)-$
Write the structure of the mono-alkylation product in the following:	

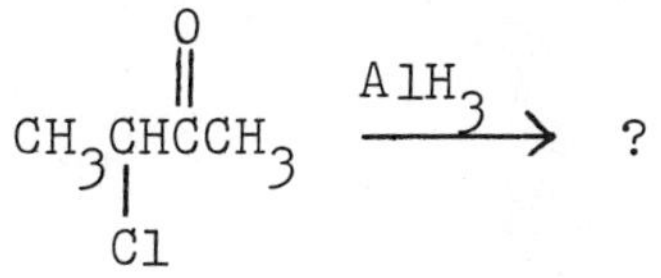

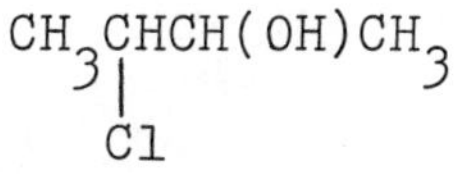

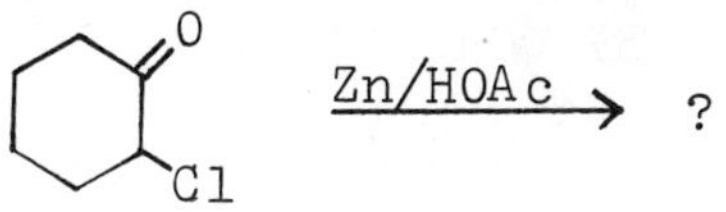

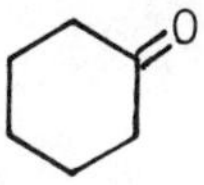

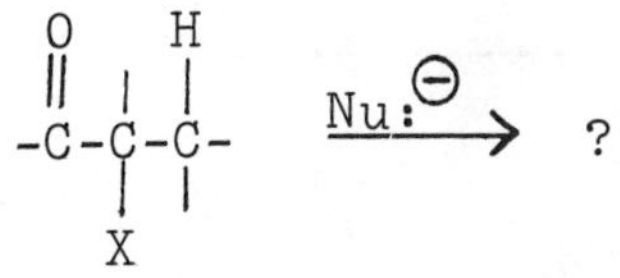

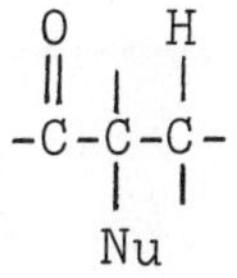

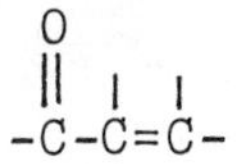

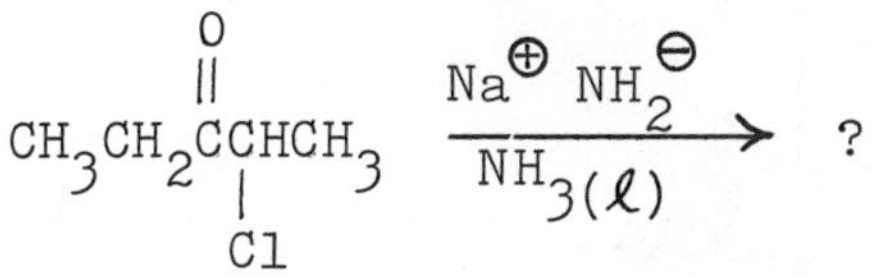

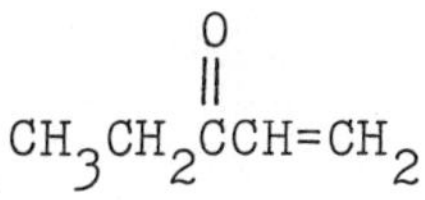

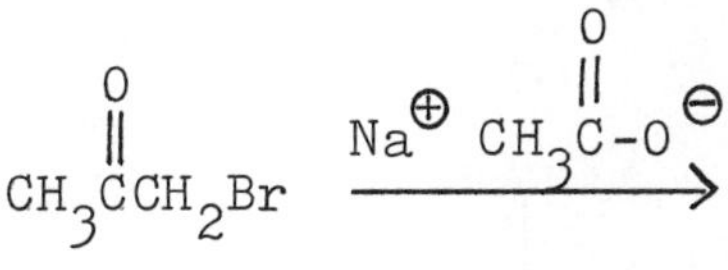

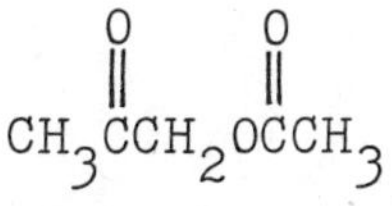

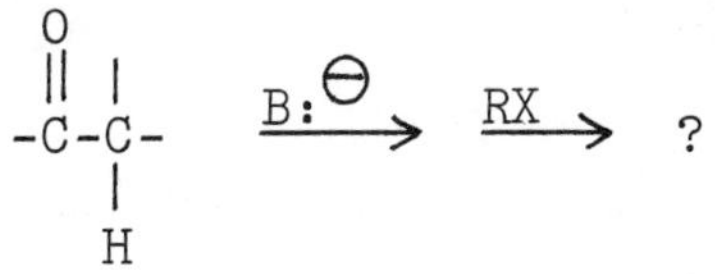

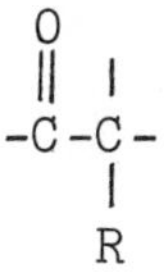

(continued next page)

a) cyclopentanone $\xrightarrow{Na^{\oplus}\ NH_2^{\ominus}}$ $\xrightarrow{CH_3I}$?	a) 2-methylcyclopentanone (cyclopentanone with CH_3 on the α-carbon)
b) $CH_3\overset{O}{\overset{\Vert}{C}}CH_2\overset{O}{\overset{\Vert}{C}}CH_3$ $\xrightarrow{Na^{\oplus}\ OR^{\ominus}}$ $\xrightarrow{CH_3CH_2Cl}$?	b) $CH_3\overset{O}{\overset{\Vert}{C}}\underset{CH_2CH_3}{\underset{\vert}{C}H}\overset{O}{\overset{\Vert}{C}}CH_3$
7. 2 $-\overset{O}{\overset{\Vert}{C}}-\overset{H}{\overset{\vert}{\underset{H}{\underset{\vert}{C}}}}-$ $\xrightarrow[H_2O]{OH^{\ominus}}$?	7. $-\overset{H}{\overset{\vert}{\underset{H}{\underset{\vert}{C}}}}-\overset{OH}{\overset{\vert}{\underset{\vert}{C}}}-\underset{\vert}{C}H-\overset{O}{\overset{\Vert}{C}}-$
	(aldol condensation)
Write the structure of the major product when $-\overset{O}{\overset{\Vert}{C}}-CH_2-$ is:	
a) $2CH_3CH_2CH_2CHO$	a) $CH_3CH_2CH_2\overset{OH}{\overset{\vert}{C}H}\underset{CH_2CH_3}{\underset{\vert}{C}H}CHO$
b) benzaldehyde ($C_6H_5\overset{O}{\overset{\Vert}{C}}H$) + $CH_3CH_2\overset{O}{\overset{\Vert}{C}}H$	b) $\left[C_6H_5-\overset{OH}{\overset{\vert}{C}H}\underset{CH_3}{\underset{\vert}{C}H}CHO\right]$ $\xrightarrow{-H_2O}$ $C_6H_5-CH=\underset{CH_3}{\underset{\vert}{C}}CHO$

8. What is the product of: $CH_3CH_2CH_2CH(OH)CH(CH_2CH_3)CHO \xrightarrow[\text{heat}]{OH^{\ominus}/H_2O}$?	8. $CH_3CH_2CH_2CH{=}C(CH_2CH_3)CHO$
9. $2\ {-}\overset{O}{\overset{\Vert}{C}}{-}\overset{H}{\underset{H}{C}}{-} \xrightarrow{H_3O^{\oplus}}$? Write the structure of the major product when ${-}\overset{O}{\overset{\Vert}{C}}{-}CH_2{-}$ is: a) $CH_3CH_2CH_2CHO$ b) cyclohexanone	9. ${-}\overset{H}{\underset{H}{C}}{-}\overset{\vert}{C}{=}\overset{\vert}{C}{-}\overset{O}{\overset{\Vert}{C}}{-}\ +\ H_2O$ a) $CH_3CH_2CH_2CH{=}C(CH_2CH_3)CHO$ b) 2-cyclohexylidenecyclohexanone

CHAPTER 19

CARBOXYLIC ACIDS AND THEIR DERIVATIVES

LEARNING OBJECTIVES

When you have completed this chapter, you should be able to:

1. name or write structures for carboxylic acids and derivatives (problems 1, 2, 37, 38);
2. account for properties of the phenyl anion and phenyl group (problems 3, 4);
3. compare the acidic or basic properties of carboxylic acids and derivatives (problems 5, 39, 43, 48 and related problems 42, 44);
4. draw resonance structures for a protonated carboxylic acid (problem 10);
5. write equations for the synthesis of carboxylic acids (problems 6, 7, 14);
6. write mechanisms for ketene reactions (problems 8, 9, 49);
7. write mechanisms for ester formation and the reverse reaction ester hydrolysis (problems 11, 12, 13, 15, 17, 20, 21, 55 and related problems 18, 19, 22, 59, 64);
8. write equations for the synthesis of amides (problem 23);
9. write mechanisms for the formation of amides and hydrolysis of amides (problems 24, 25, 28 and related problems 26, 27, 30, 46);
10. predict products of amide hydrolysis (problems 29, 31, 32);
11. write mechanisms for the formation of and reactions of acid chlorides and anhydrides (problems 33, 34, 47, 50, 51, 53, 54, 58);
12. explain the steps in the separation of a carboxylic acid from an ester (problem 41);
13. explain the properties of or identify carboxylic acids and derivatives by use of IR and NMR spectroscopy (problems 35, 36, 63, 64, 67);
14. explain how IR and NMR can be used to distinguish compounds (problems 65, 66, 68);
15. write mechanisms for miscellaneous reactions which involve acids and derivatives (problems 52, 60, 61, 62);
16. perform calculations which involve free energy and equilibrium constants (problems 56, 57).

ANSWERS TO QUESTIONS

1. (a) 2-methylbutanoic acid (α-methylbutyric acid)
   (b) 4-methylpentanoic acid (γ-methylvaleric acid)
   (c) same as (a)
   (d) 4-bromo-2-methylbutanoic acid (α-methyl-γ-bromobutyric acid)
   (e) 3-(2-bromoethyl)-hexanoic acid
   (f) p-nitrobenzoic acid
   (g) 3-(4-chlorophenyl)-4-methylpentanoic acid

2. (a) calcium acetate
   (b) sodium formate
   (c) 1,4-benzenedicarbonitrile
   (d) phenyl acetate
   (e) 3-pyridinecarboxamide (nicotinamide)
   (f) benzoic anhydride
   (g) benzyl benzoate
   (h) benzyl phenyl carbonate
   (i) diphenylmethyl α-furoate
   (j) α-aminopropionic acid (2-aminopropanoic acid, alanine)

3. The nonbonding electron pair is in an orbital which is perpendicular to the p-orbitals of the π-network.

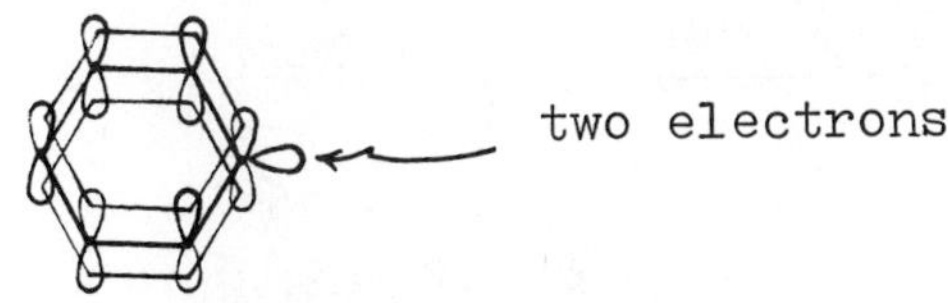

4. The phenyl group is electron withdrawing.

5. (a) $pK_{(o\text{-nitro})} = 2.17$; $pK_{(m\text{-nitro})} = 3.49$

   (b) $Ka_{(p\text{-nitrobenzoic acid})} = 3.98 \times 10^{-4}$

   (c and d) The acidities of m- and p-nitrobenzoic acids are almost identical. This suggests that the conjugate bases of the two acids have about the same stability. Resonance stabilization because of the nitro group is relatively unimportant.

6. (a) $CH_2{=}CH_2 \xrightarrow[H_3O^{\oplus}]{MnO_4^{\ominus}} \xrightarrow{heat} 2HCOOH$ (or $CH_2{=}CH_2 \xrightarrow{O_3} \xrightarrow{H_2O_2/OH^{\ominus}}$)

   (b) $CH_2{=}CH_2 \xrightarrow{H_2O/H^{\oplus}} \xrightarrow[heat]{CrO_3/H^{\oplus}} CH_3COOH$

   (c) $CH_3CH{=}C(CH_3)_2 \xrightarrow{HBr} \xrightarrow[ether]{Mg} \xrightarrow{CO_2} \xrightarrow{H_3O^{\oplus}} CH_3CH_2C(CH_3)_2(COOH)$

(continued next page)

(d) $C_6H_5CH{=}CHC_6H_5 \xrightarrow[\text{heat}]{MnO_4^{\ominus}/OH^{\ominus}} \xrightarrow{H_3O^{\oplus}} 2C_6H_5COOH$

(or $C_6H_5CH{=}CHC_6H_5 \xrightarrow{O_3} \xrightarrow{H_2O_2/H^{\oplus}}$)

(e) cyclohexene $\xrightarrow[\text{heat}]{MnO_4^{\ominus}/OH^{\ominus}} \xrightarrow{H_3O^{\oplus}} HO_2C(CH_2)_4CO_2H$

(or cyclohexene $\xrightarrow{O_3} \xrightarrow{H_2O_2/H^{\oplus}}$)

7. $Br(CH_2)_4Br \xrightarrow[\text{DMSO}]{Na^{\oplus}\ CN^{\ominus}} NC(CH_2)_4CN \xrightarrow[\text{heat}]{H^{\oplus}/H_2O} HO_2C(CH_2)_4CO_2H$

8. (cyclopentylidene)C=O: + $H_2\ddot{O}$ ⟶ [(cyclopentylidene)C($\ddot{O}:^{\ominus}$)–$\ddot{O}^{\oplus}H_2$ ⟷ $^{\ominus}$(cyclopentyl)–C(=O)–$\ddot{O}^{\oplus}H_2$] ⟶ (cyclopentyl, H)–C(=$\ddot{O}$:)–$\ddot{O}$–H

9. A more stable anion in which the negative charge can be delocalized on oxygen is formed by attack of water at the carbon of the C=O (see problem 8).

10. $CH_3-\overset{\oplus}{C}(\ddot{O}H)(\ddot{O}H)$ ⟷ $CH_3-C(=\overset{\oplus}{\ddot{O}}H)(\ddot{O}H)$ ⟷ $CH_3-C(\ddot{O}H)(=\overset{\oplus}{\ddot{O}}H)$

11. $C_6H_5\overset{\overset{:O:}{\|}}{C}-OH \underset{}{\overset{H^{\oplus}}{\rightleftharpoons}} \left[C_6H_5\overset{\overset{\oplus}{:OH}}{\overset{\|}{C}}-\ddot{O}H \longleftrightarrow C_6H_5\overset{\overset{:\ddot{O}-H}{|}}{\underset{\oplus}{C}}-\ddot{O}H \right] \overset{H\ddot{O}CH_2CH_3}{\rightleftharpoons}$

$C_6H_5\overset{\overset{\oplus\,:O-H}{\|}}{C}-\ddot{O}CH_2CH_3 \underset{}{\overset{(-H_2O)}{\rightleftharpoons}} C_6H_5-\overset{\overset{:\ddot{O}-H}{|}}{\underset{\underset{:\ddot{O}CH_2CH_3}{|}}{C}}-\overset{\oplus}{O}H_2 \rightleftharpoons C_6H_5-\overset{\overset{:\ddot{O}-H}{|}}{\underset{\underset{H\,\oplus\,CH_2CH_3}{\overset{|}{O:}}}{C}}-\ddot{O}H$

(curved arrow: $H\ddot{O}CH_2CH_3$ removes H from the protonated ester)

$\rightleftharpoons$

$C_6H_5\overset{\overset{:O:}{\|}}{C}-\ddot{O}CH_2CH_3 + CH_3CH_2\overset{\oplus}{O}H_2$

12. (a) ^{18}O is found in the ester.

(b) It is not found in the acyl oxygen.

The product would be $C_6H_5\overset{\overset{O}{\|}}{C}-{}^{18}OCH_2CH_3$.

(Follow through the mechanism in problem 11 using $CH_3CH_2-{}^{18}OH$).

13. (a) $CH_3-\overset{\overset{:O:}{\|}}{C}-Cl + CH_3\ddot{O}H \longrightarrow CH_3-\overset{\overset{:\ddot{O}:^{\ominus}}{|}}{\underset{\underset{CH_3\overset{\oplus}{O}H}{|}}{C}}-Cl \xrightarrow{-Cl^{\ominus}} CH_3-\overset{\overset{:O:}{\|}}{C}-\overset{\oplus}{O}(H)CH_3 \xrightarrow{:\ddot{Cl}:^{\ominus}}$

$HCl + CH_3-\overset{\overset{:O:}{\|}}{C}-\ddot{O}CH_3$

(b) The acetyl chloride reaction should be run in a hood since it produces hydrogen chloride gas.

(continued next page)

(c) Pyridine, a base, reacts with the hydrogen chloride as it is formed.

$$\text{C}_5\text{H}_5\ddot{\text{N}} + \text{HCl} \longrightarrow \text{C}_5\text{H}_5\text{N}^{\oplus}\text{-H} \quad \text{Cl}^{\ominus}$$

14. Syntheses other than those shown are possible in some questions.

(a) $H_2C{=}O \xrightarrow{D_2/Pt} CH_2DOH \xrightarrow{CH_3C(=O)\text{-}Cl} CH_2DOC(=O)CH_3$

(b) $HCOOH \xrightarrow{LiAlD_4} \xrightarrow{H_2O} HCD_2OH \xrightarrow{CH_3C(=O)Cl} CHD_2OC(=O)CH_3$

(c) CH_2DOH (from a) $\xrightarrow{SOCl_2} CH_2DCl \xrightarrow[\text{ether}]{Mg} \xrightarrow{CO_2} \xrightarrow{H_3O^{\oplus}}$

CH_2DCO_2H

15. Only mechanism A supports retention of configuration.

16. $(CH_3)_3C^{\oplus} + :\ddot{O}H_2 \rightleftharpoons (CH_3)_3C\text{-}\overset{\oplus}{O}H_2 \rightleftharpoons (CH_3)_3C\text{-}\ddot{O}\text{-}H + H_3O^{\oplus}$ (proton removed by $:\ddot{O}H_2$)

17.

$$(CH_3)_3C\ddot{O}H \underset{}{\overset{H^{\oplus}}{\rightleftharpoons}} (CH_3)_3C\text{-}\overset{\oplus}{O}H_2 \underset{}{\overset{-H_2O}{\rightleftharpoons}} (CH_3)_3C^{\oplus} \overset{C_6H_5C(=O)\text{-}\ddot{O}H}{\rightleftharpoons}$$

$$(CH_3)_3C\text{-}\ddot{O}\text{-}C(=O)C_6H_5 \underset{}{\overset{-H^{\oplus}}{\rightleftharpoons}} (CH_3)_3C\text{-}\overset{\oplus}{O}(H)\text{-}C(=O)C_6H_5$$

18. (d,l)-$CH_3CH_2CH_2C(CH_3)CH_2CH_3$ (with OH on the C)

19. Yes. Hydrolysis of this ester produces the stable cation shown below which reacts with $H_2{}^{18}O$.

$$(C_6H_5)_2\overset{\oplus}{C}H \underset{}{\overset{H_2{}^{18}O}{\rightleftharpoons}} (C_6H_5)_2CH{}^{18}\overset{\oplus}{O}H_2 \overset{-H^{\oplus}}{\rightleftharpoons} (C_6H_5)_2CH{}^{18}OH$$

20. (a)

$$CH_3\overset{O}{\overset{\|}{C}}OCH_2CH_3 \overset{H^{\oplus}}{\rightleftharpoons} \left[CH_3\overset{\oplus OH}{\overset{\|}{C}}OCH_2CH_3 \leftrightarrow CH_3\text{-}\overset{OH}{\overset{|}{\underset{\oplus}{C}}}\text{-}OCH_2CH_3\right] \overset{CH_3OH}{\rightleftharpoons}$$

$$H^{\oplus} + CH_3\text{-}\overset{O}{\overset{\|}{C}}\text{-}OCH_3 + CH_3CH_2OH \rightleftharpoons CH_3\text{-}\overset{O\text{-}H}{\underset{OCH_3}{C}}\text{-}\overset{H}{\overset{|}{\underset{\oplus}{O}}}\text{-}CH_2CH_3 \rightleftharpoons CH_3\text{-}\overset{OH}{\underset{\overset{\oplus}{O}(H)CH_3}{C}}\text{-}OCH_2CH_3$$

(b)

$$CH_3\overset{O}{\overset{\|}{C}}\text{-}O\text{-}C(CH_3)_3 \overset{H^{\oplus}}{\rightleftharpoons} CH_3\overset{\oplus OH}{\overset{\|}{C}}\text{-}O\text{-}C(CH_3)_3 \rightleftharpoons$$

$$CH_3\overset{OH}{\underset{\overset{\oplus}{O}(H)CH_3}{C}}\text{-}OH \rightleftharpoons CH_3OH \quad CH_3\overset{\oplus OH}{\overset{\|}{C}}\text{-}OH \overset{H^{\oplus}}{\rightleftharpoons} CH_3\overset{O}{\overset{\|}{C}}OH + (CH_3)_3C^{\oplus}$$

$$\rightleftharpoons \quad CH_3\text{-}\overset{O\text{-}H}{\underset{OCH_3}{C}}\text{-}\overset{\oplus}{O}H_2 \quad \overset{-H_3O^{\oplus}}{\rightleftharpoons} \quad CH_3\overset{O}{\overset{\|}{C}}\text{-}OCH_3$$

$$(CH_3)_3C^{\oplus} \overset{CH_3OH}{\rightleftharpoons} (CH_3)_3C\text{-}\overset{\oplus}{\underset{H}{O}}\text{-}CH_3 \overset{-H^{\oplus}}{\rightleftharpoons} (CH_3)_3COCH_3$$

21. Acid-catalyzed transesterification produces cation A which dissociates into benzoic acid and t-butyl cation. The latter reacts with methanol to form t-butyl methyl ether.

$$\underset{A}{C_6H_5\overset{\oplus\ddot{O}H}{\overset{\|}{C}}-\ddot{O}-C(CH_3)_3} \longrightarrow C_6H_5CO_2H + (CH_3)_3C^{\oplus}$$

$$\downarrow CH_3\ddot{O}H$$

$$(CH_3)_3C\ddot{O}CH_3 \xleftarrow{-H^{\oplus}} (CH_3)_3C-\overset{H}{\underset{\oplus}{\ddot{O}}}-CH_3$$

In alkaline solution, cation A is not formed, and a similar dissociation requires formation of a cation and an anion ($C_6H_5COO^{\ominus}$ is a poorer leaving group than C_6H_5COOH).

$$C_6H_5\overset{:O:}{\overset{\|}{C}}-\ddot{O}-C(CH_3)_3 \not\longrightarrow C_6H_5COO^{\ominus} + (CH_3)_3C^{\oplus}$$

22. Yes. No reaction occurs at the chiral center. The mechanism (below) gives retention of configuration in the alcohol.

$$CH_3\ddot{O}H + B:^{\ominus} \rightleftharpoons CH_3\ddot{O}:^{\ominus} + HB$$

$$CH_3CH_2\overset{:O:}{\overset{\|}{C}}-OCH(CH_3)CH_2CH_3 \ (R) + CH_3\ddot{O}:^{\ominus} \longrightarrow CH_3CH_2-\overset{:\ddot{O}:^{\ominus}}{\underset{:\ddot{O}CH_3}{C}}-\ddot{O}CH(CH_3)CH_2CH_3$$

$$\downarrow$$

$$B:^{\ominus} + H\ddot{O}CH(CH_3)CH_2CH_3 \ (R) \xleftarrow{B-H} {}^{\ominus}:\ddot{O}-CH(CH_3)CH_2CH_3$$

$$+ CH_3CH_2\overset{:O:}{\overset{\|}{C}}\ddot{O}CH_3$$

23. Amides are best formed from an acid chloride and ammonia or an amine.

(a) $CH_3\overset{O}{\overset{\|}{C}}Cl + NH_3$

(b) $CH_3CH_2\overset{O}{\overset{\|}{C}}Cl + CH_3NH_2$

(c) $CH_3CH_2\overset{O}{\overset{\|}{C}}Cl + (CH_3)_2NH$

(d) $CH_3\overset{O}{\overset{\|}{C}}Cl$ + pyrrolidine (CH_2–CH_2–CH_2–CH_2 ring closed through N–H)

(e) $2CH_3\overset{O}{\overset{\|}{C}}Cl + NH_2(CH_2)_4NH_2$

(f) $Cl\text{-}\overset{O}{\overset{\|}{C}}\text{-}C_6H_4\text{-}\overset{O}{\overset{\|}{C}}\text{-}Cl + 2CH_3NH_2$

24. 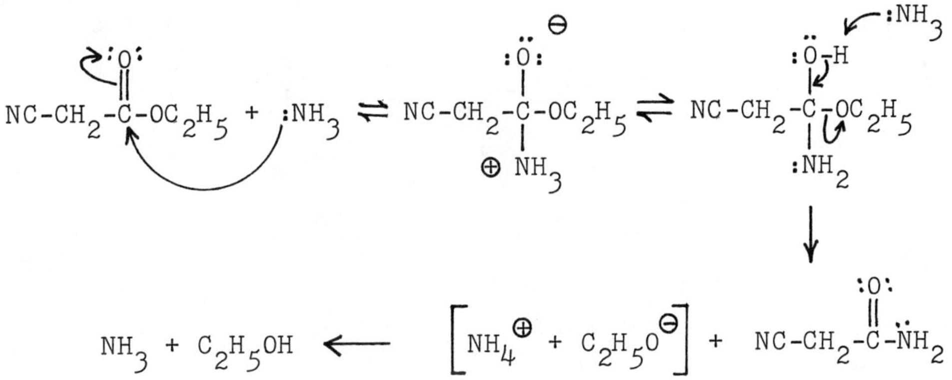 $NC\text{-}CH_2\text{-}\overset{:O:}{\overset{\|}{C}}\text{-}OC_2H_5 + :NH_3 \rightleftharpoons NC\text{-}CH_2\text{-}\overset{:\ddot{O}:^{\ominus}}{\overset{|}{C}}\text{-}OC_2H_5\ (\text{with } \oplus NH_3 \text{ on C}) \rightleftharpoons NC\text{-}CH_2\text{-}\overset{:\ddot{O}\text{-}H}{\overset{|}{C}}\text{-}OC_2H_5\ (\text{with } :NH_2 \text{ on C};\ :NH_3 \text{ attacks H})$

$\downarrow$

$NH_3 + C_2H_5OH \longleftarrow \left[NH_4^{\oplus} + C_2H_5O^{\ominus}\right] + NC\text{-}CH_2\text{-}\overset{:O:}{\overset{\|}{C}}\text{-}\ddot{N}H_2$

25. 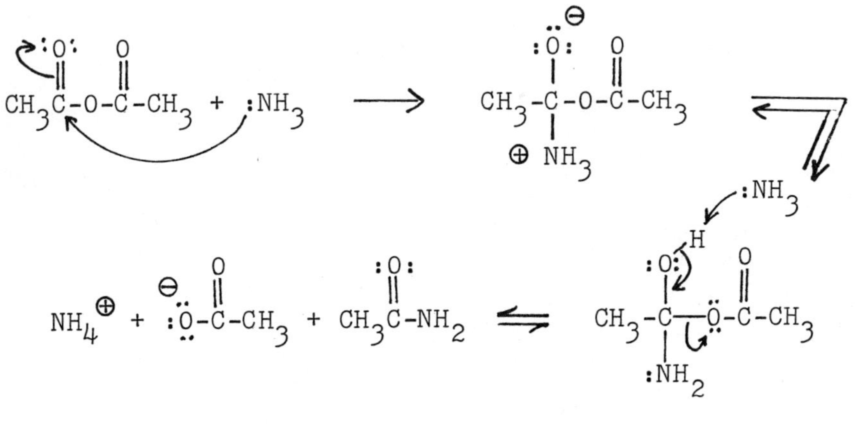 $CH_3\overset{:O:}{\overset{\|}{C}}\text{-}O\text{-}\overset{O}{\overset{\|}{C}}\text{-}CH_3 + :NH_3 \longrightarrow CH_3\text{-}\overset{:\ddot{O}:^{\ominus}}{\overset{|}{C}}\text{-}O\text{-}\overset{O}{\overset{\|}{C}}\text{-}CH_3\ (\text{with } \oplus NH_3 \text{ on C}) \rightleftharpoons$

$CH_3\text{-}\overset{:O\text{-}H}{\overset{|}{C}}\text{-}\ddot{O}\text{-}\overset{O}{\overset{\|}{C}}\text{-}CH_3\ (\text{with } :NH_2 \text{ on C};\ :NH_3 \text{ attacks H}) \rightleftharpoons NH_4^{\oplus} + {}^{\ominus}:\ddot{O}\text{-}\overset{O}{\overset{\|}{C}}\text{-}CH_3 + CH_3\overset{:O:}{\overset{\|}{C}}\text{-}NH_2$

26. (a) [Resonance structures: RC(=O)–O⊕(H)–C(=N–cyclohexyl)–N⊖–cyclohexyl ⟷ RC(=O)–O⊕(H)–C(=N–cyclohexyl)–N⊖–cyclohexyl]

(b) In both systems, the negative charge is stabilized by resonance.

(c) In this anion both contributors have the negative charge on the same element (nitrogen).

27. The anion below is resonance stabilized, an alkoxide ion is not.

$$R\text{-}\ddot{N}{=}C(\text{-}\ddot{O}:^{\ominus})\text{-}N(H)R \longleftrightarrow R\text{-}\ddot{N}^{\ominus}\text{-}C(=\ddot{O})\text{-}N(H)R$$

28. (a)

$$CH_3\overset{\overset{:O:}{\|}}{C}\text{-}NH_2 \underset{}{\overset{H^{\oplus}}{\rightleftharpoons}} \left[CH_3\text{-}\overset{\overset{\oplus}{:O\text{-}H}}{\overset{\|}{C}}\text{-}NH_2 \longleftrightarrow CH_3\text{-}\overset{:\ddot{O}\text{-}H}{\underset{\oplus}{C}}\text{-}NH_2 \right] \overset{H_2\ddot{O}:}{\rightleftharpoons}$$

$$CH_3\text{-}\overset{:O:}{\overset{\|}{C}}\text{-}\ddot{O}H \overset{-NH_4^{\oplus}}{\rightleftharpoons} CH_3\text{-}C(:\ddot{O}\text{-}H)(:\ddot{O}H)\text{-}\overset{\oplus}{N}H_3 \rightleftharpoons CH_3\text{-}C(:\ddot{O}\text{-}H)(\overset{\oplus}{O}H_2)\text{-}NH_2$$

(b)

$$H_2N\text{-}\overset{:O:}{\overset{\|}{C}}\text{-}\ddot{N}H_2 \xrightarrow{H^{\oplus}} H_2N\text{-}\overset{\oplus\ddot{O}H}{\overset{\|}{C}}\text{-}\ddot{N}H_2 \xrightarrow{:\ddot{O}H_2} H_2N\text{-}C(:\ddot{O}H)(\overset{\oplus}{O}H_2)\text{-}\ddot{N}H_2$$

$$\rightleftharpoons H_2N\text{-}C(:\ddot{O}\text{-}H)(:\ddot{O}\text{-}H)\text{-}\overset{\oplus}{N}H_3 \xrightarrow{-NH_4^{\oplus}} H_2N\text{-}\overset{:O:}{\overset{\|}{C}}\text{-}\ddot{O}H \quad \text{repeat}$$

(continued next page)

$$HO\text{-}\overset{\overset{\displaystyle O}{\|}}{C}\text{-}OH \longrightarrow CO_2 + H_2O$$

29. (a) $C_6H_5COO^{\ominus}Na^{\oplus}$ + NH_3

(b) $C_6H_5COO^{\ominus}Na^{\oplus}$ + $CH_3CH_2NH_2$

(c) $CH_3COO^{\ominus}Na^{\oplus}$ + $C_2H_5(CH_3)NH$

(d) cyclohexyl-$COO^{\ominus}Na^{\oplus}$ + cyclohexyl-NH-(2-oxo-3-NH_2-cyclohexyl)

(e) $CH_3COO^{\ominus}Na^{\oplus}$ + $H_2N\overset{C_2H_5}{\underset{H}{C}}\text{-}COO^{\ominus}Na^{\oplus}$

(f) $H_2NCH_2COO^{\ominus}Na^{\oplus}$ + $H_2N\text{-}\overset{CH_3}{CH}\text{-}COO^{\ominus}Na^{\oplus}$

30. An amide has the structure $\text{-}\overset{\overset{O}{\|}}{C}\text{-}\ddot{N}\text{-}$. Formation of trimethylamin would require the structure $\text{-}\overset{\overset{O}{\|}}{C}\text{-}\overset{\oplus}{N}(CH_3)_3$.

31. (a) $H_2NCH_2CH_2CH_2COO^{\ominus}$

(b) $CH_3NHCH_2CH_2CH_2CH_2COO^{\ominus}$

(c) $CH_3\overset{\oplus}{N}H_2CH_2CH_2CH_2CH_2COOH$

32. (a) two

(continued next page)

(b) 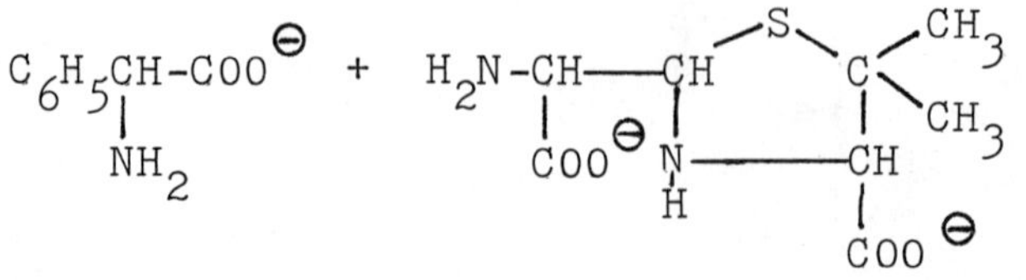 $C_6H_5CH(NH_2)-COO^{\ominus}$ + $H_2N-CH(COO^{\ominus})-CH$ (thiazolidine ring: $-CH-S-C(CH_3)_2-CH(COO^{\ominus})-NH-$)

33. 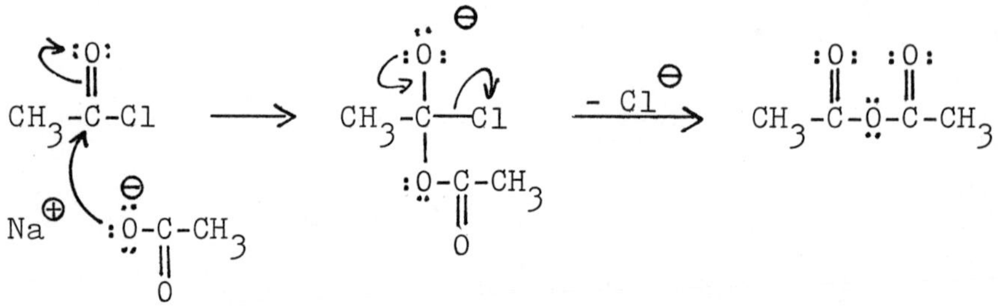 $CH_3-C(=O)-Cl$ + $Na^{\oplus}$ $^{\ominus}:\ddot{O}-C(=O)-CH_3$ ⟶ $CH_3-C(-\ddot{O}:^{\ominus})(-Cl)-O-C(=O)-CH_3$ $\xrightarrow{-\ Cl^{\ominus}}$ $CH_3-C(=O)-\ddot{O}-C(=O)-CH_3$

34. (a) $HOOC(CH_2)_3COOH$

(b) $HOOC(CH_2)_3COOCH_3$

(c) 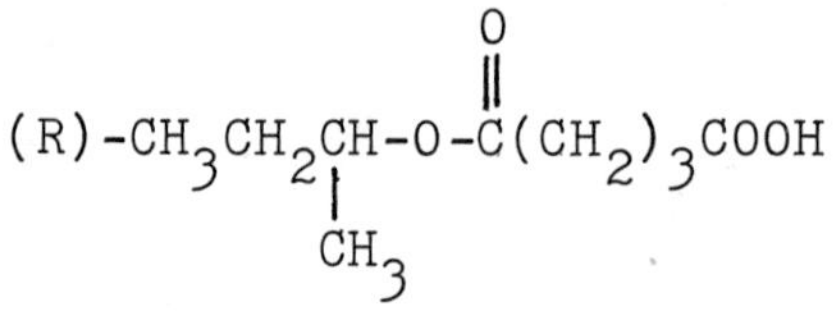 (R)-$CH_3CH_2CH(CH_3)-O-C(=O)(CH_2)_3COOH$

(d) 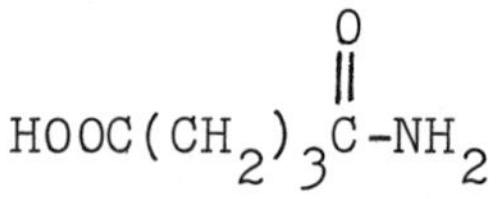 $HOOC(CH_2)_3C(=O)-NH_2$

(e) 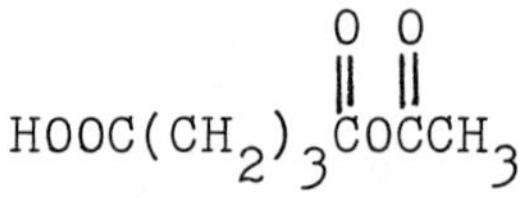 $HOOC(CH_2)_3C(=O)OC(=O)CH_3$

(f) 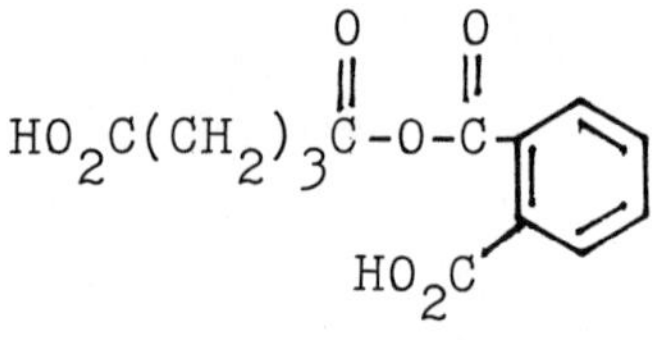 $HO_2C(CH_2)_3C(=O)-O-C(=O)-C_6H_4-CO_2H$ (ortho)

35. (a) Bonding an unsaturated group to the carbonyl carbon decreases the double bond character as shown by resonance form B.

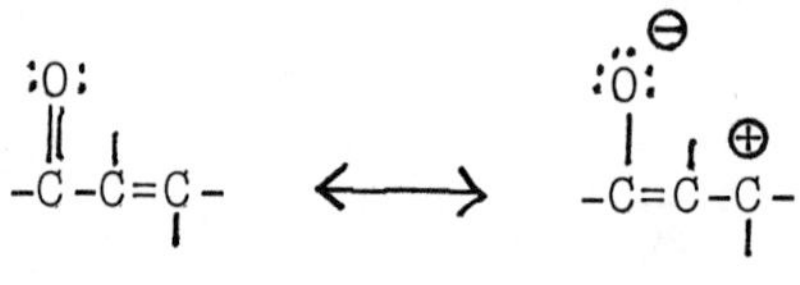

$$-C(=\ddot{O}:)-C=C- \longleftrightarrow -C(-\ddot{O}:^{\ominus})=C-C^{\oplus}-$$

A B

(b) The presence of an unsaturated group attached to the -O- of an ester reduces the resonance interaction of the -O- with the carbonyl and, as a result, increases the double bond character of the carbonyl.

A $-\ddot{O}-C(=\ddot{O}:)- \longleftrightarrow -\overset{\oplus}{O}=C(-\ddot{O}:^{\ominus})-$

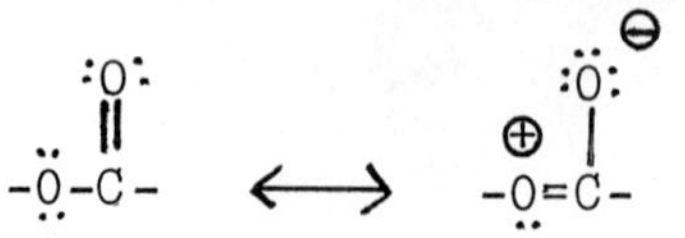

(continued next page)

B $\quad -\overset{|}{C}=\overset{|}{C}-\ddot{\underset{..}{O}}-\overset{\overset{:O:}{||}}{C}- \longleftrightarrow -\underset{\ominus}{\overset{|}{\ddot{C}}}-\overset{|}{C}=\underset{\oplus}{\ddot{O}}-\overset{\overset{:O:}{||}}{C}- \longleftrightarrow -\overset{|}{C}=\overset{|}{C}-\underset{\oplus}{\ddot{O}}=\overset{\overset{:\ddot{O}:^{\ominus}}{|}}{C}-$

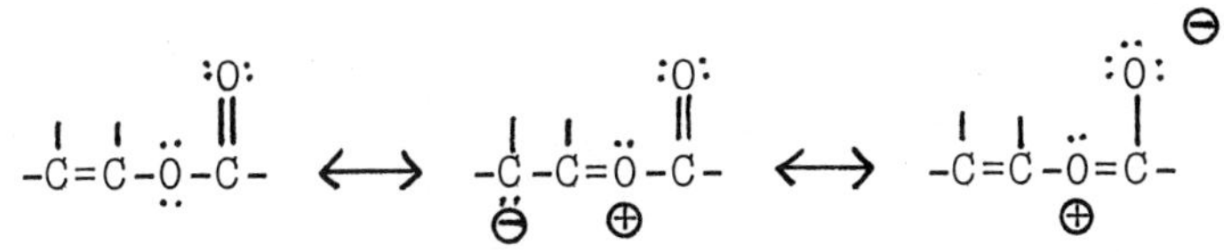

greater double bond character in the carbonyl of B

36. (a) CH_3CO_2H: $-CO_2H$, singlet 10-12 δ
$CH_3C(O)NH_2$: $-NH_2$, singlet 5-6.5 δ

(b) CH_3CHO: $-CHO$, singlet 9-10 δ
CH_3CO_2H: $-CO_2H$, singlet 10-12 δ

(c) $CH_3CO_2CH_3$: two singlets (nonequivalent CH_3 groups)
$CH_3C(O)CH_3$: one singlet (equivalent CH_3 groups)

(d) $CH_3CH_2CO_2CH_3$: CH_3CH_2-; triplet, quartet. CH_3O-; singlet
$CH_3CH_2CH_2CO_2H$: $CH_3CH_2CH_2-$; triplet multiplet, triplet
$-CO_2H$; singlet 10-12 δ

(e) $CH_3C(O)OC(O)CH_3$: one singlet (equivalent CH_3 groups)
CH_3CO_2H: $-CH_3$, singlet; $-CO_2H$, singlet 10-12 δ

37. (a) C-C-C-C-C-CO_2H

hexanoic acid

(b) 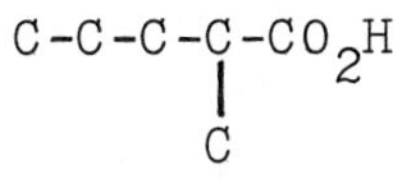 C-C-C-C(-C)-CO_2H

2-methylpentanoic acid

(c) 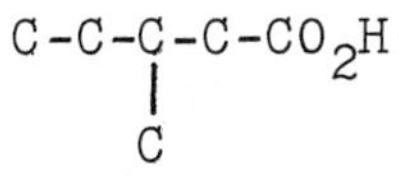 C-C-C(-C)-C-CO_2H

3-methylpentanoic acid

(d) 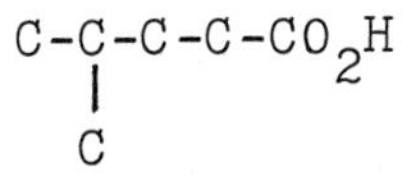 C-C(-C)-C-C-CO_2H

4-methylpentanoic acid

(continued next page)

(e)
```
        C
        |
  C-C-C-CO2H
        |
        C
```
2,2-dimethylbutanoic acid

(f)
```
      C
      |
  C-C-C-CO2H
      |
      C
```
3,3-dimethylbutanoic acid

(g)
```
     C C
     | |
   C-C-C-CO2H
```
2,3-dimethylbutanoic acid

(h)
```
  C-C-C-CO2H
      |
      C-C
```
2-ethylbutanoic acid

38. (a) trifluoroacetic acid
(b) trifluoroacetic anhydride
(c) trifluoroacetamide
(d) trifluoroacetonitrile
(e) α-furoyl chloride
(f) phenyl acetate
(g) methyl benzoate
(h) benzyl benzoate
(i) 4-chlorobenzamide
(j) N-methyl-4-chlorobenzamide
(k) N-ethyl-N-methyl-4-chlorobenzamide
(l) dimethylmalonic acid
(m) diethyl malonate (malonic ester)
(n) 3-chloro-3-phenylpropanoic acid
(o) monoperoxyphthalic acid
(p) trans-cyclohexene carbonate
(q) cis-cyclohexene carbonate
(r) o-hydroxybenzoic acid (salicylic acid)
(s) methyl o-hydroxybenzoate (methyl salicylate)
(t) trans-2-chlorocyclohexanecarboxylic acid
(u) cyclohexyl cyanide (cyclohexane carbonitrile)
(v) peroxyformic acid

(continued next page)

(w) (E)-3-chloro-2-methyl-2-pentenoic acid

(x) diazoacetone

(y) ketene

(z) cyclohexane-1,1-dicarboxylic acid

(Text S. 19.2)

39. pentane < acetylene < 1-propanol < propanoic acid < 3-chloropropanoic acid < 2-chloropropanoic acid < hydrochloric acid

(Text S. 19.3)

40. deuterium < chloro < methoxy < hydroxy < carboxy

41. (a) (i) extraction with bicarbonate converts the acid to a water soluble salt. (ii) acidification regenerates the acid (now freed of ester). (iii) washing removes any excess hydrochloric acid. No unnecessary steps.

(b) (i) sodium chloride would not convert the acid to its sodium salt (chloride ion is a weaker base than bicarbonate ion). (ii) carboxylic acids are strong enough to produce salts with bicarbonate ion. Hydroxide ion is unnecessary.

42. (a) H_2 (b) $HSO_4^{\ominus}$ (c) CF_3COOD (d) $HSO_4^{\ominus}$

(e) $CH_3\overset{\oplus}{O}H_2$ (f) $Cl_2CHCO_2^{\ominus}$ (g) $CH_3CO_2H_2^{\oplus}$ (h) HCl

(i) NH_3 (j) $H_3O^{\oplus}$ (k) $OH^{\ominus}$

43. The more potent (stronger) base has the weaker conjugate acid. The conjugate acid of benzamidine is weaker (more stable) because of greater resonance stabilization. This can be predicted by noting that the two resonance forms of benzamidine are of equal energy whereas those of benzamide are not.

$C_6H_5-C(=\overset{\oplus}{N}H_2)-\ddot{N}H_2 \longleftrightarrow C_6H_5-C(-:NH_2)=\overset{\oplus}{N}H_2$ and

equal energy

$C_6H_5-C(=\overset{\oplus}{:OH})-\ddot{N}H_2 \longleftrightarrow C_6H_5-C(-:\ddot{O}H)=\overset{\oplus}{N}H_2$

unequal energy

$$K_{b(\text{benzamidine})} = 10^{-2.4} = 10^{0.6} \times 10^{-3} = 4 \times 10^{-3}$$

$$\frac{K_{b(\text{benzamidine})}}{K_{b(\text{benzamide})}} = \frac{4 \times 10^{-3}}{K_{a(\text{benzamide})}} = \frac{10^{14}}{1}$$

$$K_{b(\text{benzamide})} = 4 \times 10^{-17}; \quad pK_{b(\text{benzamide})} = 16.4$$

44. The actual preservative is benzoic acid

$$C_6H_5COO^{\ominus} + H^{\oplus} \rightleftarrows C_6H_5COOH$$

45. Syntheses other than those shown below may be possible in some questions.

(a) $CH_3COOH \xrightarrow[\text{ether}]{SOCl_2} CH_3COCl$

(continued next page)

(b) $CH_3COOH \xrightarrow{NaOH} CH_3COO^{\ominus}Na^{\oplus} \xrightarrow[\text{(from a)}]{CH_3COCl} CH_3\overset{O}{\overset{\|}{C}}\text{-O-}\overset{O}{\overset{\|}{C}}CH_3$

(c) $C_6H_6 \xrightarrow[AlCl_3]{CH_3COCl} C_6H_5\overset{O}{\overset{\|}{C}}CH_3$

(d) $C_6H_5COOH \xrightarrow{LiAlH_4} \xrightarrow{H_2O} C_6H_5CH_2OH \xrightarrow[(CH_3\overset{\|}{\underset{O}{C}})_2O]{CrO_3} \xrightarrow{H_2O}$

C_6H_5CHO

(e) $C_6H_5COOH \xrightarrow{SOCl_2} C_6H_5\overset{O}{\overset{\|}{C}}\text{-Cl}$

(f) $C_6H_5\overset{O}{\overset{\|}{C}}Cl$ (from e) $\xrightarrow[\text{excess}]{NH_3} C_6H_5\overset{O}{\overset{\|}{C}}NH_2$

(g) $CH_3COOH \xrightarrow{SOCl_2} \xrightarrow{CH_2N_2} \xrightarrow{Ag^{\oplus}} CH_3CH_2COOH$ (Arndt-Eistert)

(h) $CH_3CH_2COOH \xrightarrow{SOCl_2} \xrightarrow{NH_3} CH_3CH_2\overset{O}{\overset{\|}{C}}NH_2 \xrightarrow[200^{\circ}]{P_2O_5} CH_3CH_2CN$

(i) $C_6H_5CH_2OH$ (from d) $\xrightarrow{SOCl_2} \xrightarrow[DMSO]{NaCN} C_6H_5CH_2CN \xrightarrow[\text{heat}]{H_3O^{\oplus}}$

$C_6H_5CH_2COOH$

(j) $CH_3COOH \xrightarrow[\text{ether}]{LiAlH_4} \xrightarrow{H_2O} CH_3CH_2OH$

(continued next page)

(k) $C_6H_5CH_2OH$ (from d) $\xrightarrow{ClCCl\ (C{=}O)}$ $(C_6H_5CH_2O)_2C{=}O$

(l) $C_6H_5CH_2OH$ (from d) $\xrightarrow[\text{(from a)}]{CH_3CCl\ (C{=}O)}$ $C_6H_5CH_2OCCH_3$ (C=O)

(m) CH_3COOH $\xrightarrow[\text{(from j)}]{CH_3CH_2OH}$ $\xrightarrow[\text{heat}]{H^{\oplus}}$ $CH_3C(=O){-}OC_2H_5 + H_2O$ (remove as formed)

(n) CH_3CH_2COOH (from g) $\xrightarrow[H_2O]{NaHCO_3}$ $CH_3CH_2COO^{\ominus}Na^{\oplus}$

(o) C_6H_5CCl (C=O) (from e) $\xrightarrow[\text{excess}]{NH_3}$ $\xrightarrow[200^{\circ}]{P_2O_5}$ $C_6H_5C{\equiv}N$

(p) $C_6H_5CH_2OH$ (from d) $\xrightarrow{H_2/Pt}$ $C_6H_{11}CH_2OH$

(q) CH_3COOH $\xrightarrow{Ag_2O}$ CH_3COOAg $\xrightarrow[CCl_4]{Br_2}$ CH_3Br (Hunsdiecker)

(r) $C_6H_5CH_2OH$ (from d) $\xrightarrow{PBr_3}$ $C_6H_5CH_2Br$

(s) C_6H_5COOH $\xrightarrow[\text{heat}]{\text{Cu/quinoline}}$ C_6H_6

$C_6H_5CH_2COOH$ (from i) $\xrightarrow[HF]{C_6H_6}$ $C_6H_5CCH_2C_6H_5$ (C=O)

46. Entropy favors ring (thus, imide) formation

47. (a) $HOCH_2CH_2CH_2COOH \longrightarrow$ (five-membered lactone: CH_2–CH_2–CH_2–O–C(=O)) (inert to cold base)

(b) $CH_3CH(OH)CH_2COOH \longrightarrow$ (four-membered lactone: CH_2—C=O, CH(—CH_3)—O) $\xrightarrow{OH^{\ominus}}$ $CH_3CH(OH)CH_2COO^{\ominus}$

water soluble

(c) $2CH_3CH(OH)COOH \longrightarrow$ (six-membered dilactone: O=C–O–$CHCH_3$–C(=O)–O–CH_3HC) (inert to cold base)

Excess strain prohibits formation of a three-membered lactone in c. Ring strain in the lactone produced in b accelerates ring opening (see problem 59).

48. Amides (–C(=O)–NH–) and imides (–C(=O)–N(H)–C(=O)–) form resonance-stabilized

anions (conjugate bases) with the imide anion having lower energy because of distribution of the charge to two oxygens.

$$\left[\; -\overset{:O:}{\overset{\|}{C}}-\ddot{N}^{\ominus}- \;\leftrightarrow\; -\overset{:\ddot{O}:^{\ominus}}{\overset{|}{C}}=\ddot{N}- \;\right] \quad \text{vs} \quad \left[\; -\overset{:O:}{\overset{\|}{C}}-\ddot{N}^{\ominus}-\overset{:O:}{\overset{\|}{C}}- \;\leftrightarrow\; -\overset{:\ddot{O}:^{\ominus}}{\overset{|}{C}}=\ddot{N}-\overset{:O:}{\overset{\|}{C}}- \;\leftrightarrow\; -\overset{:O:}{\overset{\|}{C}}-\ddot{N}=\overset{:\ddot{O}:^{\ominus}}{\overset{|}{C}}- \;\right]$$

49. Treat H_2O, ROH and RCO_2H as ZOH, where Z = H, R, and RC=O

The first three then are:

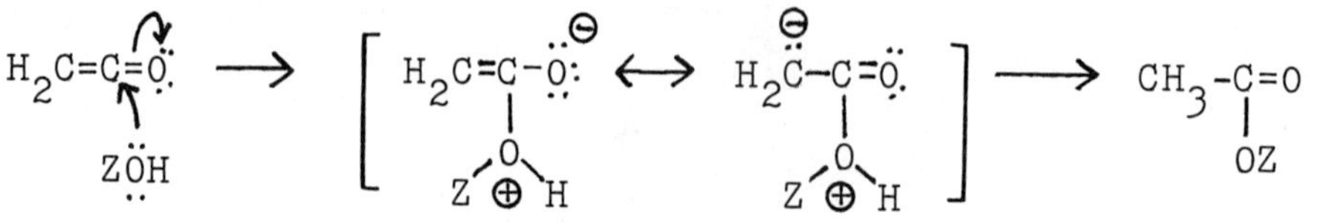

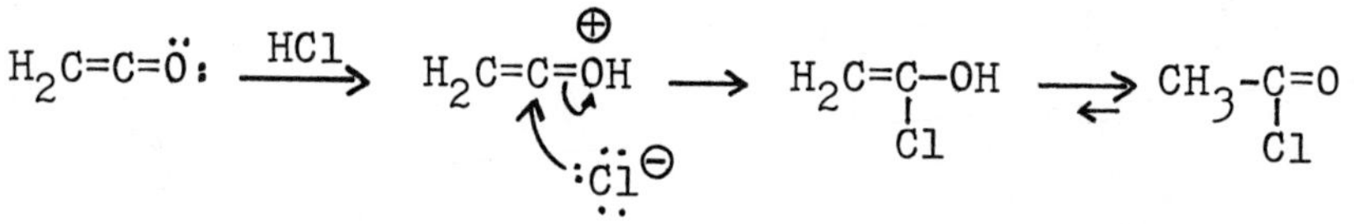

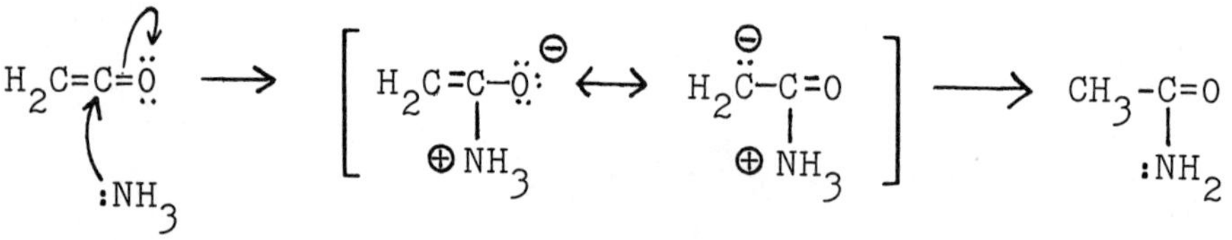

$H_2C{=}C{=}\ddot{O}$: + $Z\ddot{O}H$ → [$H_2C{=}C(-\ddot{O}:^{\ominus})-\overset{\oplus}{O}(Z)H$ ↔ $H_2\overset{\ominus}{\ddot{C}}-C(=\ddot{O})-\overset{\oplus}{O}(Z)H$] → $CH_3-C(=O)-OZ$

$H_2C{=}C{=}\ddot{O}$: $\xrightarrow{HCl}$ $H_2C{=}C{=}\overset{\oplus}{O}H$ + :$\ddot{Cl}^{\ominus}$ → $H_2C{=}C(Cl)-OH$ ⇌ $CH_3-C(=O)-Cl$

$H_2C{=}C{=}\ddot{O}$ + :NH_3 → [$H_2C{=}C(-\ddot{O}:^{\ominus})-\overset{\oplus}{N}H_3$ ↔ $H_2\overset{\ominus}{\ddot{C}}-C(=O)-\overset{\oplus}{N}H_3$] → $CH_3-C(=O)-:NH_2$

50. [Mechanism scheme: cyclic anhydride + $H_2\ddot{N}-CH(CO_2H)-(CH_2)_2-CO_2H$ → tetrahedral intermediate ($-O^{\ominus}$, $\overset{\oplus}{N}H_2-C(H)(CO_2H)(CH_2)_2CO_2H$) → ring-opened carboxylate $-C(=O)-\ddot{O}:^{\ominus}$ / $-C(=O)-\overset{\oplus}{N}H_2-CH(CO_2H)(CH_2)_2CO_2H$ ⇌ $-C(=O)-OH$ / $-C(=O)-\ddot{N}H-C(H)(CO_2H)(CH_2)_2CO_2H$ → cyclic tetrahedral intermediate ($:\ddot{O}:^{\ominus}$, $\ddot{O}H$, $\overset{\oplus}{N}H-C(H)(CO_2H)(CH_2)_2CO_2H$) → imide: ring $-C(=O)-N(-C(H)(CO_2H)(CH_2)_2CO_2H)-C(=O)-$]

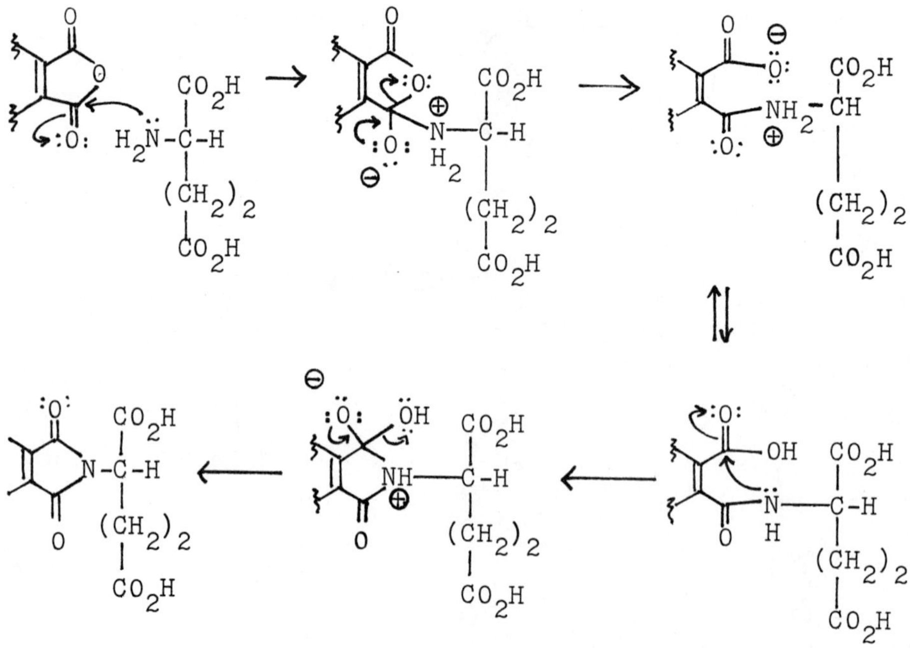

(continued next page)

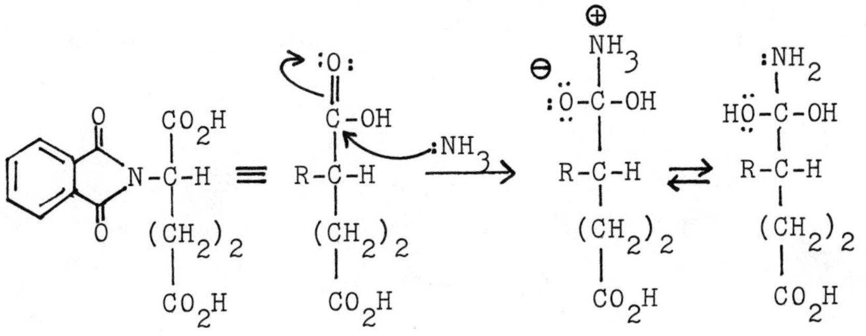

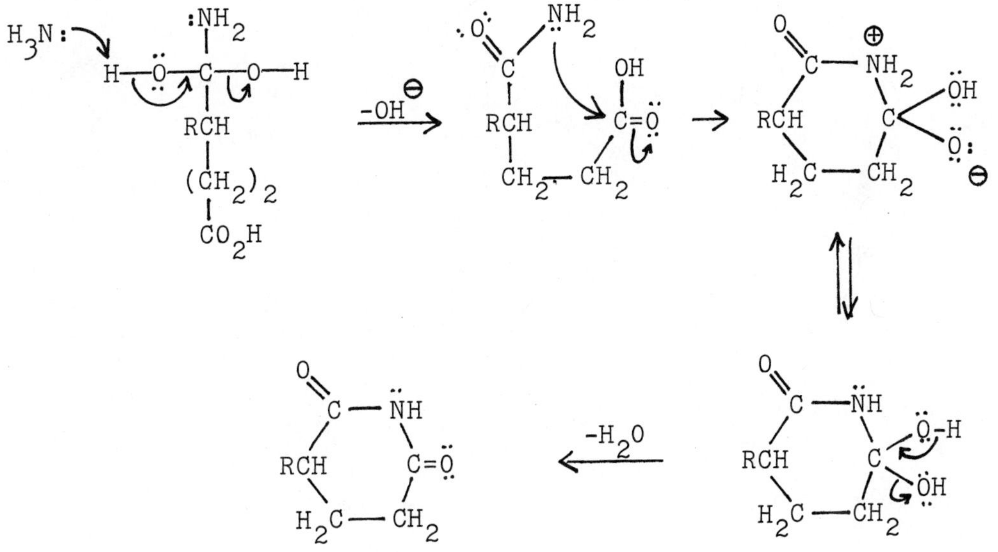

R-C-H, $(CH_2)_2$, CO_2H; N-C-H; C-OH; :NH_3; $NH_3^{\oplus}$; $:\ddot{O}$-C-OH; :NH_2; $H\ddot{O}$-C-OH

H_3N:; H-Ö-C-O-H; RCH; $-OH^{\ominus}$; C=O; CH_2; NH_2; OH; H_2C—CH_2; $\ddot{N}H$; $-H_2O$

51. 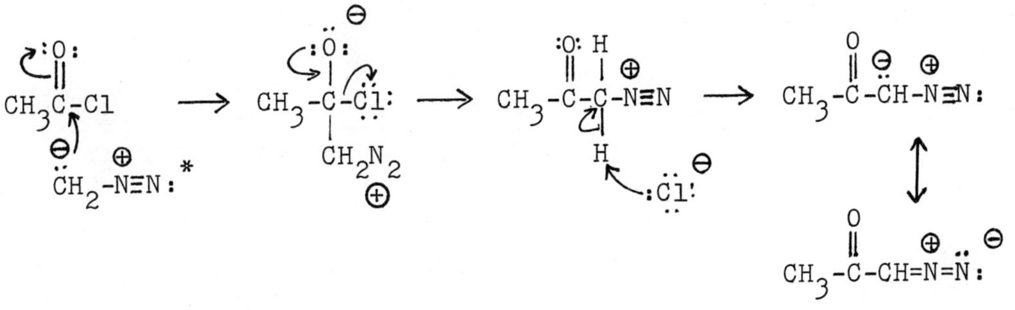 CH_3C-Cl ; $\ddot{C}H_2$-N≡N:* → CH_3-C-Cl: ($CH_2N_2^{\oplus}$) → CH_3-C-C-N≡N (H, :Cl:$^{\ominus}$) → CH_3-C-CH-N≡N: ↔ CH_3-C-CH=N=N:

*one of resonance forms: $\ddot{C}H_2^{\ominus}$-N≡N: ↔ CH_2=N=$\ddot{N}$:$^{\ominus}$

52. (a) The need for a proton on the γ carbon so that isomerization to the β,γ - unsaturated acid may occur. The mechanism is:

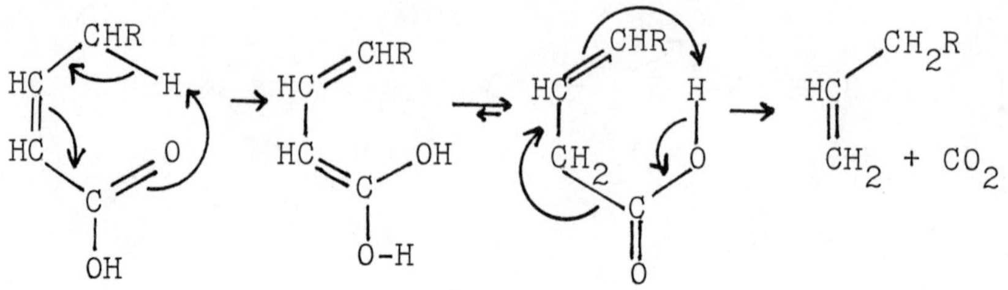

(b) The smaller ring can not form a cyclic activated complex required for decarboxylation, but the larger ring can (a model will be helpful).

53. (a) -O~~~O-(C(=O)-O~~~O)$_n$-C(=O)-O~~~O-C(=O)-

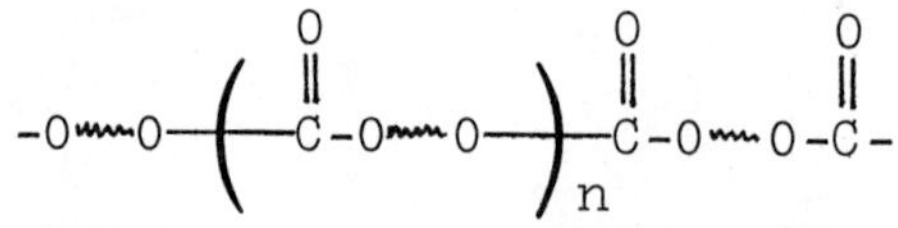

(b)

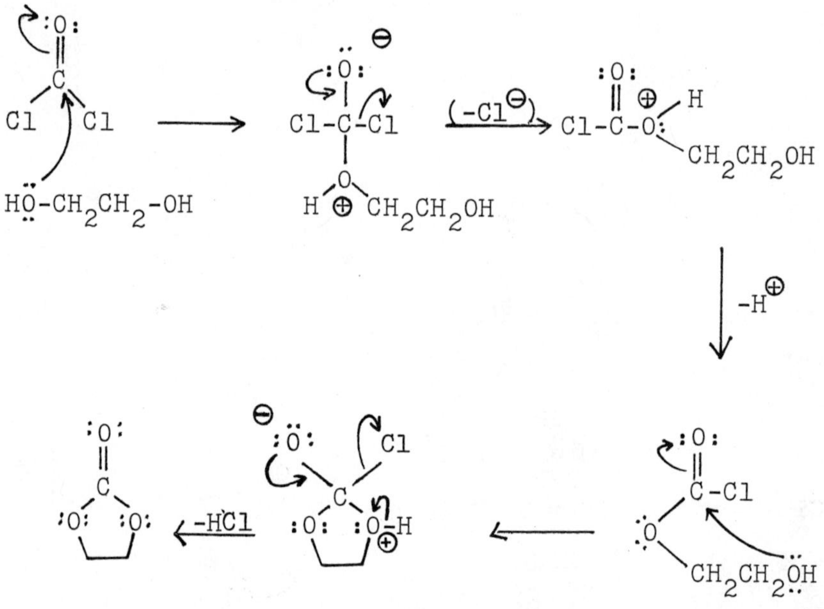

(c) Carrying out the reaction using a large excess of one of the reagents (and in dilute solution) should favor formation of the cyclic carbonate.

54. The "amide" which it forms has a good leaving group ($-\overset{\oplus}{N}(CH_3)_3$) and is easily hydrolyzed.

$$-\overset{O}{\overset{\|}{C}}-Cl + :N(CH_3)_3 \longrightarrow -\overset{O^{\ominus}}{\overset{|}{C}}(Cl)-\overset{\oplus}{N}(CH_3)_3 \xrightarrow{-:Cl^{\ominus}} -\overset{O}{\overset{\|}{C}}-\overset{\oplus}{N}(CH_3)_3$$

$$\downarrow H_2\ddot{O}:$$

$$-\overset{O}{\overset{\|}{C}}-\ddot{O}H \xleftarrow{-(CH_3)_3\overset{\oplus}{N}H} -\overset{O}{\overset{\|}{C}}-\overset{\oplus}{O}H_2 \;\; ((CH_3)_3N:) \longleftarrow -\overset{O^{\ominus}}{\overset{|}{C}}(\overset{\oplus}{O}H_2)-\overset{\oplus}{N}(CH_3)_3$$

55. The stability of the triphenylmethyl cation permits an S_N1-type reaction.

$$C_6H_5\overset{O}{\overset{\|}{C}}-\ddot{O}-C(C_6H_5)_3 \longrightarrow C_6H_5\overset{O}{\overset{\|}{C}}-\ddot{O}:^{\ominus} + \overset{\oplus}{C}(C_6H_5)_3 \xrightarrow{H_2\ddot{O}:}$$

$$C_6H_5COOH + (C_6H_5)_3COH \longleftarrow H_2\overset{\oplus}{\ddot{O}}-C(C_6H_5)_3 \;\; (C_6H_5CO\ddot{O}:^{\ominus})$$

56. $\Delta G = -2.3\ RT \log K_a$

since $-\log K_a = pK_a$

$\Delta G = 2.3RT\ pK_a$; $\Delta\Delta G = 1.4$

$\Delta pK_a = 1.03$

57. $\Delta G = -2.3RT \log K$

$\Delta G = -2.3\ (1.99 \times 10^{-3})\ (298) \log 4$

$\Delta G = -0.82$

58.

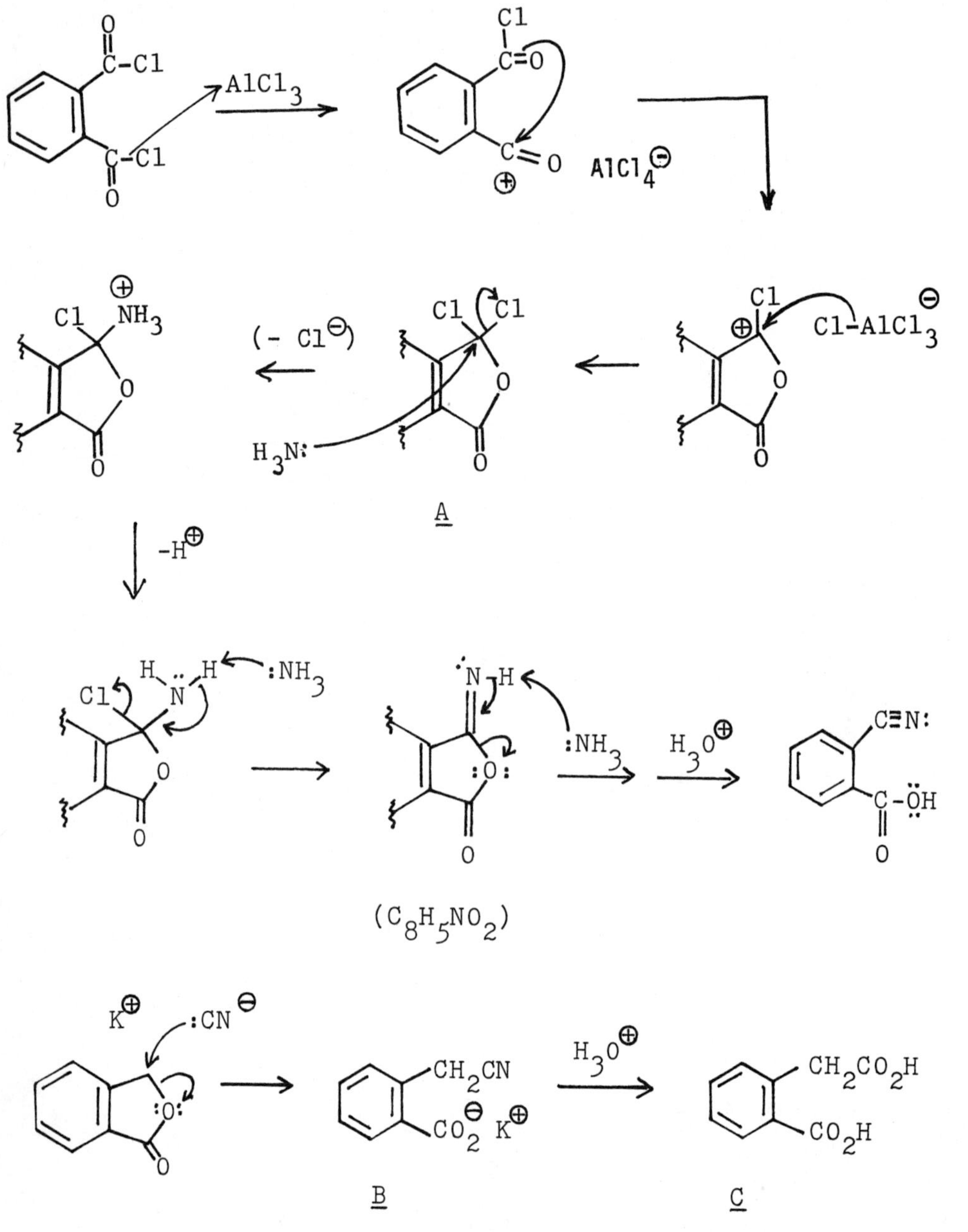
AlCl3
AlCl4⊖
Cl-AlCl3
(- Cl⊖)
H3N:
A
-H⊕
:NH3
H3O⊕
(C8H5NO2)
K⊕
:CN⊖
B
C

(See text p. 779 for mechanism of nitrile hydrolysis)

59. The reaction is an S_N2 on carbon as shown below.

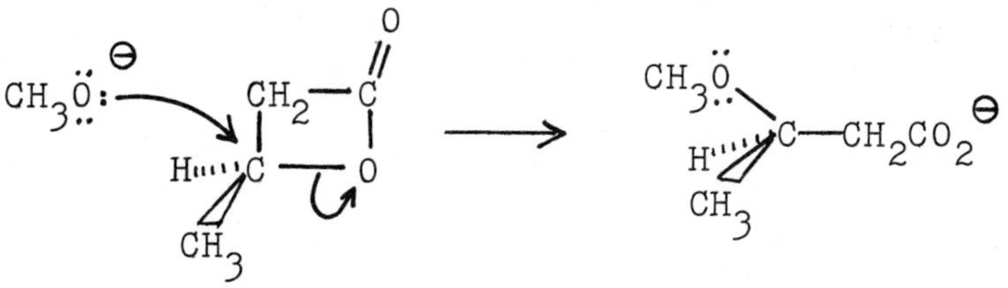

$$CH_3\ddot{O}:^{\ominus} + \text{(β-lactone: } CH_2-C(=O)-O-C(H)(CH_3)) \longrightarrow CH_3\ddot{O}-C(H)(CH_3)-CH_2CO_2^{\ominus}$$

60. $Cl_2 \xrightarrow{light} 2Cl\cdot$

$C_6H_5CH_3 + Cl\cdot \longrightarrow C_6H_5CH_2\cdot + HCl$

$C_6H_5CH_2\cdot + Cl_2 \longrightarrow C_6H_5CH_2Cl + Cl\cdot$

repeat, replacing all three methyl hydrogens $\longrightarrow$ $C_6H_5CCl_3$

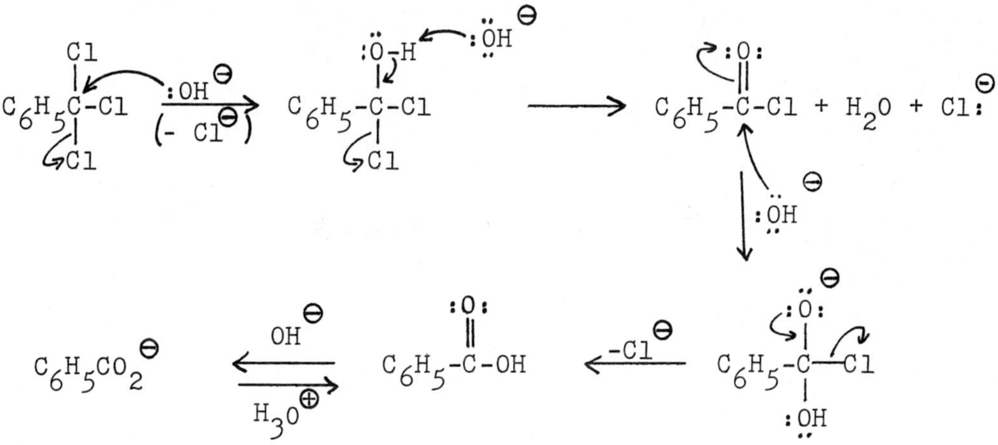

$$C_6H_5CCl_3 \xrightarrow[(-\,Cl^{\ominus})]{:OH^{\ominus}} C_6H_5-C(OH)Cl_2 \xrightarrow{:\ddot{O}H^{\ominus}} C_6H_5-C(=O)-Cl + H_2O + Cl:^{\ominus}$$

$$C_6H_5-C(=O)-Cl \xrightarrow{:\ddot{O}H^{\ominus}} C_6H_5-C(O^{\ominus})(OH)-Cl \xrightarrow{-Cl^{\ominus}} C_6H_5-C(=O)-OH \underset{H_3O^{\oplus}}{\overset{OH^{\ominus}}{\rightleftharpoons}} C_6H_5CO_2^{\ominus}$$

61. <u>step 1</u>: $C_6H_5C(=O)-C(=O)-C_6H_5 \xrightarrow{H^{\oplus}} C_6H_5C(=O)-C(=\overset{\oplus}{O}H)-C_6H_5 \xrightarrow{H_2\ddot{N}-NH_2} C_6H_5C(=O)-C(\ddot{O}H)(\overset{\oplus}{N}H_2-NH_2)-C_6H_5$

$\rightleftharpoons$

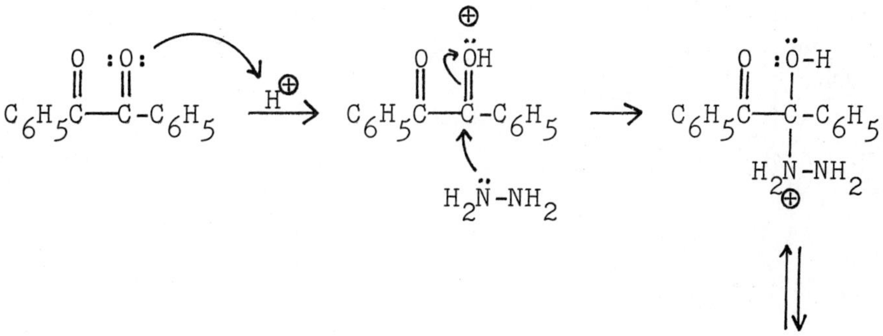

(continued next page)

$$C_6H_5C(=O)\text{-}C(=N\text{-}NH_2)\text{-}C_6H_5 \leftarrow C_6H_5C(=O)\text{—}C(=\overset{\oplus}{N}H\text{-}NH_2)\text{-}C_6H_5 \;(H_2O:) \leftarrow C_6H_5C(=O)\text{—}C(\overset{\oplus}{O}H_2)(:NH\text{-}NH_2)\text{-}C_6H_5$$

step 2: $C_6H_5\text{-}C(=O)\text{—}\overset{\ominus}{C}(N_2^{\oplus})\text{-}C_6H_5 \longrightarrow O{=}C{=}C(C_6H_5)_2 + N_2$

62. A possible mechanism is: (^{18}O is represented by O^*)

$$H\text{-}O^*\text{—}O^*\text{—}H + OH^{\ominus} \rightleftarrows H_2O + {}^{\ominus}:O^*\text{—}O^*\text{—}H$$

$$R\text{-}C{\equiv}N: + {}^{\ominus}:O^*\text{—}O^*\text{-}H \longrightarrow R\text{-}C(=N:^{\ominus})(:O^*\text{—}O^*\text{-}H) \xrightarrow{H_2O} R\text{-}C(=N\text{-}H)(:O^*\text{—}O^*\text{-}H)$$

$$\downarrow (-\ O^*\text{-}H^{\ominus})$$

$$R\text{-}C(=N\text{-}H)(:O^*\text{—}O^*\text{—}O^*\text{-}H) \xleftarrow{{}^{\ominus}:O^*\text{—}O^*\text{-}H} R\text{-}C(=N\text{-}H)(:O^*:^{\oplus})$$

$$\downarrow (-\ O_2),\ :OH^{\ominus}$$

$$R\text{-}C(=N\text{-}H)(:O:^{\ominus}) \xrightarrow{H_2O} R\text{-}C(=N\text{-}H)(:O\text{-}H) \rightleftarrows R\text{-}C(=O)\text{—}NH_2$$

63.

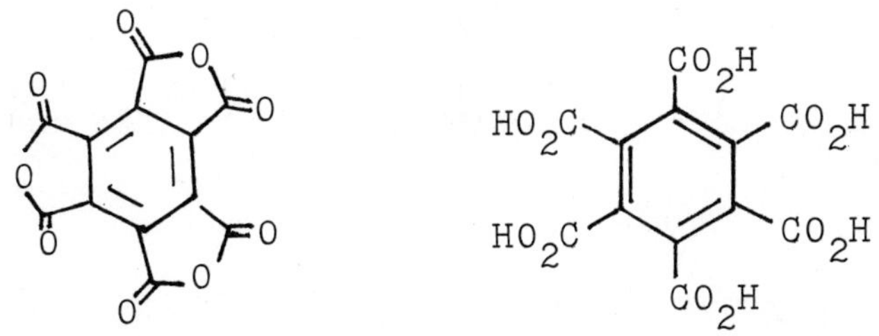

mellitic anhydride   mellitic acid

(Note that mellitic anhydride is an oxide of carbon!)

64. 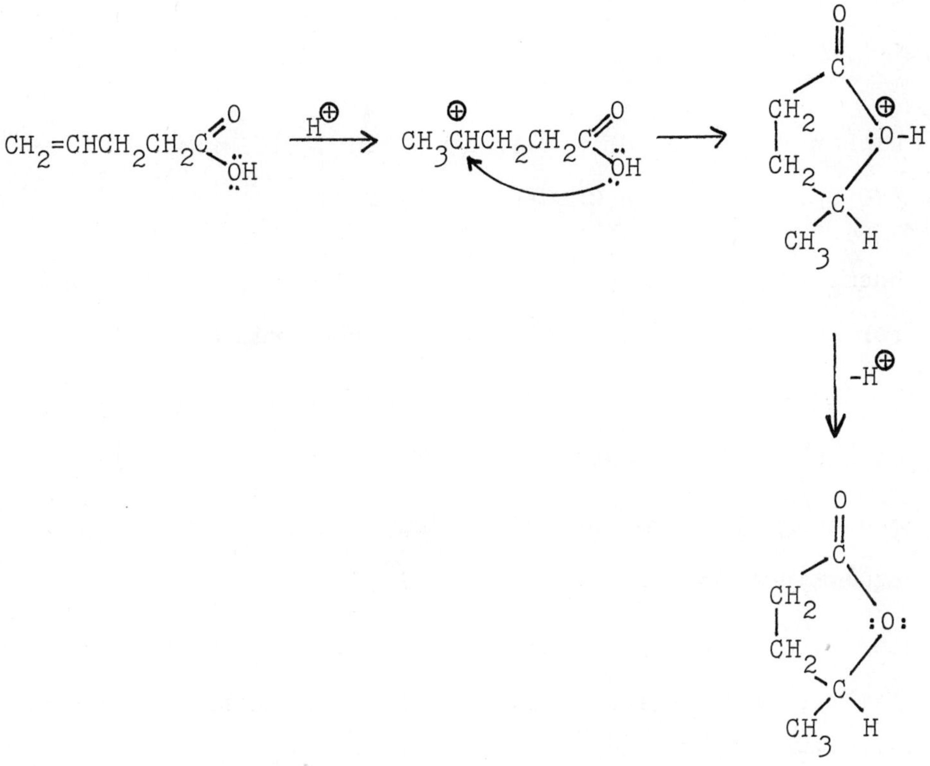 $CH_2{=}CHCH_2CH_2C(=O)\ddot{O}H \xrightarrow{H^{\oplus}} CH_3\overset{\oplus}{C}HCH_2CH_2C(=O)\ddot{O}H \longrightarrow$ (cyclic protonated lactone) $\xrightarrow{-H^{\oplus}}$ (lactone)

65. (a → d) In a thru d, acetic acid is an example which can be distinguished from other compounds by the presence of OH stretch of $-\overset{O}{\overset{\|}{C}}-OH$ (3300-2500 cm^{-1})

(e) C=O stretch (1720-1710 cm^{-1}) in ir of CH_3CO_2H

(continued next page)

(f) OH stretch (3300-2500 cm^{-1}) in ir of CH_3CO_2H

(g) two C-O stretches in ir of $CH_3C(O)OCH_3$ at 1275-1185 and 1160-1050 cm^{-1}

(h) two C=O stretches (symmetric and asymmetric) in the ir of $CH_3C(O)OC(O)CH_3$ near 1820 and 1760 cm^{-1}

(i) N-H stretches (symmetric and asymmetric) in the ir of $CH_3C(O)NH_2$ near 3300 cm^{-1}

(j) N-H stretches in ir of $CH_3C(O)NH_2$

(k) higher frequency of C=O in ir of $CH_3C(O)Cl$ near 1810 cm^{-1} (1670-1630 cm^{-1} in $CH_3C(O)N(CH_3)_2$)

66. (a) chemical shift of OH protons: 10-12 δ for -COOH, 2-4 δ for $-CH_2OH$. Only two singlets in spectrum of CH_3CO_2H

(b) two singlets for CH_3CO_2H ($-CH_3$ and -COOH); one singlet for equivalent CH_3 groups of $CH_3C(O)CH_3$

(c) one singlet in nmr of $CH_3C(O)CH_3$; two singlets for non-equivalent CH_3 groups in $CH_3C(O)OCH_3$

(d) three singlets in nmr of $C_6H_5CH_2C(O)CH_3$; quartet and triplet for CH_3CH_2- and split signal for aromatic protons (electronegative -COO- attached to ring) in $C_6H_5CO_2CH_2CH_3$

(e) absorption 10-12 δ for proton of $-CO_2H$ in $(CH_3)_2CHCO_2H$

(f) absorption 6-8.2 δ for proton of -NH- in $CH_3CONHCH_3$

(g) two singlets of 1:1 ratio in nmr of $CH_3CO_2CH_3$; two singlets of 1:2 ratio in nmr of $CH_3CON(CH_3)_2$

67.

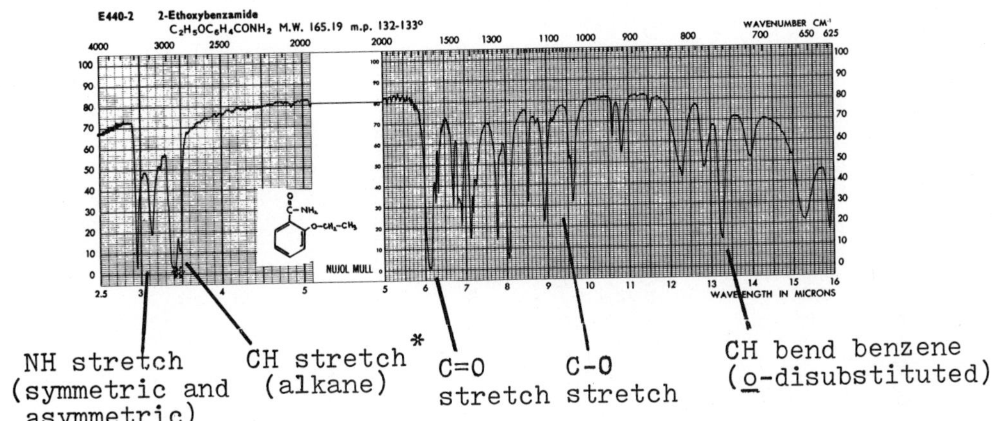

(also: CH stretch, benzene just above 3000 cm^{-1} and C=C stretch benzene near 1600 and 1500 cm^{-1})

* absorption just below 3000 cm^{-1} primarily that of C-H stretch from nujol.

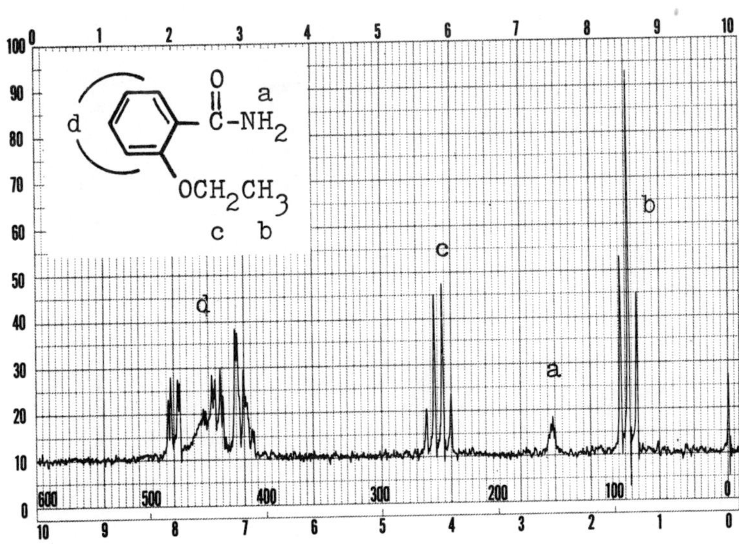

68.

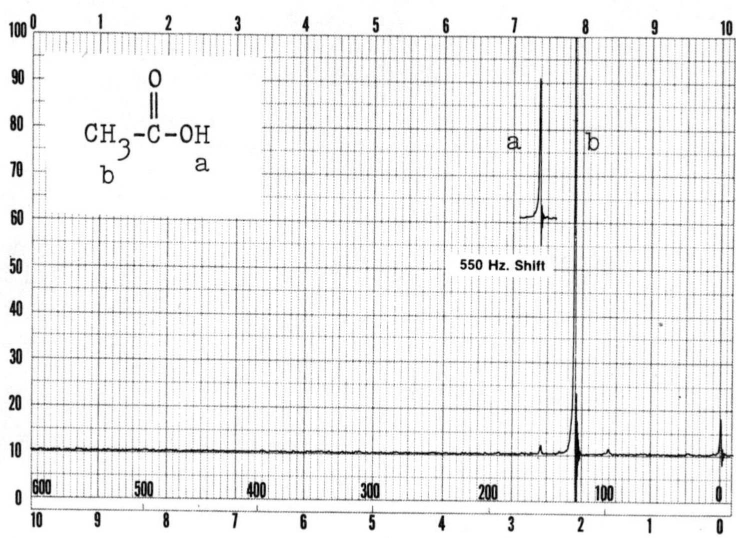

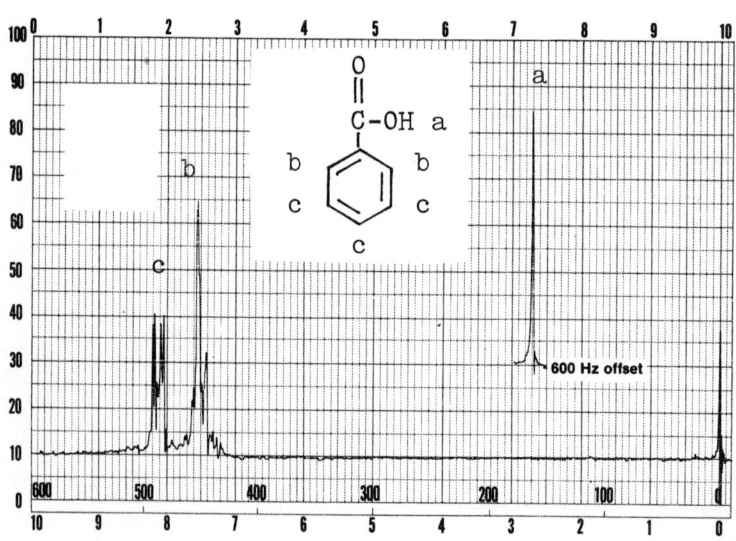

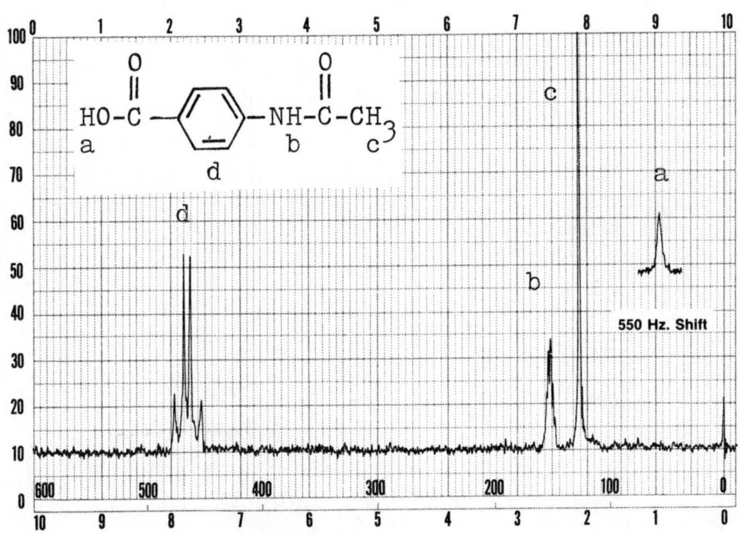
HO-C(=O)-C6H4-NH-C(=O)-CH3
a
d
b
c
550 Hz. Shift

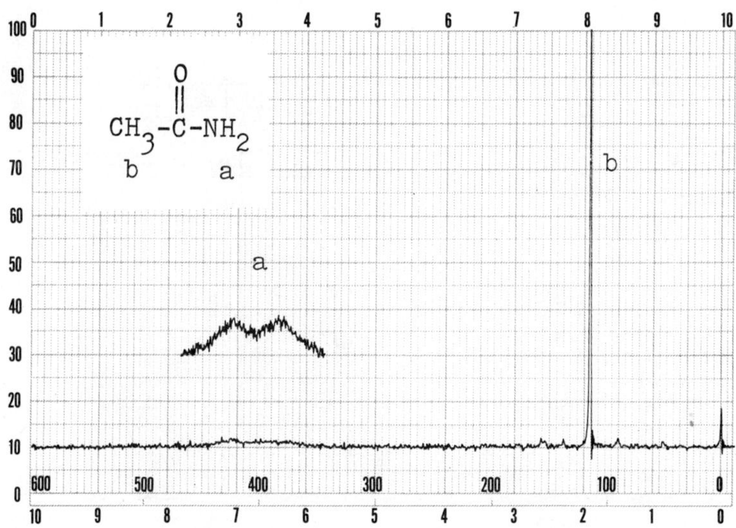
CH3-C(=O)-NH2
b
a

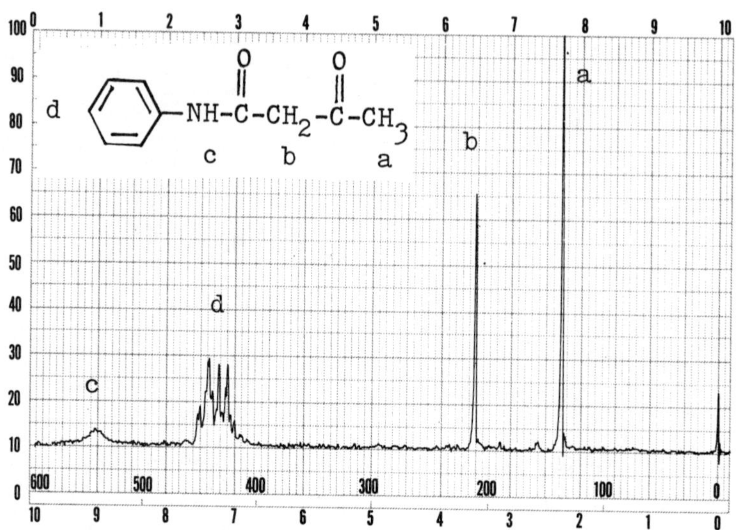
d
NH-C-CH2-C-CH3
c b a
a
b
d
c

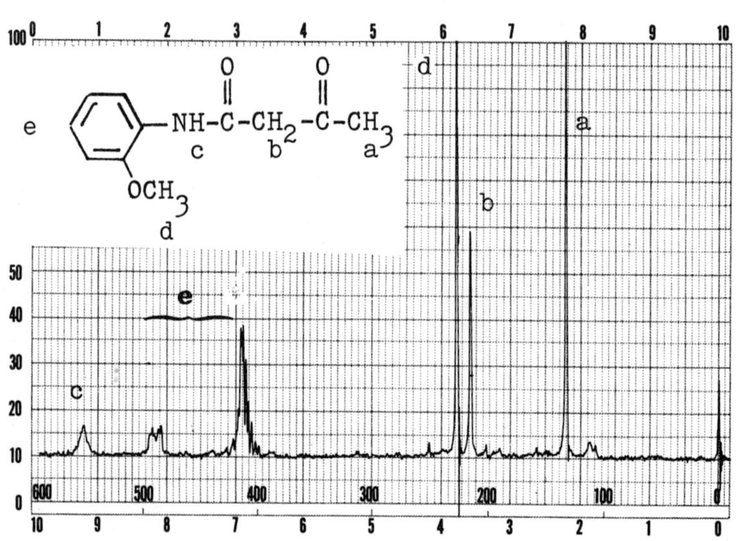
e
NH-C-CH2-C-CH3
c b a
OCH3
d
d
a
b
e
c

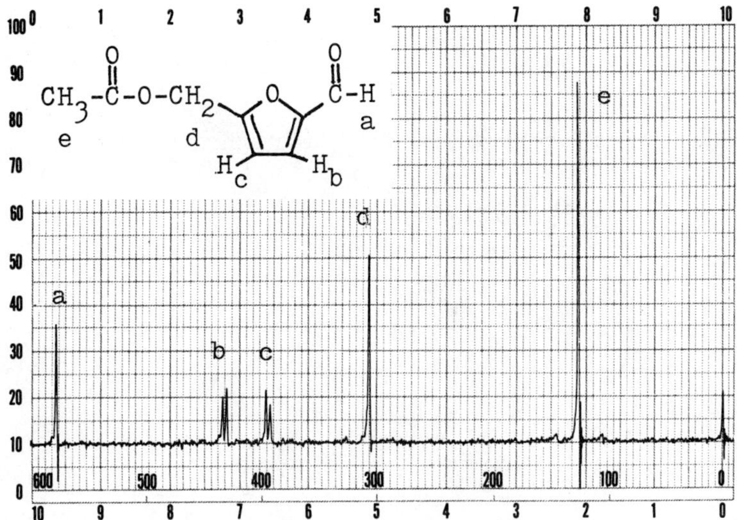
CH3-C-O-CH2
e
d
C-H
a
Hc
Hb
a
b
c
d
e
600
500
400
300
200
100
0

SUMMARY OF REACTIONS

1. Synthesis of Carboxylic Acids

(a) Review of methods previously covered, Figure 19-3, p. 778.

(b) $RMgX \xrightarrow{CO_2} \xrightarrow{H_2O} RC(=O){-}OH$ (Grignard synthesis, p. 778)

(c) $R{-}C{\equiv}N^{*} \xrightarrow{H_3O^{\oplus}} RC(=O){-}OH$ (nitrile hydrolysis, p. 779)

*from: (1) $R{-}X \xrightarrow{CN^{\ominus}} RCN + X^{\ominus}$ (S_N2)

(2) $R{-}C(=O)NH_2 \xrightarrow{P_2O_5} RCN$ (p. 557)

(d) $RC(=O){-}OH \xrightarrow{SOCl_2} RC(=O){-}Cl \xrightarrow{CH_2N_2} R{-}C(=O){-}\overset{\ominus}{\ddot{C}}H\overset{\oplus}{N}_2 \xrightarrow[H_2O]{Ag_2O} RCH_2C(=O)OH$

(Arndt-Eistert, p. 780)

2. Reactions of Carboxylic Acids

$R{-}C(=O){-}OH$ →

- $\xrightarrow{R'OH/H^{\oplus}} RC(=O){-}O{-}R' + H_2O$ (ester, p. 783)
- $\xrightarrow{SOCl_2} \xrightarrow{R'OH} RC(=O){-}O{-}R' + HCl$ (ester, p. 784)
- $\xrightarrow{CH_2N_2} RC(=O)OCH_3$ (methyl ester, p. 785)
- $SOCl_2$, PCl_3, PCl_5, PBr_3 $\longrightarrow R{-}C(=O){-}X$ (acid halide, p. 800)
- $\xrightarrow{SOCl_2} \xrightarrow{2NH_3} R{-}C(=O){-}NH_2 + NH_4Cl$ (amide, p. 792)
- $\xrightarrow{SOCl_2} \xrightarrow{R'COO^{\ominus}} RC(=O)OC(=O)R + Cl^{\ominus}$ (anhydride, p. 800)

3. Reactions of Dicarboxylic Acids

(a) Decarboxylation (p. 801)

$$HO\overset{O}{\overset{\|}{C}}-\overset{O}{\overset{\|}{C}}OH \longrightarrow CO_2 + HCOOH$$

$$HO-\overset{O}{\overset{\|}{C}}-CH_2\overset{O}{\overset{\|}{C}}OH \xrightarrow{heat} CO_2 + CH_3COOH$$

(b) Formation of anhydrides (p. 804)

$$\text{(cyclic) } \begin{matrix} COOH \\ COOH \end{matrix} \xrightarrow{heat} \text{(cyclic anhydride) } \begin{matrix} \overset{O}{\overset{\|}{C}} \\ \quad O \\ \underset{O}{\underset{\|}{C}} \end{matrix} + H_2O$$

γ and δ dicarboxylic acids

4. Decarboxylation (see also 3a)

(a) β-ketoacids

$$-\overset{O}{\overset{\|}{C}}-\overset{|}{\underset{|}{C}}-\overset{O}{\overset{\|}{C}}-OH \xrightarrow{heat} CO_2 + -\overset{O}{\overset{\|}{C}}-\overset{|}{\underset{|}{C}}-H \qquad \text{(p. 801)}$$

(b) β,γ-unsaturated acids

$$-\overset{|}{C}=\overset{|}{C}-\overset{|}{\underset{|}{C}}-\overset{O}{\overset{\|}{C}}-OH \xrightarrow{heat} -\overset{|}{\underset{H}{C}}-\overset{|}{C}=\overset{|}{C}- + CO_2 \text{ (p. 801)}$$

5. Ester Hydrolysis

$$R-\overset{O}{\overset{\|}{C}}-O-R' \xrightarrow{H_3O^{\oplus}/heat} R-\overset{O}{\overset{\|}{C}}-OH + R'OH \qquad \text{(p. 786)}$$

$$R-\overset{O}{\overset{\|}{C}}-O-R' \xrightarrow{OH^{\ominus}/H_2O} R-\overset{O}{\overset{\|}{C}}-O^{\ominus} + R'OH \qquad \text{(p. 789)}$$

6. Transesterification

$$R\text{-}\overset{O}{\overset{\|}{C}}\text{-}O\text{-}R' + R''OH \xrightleftharpoons{H^{\oplus}} R\text{-}\overset{O}{\overset{\|}{C}}\text{-}O\text{-}R'' + R'OH$$ (p. 790)

7. Amide Hydrolysis

$$R\text{-}\overset{O}{\overset{\|}{C}}\text{-}NH_2 \xrightarrow{H_3O^{\oplus}/\text{heat}} R\overset{O}{\overset{\|}{C}}OH + NH_4^{\oplus}$$ (p. 797)

$$R\text{-}\overset{O}{\overset{\|}{C}}\text{-}NH_2 \xrightarrow[\text{heat}]{OH^{\ominus}/H_2O} R\overset{O}{\overset{\|}{C}}\text{-}O^{\ominus} + NH_3$$ (p. 798)

8. Reduction of Carboxylic Acids and Derivatives (p. 813)

$$R\text{-}\overset{O}{\overset{\|}{C}}OH \xrightarrow{LiAlH_4} \xrightarrow{H_2O} RCO_2H$$

$$R\text{-}\overset{O}{\overset{\|}{C}}OH \xrightarrow{BH_3} \xrightarrow{H_3O^{\oplus}} RCO_2H$$

$$R\text{-}\overset{O}{\overset{\|}{C}}\text{-}OR' \xrightarrow{LiAlH_4} \xrightarrow{H_2O} RCH_2OH + R'OH$$

$$R\text{-}\overset{O}{\overset{\|}{C}}\text{-}Cl \xrightarrow{LiAlH_4} \xrightarrow{H_2O} RCH_2OH$$

$$R\text{-}\overset{O}{\overset{\|}{C}}\text{-}NR'_2 \xrightarrow{LiAlH_4} \xrightarrow{H_2O} RCH_2NR_2$$ (an amine)

REACTION REVIEW

A. Reactions from Chapter 19

Questions	Answers
1. $RMgX \xrightarrow{CO_2} \xrightarrow{H_2O}$?	1. $R\overset{O}{\overset{\|}{C}}OH$

(continued next page)

Write the structure of the product when RMgX is:

a) $(CH_3)_2CHCH_2MgX$

b) $CH_3-C_6H_4-MgX$ (para)

a) $(CH_3)_2CHCH_2CO_2H$

b) $CH_3-C_6H_4-CO_2H$ (para)

2. $R-C\equiv N \xrightarrow{H_3O^{\oplus}}$?

Write the structure of the product when $RC\equiv N$ is:

a) $CH_2=CHCH_2CN$

b) $Cl-C_6H_4-CH_2CN$ (para)

2. $RCOH$ (with C=O, i.e. RCO_2H)

a) $CH_2=CHCH_2CO_2H$

b) $Cl-C_6H_4-CH_2CO_2H$ (para)

3. $R-\overset{O}{\overset{\|}{C}}Cl \xrightarrow{CH_2N_2} \xrightarrow[H_2O]{Ag_2O}$?

Write the structure of the product when $R\overset{O}{\overset{\|}{C}}Cl$ is:

a) $CH_3CH_2\overset{O}{\overset{\|}{C}}Cl$

b) (R)-$C_6H_5\underset{CH_3}{\underset{|}{C}}H\overset{O}{\overset{\|}{C}}Cl$

3. $RCH_2\overset{O}{\overset{\|}{C}}OH$

a) $CH_3CH_2CH_2CO_2H$

b) (S)-$C_6H_5\underset{CH_3}{\underset{|}{C}}HCH_2CO_2H$

retention of configuration and a change in priority of groups

4. $RCOH + R'OH \xrightarrow{H^{\oplus}}$? (with C=O on C)	4. $RCOR'$ (with C=O on C)
Write the structure of the ester formed in the following:	
a) $C_6H_5CO_2H + CH_3CH_2OH \xrightarrow{H^{\oplus}}$?	a) $C_6H_5COCH_2CH_3$ (with C=O on C)
b) $CH_3CH_2CO_2H$ + cyclopentanol (OH) $\xrightarrow{H^{\oplus}}$?	b) CH_3CH_2CO–cyclopentyl (with C=O on C)
5. $RCOH \xrightarrow{SOCl_2}$? (with C=O on C)	5. $RCCl$ (with C=O on C)
Write the structure of the product when RCO_2H is:	
a) $CH_3CH_2CO_2H$	a) CH_3CH_2CCl (with C=O on C)
b) CH_3–C_6H_4–CO_2H	b) CH_3–C_6H_4–CCl (with C=O on C)
6. $RCOH \xrightarrow{PX_3 \text{ or } PX_5}$? (with C=O on C)	6. RCX (with C=O on C)
Write the structure of the product in the following:	
a) $CH_3CO_2H \xrightarrow{PCl_3}$?	a) CH_3CCl (with C=O on C)
	(continued next page)

b) $CH_3CO_2H \xrightarrow{PI_3}$?

b) $CH_3\overset{O}{\overset{\|}{C}}I$

c) $CH_3CO_2H \xrightarrow{PCl_5}$?

c) $CH_3\overset{O}{\overset{\|}{C}}Cl$

7. $R\overset{O}{\overset{\|}{C}}Cl \xrightarrow{HNu:}$?

Write the structure of the product in the following:

a) $CH_3\overset{O}{\overset{\|}{C}}Cl \xrightarrow{NH_3}$?

b) $C_6H_5\overset{O}{\overset{\|}{C}}Cl \xrightarrow{C_2H_5OH}$

c) $(CH_3)_2CH\overset{O}{\overset{\|}{C}}Cl \xrightarrow[H_2O]{OH^{\ominus}} \xrightarrow{H_3O^{\oplus}}$?

d) $CH_3CH_2\overset{O}{\overset{\|}{C}}Cl \xrightarrow{CH_3COO^{\ominus}Na^{\oplus}}$?

7. $R\overset{O}{\overset{\|}{C}}Nu + HCl$

a) $CH_3\overset{O}{\overset{\|}{C}}\text{-}NH_2$

b) $C_6H_5\overset{O}{\overset{\|}{C}}OC_2H_5$

c) $(CH_3)_2CH\overset{O}{\overset{\|}{C}}\text{-}OH$

d) $CH_3CH_2\overset{O}{\overset{\|}{C}}O\overset{O}{\overset{\|}{C}}CH_3$

8. $R\overset{O}{\overset{\|}{C}}\text{-}CH_2\text{-}\overset{O}{\overset{\|}{C}}OH \xrightarrow{heat}$?

Write the structure of the product when the acid is:

a) $CH_3\overset{O}{\overset{\|}{C}}\text{-}CH_2\text{-}\overset{O}{\overset{\|}{C}}OH$

8. $R\overset{O}{\overset{\|}{C}}CH_3 + CO_2$

a) $CH_3\overset{O}{\overset{\|}{C}}CH_3 + CO_2$

(continued next page)

b) (2-oxocyclohexanecarboxylic acid: cyclohexanone ring with C=O and $\overset{O}{\overset{\parallel}{C}}OH$)	b) (cyclohexanone) + CO_2
9. $RCH{=}CH{-}\underset{\mid}{\overset{\mid}{C}}{-}\overset{O}{\overset{\parallel}{C}}{-}OH \xrightarrow{heat}$? Write the structure of the product when the acid is: $CH_3CH_2CH{=}CHCH_2\overset{O}{\overset{\parallel}{C}}OH$	9. $RCH_2CH{=}\overset{\mid}{C}{-}$ + CO_2 $CH_3CH_2CH_2CH{=}CH_2$ + CO_2
10. $R\overset{O}{\overset{\parallel}{C}}OR' \xrightarrow[heat]{H_3O^{\oplus}}$? (or $\xrightarrow[heat]{OH^{\ominus}/H_2O} \xrightarrow{H_3O^{\oplus}}$?) Write the structure of the products in the following: a) $CH_3\overset{O}{\overset{\parallel}{C}}OCH_2CH_3 \xrightarrow[heat]{OH^{\ominus}/H_2O}$ $\xrightarrow{H_3O^{\oplus}}$? b) $C_6H_5\overset{O}{\overset{\parallel}{C}}{-}^{18}OCH_3 \xrightarrow[heat]{H_3O^{\oplus}}$	10. $R\overset{O}{\overset{\parallel}{C}}OH + R'OH$ a) $CH_3\overset{O}{\overset{\parallel}{C}}OH + CH_3CH_2OH$ b) $C_6H_5\overset{O}{\overset{\parallel}{C}}OH + CH_3{}^{18}OH$
11. $R\overset{O}{\overset{\parallel}{C}}OR' + R''OH \xrightarrow{H^{\oplus}}$?	11. $R\overset{O}{\overset{\parallel}{C}}OR'' + R'OH$ (continued next page)

Write the structure of the products in the following:

a) $C_6H_5COOCH_3 \xrightarrow[H^{\oplus}]{CH_3CH_2OH}$?

a) $C_6H_5COOCH_2CH_3 + CH_3OH$

b) $CH_3COOCH_2CH_3 \xrightarrow[H^{\oplus}]{\text{cyclopentanol (C}_5H_9OH)}$?

b) $CH_3COO\text{-}C_5H_9$ (cyclopentyl acetate) $+ CH_3CH_2OH$

12. $RCONH_2$ (a) $\xrightarrow[\text{heat}]{H_3O^{\oplus}}$?

(b) $\xrightarrow[\text{heat}]{OH^{\ominus}/H_2O}$?

12. (a) $RCOOH + NH_4^{\oplus}$

(b) $RCOO^{\ominus} + NH_3$

Write the structure of the product in the following:

c) $CH_3CONH_2 \xrightarrow[\text{heat}]{H_3O^{\oplus}}$?

d) $C_6H_5C(=O)\text{-}NHCH_3 \xrightarrow[\text{heat}]{OH^{\ominus}/H_2O}$?

c) $CH_3COOH + NH_4^{\oplus}$

d) $C_6H_5C(=O)\text{-}O^{\ominus} + CH_3NH_2$

13. $RCOOH \xrightarrow{LiAlH_4} \xrightarrow{H_2O}$?

13. RCH_2OH

(continued next page)

Write the structure of the product when RCO_2H is: a) $C_6H_5-CO_2H$ (benzene ring–CO_2H) b) $HO_2CCH_2CH_2CO_2H$	a) $C_6H_5-CH_2OH$ (benzene ring–CH_2OH) b) $HOCH_2CH_2CH_2CH_2OH$
14. $RC(=O)-Nu \xrightarrow{LiAlH_4} \xrightarrow{H_2O}$ (-Nu: -Cl, -OR, $-N\langle$) Write the structure of the product when $RC(=O)Nu$ is: a) $CH_3C(=O)Cl$ b) $CH_3-C_6H_4-C(=O)OCH_3$ (para-substituted benzene ring) c) $CH_3CH_2C(=O)NHCH_3$	14. RCH_2OH when -Nu is -Cl or -OR $RCH_2N\langle$ when -Nu is $-N\langle$ a) CH_3CH_2OH b) $CH_3-C_6H_4-CH_2OH$ (para-substituted benzene ring) c) $CH_3CH_2CH_2NHCH_3$

B. Write a structure for each letter.

Question	Answer
1. $CH_3CH_3 \xrightarrow[h\nu]{Cl_2}$ A	1. A: CH_3CH_2Cl

(continued next page)

$\xrightarrow{Na^{\oplus}\ CN^{\ominus}}$ B $\xrightarrow[heat]{H_3O^{\oplus}}$ C $\xrightarrow[H^{\oplus}]{CH_3CH_2OH}$ D	B: CH_3CH_2CN C: $CH_3CH_2CO_2H$ D: $CH_3CH_2\overset{O}{\overset{\Vert}{C}}OCH_2CH_3$
2. C_6H_5Br (bromobenzene) $\xrightarrow[ether]{Mg}$ A $\xrightarrow{CO_2}$ $\xrightarrow{H_3O^{\oplus}}$ B $\xrightarrow{SOCl_2}$ C $\xrightarrow{CH_3NH_2}$ D	2. A: C_6H_5MgBr B: $C_6H_5CO_2H$ C: $C_6H_5\overset{O}{\overset{\Vert}{C}}Cl$ D: $C_6H_5\overset{O}{\overset{\Vert}{C}}NHCH_3$
3. $CH_3CH_2CH_2OH$ $\xrightarrow[H^{\oplus}]{CrO_3}$ A $\xrightarrow{PCl_3}$ B $\xrightarrow{CH_3OH}$ C	3. A: $CH_3CH_2CO_2H$ B: $CH_3CH_2\overset{O}{\overset{\Vert}{C}}Cl$ C: $CH_3CH_2\overset{O}{\overset{\Vert}{C}}OCH_3$
4. $C_6H_{11}Cl$ (chlorocyclohexane) $\xrightarrow[heat]{KOH/alcohol}$ A $\xrightarrow{O_3}$ $\xrightarrow[H^{\oplus}]{H_2O_2}$ B	4. A: cyclohexene B: $HO\overset{O}{\overset{\Vert}{C}}(CH_2)_4\overset{O}{\overset{\Vert}{C}}OH$ (continued next page)

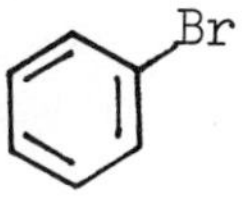

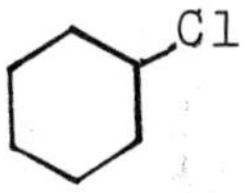

$\xrightarrow{LiAlH_4}$ $\xrightarrow{H_2O}$ C	C: $HOCH_2(CH_2)_4CH_2OH$
5. (toluene, $C_6H_5CH_3$) $\xrightarrow[OH^{\ominus}/heat]{KMnO_4}$ $\xrightarrow{H_3O^{\oplus}}$ A $\xrightarrow[FeBr_3]{Br_2}$ B	5. A: $C_6H_5CO_2H$ (benzoic acid) B: m-$BrC_6H_4CO_2H$ (3-bromobenzoic acid)
6. $CH_2{=}CHCH_3$ $\xrightarrow{NBS}$ A $\xrightarrow{Li(CH_2C_6H_5)_2Cu}$ B $\xrightarrow{BH_3}$ $\xrightarrow[OH^{\ominus}]{H_2O_2}$ C $\xrightarrow[H^{\oplus}]{K_2Cr_2O_7}$ D	6. A: $CH_2{=}CHCH_2Br$ B: $CH_2{=}CHCH_2CH_2C_6H_5$ C: $HOCH_2CH_2CH_2CH_2C_6H_5$ D: $HO_2CCH_2CH_2CH_2C_6H_5$
7. (1-methylcyclopentene) $\xrightarrow{O_3}$ $\xrightarrow[H^{\oplus}]{H_2O_2}$ A $\xrightarrow{SOCl_2}$ B $\xrightarrow{CH_2N_2}$ $\xrightarrow[H_2O]{Ag_2O}$ C	7. A: $CH_3\overset{O}{\overset{\Vert}{C}}(CH_2)_3\overset{O}{\overset{\Vert}{C}}OH$ B: $CH_3\overset{O}{\overset{\Vert}{C}}(CH_2)_3\overset{O}{\overset{\Vert}{C}}Cl$ C: $CH_3\overset{O}{\overset{\Vert}{C}}(CH_2)_3CH_2\overset{O}{\overset{\Vert}{C}}OH$

CHAPTER 20

SYNTHESIS OF CARBON-CARBON BONDS EMPLOYING ESTERS AND OTHER ACID DERIVATIVES

LEARNING OBJECTIVES

When you have completed this chapter, you should be able to:

1. write mechanisms for the halogenation of carboxylic acids (problems 1, 3);
2. write equations which illustrate the synthesis and reactions of α-halo acids (problem 4);
3. predict reactants or products or write mechanisms for condensation reactions:
   a) acetoacetic ester and related compounds (problems 8, 13, 26, 26);
   b) Claisen and crossed-Claisen (problem 9);
   c) Diekmann (problems 10, 11, 32);
   d) Knoevenagel and Perkin (problem 12);
   e) malonic ester (problem 14);
   f) Michael additions (problems 16, 17, 27, 28);
   g) Wittig (problems 18, 39);
   h) miscellaneous (problems 29, 30, 31, 34, 35, 36, 37);
4. write equations for the multistep synthesis of compounds using reactions from this and previous chapters (problems 22, 23, 25);
5. identify or distinguish compounds by chemical or spectral methods (problems 24, 40, 42);
6. identify absorptions in IR and NMR spectra (problem 41).

ANSWERS TO QUESTIONS

1. (a) The mechanism is free-radical chlorination. For example:

$$Cl_2 \xrightarrow{270^o} 2Cl\cdot$$

$$\text{C}_6\text{H}_{11}\text{(H)CO}_2\text{H} + Cl\cdot \longrightarrow \text{C}_6\text{H}_{10}(\cdot)\text{CO}_2\text{H} + HCl$$

$$\text{C}_6\text{H}_{10}(\cdot)\text{CO}_2\text{H} + Cl_2 \longrightarrow \text{C}_6\text{H}_{10}(\text{Cl})\text{CO}_2\text{H} + Cl\cdot$$

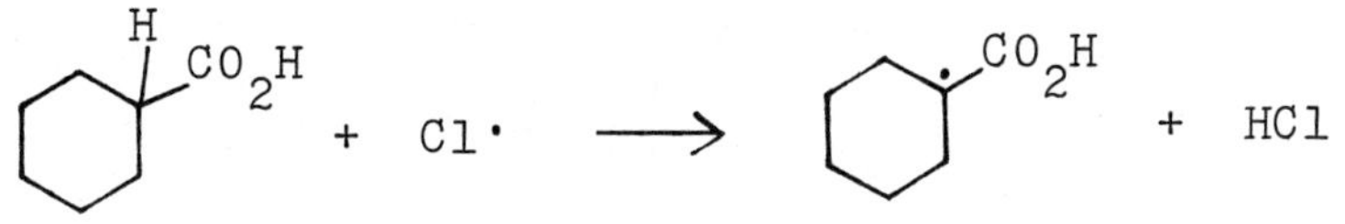

etc.

(continued next page)

(b) Possible products are:

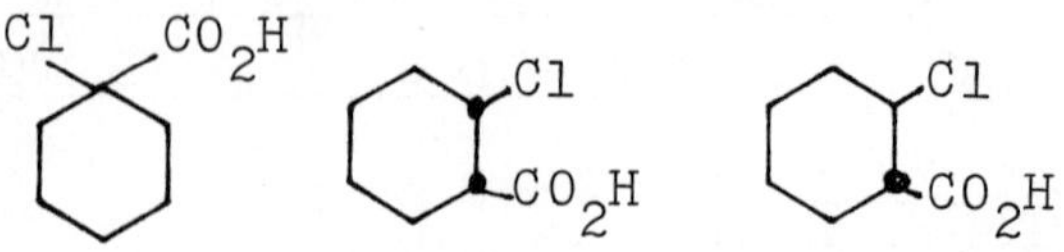

resolvable resolvable

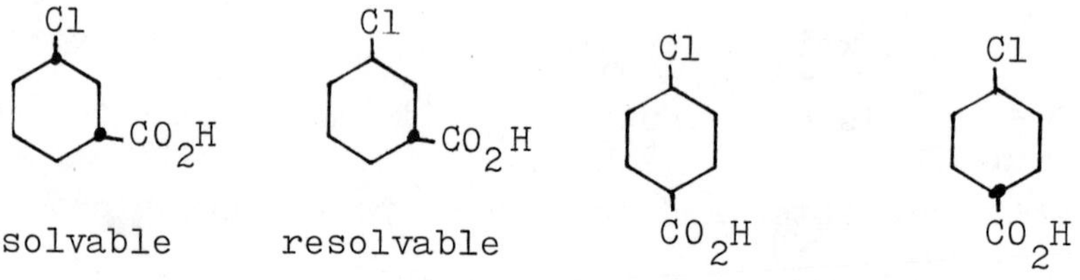

resolvable resolvable

(Text: free-radical chlorination S. 3.7)

2. Diminished delocalization in A compared to B permits greater enolization in A compared to B.

A: -C(=O)-Cl ⟷ -C(-O⁻)=Cl⁺   B: -C(=O)-O-H ⟷ -C(-O⁻)=O⁺-H

3. 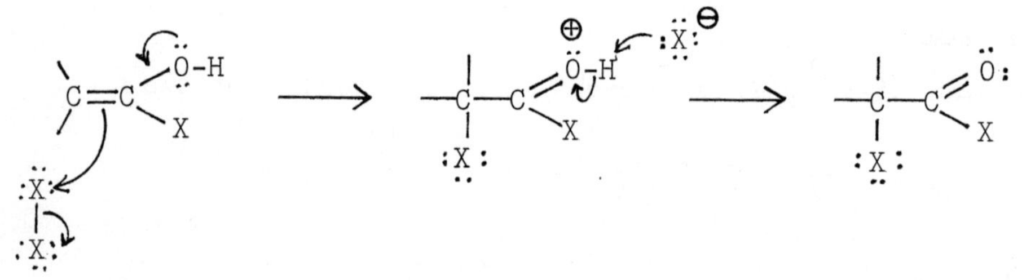 C=C(OH)X + X–X ⟶ X–C–C(=O⁺H)X + X⁻ ⟶ X–C–C(=O)X

4.

Starting material: Cl–C(H)–COOH, i.e. $-\overset{Cl}{C}(H)-COOH$

$\xrightarrow{NH_3} \xrightarrow{H_3O^{\oplus}}$ $-\overset{H}{C}(NH_2)-CO_2$ (S_N2)

$\xrightarrow{CN^{\ominus}}$ $-C(CN)-CO_2H$ (S_N2)

$\xrightarrow{SOCl_2}$ $-\overset{Cl}{C}(H)-\overset{O}{C}-Cl$ (SNi)

$\xrightarrow{CH_2N_2}$ $-\overset{Cl}{C}(H)-\overset{O}{C}-OCH_3$

$\xrightarrow{OH^{\ominus}, H_2O}$ $-C(OH)-\overset{O}{C}-O^{\ominus}$ ⟶ (cyclic dilactone: six-membered ring with two ring O and two C=O)

$-\overset{H}{C}-\overset{Cl}{C}-CO_2H \xrightarrow{R_3N:} \xrightarrow{H_3O^{\oplus}} >C=C-COOH$ (E2)

See Figure 20-1 for additional examples.

5. $CH_3I \xrightarrow[ether]{Mg} CH_3MgI$

CH_2-CH_2 (ethylene oxide, O bridging) $\xrightarrow{CH_3MgI} \xrightarrow{H_2O} CH_3CH_2CH_2OH \xrightarrow{Cr_2O_7^{2-}/H^{\oplus}}$

(continued next page)

$$CH_3CH_2COOH \xrightarrow{Cl_2/PCl_3} \xrightarrow{H_3O^{\oplus}} CH_3\underset{\large Cl}{\underset{|}{C}}HCOOH \xrightarrow{NH_3} \xrightarrow{H_3O^{\oplus}}$$

$$CH_3\underset{NH_2}{\underset{|}{C}}HCOOH$$

6. The reaction shown below occurs in preference to the reaction shown in the problem because of the resonance stabilization of anion A.

$$CH_3\text{-}\overset{O}{\overset{\|}{C}}\text{-}CH_2\text{-}\overset{O}{\overset{\|}{C}}OC_2H_5 + \overset{\ominus}{\ddot{C}}H_2\text{-}\overset{O}{\overset{\|}{C}}OC_2H_5 \rightarrow \left[\begin{array}{c} CH_3\overset{:O:}{\overset{\|}{C}}\text{-}\overset{\ominus}{\ddot{C}}H\text{-}\overset{:O:}{\overset{\|}{C}}OC_2H_5 \\ \updownarrow \\ CH_3\overset{\ominus\,:\ddot{O}:}{\overset{|}{C}}=CH\text{-}\overset{:O:}{\overset{\|}{C}}OC_2H_5 \\ \updownarrow \\ CH_3\text{-}\overset{:O:}{\overset{\|}{C}}\text{-}CH=\overset{:\ddot{O}:\,\ominus}{\overset{|}{C}}OC_2H_5 \end{array} \right] + CH_3\overset{O}{\overset{\|}{C}}OC_2H_5$$

A

7. Refer to the mechanism p. 840. In general, with bases such as ethoxide ion, the only equilibrium step favorable to product formation is the last one, i.e.,

$$\text{-}\overset{O}{\overset{\|}{C}}\text{-}CH_2\text{-}\overset{O}{\overset{\|}{C}}\text{-} + B:^{\ominus} \rightleftarrows \text{—}\overset{O}{\overset{\vdots}{C}}\cdots\overset{\ominus}{CH}\cdots\overset{O}{\overset{\vdots}{C}}\text{—} + HB$$

Since this reaction is not possible for ethyl 2,2,4-trimethyl-3-oxopentanoate, a stronger base must be used to make the initial condensation step irreversible.

8. $$CH_3\text{-}\overset{O}{\overset{\|}{C}}\text{-}CH_3 + :\ddot{O}Et^{\ominus} \rightleftarrows EtOH + CH_3\text{-}\overset{:O:}{\overset{\|}{C}}\text{-}\overset{\ominus}{\ddot{C}}H_2 \longleftrightarrow CH_3\text{-}\overset{\ominus\,:\ddot{O}:}{\overset{|}{C}}=CH_2$$

(continued next page)

$CH_3\text{-}\overset{O}{\overset{\|}{C}}\text{-}OC_2H_5 + {:}\overset{\ominus}{C}H_2\text{-}\overset{O}{\overset{\|}{C}}CH_3 \longrightarrow CH_3\text{-}\underset{OC_2H_5}{\overset{O^{\ominus}}{C}}\text{-}CH_2\text{-}\overset{O}{\overset{\|}{C}}\text{-}CH_3 \rightleftharpoons CH_3\text{-}\overset{O}{\overset{\|}{C}}\text{-}CH_2\overset{O}{\overset{\|}{C}}CH_3 + {:}\ddot{O}Et^{\ominus}$

$\longrightarrow CH_3\text{-}\overset{O}{\overset{\|}{C}}\text{=}\overset{\ominus}{C}H\text{=}\overset{O}{\overset{\|}{C}}\text{-}CH_3 + EtOH \xrightarrow{HCl} CH_3\overset{O}{\overset{\|}{C}}CH_2\overset{O}{\overset{\|}{C}}CH_3$

9. (a) $C_2H_5O\overset{O}{\overset{\|}{C}}CH_2CH_2\overset{O}{\overset{\|}{C}}OC_2H_5 + H\overset{O}{\overset{\|}{C}}OC_2H_5 \xrightarrow[C_2H_5OH]{Na^{\oplus}\,{}^{\ominus}OC_2H_5}$

(b) $C_6H_5\overset{O}{\overset{\|}{C}}OC_2H_5 + CH_3\overset{O}{\overset{\|}{C}}OC_2H_5 \xrightarrow[C_2H_5OH]{Na^{\oplus}\,{}^{\ominus}OC_2H_5}$

10. $C_2H_5O\overset{O}{\overset{\|}{C}}(CH_2)_4\overset{O}{\overset{\|}{C}}OC_2H_5 \xrightarrow{C_2H_5\ddot{O}{:}^{\ominus}} C_2H_5O\overset{O}{\overset{\|}{C}}\text{-}\overset{\ominus}{C}H(CH_2)_3\overset{O}{\overset{\|}{C}}OC_2H_5$ *

$\longrightarrow$ [cyclopentane with $C(O^{\ominus})(OC_2H_5)$ and adjacent $CH(\overset{O}{\overset{\|}{C}}OC_2H_5)$] $\longrightarrow$ 2-carbethoxycyclopentanone (H, $\overset{O}{\overset{\|}{C}}OC_2H_5$) $\xrightarrow{C_2H_5\ddot{O}{:}^{\ominus}}$ enolate of 2-carbethoxycyclopentanone ($\overset{O}{\overset{\|}{C}}OC_2H_5$) *

$\downarrow H^{\oplus}$

2-carbethoxycyclopentanone (O, H, $\overset{O}{\overset{\|}{C}}OC_2H_5$)

*resonance-stabilized

11. CH_3-substituted cycloheptanone: 2-methyl-7-($\overset{O}{\overset{\|}{C}}-OC_2H_5$)cycloheptanone

12. (a) $CH_3CH_2C(=O)H$ + $(C_2H_5O\overset{O}{\overset{\|}{C}})_2CH_2$ $\xrightarrow{\text{piperidine/heat}}$

(b) $C_6H_5C(=O)H$ + $(CH_3O\overset{O}{\overset{\|}{C}})_2CH_2$ $\xrightarrow{\text{piperidine/heat}}$

(c) $(CH_3CH_2)_2C{=}O$ + $C_2H_5O\overset{O}{\overset{\|}{C}}CH_2CN$ $\xrightarrow{\text{piperidine/heat}}$

(d) $C_6H_5C(=O)H$ + $CH_3\overset{O}{\overset{\|}{C}}CH_2\overset{O}{\overset{\|}{C}}OC_2H_5$ $\xrightarrow{\text{piperidine/heat}}$

13. (a) $CH_3\overset{:O:}{\overset{\|}{C}}-CH_2-\overset{O}{\overset{\|}{C}}OC_2H_5$ $\xrightarrow{:\ddot{O}H^{\ominus}}$ $CH_3\overset{:\ddot{O}:^{\ominus}}{\overset{|}{\underset{OH}{\underset{|}{C}}}}-CH_2-\overset{O}{\overset{\|}{C}}OC_2H_5$

$\upharpoonleft\downharpoonright$

$CH_3\overset{O}{\overset{\|}{C}}OH$ + $\left[\overset{\ominus}{\ddot{C}}H_2-\overset{O}{\overset{\|}{C}}OC_2H_5 \longleftrightarrow CH_2{=}\overset{:\ddot{O}:^{\ominus}}{\overset{|}{C}}OC_2H_5\right]$

$\downarrow OH^{\ominus}$ $\qquad\qquad$ $\downarrow H_2O$

$CH_3CO_2^{\ominus}$ + H_2O $\qquad\qquad$ $CH_3-\overset{:O:}{\overset{\|}{C}}OC_2H_5$

$\downarrow :\ddot{O}H^{\ominus}$

(continued next page)

$$\downarrow$$

$$C_2H_5\ddot{O}:^{\ominus} + CH_3\overset{O}{\overset{\|}{C}}OH \rightleftharpoons CH_3-\underset{OH}{\overset{:\ddot{O}:^{\ominus}}{C}}-\ddot{O}C_2H_5$$

$$C_2H_5OH + CH_3CO_2^{\ominus}$$

(b) $$CH_3\overset{O}{\overset{\|}{C}}CH_2\overset{O}{\overset{\|}{C}}OEt \xrightarrow{Na^{\oplus}\ OEt^{\ominus}} \xrightarrow{CH_3CH_2CH_2Br} CH_3\overset{O}{\overset{\|}{C}}\underset{CH_2CH_2CH_3}{CH}\overset{O}{\overset{\|}{C}}OEt$$

$$\downarrow \text{conc. } OH^{\ominus}/H_2O$$

$$CH_3CO_2H + CH_3CH_2CH_2CH_2CO_2H \xleftarrow{H_3O^{\oplus}} CH_3CO_2^{\ominus} + CH_3CH_2CH_2CH_2CO_2 + EtOH$$

14. A: $(C_2H_5O\overset{O}{\overset{\|}{C}})_2\ddot{C}H^{\ominus}\ Na^{\oplus}$

B: $(C_2H_5O\overset{O}{\overset{\|}{C}})_2CH\underset{CH_3}{CH}CO_2C_2H_5$

C: $(Na^{\oplus}\ {}^{\ominus}O_2C)_2CH\underset{CH_3}{CH}CO_2^{\ominus}\ Na^{\oplus}$

D: $(HO_2C)_2CH\underset{CH_3}{CH}CO_2H$

E: $HO_2CCH_2\underset{CH_3}{CH}CO_2H$

15.

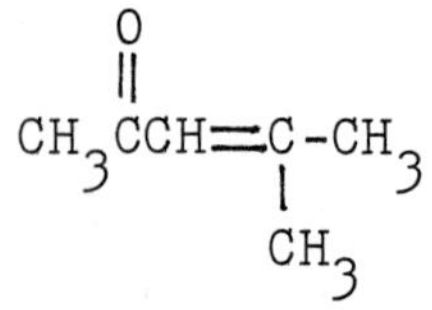

A

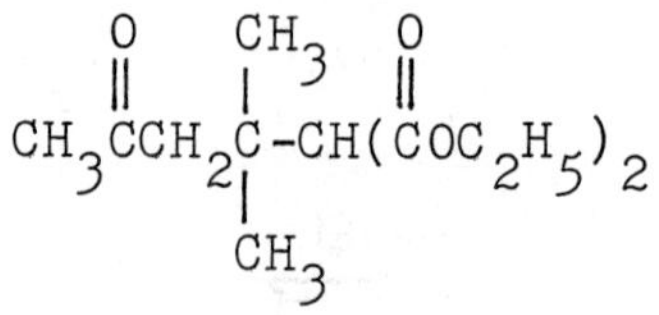

B

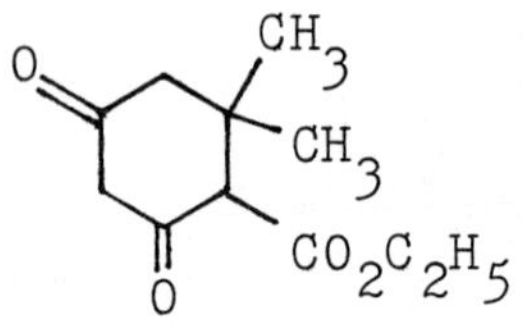

C

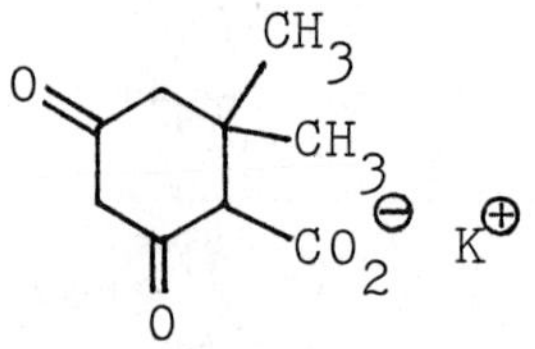

D

16. Part A

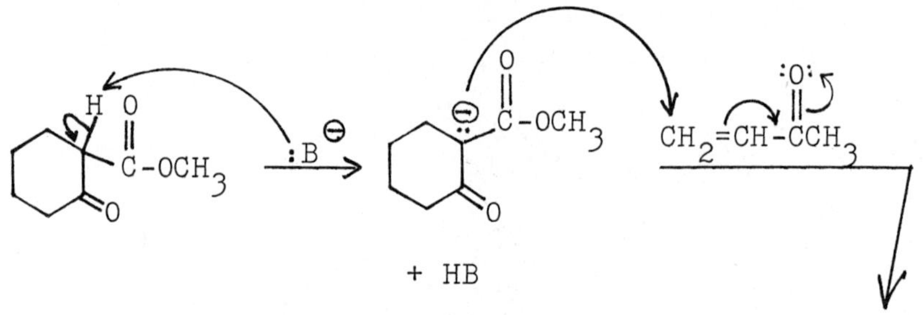

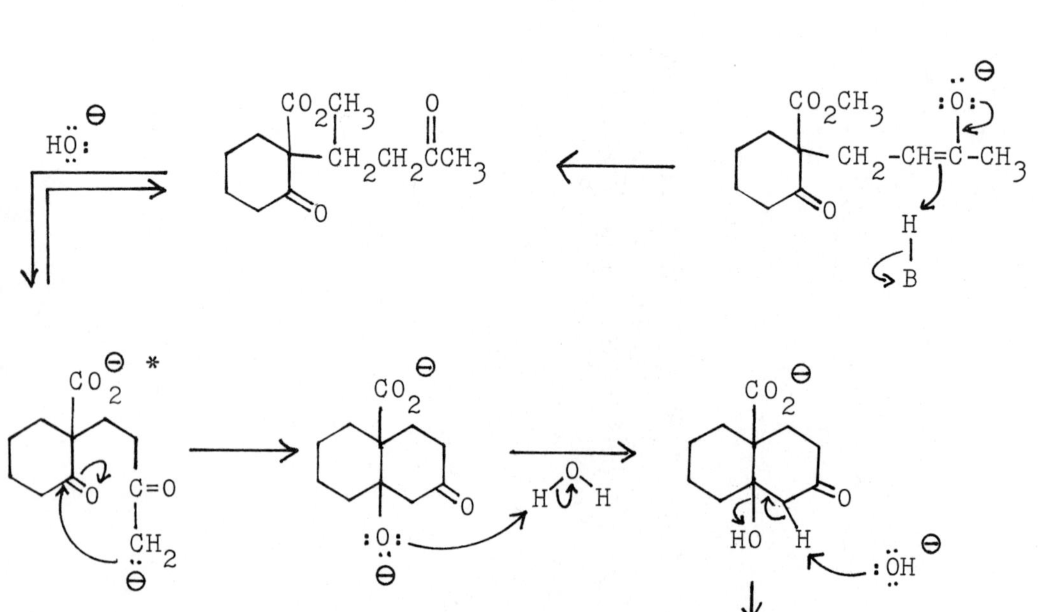

(continued next page)

$-CO_2$ $H_3O^{\oplus}$ CO_2 CO_2H

* $CO_2CH_3 \rightarrow CO_2^{\ominus}$ alkaline hydrolysis of an ester

17. CH_3 H $CH_2{=}CHC{\equiv}N:$ CH_3 CH_2-$\ddot{C}H^{\ominus}$-$C{\equiv}N:$ $\leftrightarrow$ $-CH{=}C{=}\ddot{N}^{\ominus}:$ H

EtOH

CH_3 $CH_2CH_2C\ N$ H CH_3 $CH_2CH_2C\ N$ H H $:\ddot{O}Et^{\ominus}$

$H_3O^{\oplus}$

CH_3 CH_2CH_2CN H H $:\ddot{O}H_2$ H H N-H H

CH_3 O CH_2CH_2CN H $-H^{\oplus}$ $\ddot{O}$-H

18. (a) $CH_3CH_2\overset{O}{\overset{\|}{C}}\text{-}H + (C_6H_5)_3\overset{\oplus}{P}\text{-}\ddot{C}H_2^{\ominus}$

(b) $CH_3\overset{O}{\overset{\|}{C}}\text{-}H + (C_6H_5)_3\overset{\oplus}{P}\text{-}\overset{\ominus}{\ddot{C}}HCH_3$

(c) cyclohexanone $+ (C_6H_5)_3\text{-}\overset{\oplus}{P}\text{-}\overset{\ominus}{\ddot{C}}HCH_3$

(d) $(C_6H_5)_2C{=}O + (C_6H_5)_3\overset{\oplus}{P}\text{-}\overset{\ominus}{\ddot{C}}H_2$

(e) $C_6H_5\overset{O}{\overset{\|}{C}}\text{-}H + (C_6H_5)_3\overset{\oplus}{P}\text{-}\overset{\ominus}{\ddot{C}}HC_6H_5$

(f) $(CH_3)_2C{=}O + (CH_3CH_2O)_2\overset{\oplus}{P}(O^{\ominus})\text{-}\overset{\ominus}{\ddot{C}}H\overset{O}{\overset{\|}{C}}\text{-}OCH_3$

19. Thiol esters are not stabilized by resonance to the extent that esters are because sulfur's outer electrons are in the third main quantum level, and overlap between sulfur and carbon orbitals is less effective than between oxygen and carbon orbitals.

20. The subscript numbers indicate the number of double bonds present.

21. (a) $2CH_3CH_2\overset{O}{\overset{\|}{C}}OCH_3 \xrightarrow{^{\ominus}OCH_3} \xrightarrow{\text{dil HCl}} CH_3\text{-}CH(\text{-}C(=O)\text{-}CH_2CH_3)\text{-}COOCH_3$

(b) $Cl\text{-}C_6H_4\text{-}CH{=}O \xrightarrow[CH_3CH_2COO^{\ominus}Na^{\oplus}]{(CH_3CH_2\overset{O}{\overset{\|}{C}})_2O} Cl\text{-}C_6H_4\text{-}CH{=}C(CH_3)COOH$

(continued next page)

(c) $2CH_3CH_2\overset{O}{\overset{\|}{C}}H \xrightarrow{OH^{\ominus}} CH_3CH_2\overset{OH}{\overset{|}{C}}H\underset{CH_3}{\underset{|}{C}}HCHO$

(d) $CH_3\overset{O}{\overset{\|}{C}}H + CH_3CH_2O\overset{O}{\overset{\|}{C}}CH_2\overset{O}{\overset{\|}{C}}OCH_2CH_3 \xrightarrow{\text{piperidine}} (CH_3)(H)C{=}C(CO_2CH_2CH_3)_2$

Hydrolysis followed by heating may lead to an α,β-unsaturated acid.

$CH_3CH{=}C(CO_2CH_2CH_3)_2 \xrightarrow[\text{heat}]{H_3O^{\oplus}} CH_3CH{=}CHCOOH + CO_2 + CH_3CH_2OH$

(e) $CH_3O\overset{O}{\overset{\|}{C}}CH_3 + CH_2{=}CH\overset{O}{\overset{\|}{C}}OCH_3 \xrightarrow{^{\ominus}OCH_3} \xrightarrow{H_3O^{\oplus}} CH_3O\overset{O}{\overset{\|}{C}}CH_2CH_2CH_2\overset{O}{\overset{\|}{C}}OCH_3$

(f) $\text{(2-furyl)}\overset{O}{\overset{\|}{C}}\text{-O-}CH_3 + C_6H_5CH_2COOCH_3 \xrightarrow{^{\ominus}OCH_3} \text{(2-furyl)-}\overset{O}{\overset{\|}{C}}\underset{C_6H_5}{\underset{|}{C}}HCOOCH_3$

(g) $CH_3\overset{O}{\overset{\|}{C}}CH_2\overset{O}{\overset{\|}{C}}OC_2H_5 \xrightarrow{^{\ominus}OC_2H_5} \xrightarrow{CH_3I} CH_3\overset{O}{\overset{\|}{C}}\underset{CH_3}{\underset{|}{C}}H\overset{O}{\overset{\|}{C}}OC_2H_5 \xrightarrow[\text{heat}]{H_3O^{\oplus}}$

$CH_3\overset{O}{\overset{\|}{C}}CH_2CH_3 + CO_2 + C_2H_5OH$

(h) $C_2H_5O\overset{O}{\overset{\|}{C}}CH_2\overset{O}{\overset{\|}{C}}OC_2H_5 \xrightarrow{^{\ominus}OC_2H_5} \xrightarrow{CH_3I} C_2H_5O\overset{O}{\overset{\|}{C}}\underset{CH_3}{\underset{|}{C}}H\overset{O}{\overset{\|}{C}}OC_2H_5 \xrightarrow[\text{heat}]{H_3O^{\oplus}}$

$CH_3CH_2CO_2H + CO_2 + C_2H_5OH$

(continued next page)

(i) $C_2H_5O\overset{O}{\overset{\|}{C}}(CH_2)_5\overset{O}{\overset{\|}{C}}OC_2H_5 \xrightarrow{^{\ominus}OC_2H_5}$ 2-oxocyclohexane with H and $COOC_2H_5$ on the carbon next to C=O

(j) $CH_3O\overset{O}{\overset{\|}{C}}(CH_2)_8\overset{O}{\overset{\|}{C}}OCH_3 \xrightarrow{Na} \xrightarrow{H_2O}$ ring of $(CH_2)_8$ closed by $C{=}O$ and $C(H)(OH)$

(k) 2-($\overset{O}{\overset{\|}{C}}OCH_3$)cyclohexanone $\xrightarrow{^{\ominus}OCH_3}$ 2-($\overset{O}{\overset{\|}{C}}OCH_3$)cyclohexanone $\xrightarrow{CH_3CH{=}CH\overset{O}{\overset{\|}{C}}CH_3}$

2-($COOCH_3$)-2-[$CH(CH_3)CH_2\overset{O}{\overset{\|}{C}}CH_3$]cyclohexanone $\xrightarrow[\text{aldol}]{OH^{\ominus}} \xrightarrow{-H_2O}$ bicyclic enone with $^{\ominus}OOC$ at the ring-fusion carbon, CH_3 on the neighbouring ring carbon, and C=O

$\xrightarrow[CO_2]{H^{\oplus}/\text{heat}}$ bicyclic enone with CH_3 and C=O (no carboxylate)

(l) $(C_6H_5)_3\overset{\oplus}{P}-\overset{\ominus}{\ddot{C}}H_2 + CH_3\overset{O}{\overset{\|}{C}}-H \longrightarrow CH_3CH{=}CH_2 + (C_6H_5)_3PO$

(m) $BrZnCH_2\overset{O}{\overset{\|}{C}}OCH_3 \xrightarrow{CH_3\overset{O}{\overset{\|}{C}}CH_3} \xrightarrow{H_3O^{\oplus}} (CH_3)_2\underset{OH}{\underset{|}{C}}CH_2COOCH_3$

22. (a) $CH_3CO_2C_2H_5 \xrightarrow[\text{heat}]{H_3O^{\oplus}} CH_3COOH + C_2H_5OH$

(continued next page)

(b) see a

(c) C_2H_5OH (from a) $\xrightarrow[\text{heat}]{H_2SO_4}$ $CH_2{=}CH_2$

(d) $CH_2{=}CH_2$ (from c) $\xrightarrow{O_3}$ $\xrightarrow{BH_3}$ CH_3OH

(e) CH_3CO_2H (from a) $\xrightarrow{SOCl_2}$ $\xrightarrow{CH_2N_2}$ $\xrightarrow[H_2O]{Ag^{\oplus}}$ CH_3CH_2COOH

(f) $CH_3CO_2C_2H_5$ $\xrightarrow{^{\ominus}OC_2H_5}$ $CH_3\overset{O}{\overset{\|}{C}}CH_2\overset{O}{\overset{\|}{C}}{-}OC_2H_5$

(g) $CH_3\overset{O}{\overset{\|}{C}}CH_2\overset{O}{\overset{\|}{C}}O_2C_2H_5$ (from f) $\xrightarrow[H_2O]{NaBH_4}$ $CH_3\underset{OH}{\underset{|}{C}}HCH_2\overset{O}{\overset{\|}{C}}OC_2H_5$

(h) $CH_3CH_2CO_2H$ (from e) + Br_2 $\xrightarrow{PBr_3}$ $CH_3\underset{Br}{\underset{|}{C}}HCOOH$

(i) $CH_3\overset{O}{\overset{\|}{C}}CH_2CO_2C_2H_5$ (from f) $\xrightarrow{LiAlH_4}$ $\xrightarrow{H_2O}$ $CH_3\underset{OH}{\underset{|}{C}}HCH_2CH_2OH$

(j) $CH_3CO_2C_2H_5$ $\xrightarrow{Cl_2/PCl_3}$ $ClCH_2COOH$ $\xrightarrow[DMSO]{NaCN}$ $\xrightarrow[\text{heat}]{H_3O^{\oplus}}$

$HOOCCH_2COOH$ $\xrightarrow{SOCl_2}$ $\xrightarrow{C_2H_5OH}$ $C_2H_5O\overset{O}{\overset{\|}{C}}CH_2\overset{O}{\overset{\|}{C}}OC_2H_5$

(k) $(C_2H_5O\overset{O}{\overset{\|}{C}})_2CH_2$ (from j) $\xrightarrow{^{\ominus}OC_2H_5}$ $\xrightarrow{CH_2{=}CHCN}$ $\xrightarrow{H_3O^{\oplus}}$

(continued next page)

$$(C_2H_5O\overset{O}{\overset{\|}{C}})_2CHCH_2CH_2CN \xrightarrow{H_3O^{\oplus}/heat} \xrightarrow[-CO_2]{heat} HOOCCH_2CH_2CH_2CN$$

(l) $$(C_2H_5O\overset{O}{\overset{\|}{C}})_2CHCH_2CH_2CN \text{ (from k) } \xrightarrow{^{\ominus}OC_2H_5} \xrightarrow{CH_3I}$$

$$(C_2H_5O\overset{O}{\overset{\|}{C}})_2\underset{\underset{CH_3}{|}}{C}CH_2CH_2CN \xrightarrow{H_3O^{\oplus}/heat} \xrightarrow[-CO_2]{heat} HOOC\underset{\underset{CH_3}{|}}{C}HCH_2CH_2CN$$

(m) $$HOOC\underset{\underset{CH_3}{|}}{C}HCH_2CH_2CN \text{ (from l) } \xrightarrow[reflux]{H_3O^{\oplus}} HOOC\underset{\underset{CH_3}{|}}{C}HCH_2CH_2COOH$$

(n) $$CH_3CH_2O\overset{O}{\overset{\|}{C}}CH_3 \xrightarrow{^{\ominus}OC_2H_5} \xrightarrow{D_2O} CH_3CH_2O\overset{O}{\overset{\|}{C}}CH_2D$$

(o) $$(C_2H_5O\overset{O}{\overset{\|}{C}})_2CH_2 \xrightarrow{^{\ominus}OC_2H_5} \xrightarrow{Br_2} (C_2H_5O\overset{O}{\overset{\|}{C}})_2CHBr \xrightarrow{(C_6H_5)_3P}$$

$$\xrightarrow{CH_3CH_2CH_2\overset{O}{\overset{\|}{C}}CH_3} CH_3CH_2CH_2\underset{\underset{CH_3}{|}}{C}=C(\overset{O}{\overset{\|}{C}}OC_2H_5)_2$$

(p) $$(C_2H_5O\overset{O}{\overset{\|}{C}})_2CH_2 \xrightarrow{^{\ominus}OC_2H_5} \xrightarrow{CH_3CH_2Br} \xrightarrow[heat]{H_3O^{\oplus}} CH_3CH_2CH_2COOH$$

(q) $$CH_3\overset{O}{\overset{\|}{C}}OC_2H_5 \xrightarrow[ether, -70^{o}]{CH_3MgI} \xrightarrow{H_2O} CH_3\overset{O}{\overset{\|}{C}}CH_3$$

23. (a) $ClCH_2CO_2H \xrightarrow{CH_2N_2} \xrightarrow[C_6H_6/\text{ether}]{Zn} ClZnCH_2COOCH_3 \xrightarrow{CH_3CHO}$

$\xrightarrow{H_3O^{\oplus}} CH_3\overset{OH}{\overset{|}{C}}HCH_2\overset{O}{\overset{\|}{C}}OCH_3$

(b) $(C_2H_5O\overset{O}{\overset{\|}{C}})_2CH_2 \xrightarrow{{}^{\ominus}OC_2H_5} \xrightarrow{CH_3I} \xrightarrow{{}^{\ominus}OC_2H_5} \xrightarrow{(CH_3)_2CHCH_2CH_2Cl}$

$\xrightarrow{H_3O^{\oplus}/\text{heat}} \xrightarrow{\text{heat}(-CO_2)} (CH_3)_2CHCH_2CH_2\underset{CH_3}{\underset{|}{C}}HCO_2H$

(c) $(HO_2C)_2CH_2 \xrightarrow{2CH_2N_2} \xrightarrow{{}^{\ominus}OCH_3} \xrightarrow{(CH_3)_2CHCl} \xrightarrow{H_3O^{\oplus}/\text{heat}}$

$\xrightarrow{\text{heat}(-CO_2)} (CH_3)_2CHCH_2CH_2COOH \xrightarrow[PBr_3]{Br_2} \xrightarrow[-HBr]{(CH_3)_3N}$

$\xrightarrow{H_3O^{\oplus}} (CH_3)_2CHCH{=}CHCO_2H$

(or $CH_2(CO_2H)_2 + (CH_3)_2CHCHO \xrightarrow[\text{heat}]{\text{piperidine}} (CH_3)_2CHCH{=}CHCO_2H$)

(d) $HC{\equiv}CH \xrightarrow{NaH} \xrightarrow{\text{ethylene oxide}} \xrightarrow{H_3O^{\oplus}} \xrightarrow{\text{repeat}} HOCH_2CH_2C{\equiv}CCH_2CH_2OH$

$\xrightarrow{H_2/Pt} \xrightarrow{CrO_3/H_3O^{\oplus}} HO_2CCH_2CH_2CH_2CH_2CO_2H$

(continued next page)

(e) $C_6H_5CHO \xrightarrow[2)\ H_3O^{\oplus}]{1)\ BrZnCH_2CO_2C_2H_5} C_6H_5CH(OH)CH_2CO_2C_2H_5 \xrightarrow{HBr}$

$C_6H_5CH(Br)CH_2CO_2C_2H_5 \xrightarrow{CN^{\ominus}} \xrightarrow{H_3O^{\oplus}} C_6H_5CH(COOH)CH_2COOH$

(f) $(C_2H_5O_2C)_2CH_2 \xrightarrow{H_2C=O/\text{piperidine}} H_2C=C(CO_2C_2H_5)$

$\xrightarrow{(C_2H_5OC(=O))_2CH_2} (C_2H_5O_2C)_2CHCH_2CH(CO_2C_2H_5) \xrightarrow{H_3O^{\oplus}/\text{heat}}$

$\xrightarrow{\text{heat}(-CO_2)} HO_2C(CH_2)_3CO_2H$

(g) $C_2H_5O_2C(CH_2)_4CO_2C_2H_5 \xrightarrow{^{\ominus}OC_2H_5}$ 2-(ethoxycarbonyl)cyclopentanone (ring C=O, with H and $CO_2C_2H_5$ on the adjacent carbon) $\xrightarrow{H_3O^{\oplus}/\text{heat}}$

$\xrightarrow[-CO_2]{\text{heat}}$ cyclopentanone $\xrightarrow[H_2O]{NaBH_4}$ cyclopentanol (H, OH)

(h) $C_2H_5O_2C(CH_2)_6CO_2C_2H_5 \xrightarrow{Na} \xrightarrow{H_2O}$ 2-hydroxycyclooctanone (=O, OH) $\xrightarrow[\text{heat}]{H_2SO_4}$

$\xrightarrow{H_2/Ni}$ cyclooctanol (OH)

(continued next page)

(i) $CH_2(CO_2C_2H_5)_2 \xrightarrow{^{\ominus}OC_2H_5} \xrightarrow{BrCH_2CH_2Br}$ [cyclopropane with $CO_2C_2H_5$, $CO_2C_2H_5$]

$\xrightarrow[heat]{H_3O^{\oplus}}$ [cyclopropane with CO_2H, CO_2H]

(j) $BrCH_2CH_2CHO + HOCH_2CH_2OH \xrightarrow{H^{\oplus}} BrCH_2CH_2CH$[1,3-dioxolane ring] (A)

$C_2H_5O_2CCH_2CN \xrightarrow{NaH} C_2H_5O_2C\overset{\ominus}{\ddot{C}}HCN \xrightarrow{(A)}$

$C_2H_5O_2C\underset{CN}{CH}CH_2CH_2\text{-}CH$[1,3-dioxolane ring] $\xrightarrow{H_3O^{\oplus}} C_2H_5O_2C\underset{CN}{CH}CH_2CH_2CHO$

(k) $CH_3CH_2CO_2H \xrightarrow[Br_2]{PBr_3} \xrightarrow{(C_2H_5)_3N} \xrightarrow{H_3O^{\oplus}} CH_2{=}CHCO_2H$

(l) $CH_3CO_2C_2H_5 \xrightarrow{NaH} \xrightarrow{C_6H_5CHO} \xrightarrow{H_2O} C_6H_5\underset{OH}{CH}CH_2CO_2C_2H_5$

(m) $CH_3\overset{O}{\overset{\|}{C}}CH_3$ + [ethylene carbonate, cyclic O–$C(=O)$–O] $\xrightarrow{NaH} CH_3\overset{O}{\overset{\|}{C}}CH_2\overset{O}{\overset{\|}{C}}OCH_2CH_2OH \xrightarrow{CH_3\overset{O}{\overset{\|}{C}}Cl}$

$CH_3\overset{O}{\overset{\|}{C}}CH_2\overset{O}{\overset{\|}{C}}OCH_2CH_2O\overset{O}{\overset{\|}{C}}CH_3$

(continued next page)

24. Most of the pairs can be distinguished by boiling or melting points.

(a) Acetic acid turns blue litmus red, ethyl acetate does not. The IR spectrum of acetic acid contains an OH stretch (3200-3600 cm^{-1}).

(b) The IR spectrum of acetoacetic ester should show two C=O stretches while that of ethyl acetate shows only one. The NMR spectrum of acetoacetic ester shows two singlets; that of ethyl acetate shows only one.

(c) The IR spectrum of acetic acid shows an OH stretch (3200-3600 cm^{-1}); that of acetic anhydride does not.

(d) Sodioacetic ester forms acetoacetic ester in water solution which can be extracted by organic solvents. The NMR spectrum of sodioacetic ester is much more complex (singlet for $CH_3\overset{O}{\overset{\|}{C}}-$, triplet and quartet for CH_3CH_2- etc) than that for sodium acetate (one singlet for CH_3-).

(e) 2-Carbethoxycyclohexanone should show two C=O stretching vibrations in the IR; the IR spectrum of cyclohexanone shows only one. The triplet-quartet combination for $-OCH_2CH_3$ in the NMR spectrum of 2-carbethoxycyclohexanone will also distinguish the compounds.

(f) Since caprylic acid contains two more CH_2 groups, the two acids could be distinguished by comparing the integration of peak areas.

(continued next page)

(g) Acetic anhydride is a liquid; succinic anhydride is a solid. Each compounds shows a single C=O stretch in the IR and a single absorption in NMR. However the singlet in the NMR spectrum of acetic anhydride ($CH_3\overset{O}{\overset{\|}{C}}-$) will be further upfield than the singlet in the spectrum of succinic anhydride ($-CH_2-\overset{O}{\overset{\|}{C}}-$).

(h) The NMR spectrum of ethylene carbonate consists of a single line while that of propylene carbonate is more complex.

(i) The NMR spectrum of pivaldehyde, $(CH_3)_3CHO$ consists of a single upfield absorption. The spectrum of 3,3-dimethylbutanol is more complex.

(j) [structure: bicyclic enone with O and vinyl H] shows a vinyl resonance (~5δ); its isomer does not.

(k) [structure: 2-oxocyclohexane bearing $CO_2C_2H_5$ and CH_2CH_2CN] will liberate CO_2 after the ester is hydrolyzed to the acid (β-Keto acid) and heated; the other compound will not. CO_2 is detected by bubbling it into limewater ($Ca(OH)_2$). A precipitate of $CaCO_3$ forms.

25. $CH_3\overset{O}{\overset{\|}{C}}CH_2CO_2Et$ is acetoacetic ester (AE)

(a) AE $\xrightarrow[EtOH]{Na^{\oplus}\ OEt^{\ominus}}$ $CH_3\overset{O}{\overset{\|}{C}}-\overset{\ominus}{\ddot{C}}H-CO_2Et$ * $\xrightarrow{CH_3CH_2Br}$

$CH_3\overset{O}{\overset{\|}{C}}\underset{CH_2CH_3}{\underset{|}{C}}HCO_2Et$ $\xrightarrow{H_3O^{\oplus}}$ $CH_3\overset{O}{\overset{\|}{C}}\underset{CH_2CH_3}{\underset{|}{C}}HCOOH$ $\xrightarrow[-CO_2]{heat}$ $CH_3\overset{O}{\overset{\|}{C}}CH_2CH_2CH_3$

*(resonance stabilized)

(b) $CH_3\overset{O}{\overset{\|}{C}}CH_2CO_2Et$ $\xrightarrow[C_2H_5OH]{C_2H_5O^{\ominus}}$ $\xrightarrow{H_3O^{\oplus}}$ $2CH_3CO_2C_2H_5$
(retro Claisen)

$CH_3\overset{O}{\overset{\|}{C}}OC_2H_5$ $\xrightarrow[\text{low temp.}]{(CH_3)_3CMgCl}$ $CH_3\overset{O}{\overset{\|}{C}}C(CH_3)_3$

(c) $CH_3\overset{O}{\overset{\|}{C}}OC_2H_5$ (from b) $\xrightarrow[\text{low temperature}]{C_6H_5CH_2MgCl}$ $CH_3\overset{O}{\overset{\|}{C}}CH_2C_6H_5$

(d) $CH_3\overset{O}{\overset{\|}{C}}CH_2CH_2CH_3$ (from a) $\xrightarrow{LiAlH_4}$ $\xrightarrow{H_2O}$ $CH_3CH(OH)CH_2CH_2CH_3$

(e) AE $\xrightarrow[C_2H_5OH]{Na^{\oplus}\ OC_2H_5^{\ominus}}$ $\xrightarrow[\text{heat }(-CO_2)]{H_3O^{\oplus}}$ $CH_3\overset{O}{\overset{\|}{C}}CH_3$ $\xrightarrow[OH^{\ominus}]{CH_2O}$

$CH_3\overset{O}{\overset{\|}{C}}CH_2CH_2OH$ $\xrightarrow{H_2/Pt}$ $CH_3CH(OH)CH_2CH_2OH$

(continued next page)

(f) AE $\xrightarrow[C_2H_5OH]{Na^{\oplus}\ ^{\ominus}OC_2H_5}$ $\xrightarrow{BrCH_2\overset{O}{\overset{\|}{C}}CH_3}$ $CH_3\overset{O}{\overset{\|}{C}}\underset{CH_2\underset{\underset{O}{\|}}{C}CH_3}{\underset{|}{C}H}CO_2C_2H_5$

$\xrightarrow[H_2O]{OH^{\ominus}}$ $\xrightarrow{H_3O^{\oplus}}$ (acid) $\xrightarrow{SOCl_2}$ $\xrightarrow{CH_3OH}$ $CH_3\overset{O}{\overset{\|}{C}}\underset{CH_2\underset{\underset{O}{\|}}{C}CH_3}{\underset{|}{C}H}CO_2CH_3$

(g) $CH_3\overset{O}{\overset{\|}{C}}\underset{CH_2\underset{\underset{O}{\|}}{C}CH_3}{\underset{|}{C}H}CO_2H$ (from f) $\xrightarrow[-CO_2]{heat}$ $CH_3\overset{O}{\overset{\|}{C}}CH_2CH_2\overset{O}{\overset{\|}{C}}CH_3$

(h) $CH_3\overset{O}{\overset{\|}{C}}CH_2CH_2\overset{O}{\overset{\|}{C}}CH_3$ (from g) $\xrightarrow{CH_3MgI}$ $\xrightarrow{H_2O}$

$(CH_3)_2\underset{OH}{\underset{|}{C}}CH_2CH_2\underset{OH}{\underset{|}{C}}(CH_3)_2$

(i) AE $\xrightarrow[C_2H_5OH]{Na^{\oplus}\ ^{\ominus}OC_2H_5}$ $\xrightarrow{CH_3CH_2Cl}$ $CH_3\overset{O}{\overset{\|}{C}}\underset{CH_2CH_3}{\underset{|}{C}H}CO_2C_2H_5$ $\xrightarrow{LiAlH_4}$

$\xrightarrow{H_2O}$ $CH_3CH(OH)\underset{CH_2CH_3}{\underset{|}{C}H}CH_2OH$

26. (a) $^{\ominus}:CH_2-\overset{:O:}{\overset{\|}{C}}-\overset{\ominus}{\ddot{C}}H-\overset{:O:}{\overset{\|}{C}}OC_2H_5 \longleftrightarrow CH_2=\overset{:\ddot{O}:^{\ominus}}{\overset{|}{C}}-CH-\overset{:O:}{\overset{\|}{C}}OC_2H_5$

$\updownarrow$

etc. $\longleftrightarrow CH_2=\overset{^{\ominus}:\ddot{O}:}{\overset{|}{C}}-CH=\overset{:\ddot{O}:^{\ominus}}{\overset{|}{C}}OC_2H_5$

(b) Initial reaction at the terminal carbon leads to a more stable monoanion.

$$\overset{\ominus}{\ddot{C}}H_2\text{-}\overset{O}{\overset{\|}{C}}\text{-}\overset{\ominus}{\ddot{C}}H\text{-}\overset{O}{\overset{\|}{C}}\text{-}OC_2H_5 \xrightarrow{CH_3Cl} \overset{\ominus}{C}H_2\text{=}\overset{O}{\overset{\|}{C}}\text{-}\underset{CH_3}{\underset{|}{C}H}\text{-}\overset{O}{\overset{\|}{C}}OC_2H_5$$

$$\xrightarrow{CH_3Cl} CH_3CH_2\overset{O}{\overset{\|}{C}}\text{=}\overset{\ominus}{C}H\text{=}\overset{O}{\overset{\|}{C}}\text{-}OC_2H_5$$

(more stable)

$$\downarrow CH_3COCl$$

$$CH_3CH_2\overset{O}{\overset{\|}{C}}\underset{O=C-CH_3}{\underset{|}{C}H}\overset{O}{\overset{\|}{C}}OC_2H_5$$

27. $(NC)_2C{=}C(CN)_2 + H\ddot{O}H \longrightarrow NC\text{-}\underset{H_2\overset{\oplus}{O}}{\underset{|}{\overset{CN}{\overset{|}{C}}}}\text{-}\underset{CN}{\underset{|}{\overset{\ominus}{\ddot{C}}}}\text{-}CN$ * $\xrightarrow{-CN^{\ominus}} \underset{H_2\overset{\oplus}{O}}{NC}\!>\!C{=}C\!<\!(CN)_2 \xrightarrow{-H^{\oplus}} \underset{HO}{NC}\!>\!C{=}C\!<\!(CN)_2$

stabilized enol

*resonance stabilized anion

$$-\overset{\ominus}{\ddot{C}}\text{-}C{\equiv}N: \longleftrightarrow -C{=}C{=}\overset{\ominus}{\ddot{N}}: \longleftrightarrow \text{etc.}$$
$$\quad\ \underset{C\equiv N:}{|} \qquad\qquad\quad \underset{C\equiv N:}{|}$$

28. $(CH_3CH_2O)_2\overset{\oplus}{P}(-S^{\ominus})-\ddot{\underset{..}{S}}:^{\ominus}$ + $C_2H_5O_2CCH{=}CHCO_2C_2H_5$

29. In each of the following a resonance-stabilized anion is formed (*). In answers a, b, d, e, f, g and i the stabilization is of the type $-\overset{:O:}{\overset{\|}{C}}-\overset{\ominus}{\ddot{C}}- \longleftrightarrow -\overset{:\ddot{O}:^{\ominus}}{\overset{|}{C}}=C-$. In answers c and h it is of the type $-\overset{:O:}{\overset{\|}{C}}-\overset{\ominus}{\ddot{C}}-\overset{:O:}{\overset{\|}{C}}- \longleftrightarrow -\overset{^{\ominus}:\ddot{O}:}{\overset{|}{C}}=C-\overset{:O:}{\overset{\|}{C}}- \longleftrightarrow -\overset{:O:}{\overset{\|}{C}}-C=\overset{:\ddot{O}:^{\ominus}}{\overset{|}{C}}-$. For convenience, the resonance structure with the charge on carbon is used in the examples below.

(a) $R\overset{O}{\overset{\|}{C}}CH_3 + H:^{\ominus} \longrightarrow R\overset{O}{\overset{\|}{C}}\ddot{C}H_2^{\ominus} + H_2$

$C_2H_5O\overset{:O:}{\overset{\|}{C}}-OC_2H_5$ + $^{\ominus}:CH_2\overset{O}{\overset{\|}{C}}R$ — $C_2H_5O\overset{:\ddot{O}:^{\ominus}}{\overset{|}{C}}(CH_2\overset{O}{\overset{\|}{C}}R)-OC_2H_5 \longrightarrow C_2H_5O\overset{O}{\overset{\|}{C}}CH_2\overset{O}{\overset{\|}{C}}R + C_2H_5O^{\ominus}$

(b) $C_2H_5O\overset{O}{\overset{\|}{C}}CH_3 \xrightarrow{C_2H_5O^{\ominus}} C_2H_5O\overset{O}{\overset{\|}{C}}\ddot{C}H_2^{\ominus}$ *

$H-\overset{:O:}{\overset{\|}{C}}-OC_2H_5$ + $^{\ominus}:CH_2\overset{O}{\overset{\|}{C}}OC_2H_5 \longrightarrow H-\overset{:\ddot{O}:^{\ominus}}{\overset{|}{C}}(CH_2\overset{O}{\overset{\|}{C}}OC_2H_5)-OC_2H_5 \longrightarrow H\overset{O}{\overset{\|}{C}}CH_2\overset{O}{\overset{\|}{C}}OC_2H_5 + C_2H_5O^{\ominus}$

(continued next page)

(c) $C_2H_5O\overset{O}{\overset{\|}{C}}CH_2\overset{O}{\overset{\|}{C}}OC_2H_5 \xrightarrow{C_2H_5O^{\ominus}} C_2H_5O_2C\overset{\ominus}{\ddot{C}}HCO_2C_2H_5$ * $Br-CH_2CH_2-Br$ → $C_2H_5O_2CCHCO_2C_2H_5$ (with CH_2CH_2Br substituent) $\xrightarrow{C_2H_5O^{\ominus}}$ $C_2H_5O_2C-\overset{\ominus}{\ddot{C}}-CO_2C_2H_5$ (with CH_2CH_2-Br substituent) → $(C_2H_5O_2C)_2C$ with CH_2—CH_2 ring

(d) 1-pyrrolidinylcyclohexene + $CH_2=C(CH_3)CO_2CH_3$ → iminium ion ($N^{\oplus}$), H, H, $CH_2-\overset{\ominus}{\ddot{C}}-CO_2C_2H_5$ with CH_3 * → enamine with $CH_2CHCO_2C_2H_5$, CH_3

(e) $CH_3\overset{O}{\overset{\|}{C}}CH_2CH_2CHCO_2C_2H_5$ (with $C=O$, OC_2H_5 substituent) $\xrightarrow{C_2H_5O^{\ominus}}$ $\overset{\ominus}{\ddot{C}}H_2-\overset{O}{\overset{\|}{C}}CH_2CH_2CHCO_2C_2H_5$ (with $C=\ddot{O}$, OC_2H_5 substituent) * →

(continued next page)

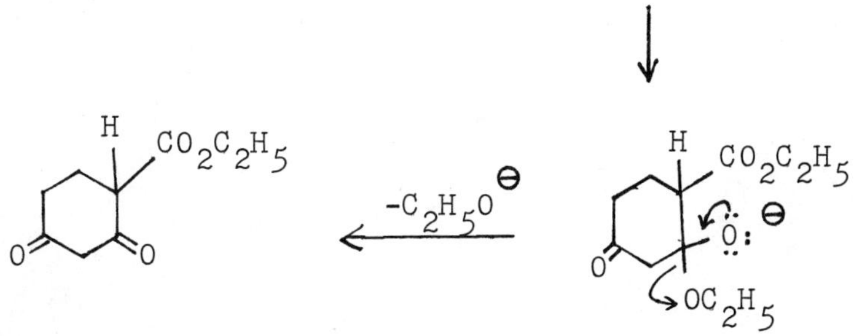

$-C_2H_5O^{\ominus}$

(f)

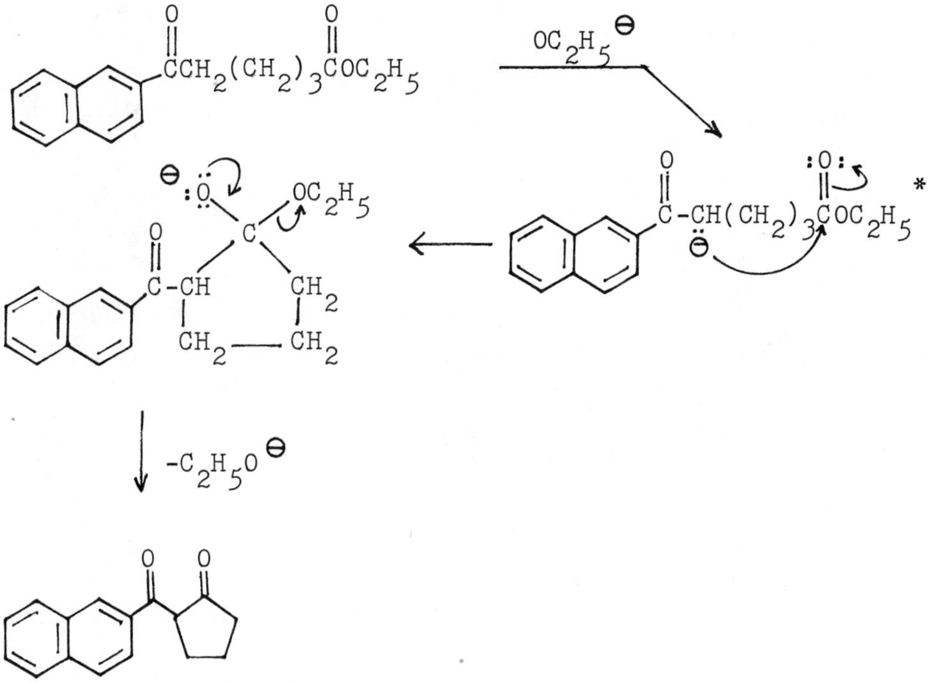

$OC_2H_5^{\ominus}$

$C_{10}H_7\text{-}\overset{O}{\overset{\|}{C}}CH_2(CH_2)_3\overset{O}{\overset{\|}{C}}OC_2H_5$

$C_{10}H_7\text{-}\overset{O}{\overset{\|}{C}}\text{-}\ddot{C}H^{\ominus}(CH_2)_3\overset{:O:}{\overset{\|}{C}}OC_2H_5$ *

$-C_2H_5O^{\ominus}$

(g)

$B:^{\ominus}$

$CH_2{=}CH\text{-}\overset{O}{\overset{\|}{C}}CH_3$

CH_3 $-CH_2\text{-}\ddot{C}H^{\ominus}\text{-}\overset{O}{\overset{\|}{C}}\text{-}CH_3$

HB

$B:^{\ominus}$

CH_3 $CH_2\text{-}CH_2$ $C{=}O$ $:CH_2^{\ominus}$

CH_3 O $O^{\ominus}$

CH_3 HO H $:B^{\ominus}$

CH_3 O

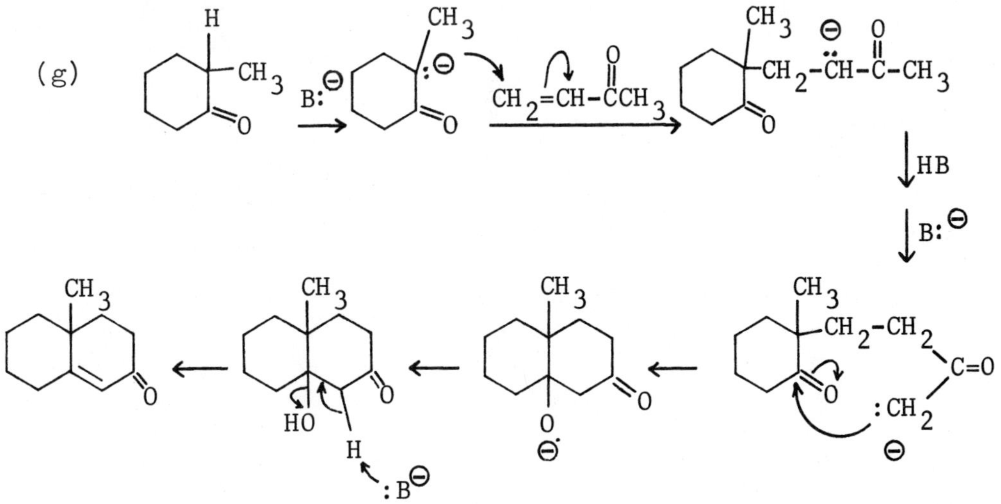

(continued next page)

(h) [cyclobutane]$(CH_2Br)(CH_2{-}Br)$ $\quad C_2H_5O_2C\ddot{C}HCO_2C_2H_5$ (⊖) * $\longrightarrow$ [cyclobutane]$(CH_2{-}Br)(CH_2{-}CH(CO_2C_2H_5)_2)$

$\downarrow C_2H_5O^{\ominus}$

[cyclobutane]$(CH_2{-}Br)(CH_2\ddot{C}^{\ominus}(CO_2C_2H_5)_2)$ $\longrightarrow$ [spiro]$(CH_2)_2C(CO_2C_2H_5)_2$ $\xrightarrow{H_3O^{\oplus}}$

[spiro[3.3]heptane]$(C(=O){-}OH)(C{=}O(OH))$ $\longrightarrow$ [spiro[3.3]heptane]$=C(OH)_2$ + CO_2 $\longrightarrow$ [spiro[3.3]heptane]$-C(=O){-}OH$

(i) $C_6H_5CH{=}CHC(=O)CH{=}CHC_6H_5$ + $^{\ominus}:CH(CO_2C_2H_5)_2$ $\longrightarrow$ $C_6H_5CH(CH(CO_2C_2H_5)_2){-}\ddot{C}^{\ominus}HC(=O)CH{=}CHC_6H_5$

$\downarrow H_3O^{\oplus}$

$C_6H_5CH(CH(CO_2C_2H_5)_2){-}CH_2C(=O)CH{=}CHC_6H_5$ $\xrightarrow{C_2H_5O^{\ominus}}$ $C_6H_5CH((C_2H_5O_2C)_2C:^{\ominus})CH_2C(=O)CH{=}CHC_6H_5$

$\downarrow H_3O^{\oplus}$

[cyclohexanone with C_6H_5, C_6H_5 and $(CO_2C_2H_5)_2$ substituents]

30. (a) The nitro group stabilizes the carbanion.

$$CH_3-\overset{\oplus}{N}(=O)-O^{\ominus} \xrightarrow[-HOD]{OD^{\ominus}} \left[:\overset{\ominus}{C}H_2-\overset{\oplus}{N}(=\ddot{O})-\ddot{O}:^{\ominus} \longleftrightarrow CH_2=\overset{\oplus}{N}(-\ddot{O}:^{\ominus})-\ddot{O}:^{\ominus} \right]$$

$$D-O-D \quad CH_2=\overset{\oplus}{N}(-\ddot{O}:^{\ominus})-\ddot{O}:^{\ominus} \longrightarrow DO^{\ominus} + DCH_2-\overset{\oplus}{N}(=\ddot{O})-\ddot{O}:^{\ominus} \xrightarrow[\text{twice}]{\text{repeat}} CD_3-NO_2$$

The aci form of nitromethane is $CH_2=\overset{\oplus}{N}(-OH)-O^{\ominus}$

(b)

(i) $$C_6H_5CH{=}O + :\overset{\ominus}{C}H_2NO_2 \longrightarrow C_6H_5CH(\ddot{O}:^{\ominus})-CH_2NO_2 \xrightarrow{H_2O} C_6H_5-CH(OH)-CH_2-NO_2 \xrightarrow{:\ddot{O}H^{\ominus}} C_6H_5CH{=}CHNO_2$$

(ii) $$C_6H_5CH_2NO_2 \xrightarrow{OH^{\ominus}} C_6H_5\overset{\ominus}{\ddot{C}}HNO_2 \xrightarrow{Br-Br} C_6H_5CHBrNO_2 + Br^{\ominus}$$

(iii) $$(CH_3)_2CHNO_2 \xrightarrow{OH^{\ominus}} (CH_3)_2\overset{\ominus}{\ddot{C}}NO_2 \xrightarrow{CH_2=CHCN} \xrightarrow{H_2O} (CH_3)_2C(NO_2)CH_2CH_2CN$$

31. $$C_6H_4(CH_2-C{\equiv}N)_2 \xrightarrow{OC_2H_5^{\ominus}} C_6H_4(CH_2-C{\equiv}N:)(\overset{\ominus}{\ddot{C}}H-C{\equiv}N) \longrightarrow C_6H_4[CH_2-C(=\ddot{N}:^{\ominus})-CH(CN)] \xrightarrow{HOC_2H_5}$$

(continued next page)

H_2O: ; $H_3O^{\oplus}$; $C=\ddot{N}H$; A ; $+ C_2H_5O^{\ominus}$; $=NH_2^{\oplus}$; H CN ; $OH_2^{\oplus}$; NH_2 ; $O-H$; $:\ddot{O}H_2$; $\overset{\oplus}{N}H_3$; $=O$

32. A can be formed from $C_2H_5OOC(CH_2)_4\underset{CH_3}{CHCOOC_2H_5}$

A $\xrightarrow{OC_2H_5^{\ominus}}$; CH_3 ; $CO_2C_2H_5$; $\ddot{O}:^{\ominus}$; OC_2H_5 ; $C-CO_2C_2H_5$; OC_2H_5 ; H H O ; H_3C ; $:O:$; COC_2H_5 ; OC_2H_5 ; H O ; CH_3 ; $\ddot{O}^{\ominus}$; OC_2H_5 ; COC_2H_5 ; O ; CH_3 ; O ; COC_2H_5 ; O

•(see problem 11, p. 844)

33. $C_6H_5-\underset{H}{\overset{H}{C}}-\underset{C_6H_5}{\overset{CN}{C}}-C_6H_5$; $H_2\ddot{N}:^{\ominus}$ $\longrightarrow$ $(C_6H_5)(H)C=C(C_6H_5)_2$ + $CN^{\ominus}$ + $:NH_3$

34. $C_6H_5CH_2CN \xrightarrow{CN^{\ominus}} C_6H_5\overset{\ominus}{C}HCN \xrightarrow{C_6H_5C(=O)-H} C_6H_5-C(H)(CN)-C(H)(O^{\ominus})-C_6H_5$

1. HCN
2. $-H_2O$

$C_6H_5(NC)C=C(H)C_6H_5$ + $:N\equiv C:^{\ominus}$ ⟶ $C_6H_5-\overset{\ominus}{C}(NC)-C(H)(CN)-C_6H_5$ $\xrightarrow{H_2O}$ $C_6H_5-C(H)(NC)-C(H)(CN)-C_6H_5$

Addition of $CN^{\ominus}$ occurs at the carbon of the carbon-carbon double which leads to the lowest energy anion, i.e., the anion resonance stabilized by the other -CN group.

$C_6H_5-\overset{\ominus}{C}(NC)-C(H)(CN)-C_6H_5$ more stable than $C_6H_5-C(NC)(NC)-\overset{\ominus}{C}(H)-C_6H_5$

35. $C_2H_5OC(=O)-\overset{\ominus}{C}H-CH_2-C(=O)OC_2H_5$ + $C_2H_5O-C(=O)-CH_2CH_2C(=O)OC_2H_5$ ⟶ $C_2H_5OC(=O)-CH(-C(O^{\ominus})(OC_2H_5)-CH_2-CH_2-C(=O)OC_2H_5)-CH_2C(=O)OC_2H_5$

$\xrightarrow{-C_2H_5O^{\ominus}}$ $C_2H_5OC(=O)-CH(-C(=O)-CH_2-CH_2-C(=O)OC_2H_5)-CH_2-C(=O)OC_2H_5$

$\xrightarrow{OC_2H_5^{\ominus}}$ $C_2H_5OC(=O)-CH(-C(=O)-CH_2-\overset{\ominus}{C}H-C(=O)OC_2H_5)-CH_2-C(=O)OC_2H_5$

↓

(continued next page)

$$\xrightarrow{-C_2H_5O^{\ominus}}$$

$C_2H_5O_2C$–(cyclohexane ring with $:\ddot{O}:^{\ominus}$, OC_2H_5)–$CO_2C_2H_5$, =O $\longrightarrow$ $C_2H_5O_2C$–(cyclohexane-1,4-dione)–$CO_2C_2H_5$

36. $NC\text{-}CH\text{-}CN$ (H abstracted by $C_5H_5N:$) $\longrightarrow$ $C_6H_5\overset{\oplus}{N}H$ + $NC\text{-}\overset{\ominus}{\ddot{C}H}\text{-}CN$ (resonance stabilized)

$\downarrow$ $C_6H_5CH=\ddot{O}:$

$NC\text{-}CH\text{-}CN$, C_6H_5CH, $:\ddot{O}:^{\ominus}$ $\xleftarrow{C_5H_5\overset{\oplus}{N}\text{-}H}$ $NC\text{-}C(H)\text{-}CN$, C_6H_5CH, $:\ddot{O}H$ (H removed by $:NC_6H_5$)

$\downarrow$

$C_6H_5CH=C(CN)_2$

37. $C_6H_5\text{-}C(C_6H_5)(H)\text{-}CN$ + $H_2\ddot{N}:^{\ominus}$ $\longrightarrow$ $C_6H_5\text{-}\overset{\ominus}{\ddot{C}}(C_6H_5)\text{-}CN$ + $C_6H_5CH_2\text{-}Br$ $\longrightarrow$ $C_6H_5\text{-}C(C_6H_5)(CH_2C_6H_5)\text{-}CN$

(also see problem 33)

38. $3C_6H_6 + PCl_3 \xrightarrow{AlCl_3} (C_6H_5)_3P + 3HCl$

39. $CH_3\text{-}\overset{O}{\overset{\|}{C}}\text{-}CH_3 + (C_6H_5)_3\overset{\oplus}{P}\text{-}\overset{\ominus}{\ddot{C}}H_2 \longrightarrow (CH_3)_2C=CH_2$

There is only one monoalkene.

40.

A

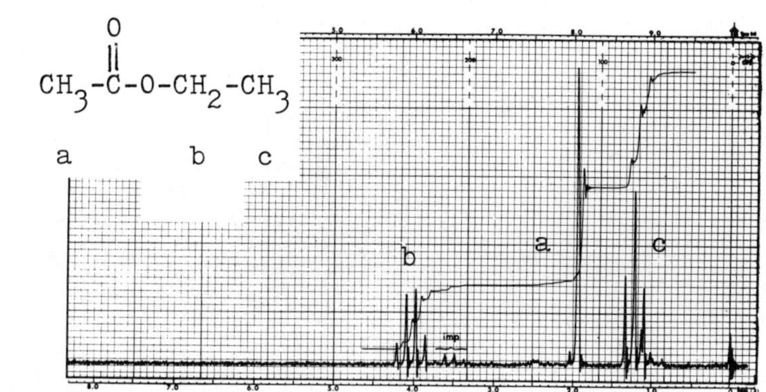

B

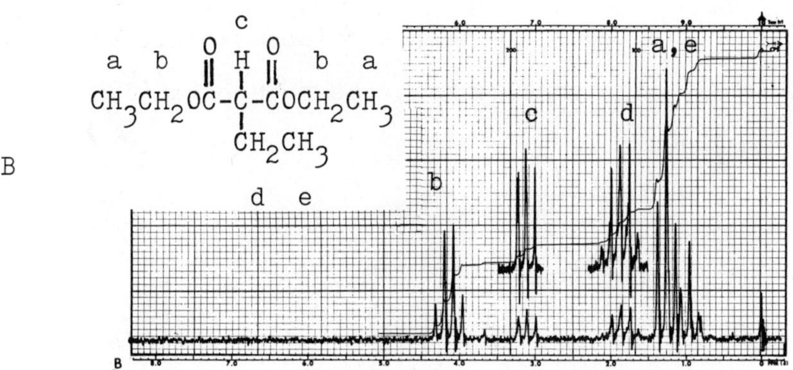

C

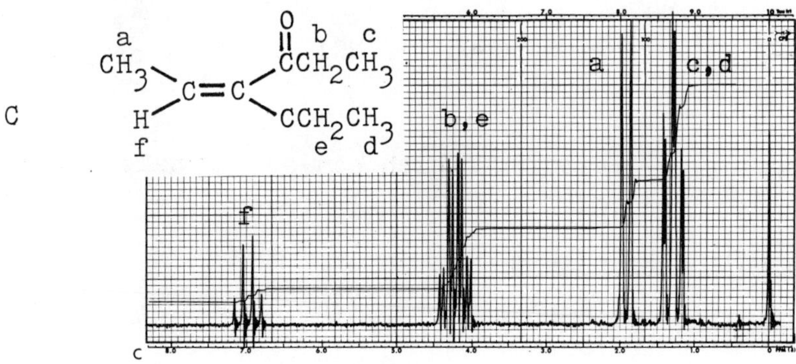

(continued next page)

D

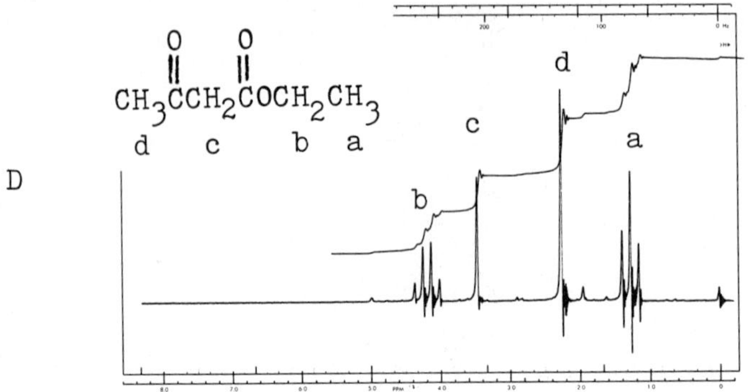

Chemical means of distinguishing these compounds. Only diethylethylidenemalonate will decolorize Br_2/CCl_4. Only ethyl acetate will <u>not</u> decarboxylate following hydrolysis to the acid (evolution of CO_2). A sample of acetoacetic ester (two acidic hydrogens) will consume twice as much base ($C_2H_5O^{\ominus}$) as diethyl ethylmolonate.

41. $$CH_3\text{-}\overset{O}{\overset{\|}{C}}\text{-}CH_2\text{-}\overset{O}{\overset{\|}{C}}\text{-}O\text{-}CH_3$$

2.12δ 3.35δ 3.62δ

<u>SUMMARY OF REACTIONS</u>

1. <u>α-halo acids</u> (Hell-Volhard-Zelensky reaction, p. 836)

$$RCH_2\overset{O}{\overset{\|}{C}}OH \xrightarrow{X_2/PX_3} \underset{\underset{X}{|}}{RCH}\overset{O}{\overset{\|}{C}}\text{-}OH + HCl \quad (X = Cl \text{ or } Br)$$

see summary of reactions which convert α-halo acids into other acids and esters, p. 838-9.

2. <u>Condensation Reactions</u>

(a) <u>Claisen</u> (p. 840)

$$2RCH_2\overset{O}{\overset{\|}{C}}OC_2H_5 \xrightarrow[C_2H_5OH]{Na^{\oplus}\ {}^{\ominus}OC_2H_5} \xrightarrow{H_3O^{\oplus}} RCH_2\overset{O}{\overset{\|}{C}}\underset{\underset{R}{|}}{CH}\overset{O}{\overset{\|}{C}}OC_2H_5$$

(continued next page)

(b) Crossed Claisen (p. 842)

$$RCH_2\overset{O}{\overset{\|}{C}}OC_2H_5 + -\underset{|}{\overset{|}{C}}\overset{O}{\overset{\|}{C}}OC_2H_5 \xrightarrow[C_2H_5OH]{Na^{\oplus}\ {}^{\ominus}OC_2H_5} \xrightarrow{H_3O^{\oplus}} -\underset{|}{\overset{|}{C}}\overset{O}{\overset{\|}{C}}\underset{R}{\underset{|}{C}}H\overset{O}{\overset{\|}{C}}OC_2H_5$$

no hydrogens to C=O

(c) Dieckmann (p. 844)

$$C_2H_5O\overset{O}{\overset{\|}{C}}(CH_2)_n\overset{O}{\overset{\|}{C}}OC_2H_5 \xrightarrow[C_2H_5OH]{Na^{\oplus}\ {}^{\ominus}OC_2H_5} \xrightarrow{H_3O^{\oplus}} (CH_2)_{n-1}\ \text{ring with } C{=}O \text{ and } C\text{-}H(CO_2C_2H_5)$$

(d) Perkin (p. 844)

$$Ar\overset{O}{\overset{\|}{C}}\text{-}H + RCH_2\overset{O}{\overset{\|}{C}}O\overset{O}{\overset{\|}{C}}CH_2R \xrightarrow[heat]{RCH_2CO_2^{\ominus}} ArCH{=}\underset{R}{\underset{|}{C}}\overset{O}{\overset{\|}{C}}OH$$

(e) Knoevenagel (p. 845)

$$RO\overset{O}{\overset{\|}{C}}CH_2\overset{O}{\overset{\|}{C}}OR + R'\text{-}\overset{O}{\overset{\|}{C}}\text{-}H \xrightarrow{\text{piperidine (N-H)}} R'HC{=}C(\overset{O}{\overset{\|}{C}}OR)_2$$

3. Alkylation of Anions

(a) Acetoacetic ester synthesis (p. 847)

$$CH_3\overset{O}{\overset{\|}{C}}CH_2\overset{O}{\overset{\|}{C}}OC_2H_5 \xrightarrow[C_2H_5OH]{Na^{\oplus}\ {}^{\ominus}OC_2H_5} \xrightarrow{R\text{-}X} CH_3\overset{O}{\overset{\|}{C}}\underset{R}{\underset{|}{C}}H\overset{O}{\overset{\|}{C}}OC_2H_5$$

$$\xrightarrow{Na^{\oplus}\ {}^{\ominus}OC_2H_5} \xrightarrow{R'\text{-}X} CH_3\overset{O}{\overset{\|}{C}}\text{-}\underset{R}{\underset{|}{\overset{R'}{\overset{|}{C}}}}\text{—}\overset{O}{\overset{\|}{C}}OC_2H_5 \xrightarrow{H_3O^{\oplus}}$$

(monoalkylated product) $\xrightarrow{H_3O^{\oplus}}$

(continued next page)

$$\downarrow$$

$$CH_3\overset{O}{\overset{\|}{C}}CRR'\overset{O}{\overset{\|}{C}}OH \xrightarrow{\text{heat}} CH_3\overset{O}{\overset{\|}{C}}CHRR' + CO_2$$

$$\downarrow$$

$$CH_3\overset{O}{\overset{\|}{C}}\underset{R}{\underset{|}{C}}H\overset{O}{\overset{\|}{C}}OH \xrightarrow{\text{heat}} CH_3\overset{O}{\overset{\|}{C}}CH_2R + CO_2$$

(b) Malonic ester synthesis (p. 848)

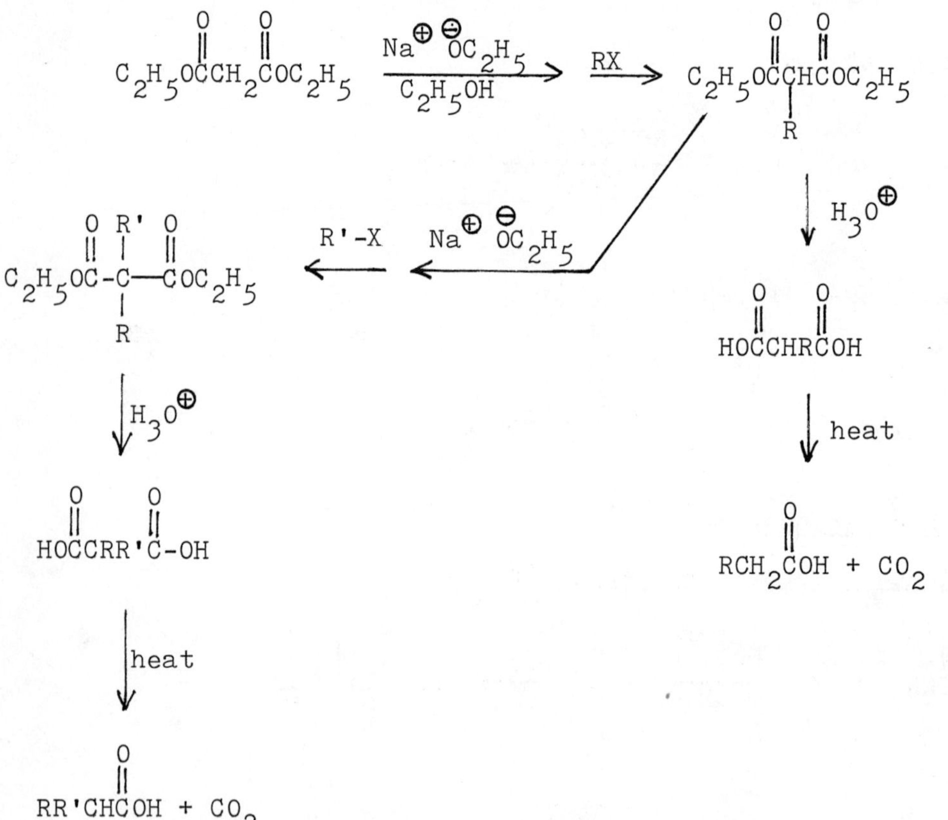

4. Michael Additions (p. 849)

General: $Nu:^{\ominus} + \rangle C=\overset{|}{C}-\overset{|}{C}=O \longrightarrow \xrightarrow{H_3O^{\oplus}} Nu-\overset{|}{\underset{|}{C}}-\overset{|}{\underset{H}{C}}-\overset{|}{C}=O$

(a) $-\overset{O}{\overset{\|}{C}}-CH_2-\overset{O}{\overset{\|}{C}}- + \rangle C=\overset{|}{C}-\overset{|}{C}=O \xrightarrow{B^{\ominus}} \xrightarrow{H_3O^{\oplus}} (-\overset{O}{\overset{\|}{C}}\text{-})_2CH-\overset{|}{\underset{|}{C}}-\overset{|}{\underset{H}{C}}-\overset{|}{C}=O$

(b) cyanoethylation (see p. 853 for summary)

$Nu:^{\ominus} + \rangle C=\overset{|}{C}-C\equiv N \longrightarrow \xrightarrow{H_3O^{\oplus}} Nu-\overset{|}{\underset{|}{C}}-\overset{|}{\underset{H}{C}}-C\equiv N$

(c) Robinson annelation (p. 850, Michael + aldol condensations)

cyclohexanone $\xrightarrow[\text{base}]{H_2C=CH\overset{O}{\overset{\|}{C}}CH_3}$ 2-(3-oxobutyl)cyclohexanone (C=O, CH_3) $\xrightarrow{\text{base}}$ bicyclic ketol (OH, =O) $\xrightarrow{-H_2O}$ bicyclic enone (=O)

Michael adduct

5. Reformatsky reaction (p. 851)

$BrCH_2\overset{O}{\overset{\|}{C}}OC_2H_5 \xrightarrow[\text{ether}]{Zn} BrZnCH_2\overset{O}{\overset{\|}{C}}OC_2H_5 \xrightarrow{R-\overset{O}{\overset{\|}{C}}-R} R-\overset{R}{\overset{|}{\underset{OZnBr}{\underset{|}{C}}}}-CH_2\overset{O}{\overset{\|}{C}}OC_2H_5$

$\downarrow H_2O$

$R-\overset{R}{\overset{|}{\underset{OH}{\underset{|}{C}}}}-CH_2\overset{O}{\overset{\|}{C}}OC_2H_5$

6. Wittig reaction (p. 854)

$$(C_6H_5)_3P + RCH_2X \longrightarrow (C_6H_5)_3\overset{\oplus}{P}CH_2R\ X^{\ominus} \xrightarrow{\text{base}} (C_6H_5)_3\overset{\oplus}{P}-\overset{\ominus}{\ddot{C}}HR$$

$$\downarrow R'CHO$$

$$(C_6H_5)_3PO + R'CH=CHR$$

7. Acyloin condensation (p. 857)

$$(CH_2)_n\text{-ring with } C(=O)OC_2H_5 \text{ and } C(=O)OC_2H_5 \xrightarrow{2Na} (CH_2)_n\text{-ring with } C\text{-}O^{\ominus} = C\text{-}O^{\ominus} \xrightarrow{H_2O} (CH_2)_n\text{-ring with } C=O \text{ and } C\text{-}OH$$

CHAPTER 21 - AMINES

LEARNING OBJECTIVES

When you have completed this chapter, you should be able to:

1. provide names or structures for amines (problems 1, 25, 26);
2. demonstrate the configurational instability of amines (problem 3);
3. predict products in reactions of amines (problems 4, 42, 48);

(continued next page)

4. write equations which illustrate the synthesis of amines (problems 6, 37, 38, 39);
5. write equations for the multistep synthesis of various compounds based on reactions in this and previous chapters (problems 32, 33, 34, 36);
6. write mechanisms for reactions of amines and other compounds (problems 8, 9, 17, 23, 41, 43, 45, 46, 47, 49);
7. suggest reactants in the Bischler-Napieralski reaction (problem 10);
8. draw resonance structures for some compounds containing nitrogen (problems 11, 14);
9. suggest reactants and products in enamine reactions (problems 15, 16, 19);
10. write mechanisms for enamine reactions (problems 17, 20);
11. account for products in syn-elimination reactions (problems 21, 22);
12. suggest tests and procedures for identifying amines and for distinguishing amines from other compounds (problems 27, 30, 31);
13. identify products in the Hofmann elimination reaction (problem 40);
14. make assignments in the NMR and IR spectra of amines (problem 50);
15. identify or distinguish compounds based on their spectra (problems 51, 52, 53, 54).

ANSWERS TO QUESTIONS

1. (a) methylamine
   (b) dimethylammonium chloride
   (c) N-ethylaniline
   (d) dimethyldiphenylammonium hydroxide
   (e) 2-aminopyridine
   (f) 1-chloro-2,2-dimethylaziridine
   (g) N-methyl-p-toluidine

2. Azanonatetraene is aromatic. The 10 π electrons in the system include the nonbonding pair on nitrogen. The electron density in the ring of the amide is reduced because of resonance with the $-COOC_2H_5$ group decreasing the aromaticity (see resonance structure B).

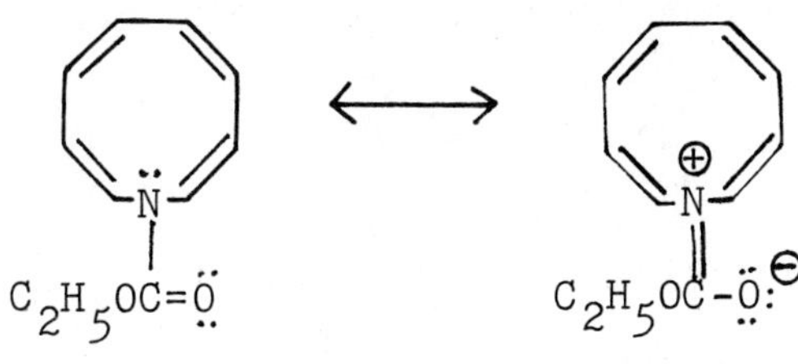

3. (a) If the amine is configurationally stable, the methylene protons of $C_6H_5CH_2$- should be chemical-shift nonequivalent and constitute an AB system. If configurationally unstable, they should appear as a single line.

(b) Inversion at nitrogen

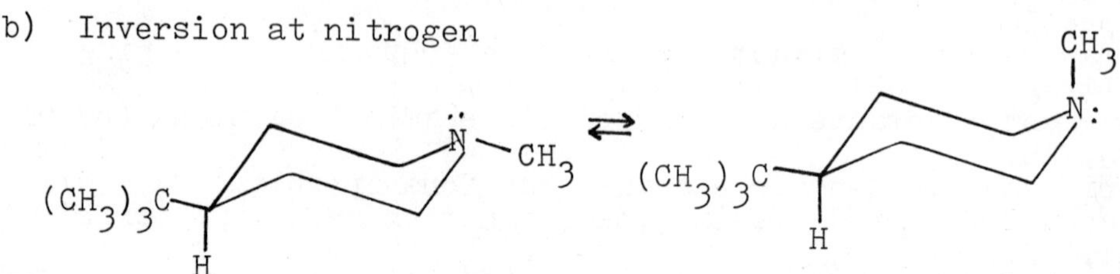

4. (a) $CH_3\text{-}CH(C_2H_5)\text{-}CH_2\text{-}NH_3^{\oplus}$ $\quad$ $CH_3CH(OH)COO^{\ominus}$

$CH_3\text{-}CH(C_2H_5)\text{-}CH_2\text{-}NH_3^{\oplus}$	$CH_3CH(OH)COO^{\ominus}$
R	R
S	R

(b) $CH_3CH(C_2H_5)\text{-}CH_2\text{-}\overset{\oplus}{N}H(CH_3)C_2H_5$ $\quad$ $CH_3CH(OH)COO^{\ominus}$

$CH(C_2H_5)$	$N(CH_3)$	$CH_3CH(OH)COO^{\ominus}$
R	R	R
S	R	R
R	S	R
S	S	R

5. No. The potassium salt of the imide is (maleimide anion: ring with two C=O and $:\overset{\ominus}{N}:$) $K^{\oplus}$ containing $4\ \pi\ e^{\ominus}$ (consider the resonance form (ring with two $\overset{\oplus}{C}\text{-}\ddot{O}:^{\ominus}$ groups and $:N$)).

6. (a) $\xrightarrow[H_2/Ni]{NH_3}$

(b) (1) $\xrightarrow{PBr_3}$ CH_3Br $\xrightarrow{\text{(phthalimide: O, O, } N^{\ominus}\!: K^{\oplus})}$ $\xrightarrow{OH^{\ominus}/H_2O}$

(2) CH_3OH $\xrightarrow{CrO_3/\text{pyridine}}$ CH_2O $\xrightarrow[H_2/Ni]{NH_3}$

(c) $\xrightarrow{PBr_3}$ CH_3Br $\xrightarrow{CH_3CH_2NH_2}$ $CH_3CH_2\overset{\oplus}{N}H_2CH_3\,Br^{\ominus}$ $\xrightarrow{OH^{\ominus}/H_2O}$

(d) $\xrightarrow{SOCl_2}$ $CH_3\overset{O}{\overset{\|}{C}}Cl$ $\xrightarrow{NH_3}$

(e) $\xrightarrow{C_6H_5\overset{O}{\overset{\|}{C}}\text{-}Cl}$

(f) (1) $\xrightarrow{SOCl_2}$ $CH_3CH_2CH_2\overset{O}{\overset{\|}{C}}Cl$ $\xrightarrow{NH_3}$ $\xrightarrow{Br_2/H_2O/OH^{\ominus}}$

(2) $CH_3CH_2CH_2\overset{O}{\overset{\|}{C}}Cl$ $\xrightarrow{Na^{\oplus}\ N_3^{\ominus}}$ $\xrightarrow{\text{heat}}$ $\xrightarrow{H_2O}$

(g) $CH_3CH_2CH_2CH_2OH$ $\xrightarrow[H^{\oplus}]{KMnO_4}$ $CH_3CH_2CH_2CO_2H$, then as in f

(h) $\xrightarrow{Na^{\oplus}\ CN^{\ominus}}$ CH_3CN $\xrightarrow[\text{(excess)}]{LiAlH_4}$ $\xrightarrow{H_2O}$ (or $\xrightarrow{H_2/Pt}$)

(i) $\xrightarrow[\text{(excess)}]{LiAlH_4}$ $\xrightarrow{H_2O}$

7. A possible mechanism is:

EtO2C H H CN * EtO2C H H CN ⊖ CH3 NH2 NH2 CH3 CH3 NH2 ⊕ CH3 NH2

EtO2C H H CN EtO2C H H CN ⊖ CH3 N CH3 CH3 NH CH3 H H3N: NH3 ⊕

EtO2C H H CN # ArCHO H3N: EtO2C H H CN CH-Ar OH CH3 N ⊖ CH3 CH3 N CH3

H Ar H EtO2C CN Ar EtO2C H CN tautomerize CH3 N CH3 CH3 N CH3

(continued next page)

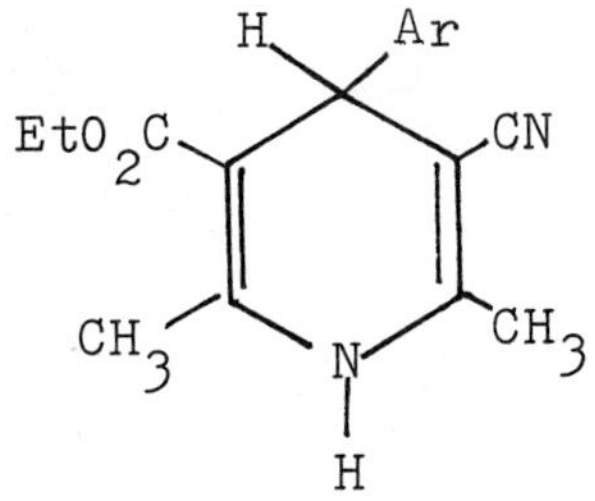

H Ar

EtO_2C CN

CH_3 N CH_3

H

*resonance stabilized

$-\overset{|}{\underset{\ominus}{C}}-C{\equiv}N: \longleftrightarrow -\overset{|}{C}=C=\underset{\ominus}{N}:$

#resonance stabilized

$-\underset{\ominus}{N}-\overset{|}{C}=\overset{|}{C}-C{\equiv}N: \longleftrightarrow -N=\overset{|}{C}-\overset{|}{C}=C=\underset{\ominus}{N}:$

8. The carbonyl group undergoes a facile addition-elimination to regenerate the starting material.

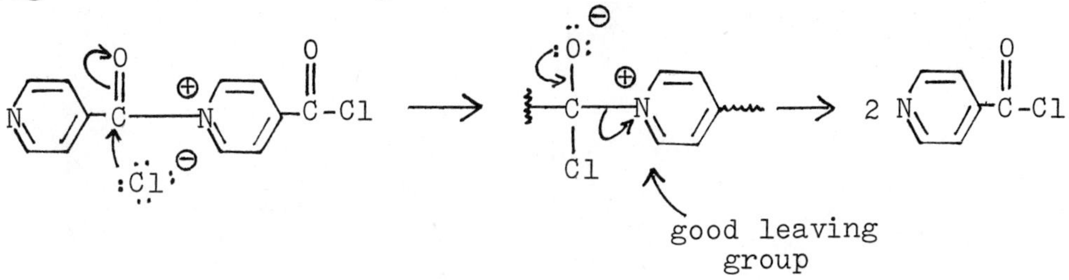

good leaving group

9. 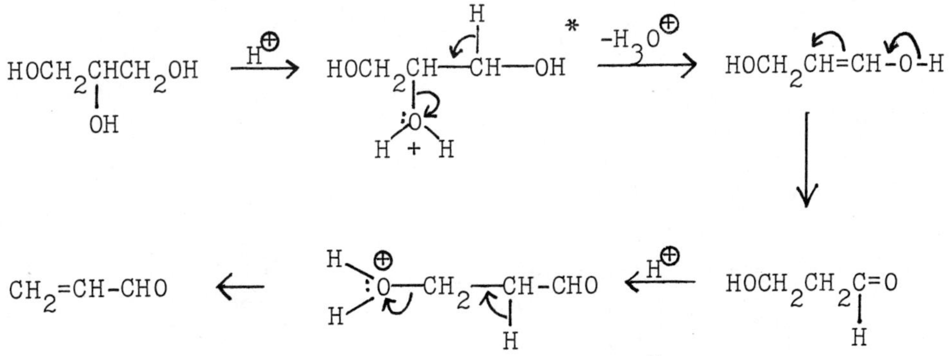 $HOCH_2CHCH_2OH$ (OH) $\xrightarrow{H^{\oplus}}$ $HOCH_2CH—CH—OH$ (H, $:\overset{\oplus}{O}H_2$) * $\xrightarrow{-H_3O^{\oplus}}$ $HOCH_2CH=CH-O-H$ $\longrightarrow$ $HOCH_2CH_2C=O$ (H) $\xrightarrow{H^{\oplus}}$ $H_2\overset{\oplus}{O}-CH_2-CH-CHO$ (H) $\longrightarrow$ $CH_2=CH-CHO$

*or $HOCH_2CH-CH_2OH$ ($\oplus OH_2$) $\xrightarrow{-H_2O}$ $HOCH_2\overset{\oplus}{C}H—CHOH$ (H $\leftarrow :\ddot{O}H_2$) $\xrightarrow{etc}$

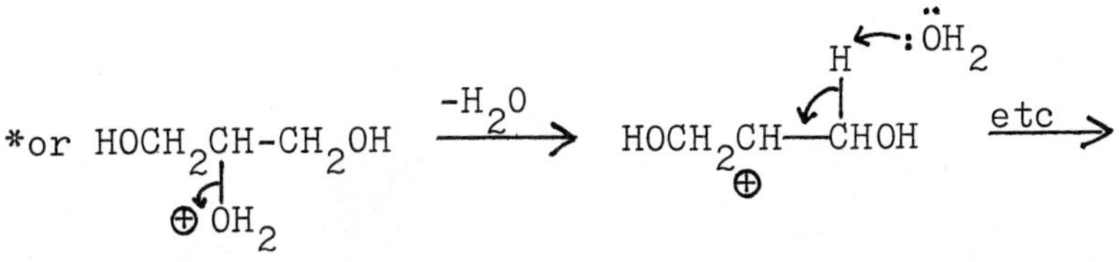

10. (a) $C_6H_6 \xrightarrow[Fe]{Br_2} C_6H_5Br \xrightarrow[ether]{Mg} C_6H_5MgBr \xrightarrow{H-\overset{O}{\overset{\|}{C}}-H} \xrightarrow{H_3O^{\oplus}}$

$C_6H_5CH_2OH \xrightarrow{PCl_3} C_6H_5CH_2Cl \xrightarrow[DMSO]{Na^{\oplus}\ CN^{\ominus}} C_6H_5CH_2CN$

$\xrightarrow[AlCl_3]{LiAlH_4} \xrightarrow{H_2O} C_6H_5CH_2CH_2NH_2 \xrightarrow{CH_3\overset{O}{\overset{\|}{C}}-Cl} C_6H_5CH_2CH_2NH\overset{O}{\overset{\|}{C}}CH_3$

(b) (i) $3\text{-}CH_3C_6H_4CH_2CH_2N(CH_3)-HC{=}O$ (ii) $C_6H_5CH(CH_3)CH_2N(CH_2C(CH_3)_3)-HC{=}O$

11. $(CH_3)_2\ddot{N}-\ddot{N}=\ddot{O}: \longleftrightarrow (CH_3)_2\overset{\oplus}{N}=\ddot{N}-\ddot{O}:^{\ominus}$

12. (a) $R_3N: + \overset{\oplus}{N}=O \longrightarrow R_3\overset{\oplus}{N}-N=O$

3° amine

(b) $R_2HN: + \overset{\oplus}{N}=O \longrightarrow R_2\overset{\oplus}{N}(H)-N=O \longrightarrow R_2\ddot{N}-N=O$ (H removed by $H_2O:$)

2° amine

13. (a) $R-\ddot{N}(H)-\ddot{N}=\ddot{O}: \overset{H^+}{\rightleftharpoons} R-\ddot{N}(H)-\ddot{N}=\overset{\oplus}{\ddot{O}}H \longrightarrow R-\ddot{N}=\ddot{N}-\ddot{O}H$ (H removed by $H_2\ddot{O}:$)

(A) (B)

(continued next page)

(b) 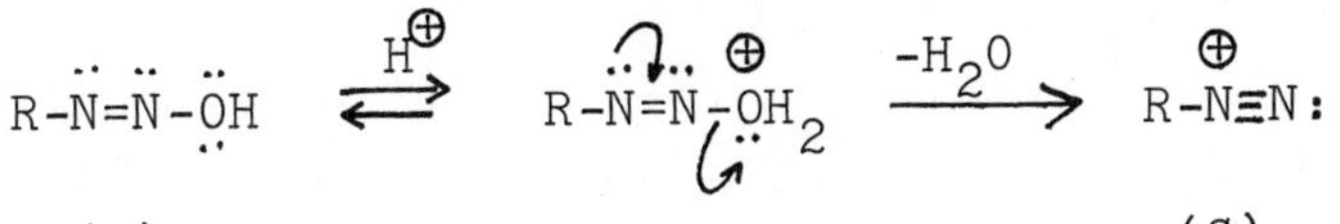 $R{-}\ddot{N}{=}\ddot{N}{-}\ddot{O}H \underset{}{\overset{H^{\oplus}}{\rightleftharpoons}} R{-}\ddot{N}{=}\ddot{N}{-}\overset{\oplus}{O}H_2 \xrightarrow{-H_2O} R{-}\overset{\oplus}{N}{\equiv}N:$

(B) (C)

14\. (a) (1) conversion of carboxylic acids to methyl esters;

(2) synthesis of cyclopropanes;

(3) reaction with acyl halides in the Arndt-Eistert reaction.

(b) 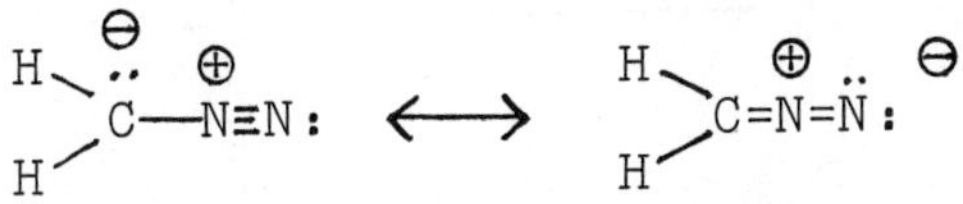 $H_2\overset{\ominus}{\ddot{C}}{-}\overset{\oplus}{N}{\equiv}N: \longleftrightarrow H_2C{=}\overset{\oplus}{N}{=}\overset{\ominus}{\ddot{N}}:$

15\. (a) $CH_3CH_2CH{=}O$ and $C_6H_5NHCH_3$

(b) $(CH_3)_2CHCH{=}O$ and piperidine (NH)

(c) 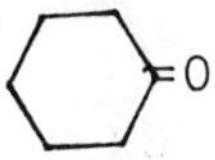 cyclohexanone (=O) and piperidine (NH)

16\. $H^{\oplus}$ + 1-(dimethylamino)cyclohexene (N, CH_3, CH_3) ⟶ cyclohexylidene$=\overset{\oplus}{N}(CH_3)_2$ + $:\ddot{O}H_2$ ⟶ 1-$\ddot{N}(CH_3)_2$-1-$\overset{\oplus}{O}H_2$-cyclohexane ⇅ 1-$\overset{\oplus}{N}H(CH_3)_2$-1-O-H-cyclohexane + $H_2\ddot{O}:$ ⟶ $(CH_3)_2\ddot{N}H$ + cyclohexanone (=O)

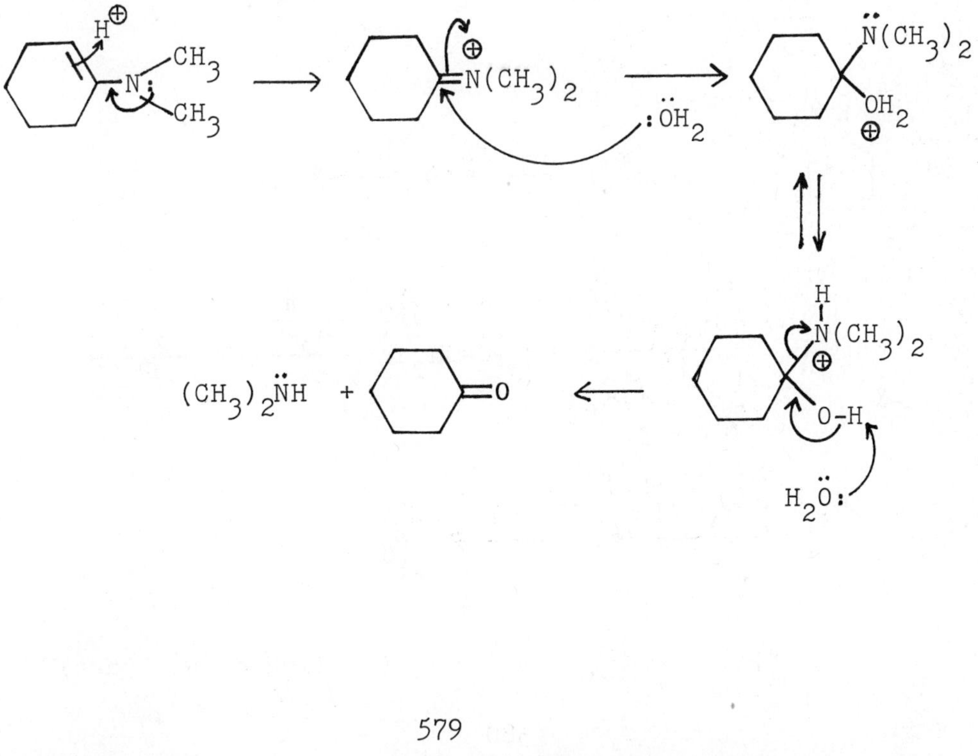

17. CH_3CH_2Cl → $Cl^{\ominus}$ CH_2CH_3

18. (a) $\xrightarrow{LiAlH_4}$ $\xrightarrow{H_2O}$ (or $\xrightarrow{H_2/Pt}$)

(b) $\xrightarrow{CH_3MgX}$ $\xrightarrow{H_2O}$

(c) $\xrightarrow[H^{\oplus}]{}$ $\xrightarrow{CH_3I}$ $I^{\ominus}$ CH_3

$\xrightarrow{H_2O}$

(d) CH_3 (from c) $\xrightarrow{LiAlH_4}$ $\xrightarrow{H_2O}$

19. (a) $CH_3\overset{O}{\overset{\|}{C}}CH_3$ $\xrightarrow[H^{\oplus}]{(CH_3)_2NH}$ $CH_3\text{-}\overset{N(CH_3)_2}{\overset{|}{C}}\text{=}CH_2$ $\xrightarrow{CH_3\overset{O}{\overset{\|}{C}}Cl}$ $\xrightarrow[H^{\oplus}]{H_2O}$

(b) $\xrightarrow[H^{\oplus}]{(CH_3)_2NH}$ $\xrightarrow{CH_3\overset{O}{\overset{\|}{C}}Cl}$ $\xrightarrow[H^{\oplus}]{H_2O}$

(continued next page)

(c) $(CH_3)_2CHCCH_2CH_3$ (C=O) $\xrightarrow[H^{\oplus}]{\text{pyrrolidine (N-H)}}$ $(CH_3)_2CH\text{-}C\text{=}CHCH_3$ (C bearing pyrrolidin-1-yl) $\xrightarrow{C_6H_5CH_2CCl\ (C=O)}$

(minor)

$\xrightarrow[H^{\oplus}]{H_2O}$

(d) cyclohexanone $\xrightarrow[H^{\oplus}]{(CH_3)_2NH}$ $\xrightarrow{\text{cyclohexyl-CCl (C=O)}}$ $\xrightarrow[H^{\oplus}]{H_2O}$ 2-(cyclohexylcarbonyl)cyclohexanone

$\xrightarrow{EtO^{\ominus}/EtOH}$ $\xrightarrow{CH_3Cl}$

20. $\text{-}\ddot{N}\text{—}C\text{=}C\text{-}$ (α, β) C_β is nucleophilic. Nitrogen provides the electrons,

i.e., $\text{-}\ddot{N}\text{—}C\text{=}C\text{-}$ $E^{\oplus}$

21. The groups being eliminated are removed from the same side (eclipsed in the transition state).

$\ominus$ $:\ddot{O}$ $N^{\oplus}$ H C—C

22. (a) The ester pyrolysis, like the Cope reaction, is a stereospecific syn (cis)-elimination. The reaction is concerted and proceeds without isomerization. For example:

CH_3, O=C, O, H, C—C, CH_3, CH_3, CD_3, CD_3 $\longrightarrow$ CH_3, CH_3, C=C, CD_3, CD_3 + CH_3COOH

(continued next page)

(b) Dehydration of the corresponding alcohol involves formation of a carbocation intermediate. From this cation both cis and trans products are formed (in general, rearrangement may occur).

23. (a) and (c)

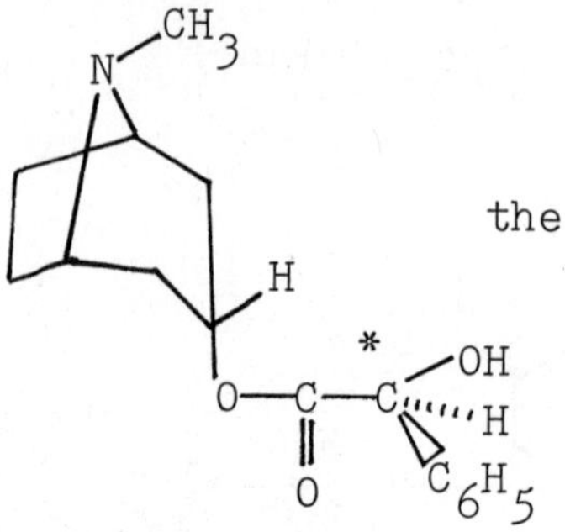

N-CH3 (tropane ester structure): O—C(=O)—C*(OH)(H)(C6H5)

the chiral center is starred

(b) ~C(=O)-C(OH)(H)-C6H5 (R or S) $\xrightarrow{B:^{\ominus}}$ [~C(=O)-C⁻(OH)-C6H5 ⟷ ~C(O⁻)=C(OH)-C6H5] achiral

$\downarrow$ HB

~C(=O)-CH(OH)-C6H5 (R and S)

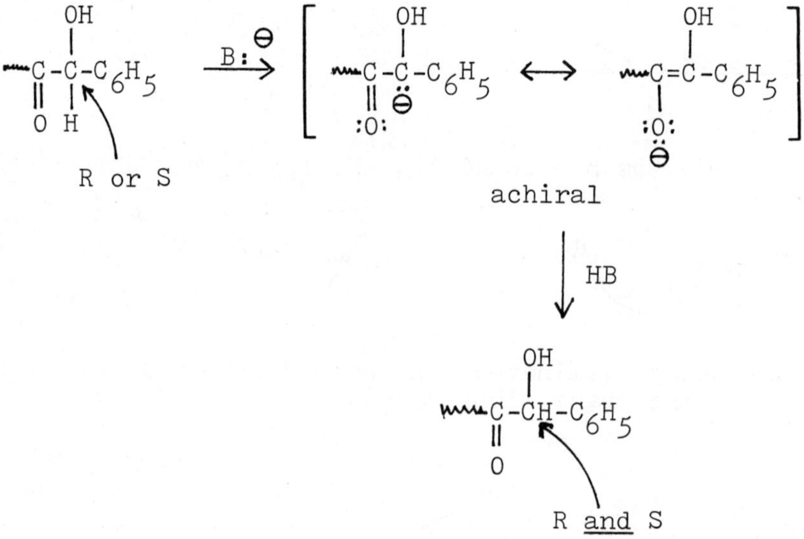

24. 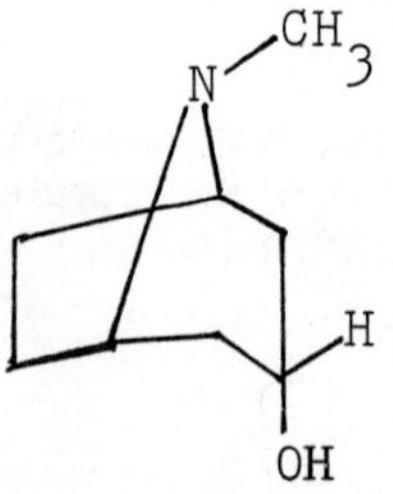 tropine and $C_6H_5CHCO_2H$ (with OH) (±)-tropic acid

25. (a) allylamine, 1°
(b) allylmethylamine, 2°
(c) methyl-n-propylamine, 2°
(d) aniline, 1°
(e) aminoacetic acid, 1°
(f) dimethylisopropylamine, 3°
(g) tri-t-butylamine, 3°
(h) 1-azaanthracene, 3°
(i) N-methyl-N-phenyl-3-aminopropanenitrile, 3°
(j) 3-phenylpyridine, 3°
(k) 1-methylaziridine, 3°
(l) 2-methylaziridine, 2°
(m) N-methylmorpholine, 3°
(n) N-methyl-m-nitroaniline, 2°
(o) N-methylpiperidine, 3°
(p) 10-methylacridine-N-oxide

26. (a) 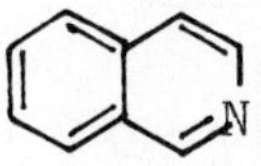 N

(b) 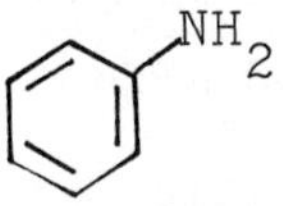 NH_2

(c) 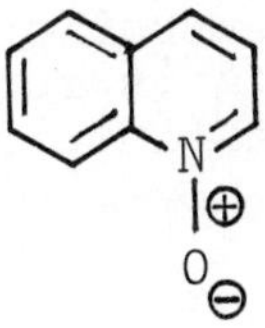 N ⊕ O ⊖

(d) 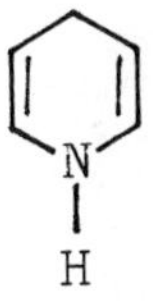 N H

(e) $CH_3CH_2C(CH_3)_2-C(=O)-NH_2$

(f) 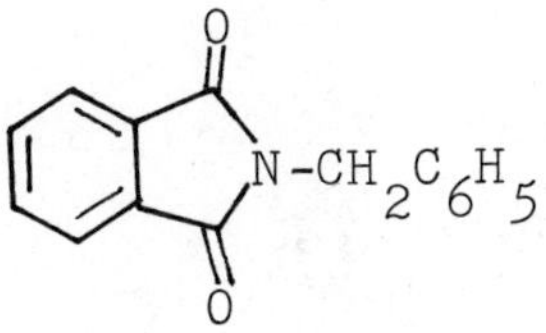 O O $N-CH_2C_6H_5$

(g) $(C_6H_5)_2NCH_2CH_2CH=CHC(=O)-OC_2H_5$

(h) $C_6H_5N=C=O$

(i) $CH_3-C(CH_3)_2-NH_2$

(j) $CH_3CH_2CH_2CH_2CH_2NH_2$

(k) $CH_3-NH-CH(C_2H_5)CH_2CH_3$

(l) $(C_2H_5)_3N$

(continued next page)

(m) $H_2NCH_2CH_2NH_2$

(n) $CH_3CH_2-\overset{\oplus}{N}(CH_3)_2$ with C_6H_5 on N, $OH^{\ominus}$

27. The Hinsberg test is conducted by shaking a sample of amine with benzenesulfonyl chloride in dilute aqueous sodium hydroxide solution.

1° amine:

$$R\text{-}NH_2 \xrightarrow[NaOH]{C_6H_5SO_2Cl} [C_6H_5SO_2NHR] \xrightarrow{NaOH} C_6H_5SO_2\overset{\ominus}{N}R\ Na^{\oplus} \quad (A)$$

soluble

$\downarrow H_3O^{\oplus}$

$$C_6H_5SO_2NHR \quad (B)$$

insoluble

Observations: When the amine is shaken with benzenesulfonyl chloride a clear solution is formed (soluble salt A is formed). Acidification of this solution gives a precipitate (insoluble sulfonamide B).

2° amine:

$$R_2NH \xrightarrow[NaOH]{C_6H_5SO_2Cl} C_6H_5SO_2NR_2 \xrightarrow{NaOH} NR$$

insoluble

Observations: When the amine is shaken with benzenesulfonyl chloride an insoluble layer separates (an insoluble sulfonamide is formed - this sulfonamide lacks an acidic hydrogen or nitrogen and does not form a soluble sodium salt).

(continued next page)

3° amine:

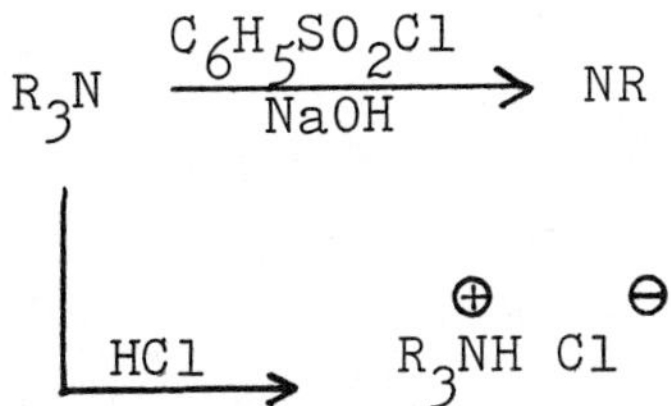

Observations: When the amine is shaken with benzenesulfonyl chloride an insoluble layer separates (the starting material - no reaction occurs). The insoluble amine will form a soluble salt in acid.

28. (a) Skraup: quinoline synthesis from aniline, glycerol and nitrobenzene.

 (b) Hofmann hypohalite: 1° amines from alkyl and aryl carboxamides + halogen.

 (c) Hantzsch: synthesis of dihydropyridines from an unsaturated carbonyl compound, an aldehyde and ammonia.

 (d) Curtius rearrangement: Conversion of an acyl azide to an isocyanate via a nitrene intermediate. Hydrolysis of the isocyanate leads to a 1° amine.

 (e) Gabriel: 1° amine synthesis via alkylation of potassium phthalimide followed by hydrolysis.

 (f) Cope: synthesis of alkenes by pyrolysis of amine oxides.

29. more $(CH_3)_2N\text{-}NH_2$. The first step produces $CH_3NH\text{-}NH_2$ and CH_3NH- is a stronger nucleophile than $\text{-}NH_2$.

30. (a) Bubble hydrogen chloride into mixture, ethyl amine hydrochloride precipitates; filter to remove solid; add sodium hydroxide solution to salt to liberate ethylamine (in a distillation apparatus or flask with a delivery tube), allow ethylamine to diffuse (b.p. 17°) into a cold trap.

 (b) Separate and isolate ethylamine as in (a). Separate ether (b.p. 35°) and ethyl alcohol (78°) by distillation.

(continued next page)

(c) Separate and isolate ethylamine as in (a). Separate ether (b.p. 35°), ethanol (78°) and ethylene glycol (b.p. 197°) by distillation.

(d) Shake mixture with dilute hydrochloric acid (amine dissolves as hydrochloride salt). Separate layers. Make aqueous layer basic and extract amine with ether. Remove ether by distillation. Shake organic layer with dilute sodium hydroxide (dissolves acid as the salt). Extract heptanol with ether and remove ether by distillation. Regenerate free heptanoic acid by acidifying mixture with dil. hydrochloric acid. Extract acid with ether and remove ether by distillation.

(e) Use Hinsberg procedure.
Shake mixture with alkaline benzenesulfonyl chloride (amide from 1° amine dissolves; amides from dibutylamine and triethylamine - which "does not react", see p. 911 - form a lower insoluble layer). Separate layers. Precipitate amide of 1° amine from aqueous layer by acidifying mixture. Regenerate 1° amine by hydrolysis. Extract amine from water using ether and remove ether by distillation. Dissolve 3° amine (in other layer) in dil. hydrochloric acid. Separate layers (salt of 3° amine in aqueous layer; amide of 2° amine is the other layer). Regenerate 3° amine by making solution basic. Extract 3° amine with ether and remove ether by distillation. The 2° amide is recovered by hydrolyzing the amide, extracting the amine with ether and removing the ether by evaporation.

(f) Shake mixture with concentrated aqueous sodium bisulfite. Filter solid bisulfite addition product of 2-hexanone. Regenerate 2-hexanone by addition of acid. Extract ketone with ether. Remove ether by distillation. Make mixture containing alcohol and amine acid to litmus. Extract alcohol with ether. Remove ether by distillation. Liberate amine from aqueous mixture by making it basic to litmus. Extract amine with ether. Remove ether by distillation.

31. (a) Ethylamine reacts with nitrous acid with evolution of nitrogen, ethanol does not. Ethylamine "fumes" with HCl (forms the hydrochloride salt), ethanol does not. (Although not a chemical test, it might be noted that at room temperature ethylamine is a gas, ethanol is a liquid).

(b) Hinsberg test: ethylamine (1°), diethylamine (2°). The ir of ethylamine should show a doublet N-H stretch (symmetric and asymmetric stretches of NH_2), the spectrum of diethylamine will have a single N-H band.

(c) Hinsberg test: ethylamine (1°), triethylamine (3°). The ir spectrum of ethylamine will have an N-H stretch, the spectrum of triethylamine will not.

(d) An aqueous solution of tetraethylammonium chloride will give a precipitate (AgCl) with silver nitrate. Because of the positive charge on nitrogen, the nmr signals in tetraethylammonium chloride appear further downfield.

(e) Hinsberg test: isoamylamine (isopentylamine) (1°), dihexylamine (2°). The ir spectrum of isoamylamine should show doublet for N-H stretches of NH_2 (symmetric and asymmetric).

(f) Hinsberg test: dipropylamine (2°), aniline (1°). The nmr spectrum of aniline contains absorption near 7δ for benzene protons; there are no absorptions in that region for dipropylamine.

(g) Hexylamine reacts with nitrous acid with evolution of nitrogen <u>or</u> hexylamine forms a hydrochloride salt which is water soluble; 2-hexene reacts with HCl to give a water insoluble alkyl chloride. The nmr spectrum of 2-hexene will have alkene hydrogen absorption near 5δ.

(h) Morpholine gives a Hinsberg test (2°), pyridine does not. The nmr spectrum of pyridine contains absorptions in the aromatic proton region (near 7δ).

(i) same as (h) (piperidine, like morpholine, is a 2° amine).

(for discussion of the Hinsberg test see Text S. 21.8 and answer 27).

32. (a) $CH_3COOH \xrightarrow{SOCl_2} \xrightarrow{NH_3} CH_3\overset{O}{\overset{\|}{C}}NH_2$

(b) $CH_3COOH \xrightarrow{LiAlH_4} \xrightarrow{H_2O} CH_3CH_2OH$

(c) $CH_3\overset{O}{\overset{\|}{C}}NH_2$ (from a) $\xrightarrow{LiAlH_4} \xrightarrow{H_2O} CH_3CH_2NH_2$

(d) $CH_3\overset{O}{\overset{\|}{C}}NH_2$ (from a) $\xrightarrow[H_2O/OH^{\ominus}]{Br_2} CH_3NH_2$

(e) CH_3CH_2OH (from b) $\xrightarrow{PCl_3} CH_3CH_2Cl$

(f) CH_3CH_2Cl (from e) $\xrightarrow[ether]{Mg} CH_3CH_2MgCl \xrightarrow{CO_2} \xrightarrow{H_2O}$

CH_3CH_2COOH (or $CH_3CO_2H \xrightarrow{SOCl_2} \xrightarrow{CH_2N_2} \xrightarrow[H_2O]{Ag_2O}$)

(g) CH_3CN (from c) $\xrightarrow{CH_3MgI} \xrightarrow{H_2O} CH_3\overset{O}{\overset{\|}{C}}CH_3 \xrightarrow[H_2/Pt]{NH_3} CH_3\underset{NH_2}{\underset{|}{C}}HCH_3$

(h) $CH_3CH_2OH \xrightarrow[heat]{H_2SO_4} CH_2{=}CH_2 \xrightarrow{CH_3\overset{O}{\overset{\|}{C}}OOH}$ CH_2—CH_2 (epoxide, ring closed by O)

(i) CH_2—CH_2 (epoxide, ring closed by O) (from h) $\xrightarrow{NH_3} HOCH_2CH_2NH_2$

(continued next page)

(j) $\underset{\diagdown O \diagup}{CH_2—CH_2}$ (from h) + CH_3CH_2MgCl (from e) $\longrightarrow$ $\xrightarrow{H_2O}$

$CH_3CH_2CH_2CH_2OH$

(k) $CH_3(CH_2)_2CH_2OH$ (from j) $\xrightarrow{CrO_3/H^{\oplus}}$ $CH_3(CH_2)_2COOH$

(l) CH_3COOH $\xrightarrow{SOCl_2}$ $CH_3\overset{O}{\overset{\|}{C}}\text{-}Cl$ $\xrightarrow{CH_3NH_2\text{(from d)}}$ $CH_3\overset{O}{\overset{\|}{C}}\text{-}NH\text{-}CH_3$

$\xrightarrow{LiAlH_4}$ $\xrightarrow{H_3O^{\oplus}}$ $CH_3CH_2NHCH_3$

(m) $CH_3\overset{O}{\overset{\|}{C}}\text{-}OH$ $\xrightarrow{SOCl_2}$ $\xrightarrow{C_2H_5OH}$ $CH_3\overset{O}{\overset{\|}{C}}\text{-}OC_2H_5$

(n) $CH_3\overset{O}{\overset{\|}{C}}\text{-}OC_2H_5$ (from m) $\xrightarrow{Na^{\oplus}\ \overset{\ominus}{O}C_2H_5}$ $CH_3\overset{O}{\overset{\|}{C}}CH_2\overset{O}{\overset{\|}{C}}OC_2H_5$

(o) CH_3CH_2OH (from b) $\xrightarrow[heat]{H_2SO_4}$ $CH_2{=}CH_2$

(p) $CH_2{=}CH_2$ (from o) $\xrightarrow{O_3}$ $\xrightarrow{Zn/CH_3COOH}$ 2 $H_2C{=}O$

(q) 2 $CH_3\overset{O}{\overset{\|}{C}}CH_2CO_2C_2H_5$ (from n) + HCHO (from p) $\xrightarrow{NH_3}$

(continued next page)

$$\text{(1,4-dihydropyridine: EtO}_2\text{C, CO}_2\text{Et at 3,5; CH}_3\text{ at 2,6; N–H)} \xrightarrow[H_2SO_4]{HNO_3} \xrightarrow{KOH/EtOH} \xrightarrow[heat]{CaO}$$ (see p. 890)

33. (a) HCl

(b) $CH_3\overset{O}{\overset{\|}{C}}Cl$ or $(CH_3\overset{O}{\overset{\|}{C}})_2O$

(c) $CH_3CH{=}O/H_2/Ni$ or CH_3CH_2Br (less desirable)

(d) $CH_3CH_2CH{=}O/H_2/Ni$ or $CH_3CH_2CH_2Br$ (less desirable)

(e) $(CH_3)_2C{=}O$

(f) $(CH_3)_2C{=}NCH_3$ (from e) + H_2/Pt (or $LiAlH_4$ then H_2O)

34. (a) $CH_3CH_2CH_2CH_2Cl \xrightarrow{Na^{\oplus}\ CN^{\ominus}} \xrightarrow{H_2/Pt}$ ($\xrightarrow{\text{or } LiAlH_4} \xrightarrow{H_2O}$)

(b) $ClCH_2CH_2CH_2CH_2Cl \xrightarrow{2Na^{\oplus}\ CN^{\ominus}} \xrightarrow{LiAlH_4} \xrightarrow{H_2O}$

(c) $CH_3CH_2CO_2H \xrightarrow{SOCl_2} \xrightarrow{CH_3CH_2CH_2CH_2NH_2}$

(d) $CH_3CH_2\overset{O}{\overset{\|}{C}}NH(CH_2)_3CH_3$ (from c) $\xrightarrow{LiAlH_4} \xrightarrow{H_2O}$

(e) $2CH_3\overset{O}{\overset{\|}{C}}OC_2H_5 \xrightarrow[C_2H_5OH]{Na^{\oplus}\ OC_2H_5^{\ominus}}$ (Claisen condensation)

(f) $2CH_3\overset{O}{\overset{\|}{C}}\text{-}CH_2CO_2C_2H_5$ (from e) + NH_3 + $H\overset{O}{\overset{\|}{C}}H$ $\longrightarrow$ (1,4-dihydropyridine: $C_2H_5\overset{O}{\overset{\|}{C}}O$ and $O\overset{O}{\overset{\|}{C}}C_2H_5$ at 3,5; CH_3 at 2,6; N–H)

(continued next page)

$\xrightarrow[H_2SO_4]{HNO_3}$ 3,5-di(propanoyloxy)-2,6-dimethylpyridine ($C_2H_5\overset{O}{\overset{\|}{C}}O$, $O\overset{O}{\overset{\|}{C}}C_2H_5$, CH_3, N, CH_3) $\xrightarrow[-CO_2]{CaO/heat}$ 2,6-dimethylpyridine (CH_3, N, CH_3)

(g) $CH_3\overset{O}{\overset{\|}{C}}CH_3 + CH_3NHCH_2CH_3 \xrightarrow{H^{\oplus}}$ (enamine synthesis)

(h) $CH_2{=}CH(CH_3)N(CH_3)CH_2CH_3$ (from g) $\xrightarrow{H_2/Pt}$

(i) $(CH_3)_2NH \xrightarrow{CH_2=CH-CN}$ (Michael Addition)

(j) $N{\equiv}CCH_2CH_2OH \xrightarrow{CH_2=CHCN} NCCH_2CH_2OCH_2CH_2CN \xrightarrow{H_2/Pt}$

35. Any Grignard reagent which forms will abstract a proton from the amino group of another molecule.

$H_2N-C_6H_4-MgBr + H_2\ddot{N}-C_6H_4-Br \rightarrow H_2N-C_6H_4-H + BrMg(H)\ddot{N}-C_6H_4-Br$

(review reactions of Grignard reagents with proton donors p. 229)

36. (a) $CH_3CH_2OH \xrightarrow[heat]{H_2SO_4} CH_2{=}CH_2 \xrightarrow{CH_3CO_3H}$ CH_2—CH_2 (ring closed by O) $\xrightarrow{NH_3}$

$HOCH_2CH_2NH_2 \xrightarrow{HCl} HOCH_2CH_2\overset{\oplus}{N}H_3\overset{\ominus}{Cl} \xrightarrow{PCl_3} ClCH_2CH_2\overset{\oplus}{N}H_3\overset{\ominus}{Cl}$

$\xrightarrow{OH^{\ominus}}$ CH_2—CH_2 (ring closed by N–H)

(b) Products:

(a) $H_2NCH_2CH_2OC_2H_5$ (b) $H_2NCH_2CH_2OH$ (c) $H_2NCH_2CH_2SH$

(continued next page)

(d) $H_2NCH_2CH_2NHCH_3$ (e) $\triangleright$N-CH_2CH_2CN (f) $\triangleright$N-$\underset{\displaystyle CH_3}{CH}CH_2CO_2C_2H_5$

37. (a) $CH_3\overset{O}{\overset{\|}{C}}CH_3 \xrightarrow{NH_3} \xrightarrow{H_2/Ni} (CH_3)_2CHNH_2$

(b) $CH_3\overset{O}{\overset{\|}{C}}CH_3 \xrightarrow{CH_3MgI} \xrightarrow{H_3O^{\oplus}} (CH_3)_3COH \xrightarrow{HCl} (CH_3)_3CCl$

$\xrightarrow[ether]{Mg} \xrightarrow{CO_2} \xrightarrow{H_2O} \xrightarrow{SOCl_2} (CH_3)_3C\text{-}\overset{O}{\overset{\|}{C}}\text{-}Cl \xrightarrow{NH_3}$

$\xrightarrow{LiAlH_4} \xrightarrow{H_3O^{\oplus}} (CH_3)_3CCH_2NH_2$

(c) $(CH_3)_2CHNH_2$ (from a) $\xrightarrow{\text{ethylene oxide}} (CH_3)_2CHNHCH_2CH_2OH$

(d) $CH_3\text{-}\overset{O}{\overset{\|}{C}}\text{-}O\text{-}C_2H_5 \xrightarrow[C_2H_5OH]{NH_3} CH_3\text{-}\overset{O}{\overset{\|}{C}}\text{-}NH_2 \xrightarrow{P_2O_5} CH_3C{\equiv}N$

$\xrightarrow{2H\text{-}C{\equiv}C\text{-}MgBr} \xrightarrow{H_2O} CH_3\text{-}\underset{\displaystyle NH_2}{C}(C{\equiv}CH)_2$

(e) $CH_2{=}CH_2 \xrightarrow{CH_3CO_3H} \xrightarrow[H_2O]{Na^{\oplus}\ CN^{\ominus}} HOCH_2CH_2CN \xrightarrow{H_2/Ni}$

$HOCH_2CH_2CH_2NH_2$

(f) $CH_3CH_2NH_2 \xrightarrow{3CH_3I} CH_3CH_2\overset{\oplus}{N}(CH_3)_3I^{\ominus} \xrightarrow[heat]{Na^{\oplus}\ OH^{\ominus}} CH_2{=}CH_2$

(continued next page)

(g) $HOCH_2CH_2NH_2 \xrightarrow{TsCl} TsOCH_2CH_2NH_2 \xrightarrow{NH_2CH_2CH_2OH} \xrightarrow{P_2O_5}$ morpholine

$HO\text{-}CH_2CH_2\text{-}N(H)\text{-}CH_2\text{-}CH_2\text{-}OH \xrightarrow{-H_2O}$ morpholine

(h) $CH_3CH_3 \xrightarrow[light]{Br_2} CH_3CH_2\text{-}Br \xrightarrow{\text{potassium phthalimide}}$

N-ethylphthalimide ($N\text{-}CH_2CH_3$) $\xrightarrow[H_2O]{OH^{\ominus}}$ $C_6H_4(COO^{\ominus})_2$ + $CH_3CH_2NH_2$

(i) $CH_3CH{=}CH_2 \xrightarrow{HBr} \xrightarrow{NaCN} CH_3CH(CN)CH_3 \xrightarrow{H_2/Ni} CH_3CH(CH_2NH_2)CH_3$

(j) $CH_3CH{=}CH_2 \xrightarrow{BH_3} \xrightarrow[OH^{\ominus}]{H_2O_2} CH_3CH_2CH_2OH \xrightarrow{PBr_3} \xrightarrow{Na^{\oplus}\ CN^{\ominus}}$

$\xrightarrow{LiAlH_4} \xrightarrow{H_2O} CH_3CH_2CH_2CH_2NH_2$

(k) $C_2H_5O\overset{O}{\overset{\|}{C}}CH_2\overset{O}{\overset{\|}{C}}OC_2H_5 \xrightarrow{Na^{\oplus}\ OC_2H_5^{\ominus}} \xrightarrow{C_2H_5Br} C_2H_5O\text{-}\overset{O}{\overset{\|}{C}}\text{—}C(CH_2CH_3)_2\text{—}\overset{O}{\overset{\|}{C}}\text{-}OC_2H_5$

$\xrightarrow[heat]{H_3O^{\oplus}} \xrightarrow[-CO_2]{heat} CH_3CH_2\text{-}CH(COOH)\text{-}CH_2CH_3 \xrightarrow{SOCl_2} \xrightarrow{NH_3}$

(continued next page)

$$CH_3CH_2\underset{\underset{O=C-NH_2}{|}}{C}HCH_2CH_3 \xrightarrow[OH^{\ominus}]{Br_2} CH_3CH_2\underset{\underset{NH_2}{|}}{C}HCH_2CH_3$$

38. $$CH_3CH_2CH_2CH_2\overset{O}{\overset{\|}{C}}CH_3 \xrightarrow{Cl_2/OH^{\ominus}} \xrightarrow{H_3O^{\oplus}} CH_3CH_2CH_2CH_2\overset{O}{\overset{\|}{C}}\text{-}OH \xrightarrow{SOCl_2}$$

$$\xrightarrow{NH_3} CH_3CH_2CH_2CH_2\overset{O}{\overset{\|}{C}}\text{-}NH_2 \xrightarrow{LiAlH_4} \xrightarrow{H_2O.} CH_3CH_2CH_2CH_2CH_2NH_2$$

39. (a) (1) $\xrightarrow{\text{potassium phthalimide } (:N:^{\ominus}\ K^{\oplus})} \xrightarrow{H_2O/OH^{\ominus}}$

(2) $\xrightarrow[\text{(excess)}]{NH_3}$

(b) $\xrightarrow{Na^{\oplus}\ CN^{\ominus}} \xrightarrow{LiAlH_4} \xrightarrow{H_2O}$

(c) $\xrightarrow[H_2/Pt]{NH_3}$

(d) $\xrightarrow[H_2O/OH^{\ominus}]{Br_2}$

(e) $\xrightarrow{CrO_3/\text{pyridine}} \xrightarrow[H_2/Pt]{NH_3}$

(f) $\xrightarrow{K_2Cr_2O_7/H^{\oplus}} CH_3CH_2CH_2CH_2CO_2H \xrightarrow{SOCl_2} \xrightarrow{NH_3}$

$$CH_3CH_2CH_2CH_2\overset{O}{\overset{\|}{C}}NH_2 \xrightarrow[OH^{\ominus}/H_2O]{Br_2}$$

40. (a) 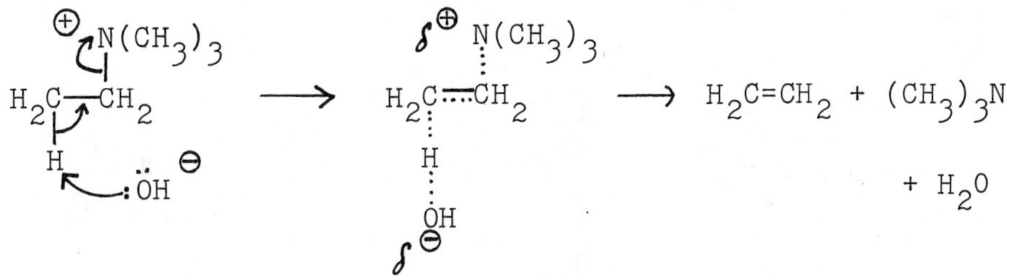 $H_2C—CH_2$ with $\overset{\oplus}{N}(CH_3)_3$ and H, $:\ddot{O}H^{\ominus}$ $\longrightarrow$ $H_2C\cdots CH_2$ with $\delta^{\oplus}$ $N(CH_3)_3$, H, $\ddot{O}H$ $\delta^{\ominus}$ $\longrightarrow$ $H_2C{=}CH_2 + (CH_3)_3N + H_2O$

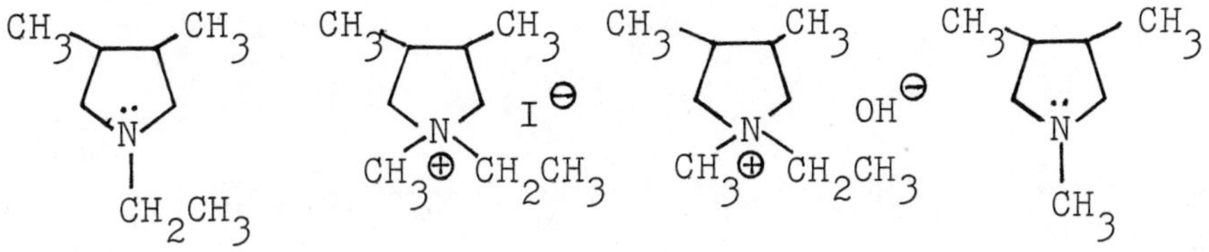

(b) A (cis or trans), B, C, D, E, F, G, H, 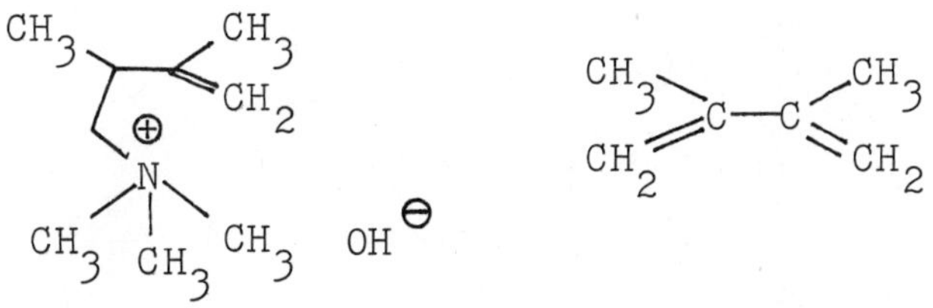 I, J

A: CH_3, CH_3, $\ddot{N}$, CH_2CH_3

B: CH_3, CH_3, N, $CH_3\oplus$ CH_2CH_3 $I^{\ominus}$

C: CH_3, CH_3, N, $CH_3\oplus$ CH_2CH_3 $OH^{\ominus}$

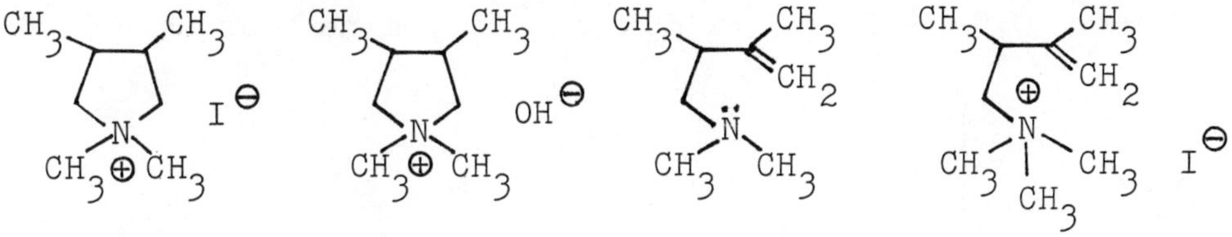

D: CH_3, CH_3, $\ddot{N}$, CH_3

E: CH_3, CH_3, N, $CH_3\oplus$ CH_3 $I^{\ominus}$

F: CH_3, CH_3, N, $CH_3\oplus$ CH_3 $OH^{\ominus}$

G: CH_3, CH_3, CH_2, $\ddot{N}$, CH_3 CH_3

H: CH_3, CH_3, CH_2, $\overset{\oplus}{N}$, CH_3 CH_3 CH_3 $I^{\ominus}$

I: CH_3, CH_3, CH_2, $\overset{\oplus}{N}$, CH_3 CH_3 CH_3 $OH^{\ominus}$

J: CH_3, $CH_2{=}C—C{=}CH_2$, CH_3

41. 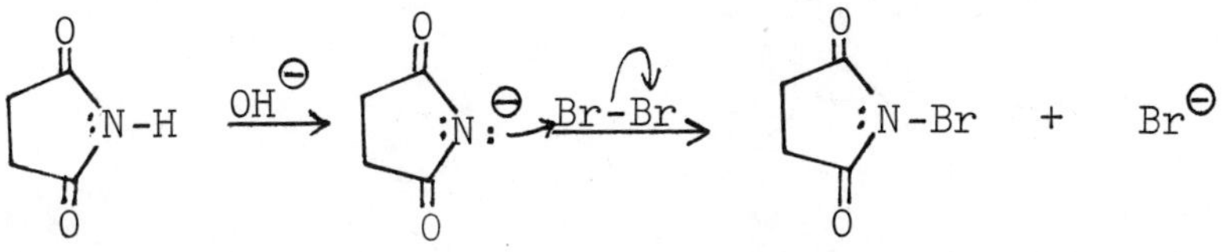 O, O, $:N-H$ $\xrightarrow{OH^{\ominus}}$ O, O, $:N:^{\ominus}$ $\xrightarrow{Br-Br}$ O, O, $:N-Br$ + $Br^{\ominus}$

(continued next page)

$$\text{N-bromosuccinimide} \xrightarrow{:OH^{\ominus}} \text{tetrahedral intermediate} \rightarrow HO_2CCH_2CH_2\overset{O}{\overset{\|}{C}}\text{-}\ddot{N}^{\ominus}\text{-}Br$$

$$\downarrow -Br^{\ominus}$$

$$HO_2CCH_2CH_2\text{-}N{=}C{=}O \leftarrow HO_2CCH_2CH_2\text{-}\overset{O}{\overset{\|}{C}}\text{-}\ddot{N}$$

hydrolysis

$HO_2CCH_2CH_2NH_2$ (actually ${}^{\ominus}O_2CCH_2CH_2NH_2$ in $OH^{\ominus}$ solution)

42. $H_2N\text{-}C_6H_4\text{-}\overset{O}{\overset{\|}{C}}\text{-}O\text{-}CH_2CH_2\text{-}\overset{H}{\underset{\oplus}{N}}(C_2H_5)_2\ Cl^{\ominus}$ (procaine hydrochloride)

$$\downarrow NaOH$$

$$H_2N\text{-}C_6H_4\text{-}\overset{O}{\overset{\|}{C}}\text{-}O\text{-}CH_2CH_2\text{-}N(C_2H_5)_2 + NaCl + H_2O$$

$$\downarrow NaOH/reflux$$

$$\downarrow H_3O^{\oplus}$$

$$H_2N\text{-}C_6H_4\text{-}CO_2H + HOCH_2CH_2N(C_2H_5)_2$$

A (excess HA) ↓

B: $CH_3\overset{O}{\overset{\|}{C}}Cl$ → D; HCl → C

D C (continued next page)

$H_3\overset{\oplus}{N}-C_6H_4-CO_2H\ A^{\ominus}$ (soluble in H_2O)

$CH_3\overset{O}{\overset{\|}{C}}OCH_2CH_2\overset{H}{\underset{\oplus}{N}}(C_2H_5)_2\ Cl^{\ominus}$ D

$HOCH_2CH_2\overset{\oplus}{N}H(C_2H_5)_2\ Cl^{\ominus}$ C

$$\downarrow OH^{\ominus}/H_2O$$

$CH_3\overset{O}{\overset{\|}{C}}OCH_2CH_2N(C_2H_5)_2$

$(C_8H_{17}O_2N)$

43. (a) $(CH_3)_2CHCH_2NH_2 \xrightarrow{HONO} CH_3\underset{CH_3}{\underset{|}{C}}H-CH_2-\overset{\oplus}{N}_2 \longrightarrow CH_3\overset{\oplus}{C}HCH_2CH_3$

$CH_3\overset{\oplus}{C}HCH_2CH_3 \xrightarrow{-H^{\oplus}} CH_3CH=CHCH_3$

$CH_3\overset{\oplus}{C}HCH_2CH_3 \xrightarrow{H_2O} CH_3\underset{\oplus OH_2}{\underset{|}{C}}HCH_2CH_3 \xrightarrow{-H^{\oplus}} CH_3\underset{OH}{\underset{|}{C}}HCH_2CH_3$

(b) $CH_3-\underset{\oplus N_2}{\overset{CH_3}{C}}-\underset{OH}{\overset{CH_3}{C}}-CH_3 \xrightarrow{-N_2} (CH_3)_2\overset{\oplus}{C}-\underset{O-H}{\overset{CH_3}{C}}-CH_3 \longrightarrow (CH_3)_3C\overset{\|}{\underset{O}{C}}CH_3$

(with $:\ddot{O}H_2$ removing the proton from O-H)

(c) 1-hydroxycyclohexyl-$CH_2\overset{\oplus}{N}_2$ $\xrightarrow{-N_2}$ cycloheptyl cation* bearing O-H (with $:\ddot{O}H_2$) $\xrightarrow[(-H_3O^{\oplus})]{}$ cycloheptanone

*resonance stabilized

44. (a) [2-methylpyridine] $\xrightarrow{CH_3I}$ [N-methyl-2-methylpyridinium iodide, $I^{\ominus}$] A $\xrightarrow[\text{morpholine}]{C_6H_5CHO}$ [N-methylpyridinium with 2-CH_2-CH(OH)C_6H_5, $I^{\ominus}$] $\xrightarrow{-H_2O}$ [N-methylpyridinium with 2-CH=CH-C_6H_5, I] B

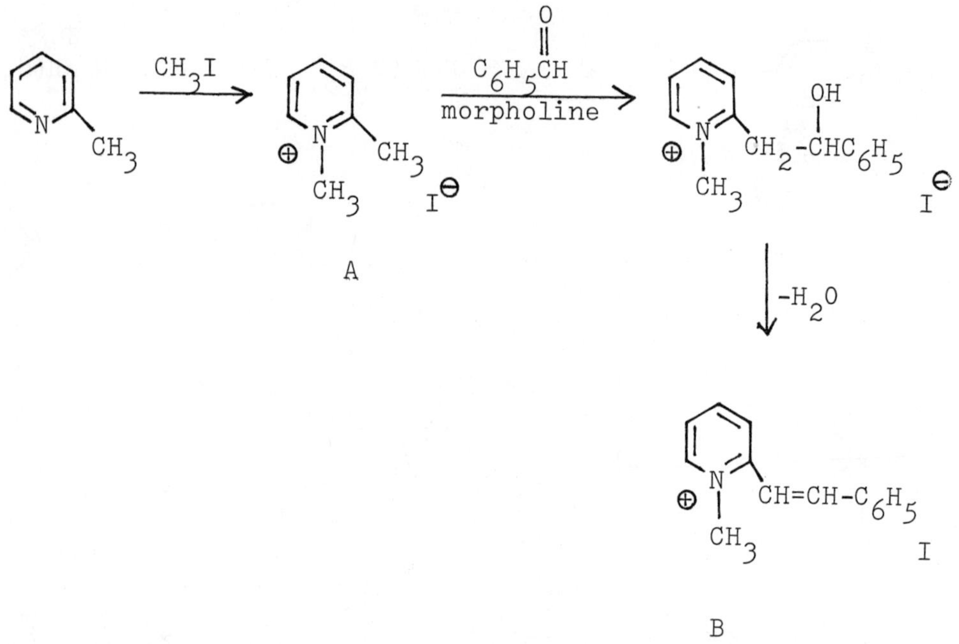

(b) Morpholine serves as a base to remove an acidic proton.

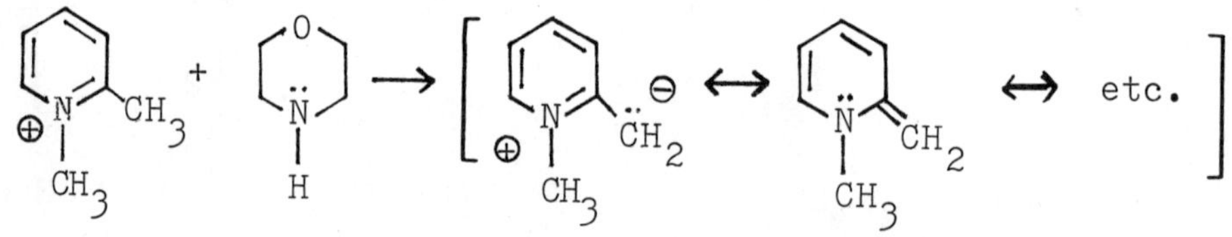

[N-methyl-2-methylpyridinium] + [morpholine] $\longrightarrow$ [[2-$\ddot{C}H_2^{\ominus}$ N-methylpyridinium] $\longleftrightarrow$ [N-methyl-1,2-dihydro-2-methylenepyridine] $\longleftrightarrow$ etc.]

(c) The analogous reaction using α-picoline could lead to reaction of benzaldehyde on carbon or nitrogen. In addition, the anion is less likely to form from α-picoline which lacks a positive charge on nitrogen.

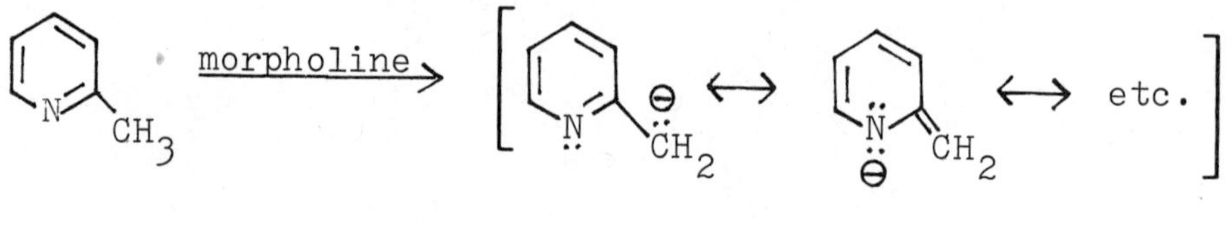

[2-methylpyridine] $\xrightarrow{\text{morpholine}}$ [[2-pyridyl-$\ddot{C}H_2^{\ominus}$] $\longleftrightarrow$ [2-methylene pyridine N anion] $\longleftrightarrow$ etc.]

45. nitration: The nitrogen-containing pyridine ring is deactivated, and electrophilic aromatic substitution occurs in the other ring.

nucleophilic attack: Nucleophilic attack occurs at the >C=N- moiety, e.g., in quinoline:

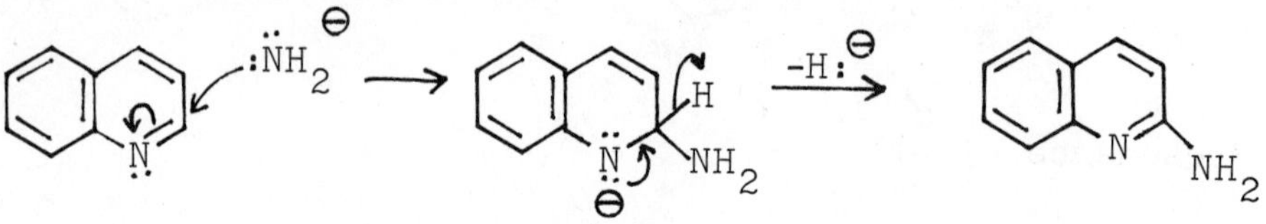

[quinoline] + $:\ddot{N}H_2^{\ominus}$ $\longrightarrow$ [2-amino-1,2-dihydroquinoline anion] $\xrightarrow{-H:^{\ominus}}$ [2-aminoquinoline]

46. (a) a cation rearrangement reaction.

(b) $R_2C{=}\ddot{N}{-}OH \xrightarrow{H^{\oplus}} {}^{*}R_2C{=}\ddot{N}{-}OH_2^{\oplus} \xrightarrow{-H_2O} R{-}\overset{\oplus}{C}{=}\ddot{N}{-}R$

$R{-}\overset{\oplus}{C}{=}\ddot{N}{-}R + H_2\ddot{O}: \longrightarrow R{-}C(\overset{\oplus}{O}H_2){=}\ddot{N}{-}R \xrightarrow{H^{\oplus}} R{-}C(={}^{\oplus}\!:O{-}H){-}NH{-}R \xrightarrow{(-H^{\oplus})} R{-}\underset{\underset{:O:}{\|}}{C}{-}NHR$

*or the concerted reaction $R_2C{=}N{-}\overset{\oplus}{O}H_2 \longrightarrow H_2\overset{\oplus}{O}{-}C(R){=}\ddot{N}{-}R$

(c) $C_6H_5\overset{O}{\overset{\|}{C}}C_6H_5 \xrightarrow{NH_2OH} (C_6H_5)_2C{=}N{-}OH \xrightarrow{H^{\oplus}} C_6H_5\overset{O}{\overset{\|}{C}}NH{-}C_6H_5$

$\xrightarrow{OH^{\ominus}/H_2O/\text{heat}} C_6H_5COO^{\ominus} + C_6H_5NH_2$

(d) cyclohexanone oxime: $C_6H_{10}{=}N{-}OH$

47. (a) pyridine $\xrightarrow{Li^{\oplus}\ :NH_2^{\ominus}}$ 2-amino-1-lithio-1,2-dihydropyridine (A) $\xrightarrow{H_2O}$ 2-amino-1,2-dihydropyridine (N–H) $\longrightarrow$ 2-aminopyridine (NH_2) + H_2

A

(continued next page)

(b) Attack of amide ion at the 3-position gives an anion which is not stabilized by the nitrogen (nitrogen can accommodate the negative charge more effectively).

NH2 H ⟷ NH2 H ⟷ NH2 H

(c) C_6H_5

48. $CH_3OCH_2CH_2NH_2 \xrightarrow{HCl} CH_3OCH_2CH_2\overset{\oplus}{N}H_3\ Cl^{\ominus}$

A B

$\downarrow C_6H_5SO_2Cl$

$CH_3OCH_2CH_2\ddot{N}HSO_2C_6H_5$

$\downarrow NaOH/H_2O$

$CH_3OCH_2CH_2\ddot{N}^{\ominus}SO_2C_6H_5\ Na^{\oplus}$

A $\xrightarrow{CH_3\overset{O}{\overset{\|}{C}}Cl}$ $CH_3OCH_2CH_2NH\overset{O}{\overset{\|}{C}}CH_3$ C

A $\xrightarrow{HI}$ $CH_3I + HOCH_2CH_2NH_2$

49. A. To account for the rearranged carbon skeleton which comes from the 3° amine, a cyclic ammonium intermediate has been proposed.

(continued next page)

(1) $CH_3CHCH_2\ddot{N}(CH_3)_2$ (with Cl on C-2) $\longrightarrow$ $CH_3\text{-}CH\text{—}CH_2$ (three-membered ring with $\overset{\oplus}{N}(CH_3)_2$) $Cl^{\ominus}$

(2) $(C_6H_5)_2CHCN \quad + \; B:^{\ominus} \longrightarrow \quad (C_6H_5)_2\ddot{C}^{\ominus}CN \quad + \; HB$

(resonance stabilized)

The anion formed in step (2) attacks the less sterically hindered carbon of the cyclic intermediate shown in step (1).

$CH_3\text{-}CH\text{—}CH_2$ (ring with $\overset{\oplus}{N}(CH_3)_2$) $\xrightarrow{(C_6H_5)_2\ddot{C}^{\ominus}CN}$ $CH_3CH(N(CH_3)_2)CH_2C(C_6H_5)_2CN$

B. $N{\equiv}C\text{—}C(C_6H_5)_2\text{~~~}$ $\xrightarrow{C_2H_5\text{-}MgCl}$ $ClMgN{=}C(C_2H_5)\text{-}C\text{~~~}$

$\xrightarrow{H^{\oplus}}$ $HN{=}C(C_2H_5)\text{-}C\text{~~~}$ $\xrightarrow{H^{\oplus}}$ $H_2\overset{\oplus}{N}{=}C(C_2H_5)\text{-}C\text{~~~}$ $\xrightarrow{H_2\ddot{O}:}$ $H_2N\text{-}C(\overset{\oplus}{O}H_2)(C_2H_5)\text{-}C\text{~~~}$

$\rightleftharpoons$

$H_3\overset{\oplus}{N}\text{-}C(OH)(C_2H_5)\text{-}C\text{—}$ $\longrightarrow$ $C_2H_5\text{-}\overset{O}{\overset{\|}{C}}\text{-}C(C_6H_5)_2\text{-}CH_2\text{-}CH(CH_3)\text{-}CH_2\text{-}N(CH_3)_2$

$+ \; NH_4^{\oplus}$

50. (a)

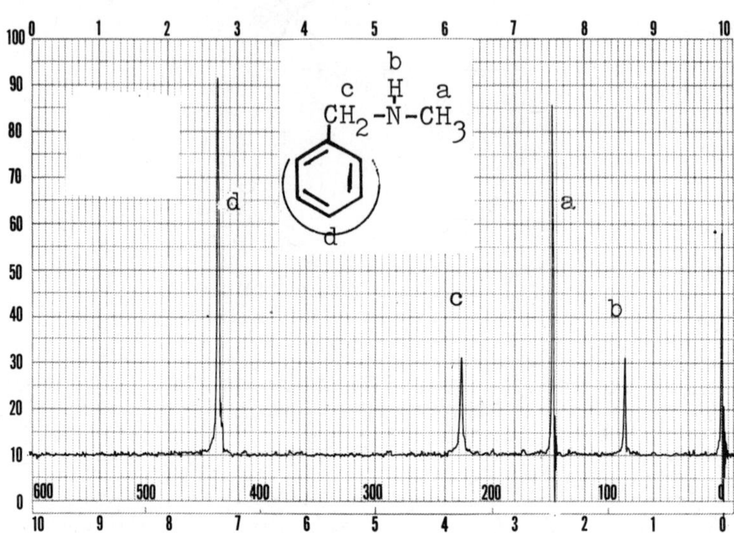

(b)

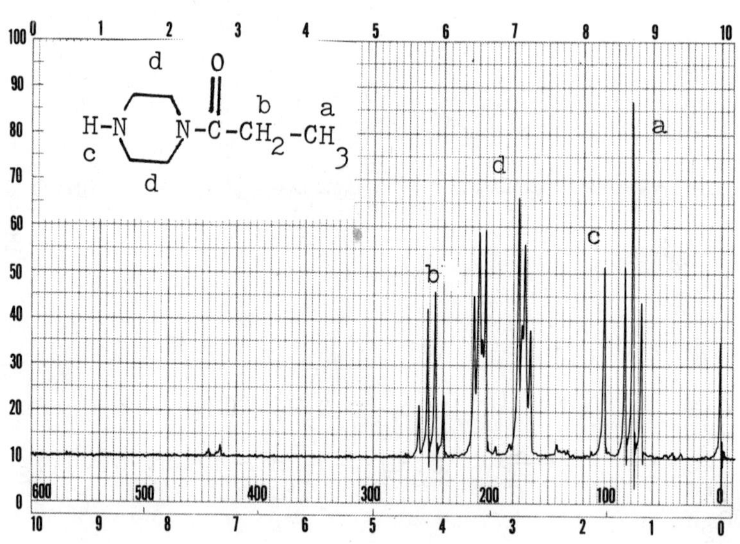

(c)

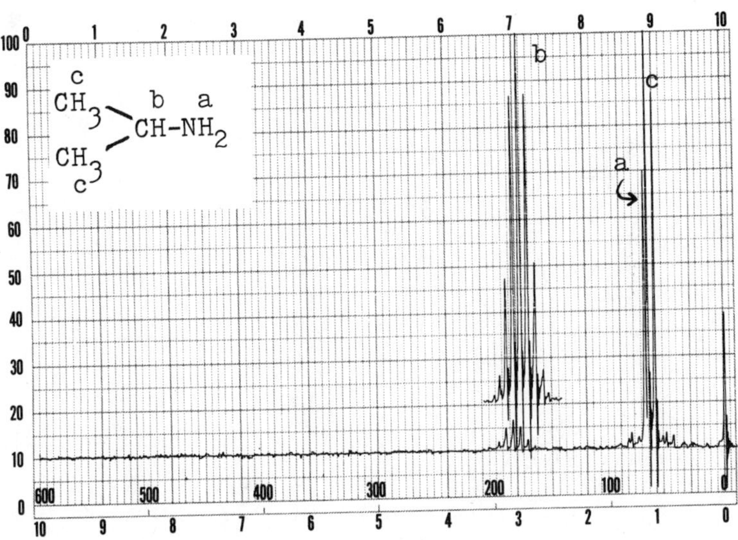

a, singlet
c, doublet

b, multiplet

(d)

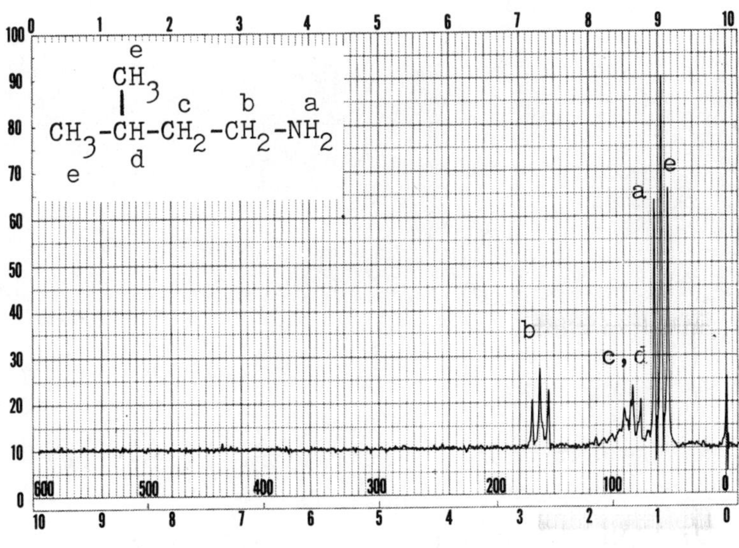

a, singlet
e doublet

c,d, complex of multiplets

(e)

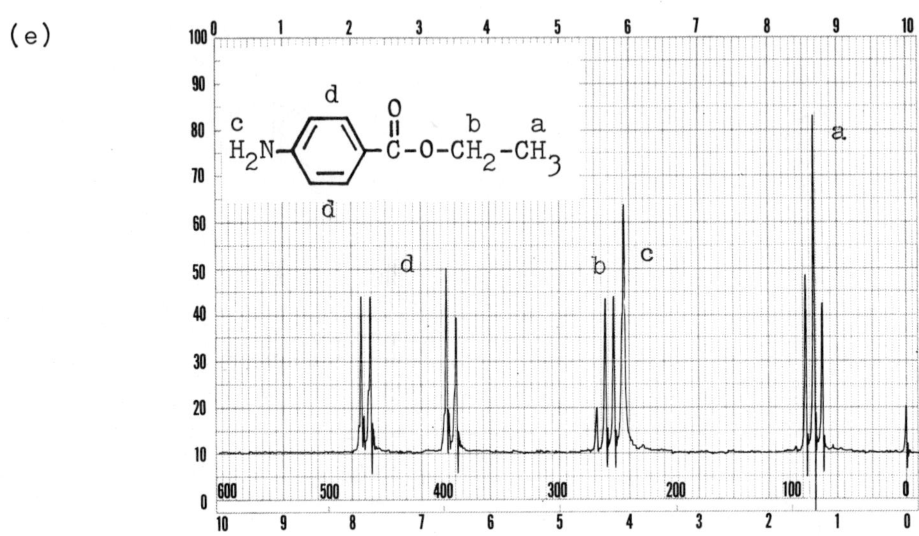

51. The NMR spectrum is that of p-toluidine.

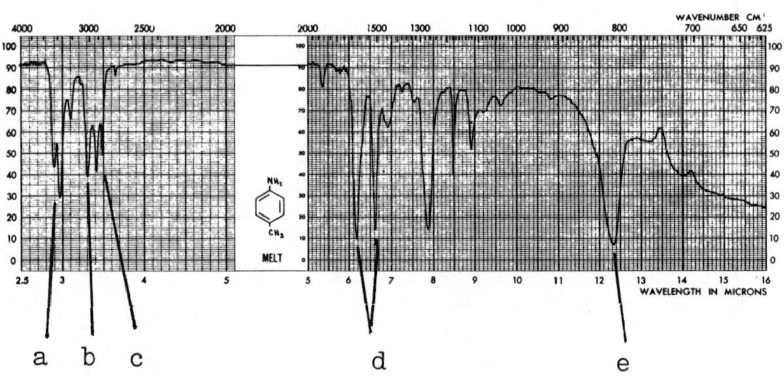

a: N-H stretch (symmetric and asymmetric)

b: =C-H stretch (benzene)

c: C-H stretch (alkyl)

d: C=C stretch (benzene)

e: =C-H out-of-plane bending (p-disubstituted benzene)

52. A is N,N-diethylaniline $(CH_3CH_2)_2N-C_6H_5$

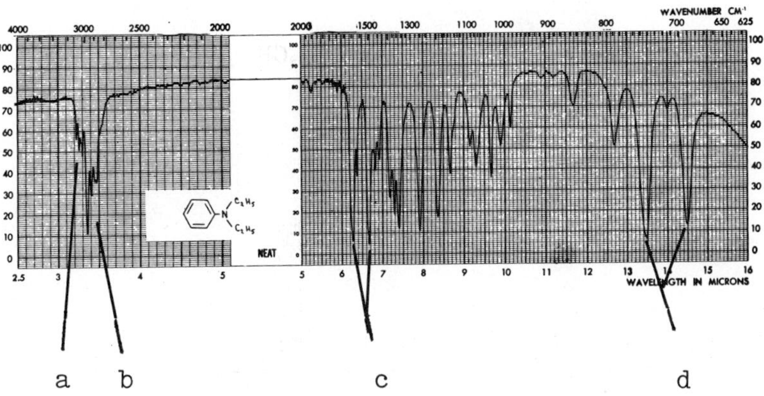

a: =C-H stretch (benzene)

b: C-H stretch (alkyl)

c: C=C stretch (benzene)

d: =C-H out-of-plane bending (monosubstituted benzene)

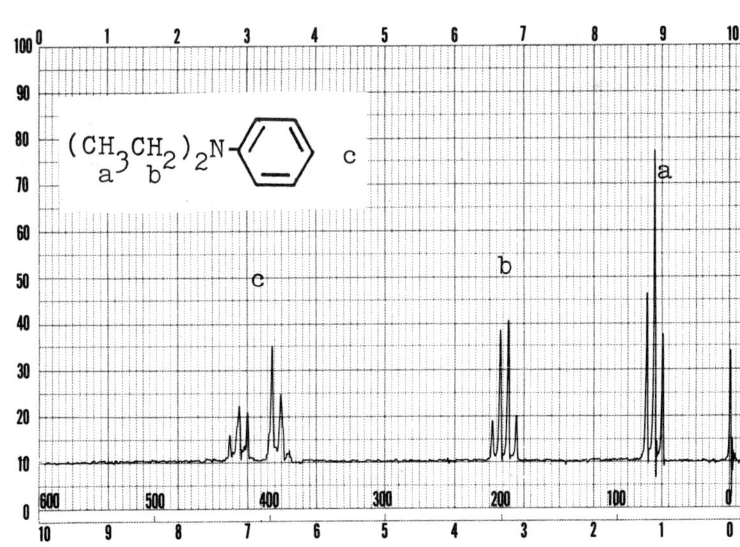

B is 3-diethylaminopropylamine $(CH_3CH_2)_2NCH_2CH_2NH_2$

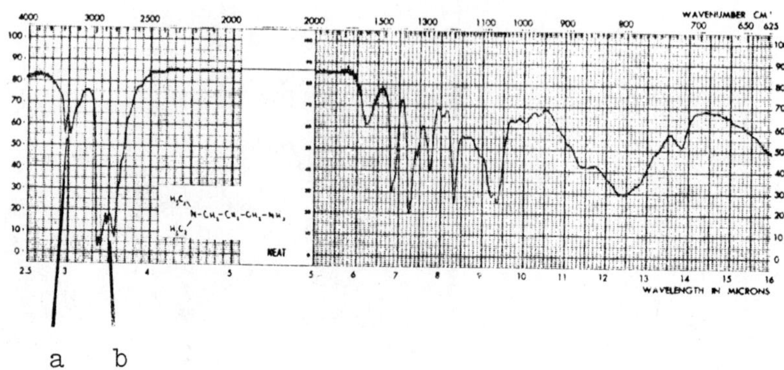

a: N-H stretch (symmetric and asymmetric), 1° amine

b: C-H stretch (alkyl)

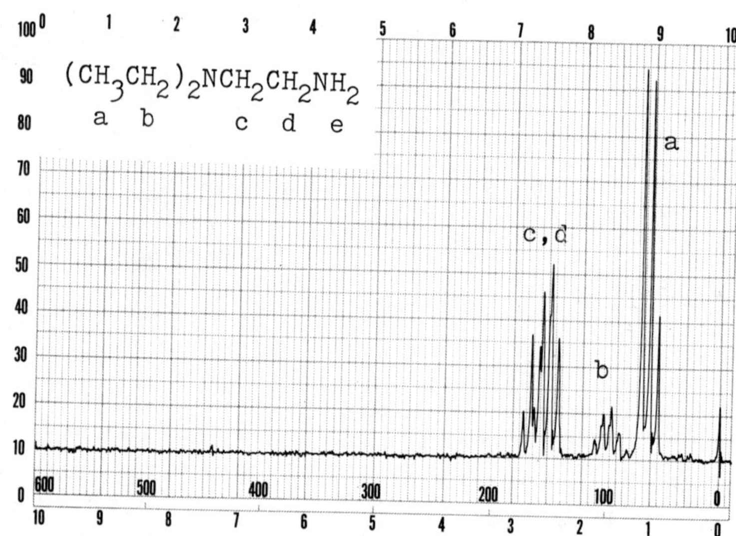

C is diisobutylamine $(CH_3)_2CHCH_2NHCH_2CH(CH_3)_2$

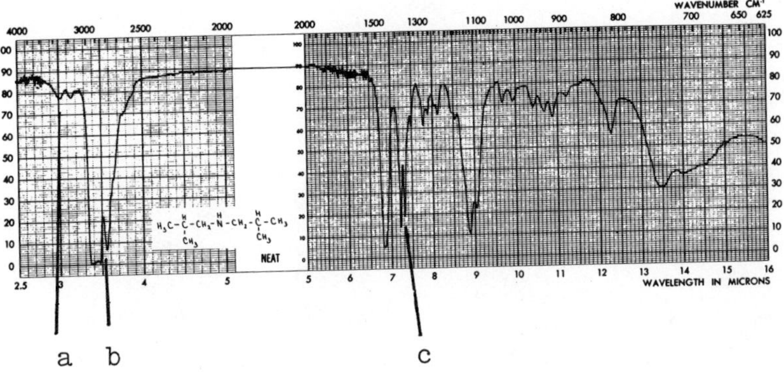

a: N-H stretch, 2° amine

b: C-H stretch (alkyl)

c: C-H bending (doublet, isopropyl)

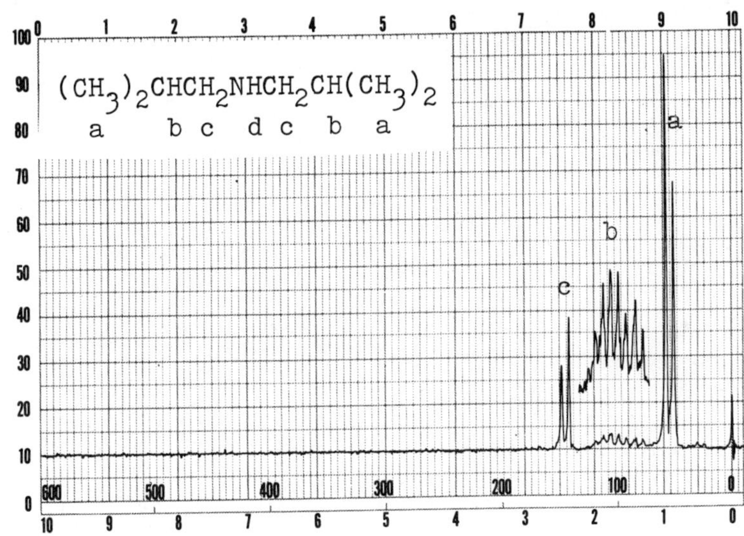

N-H absorption hidden in other absorptions

53. Cyclohexanone $\xrightarrow{\text{conc. } HNO_3}$ $HO_2C(CH_2)_4CO_2H$ (A) $\xrightarrow{SOCl_2}$ $ClC(=O)(CH_2)_4C(=O)Cl$ (B)

$\xrightarrow{NH_3}$ $H_2NC(=O)(CH_2)_4C(=O)NH_2$ (C) $\xrightarrow[H_2O]{Br_2/OH^{\ominus}}$ $H_2N(CH_2)_4NH_2$ (D)

54. (a) C_6H_5-CH_2-NH-CH_2-C_6H_5

(b) $C_6H_5C(=O)$-H $\xrightarrow{NH_3}$ [C_6H_5-CH=NH] $\xrightarrow{H_2/Pt}$

$C_6H_5CH_2NH_2$ + C_6H_5-C(=O)-H $\longrightarrow$ C_6H_5-C(H)($O^{\ominus}$)-$\overset{\oplus}{N}H_2$-$CH_2C_6H_5$

$\rightleftharpoons$ C_6H_5-CH(OH)-N(H)$CH_2C_6H_5$ $\xrightarrow{-H_2O}$ $C_6H_5CH=NCH_2C_6H_5$

$\xrightarrow{H_2/Pt}$ $C_6H_5CH_2NHCH_2C_6H_5$

SUMMARY OF REACTIONS

1. Synthesis of Amines

(a) Ammonolysis of alkyl halides (p. 882)

$$R\text{-}X \xrightarrow{NH_3} R\overset{\oplus}{N}H_3\overset{\ominus}{X} \xrightarrow{NH_3} RNH_2 + NH_4^{\oplus}X^{\ominus}$$

(b) Gabriel synthesis (p. 883)

$$\text{(phthalimide)}\,\overset{\ominus}{:N}:\overset{\oplus}{K} \xrightarrow{R\text{-}X} \text{(phthalimide)}\,:N\text{-}R \xrightarrow[OH^{\ominus}]{H_2O} C_6H_4(COO^{\ominus})_2 + R\ddot{N}H_2$$

(c) Reduction of nitriles (p. 885)

$$RC\equiv N \xrightarrow{\begin{array}{c}(1)\ H_2/Ni \\ (2)\ LiAlH_4 \\ AlCl_3 \\ (3)\ B_2H_6\end{array}} RCH_2NH_2$$

(d) Reductive Amination (p. 885)

$$\text{R-}\overset{\text{O}}{\overset{\|}{\text{C}}}\text{-H(R)} + \text{NH}_3 \xrightarrow[\text{(or NaBH}_3\text{CN)}]{\text{H}_2\text{/Ni}} \text{R-}\underset{\text{H}}{\underset{|}{\overset{\text{NH}_2}{\overset{|}{\text{C}}}}}\text{-H(R)}$$

1° amine

$$\text{R-}\overset{\text{O}}{\overset{\|}{\text{C}}}\text{-H(R)} + \text{R'NH}_2 \xrightarrow[\text{(or NaBH}_3\text{CN)}]{\text{H}_2\text{/Ni}} \text{R-}\underset{\text{H}}{\underset{|}{\overset{\text{NHR'}}{\overset{|}{\text{C}}}}}\text{-H(R)}$$

2° amine

(e) Hofmann hypohalite reaction (p. 886)

$$\text{R}\overset{\text{O}}{\overset{\|}{\text{C}}}\text{-NH}_2 \xrightarrow[\text{OH}^\ominus\text{/H}_2\text{O}]{\text{Br}_2} \text{R-NH}_2$$

(f) Curtius rearrangement (p. 888)

$$\text{R-}\overset{\text{O}}{\overset{\|}{\text{C}}}\text{-Cl} \xrightarrow{\text{Na}^\oplus\text{N}_3^\ominus} \text{R-}\overset{\text{O}}{\overset{\|}{\text{C}}}\text{-}\underset{\ominus}{\ddot{\text{N}}}\text{-}\underset{\oplus}{\text{N}}\equiv\text{N:} \xrightarrow[\text{-N}_2]{\text{heat}} \text{R-N=C=O} \xrightarrow{\text{H}_2\text{O}} \text{R-NH}_2 + \text{H}_2\text{O}$$

(g) Reduction of Amides (p. 888)

$$\text{R-}\overset{\text{O}}{\overset{\|}{\text{C}}}\text{-NH-} \xrightarrow{\text{LiAlH}_4} \xrightarrow{\text{H}_2\text{O}} \text{R-CH}_2\text{-NH-}$$

(h) Reduction of aryl nitro compounds (p. 888)

$$\text{Ar-NO}_2 \xrightarrow{\text{Sn/HCl}} \text{Ar}\overset{\oplus}{\text{N}}\text{H}_3\overset{\ominus}{\text{Cl}} \xrightarrow{\text{OH}^\ominus\text{/H}_2\text{O}} \text{ArNH}_2$$

2. Synthesis of Heterocycles

(a) pyridines (Hantzsch synthesis) p. 889

(b) quinolines (Skraup synthesis) p. 892

(c) isoquinolines (Bischler-Napieralski synthesis) p. 893

3. Reactions of Amines

(a) acidity (p. 894)

$$-\ddot{N}-H \xrightarrow[\text{liq. } NH_3]{Na} -\ddot{N}:^{\ominus} \; Na^{\oplus}$$

(b) acylation (p. 894)

$$-N-H \xrightarrow[OH^{\ominus}]{R-\overset{O}{\overset{\|}{C}}-Cl} -N-\overset{O}{\overset{\|}{C}}-R + H_2O + Cl^{\ominus}$$

(c) nitrosation (p. 895)

1° arylamines

$$ArNH_2 \xrightarrow[HA \quad 0^o]{Na^{\oplus}NO_2^{\ominus}} ArN_2^{\oplus} \; A^{\ominus}$$

benzene diazonium salt

2° amines

$$R_2NH \xrightarrow[HCl]{Na^{\oplus}NO_2^{\ominus}} R_2N-N=O$$

(or aryl) nitrosamines

3° arylamines

$$R_2N-C_6H_5 \xrightarrow[HCl]{Na^{\oplus}NO_2^{\ominus}} R_2N-C_6H_4-N=O$$

reactions of HONO with other amines are of little synthetic value.

4. Imine Formation (p. 899)

$$R'-NH_2 + R-\overset{O}{\overset{\|}{C}}-R \rightarrow R'-NH-\overset{OH}{\overset{|}{C}}R_2 \xrightarrow{-H_2O} R'\ddot{N}=CR_2 \text{ (ketimine)}$$

$$R'-NH_2 + R-\overset{O}{\overset{\|}{C}}-H \rightarrow R'-NH-\overset{OH}{\overset{|}{C}}HR \xrightarrow{-H_2O} R'\ddot{N}=C(H)R \text{ (aldimine)}$$

5. Enamine Synthesis and Alkylation of Enamines (p. 899)

$$\underset{2^\circ}{:NH} + O=C(-CH-)(H(R)) \xrightarrow{H^\oplus} \underset{\text{enamine}}{:N-C(=C-)(H(R))} \xrightarrow{R'-X} \oplus N=C(-C-R')(H(R))$$

$$\xrightarrow{H_3O^\oplus} :NH + O=C(-CR')(H(R))$$

6. Oxidation, Cope Elimination (p. 903)

$$-\underset{H}{C}-\underset{N(CH_3)_2}{C}- \xrightarrow{H_2O_2} -\underset{H}{C}-\underset{\ominus O-\overset{\oplus}{N}(CH_3)_2}{C}- \xrightarrow{heat} \rangle C=C\langle + (CH_3)_2NOH$$

REACTION REVIEW

A. Reactions from Chapter 21

Questions	Answers
1. $RX + NH_3 \longrightarrow$? (excess)	1. $R\overset{\oplus}{N}H_3\overset{\ominus}{X} \xrightarrow{NH_3} RNH_2 + NH_4^{\oplus}X^{\ominus}$

Write the structure of the amine formed in the following: a) $CH_3CH_2Cl + NH_3$(excess) $\rightarrow$? b) $(CH_3)_2CHBr + CH_3NH_2$ (excess) $\rightarrow$? (c) $CH_3CH_2Br + NH_3$ (large excess) $\rightarrow$?	a) $CH_3CH_2NH_2$ b) $(CH_3)_2CHNHCH_3$ c) $(CH_3CH_2)_4\overset{\oplus}{N}\ \overset{\ominus}{Br}$
2. Potassium phthalimide ($:\overset{\ominus}{N}:K^{\oplus}$) $\xrightarrow{RX}$ $\xrightarrow[H_2O]{OH^{\ominus}}$? Write the structure of the amine formed when RX is: a) $CH_3CH_2CH_2Br$ b) Cl–C_6H_4–CH_2Cl	2. RNH_2 + $C_6H_4(CO_2^{\ominus})_2$ a) $CH_3CH_2CH_2NH_2$ b) Cl–C_6H_4–CH_2NH_2
3. $RC\equiv N \xrightarrow{H_2/Pt}$? Write the structure of the product when $RC\equiv N$ is: a) $CH_3CH_2C\equiv N$ b) 1-cyclohexenyl–$C\equiv N$	3. RCH_2NH_2 a) $CH_3CH_2CH_2NH_2$ b) cyclohexyl–CH_2NH_2

4. $RC\equiv N \xrightarrow[AlCl_3]{LiAlH_4} \xrightarrow{H_2O}$? (or $\xrightarrow[H_2O]{B_2H_6}$) Write the structure of the product when $RC\equiv N$ is: a) $CH_2{=}CH{-}CH_2{-}C\equiv N$ b) $C_6H_5{-}C\equiv N$ (benzene ring–C≡N)	4. RCH_2NH_2 a) $CH_2{=}CHCH_2CH_2NH_2$ b) $C_6H_5{-}CH_2NH_2$ (benzene ring–CH_2NH_2)
5. $R\overset{O}{\overset{\parallel}{C}}NH_2 \xrightarrow[OH^{\ominus}/H_2O]{Br_2}$? Write the structure of the product when $R\overset{O}{\overset{\parallel}{C}}NH_2$ is: a) $CH_3CH_2\overset{O}{\overset{\parallel}{C}}NH_2$ b) cyclopentyl–$\overset{O}{\overset{\parallel}{C}}{-}NH_2$	5. RNH_2 a) $CH_3CH_2NH_2$ b) cyclopentyl–NH_2
6. $R{-}\overset{O}{\overset{\parallel}{C}}Cl \xrightarrow[2)\ heat]{1)\ Na^{\oplus}\ N_3^{\ominus}} \xrightarrow{H_2O}$? Write the structure of the product when $R\overset{O}{\overset{\parallel}{C}}Cl$ is:	6. RNH_2 (continued next page)

a) $CH_3CH_2\overset{O}{\overset{\Vert}{C}}Cl$	a) $CH_3CH_2NH_2$
b) cyclopentyl–$\overset{O}{\overset{\Vert}{C}}$-Cl	b) cyclopentyl–NH_2
7. $R\overset{O}{\overset{\Vert}{C}}H(R) + NH_3 \xrightarrow{H_2/Ni\ *} ?$ *(or $NaBH_3CN$) Write the structure of the product in the following: a) $CH_3CH_2\overset{O}{\overset{\Vert}{C}}CH_3 + NH_3 \xrightarrow{H_2/Ni} ?$ b) cyclopentanone $+ CH_3NH_2 \xrightarrow{H_2/Ni} ?$	7. $RCH\text{-}NH_2$ (with H(R) on the C) a) $CH_3CH_2\underset{}{\overset{NH_2}{\overset{\vert}{C}}}HCH_3$ b) cyclopentyl–$NHCH_3$
8. $R\text{-}\overset{O}{\overset{\Vert}{C}}\text{-}NH\text{-} \xrightarrow{LiAlH_4} \xrightarrow{H_2O} ?$ Write the structure of the product when $R\overset{O}{\overset{\Vert}{C}}NH\text{-}$ is: a) $CH_3CH_2\overset{O}{\overset{\Vert}{C}}\text{-}NH_2$ b) 2-piperidinone (six-membered ring with C=O and NH)	8. $RCH_2NH\text{-}$ a) $CH_3CH_2CH_2NH_2$ b) piperidine (six-membered ring with NH)

Question	Answer
9. Nitrobenzene ($C_6H_5NO_2$) $\xrightarrow[HCl]{Sn}$ $\xrightarrow[H_2O]{OH^{\ominus}}$?	9. Aniline ($C_6H_5NH_2$)
10. RNH (with H(R) on N) $\xrightarrow{R'\overset{O}{\overset{\|}{C}}Cl}$? Write the structure of the products in the following: a) $CH_3CH_2NH_2 \xrightarrow{CH_3\overset{O}{\overset{\|}{C}}Cl}$? b) $CH_3CH_2NHCH_3 \xrightarrow{C_6H_5\overset{O}{\overset{\|}{C}}Cl}$?	10. $RN\overset{O}{\overset{\|}{C}}R' + HCl$ (with H(R) on N) a) $CH_3CH_2NH\overset{O}{\overset{\|}{C}}CH_3$ b) $CH_3CH_2N\overset{O}{\overset{\|}{C}}C_6H_5$ (with CH_3 on N)
11. a) Aniline ($C_6H_5NH_2$) b) RNHR(Ar) c) $C_6H_5NR_2$ $\xrightarrow[HCl]{Na^{\oplus}NO_2^{\ominus}}$? Write the structure of the products in the following: d) $CH_3-C_6H_4-NHCH_3$ (para)	11. a) $C_6H_5\overset{\oplus}{N}\equiv N\ Cl^{\ominus}$ b) R-NR(Ar) with N=O on N c) $O=N-C_6H_4-NR_2$ (para) d) $CH_3-C_6H_4-N(CH_3)-N=O$ (para) (continued next page)

(e) $C_6H_5-N(CH_3)_2$	(e) $O=N-C_6H_4-N(CH_3)_2$
(f) $CH_3-C_6H_4-NH_2$	(f) $CH_3-C_6H_4-\overset{\oplus}{N}\equiv N\ Cl^{\ominus}$
12. $\rangle NH + RCH_2\overset{O}{\overset{\parallel}{C}}H(R) \xrightarrow{H^{\oplus}} ?$ (2°) Write the structure of the enamine formed in the following: a) pyrrolidine (N–H) + $CH_3CH_2\overset{O}{\overset{\parallel}{C}}CH_2CH_3 \xrightarrow{H^{\oplus}} ?$ b) morpholine (N–H) + cyclohexanone $\xrightarrow{H^{\oplus}} ?$	12. $\rangle N-\underset{H(R)}{\underset{\vert}{C}}=CHR$ a) pyrrolidinyl–$N-\underset{CH_2CH_3}{\underset{\vert}{C}}=CHCH_3$ b) morpholinyl (O, N)–N–cyclohexenyl
13. $:\!\!\rangle N-\overset{\vert}{C}=C\langle \xrightarrow{RX} \xrightarrow{H_3O^{\oplus}} ?$ enamine Write the structure of the product in the following: a) $(CH_3)_2N$–cyclopentenyl $\xrightarrow{CH_3I}$ $\xrightarrow{H_3O^{\oplus}} ?$	13. $O=\overset{\vert}{C}-\overset{\vert}{\underset{\vert}{C}}-R$ a) 2-methylcyclopentanone ($C=O$, CH_3) (continued next page)

b) 4-(1-cyclohexenyl)morpholine $\xrightarrow{CH_3\overset{O}{\overset{\|}{C}}Cl}$ $\xrightarrow{H_3O^{\oplus}}$?	b) 2-acetylcyclohexanone (C_6H_9O–$\overset{O}{\overset{\|}{C}}CH_3$)
14. $-\underset{H}{\overset{\mid}{\underset{\mid}{C}}}-\underset{N(CH_3)_2}{\overset{\mid}{\underset{\mid}{C}}}-$ $\xrightarrow{H_2O_2}$ $\xrightarrow{heat}$? Write the structure of the products when $-\underset{H}{\overset{\mid}{\underset{\mid}{C}}}-\underset{N(CH_3)_2}{\overset{\mid}{\underset{\mid}{C}}}-$ is: a) cyclohexyl–$CH_2N(CH_3)_2$ b) CH_3, H, CH_3CH_2 on C—C bearing $N(CH_3)_2$, H, CH_3	14. $\rangle C=C\langle$ a) cyclohexane$=CH_2$ b) $CH_3CH_2(CH_3)C=C(H)CH_3$

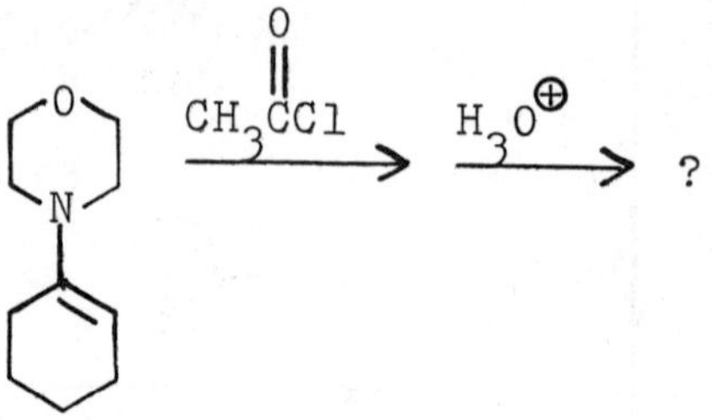

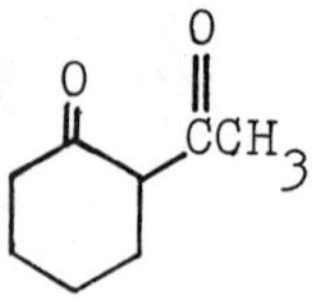

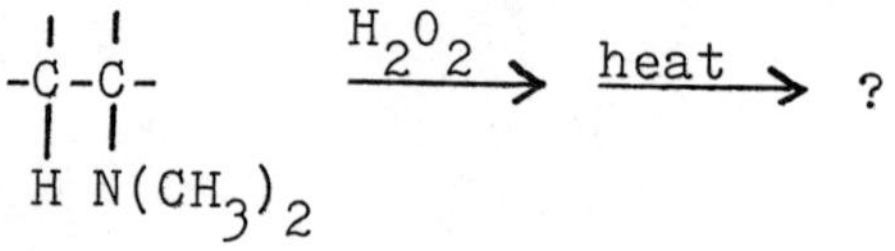

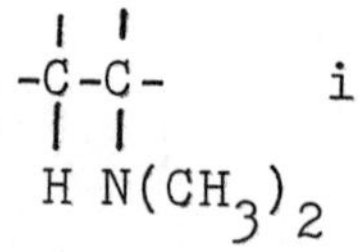

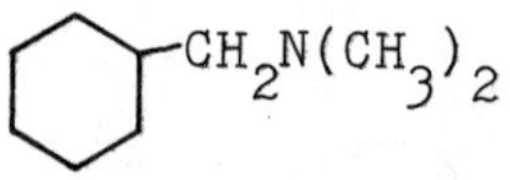

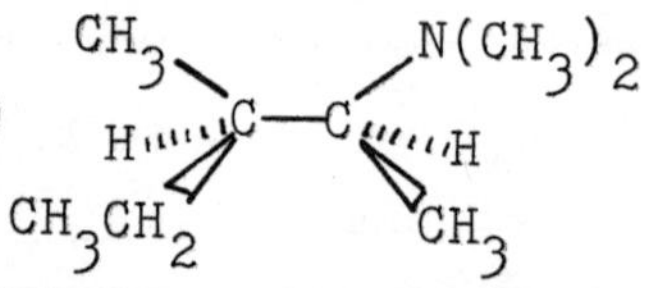

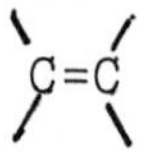

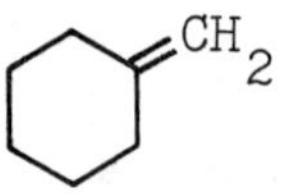

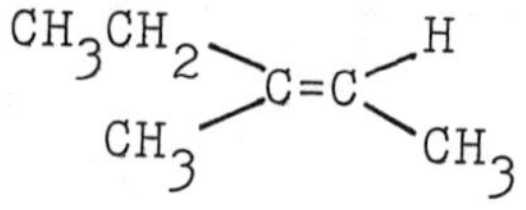

CHAPTER 22 - ARYL NITROGEN COMPOUNDS

ANSWERS TO QUESTIONS

1. (a)

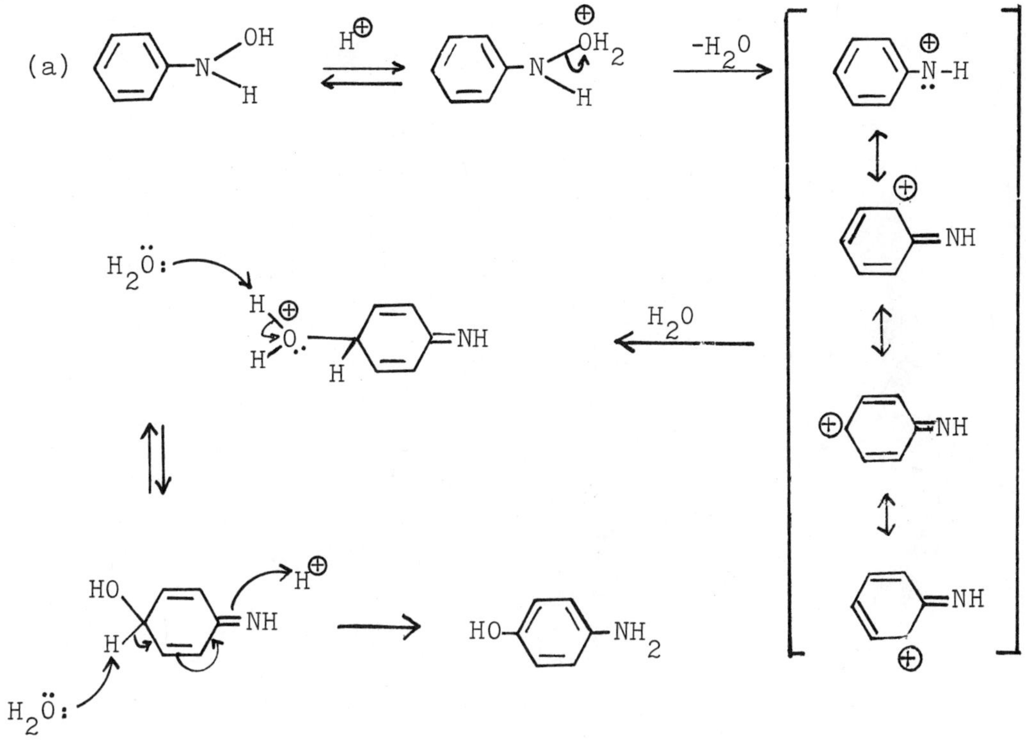

(b)

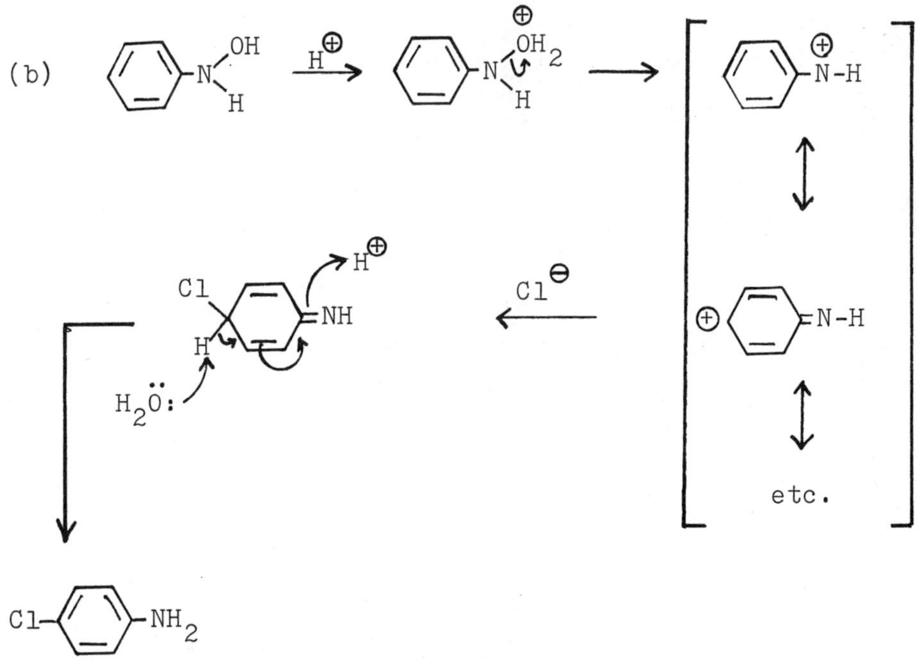

2. The increased stability may be attributed to additional resonance stability afforded by the nonbonding electron pair on nitrogen of the dimethylamino group.

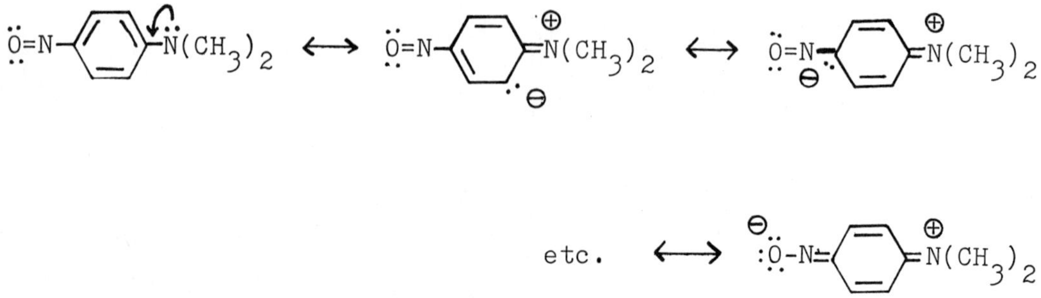

$\ddot{O}$=N–C_6H_4–$\ddot{N}(CH_3)_2$ ⟷ $\ddot{O}$=N–C_6H_4=$\overset{\oplus}{N}(CH_3)_2$ ($\ominus$ on ring) ⟷ $\ddot{O}$=N$^{\ominus}$–C_6H_4=$\overset{\oplus}{N}(CH_3)_2$

etc. ⟷ $\overset{\ominus}{:\ddot{O}}$–N=$C_6H_4$=$\overset{\oplus}{N}(CH_3)_2$

3. (a) 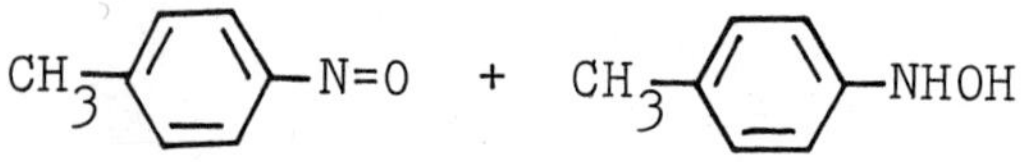 CH_3–C_6H_4–N=O + CH_3–C_6H_4–NHOH

(b) (m-Cl)C_6H_4–N=O + CH_3–C_6H_4–NHOH

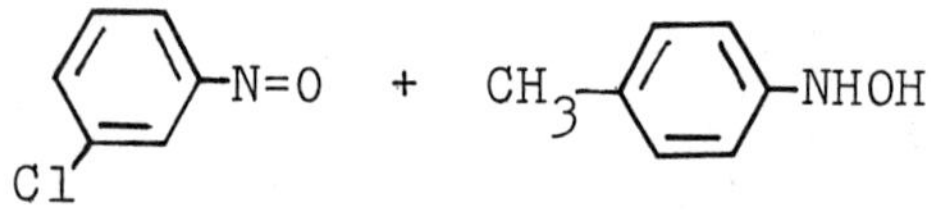

Cl

(c) CH_3–C_6H_4–N=O + (m-Cl)C_6H_4–NHOH

Cl

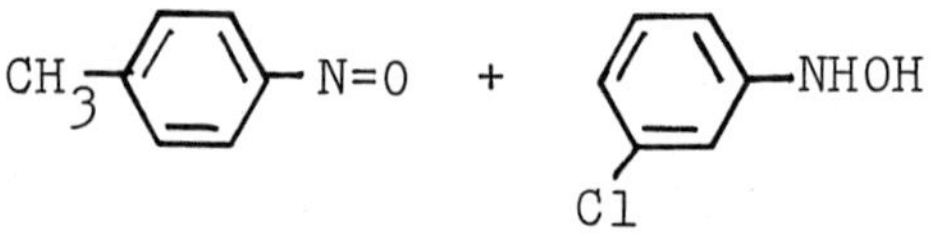

4. (a) 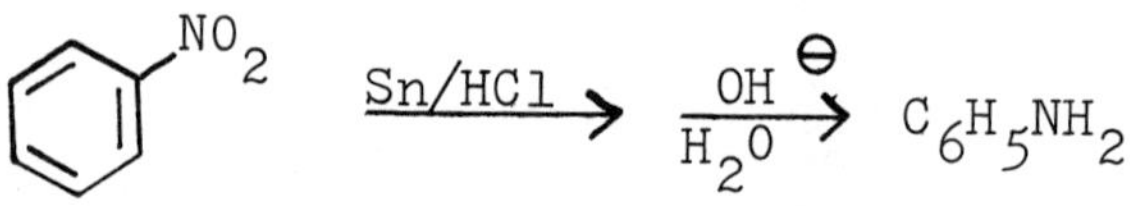 $C_6H_5NO_2$ $\xrightarrow{Sn/HCl}$ $\xrightarrow[H_2O]{OH^{\ominus}}$ $C_6H_5NH_2$

(b) $C_6H_5NO_2$ $\xrightarrow[C_2H_5OH/\ KOH]{NH_2NH_2/Ru/C}$ $(C_6H_5NH)_2$

(c) $C_6H_5NO_2$ $\xrightarrow[C_2H_5OH]{Zn/NaOH}$ $C_6H_5N{=}NC_6H_5$

(d) $C_6H_5NO_2$ $\xrightarrow[H_2O]{Zn/NH_4Cl}$ C_6H_5NHOH

(continued next page)

(e) $C_6H_5NH_2$ (from a) $\xrightarrow{(CH_3\overset{O}{\overset{\|}{C}})_2O}$ $\xrightarrow[AlCl_3]{CH_3Cl}$ $CH_3-C_6H_4-NH-\overset{O}{\overset{\|}{C}}CH_3$

$\xrightarrow[heat]{OH^{\ominus}/H_2O}$ $CH_3-C_6H_4-NH_2$

(f) $C_6H_5NO_2$ $\xrightarrow{LiAlD_4}$ $\xrightarrow{D_2O}$ $C_6H_5ND_2$

(Text S. 21.1)

5. $(CH_3)_2\underset{CN}{\underset{|}{C}}-Br$ from the reaction:

AlBN $\xrightarrow{heat}$ $2(CH_3)_2\underset{CN}{\underset{|}{C}}\cdot$ + N_2

$(CH_3)_2\underset{CN}{\underset{|}{C}}\cdot$ + Br_2 $\longrightarrow$ $(CH_3)_2\underset{CN}{\underset{|}{C}}-Br$ + $Br\cdot$ etc.

6. $Cl-C_6H_4-\overset{\oplus}{N}\equiv N:$ + $C_6H_5-\ddot{N}(CH_3)_2$ $\longrightarrow$ $Cl-C_6H_4-\ddot{N}=\ddot{N}-C_6H_4(H)=\overset{\oplus}{N}(CH_3)_2$ $\xrightarrow{H_2\ddot{O}:}$

$Cl-C_6H_4-\ddot{N}=\ddot{N}-C_6H_4-\ddot{N}(CH_3)_2$

7. (a) $CH_3-C_6H_5$ $\xrightarrow[H_2SO_4]{HNO_3}$ $CH_3-C_6H_4-NO_2$ $\xrightarrow[HCl]{Sn}$ $\xrightarrow[H_2O]{OH^{\ominus}}$

(continued next page)

(b) $CH_3-C_6H_4-NO_2$ (from a) $\xrightarrow[\text{EtOH}]{\text{Zn/NaOH}}$

(c) $CH_3-C_6H_4-NO_2$ $\xrightarrow[\text{heat}]{As_2O_3/OH^{\ominus}/H_2O}$

8. (a) (i) $C_6H_5^{\oplus} + BF_4^{\ominus} \longrightarrow \left[\overset{\delta+}{C_6H_5}\cdots F \cdots \overset{\delta-}{BF_3}\right] \longrightarrow C_6H_5F + BF_3$

(ii) $C_6H_5^{\oplus} + H_2O \longrightarrow \left[\overset{\delta+}{C_6H_5}\cdots \overset{\delta+}{O}(H)-H\right] \longrightarrow C_6H_5OH + H^{\oplus}$

9. $Ar-\ddot{N}=\ddot{N}-NH-CH_2R \xrightarrow{H^{\oplus}} Ar-\overset{\oplus}{N}H=N-NH-CH_2R$ (with $:\ddot{Br}:^{\ominus}$ removing H) $\longrightarrow Ar-\ddot{N}H-N=N-CH_2R$

$\downarrow H^{\oplus}$

$ArNH_2 + N_2 + RCH_2Br \longleftarrow Ar-\overset{\oplus}{N}H_2-N=N-CH_2R$ (attacked by $:\ddot{Br}:^{\ominus}$)

The use of aniline would prevent the last step since an S_N2 reaction on an sp^2 carbon would be required.

10. (a) See (i) if the reaction is accelerated by a radical initiator, (ii) if there is a decrease in rate due to the presence of a radical inhibitor, (iii) if there is a general insensitivity of the rate to changes in solvent polarity.

(b) $H-\overset{\oplus}{P}(H)(:\ddot{O}:^{\ominus})-\ddot{O}-H$ or $H-\ddot{P}(:\ddot{O}-H)-\ddot{O}-H$ (see Chapter 24)

11. 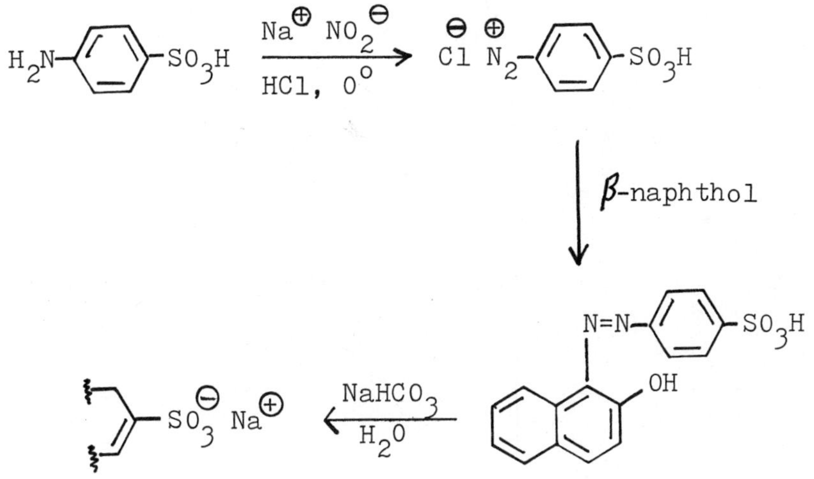 $H_2N-C_6H_4-SO_3H \xrightarrow[HCl,\ 0^o]{Na^{\oplus}\ NO_2^{\ominus}} Cl^{\ominus}\ N_2^{\oplus}-C_6H_4-SO_3H$

β-naphthol

N=N–C_6H_4–SO_3H

OH

$SO_3^{\ominus}\ Na^{\oplus}$ $\xleftarrow[H_2O]{NaHCO_3}$

12. 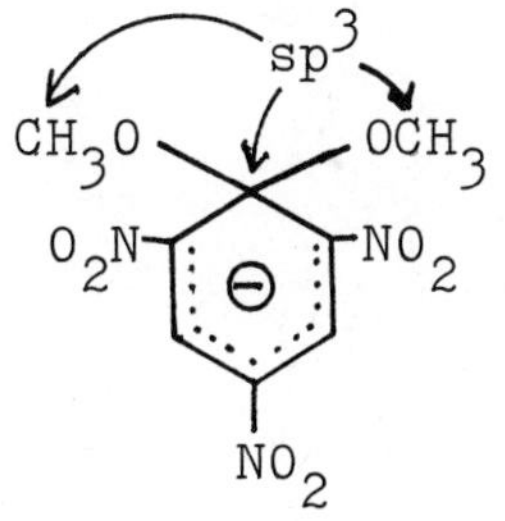 sp^3

CH_3O OCH_3

O_2N NO_2

NO_2

All others are sp^2.

13. One pair of electrons occupies an orbital which is perpendicular to the π system.

14. Aromaticity is a ground state (thermodynamic) property; reactivity (rate) is related to the difference in energy between the ground state and the activated complex.

15. 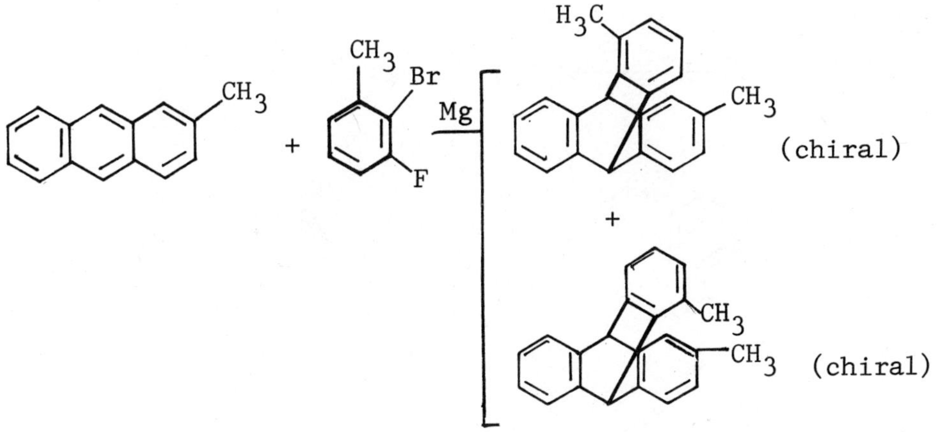 CH_3 + CH_3 Br F $\xrightarrow{Mg}$ H_3C CH_3 (chiral)

+

CH_3 CH_3 (chiral)

(continued next page)

CH_3-anthracene + CH_3, Br, F-benzene $\xrightarrow{Mg}$ [CH_3, CH_3 product (achiral) + CH_3, CH_3 product (chiral)]

16. (a) $C_6H_5N(H)-OH$ + o-$CH_3C_6H_4NO$ $\xrightarrow[H_2O]{OH^{\ominus}}$ $C_6H_5-\overset{O^{\ominus}}{\underset{\oplus}{N}}=N-C_6H_4CH_3$

(b) $C_6H_5N(H)-OH$ $\xrightarrow[0^{o}]{Cr_2O_7^{2-}/H^{\oplus}}$ C_6H_5NO

(c) $C_6H_5NO_2$ $\xrightarrow[FeCl_3]{Cl_2}$ m-$ClC_6H_4NO_2$ $\xrightarrow[H_2O/heat]{As_2O_3/OH^{\ominus}}$ m-$ClC_6H_4-\overset{O^{\ominus}}{\underset{\oplus}{N}}=N-C_6H_4Cl$-m

(d) $C_6H_5NO_2$ $\xrightarrow[HCl]{Sn}$ $\xrightarrow[H_2O]{OH^{\ominus}}$ $C_6H_5NH_2$

(e) C_6H_5NO + m-$CH_3C_6H_4NH_2$ $\longrightarrow$ $C_6H_5-N=N-C_6H_4CH_3$ $\xrightarrow{h\nu}$

C_6H_5-N=N-C_6H_4-CH_3 (cis)

(continued next page)

(f) C_6H_5NO + [3-methylaniline: benzene ring with CH_3 and NH_2 meta] $\longrightarrow$ $C_6H_5-\ddot{N}=\ddot{N}-$[3-methylphenyl, CH_3]

(g) $C_6H_5N{=}NC_6H_5 \xrightarrow[H_2O/OH^{\ominus}]{Zn} C_6H_5NHNHC_6H_5$

(h) $(C_6H_5NH)_2 \xrightarrow[H_2O]{Na_2S_2O_4} 2C_6H_5NH_2$

(i) $C_6H_5NHNHC_6H_5 \xrightarrow{O_2} C_6H_5N{=}NC_6H_5$

(j) $C_6H_5NH_2 \xrightarrow[H_2O/0^{\circ}]{HCN/NaNO_2} C_6H_5N{=}N{-}CN$

(k) $C_6H_5NH_2 \xrightarrow[0^{\circ}/H_2O]{NaNO_2/H_2SO_4} \xrightarrow{SnCl_2} C_6H_5NHNH_2$

17. (a) $C_6H_5NO_2 \xrightarrow[HCl]{Sn} \xrightarrow[H_2O]{OH^{\ominus}} \xrightarrow[H_2SO_4/H_2O/0^{\circ}]{NaNO_2} \xrightarrow{CuCN} C_6H_5CN$

(b) C_6H_5CN (from a) $\xrightarrow{LiAlH_4} \xrightarrow{H_2O} C_6H_5CH_2NH_2$

(c) $C_6H_5CN \xrightarrow[ether]{CH_3MgCl} \xrightarrow{H_3O^{\oplus}} C_6H_5\overset{O}{\overset{\|}{C}}CH_3$

(d) $C_6H_5CH_3 \xrightarrow[H_2SO_4]{HNO_3} CH_3-C_6H_4-NO_2 \xrightarrow{Sn/HCl} \xrightarrow[H_2O]{OH^{\ominus}} \xrightarrow{(CH_3\overset{O}{\overset{\|}{C}})_2O}$

(continued next page)

$CH_3-C_6H_4-NHCCH_3$ (C=O) $\xrightarrow[H_2SO_4]{HNO_3}$ (4-CH$_3$, 2-NO$_2$)$C_6H_3NHCCH_3$ (C=O) $\xrightarrow[heat]{H_2O/OH^{\ominus}}$

(4-CH$_3$, 2-NO$_2$)$C_6H_3NH_2$ $\xrightarrow[H_2SO_4/0^{o}]{NaNO_2}$ $\xrightarrow{H_3PO_2}$ (3-NO$_2$)$C_6H_4CH_3$

(e) (3-NO$_2$)$C_6H_4CH_3$ (from d) $\xrightarrow[ether]{LiAlH_4}$ $\xrightarrow{H_2O}$ (3-NH$_2$)$C_6H_4CH_3$

(f) $C_6H_5CH_3$ $\xrightarrow[H_2SO_4]{HNO_3}$ $CH_3-C_6H_4-NO_2$ $\xrightarrow[Ru/C]{H_2NNH_2}$ $CH_3-C_6H_4-NH-NH-C_6H_4-CH_3$

(g) $C_6H_5NO_2$ $\xrightarrow{H_2NNH_2/Ru/C}$ $C_6H_5-NH-NH-C_6H_5$ $\xrightarrow{H_3O^{\oplus}}$

$\xrightarrow{OH^{\ominus}}$ $H_2N-C_6H_4-C_6H_4-NH_2$

(h) $H_2N-C_6H_4-C_6H_4-NH_2$ $\xrightarrow[HCl,\ 0^{o}]{NaNO_2}$ $\xrightarrow{H_3PO_2}$ $C_6H_5-C_6H_5$

(continued next page)

(i) $H_2N-C_6H_4-C_6H_4-NH_2$ $\xrightarrow{(CH_3\overset{O}{\overset{\|}{C}})_2O}$ $\xrightarrow[FeBr_3]{Br_2}$ $\xrightarrow[heat]{OH^{\ominus}/H_2O}$

$H_2N-C_6H_3(Br)-C_6H_3(Br)-NH_2$ $\xrightarrow[HCl,\ 0^o]{NaNO_2}$ $\xrightarrow[heat]{C_2H_5OH}$ 3,3'-dibromobiphenyl (Br, Br)

(j) m-$CH_3C_6H_4NO_2$ (from d) $\xrightarrow[HCl]{Sn}$ $\xrightarrow[H_2O]{OH^{\ominus}}$ m-$CH_3C_6H_4NH_2$ $\xrightarrow[HCl,\ 0^o]{NaNO_2}$

$\xrightarrow{CuCl}$ m-$CH_3C_6H_4Cl$

(k) $C_6H_5CH_3$ $\xrightarrow{Br_2/h\nu}$ $\xrightarrow[H_2O]{OH^{\ominus}}$ $C_6H_5CH_2OH$

(l) $C_6H_5CH_2OH$ (from k) $\xrightarrow[OH^{\ominus}]{\text{ethylene oxide}}$ $\xrightarrow{H_2O}$ $C_6H_5CH_2OCH_2CH_2OH$

(m) $O_2N-C_6H_4-CH_3$ (from f) $\xrightarrow{Br_2/h\nu}$ $\xrightarrow{OH^{\ominus}/H_2O}$ $\xrightarrow[OH^{\ominus}]{\text{ethylene oxide}}$

$\xrightarrow{CH_3I}$ $O_2N-C_6H_4-CH_2OCH_2CH_2OCH_3$

(n) $C_6H_5NO_2$ $\xrightarrow[HCl]{Sn}$ $\xrightarrow[H_2O]{OH^{\ominus}}$ $\xrightarrow[HCl,\ 0^o]{NaNO_2}$ $\xrightarrow{CuBr}$ C_6H_5Br

(continued next page)

(o) 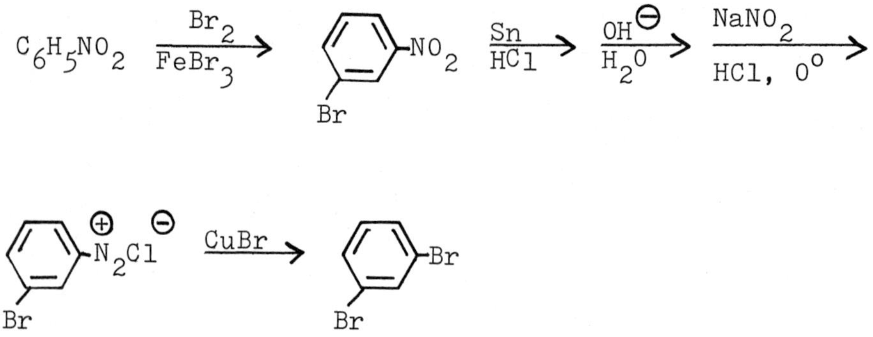 $C_6H_5NO_2 \xrightarrow[FeBr_3]{Br_2}$ (m-bromonitrobenzene: ring with NO_2 and Br) $\xrightarrow[HCl]{Sn} \xrightarrow[H_2O]{OH^\ominus} \xrightarrow[HCl,\ 0^o]{NaNO_2}$

(m-bromobenzenediazonium chloride: ring with $\overset{\oplus}{N}_2Cl^\ominus$ and Br) $\xrightarrow{CuBr}$ (m-dibromobenzene: ring with Br and Br)

(p) 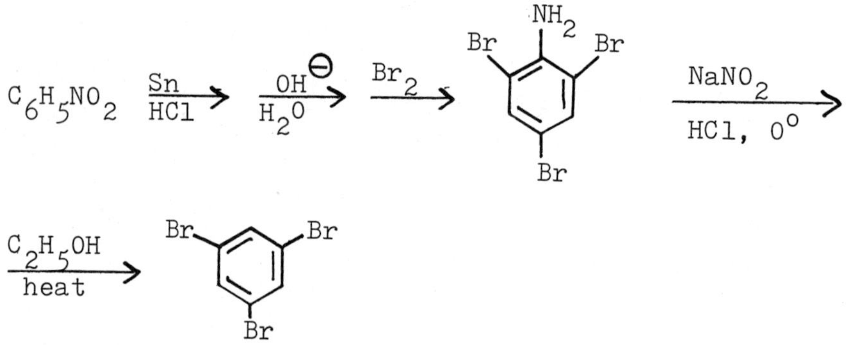 $C_6H_5NO_2 \xrightarrow[HCl]{Sn} \xrightarrow[H_2O]{OH^\ominus} \xrightarrow{Br_2}$ (2,4,6-tribromoaniline: ring with NH_2, Br, Br, Br) $\xrightarrow[HCl,\ 0^o]{NaNO_2}$

$\xrightarrow[heat]{C_2H_5OH}$ (1,3,5-tribromobenzene: ring with Br, Br, Br)

(q) $C_6H_5NO_2 \xrightarrow[HCl]{Sn} \xrightarrow[H_2O]{OH^\ominus} \xrightarrow[HCl,\ 0^o]{NaNO_2} \xrightarrow{D_3PO_2} C_6H_5D$

(r) 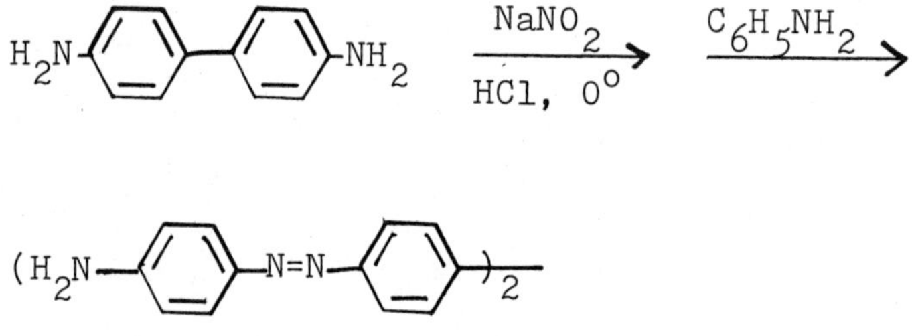 H_2N-(biphenyl)-NH_2 $\xrightarrow[HCl,\ 0^o]{NaNO_2} \xrightarrow{C_6H_5NH_2}$

$(H_2N$-(phenyl)-N=N-(phenyl)$)_2$

18. (a) 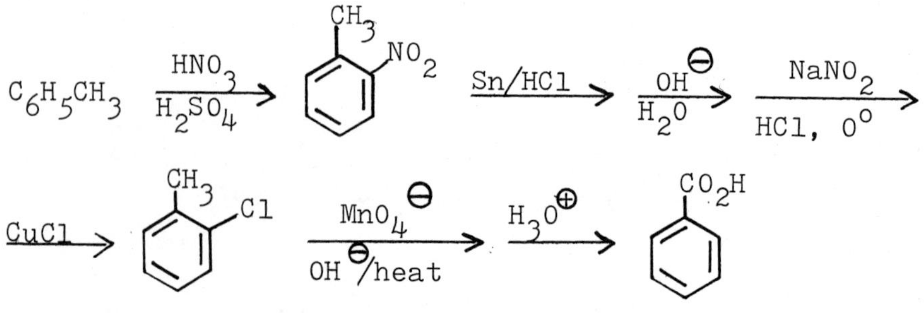 $C_6H_5CH_3 \xrightarrow[H_2SO_4]{HNO_3}$ (o-nitrotoluene: ring with CH_3 and NO_2) $\xrightarrow{Sn/HCl} \xrightarrow[H_2O]{OH^\ominus} \xrightarrow[HCl,\ 0^o]{NaNO_2}$

$\xrightarrow{CuCl}$ (o-chlorotoluene: ring with CH_3 and Cl) $\xrightarrow[OH^\ominus/heat]{MnO_4^\ominus} \xrightarrow{H_3O^\oplus}$ (ring with CO_2H)

(continued next page)

(b) 2-chlorobenzoic acid (from a) $\xrightarrow[H_2SO_4]{HNO_3}$ 2-chloro-4-nitrobenzoic acid $\xrightarrow[HCl]{Sn}$ $\xrightarrow[HCl]{NaNO_2}$ $\xrightarrow[BF_3]{HF}$ $\xrightarrow{heat}$ 2-chloro-4-fluorobenzoic acid

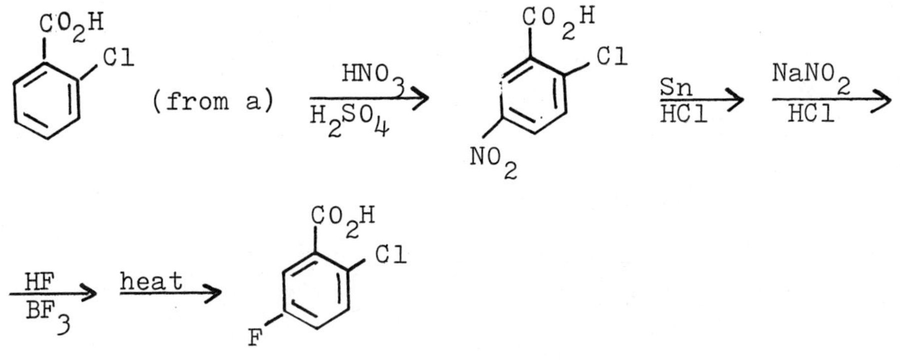

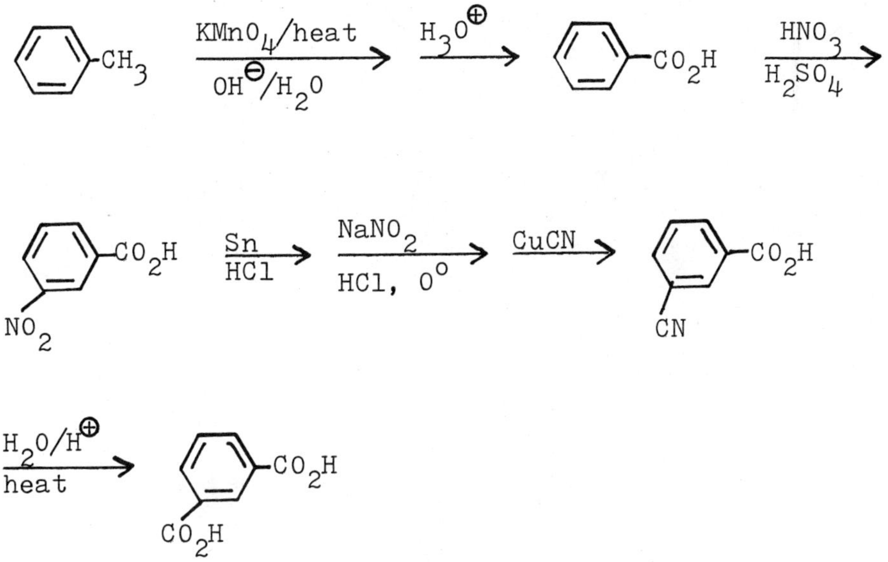

(c) $C_6H_5CH_3$ $\xrightarrow[OH^{\ominus}/H_2O]{KMnO_4/heat}$ $\xrightarrow{H_3O^{\oplus}}$ $C_6H_5CO_2H$ $\xrightarrow[H_2SO_4]{HNO_3}$ m-nitrobenzoic acid $\xrightarrow[HCl]{Sn}$ $\xrightarrow[HCl,\ 0^o]{NaNO_2}$ $\xrightarrow{CuCN}$ m-cyanobenzoic acid $\xrightarrow[heat]{H_2O/H^{\oplus}}$ isophthalic acid ($C_6H_4(CO_2H)_2$, meta)

(d) C_6H_6 $\xrightarrow[H_2SO_4]{HNO_3}$ $\xrightarrow[HCl]{Sn}$ $\xrightarrow[H_2O]{OH^{\ominus}}$ $C_6H_5NH_2$ $\xrightarrow[HCl,\ 0^o]{NaNO_2}$ $\xrightarrow{HBF_4}$ $\xrightarrow{heat}$ C_6H_5F $\xrightarrow[H_2SO_4]{HNO_3}$ $\xrightarrow[HCl]{Sn}$ $\xrightarrow[H_2O]{OH^{\ominus}}$ F–C_6H_4–NH_2 (para)

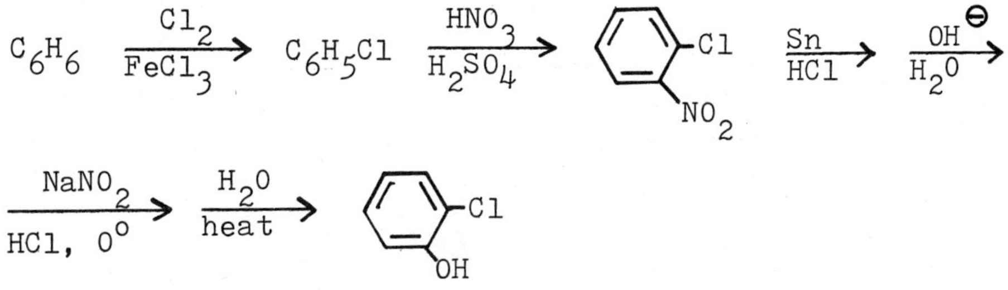

(e) C_6H_6 $\xrightarrow[FeCl_3]{Cl_2}$ C_6H_5Cl $\xrightarrow[H_2SO_4]{HNO_3}$ o-chloronitrobenzene $\xrightarrow[HCl]{Sn}$ $\xrightarrow[H_2O]{OH^{\ominus}}$ $\xrightarrow[HCl,\ 0^o]{NaNO_2}$ $\xrightarrow[heat]{H_2O}$ o-chlorophenol

(f) 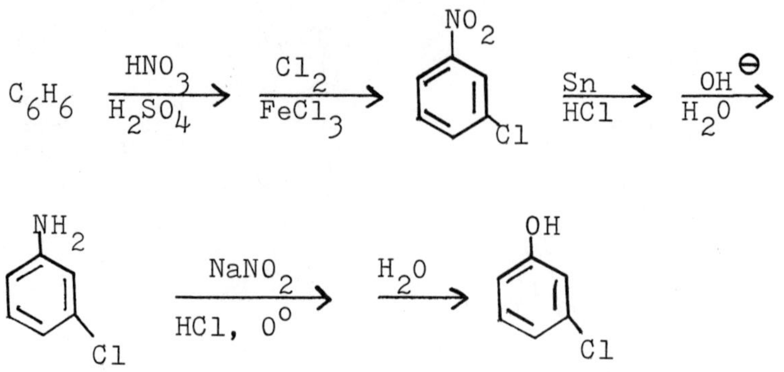 $C_6H_6 \xrightarrow[H_2SO_4]{HNO_3} \xrightarrow[FeCl_3]{Cl_2}$ m-chloronitrobenzene (NO_2, Cl) $\xrightarrow[HCl]{Sn} \xrightarrow[H_2O]{OH^{\ominus}}$

m-chloroaniline (NH_2, Cl) $\xrightarrow[HCl,\ 0^{o}]{NaNO_2} \xrightarrow{H_2O}$ m-chlorophenol (OH, Cl)

(g) 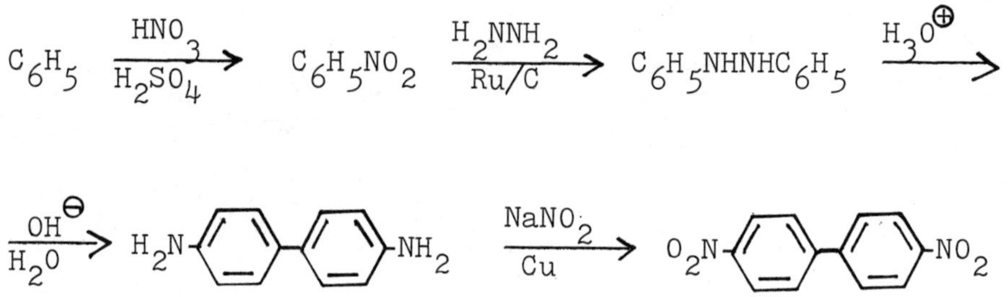 $C_6H_5 \xrightarrow[H_2SO_4]{HNO_3} C_6H_5NO_2 \xrightarrow[Ru/C]{H_2NNH_2} C_6H_5NHNHC_6H_5 \xrightarrow{H_3O^{\oplus}}$

$\xrightarrow[H_2O]{OH^{\ominus}} H_2N\text{-}C_6H_4\text{-}C_6H_4\text{-}NH_2 \xrightarrow[Cu]{NaNO_2} O_2N\text{-}C_6H_4\text{-}C_6H_4\text{-}NO_2$

19. $R_2CH\text{-}N{=}O \rightleftharpoons R_2C{=}N\text{-}OH$ (oxime)

Nitrobenzene lacks the α-hydrogen necessary for this tautomerism.

20. The inductive effect of the nitro group destabilizes the transition state leading to the aryl cation.

21. 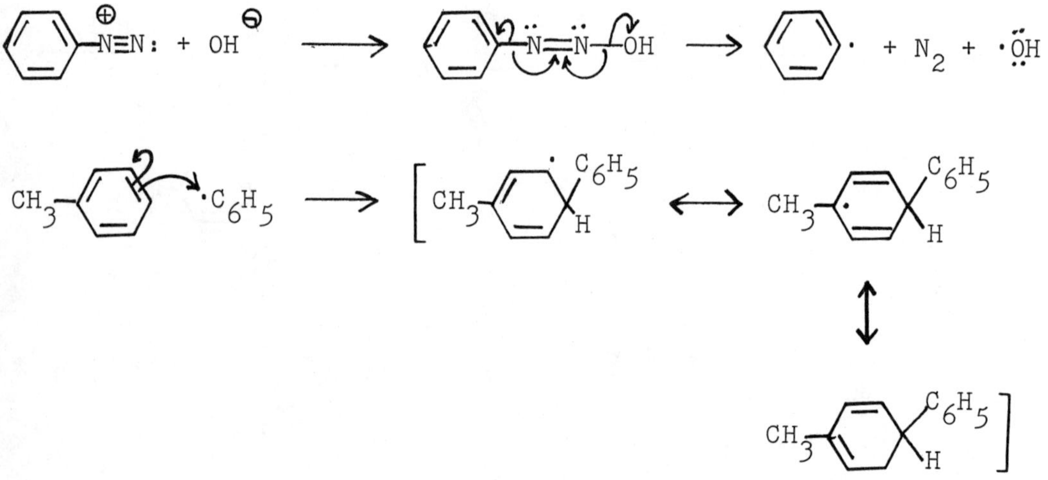 $C_6H_5\text{-}\overset{\oplus}{N}{\equiv}N: + OH^{\ominus} \longrightarrow C_6H_5\text{-}\ddot{N}{=}\ddot{N}\text{-}OH \longrightarrow C_6H_5\cdot + N_2 + \cdot\ddot{O}H$

$CH_3\text{-}C_6H_5 + \cdot C_6H_5 \longrightarrow$ [CH_3-cyclohexadienyl radical (C_6H_5, H) $\longleftrightarrow$ CH_3-cyclohexadienyl radical (C_6H_5, H) $\updownarrow$ CH_3-cyclohexadienyl radical (C_6H_5, H)]

(continued next page)

$$CH_3\text{-}C_6H_5(H)(\dot{})\ \cdot\ddot{O}H \longrightarrow CH_3\text{-}C_6H_4\text{-}C_6H_5 + H_2O$$

The same sequence (radical addition followed by H abstraction) begun at a position meta to the methyl group leads to product A.

22. $C_6H_6 \xrightarrow[H_2SO_4]{HNO_3} \xrightarrow[HCl]{Sn} \xrightarrow[H_2O]{OH^\ominus} \xrightarrow[HCl,\ 0^o]{NaNO_2} \xrightarrow[C_6H_6]{OH^\ominus} \xrightarrow{heat}$

$$C_6H_5\text{-}C_6H_5 \xrightarrow{C_6H_5N_2^\oplus} \xrightarrow{OH^\ominus/heat} C_6H_5\text{-}C_6H_4\text{-}C_6H_5$$

23. $C_6H_5\overset{O}{\overset{\|}{C}}\text{-}Cl \xrightarrow{AlCl_3} \left[C_6H_5\overset{\oplus}{C}=O \longleftrightarrow C_6H_5C\equiv\overset{\oplus}{O} \right]$

$$C_6H_6 + C_6H_5\text{-}\overset{\oplus}{C}=O \longrightarrow \text{(cation)}\ C(=O)C_6H_5\ ^* \xrightarrow{H^\oplus} \underset{(A)}{C_6H_5\overset{O}{\overset{\|}{C}}C_6H_5}$$

$$C_6H_5\text{-}\overset{\ddot{O}:}{\overset{\|}{C}}\text{-}C_6H_5 + C_6H_5\text{-}MgCl \longrightarrow C_6H_5\text{-}\overset{:\ddot{O}MgCl}{\underset{C_6H_5}{C}}\text{-}C_6H_5 \xrightarrow{H_2O} \underset{(B)}{(C_6H_5)_3COH}$$

$$(C_6H_5)_3COH \xrightarrow{HCl} (C_6H_5)_3C\text{-}\overset{\oplus}{O}H_2 \xrightarrow{-H_2O} (C_6H_5)_3C^\oplus$$

(continued next page)

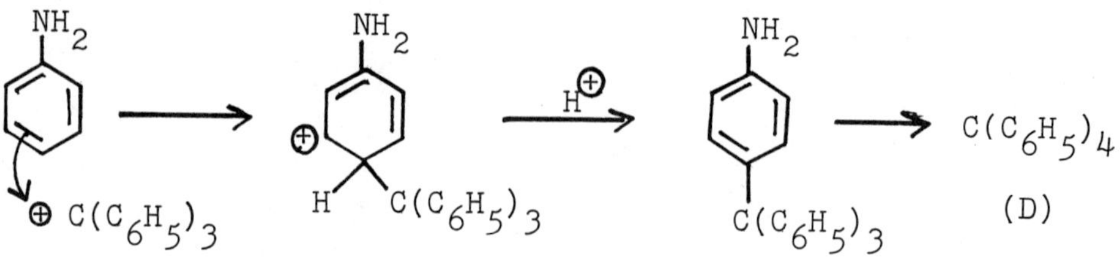

(C)

$C(C_6H_5)_4$ (D)

*resonance-stabilized cation

24. (a) $:N\equiv C-\ddot{S}:^{\ominus} \longleftrightarrow {}^{\ominus}:\ddot{N}=C=\ddot{S}:$ sulfur is more nucleophilic than nitrogen

(b) $C_6H_5SCN \xrightarrow[\text{heat}]{H_3O^{\oplus}} C_6H_5SH$

25.

A

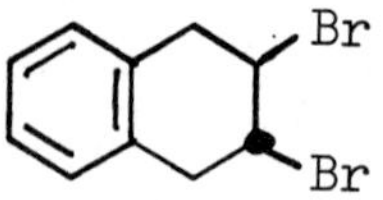

B

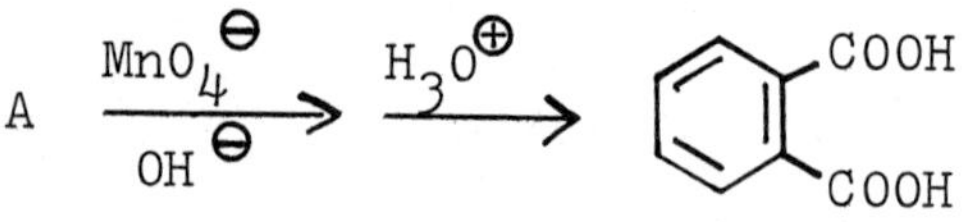

C

26. The coupling reaction requires the presence of a strong electron donating group in order that the process shown below may occur.

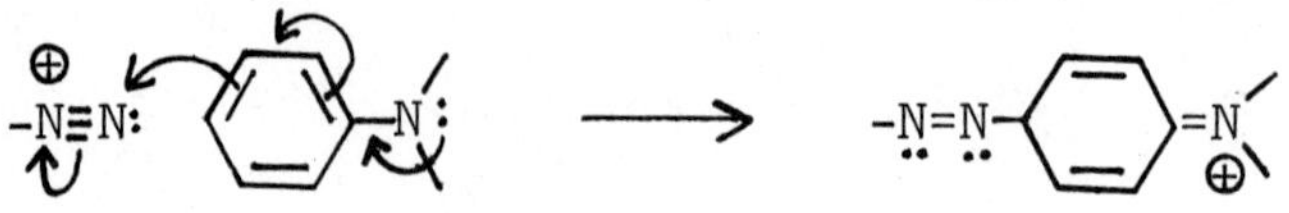

Because of crowding (steric repulsions), the dimethylamino group in N,N-2,6-tetramethylaniline is forced into a conformation which prevents the nitrogen orbital containing the electron pair from being parallel to the orbitals of the π system. Thus, donation of the nitrogen electron pair is not possible.

27. 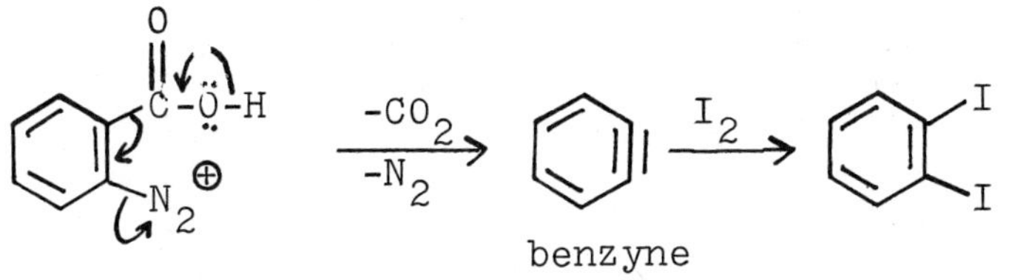 (ortho-diazonium benzoate) $\xrightarrow[-N_2]{-CO_2}$ benzyne $\xrightarrow{I_2}$ 1,2-diiodobenzene

benzyne

28. 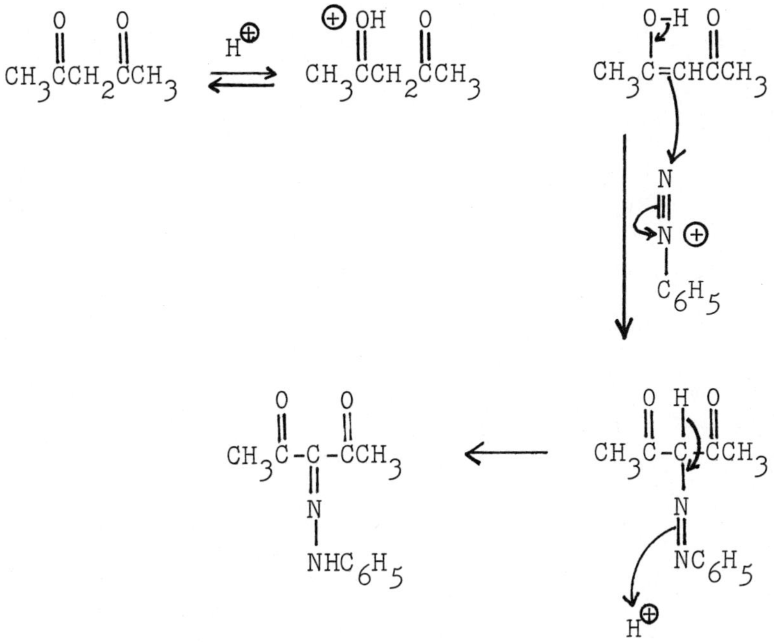 $CH_3\overset{O}{\overset{\|}{C}}CH_2\overset{O}{\overset{\|}{C}}CH_3 \overset{H^{\oplus}}{\rightleftharpoons} CH_3\overset{\oplus OH}{\overset{\|}{C}}CH_2\overset{O}{\overset{\|}{C}}CH_3$; $CH_3\overset{O-H}{\overset{|}{C}}{=}CH\overset{O}{\overset{\|}{C}}CH_3$ + $N{\equiv}N^{\oplus}-C_6H_5$

$\downarrow$

$CH_3\overset{O}{\overset{\|}{C}}-\underset{N=NC_6H_5}{\overset{H}{C}}-\overset{O}{\overset{\|}{C}}CH_3 \; (-H^{\oplus}) \longrightarrow CH_3\overset{O}{\overset{\|}{C}}-\underset{\underset{NHC_6H_5}{|}}{\overset{}{C}}{=}N-\overset{O}{\overset{\|}{C}}CH_3$

29. (1) para red

(a)  2-naphthol (OH) + $N_2^{\oplus}$–C_6H_4–NO_2 OR

(b) 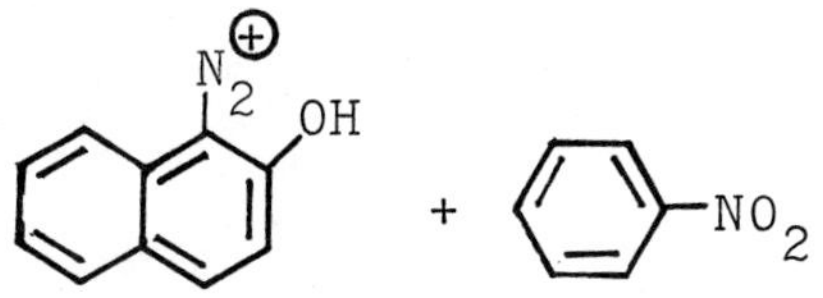 1-($N_2^{\oplus}$)-2-naphthol (OH) + C_6H_5–NO_2

Combination (a) is preferred since nitrobenzene will not undergo coupling.

(2) methyl orange

(a)  $Na^{\oplus}\,{}^{\ominus}O_3S$–$C_6H_4$–$N_2^{\oplus}$ + C_6H_5–$N(CH_3)_2$ OR

(continued next page)

(b)  $Na^{\oplus}\ {}^{\ominus}O_3S$–C₆H₅ + $\overset{\oplus}{N_2}$–C₆H₄–NO_2

Combination (a) is preferred since solium benzenesulfonate will not undergo coupling.

30. (a) Increased basicity increases the concentration of phenoxide ion which is more reactive in coupling than phenol.

(b) Amines are relatively nucleophilic and at pH's near 7 may react with a diazonium salt to form a diazoamine. This reaction is reversed by the addition of acid and, in acid solution, the coupling product is formed.

31. The greater reactivity of phenol compared to anisole is probably due to the presence of the highly reactive phenoxide ion produced in small quantity in solution.

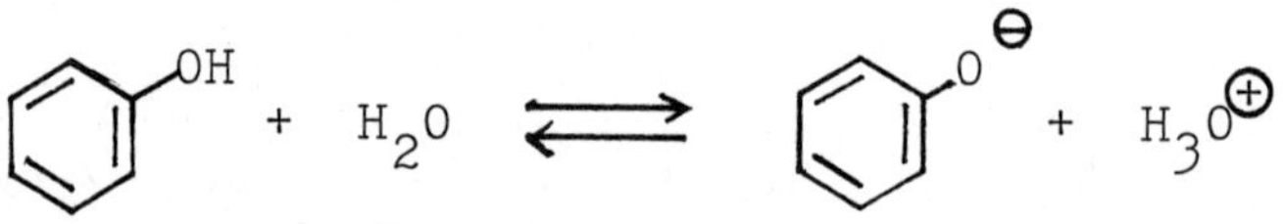

$$C_6H_5OH + H_2O \rightleftharpoons C_6H_5O^{\ominus} + H_3O^{\oplus}$$

The strongly electron withdrawing nitro groups increase the reactivity of the benzenediazonium ion as an electrophile to the extent that it reacts with the less reactive anisole.

32. (a) 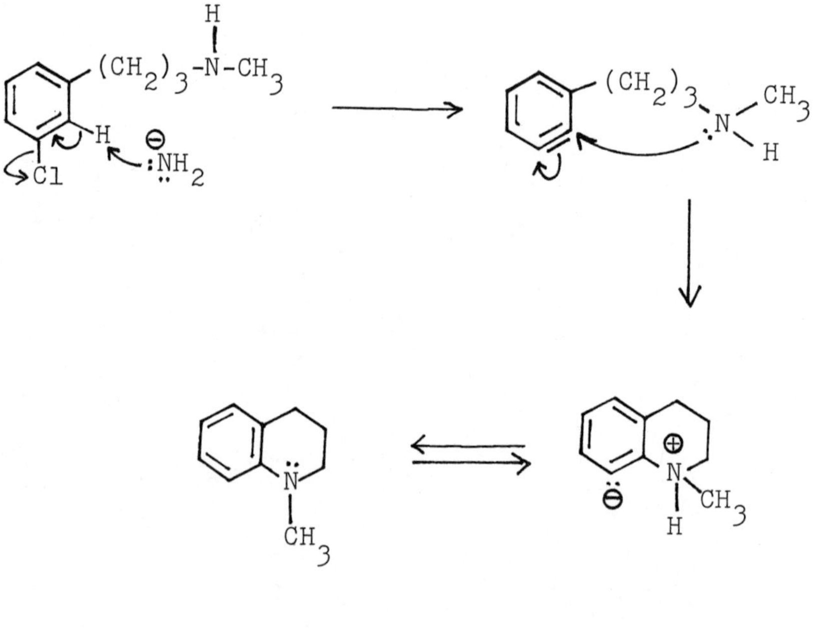 $(CH_2)_3$-N(H)-CH_3 ; H ; Cl ; $:\overset{\ominus}{NH_2}$ → $(CH_2)_3$; N ; CH_3 ; H

↓

N ; CH_3 ⇌ $\overset{\oplus}{N}$; CH_3 ; H ; $\ominus$

(continued next page)

(b) $\longrightarrow$ + SO_2 + N_2

$\longrightarrow$

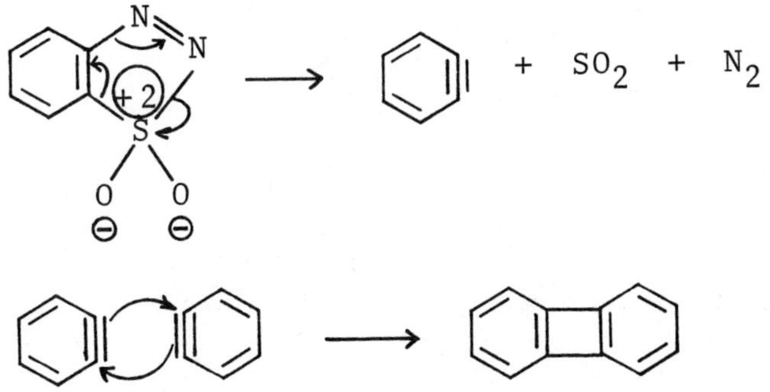

(c) $N\equiv{}^{15}N:$ $\longrightarrow$ $N={}^{15}N$ $\longrightarrow$ $:N\equiv{}^{15}N-$

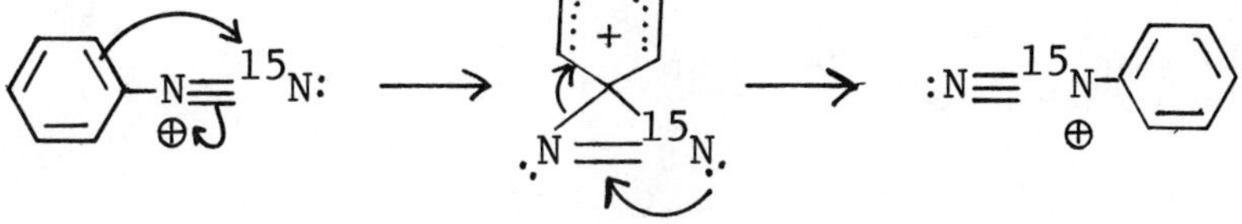

(d) CO_2H, $N_2^{\oplus}$ $\xrightarrow{Cu(I)}$ CO_2H $\xrightarrow{\text{dimerization}}$

CO_2H, HO_2C

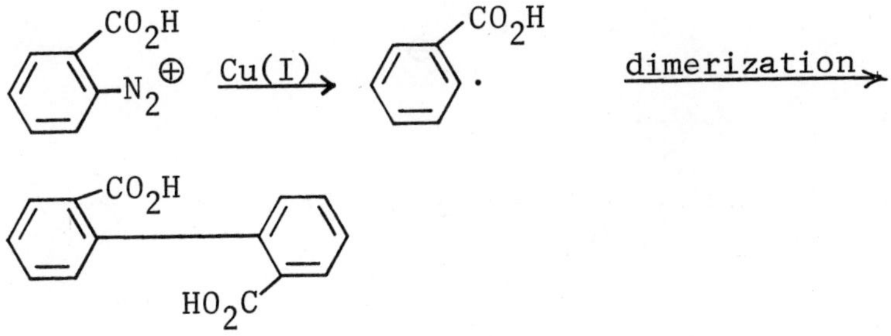

(e) $C_6H_5N_2^{\oplus}BF_4^{\ominus}$ $\longrightarrow$ $C_6H_5^{\oplus}$ + $BF_4^{\ominus}$ + N_2

$\oplus$ $:Br-C_6H_5$ $\longrightarrow$ $Br^{\oplus}$ $BF_4^{\ominus}$

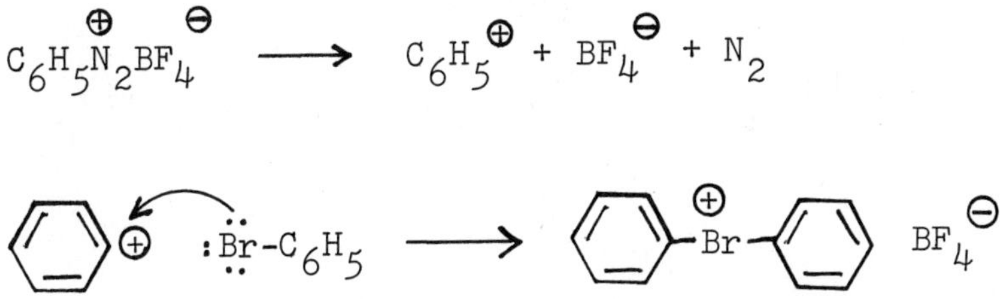

(f) $C_6H_5-\overset{\oplus}{N}\equiv N:$ $\overset{\ominus}{N}=\overset{\oplus}{N}=N^{\ominus}$ $\longrightarrow$ $C_6H_5-N=N-\overset{\ominus}{N}-\overset{\oplus}{N}\equiv N:$

$\downarrow$

$C_6H_5-N=\overset{\oplus}{N}=N^{\ominus}$ + N_2

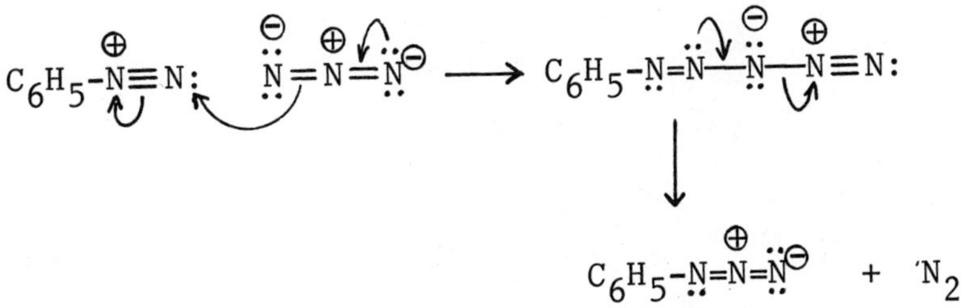

33. $C_6H_5N_2^{\oplus}A^{\ominus}$ $\xrightarrow{-N_2}$ $C_6H_5^{\oplus}$ $\xrightarrow{C_2H_5\ddot{O}H}$ $C_6H_5\overset{\oplus}{O}(H)C_2H_5$ $\xrightarrow{:A^{\ominus}}$ $C_6H_5OC_2H_5$

SUMMARY OF REACTIONS

1. Reduction of nitrobenzene (S 22.2)

$C_6H_5NO_2$ $\xrightarrow[H_2O]{Zn/NH_4^{\oplus}Cl^{\ominus}}$ $C_6H_5N(H)OH$ (phenylhydroxylamine) $\xrightarrow{H^{\oplus}}$ HO–C_6H_4–NH_2

(phenylhydroxylamine) $\xrightarrow{Zn/HCl}$ $\xrightarrow{OH^{\ominus}/H_2O}$ aniline

$C_6H_5NO_2$ $\xrightarrow{Sn/HCl}$ $\xrightarrow{OH^{\ominus}}$ $C_6H_5NH_2$ (aniline)

(or Fe/HCl)

$C_6H_5NO_2$ $\xrightarrow{LiAlH_4}$ $C_6H_5NH_2$ (aniline)

$C_6H_5NO_2$ $\xrightarrow[heat]{Ar_2O_3/OH^{\ominus}/H_2O}$ $C_6H_5\overset{\oplus}{N}(O^{\ominus})=NC_6H_5$ (azoxybenzene)

(azoxybenzene) $\xrightarrow{Zn/HCl}$ aniline

$C_6H_5NO_2$ $\xrightarrow[C_2H_5OH]{Zn/NaOH}$ $C_6H_5N=NC_6H_5$ (azobenzene)

$C_6H_5NO_2$ $\xrightarrow[C_2H_5OH/KOH]{NH_2NH_2/Ru/C}$ $C_6H_5NH-NHC_6H_5$ (hydrazobenzene)

2. Reduction of some other aryl nitrogen compounds

Azoxybenzene ($C_6H_5-N(O^{\ominus})^{\oplus}=N-C_6H_5$) → Fe/distill or $(C_2H_5O)_3P$ → $C_6H_5-N=N-C_6H_5$

(azobenzene)

azobenzene → $Zn/OH^{\ominus}/H_2O$/heat → hydrazobenzene; hydrazobenzene → air → azobenzene

$C_6H_5-NH—NH-C_6H_5$

(hydrazobenzene)

hydrazobenzene → H_2O_2, CH_3CO_2H → azoxybenzene

hydrazobenzene → $H^{\oplus}$ → $H_2N-C_6H_4-C_6H_4-NH_2$ (as a salt)

(benzidine)

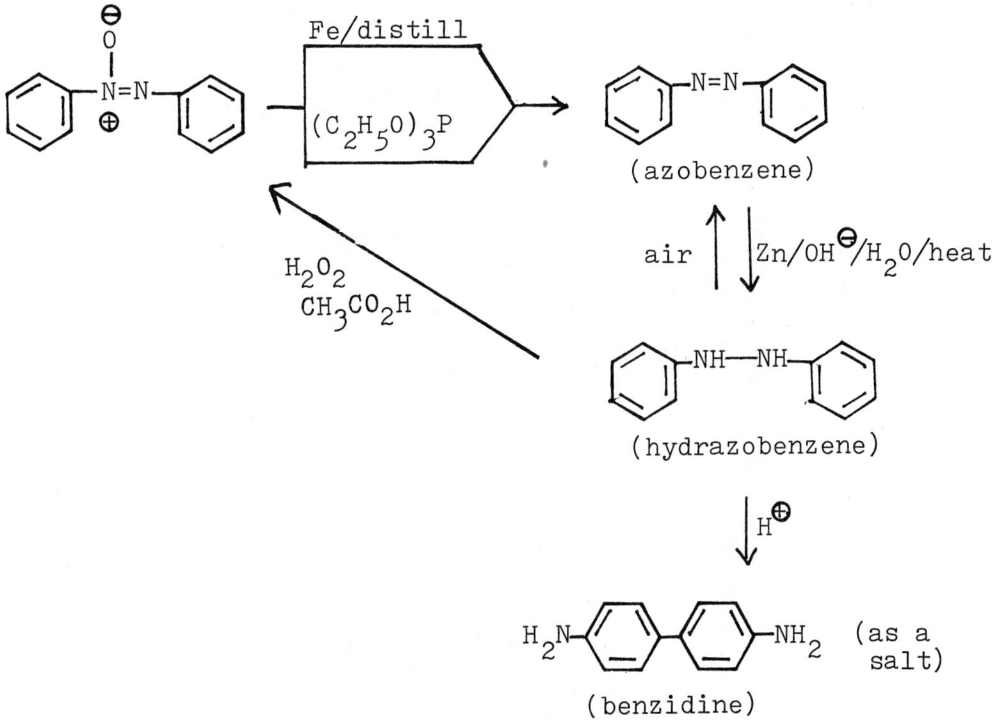

3. Conversion of aryl nitrogen compounds to aniline

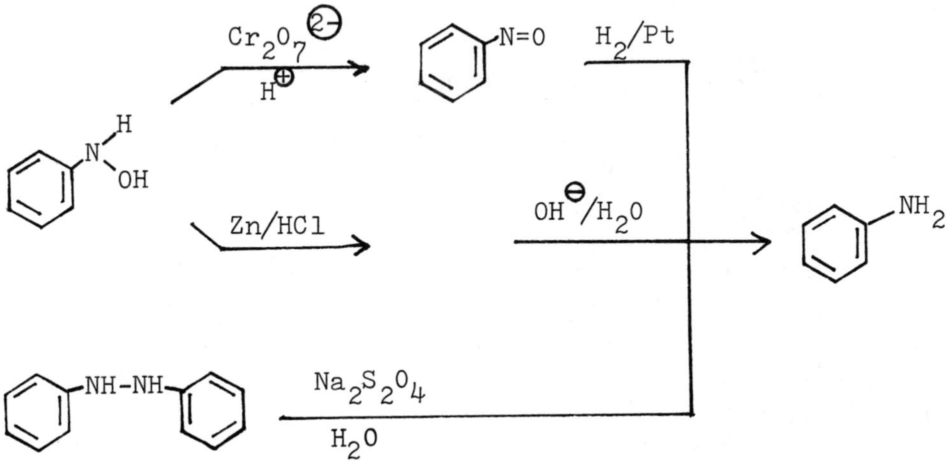

$C_6H_5-N(H)OH$ → $Cr_2O_7^{2-}$, $H^{\oplus}$ → $C_6H_5-N=O$ → H_2/Pt → $C_6H_5-NH_2$

$C_6H_5-N(H)OH$ → Zn/HCl → $OH^{\ominus}/H_2O$ → $C_6H_5-NH_2$

$C_6H_5-NH-NH-C_6H_5$ → $Na_2S_2O_4$, H_2O → $C_6H_5-NH_2$

(reduction of nitrobenzenes, see part 1)

4. Diazotization

$$ArNH_2 \xrightarrow[HA,\,0^o]{NaNO_2^{*}} Ar\overset{\oplus}{N}{\equiv}N\ \overset{\ominus}{A}$$

benzene diazonium salt

HA usually H_2SO_4 or HX* ($NaNO_2 + HA \longrightarrow HONO + NaA$)

(effect of pH on products, see Figure 22.2)

$Ar\overset{\oplus}{N}{\equiv}N\ \overset{\ominus}{A}$ reacts with:

- $\xrightarrow{HF/BF_3}$ $Ar\text{-}N_2^{\oplus}\ BF_4^{\ominus} \xrightarrow[-N_2]{heat} Ar\text{-}F$
- $\xrightarrow[-N_2]{H_2O}$ Ar-OH
- $\xrightarrow{CuX}$ Ar-X (Sandmeyer)
- $\xrightarrow{Cu/HX}$ Ar-X (Gattermann)
- $\xrightarrow{HI}$ Ar-I
- $\xrightarrow{CuCN}$ Ar-CN
- $\xrightarrow[Cu_2SO_4]{NaNO_2}$ $Ar\text{-}NO_2$
- $\xrightarrow{RCH_2NH_2}$ $Ar\text{-}N{=}NNHCH_2R \xrightarrow{HBr} ArNH_2 + RCH_2Br + N_2$
- H_3PO_2 or C_2H_5OH → Ar-H
- $SnCl_2$ or Na_2SO_3 → $ArNHNH_2$
- $\xrightarrow{Ar'\text{-}\overset{|}{N}\text{-}}$ $Ar\text{-}N{=}N\text{-}Ar'\text{-}\overset{|}{N}\text{-}$ (diazo coupling)
- $\xrightarrow{Ar'\text{-}OH}$ $Ar\text{-}N{=}N\text{-}Ar'\text{-}OH$ (diazo coupling)

5. Nucleophilic aromatic substitution (addition-elimination)

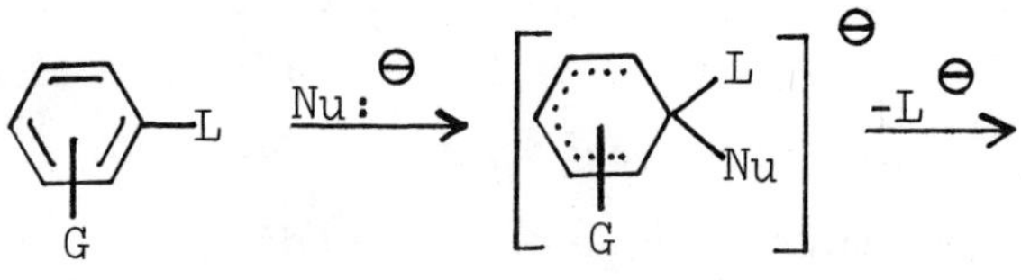

Nu:$^{\ominus}$ → -L$^{\ominus}$ → (S 22.4)

G = deactivating group

6. Dehydrobenzene (benzyne) (S 22.5)

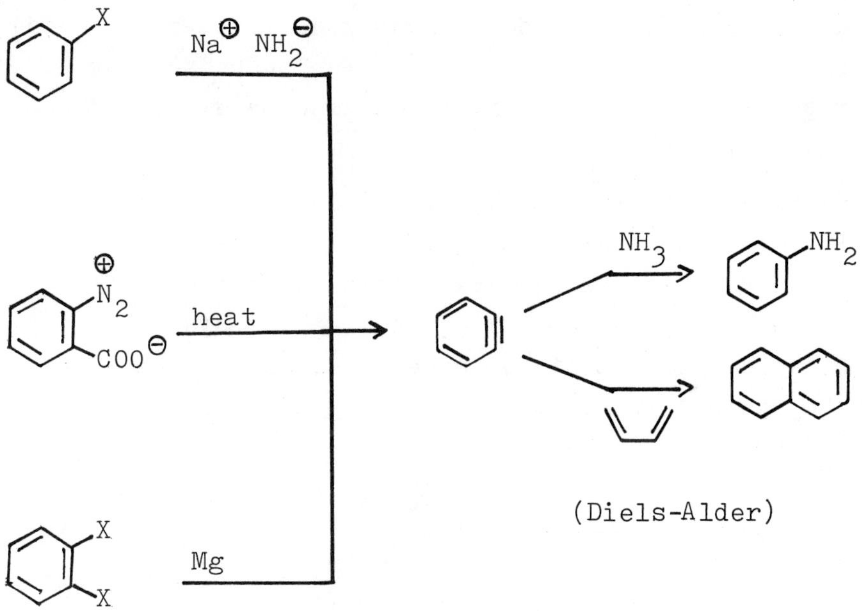

$Na^{\oplus}$ $NH_2^{\ominus}$

heat

Mg

NH_3

(Diels-Alder)

ANSWERS TO QUESTIONS

1. (a) Extract the mixture with water which dissolves acetic acid. The organic layer contains hexane (bp=68°) and cyclohexanol (bp=161°) which can be separated by distillation.

   (b) Extraction of the mixture with aqueous sodium bicarbonate followed by acidification of the aqueous phase produces p-chlorobenzoic acid. Extraction of the organic residue (hexane + phenol) with aqueous sodium hydroxide followed by acidification of the aqueous layer produces phenol. The residue is hexane.

   (c) Extraction of the mixture with water separates acetic acid. Extraction of the organic phase (hexane + p-chlorobenzoic acid) with aqueous sodium hydroxide followed by acidification of the aqueous layer produces p-chlorobenzoic acid. The residue is hexane. Hexane can also be distilled from p-chlorobenzoic acid.

   (d) Extraction of the mixture with aqueous sodium bicarbonate followed by acidification of the aqueous phase produces p-chlorobenzoic acid. Extraction of the organic residue (alcohol + phenol) with aqueous sodium hydroxide followed by acidification of the aqueous layer produces 3-octylphenol. The residue is heptanol. See Figure 23-1, text.

2. $C_6H_5-C(CH_3)_2-\ddot{O}-\ddot{O}-H \xrightarrow{H^{\oplus}} C_6H_5-C(CH_3)_2-\ddot{O}-\overset{\oplus}{O}H_2 \longrightarrow H_2\ddot{O}: + {}^{\oplus}C(CH_3)_2-\ddot{O}-C_6H_5$

$\downarrow$

$C_6H_5OH + (CH_3)_2CO \xleftarrow{-H^{\oplus}} H-\ddot{O}-C(CH_3)_2-\overset{\oplus}{O}(H)C_6H_5 \rightleftharpoons H_2\overset{\oplus}{O}-C(CH_3)_2-\ddot{O}-C_6H_5$

3. c, d and e. The Grignard reagent reacts with the other functional group on the benzene ring, e.g., (c)

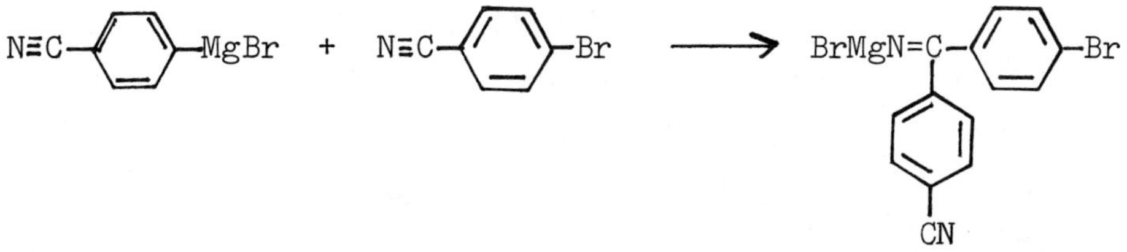

N≡C–C₆H₄–MgBr + N≡C–C₆H₄–Br ⟶ BrMgN=C(C₆H₄CN)–C₆H₄–Br

CN

(Text: Limitations of Grignard, p. 396)

4. (a) + O

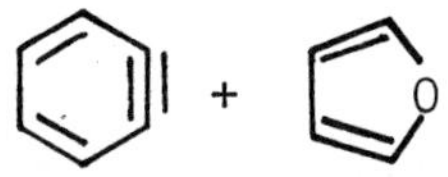

benzyne

(b) :O: $H_3O^{\oplus}$ HO⊕ ⟶ HÖ: H :ÖH₂ ⊕

:ÖH

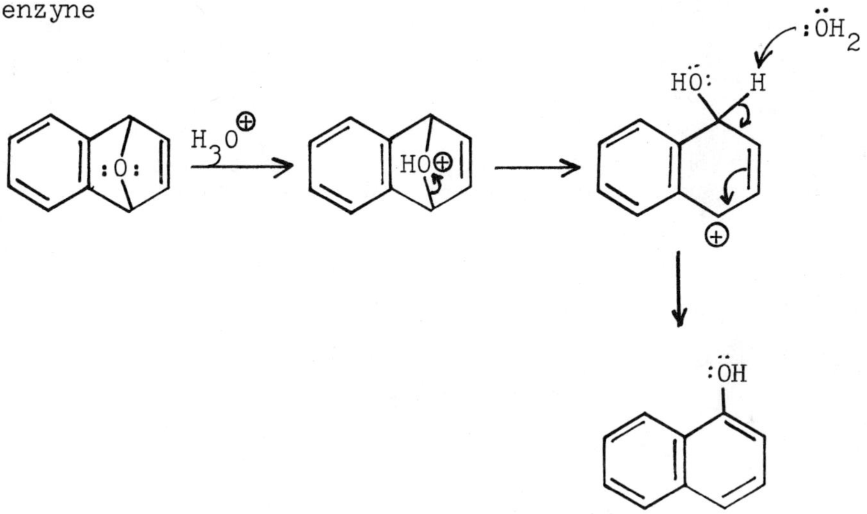

or the concerted mechanism:

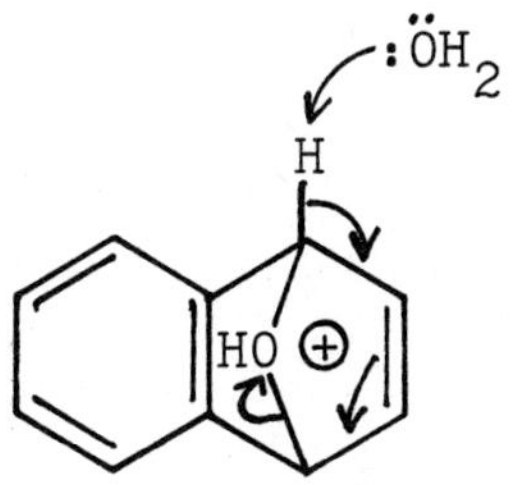

:ÖH₂ H HO ⊕

5. 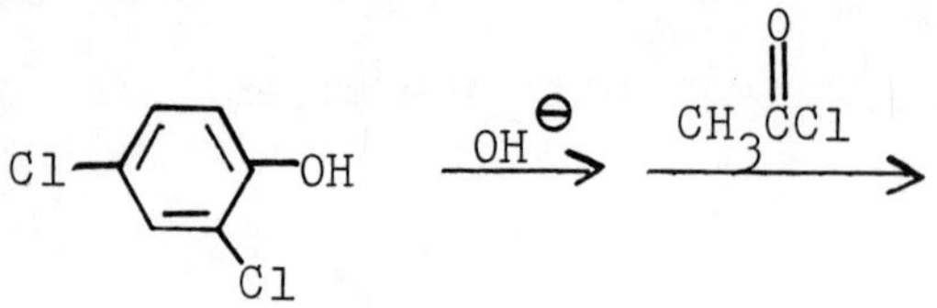 Cl–C_6H_3(Cl)–OH $\xrightarrow{OH^{\ominus}}$ $\xrightarrow{CH_3\overset{O}{\overset{\|}{C}}Cl}$

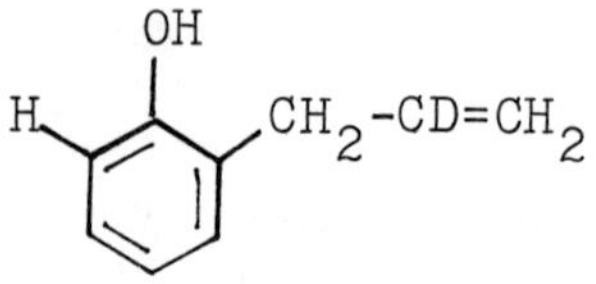

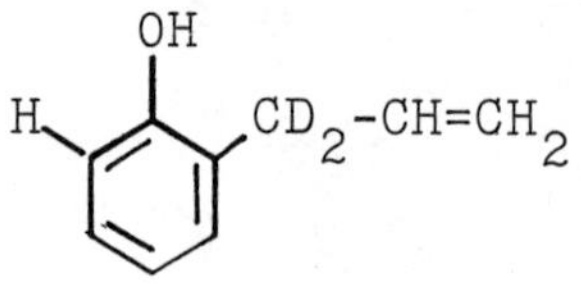

6. (a) 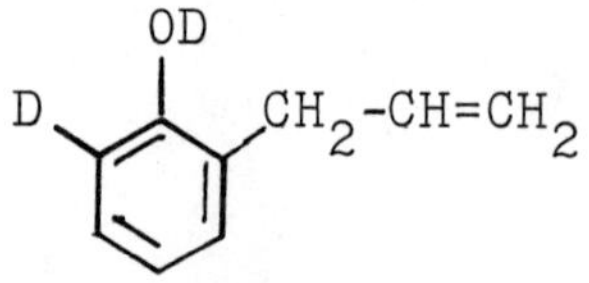 OD; D; CH_2-CH=CH_2 (c) OH; H; CH_2-CD=CH_2

(b) OH; H; CD_2-CH=CH_2 (d) OH; H; CH_2-CH=CD_2

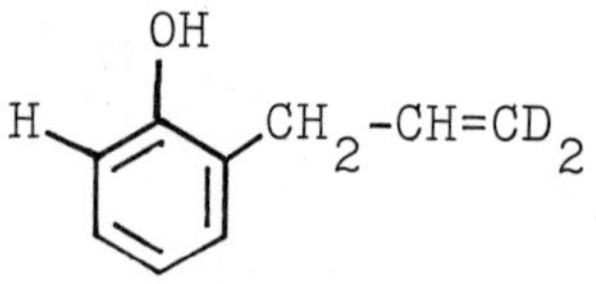

7. If an allyl cation were produced, an appropriate deuterium label would become scrambled and give rise to two products. The product mixture could be analyzed by mass spectrometry.

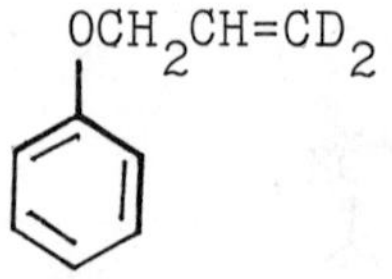

$OCH_2CH=CD_2$ (on benzene ring) does not form (OH, $CH_2CH=CD_2$ on benzene ring) even though it might be expected if the cation

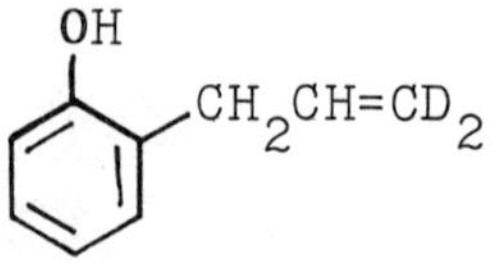

$$\left[\overset{\oplus}{C}H_2\text{-}CH\text{=}CD_2 \longleftrightarrow CH_2\text{=}CH\text{-}CD_2^{\oplus}\right]$$ is formed.

8. (a) 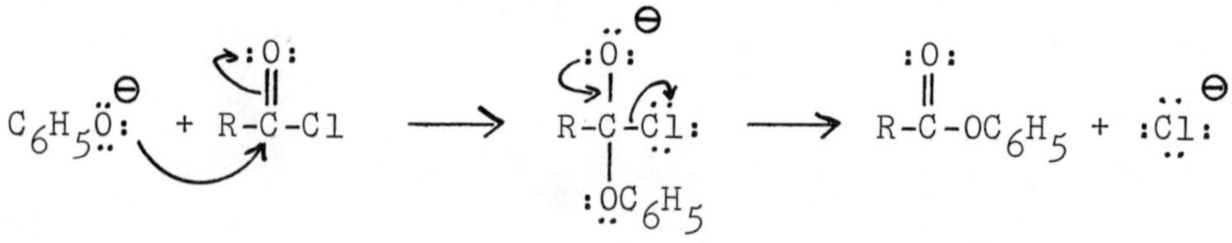 $C_6H_5\ddot{O}:^{\ominus}$ + R-$\overset{:O:}{\overset{\|}{C}}$-Cl $\longrightarrow$ R-$\overset{:\ddot{O}:^{\ominus}}{\overset{|}{C}}$-$\ddot{Cl}$: (with :$\ddot{O}C_6H_5$ below) $\longrightarrow$ R-$\overset{:O:}{\overset{\|}{C}}$-$OC_6H_5$ + $:\ddot{Cl}:^{\ominus}$

(continued next page)

(b) $C_6H_5\ddot{O}H + R-\overset{\overset{:O:}{||}}{C}-O-\overset{\overset{O}{||}}{C}-R \longrightarrow R-\overset{\overset{:\ddot{O}:^{\ominus}}{|}}{\underset{\underset{H\ \oplus\ C_6H_5}{\ddot{O}}}{C}}-O-\overset{\overset{O}{||}}{C}-R \longrightarrow R-\overset{\overset{:\ddot{O}-H}{|}}{\underset{\underset{:OC_6H_5}{|}}{C}}-\ddot{O}-\overset{\overset{O}{||}}{C}-R$

$\downarrow -RCO_2H$

$R-\overset{\overset{:O:}{||}}{C}-\ddot{O}C_6H_5$

(c) $C_6H_5\ddot{O}H + CH_2=C=\ddot{O} \longrightarrow CH_2=\overset{\overset{:\ddot{O}:^{\ominus}}{|}}{C}-\overset{\oplus}{O}(H)(C_6H_5) \longrightarrow CH_2=\overset{\overset{:\ddot{O}-H}{|}}{C}-\ddot{O}C_6H_5$

(enol)

$\rightleftharpoons$

$CH_3-\overset{\overset{:O:}{||}}{C}-\ddot{O}C_6H_5$

(d) $C_6H_5\ddot{O}H + Ar-N=C=\ddot{O} \longrightarrow Ar-N=\overset{\overset{:\ddot{O}:^{\ominus}}{|}}{C}-\overset{\oplus}{O}(H)(C_6H_5)$

$\downarrow$

$Ar-NH-\overset{\overset{:O:}{||}}{C}-\ddot{O}C_6H_5 \longleftarrow Ar-N=\overset{\overset{:\ddot{O}-H}{|}}{C}-O-C_6H_5$

9. (a) $CH_3-C_6H_4-\ddot{O}-CH_3$ + $H-\ddot{Br}:$ → $CH_3-C_6H_4-\overset{\oplus}{O}(H)-CH_3$ + $:\ddot{Br}:^{\ominus}$

↓

$CH_3-C_6H_4-OH$ + CH_3Br

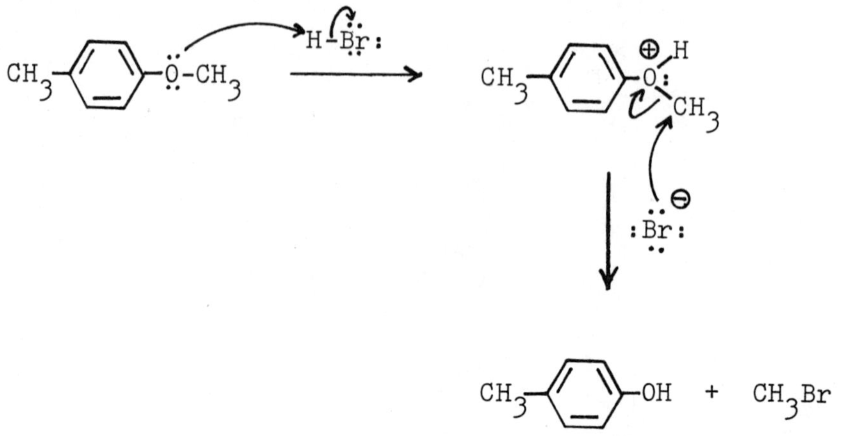

(b) Formation of p-bromotoluene would require an S_N2 displacement on an sp^2 ring carbon which does not occur (see p. 951) or formation of the very unstable phenyl cation in an S_N1 reaction which also does not occur.

10. (a) $:\ddot{O}-CrO_3H$ (ring, H) + $:\ddot{O}H_2$ → $:CrO_3H^{\ominus}$ + $:O:$ (ring) H $:\overset{\oplus}{O}H_2$

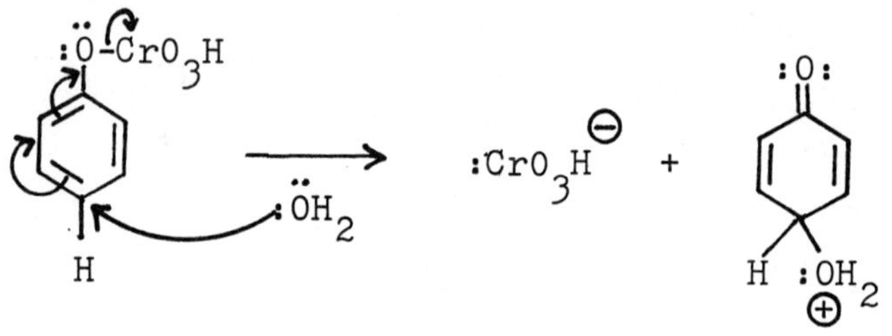

(b) $:O:$ (ring) H O-H, $H^{\oplus}$, $H_2\ddot{O}:$ ⇌ $:\ddot{O}H$ (ring) $:\ddot{O}H$

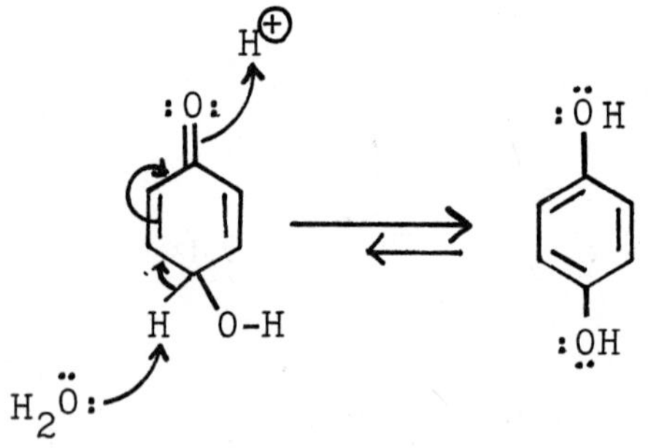

11. (a) the reduction product is $HO-C_6H_2(C(CH_3)_3)_2-CH(CH_3)_2$

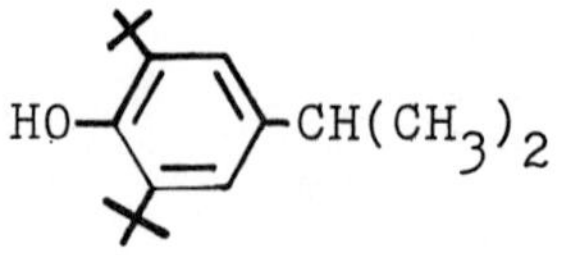

(continued next page)

(b) the oxidation product is O=⟨2,6-di-tert-butylcyclohexadienylidene⟩=$C(CH_3)_2$

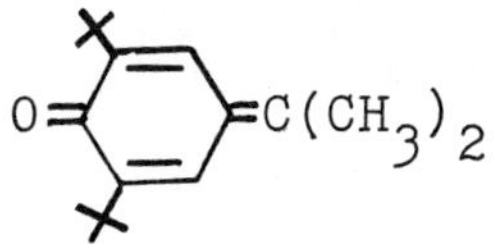

12. 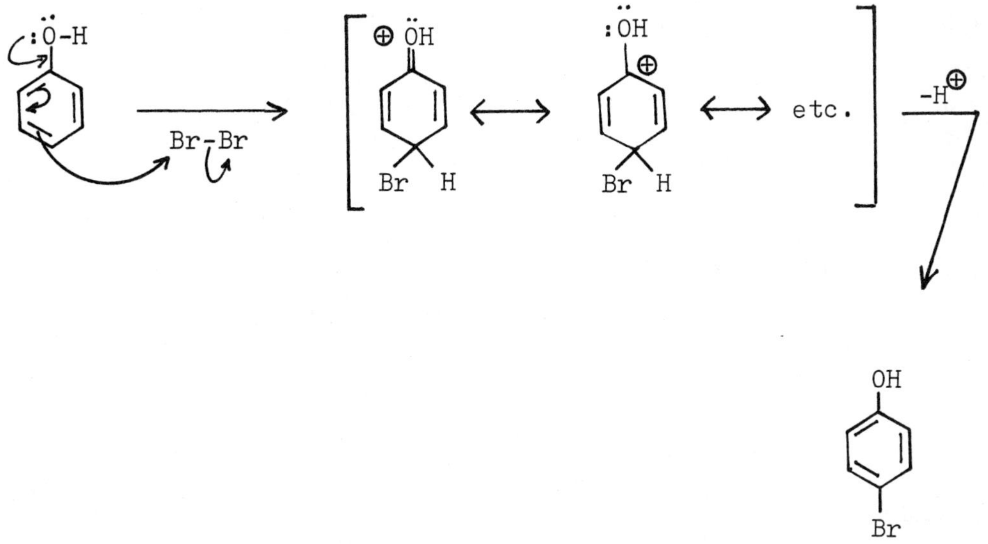 [mechanism: phenol + Br–Br → [⊕ŎH / Br H ⟷ :ŎH ⊕ / Br H ⟷ etc.] $-H^{\oplus}$ → p-bromophenol (OH, Br)]

13. Bromine is a deactivating group relative to hydrogen (electron-withdrawing *via* the inductive effect). This effect makes the conjugate base of p-bromophenol lower in energy (weaker base) than the conjugate base of phenol. The stronger acid has the weaker conjugate base.

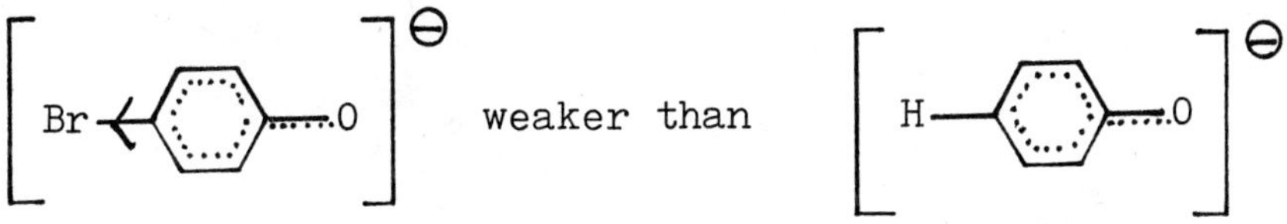

[Br–C_6H_4–O]$^{\ominus}$ weaker than [H–C_6H_4–O]$^{\ominus}$

14. (a) 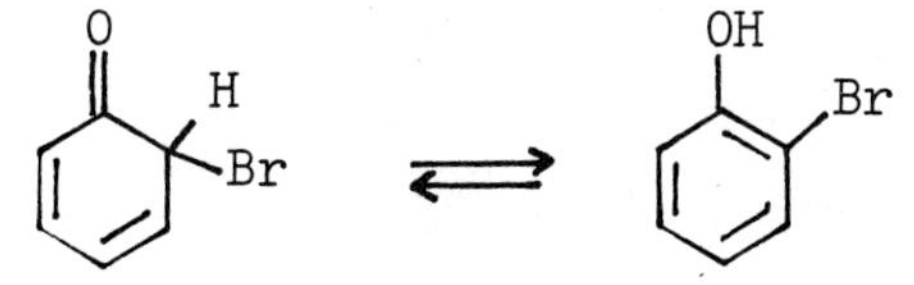 [ketone (O, H, Br) ⇌ enol (OH, Br)]

ketone enol

(b) To form an anion in step 2 of the mechanism a proton must be removed. This can be shown with the ketone; therefore it is unnecessary to show the enol form in equilibrium with the ketone.

15. Each involves anion formation followed by halogenation.

<u>bromination of phenol</u>:

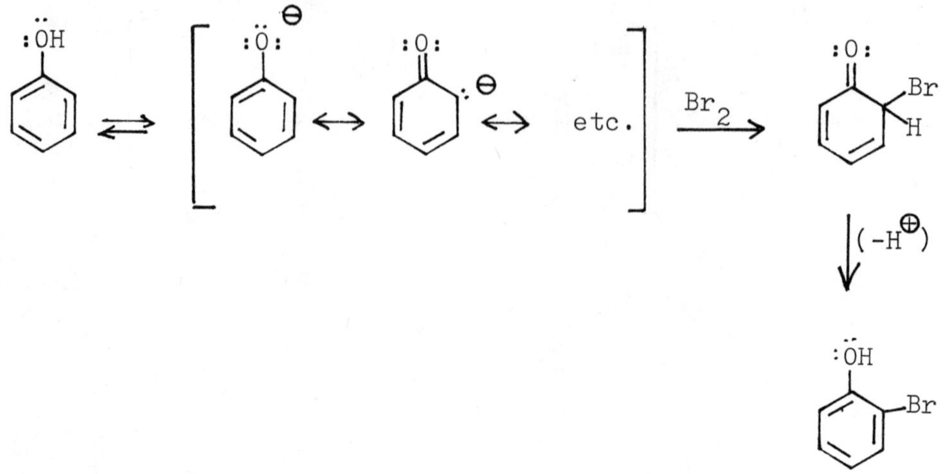

:ÖH ⇄ [:Ö:⊖ ↔ :O: ⊖ ↔ etc.] $\xrightarrow{Br_2}$:O: Br H

↓ $(-H^{\oplus})$

:ÖH Br

<u>haloform</u>:

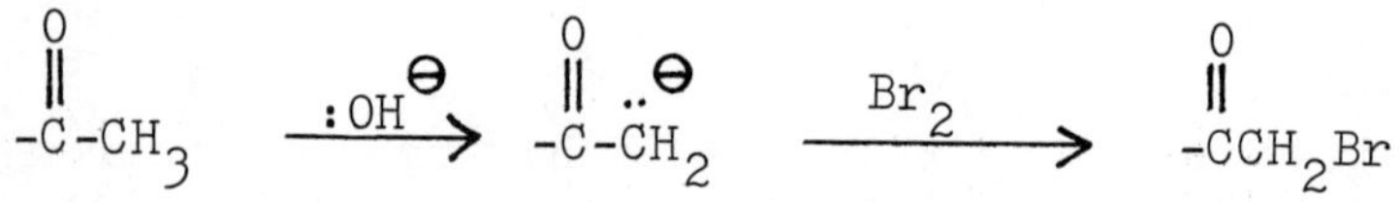

$$-\overset{O}{\overset{\|}{C}}-CH_3 \xrightarrow{:OH^{\ominus}} -\overset{O}{\overset{\|}{C}}-\ddot{C}H_2^{\ominus} \xrightarrow{Br_2} -\overset{O}{\overset{\|}{C}}CH_2Br$$

16. 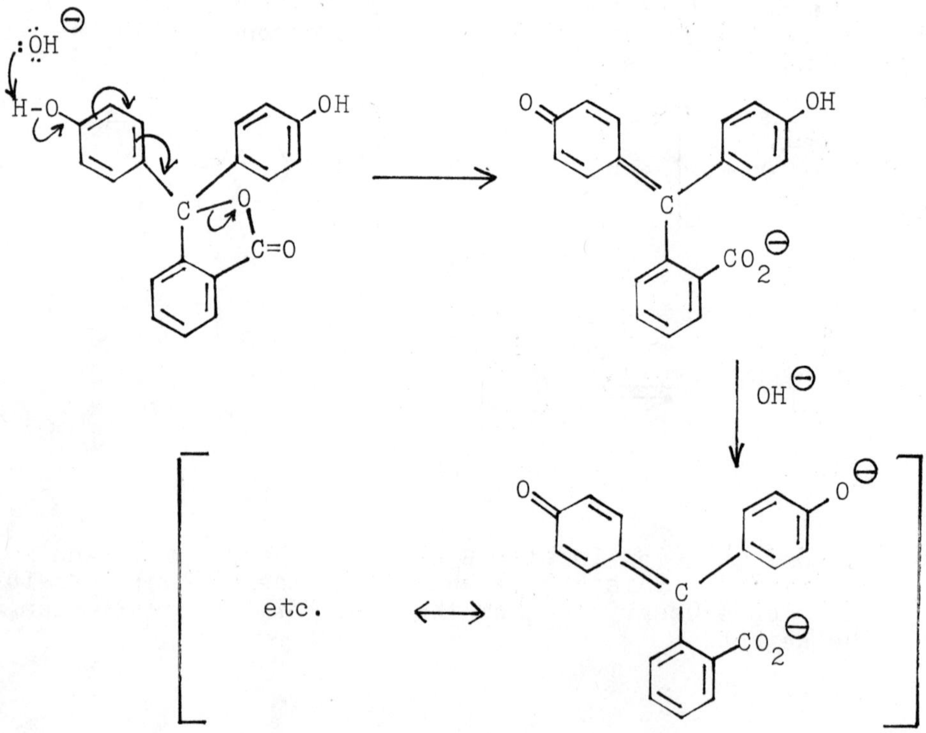 :ÖH⊖ H-O OH C O C=O ⟶ O C OH $CO_2^{\ominus}$

↓ $OH^{\ominus}$

[etc. ⟷ O C $O^{\ominus}$ $CO_2^{\ominus}$]

(see p. 979 for resonance structures)

17.

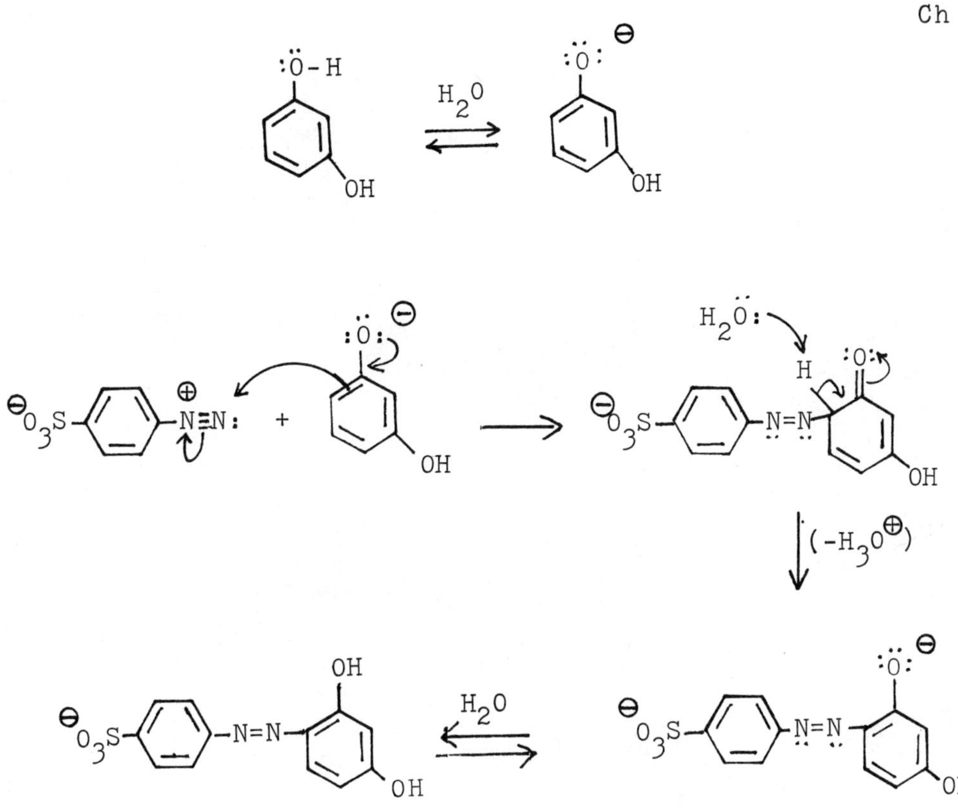

18. At high pH, some of the electrophile $ArN_2^{\oplus}$ is converted to Ar-N=N-OH which does not couple. Thus, the higher the pH, the lower the concentration of $ArN_2^{\oplus}$.

At low pH, the concentration of the more reactive phenoxide ion (in equilibrium with phenol in the solution) is decreased.

19. In the first reaction, CH_3OH reacts as a nucleophile at the electrophilic carbonyl group (of -COOH) in salicylic acid. In the second reaction, the -OH group of salicylic acid reacts as the nucleophile at the electrophilic carbonyl group of the anhydride. In general the reactions are the same.

(continued next page)

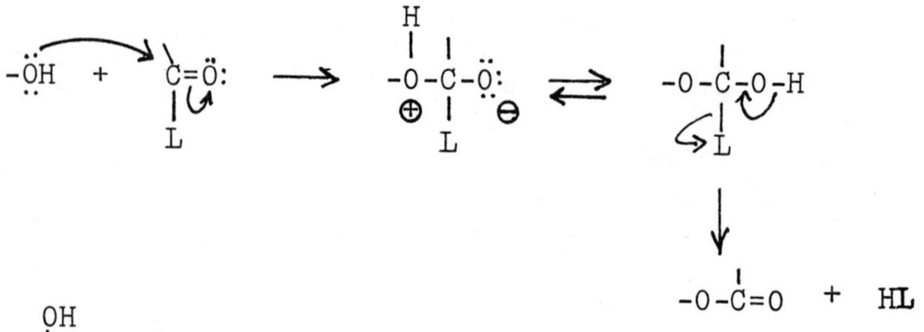

$$-\ddot{O}H + \underset{L}{\overset{\backslash}{C}}=\ddot{O}: \longrightarrow -\underset{\oplus}{\overset{H}{O}}-\underset{L}{C}-\ddot{O}:^{\ominus} \rightleftarrows -O-\underset{L}{C}-O-H \longrightarrow -O-C=O + HL$$

20. (a) 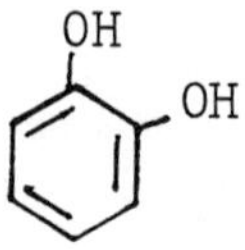 [1,2-dihydroxyphenyl ring] = Ar

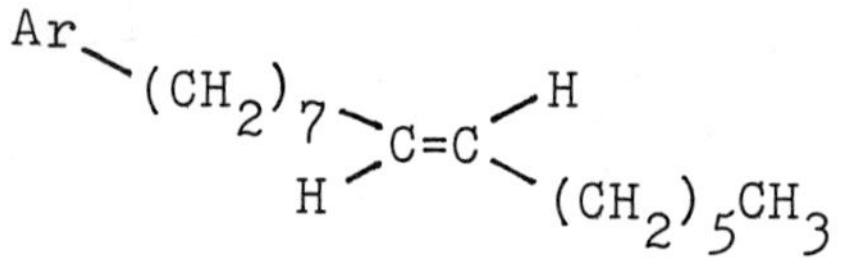

$Ar-(CH_2)_7(H)C=C(H)(CH_2)_5CH_3$ E-isomer

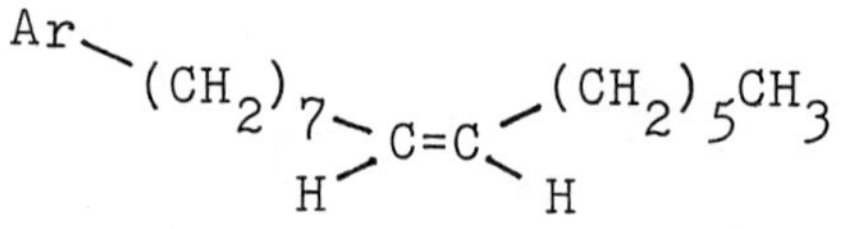

$Ar-(CH_2)_7(H)C=C(H)(CH_2)_5CH_3$ Z-isomer

(b) 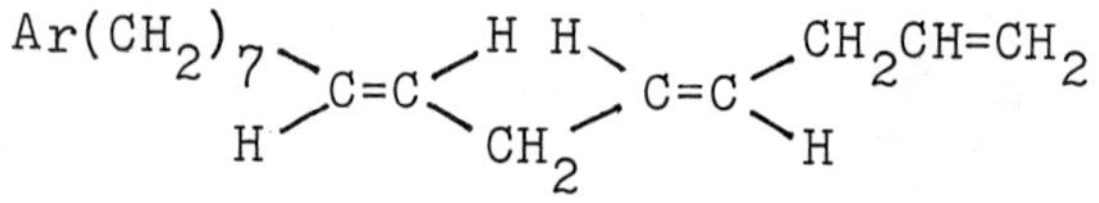 $Ar(CH_2)_7(H)C=C(H)-CH_2-(H)C=C(H)CH_2CH=CH_2$

E   E

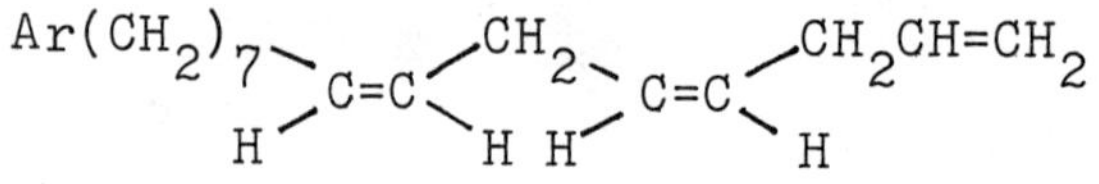

$Ar(CH_2)_7(H)C=C(H)-CH_2-(H)C=C(H)CH_2CH=CH_2$

Z   Z

$Ar(CH_2)_7(H)C=C(H)-CH_2-(H)C=C(H)CH_2CH=CH_2$

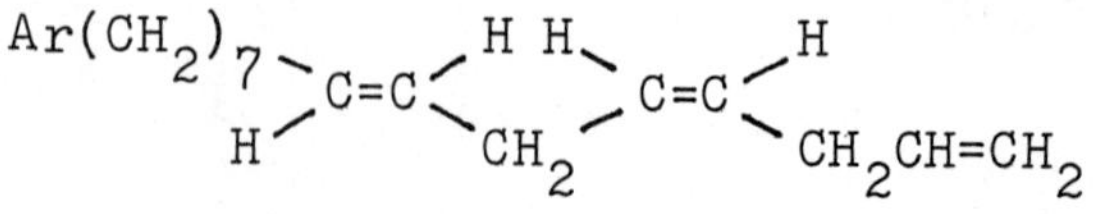

E   Z

(continued next page)

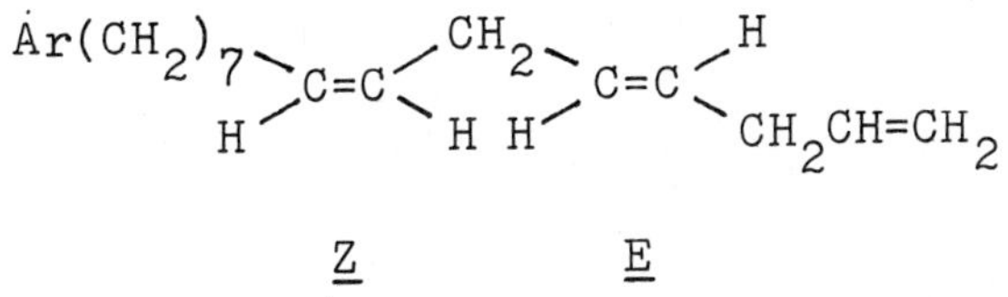

$Ar(CH_2)_7(H)C=C(H)-CH_2-(H)C=C(H)CH_2CH=CH_2$

Z E

21. 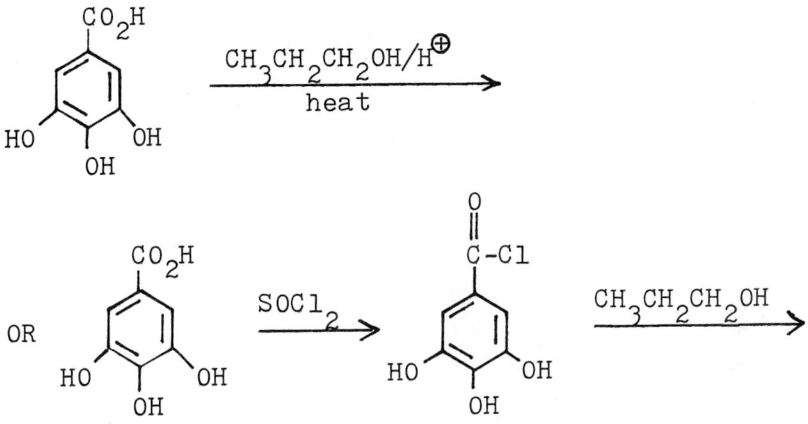 (3,4,5-trihydroxybenzoic acid) $\xrightarrow[\text{heat}]{CH_3CH_2CH_2OH/H^{\oplus}}$

OR (3,4,5-trihydroxybenzoic acid) $\xrightarrow{SOCl_2}$ (3,4,5-trihydroxybenzoyl chloride, C(=O)-Cl) $\xrightarrow{CH_3CH_2CH_2OH}$

22. Diels-Alder reactions are concerted and lead to <u>syn</u> addition to double bonds. The new chiral centers are formed with the methine hydrogens <u>trans</u> to the existing methine hydrogens because diene approaches dienophile from less-hindered face.

(Text: Review stereochemistry of Diels-Alder, p. 517)

23. 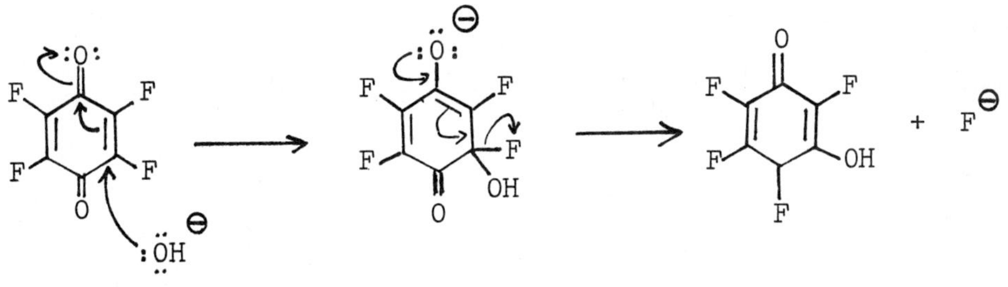 (tetrafluoro-p-benzoquinone) + $:\ddot{O}H^{\ominus}$ ⟶ (anionic intermediate, $:\ddot{O}:^{\ominus}$, F, F, F, F, OH) ⟶ (trifluoro-hydroxy-quinone, O, F, F, F, OH, F) + $F^{\ominus}$

24. 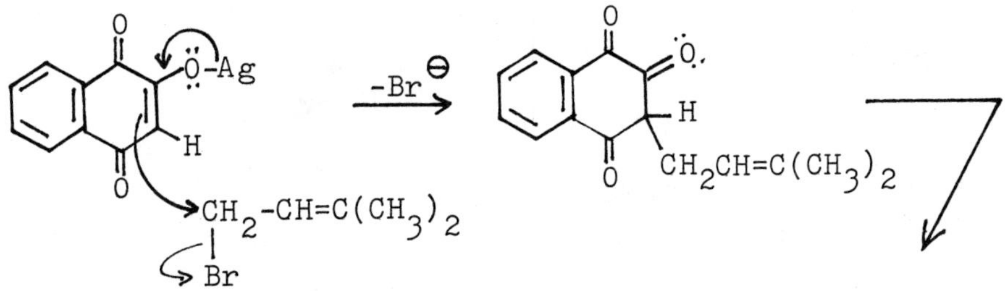 (2-hydroxy-1,4-naphthoquinone silver salt, $\ddot{O}$-Ag, H) + $CH_2-CH=C(CH_3)_2$ (Br) $\xrightarrow{-Br^{\ominus}}$ (O, O, $\ddot{O}$, H, $CH_2CH=C(CH_3)_2$)

(continued next page)

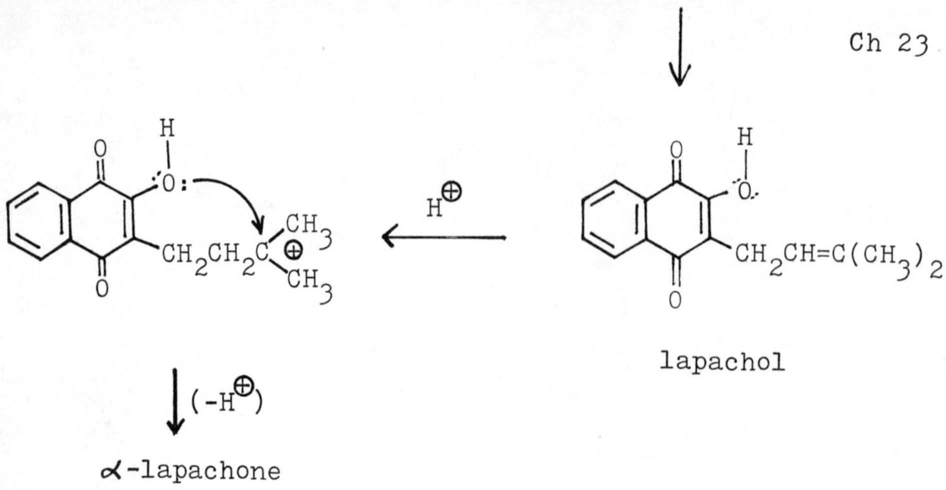

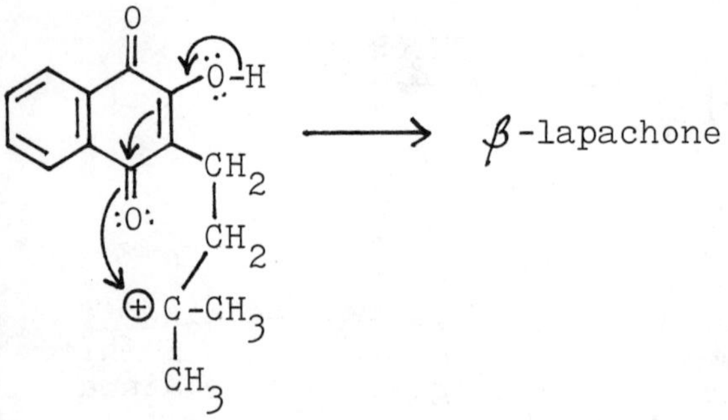

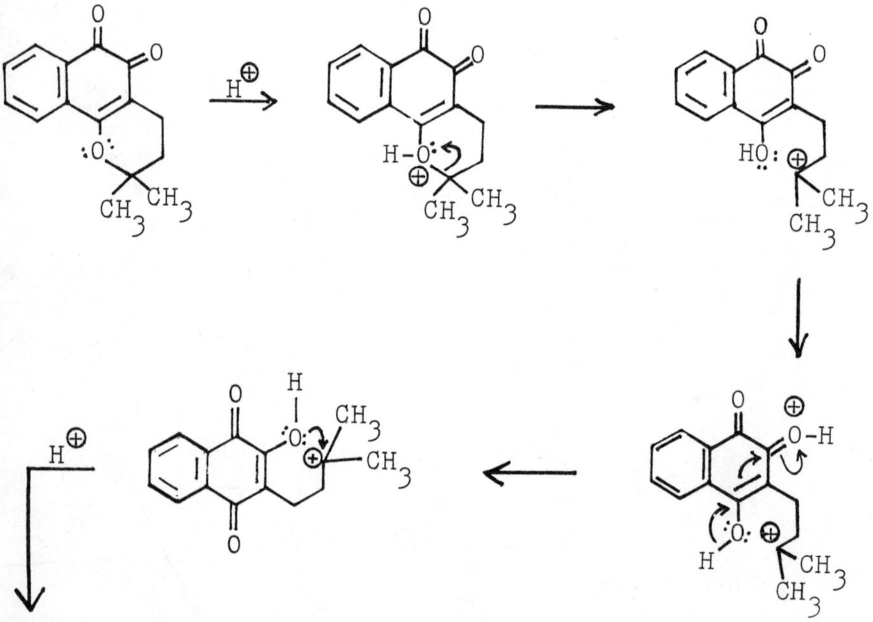

H⊕

lapachol

$CH_2CH=C(CH_3)_2$

$CH_2CH_2C^{\oplus}(CH_3)_2$

(-H⊕)

α-lapachone

β-lapachone

H⊕

(continued next page)

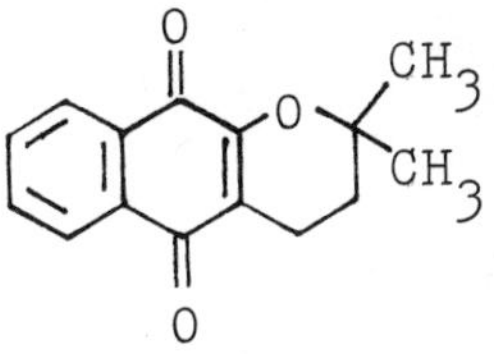

O
CH3
O
CH3
O

α-Lapachone is more stable, perhaps because of mutual carbonyl repulsion in β-lapachone.

25. (a) 2,3-dicyano-1,4-benzoquinone
(b) 1,5-dichloro-9,10-anthraquinone
(c) 1,8-dichloro-9,10-anthraquinone
(d) 2-phenyl-1,8-naphthalenediol
(e) o-methoxyphenol
(f) acetylsalicylaldehyde
(g) 5-chloro-2-hydroxybenzaldehyde
(h) (2-hydroxyphenyl) phenyl ketone (o-benzoylphenol)
(i) 2,3-anthraquinone
(j) sodium 1-hydroxy-3-naphthalenesulfonate

26. (a) $C_6H_5OH + (CH_3)_2C{=}CH_2$ (an E2 reaction)

(b) first step gives $C_6H_5O^{\ominus}$; no reaction in the second step

(c) 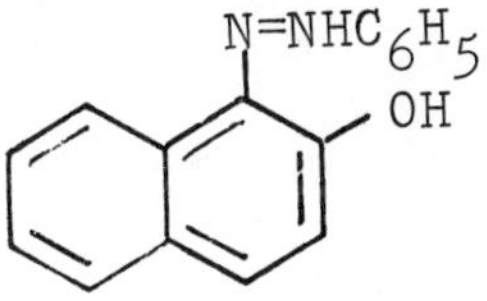 N=NHC$_6$H$_5$ OH (Coupling also occurs at position 3)

(d) 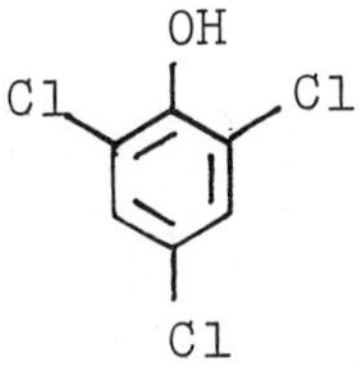 OH Cl Cl Cl

(continued next page)

(e) 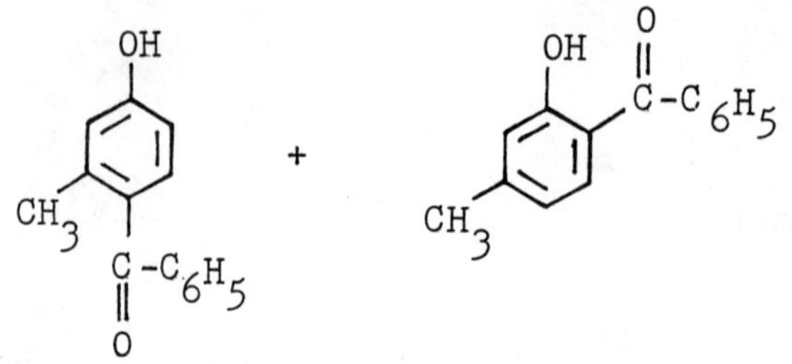

(f) 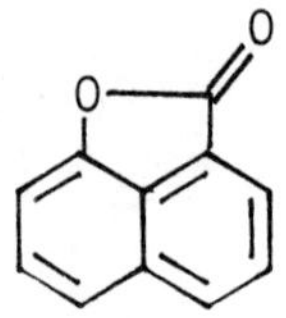

(g) 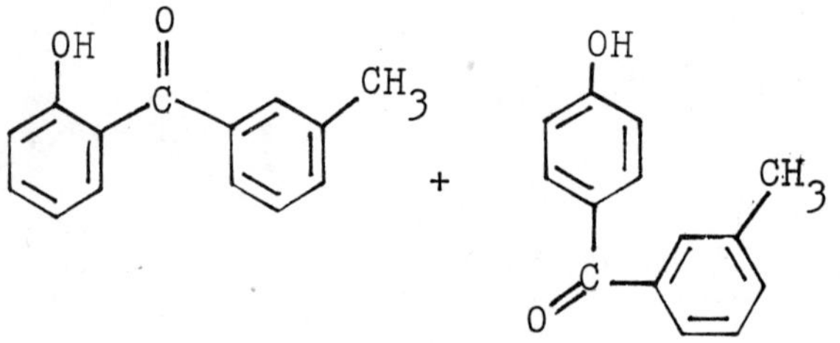

(h) 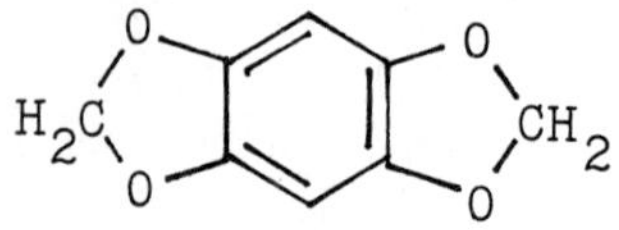

(i) $C_6H_5O^{\ominus}$ (and trace of C_6H_5OD)

(j) 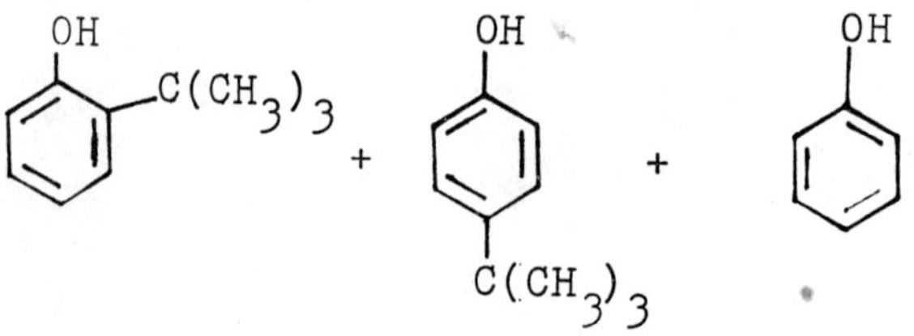 + $CH_2=C(CH_3)_2$

(k) 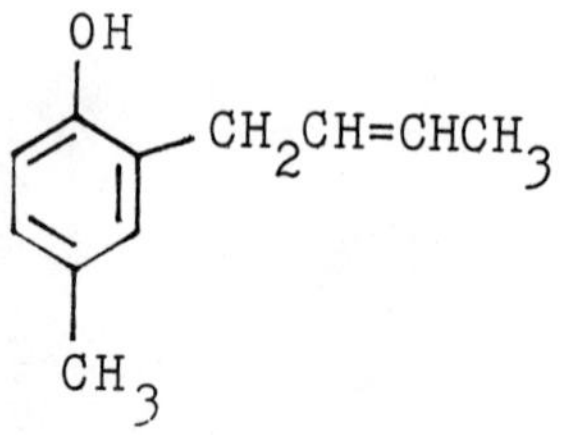

(continued next page)

(l) 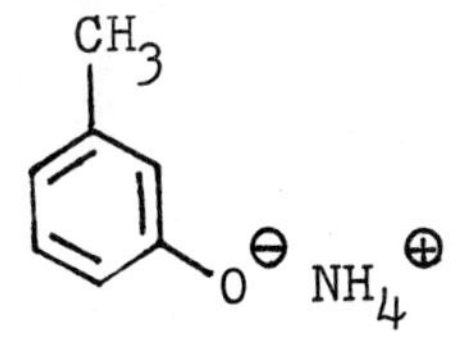

(m) 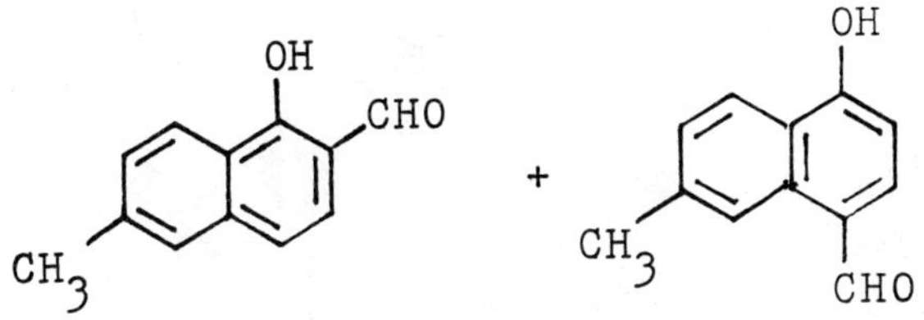

(n) 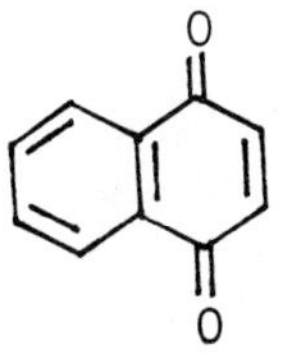

(o) 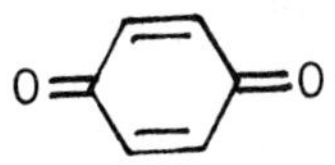

(p) 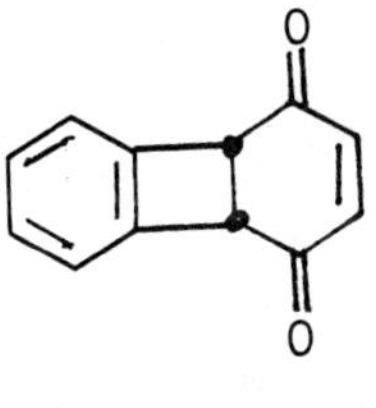

27. The IR spectrum shows, among other absorptions, C=O stretch. Integration of peak areas shows the ratio 4 aromatic: 1 vinyl: 3 aliphatic. These are in agreement with:

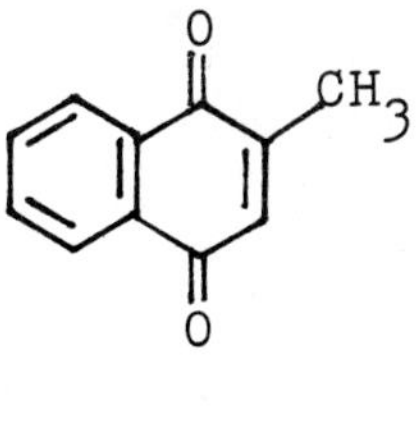

(continued next page)

An alternate synthesis is:

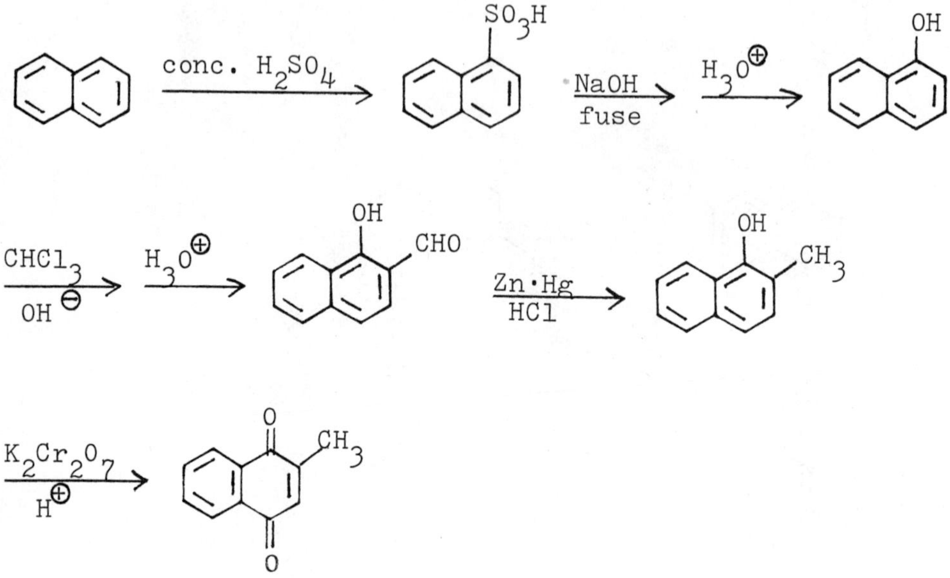

naphthalene $\xrightarrow{\text{conc. } H_2SO_4}$ 1-naphthalenesulfonic acid (SO_3H) $\xrightarrow[\text{fuse}]{NaOH}$ $\xrightarrow{H_3O^{\oplus}}$ 1-naphthol (OH)

$\xrightarrow[OH^{\ominus}]{CHCl_3}$ $\xrightarrow{H_3O^{\oplus}}$ (OH, CHO) $\xrightarrow[HCl]{Zn\cdot Hg}$ (OH, CH_3)

$\xrightarrow[H^{\oplus}]{K_2Cr_2O_7}$ (O, O, CH_3)

28. (:Ö-H, H) $\xrightarrow{CH_3O^{\ominus}}$ [(:Ö:$^{\ominus}$, H) $\longleftrightarrow$ (:O:, H, $\ominus$)] $\xrightarrow{CH_3OD}$

(:O:, H, D) $\xrightarrow{CH_3O^{\ominus}}$ [(:O:, D, $\ominus$) $\longleftrightarrow$ (:Ö:$^{\ominus}$, D)] $\xrightarrow{CH_3OD}$

(:Ö-D, D)

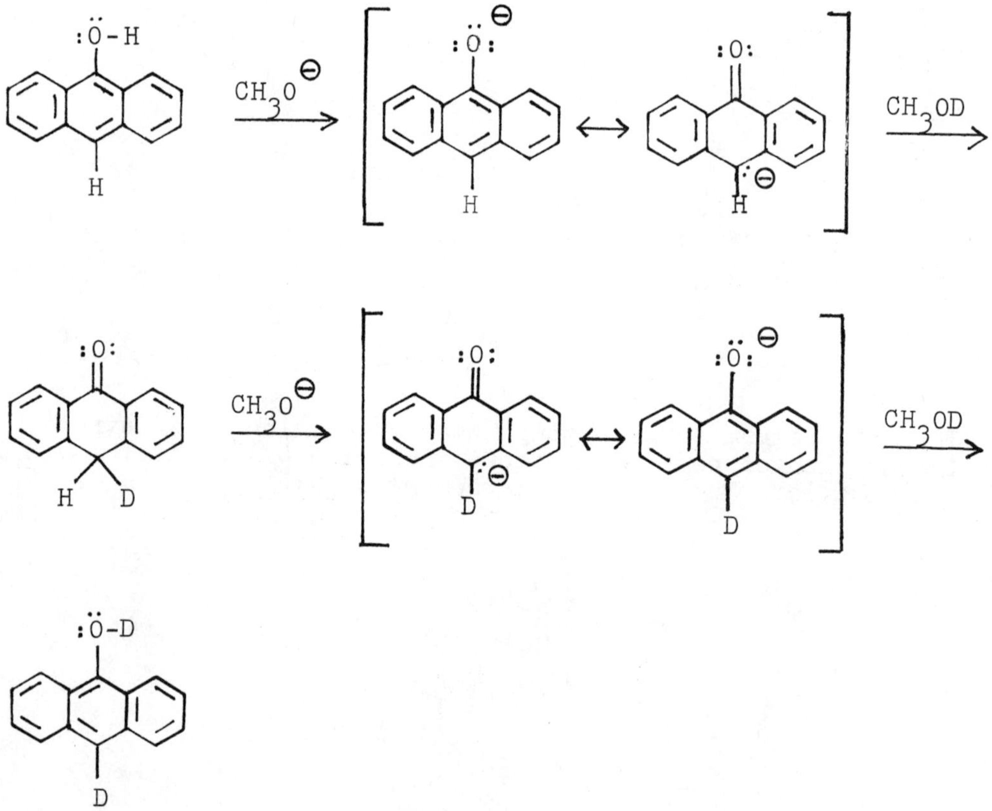

29. The strongest acids have the weakest (most stable) conjugate bases. In general, electron-releasing groups destabilize and electron-withdrawing groups stabilize the conjugate bases.

(a) phenol < *m*-nitrophenol * < *o*-nitrophenol < *p*-nitrophenol < 2,4-dinitrophenol (last three: **)

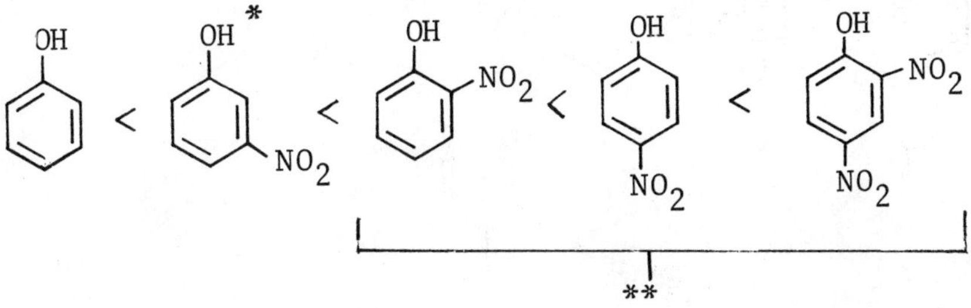

*conjugate base stabilized by the inductive effect of the NO_2 but not resonance-stabilized by NO_2.
**conjugate bases resonance-stabilized by NO_2 group(s)

(b) 4-methyl-1-naphthol < 1-naphthol < 3-nitro-1-naphthol < 2-nitro-1-naphthol

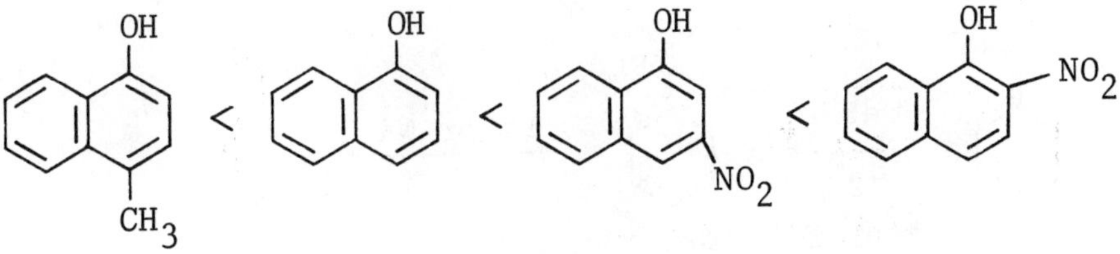

(c) *m*-cresol < catechol * < hydroquinone < *p*-cyanophenol

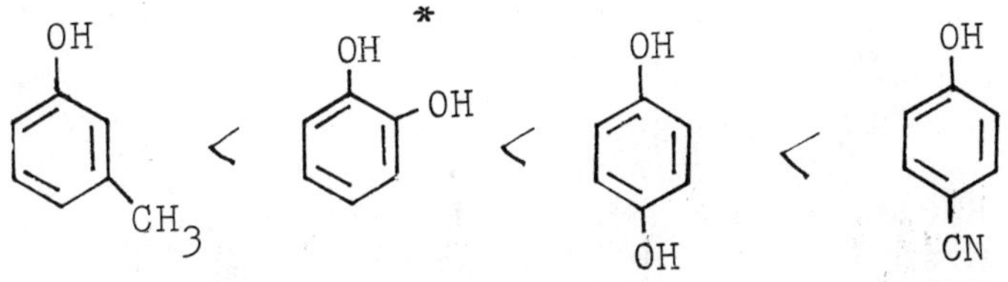

*conjugate base destabilized by resonance effect of other OH group.

30. A (2-cyano-1,4-benzoquinone) + B (hydroquinone) ⇌ C (1,4-benzoquinone) + D (2-cyanohydroquinone)

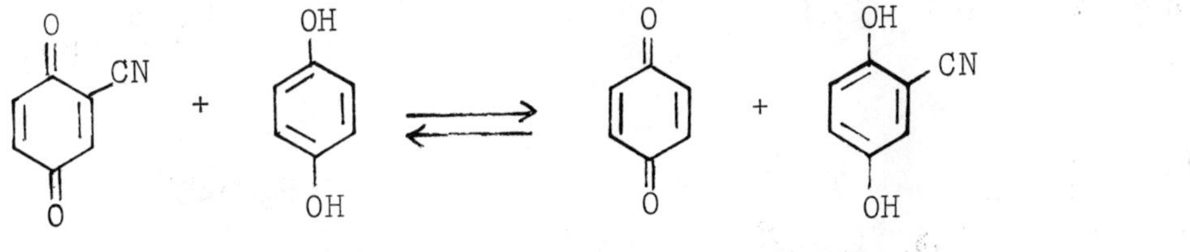

C and D are produced by a redox reaction between A and B and must be produced in equal amounts. However, the C and D pair do not have the same energy as the A and B pair, thus K_{eq} will not equal one.

31. (a)

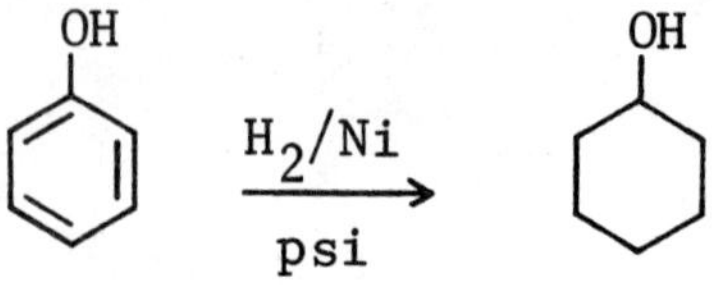
OH
H_2/Ni
psi
OH

(b)

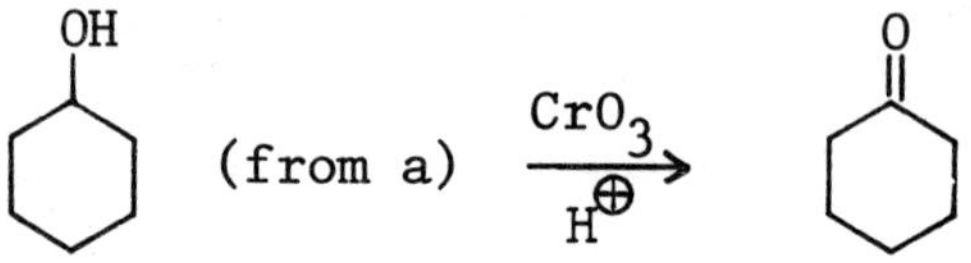
OH
(from a)
CrO_3
$H^{\oplus}$
O

(c)

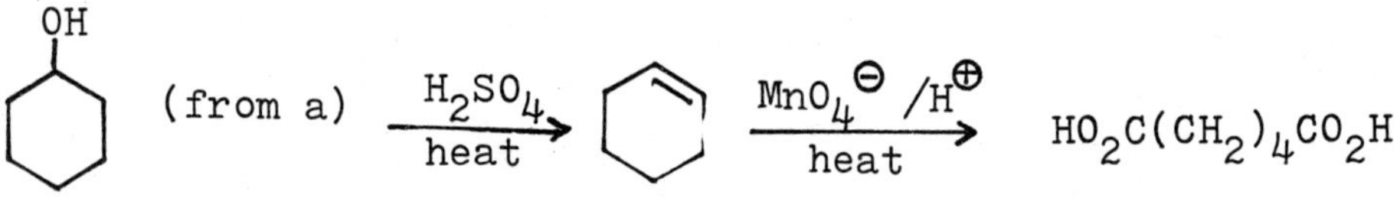
OH
(from a)
H_2SO_4
heat
$MnO_4^{\ominus}$ / $H^{\oplus}$
heat
$HO_2C(CH_2)_4CO_2H$

(d)

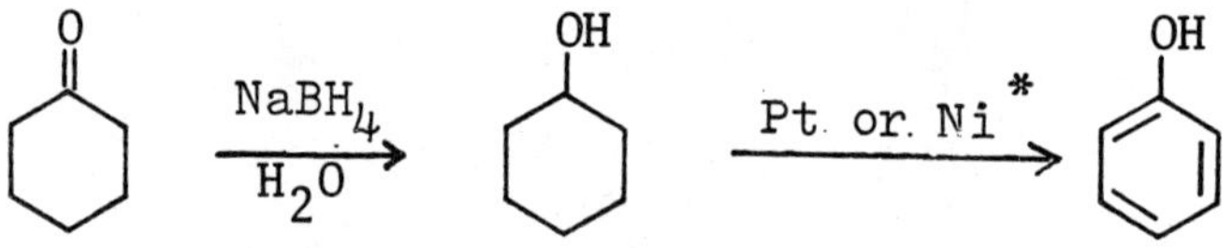
O
$NaBH_4$
H_2O
OH
Pt. or. Ni *
OH

*(or DDQ, see p. 988, text)

(e)

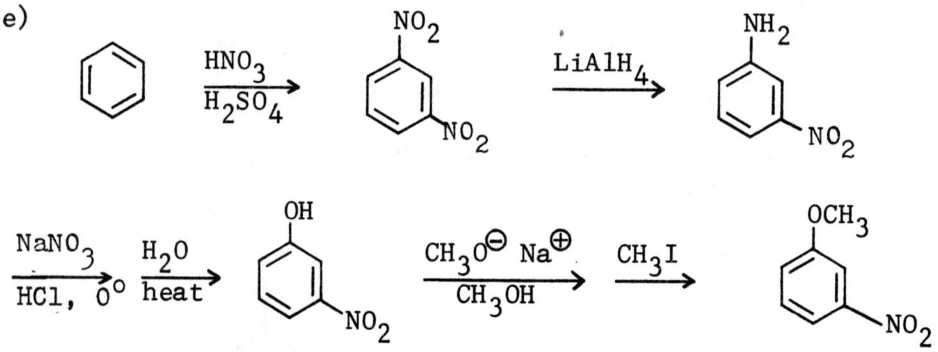
HNO_3
H_2SO_4
NO_2
NO_2
$LiAlH_4$
NH_2
NO_2
$NaNO_3$
HCl, 0°
H_2O
heat
OH
NO_2
$CH_3O^{\ominus}$ $Na^{\oplus}$
CH_3OH
CH_3I
OCH_3
NO_2

(f)

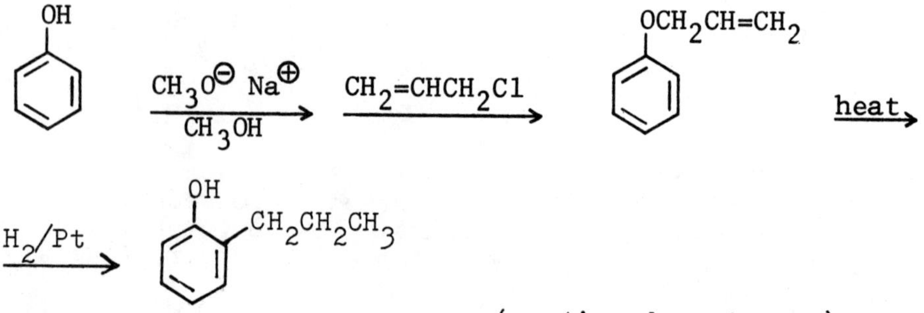
OH
$CH_3O^{\ominus}$ $Na^{\oplus}$
CH_3OH
CH_2=$CHCH_2Cl$
OCH_2CH=CH_2
heat
H_2/Pt
OH
$CH_2CH_2CH_3$

(continued next page)

(g) 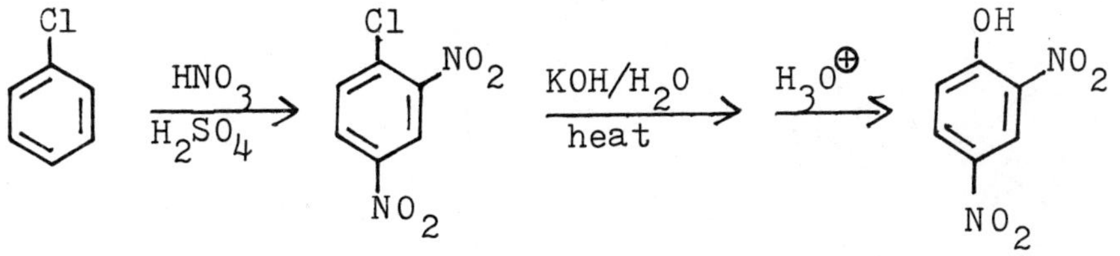 Chlorobenzene $\xrightarrow[H_2SO_4]{HNO_3}$ 1-Cl-2,4-dinitrobenzene $\xrightarrow[heat]{KOH/H_2O}$ $\xrightarrow{H_3O^{\oplus}}$ 2,4-dinitrophenol (OH, NO_2, NO_2)

(h) Benzoic acid (CO_2H) $\xrightarrow[H_2SO_4]{HNO_3}$ 3,5-dinitrobenzoic acid (CO_2H, O_2N, NO_2) $\xrightarrow{SOCl_2}$ $\xrightarrow{NH_3}$ $\xrightarrow[H_2O]{OH^{\ominus}/Br_2}$

3,5-dinitroaniline (NH_2, O_2N, NO_2) $\xrightarrow[HCl,\ 0^o]{NaNO_2}$ $\xrightarrow[heat]{H_2O}$ 3,5-dinitrophenol (OH, O_2N, NO_2) $\xrightarrow{NaOH}$ $\xrightarrow{CH_3I}$

3,5-dinitroanisole (OCH_3, O_2N, NO_2)

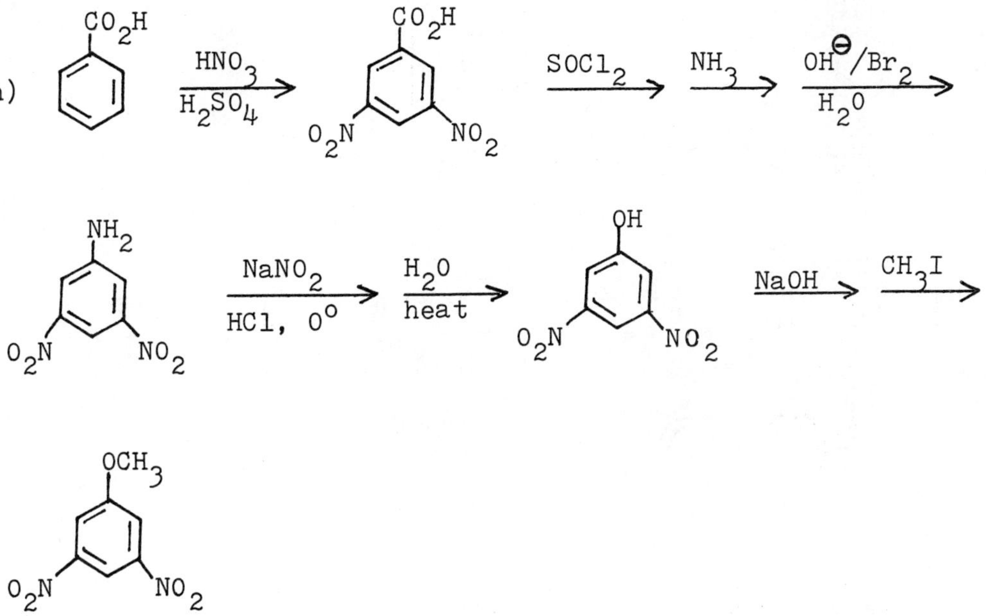

(i) 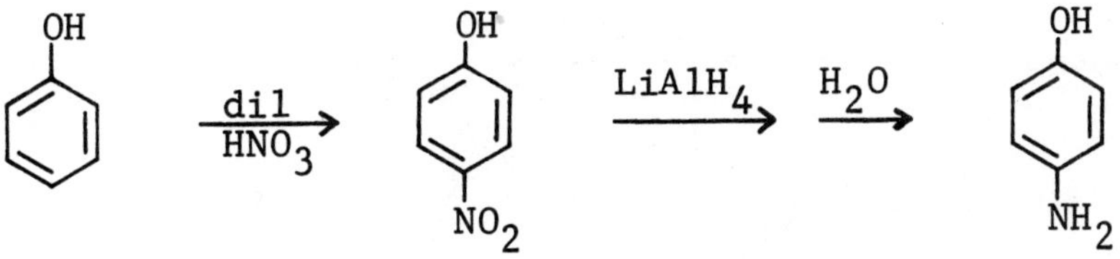 Phenol (OH) $\xrightarrow[HNO_3]{dil}$ 4-nitrophenol (OH, NO_2) $\xrightarrow{LiAlH_4}$ $\xrightarrow{H_2O}$ 4-aminophenol (OH, NH_2)

(j) 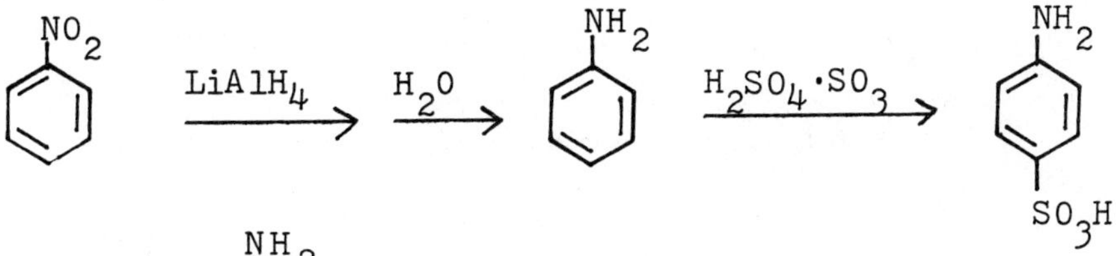 Nitrobenzene (NO_2) $\xrightarrow{LiAlH_4}$ $\xrightarrow{H_2O}$ Aniline (NH_2) $\xrightarrow{H_2SO_4 \cdot SO_3}$ 4-aminobenzenesulfonic acid (NH_2, SO_3H)

$\xrightarrow[fuse]{NaOH}$ 4-aminophenol (NH_2, OH)

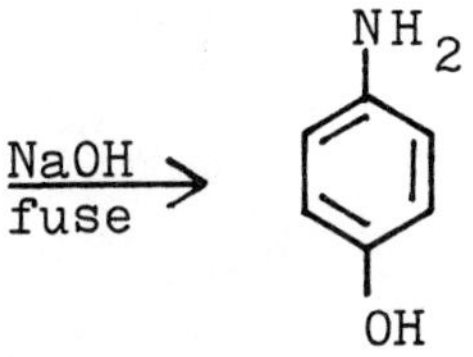

(continued next page)

(k) 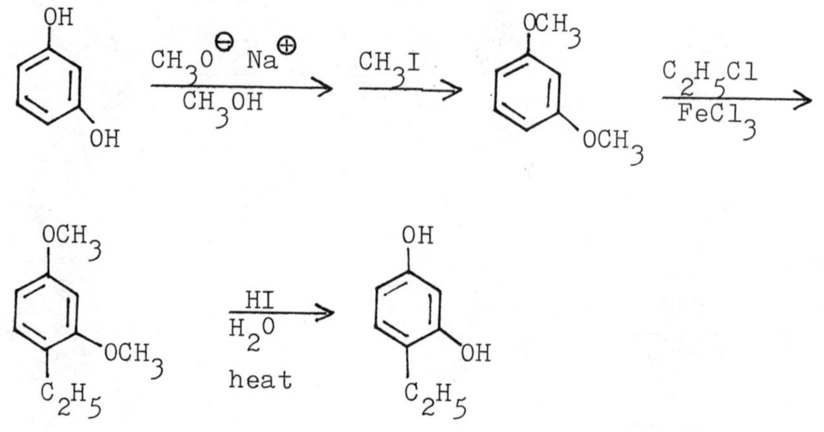 1,3-dihydroxybenzene (OH, OH) $\xrightarrow[CH_3OH]{CH_3O^{\ominus}\ Na^{\oplus}}$ $\xrightarrow{CH_3I}$ 1,3-dimethoxybenzene (OCH_3, OCH_3) $\xrightarrow[FeCl_3]{C_2H_5Cl}$

4-ethyl-1,3-dimethoxybenzene (OCH_3, OCH_3, C_2H_5) $\xrightarrow[H_2O \text{ heat}]{HI}$ 4-ethyl-1,3-dihydroxybenzene (OH, OH, C_2H_5)

(l) 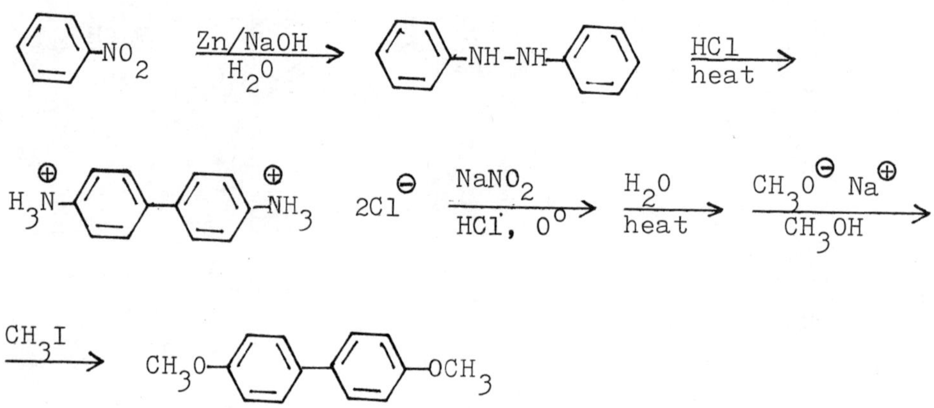 $C_6H_5NO_2$ $\xrightarrow[H_2O]{Zn/NaOH}$ C_6H_5-NH-NH-C_6H_5 $\xrightarrow[heat]{HCl}$

$H_3\overset{\oplus}{N}$-C_6H_4-C_6H_4-$\overset{\oplus}{N}H_3$ $2Cl^{\ominus}$ $\xrightarrow[HCl,\ 0^o]{NaNO_2}$ $\xrightarrow[heat]{H_2O}$ $\xrightarrow[CH_3OH]{CH_3O^{\ominus}\ Na^{\oplus}}$

$\xrightarrow{CH_3I}$ CH_3O-C_6H_4-C_6H_4-OCH_3

(m) 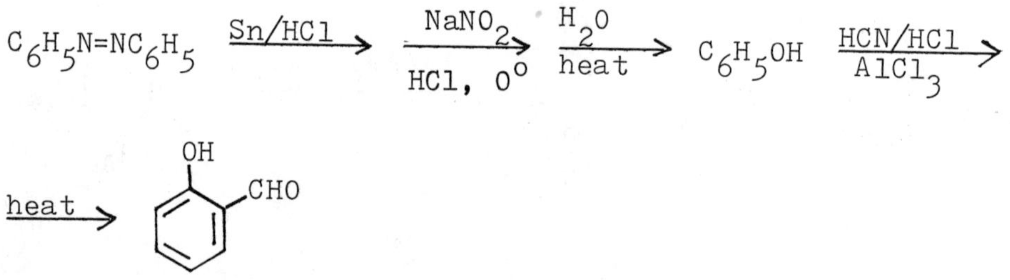 $C_6H_5N{=}NC_6H_5$ $\xrightarrow{Sn/HCl}$ $\xrightarrow[HCl,\ 0^o]{NaNO_2}$ $\xrightarrow[heat]{H_2O}$ C_6H_5OH $\xrightarrow[AlCl_3]{HCN/HCl}$

$\xrightarrow{heat}$ o-hydroxybenzaldehyde (OH, CHO)

32. (a) 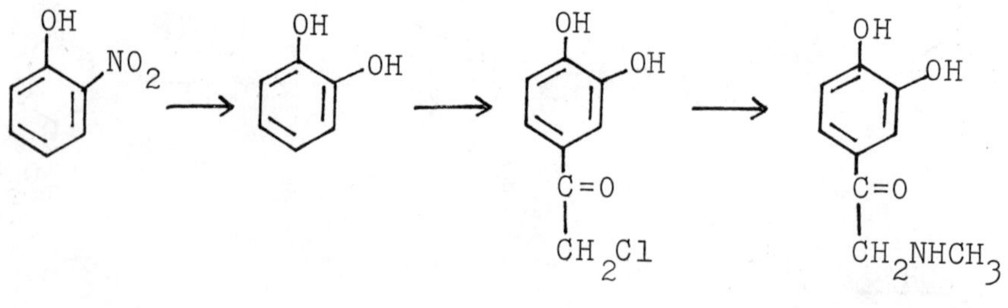 o-nitrophenol (OH, NO_2) $\longrightarrow$ o-dihydroxybenzene (OH, OH) $\longrightarrow$ 3,4-dihydroxyphenyl-C=O-CH_2Cl $\longrightarrow$ 3,4-dihydroxyphenyl-C=O-CH_2NHCH_3

A B C

(continued next page)

(b) Catecholamines are amines containing the catechol moiety.

(c) Norepinephrine can be prepared as was adrenaline except that NH_3 replaces NH_2CH_3 between B and C.

Dopamine can be prepared as follows:

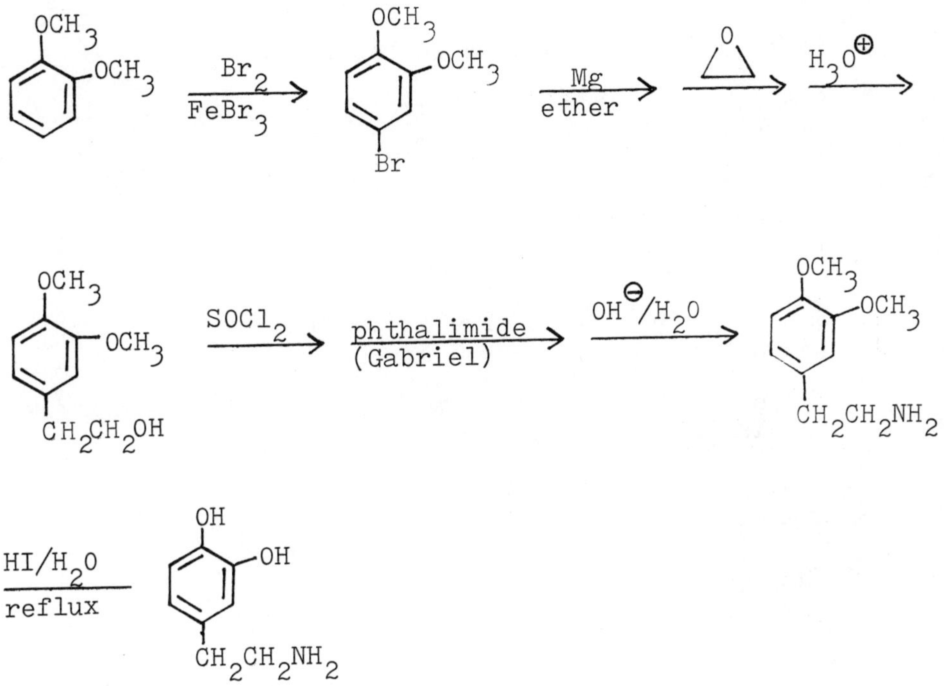

33. 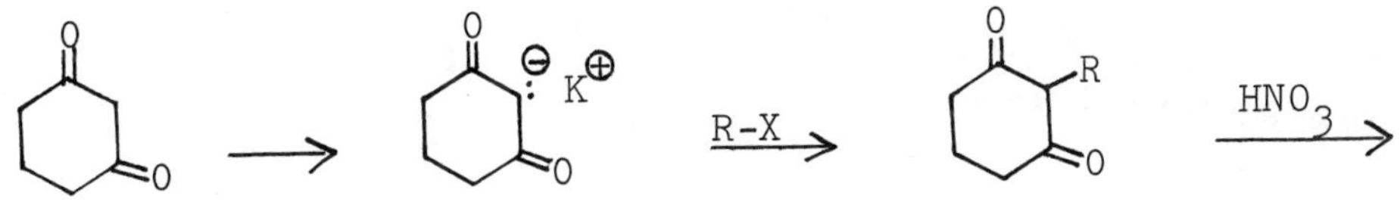

A B

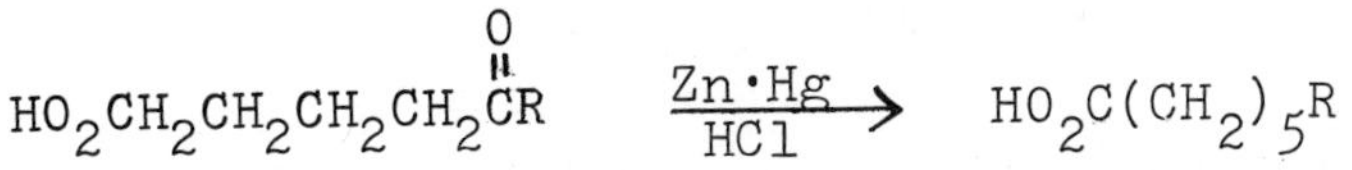

D

This reaction which is a method of lengthening a chain by six carbons, can be modified by replacing the oxidation step with a base-catalyzed cleavage.

(continued next page)

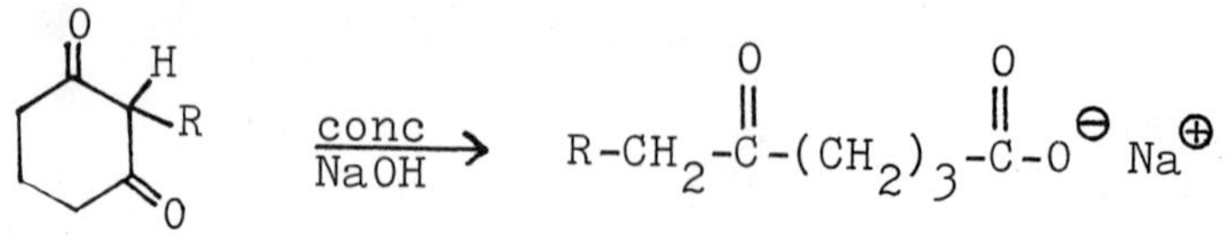

$$\xrightarrow[\text{NaOH}]{\text{conc}} R\text{-}CH_2\text{-}\overset{O}{\overset{\|}{C}}\text{-}(CH_2)_3\text{-}\overset{O}{\overset{\|}{C}}\text{-}O^{\ominus}\ Na^{\oplus}$$

34. (a) Aniline is soluble in dilute HCl.

(b) Benzoic acid is soluble in NaOH/H_2O.

(c) Phenol is soluble in water; p-hydroxybiphenyl is only slightly soluble.

(d) 2-Naphthol is soluble in NaOH solution, <u>cis</u>-4-t-butylcyclohexanol is not.

(e) p-Cresol reacts with Br_2/CCl_4 .

(f) Salicylic acid will give a positive color test with $FeCl_3$ (S. 23.11).

(g) 2,4,6-Trinitrophenol is much more acidic (see answer to problem 29a).

35.

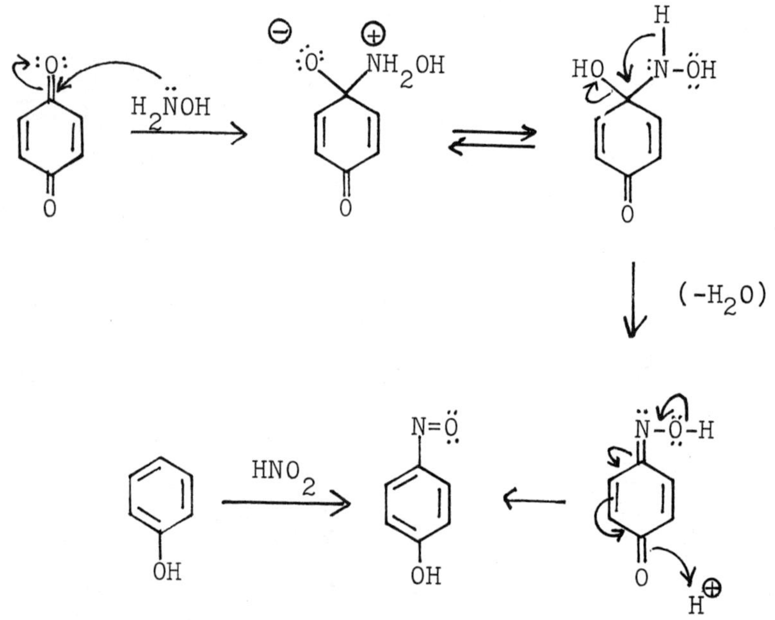

$H_2\ddot{N}OH$

$(-H_2O)$

HNO_2

36. Because of competitive disproportionation involving the formyl group (Cannizzaro reaction).

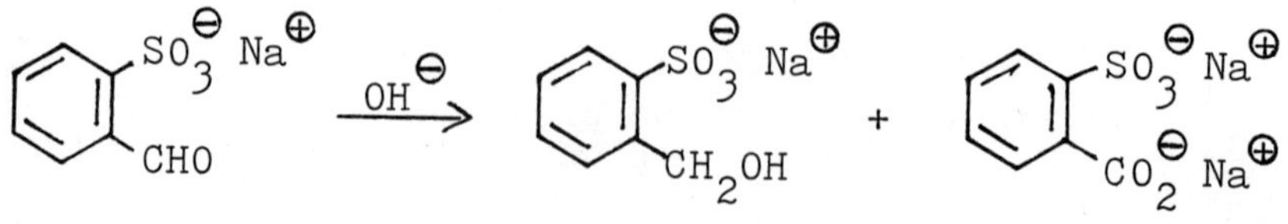

$$\xrightarrow{OH^{\ominus}}$$

37. [p-dimethoxybenzene] $\xrightarrow[H_3PO_4]{(CH_3)_3COH}$ [2-tert-butyl-1,4-dimethoxybenzene: OCH_3, $C(CH_3)_3$, OCH_3] $\xrightarrow[\text{(1 equiv)}]{HI/H_2O}$

[phenol: OH, $C(CH_3)_3$, OCH_3] + [phenol: OH, $C(CH_3)_3$, OCH_3]

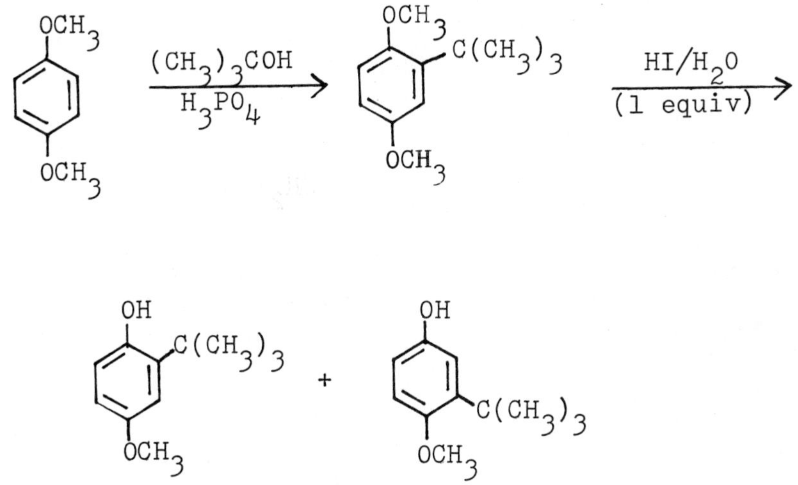

38. [2-methoxyphenol: CH_3, O, H, O] intramolecular hydrogen bonding

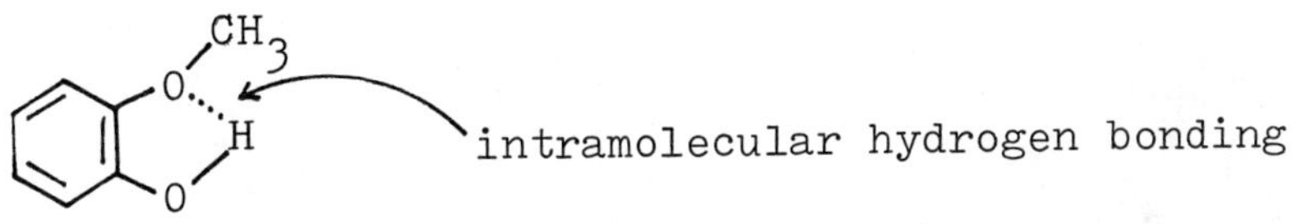

39. Syntheses other than those shown below are possible.

(a) [benzene ring with $CH_2CH_2CH_2Cl$] $\xrightarrow[H_2SO_4]{HNO_3}$ $\xrightarrow[Pt]{H_2}$ [NH_2 ring with $CH_2CH_2CH_2Cl$] $\xrightarrow[0^o]{HONO}$ $\xrightarrow[\text{heat}]{H_2O}$

$\xrightarrow{NO_2^{\oplus}\ BF_4^{\ominus}}$ [OH, NO_2 ring with $CH_2CH_2CH_2Cl$] $\xrightarrow[H^{\oplus}]{\text{dihydropyran (O)}}$

[tetrahydropyranyl–O ring with NO_2 and $CH_2CH_2CH_2Cl$] $\xrightarrow{H_2/Pt}$ [–O– ring with NH_2]

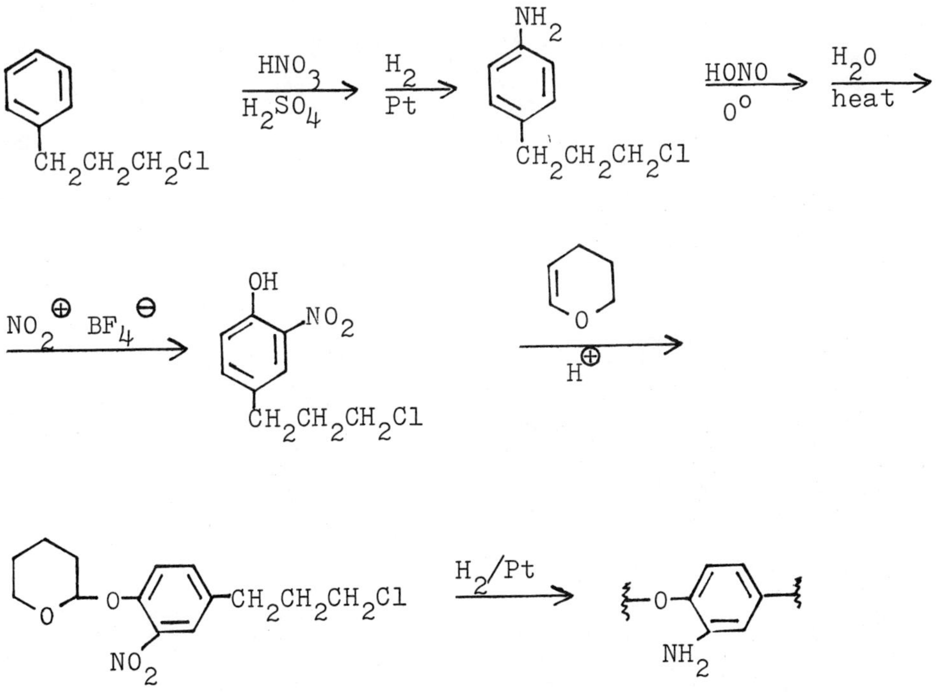

(continued next page)

$\xrightarrow{HCl}$ $\xrightarrow{CH_3CH_2ONO}$ $\xrightarrow[heat]{H_2O}$ (aryl ether, 3-OH) $\xrightarrow{NaH}$ $\xrightarrow{CH_3I}$

$\xrightarrow{H_3O^{\oplus}/heat}$ HO–C$_6$H$_3$(OCH$_3$)–$CH_2CH_2CH_2Cl$ $\xrightarrow[(CH_3)_3COH]{(CH_3)_3CO^{\ominus}\ Na^{\oplus}}$

$Na^{\oplus}\ {}^{\ominus}O$–C$_6H_3$(OCH$_3$)–$CH_2CH=CH_2$ $\xrightarrow{H_3O^{\oplus}}$ HO–C$_6$H$_3$(OCH$_3$)–CH_2-$CH=CH_2$

(b) prepare (tetrahydropyranyl)O–C$_6$H$_3$(OH)–$CH_2CH_2CH_2Cl$ as in (a) $\xrightarrow[heat]{H_2O/H^{\oplus}}$

$\xrightarrow[CH_3OH]{CH_3O^{\ominus}Na^{\oplus}}$ $\xrightarrow{CH_2I_2}$ (methylenedioxyphenyl)–$CH_2CH_2CH_2Cl$ $\xrightarrow[(CH_3)_3COH]{(CH_3)_3CO^{\ominus}Na^{\oplus}}$

(methylenedioxyphenyl)–$CH_2CH=CH_2$

(c) C$_6$H$_5$OCH$_3$ $\xrightarrow[AlCl_3]{CH_3CH_2\overset{O}{\overset{\|}{C}}Cl}$ $\xrightarrow{NaBH_4}$ $\xrightarrow[heat]{H^{\oplus}}$ p-CH_3O–C$_6$H$_4$–$CH=CHCH_3$

(continued next page)

(d) [4-allyl-2-methoxyphenol: OH, OCH_3, $CH_2CH{=}CH_2$] (from a) $\xrightarrow{H^{\oplus}}$ [OH, OCH_3, $CH{=}CH{-}CH_3$] $\xrightarrow{O_3}$ $\xrightarrow[HCl]{Zn}$

shift of the double bond

[OH, OCH_3, CHO]

40. The intermediate is: [Br, $SO_3^{\ominus}$ $Na^{\oplus}$; Br; Br; O]

41. (a) A is $C_6H_5NO_2$; B is $C_6H_5NH_2$; C is $C_6H_5\overset{\oplus}{N_2}$ $\overset{\ominus}{HSO_4}$;

D is $C_6H_5O\overset{O}{\overset{\|}{C}}CH_3$; E is C_6H_5OH ; F is CH_3CO_2H

(b) [$C_6H_5-\overset{\oplus}{N}{\equiv}N:$] $\xrightarrow{-N_2}$ [$C_6H_5^{\oplus}$] $\xrightarrow{H\ddot{O}-\overset{O}{\overset{\|}{C}}CH_3}$ [$C_6H_5-\overset{\oplus}{O}(H)-\overset{O}{\overset{\|}{C}}CH_3$] $\xrightarrow{-H^{\oplus}}$ [$C_6H_5-\ddot{O}-\overset{O}{\overset{\|}{C}}CH_3$]

(c) [$C_6H_5-O-\overset{:O:}{\overset{\|}{C}}-CH_3$] $\xrightarrow{H^{\oplus}}$ $C_6H_5-O-\overset{:\overset{\oplus}{O}H}{\overset{\|}{C}}-CH_3$ $\xrightarrow{:\ddot{O}H_2}$ $C_6H_5-O-\overset{:\ddot{O}-H}{\underset{H\overset{\oplus}{O}H}{C}}-CH_3$

(continued next page)

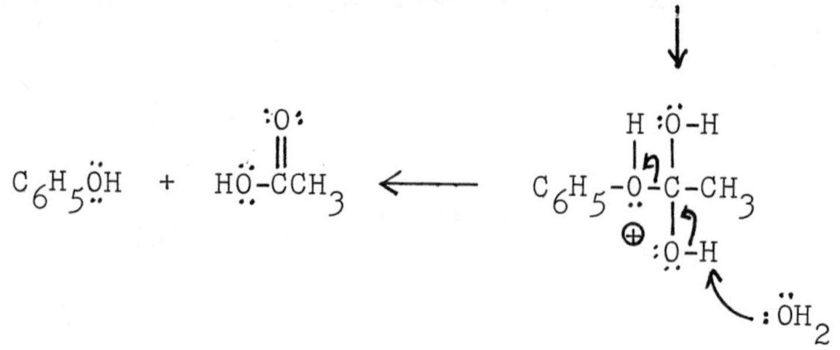

42. 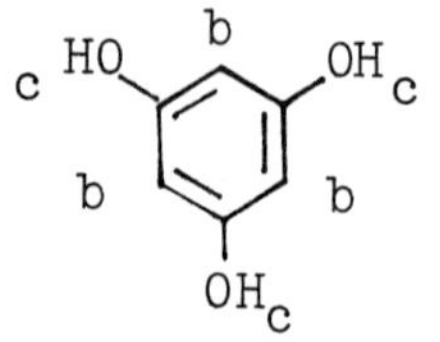

absorption "a" due to protons of water

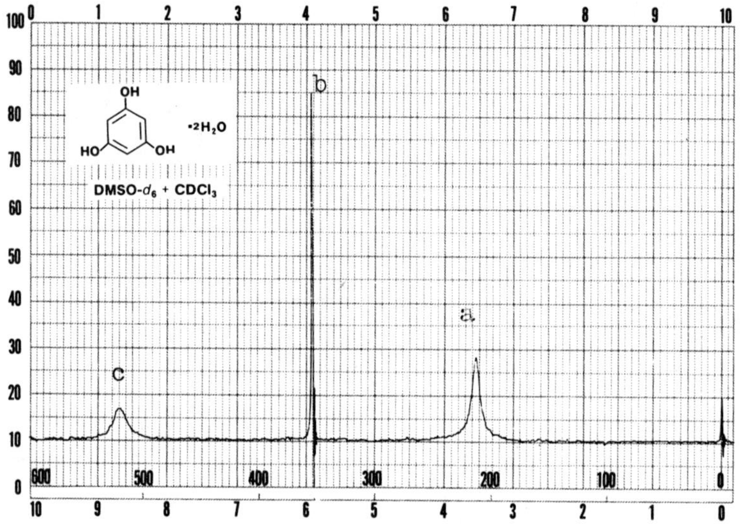

43. 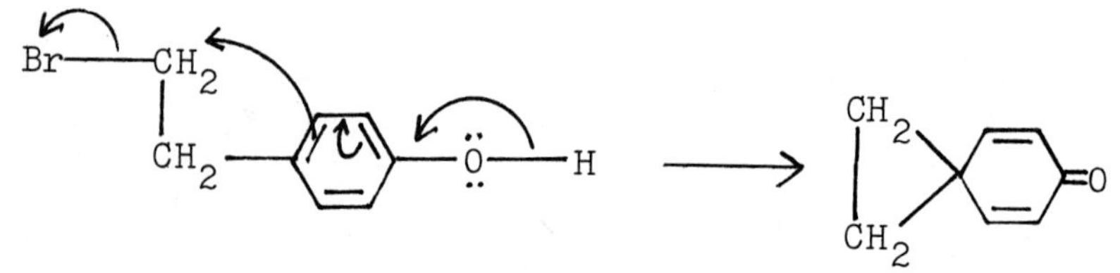

The ketone is symmetric, and both $-CH_2-$ groups become equivalent in the intermediate. Starting with Br–$^{14}CH_2CH_2$– 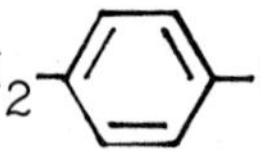 –OH, the intermediate leads to 4-(2-hydroxyethyl)phenol with scrambled ^{14}C.

(continued next page)

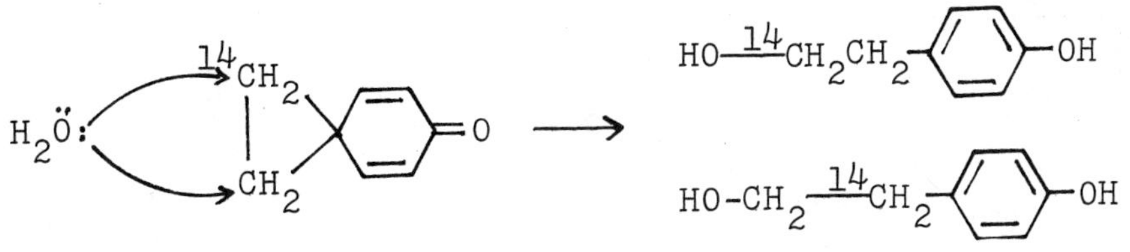

$H_2\ddot{O}:$ → $^{14}CH_2$ / CH_2 (spiro-cyclohexadienone) ⟶ $HO{-}^{14}CH_2CH_2{-}C_6H_4{-}OH$ and $HO{-}CH_2{-}^{14}CH_2{-}C_6H_4{-}OH$

44. (a)

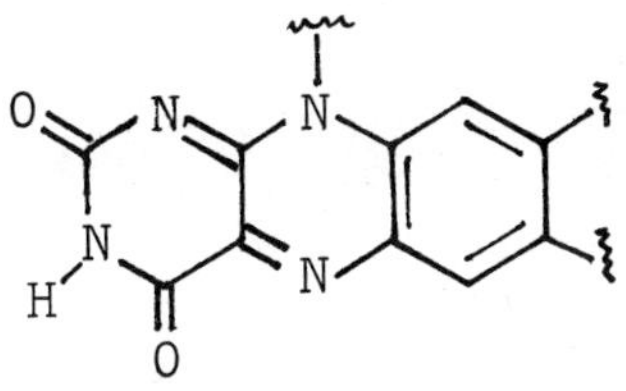

(b) 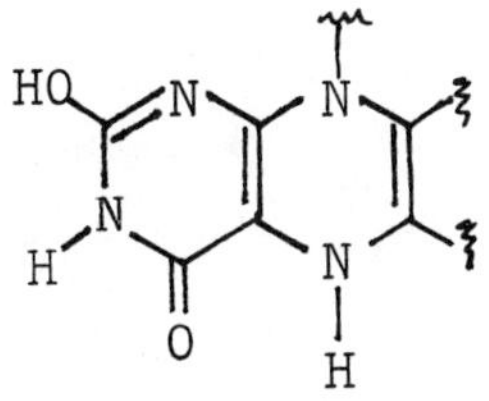 and 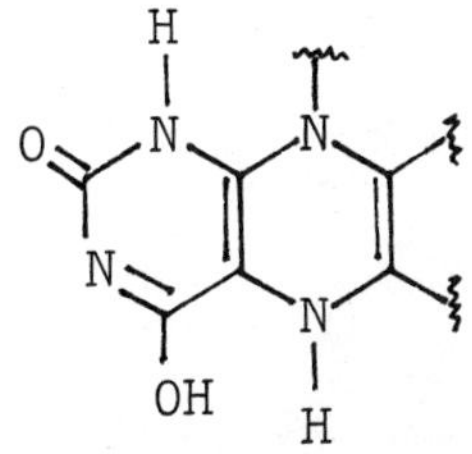 and

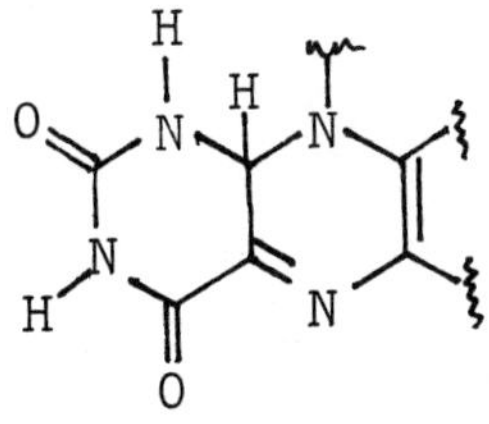

and 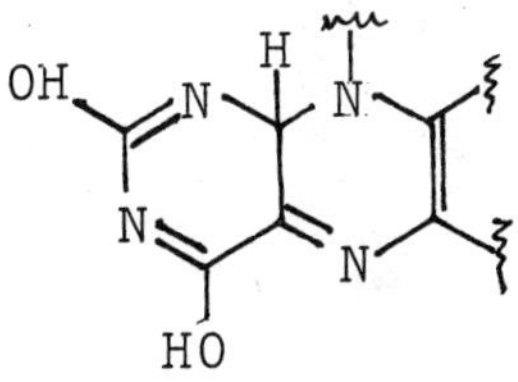 and

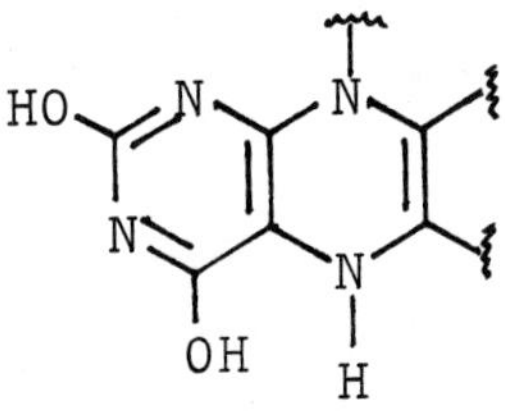

and 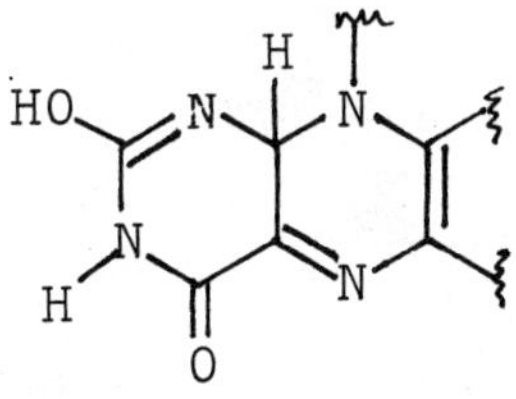 and

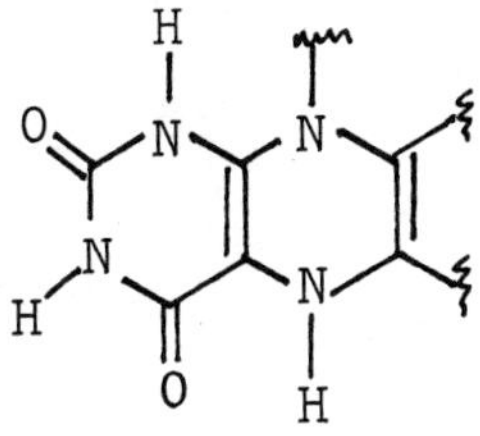

No quinone-like structures are possible

SUMMARY OF REACTIONS

1. Synthesis of phenols

(a) $Ar\text{-}SO_3H \xrightarrow[300^\circ]{NaOH/KOH} \xrightarrow{H_3O^{\oplus}} Ar\text{-}OH$ (p. 966)

(b) $G\text{-}C_6H_4\text{-}X \xrightarrow[heat]{NaOH} \xrightarrow{H_3O^{\oplus}} G\text{-}C_6H_4\text{-}OH$ (p. 966)

G = a deactivating group

(c) $ArMgX \xrightarrow[-80^\circ]{(CH_3O)_3B} ArB(OCH_3)_2 \xrightarrow{H_3O^{\oplus}} \xrightarrow[H_3O^{\oplus}]{H_2O_2} ArOH$

(d) $Ar\overset{\oplus}{N_2} \xrightarrow[heat]{H_2O} ArOH$ (p. 967)

2. Reactions of phenols (p. 968)

(a) Phenols as nucleophiles

ArOH —

- $\xrightarrow{CH_2N_2} ArOCH_3$
- $\xrightarrow{NaOH} ArO^{\ominus} Na^{\oplus} \xrightarrow{ClCH_2CO_2^{\ominus}\ Na^{\oplus}} \xrightarrow{H_3O^{\oplus}} ArOCH_2COOH$
- $\xrightarrow{OH^{\ominus}} \xrightarrow{R\text{-}X} ArOR$ *
- * $C_6H_5\text{-}O\text{-}\overset{|}{\underset{|}{C}}\text{-}\overset{|}{C}\text{=}\overset{|}{C}\text{-} \xrightarrow{heat}$ o-$(\text{-}\overset{|}{\underset{|}{C}}\text{-}\overset{|}{C}\text{=}\overset{|}{C}\text{-})C_6H_4OH$ Claisen rearrangement
- $\xrightarrow{RC(=O)\text{-}O\text{-}C(=O)\text{-}R} ArO\text{-}C(=O)\text{-}R$ #(see next page)
- $\xrightarrow{R\text{-}N=C=O} Ar\text{-}NH\text{-}C(=O)\text{-}O\text{-}R$

(continued next page)

\# Phenyl ester $C_6H_5O\text{-}C(=O)\text{-}R$ $\xrightarrow[\text{heat}]{AlCl_3}$ $\xrightarrow{H_2O}$ p-hydroxyphenyl ketone ($HO\text{-}C_6H_4\text{-}C(=O)\text{-}R$) + o-hydroxyphenyl ketone ($HO\text{-}C_6H_4\text{-}CR(=O)$)

(Fries rearrangement)

(b) Reduction

$$C_6H_5OH \xrightarrow{H_2/Ni} C_6H_{11}OH$$ (p. 971)

(c) Oxidation

$$C_6H_5OH \xrightarrow[H_3O^{\oplus}]{CrO_3} O{=}C_6H_4{=}O$$ (p. 972)

(also $C_6H_5NH_2$ $\xrightarrow[Cr_2O_7^{2-}/H^{\oplus}]{}$)

(continued next page)

(d) Ring reactions

Phenol (OH on benzene ring) reacts as follows:

$\xrightarrow[H_2O]{Br_2}$ 2,4,6-tribromophenol (OH; Br, Br, Br) $\xrightarrow[H_2O]{Br_2}$ 2,4,4,6-tetrabromocyclohexa-2,5-dienone (O; Br, Br; Br Br) (p. 974)

$\xrightarrow[CCl_4]{Br_2}$ p-bromophenol (OH; Br) (+ ortho)

$\xrightarrow[H_2O]{HNO_3}$ o-nitrophenol (OH; NO_2) (+ para) (p. 976)

$\xrightarrow{\text{conc. } H_2SO_4}$ o-hydroxybenzenesulfonic acid (OH; SO_3H) (+ para) (p. 977)

$\xrightarrow{C_6H_5\overset{\oplus}{N}{\equiv}N\ X^{\ominus}}$ $C_6H_5N{=}N{-}C_6H_4{-}OH$ (p. 979)

(coupling)

$\xrightarrow{NaOH}$ $C_6H_5O^{\ominus}Na^{\oplus}$ $\xrightarrow{CO_2}$ $o\text{-}HOC_6H_4CO_2^{\ominus}Na^{\oplus}$ (OH; $CO_2^{\ominus}$ $Na^{\oplus}$)

(Kolbe, p. 980)

$\xrightarrow[OH^{\ominus}]{CHCl_3}$ salicylaldehyde (OH; CHO) (Reimer-Tiemann, p. 981)

polyhydric phenols, p. 981
phenolic resins, p. 983

3. Quinones

(a) Synthesis

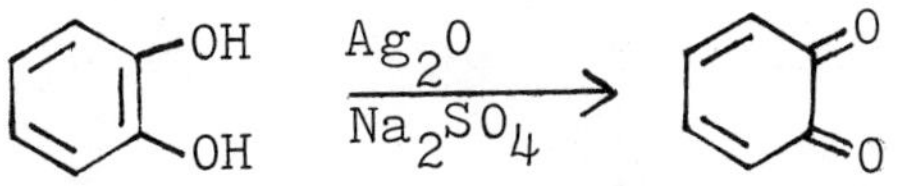

catechol $\xrightarrow[Na_2SO_4]{Ag_2O}$ o-benzoquinone (p. 697)

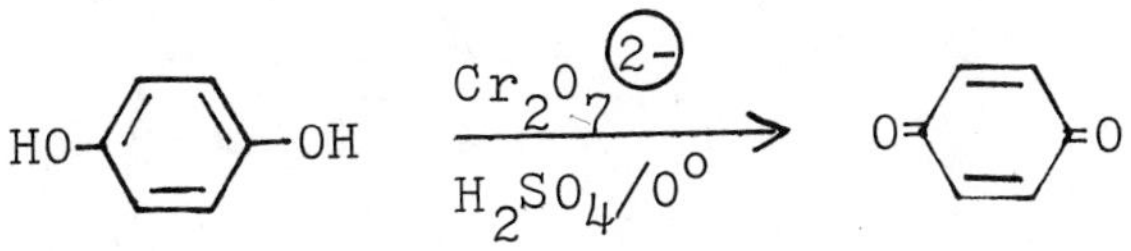

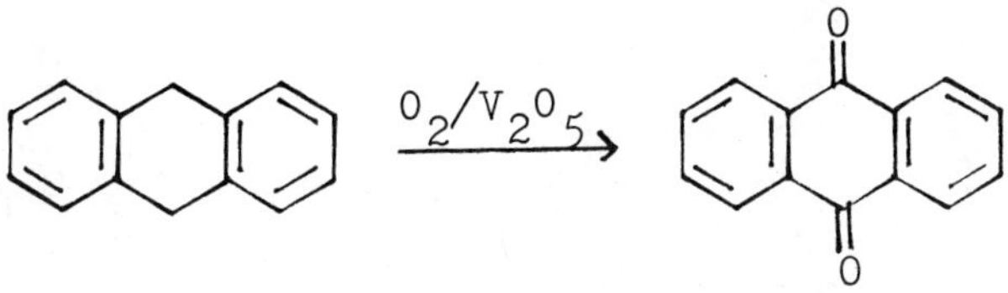

HO–C_6H_4–OH $\xrightarrow[H_2SO_4/0^o]{Cr_2O_7^{2-}}$ O=C_6H_4=O

anthracene $\xrightarrow{O_2/V_2O_5}$ anthraquinone

(b) Reactions

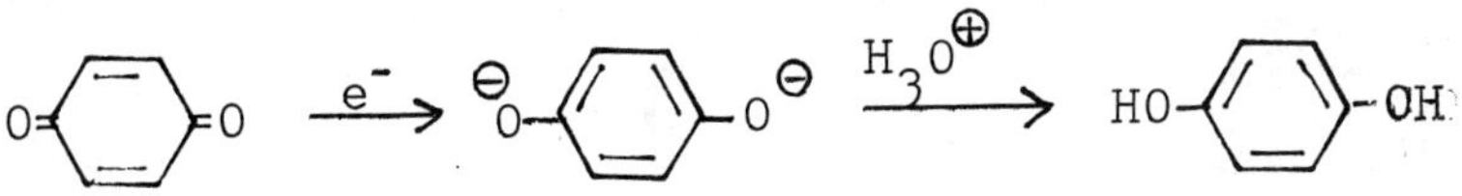

O=C_6H_4=O $\xrightarrow{e^-}$ $^{\ominus}$O–C_6H_4–O$^{\ominus}$ $\xrightarrow{H_3O^{\oplus}}$ HO–C_6H_4–OH (reduction, p. 697)

tetralin $\xrightarrow{DDQ^*}$ naphthalene

*p. 699

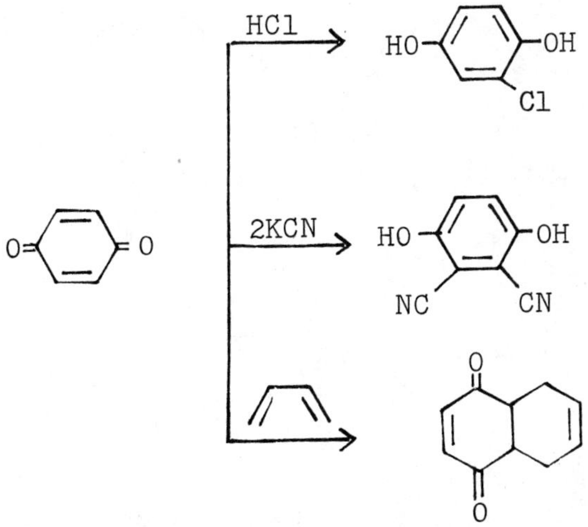

O=C_6H_4=O $\xrightarrow{HCl}$ HO–C_6H_3(Cl)–OH (addition reactions) (p. 699)

O=C_6H_4=O $\xrightarrow{2KCN}$ HO–C_6H_2(CN)(CN)–OH, NC, CN

O=C_6H_4=O + butadiene → Diels-Alder adduct (bicyclic dione) (Diels-Alder) (p. 700)

CHAPTER 24

ORGANIC COMPOUNDS OF SULFUR AND PHOSPHORUS

ANSWERS TO QUESTIONS

1. (a) 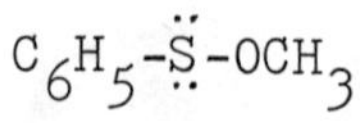 $C_6H_5\text{-}\ddot{\underset{\cdot\cdot}{S}}\text{-}OCH_3$ (b) $C_6H_5O\text{-}\ddot{\underset{\cdot\cdot}{S}}\text{-}CH_3$ (c) $C_6H_5\text{-}\overset{O^{\ominus}}{\underset{\oplus}{\overset{|}{\underset{\cdot\cdot}{S}}}}\text{-}CH_3$

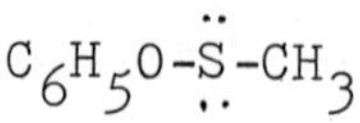

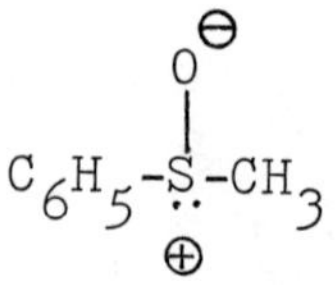

   (d) 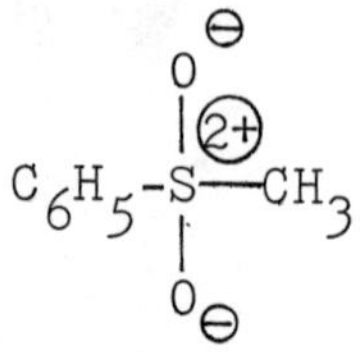 $C_6H_5\text{-}\overset{O^{\ominus}}{\underset{O^{\ominus}}{\overset{|}{\underset{|}{S^{2+}}}}}\text{—}CH_3$ (e) 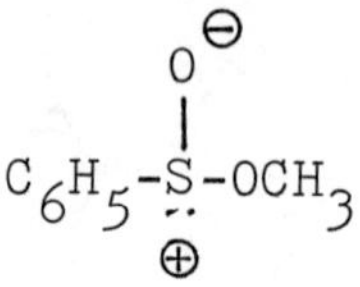 $C_6H_5\text{-}\overset{O^{\ominus}}{\underset{\oplus}{\overset{|}{\underset{\cdot\cdot}{S}}}}\text{-}OCH_3$ (f) 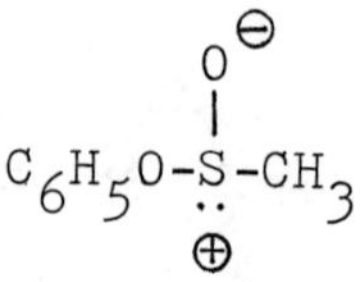 $C_6H_5O\text{-}\overset{O^{\ominus}}{\underset{\oplus}{\overset{|}{\underset{\cdot\cdot}{S}}}}\text{-}CH_3$

   (g) 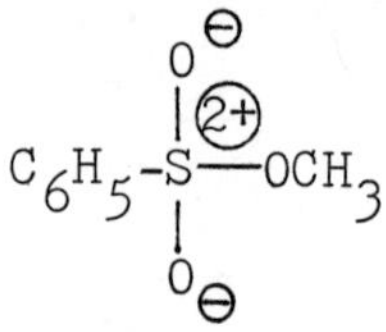 $C_6H_5\text{-}\overset{O^{\ominus}}{\underset{O^{\ominus}}{\overset{|}{\underset{|}{S^{2+}}}}}\text{—}OCH_3$ (h) 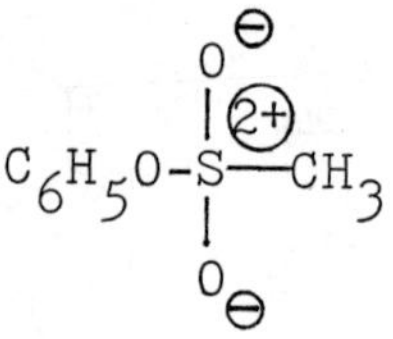 $C_6H_5O\text{-}\overset{O^{\ominus}}{\underset{O^{\ominus}}{\overset{|}{\underset{|}{S^{2+}}}}}\text{—}CH_3$ (i) 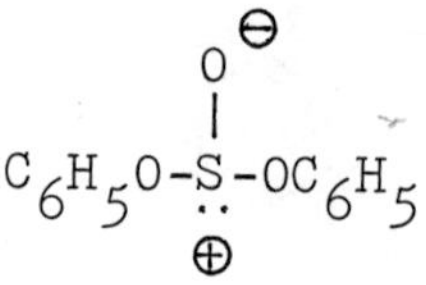 $C_6H_5O\text{-}\overset{O^{\ominus}}{\underset{\oplus}{\overset{|}{\underset{\cdot\cdot}{S}}}}\text{-}OC_6H_5$

   (j) 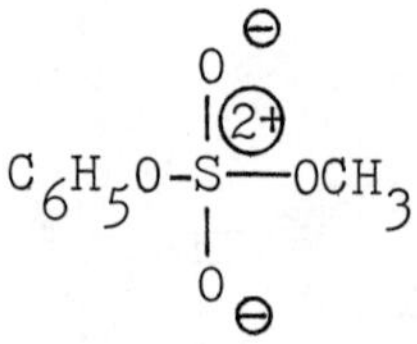 $C_6H_5O\text{-}\overset{O^{\ominus}}{\underset{O^{\ominus}}{\overset{|}{\underset{|}{S^{2+}}}}}\text{—}OCH_3$

2. (a) sulfone
   (b) sulfonate
   (c) sulfide, sulfoxide
   (d) sulfinate
   (e) sulfoxide, sulfide
   (f) sulfenate
   (g) disulfide
   (h) thiosulfinate
   (i) sulfide
   (j) thiosulfonate
   (k) thiol
   (l) sulfinyl chloride

3. (a) methyl phenyl disulfide
   (b) 1-chloro-2-naphthalenethiol
   (c) m-toluenethiol
   (d) ethyl methanesulfinate
   (e) phenyl p-toluenesulfonate
   (f) o-phenylthioanisole
   (g) 5-methyl-4,6-dithianonane

(continued next page)

(h) 2-phenyl-1,3-dithiolane (2-phenyl-1,3-dithiacyclopentane)

(i) 2-propanesulfonyl chloride (isopropylsulfonyl chloride)

(j) 3-thia-1-butanol

4. The trimethylsulfonium cation forms an ylide ($\overset{\oplus}{S}-\overset{\ominus}{C}$) in base. This ylide can react with D_2O to incorporate deuterium. Proton loss from the tetramethylammonium cation would not produce a resonance-stabilized anion (nitrogen lacks available d-orbitals).

5. No. If planar, the structure would have a plane of symmetry and could not be a pair of enantiomers. The fact that the compound has been prepared enantiomerically pure is consistent with tetrahedral geometry. The enantiomers are:

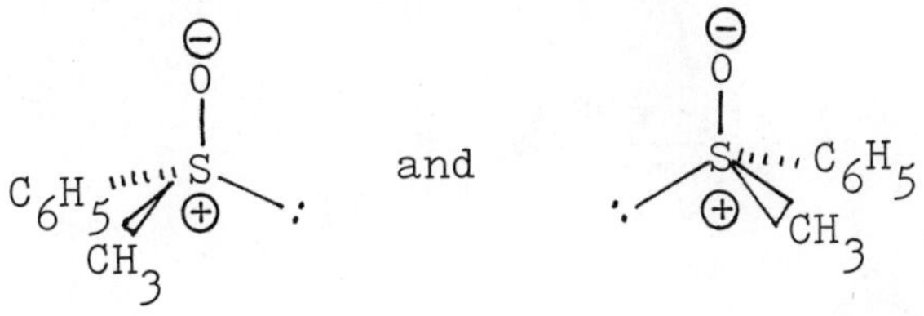

(Text: Stereochemistry of Organosulfur Compounds, p. 1030)

6. 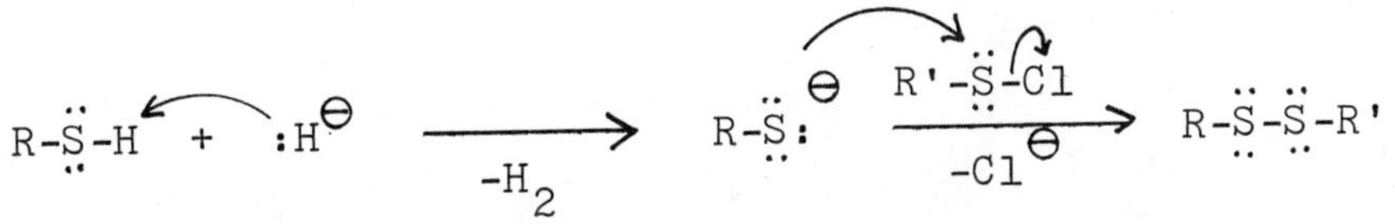

The reaction is S_N2 on sulfur.

7. (a) A is  ; B is 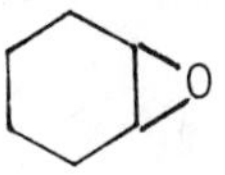 ; C is 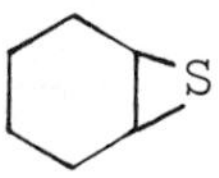

(b) D is 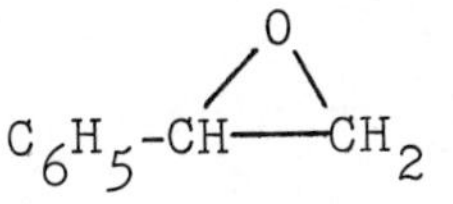 ; E is 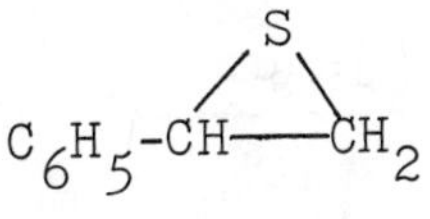

8. (a) $C_6H_5SH \xrightarrow[EtOH]{KOH} \xrightarrow{CH_3I}$

(b) $C_6H_5SH \xrightarrow[EtOH]{I_2}$ (continued next page)

(c) $C_6H_5SCH_3$ (from a) $\xrightarrow[0^{\circ}]{N_2O_4}$

(d) $C_6H_5SCH_3$ (from a) $\xrightarrow[\text{heat}]{CH_3Br/EtOH}$

(e) C_6H_5SH $\xrightarrow[EtOH]{KOH}$ $\xrightarrow[\text{excess}]{CH_3SSCH_3}$

(f) C_6H_5SH $\xrightarrow[\text{heat}]{MnO_4^{\ominus}/H^{\oplus}}$

(g) $C_6H_5SSC_6H_5$ (from b) $\xrightarrow[\text{(excess)}]{(CH_3)_3CS^{\ominus}}$

(h) $C_6H_5SSC_6H_5$ (from b) $\xrightarrow[\text{ether}]{(CH_3)_3CMgBr}$

9\. (a) $CH_3CH{=}CH_2$ $\xrightarrow[\text{peroxide}]{HBr}$ $CH_3CH_2CH_2Br$ $\xrightarrow{NH_2-\overset{S}{\overset{\|}{C}}-NH_2}$ $\xrightarrow{KOH}$

$\xrightarrow{H_3O^{\oplus}}$

(b) $CH_3CH{=}CH_2$ $\xrightarrow{HCl}$ $CH_3CHClCH_3$ $\xrightarrow{NH_2-\overset{S}{\overset{\|}{C}}-NH_2}$ $\xrightarrow{KOH}$

$CH_3\underset{S^{\ominus}K^{\oplus}}{\underset{|}{C}H}CH_3$ $\xrightarrow{CH_3CHClCH_3}$

(c) $CH_3CH{=}CH_2$ $\xrightarrow{C_6H_5CO_3H}$ $CH_3CH\overset{O}{—}CH_2$ (epoxide) $\xrightarrow{NCS^{\ominus}}$

(d) $CH_3CH\overset{S}{—}CH_2$ (from c) $\xrightarrow{CH_3MgX}$ $\xrightarrow{H_3O^{\oplus}}$

(e) $CH_2{=}CH_2$ $\xrightarrow[R\cdot]{H_2S}$ CH_3CH_2SH $\xrightarrow[EtOH]{I_2}$

(continued next page)

(f) [cyclohexene] $\xrightarrow{HBr}$ $\xrightarrow[EtOH]{Na_2S}$

(g) $CH_2=C(CH_3)_2$ $\xrightarrow[BF_3]{H_2S}$ $\xrightarrow{NaOH}$ $\xrightarrow[EtOH]{CH_3I}$

(h) $CH_3(CH_2)_6CH=CH_2$ $\xrightarrow{BH_3}$ $\xrightarrow[OH^{\ominus}]{H_2O_2}$ $\xrightarrow{SOCl_2}$ $CH_3(CH_2)_7CH_2Cl$

$\xrightarrow[ether]{Mg}$ $\xrightarrow{SO_2}$ $\xrightarrow{H_3O^{\oplus}}$

(i) $CH_3(CH_2)_7CH_2S(O)OH$ (from h) $\xrightarrow[heat]{H_2O_2}$

10. Two alternative paths are shown. Each gives the same product so that beginning with optically active starting material produces optically active product. (Note: path b is similar to path a)

[Reaction scheme: epoxide (H, Me / H, Et) attacked by $^{\ominus}SCN$ via path a or b → (a) alkoxide with NCS → rotated alkoxide → cyclic intermediate with :S:, C=N:$^{\ominus}$ → (*rotation) → thiolate with OCN → $-\ OCN^{\ominus}$ → episulfide (Me, H / H, Et)]

(*rotation)

b → $^{\ominus}O$ / H / Me / H / Et / SCN $\xrightarrow{etc.}$ episulfide (H, Me / H, Et, S)

11. (a) $O{=}C_6H_8{=}O + (CH_3)_2\overset{\oplus}{S}-\overset{\ominus}{C}H_2 \longrightarrow$ (e C-O)

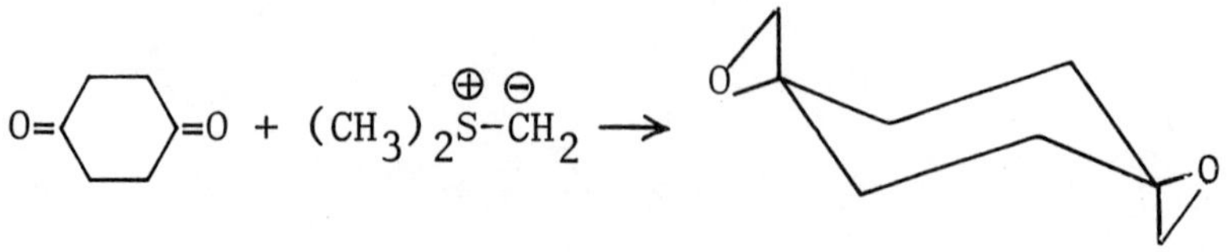

(b) $O{=}C_6H_8{=}O + (CH_3)_2\overset{\oplus}{S}(O)\overset{\ominus}{C}H_2 \longrightarrow$ (a C-O)

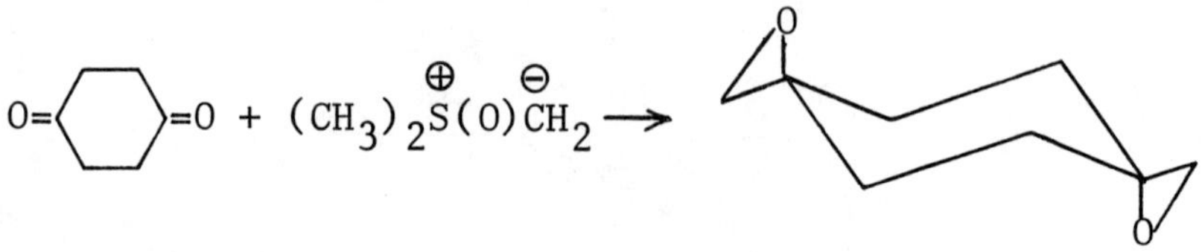

Products are diastereomers of one another

12. $CH_3S(O)_2CH_3$; dimethyl sulfone. An S_N2 attack by the carbanion on t-butyl bromide is impossible. However, this bromide will provide a proton and convert the dimsyl anion ($CH_3S(O)CH_2^{\ominus}$) to DMSO, which will be oxidized to sulfone by H_2O_2.

13. Rapid racemization occurs by formation of the achiral sulfinate anion.

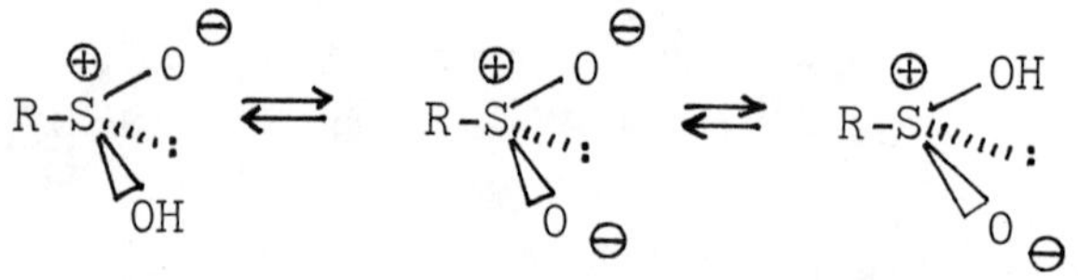

$R-\overset{\oplus}{S}(O^{\ominus})(OH)$: ⇌ $R-\overset{\oplus}{S}(O^{\ominus})(O^{\ominus})$: ⇌ $R-\overset{\oplus}{S}(OH)(O^{\ominus})$:

chiral achiral chiral

14. (c) enantiomers $CH_3, CH_3CH_2, \overset{\oplus}{S}(O^{\ominus})$: and $\overset{\oplus}{S}(O^{\ominus})$, CH_3, CH_2CH_3

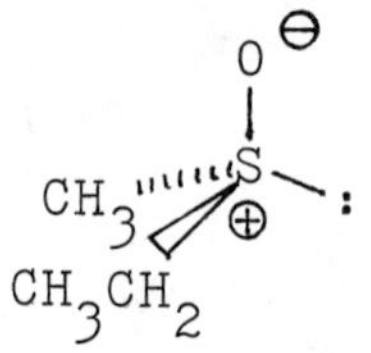

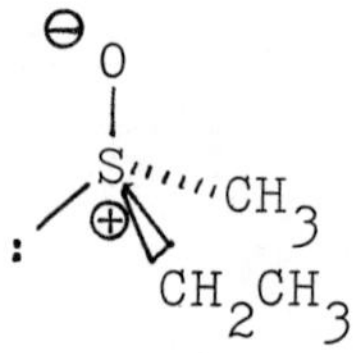

(d) $CH_3(H)C{=}C(H)-\overset{\oplus}{S}(O^{\ominus})CH_3$

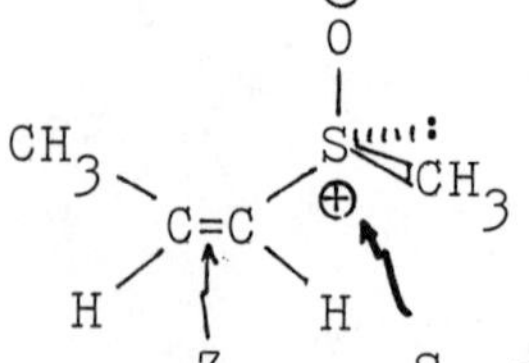

Z S and enantiomer Z,R

plus E,S and enantiomer E,R. Nonenantiomeric pairs are diastereomers.

15. Acetylating aniline decreases the basicity of the amino group and prevents it from reacting with chlorosulfonic acid or becoming protonated by chlorosulfonic acid. The latter produces $-\overset{\oplus}{N}H_3$ which deactivates the ring toward chlorosulfonation.

16. (a) phosphine
(b) phosphorus trichloride
(c) phosphorous acid
(d) phosphine oxide
(e) phosphonous acid
(f) phosphonic acid
(g) triphenyl phosphite
(h) triphenyl phosphate

17. 73 kcal/mole

18. Nonylide multiple bonds containing phosphorus are not very stable. P,P double bonds are weaker than N,N double bonds because of the greater covalent radius of P (resulting in poorer overlap). P=O and C≡P analogs of N=O and C≡N bonds are unstable because the former multiple bonds require overlap of orbitals of different main quantum levels.

19. Yes. $H^{\oplus}$ adds to the carbon of C=C already having the greater number of hydrogens.

20. $H_2C{=}\ddot{O} \xrightarrow{H^{\oplus}} H_2C{=}\overset{\oplus}{O}H$

$H_3P{:} + H_2C{=}\overset{\oplus}{O}H \longrightarrow H_3\overset{\oplus}{P}\text{-}CH_2\ddot{O}H \xrightarrow{-H^{\oplus}} H_2\ddot{P}\text{-}CH_2OH \xrightarrow[H^{\oplus}]{CH_2O} H_2\overset{\oplus}{P}(CH_2OH)_2 \xrightarrow{-H^{\oplus}} HP(CH_2OH)_2 \xrightarrow[H^{\oplus}]{CH_2O} H\overset{\oplus}{P}(CH_2OH)_3 \xrightarrow{-H^{\oplus}}$

$P(CH_2OH)_3 \xrightarrow[HCl]{CH_2O} \overset{\oplus}{P}(CH_2OH)_4\ Cl^{\ominus}$

21. A. $R\text{-}C{\equiv}C\text{-}Br + {:}\ddot{P}(C_6H_5)_2^{\ominus} \longrightarrow R\text{-}\overset{\ominus}{\ddot{C}}{=}C(Br)\text{-}\ddot{P}(C_6H_5)_2 \xrightarrow{-Br^{\ominus}} R\text{-}C{\equiv}C\text{-}\ddot{P}(C_6H_5)_2$

(continued next page)

OR

B. $R\text{-}C{\equiv}C\text{-}Br \longrightarrow \underset{(C_6H_5)_2P}{}\!\!\overset{R}{}C{=}C\overset{Br}{\underset{\ominus}{}} \longrightarrow (C_6H_5)_2P\text{-}C{\equiv}C\text{-}R$

$(C_6H_5)_2\ddot{P}{:}^{\ominus}$

OR

C. $R\text{-}C{\equiv}C\text{-}Br + {:}\ddot{P}^{\ominus}(C_6H_5)_2 \longrightarrow RC{\equiv}C{:}^{\ominus} \quad Br\text{-}\ddot{P}(C_6H_5)_2$

$\downarrow -Br^{\ominus}$

$RC{\equiv}C\text{-}\ddot{P}(C_6H_5)_2$

22. $C_6H_5CH_2OH \xrightarrow{SOCl_2} C_6H_5CH_2Cl \xrightarrow{P(C_6H_5)_3} \xrightarrow[\text{heat}]{NaOH/H_2O}$

23. $C_6H_5\ddot{C}H_2^{\ominus}$ (resonance stabilized) is a better leaving group than ${:}CH_3^{\ominus}$.

24. This reaction is favored by the strong P-O bond which is produced in the by-product.

$R\text{-}\overset{O}{\overset{\|}{C}}\text{-}\ddot{O}\text{-}H + {:}PCl_3 \longrightarrow R\text{-}\overset{:O:}{\overset{\|}{C}}\text{-}\ddot{O}^{\oplus}(H)\text{-}PCl_2 \quad {:}Cl^{\ominus} \longrightarrow R\text{-}\overset{:\ddot{O}:^{\ominus}}{\underset{Cl}{C}}\text{-}\overset{H}{\underset{\cdot\cdot}{O^{\oplus}}}\text{-}PCl_2$

$\downarrow$

$R\text{-}\overset{:O:}{\overset{\|}{C}}\text{-}Cl + H\ddot{O}\text{-}\overset{Cl}{P}\text{-}Cl$

$HOPCl_2$ reacts further with RCO_2H to produce, ultimately, $HP(OH)_2O$ and three moles of RCOCl.

25. (a) $CH_3SH \xrightarrow[C_2H_5OH/H_2O]{I_2}$

(b) $CH_3SH \xrightarrow{Na^{\oplus}\ OH^{\ominus}} CH_3S^{\ominus}Na^{\oplus} \xrightarrow{CH_3CH_2Br}$

(c) $HSCH_2CH_2SH \xrightarrow{2Na^{\oplus}\ OH^{\ominus}} Na^{\oplus}\,{}^{\ominus}SCH_2CH_2S^{\ominus}\ Na^{\oplus} \xrightarrow{2CH_3I}$

(or $2CH_3SH \xrightarrow{Na^{\oplus}\ OH^{\ominus}} 2CH_3S^{\ominus} \xrightarrow{BrCH_2CH_2Br}$)

(d) $2CH_3SH \xrightarrow[H^{\oplus}]{CH_3\overset{O}{\overset{\|}{C}}CH_3}$

(e) $C_6H_5SH \xrightarrow{Na^{\oplus}\ OH^{\ominus}} \xrightarrow{CH_3I} C_6H_5SCH_3 \xrightarrow{N_2O_4/CHCl_3}$

(f) $CH_3S(O)C_6H_5$ (from e) $\xrightarrow[heat]{H_2O_2/HOAc}$

(g) $C_6H_5SH \xrightarrow[H_3O^{\oplus}]{MnO_4^{\ominus}}$

(h) $CH_3CH_2SCH_3$ (from b) $\xrightarrow[H_2]{Raney\ Ni} CH_3CH_3 + CH_4$

(i) $CH_3SH \xrightarrow{OH^{\ominus}} \xrightarrow{CH_3I} CH_3SCH_3 \xrightarrow{CH_3Br}$

(j) $CH_3SH \xrightarrow{OH^{\ominus}} \xrightarrow{CH_3I} CH_3SCH_3 \xrightarrow[D_2]{Raney\ Ni} CH_3D$

26. (a) [cyclohexene] $\xrightarrow{C_6H_5CO_3H} \xrightarrow{NCS^{\ominus}}$

(b) [cyclohexene episulfide] (from a) $\xrightarrow{CH_3MgCl} \xrightarrow{H_3O^{\oplus}}$

(continued next page)

(c) [1-methylcyclohexane-2-thiol structure: ring with CH_3 and SH] (from b) $\xrightarrow{OH^{\ominus}}$ $\xrightarrow{CH_3I}$

(d) [cyclohexene episulfide structure: ring fused with S] (from b) $\xrightarrow{LiAlH_4}$ $\xrightarrow{H_2O}$

(e) $CH_3CH{=}CH_2 \xrightarrow[\text{peroxide}]{HBr} \xrightarrow[\text{ether}]{Mg} CH_3CH_2CH_2MgBr \xrightarrow{SO_2} \xrightarrow{H_3O^{\oplus}}$

(f) $CH_2{=}CH_2 \xrightarrow{HBr} CH_3CH_2Br \xrightarrow{NH_2C(=S)NH_2} \xrightarrow{OH^{\ominus}} \xrightarrow{H_3O^{\oplus}}$

(g) $CH_2{=}CH_2 \xrightarrow{O_3} \xrightarrow[H^{\oplus}]{Zn} H_2CO \xrightarrow{LiAlH_4} \xrightarrow{H_2O} CH_3OH \xrightarrow{SOCl_2} CH_3Cl \xrightarrow{NH_2C(=S)NH_2} \xrightarrow{OH^{\ominus}} \xrightarrow{H_3O^{\oplus}}$

27. (a) meso and d,l

(b) 4 diastereomers
(2 pairs of enantiomers)

(c) 4 diastereomers
(2 pairs of enantiomers)

(d) 4 diastereomers
(2 pairs of enantiomers)

28. $(CH_3)_2\ddot{S}: + Br{-}Br \longrightarrow (CH_3)_2\overset{\oplus}{S}{-}Br + Br^{\ominus}$

$H_2O: + (CH_3)_2\overset{\oplus}{S}{-}Br \longrightarrow H_2\overset{\oplus}{\ddot{O}}{-}S(CH_3)_2{-}Br$ (with $H_2\ddot{O}:$ removing H)

(continued next page)

$$\left[\ominus:\ddot{O}-\overset{\oplus}{S}(CH_3)_2 \longleftrightarrow \ddot{O}=S(CH_3)_2\right] \longleftarrow H-\overset{\oplus}{\ddot{O}}=S(CH_3)_2 \quad (H_2\ddot{O}:)$$

29. (a) CH_3CH_2CH (1,3-dithiane) $\xrightarrow[D_2]{\text{Raney Ni}} CH_3CH_2CHD_2$

(b) CH_3-C-CH_3 (1,3-dithiolane) $\xrightarrow[D_2]{\text{Raney Ni}} CH_3CD_2CH_3$

(c) $CH_3SCH_2CH_2CH_2SCH_3 \xrightarrow[D_2]{\text{Raney Ni}} DCH_2CH_2CH_2D\ (+\ CH_3D)$

(d) $CH_3-CH_2-\overset{O}{\overset{\|}{C}}-OCH_3 + Na^{\oplus}\ \overset{\ominus}{C}H_2\overset{\oplus}{S}(O^{\ominus})CH_3 \longrightarrow CH_3CH_2-\overset{O}{\overset{\|}{C}}-CH_2-\overset{\oplus}{S}(O^{\ominus})-CH_3 \xrightarrow[0^{\circ}]{Al\cdot Hg} CH_3CH_2-\overset{O}{\overset{\|}{C}}-CH_3$

(e) cyclobutanone $+ (CH_3)_2\underset{(+2)}{S}(O^{\ominus})-\overset{\ominus}{C}H_2 \longrightarrow$ spiro-epoxide (cyclobutane) $+ (CH_3)_2SO$

(f) $CH_3-CH=CHCH_2\overset{\oplus}{\ddot{S}}(C_6H_5)_2 \xrightarrow[\text{heat}]{OH^{\ominus}} CH_2=CH-CH=CH_2 + (C_6H_5)_2S$

(g) 2-oxocyclohexyl $O^{\ominus}Na^{\oplus} \xrightarrow{CS_2} \xrightarrow{CH_3I}$ 2-oxocyclohexyl $-O-\overset{O}{\overset{\|}{C}}-SCH_3 \xrightarrow{\text{heat}}$ cyclohexenone

(Chugaev)

(continued next page)

(h) 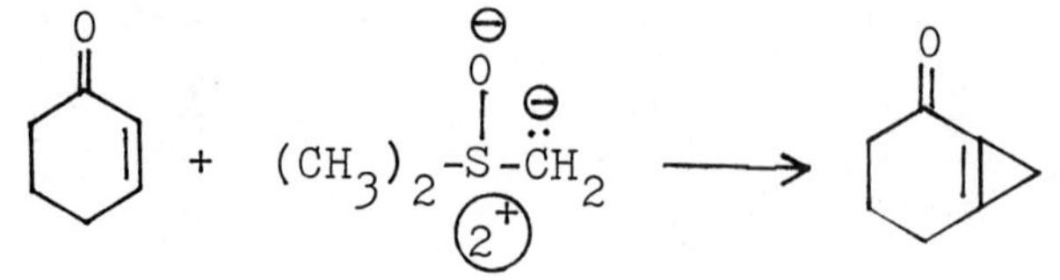 2-cyclohexenone + $(CH_3)_2\overset{\ominus O}{\underset{(2+)}{|}}S-\ddot{C}H_2^{\ominus}$ ⟶ bicyclic cyclopropane ketone

(i) 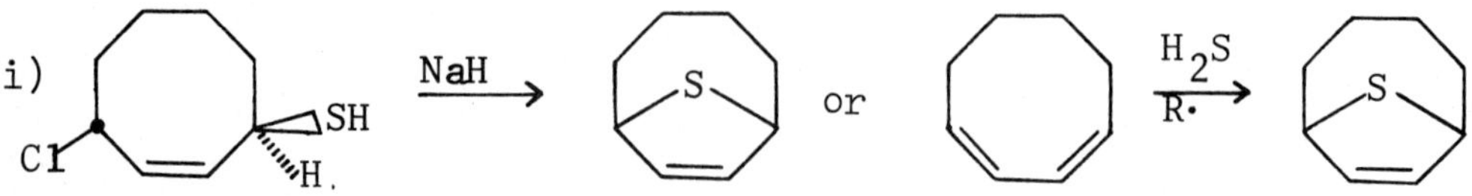 Cl-substituted cyclooctene thiol (SH, H) $\xrightarrow{NaH}$ S-bridged bicyclic alkene or cyclooctadiene $\xrightarrow[R\cdot]{H_2S}$ S-bridged bicyclic alkene

30. 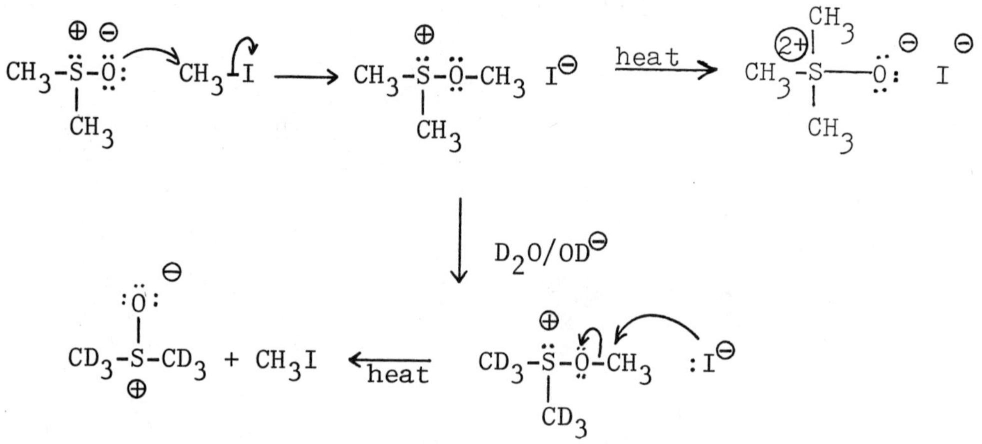 $CH_3-\overset{\oplus}{S}(CH_3)-\overset{\ominus}{O}: + CH_3-I \longrightarrow CH_3-\overset{\oplus}{S}(CH_3)-O-CH_3\ I^{\ominus} \xrightarrow{heat} CH_3-\overset{(2+)}{S}(CH_3)(CH_3)-\overset{\ominus}{O}:\ I^{\ominus}$

$\downarrow D_2O/OD^{\ominus}$

$CD_3-\underset{\oplus}{S}(\overset{\ominus}{:O:})-CD_3 + CH_3I \xleftarrow{heat} CD_3-\overset{\oplus}{S}(CD_3)-O-CH_3\quad :I^{\ominus}$

31. A produces cis and trans isomers.

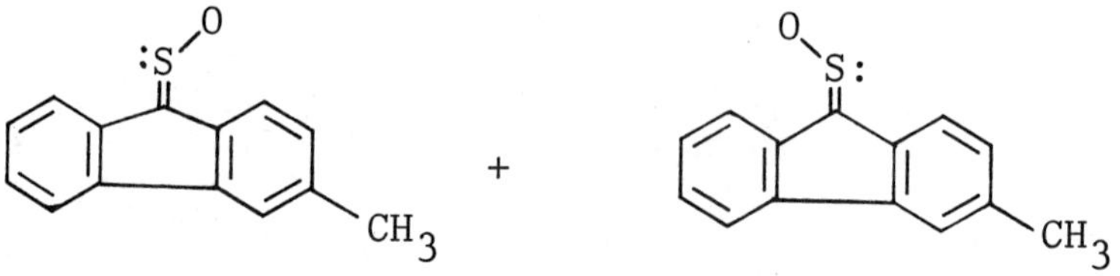

(two fluorenylidene sulfine isomers, :S=O, with CH_3) +

32. 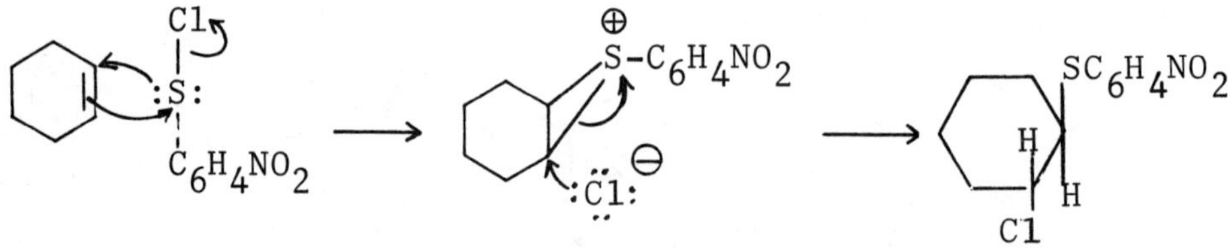 cyclohexene + $Cl-\ddot{S}-C_6H_4NO_2$ ⟶ episulfonium ion $\overset{\oplus}{S}-C_6H_4NO_2$ + $:\ddot{C}l:^{\ominus}$ ⟶ trans-1-chloro-2-($SC_6H_4NO_2$)cyclohexane (H, H, Cl)

OR

(continued next page)

$$\text{cyclohexene} + Cl\text{-}\ddot{S}\text{-}C_6H_4\text{-}NO_2 \longrightarrow \text{chloronium ion} + {}^{\ominus}:\ddot{S}\text{-}C_6H_4\text{-}NO_2 \longrightarrow \text{2-chlorocyclohexyl } p\text{-nitrophenyl sulfide}$$

33. $$R\text{-}\underset{\oplus}{\overset{\ominus :\ddot{O}:}{S}}\text{-}R + H\text{-}I \longrightarrow R\text{-}\underset{\oplus}{\overset{:\ddot{O}H}{S}}\text{-}R + :\ddot{I}:^{\ominus} \longrightarrow R\text{-}\underset{:\ddot{I}:}{\overset{:\ddot{O}H}{S}}\text{-}R \xrightarrow{HI} R\text{-}\underset{:\ddot{I}:}{\overset{:\overset{\oplus}{O}H_2}{S}}\text{-}R \xrightarrow{-H_2O} R\text{-}\underset{:\ddot{I}:}{\overset{\oplus}{\ddot{S}}}\text{-}R \xrightarrow{:\ddot{I}:^{\ominus}} R\text{-}\ddot{S}\text{-}R + I_2$$

34. The sulfinyl group possesses a lone electron pair on sulfur and can, therefore, function as an ortho, para director. The sulfonyl group lacks an electron pair on sulfur. Its inductive effect makes the sulfonyl group a meta director.

35. A concerted reaction with a five-membered cyclic activated complex. This reaction is similar to the Cope reaction (see p. 903).

$$\text{cyclic activated complex: } H\text{-}C(CH_3)(C_6H_5)\text{-}C(H)(C_6H_5)\text{-}\overset{\oplus}{S}(C_6H_5)\text{-}O^{\ominus} \longrightarrow (CH_3)(C_6H_5)C{=}C(C_6H_5)(H) + C_6H_5SOH$$

36. The increasing downfield shift of the methylene protons is related to the increasing electron withdrawing effect of the groups: $-SO_2- > -SO- > -S-$. The methylene protons of the methylene group in ethyl p-tolyl sulfoxide are diastereotopic

(continued next page)

accounting for the additional splitting in the methylene signal.

37. (a) $Cl\text{-}\ddot{S}\text{-}Cl + AlCl_3 \longrightarrow Cl\text{-}\ddot{S}^{\oplus} + AlCl_4^{\ominus}$

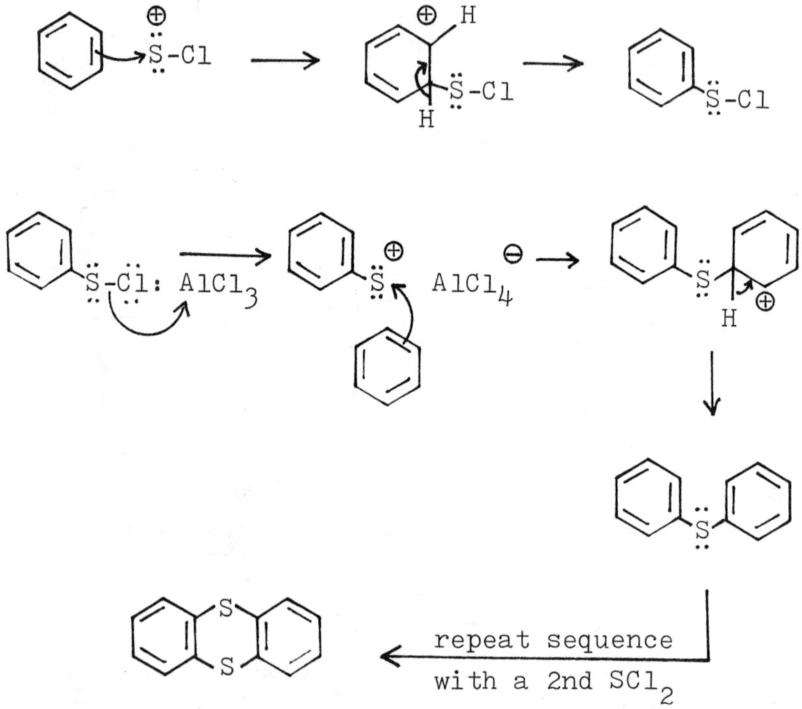

$\oplus$ H

S-Cl

S-Cl

H

S-Cl

S-Cl: $AlCl_3$

S $\oplus$ $AlCl_4$ $\ominus$

S

H $\oplus$

S

S

S

repeat sequence with a 2nd SCl_2

(b) Phenoxathiin symthesis:

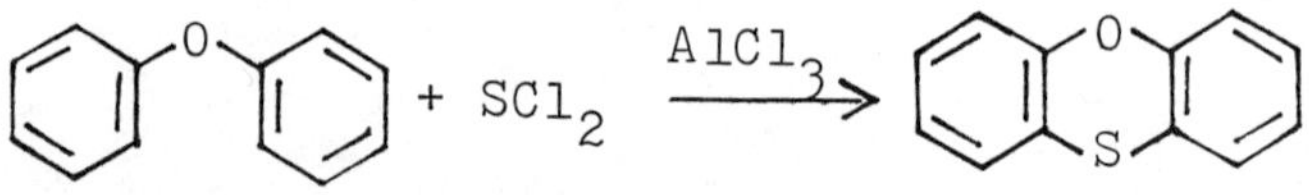

O + SCl_2 $\xrightarrow{AlCl_3}$ O S

38. 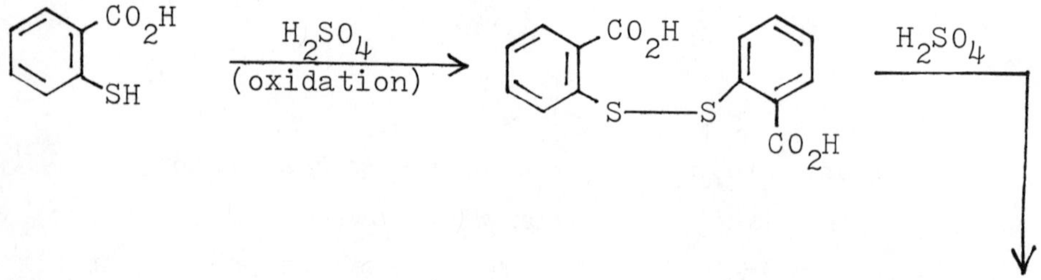 CO_2H SH $\xrightarrow[\text{(oxidation)}]{H_2SO_4}$ CO_2H S——S CO_2H $\xrightarrow{H_2SO_4}$

(continued next page)

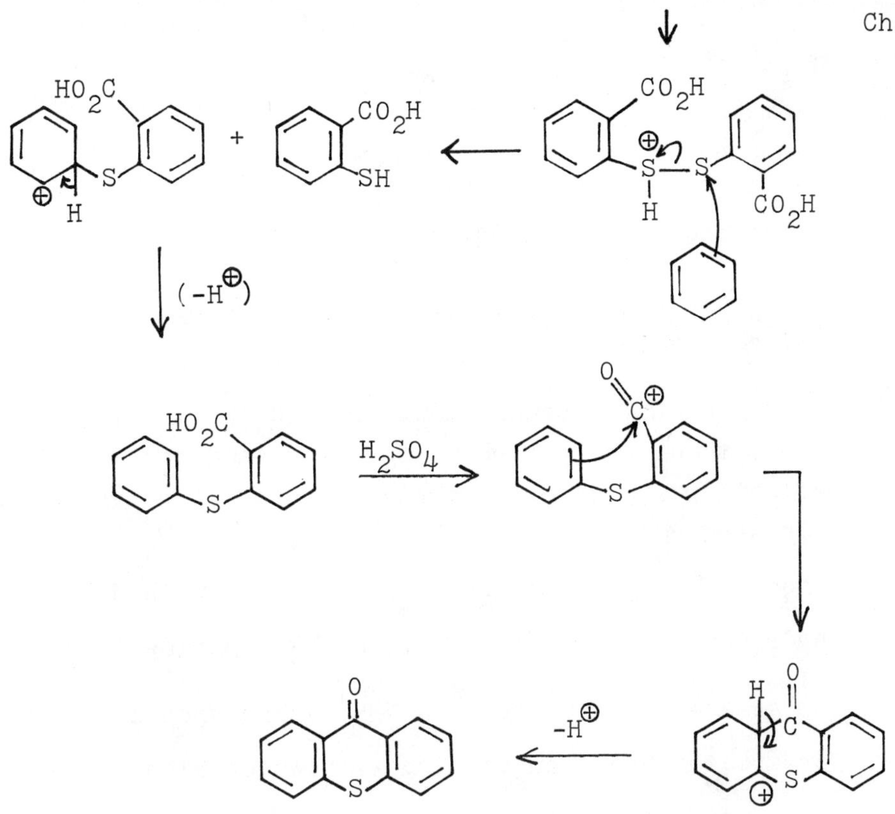

39. This suggests that the products (RCO_2H and H_2S) are more stable than the reactants (carbon-oxygen σ and π bonds are stronger than carbon-sulfur σ and π bonds).

40. (a) They are split by the other trifluoromethyl group because the two trifluoromethyl groups are non-equivalent.

(b) In rigid structure B, the two trifluoromethyl groups will be in diastereotopic environments and will demonstrate nonequivalence. A rapid equilibrium which proceeds by configurational inversion should make the trifluoromethyl groups equivalent.

41. (a) dibenzyl phosphite
(b) dibenzyl phosphonate
(c) methylenetriphenylphosphorane
(d) benzylphenylphosphinous chloride
(e) methyl dimethylphosphinite
(f) trimethylphosphine oxide
(g) ethyldimethylphosphine oxide
(h) phenyl phosphonous diamide

(continued next page)

(i) di(p-tolyl)-phosphinous amide

(j) hexamethylphosphorous triamide

(k) hexamethylphosphoric triamide

42. Phosphinic acid. It is used to convert $Ar-\overset{\oplus}{N_2}$ to Ar-H.

43. Intermolecular dipolar attraction involving the $\overset{\oplus}{P}-\overset{\ominus}{O}$ group in trimethyl phosphate.

44. (a) Ammonia hydrogen bonds to water (and to itself) more effectively than does phosphine to water (and to itself). This reflects the greater electronegativity of nitrogen compared to phosphorus.

(b) Ammonia [compare phosphine and ammonia with H_2S and H_2O (hydrides of group VI) or HCl and HF (hydrides of group VII)]

(c) The greater polarity of the N-H bonds which is responsible for greater hydrogen bonding with water also causes greater association among NH_3 molecules in the liquid state, and, as a result, a higher boiling point.

45. If the concentration of acid becomes high, the following reaction (similar to the Arbuzov reaction) may occur.

$$R-O-P(OR)_2: \underset{}{\overset{HCl}{\rightleftarrows}} R-\overset{\oplus}{O}-P(OR)_2-H \;(+\;Cl^{\ominus}) \longrightarrow R-Cl + \overset{\ominus}{O}-\overset{\oplus}{P}(OR)_2-H$$

46. (a) $PCl_3 + 2CH_3CH_2CH_2MgCl \longrightarrow (CH_3CH_2CH_2)_2PCl \xrightarrow{Na^{\oplus}\ {}^{\ominus}OCH_2CH_3} (CH_3CH_2CH_2)_2POCH_2CH_3$

(b) $C_6H_5MgBr + PCl_3 \longrightarrow C_6H_5PCl_2 \xrightarrow[ether]{LiAlH_4} C_6H_5PH_2 \xrightarrow{OH^{\ominus}}$

$\xrightarrow{CH_3I}$

(continued next page)

(c) $(CH_3)_3CMgBr \xrightarrow{PCl_2} (CH_3)_3CPCl_2 \xrightarrow{LiAlH_4} (CH_3)_3PH_2$

$\xrightarrow{OH^{\ominus}} \xrightarrow{CH_3OTs} (CH_3)_3CPHCH_3 \xrightarrow{OH^{\ominus}} \xrightarrow{CH_3OTs}$

(d) $(CH_3CH_2O)_3P + 2CH_3CH_2MgI \longrightarrow (CH_3CH_2)_2POCH_2CH_3$

$\xrightarrow{LiAlH_4} (CH_3CH_2)_2PH \xrightarrow{H_2O_2} (CH_3CH_2)_2P(O)H$

47. $R_2PH + O_2 \longrightarrow R_2\dot{P}(-O-O\cdot)-H \xrightarrow{R_2PH}$ [$R_2P(H)-O-O-P(H)R_2$] $\longrightarrow 2\left[R_2-P(=O)-H \longleftrightarrow R_2-\overset{\oplus}{P}(-O^{\ominus})-H\right]$

48. Since alkyl groups are electron donating (inductive effect), an increasing number of alkyl groups will destabilize the negative charge on phosphorus.

49. $PH_3 + R\cdot \longrightarrow \cdot PH_2 + RH$ initiation

$CH_2{=}CHR + H_2P\cdot \longrightarrow H_2P-CH_2-\dot{C}HR$

$H_2PCH_2-\dot{C}HR + PH_3 \longrightarrow H_2PCH_2-CH_2R + H_2P\cdot$

propagation

$2H_2P\cdot \longrightarrow H_2P-PH_2$

or $2H_2PCH_2\dot{C}H_2 \longrightarrow H_2PCH_2CH_2CH_2CH_2PH_2$

or $H_2P\cdot + H_2PCH_2CH_2 \longrightarrow H_2PCH_2CH_2PH_2$

termination

50. $(CH_3O)_3P: + CH_2(Br)\text{-}CO_2CH_2CH_3 \longrightarrow (CH_3O)_3\overset{\oplus}{P}\text{-}CH_2CO_2CH_2CH_3$

$\downarrow$ $Br^{\ominus}$ $(-CH_3Br)$

$Na^{\oplus}\ (CH_3O)_2\overset{\oplus}{P}(O^{\ominus})\text{-}\ddot{C}^{\ominus}HCO_2CH_2CH_3$ $\xleftarrow[-H_2]{NaH}$ $(CH_3O)_2\overset{\oplus}{P}(O^{\ominus})\text{-}CH_2CO_2CH_2CH_3$ (A)

(salt)

$\downarrow$ CH_3COCH_3 (Wittig)

$(CH_3)_2C{=}CH\text{-}CO_2CH_2CH_3$ (B)

51. $(C_6H_5)_4P^{\oplus}\ I^{\ominus}$ (A)

$(C_6H_5)_5P$ (B)

LiI (C)

52. The alcohol reacts with the phosphorus trihalide in what is potentially a two-step sequence: (1) phosphorus ester formation and (2) attack of halide ion upon the alkyl group of the ester to form an alkyl halide (like the Arbuzov reaction). If the halide ion is not a good nucleophile (e.g., $Cl^{\ominus}$ compared to $I^{\ominus}$) then one may expect ester as the main product.

53. A stable anion, similar to the phthalimide anion, is formed.

[Resonance structures: indanedione carbanion $\overset{\ominus}{C}$—CH(C_6H_5)—$\overset{\oplus}{P}(C_2H_5)_3$ $\longleftrightarrow$ enolate with :Ö:$^{\ominus}$ on upper oxygen $\longleftrightarrow$ enolate with :Ö:$^{\ominus}$ on lower oxygen]

54. (a) concerted process

$$\text{cis-2,3-dimethylthiirane} + :PR_3 \longrightarrow \text{cis-2-butene} + SPR_3$$

(b) two-step process

$$\text{cis-2,3-dimethylthiirane} + R_3P: \longrightarrow R_3\overset{\oplus}{P}\text{-CH(CH}_3\text{)-CH(CH}_3\text{)-}\ddot{S}:^{\ominus} \rightleftarrows \text{(rotated conformer)} \xrightarrow{-SPR_3} \text{trans-2-butene}$$

55. $C_6H_5\text{-}\ddot{P}(C_6H_5)\text{-}\ddot{O}\text{-}C(CH_3)_2\text{-}C{\equiv}C\text{-}H \longrightarrow C_6H_5\text{-}\ddot{P}(C_6H_5)\text{-}\ddot{O}:^{\ominus} + \left[(CH_3)_2\overset{\oplus}{C}\text{-}C{\equiv}C\text{-}H \longleftrightarrow (CH_3)_2C{=}C{=}\overset{\oplus}{C}\text{-}H\right]$

$(C_6H_5)_2\ddot{P}\text{-}\ddot{O}:^{\ominus}$

$(CH_3)_2C{=}C{=}CH\text{-}\overset{\oplus}{P}(C_6H_5)_2\text{-}\ddot{O}:^{\ominus}$ (attack by P)

56. The absence of cyclopropane derivatives suggests the absence of free carbenes as intermediates.

$$Cl_3C\text{-}\ddot{Cl}: + :PR_3 \longrightarrow Cl_3C:^{\ominus} + :\ddot{Cl}\text{-}\overset{\oplus}{P}R_3$$

(continued next page)

$R_3P: + CCl_3$ (with Cl of CCl_3 attacking $PR_3Cl^{\oplus}$) $\longrightarrow R_3\overset{\oplus}{P}-\overset{\ominus}{\ddot{C}}Cl_2 + R_3PCl_2$

or $R_3P: + Cl_3C-Cl \longrightarrow R_3\overset{\oplus}{P}-CCl_2-Cl + Cl^{\ominus}$

$R_3\overset{\oplus}{P}-CCl_3 + :PR_3 \longrightarrow Cl-\overset{\oplus}{P}R_3 + R_3P=CCl_2$

$Cl-\overset{\oplus}{P}R_3 \xrightarrow{Cl:^{\ominus}} Cl_2PR_3$

57. $CH_3OCH_2Cl \xrightarrow{(C_6H_5)_3P} \xrightarrow{B^{\ominus}} CH_3O\overset{\ominus}{\ddot{C}}HP(C_6H_5)_3$

ylide

cyclohexanone $+ CH_3O\overset{\ominus}{\ddot{C}}H\overset{\oplus}{P}(C_6H_5)_3 \longrightarrow$ (cyclohexylidene)$CHOCH_3$

$A(C_8H_{14}O)$

$\xrightarrow{H_3O^{\oplus}}$ (H)(cyclohexyl)$\overset{\oplus}{C}H-OCH_3 \xrightarrow{H_2\ddot{O}:}$ (H)(cyclohexyl)$CH(\overset{\oplus}{:O}H_2)OCH_3 \rightleftharpoons$ (H)(cyclohexyl)$CH(OH)-\overset{\oplus}{\ddot{O}}(H)CH_3$

$\longrightarrow$ (H)(cyclohexyl)$CH=O$

58. $(C_6H_5)_3P: + (NC)_2C{=}C(CN)_2 \longrightarrow (C_6H_5)_3\overset{\oplus}{P}\text{-}C(CN)_2\text{-}\overset{\ominus}{C}(CN)_2$ *

$\downarrow$ + $(NC)_2C{=}C(CN)_2$

$(C_6H_5)_3\overset{\oplus}{P}\text{-}C(CN)_2\text{-}C(CN)_2\text{-}C(CN)_2\text{-}\overset{\ominus}{C}(CN)_2$ * $\longrightarrow$ (ring: $(C_6H_5)_3P$ bonded to $(NC)_2C$—$C(CN)_2$—$C(CN)_2$—$C(NC)_2$)

*resonance stabilized $\left[-\overset{\ominus}{\ddot{C}}\text{-}C{\equiv}N: \longleftrightarrow -C{=}C{=}\overset{\ominus}{\ddot{N}}: \longleftrightarrow \text{etc.} \right]$

59. $(C_6H_5)_3P: + CH_3O_2C(H)C{=}C(H)CO_2CH_3 \longrightarrow (C_6H_5)_3\overset{\oplus}{P}\text{-}CH(CO_2CH_3)\text{-}\overset{\ominus}{\ddot{C}}H(CO_2CH_3)$

(less stable isomer) (free rotation)

$\downarrow$

$CH_3O_2C(H)C{=}C(H)CO_2CH_3$

(more stable isomer)

SUMMARY OF REACTIONS

1. Thiols (RSH, ArSH) (p. 1014)

(a) $ArH \xrightarrow{ClSO_3H} Ar\text{-}SO_2Cl \xrightarrow{Zn/HCl} Ar\text{-}SH$

(b) $Ar\overset{\oplus}{N}_2 \xrightarrow[\text{heat}]{K^{\oplus}\ {}^{\ominus}S\text{-}\overset{S}{\overset{\|}{C}}\text{-}OC_2H_5} ArS\overset{S}{\overset{\|}{C}}OC_2H_5 \xrightarrow[\text{heat}]{OH^{\ominus}/H_2O} ArS^{\ominus} \xrightarrow{H_3O^{\oplus}} ArSH$

(c) $(H_2N)_2C{=}S \xrightarrow{R\text{-}X} H_2\overset{\oplus}{N}{=}C(NH_2)\text{-}S\text{-}R\ X^{\ominus} \xrightarrow{H_2O/OH^{\ominus}} H_2N\text{-}C{\equiv}N + RS^{\ominus} \xrightarrow{H^{\oplus}} RSH$

thiourea — isothiouronium salt

(d) $RMgX \xrightarrow[\text{THF}]{S_8} \xrightarrow{H_3O^{\oplus}} RSH$

2. Disulfides (p. 1017)

$RSH \xrightarrow[C_2H_5OH/H_2O]{I_2} R\text{-}S\text{-}S\text{-}R + 2HI$

$RS^{\ominus}Na^{\oplus} \xrightarrow{Cl\text{-}S\text{-}R'} R\text{-}S\text{-}S\text{-}R' + NaCl$

3. Sulfides (p. 1017)

$R\text{-}S^{\ominus}Na^{\oplus} \xrightarrow{R'\text{-}X} R\text{-}S\text{-}R' + NaX$

$RMgX \xrightarrow{R'\text{-}S\text{-}S\text{-}R'} R\text{-}S\text{-}R' + R'SMgX$

4. Episulfides (p. 1018)

$$\text{-C—C- (epoxide, O)} \xrightarrow{NCS^{\ominus}} \text{-C—C- (episulfide, S)} + NCO^{\ominus}$$

reactions:

$$\text{episulfide} \xrightarrow{RMgX} \xrightarrow{H_3O^{\oplus}} R\text{-C-C-SH}$$

$$\text{episulfide} \xrightarrow{LiAlH_4} \xrightarrow{H_2O} H\text{-C-C-SH}$$

5. Sulfonium salts (p. 1019)

$$R'\text{-}X \xrightarrow{R\text{-}S\text{-}R} R\text{-}\overset{\oplus}{S}(R')\text{-}R \quad X^{\ominus}$$

elimination with $-S(CH_3)_2$ as a leaving group

$$H\text{-C-C-}\overset{\oplus}{S}(CH_3)_2 \quad OH^{\ominus} \xrightarrow{heat} \text{C=C} + (CH_3)_2S + H_2O$$

6. Sulfoxides (p. 1020)

$$R\text{-}S\text{-}R \xrightarrow{[O]^*} R\text{-}\overset{\oplus}{S}(O^{\ominus})\text{-}R$$

$*30\%$ H_2O_2/CH_3COOH; N_2O_4; $NaIO_4$; (m-Cl-C$_6$H$_4$-CO_3H)

$$R\text{-}\overset{\oplus}{S}(O^{\ominus})\text{-}O\text{-}R'' \xrightarrow[ether]{R'MgX} R\text{-}\overset{\oplus}{S}(O^{\ominus})\text{-}R' + R''OMgX$$

7. Sulfones (p. 1021)

$$R\text{-}S\text{-}R \quad (\text{or } R\text{-}\overset{O^{\ominus}}{\underset{\oplus}{S}}\text{-}R) \xrightarrow{H_2O_2/CH_3COOH} R\text{-}\overset{O^{\ominus}}{\underset{O^{\ominus}}{S^{2+}}}\text{-}R$$

$$ArSO_2Cl \xrightarrow[CS_2]{ArH/AlCl_3} ArSO_2Ar$$

8. Sulfonic Acids (p. 1022)

$$Ar\text{-}H \xrightarrow{\text{fuming } H_2SO_4} Ar\text{-}SO_3H$$

9. Gem-dithioethers (p. 1025)

$$\rangle C{=}O \xrightarrow[H^{\oplus}]{2RSH} RS\text{-}\overset{|}{\underset{|}{C}}\text{-}SR$$

(see p. 1026 for reactions of gem-dithioethers)

10. Sulfur ylides (p. 1027)

$$-\overset{|}{\underset{H}{C}}-\overset{|}{\underset{\oplus}{S}}- \; X^{\ominus} \xrightarrow[DMSO]{NaH} -\overset{|}{\underset{\ominus}{\ddot{C}}}-\overset{|}{\underset{\oplus}{\ddot{S}}}- \xrightarrow{\rangle C=O} -\overset{|}{C}\overset{O}{\frown}\overset{|}{C}- \text{ (epoxide)}$$

11. Phosphines (p. 1043)

$$C_6H_6 \xrightarrow[AlCl_3]{PCl_3} \xrightarrow[THF]{LiAlH_4} C_6H_5PH_2 \xrightarrow{Na} Na^{\oplus}\; C_6H_5PH^{\ominus}$$

$$PCl_3 \xrightarrow{LiAlH_4} PH_3 \xrightarrow{Na} Na^{\oplus}\; PH_2^{\ominus}$$

(see Figure 24-8 for synthesis of phosphines from $Na^{\oplus}\; \bar{C}_6H_5PH^{\ominus}$ and $Na^{\oplus}\; PH_2^{\ominus}$)

12. Nucleophilic Reactions at Phosphorus (see p. 1047)

13. Decomposition of Quaternary Phosphonium Salts (p. 1048)

$$(CH_3)_3\overset{\oplus}{P}\text{-}CH_2R \xrightarrow[H_2O]{OH^{\ominus}} (CH_3)_3\overset{\oplus}{P}\text{-}\overset{\ominus}{O} + CH_3R$$

($\cdot CH_2R$ must form a stabilized carbanion)

14. Arbuzov Reaction (p. 1049)

$$(CH_3CH_2O)_3P + RCH_2Cl \xrightarrow{S_N2} RCH_2\overset{\oplus}{P}(OCH_2CH_3)_3 \; \overset{\ominus}{}$$

$$\downarrow$$

$$CH_3CH_2Cl + \overset{\ominus}{O}\text{-}\overset{\oplus}{P}(OR)_2CH_2R$$

CHAPTER 25
AMINO ACIDS, PEPTIDES AND PROTEINS

ANSWERS TO QUESTIONS

1. [Structure: H on top; R (hashed), HO_2C (wedge), NH_2 about central carbon]

leucine	R: $-CH_2CH(CH_3)_2$
serine	R: $-CH_2OH$
valine	R: $-CH(CH_3)_2$

2. (a) $-NH_2$ is more basic, in part because $-CO_2^{\ominus}$ is resonance stabilized.

   (b) Monoprotonation of $H_2NCHRCO_2^{\ominus}$ produces $H_3\overset{\oplus}{N}CHRCO_2^{\ominus}$.

3. $R\text{-}C(H)(COH{=}O)\text{-}CO_2H \rightleftharpoons R\text{-}C(CO_2H){=}C(OH)\text{-}OH$ $\xrightarrow{(a)}$ Br–Br → $R\text{-}C(Br)(C(=O)OH)\text{-}C(=O)\text{-}O\text{-}H$ + HBr

↓ (b)

$\underset{Br}{RCH}\text{-}\overset{O}{\overset{\|}{C}}OH \rightleftharpoons R\text{-}C(Br){=}C(OH)OH$ + CO_2

4. (a) $CH_3CH_2CH(NH_3^{\oplus})CO_2^{\ominus}$ (c) $HO_2CCH_2CH_2CH(NH_3^{\oplus})CO_2^{\ominus}$

(b) $CH_3SCH_2CH_2CH(NH_3^{\oplus})CO_2^{\ominus}$

5. (a) [phthalimide]N–C(CH_3)(CH_3)–CO_2Et $\xrightarrow{H_3O^{\oplus}}$ [protonated imide, $^{\oplus}\ddot{O}H$] + $:\ddot{O}H_2$ → [tetrahedral intermediate, $H\ddot{O}$, $:\ddot{O}H_2^{\oplus}$, :N] → $H_2\ddot{O}:$ + [$H\text{-}\ddot{O}$, $\ddot{O}H$, $\overset{\oplus}{N}H$] → [:O:, $\ddot{O}H$, $\ddot{N}H$, O]

↓ repeat

(continued next page)

$\downarrow$

(phthalic acid: benzene ring bearing two $\overset{O}{\parallel}$COH groups) + $\overset{\oplus}{N}H_3$-C(CH_3)(CO_2Et)-CO_2Et

(b) $\overset{\oplus}{N}H_3$, :O: ; CH_3-C(CO_2Et)—C(=O)-OEt $\xrightarrow{H_3O^{\oplus}}$ ~~C(=$\overset{\oplus}{O}$H)-OEt

$\downarrow$ $H_2\ddot{O}$:

:$\ddot{O}H_2$; ~~C(:O-H)(O-H)-$\overset{\oplus}{O}$(Et)(H) $\rightleftarrows$ ~~C(:$\ddot{O}$-H)(-OEt)(:OH_2 $\oplus$)

$\downarrow$

~~C(=:O:)OH $\xrightarrow{\text{repeat}}$ $\overset{\oplus}{N}H_3$; CH_3-C(CO_2H)-CO_2H

(c) HO-C(=O)-C(CH_3)($H_3\overset{\oplus}{N}$)-C(=O)-O-H $\longrightarrow$ (HO)(OH)C=C(CH_3)($\overset{\oplus}{N}H_3$) + CO_2

$\rightleftarrows$

HO-C(=O)-C(H)(CH_3)($\overset{\oplus}{N}H_3$)

6. Yes. Replacement of either CO_2H group with a group other than R and H makes the central carbon a chiral center, i.e., $RCHR'CO_2H$.

7. 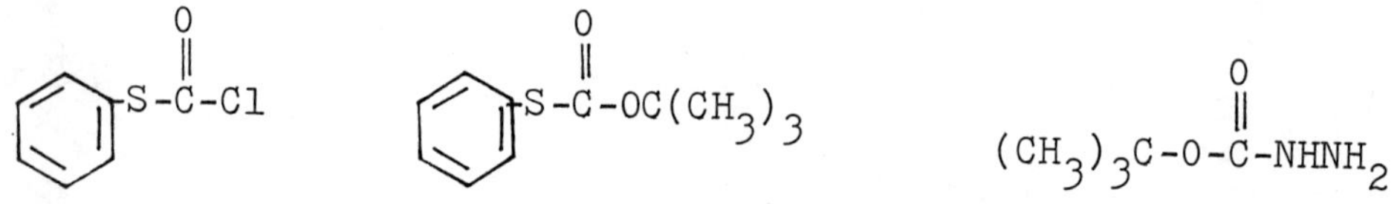 A: $C_6H_5S\text{-}\overset{O}{\overset{\|}{C}}\text{-}Cl$ B: $C_6H_5S\text{-}\overset{O}{\overset{\|}{C}}\text{-}OC(CH_3)_3$ C: $(CH_3)_3C\text{-}O\text{-}\overset{O}{\overset{\|}{C}}\text{-}NHNH_2$

A B C

8. (a) 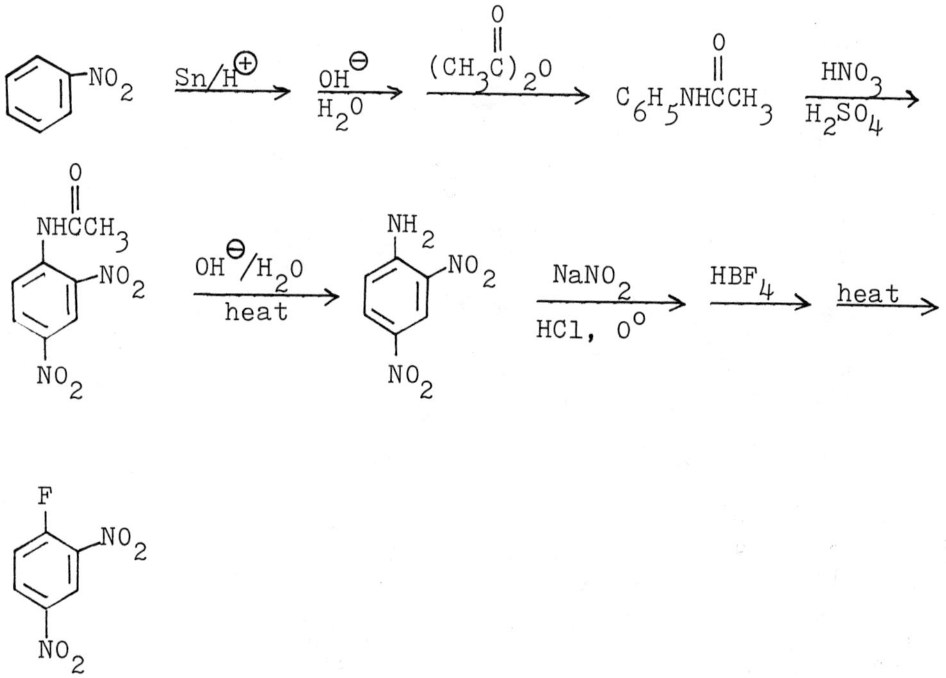 $C_6H_5NO_2 \xrightarrow{Sn/H^{\oplus}} \xrightarrow[H_2O]{OH^{\ominus}} \xrightarrow{(CH_3\overset{O}{\overset{\|}{C}})_2O} C_6H_5NH\overset{O}{\overset{\|}{C}}CH_3 \xrightarrow[H_2SO_4]{HNO_3}$ 2,4-dinitro-$C_6H_3NH\overset{O}{\overset{\|}{C}}CH_3$ $\xrightarrow[heat]{OH^{\ominus}/H_2O}$ 2,4-dinitro-$C_6H_3NH_2$ $\xrightarrow[HCl,\ 0^{o}]{NaNO_2} \xrightarrow{HBF_4} \xrightarrow{heat}$ 2,4-dinitro-C_6H_3F

(continued next page)

(b) $C_6H_5NHCOCH_3$ (from a) $\xrightarrow[AlCl_3]{CH_3Cl}$ p-$CH_3C_6H_4NHCOCH_3$ $\xrightarrow{NO_2^{\oplus}\ BF_4^{\ominus}}$

2,6-dinitro-4-methylacetanilide $\xrightarrow[heat]{H_3O^{\oplus}}$ $\xrightarrow[HCl,\ 0^{o}]{NaNO_2}$ $\xrightarrow[heat]{CH_3CH_2OH}$ 3,5-dinitrotoluene (O_2N, NO_2, CH_3)

$\xrightarrow[OH^{\ominus}/heat]{MnO_4^{\ominus}}$ $\xrightarrow{H^{\oplus}}$ $\xrightarrow{SOCl_2}$ $\xrightarrow{NH_3}$ 3,5-dinitrobenzamide (O_2N, NO_2, $\overset{O}{\overset{\|}{C}}$-$NH_2$) $\xrightarrow[OH^{\ominus}]{Br_2}$

3,5-dinitroaniline (O_2N, NO_2, NH_2) $\xrightarrow[HCl,\ 0^{o}]{NaNO_2}$ $\xrightarrow{HBF_4}$ $\xrightarrow{heat}$ 1-fluoro-3,5-dinitrobenzene (O_2N, NO_2, F)

(c) 2,4-dinitrofluorobenzene (O_2N–, –F, $N^{\oplus}(=O)O^{\ominus}$) + $:NH_2CHCH_3$ (COOH) $\longrightarrow$ $\left[\ O_2N\text{–ring}(\ominus)\text{–C}(F)(\overset{\oplus}{N}H_2CHCH_3,\ COOH),\ NO_2\ \right]^{*}$

$\downarrow -F^{\ominus}$

(continued next page)

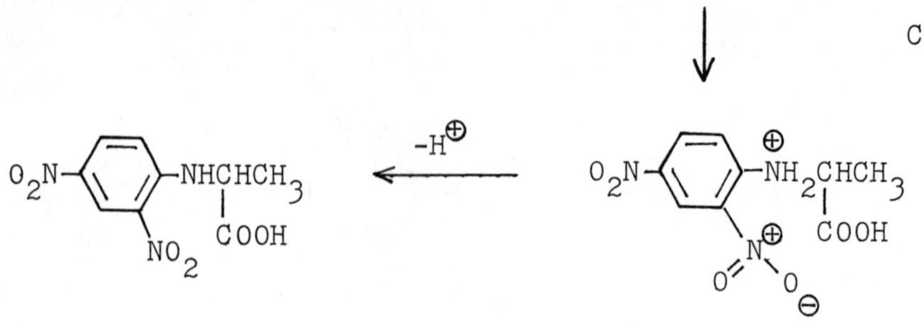

O_2N–C$_6$H$_3$(NO$_2$)–NHCHCH$_3$ (COOH) $\xleftarrow{-H^{\oplus}}$ O_2N–C$_6$H$_3$(N$^{\oplus}$(=O)O$^{\ominus}$)–N$^{\oplus}$H$_2$CHCH$_3$ (COOH)

*Resonance structures are:

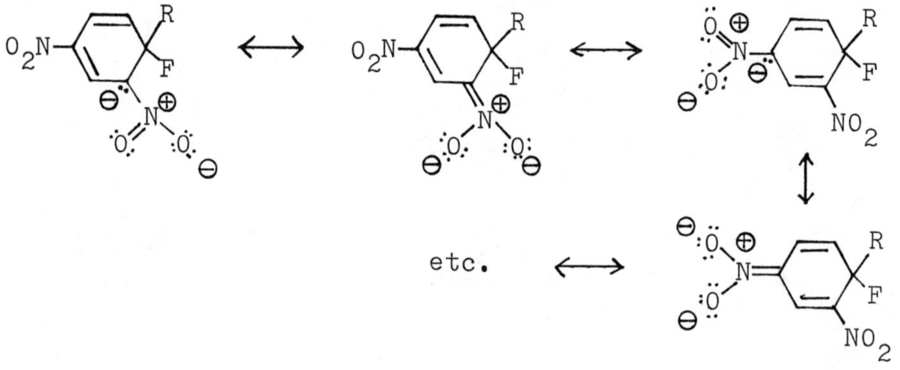

O_2N ... R, F ... $\longleftrightarrow$ O_2N ... R, F ... $\longleftrightarrow$ ... R, F, NO_2

etc. $\longleftrightarrow$ ... R, F, NO_2

9. Loss of a proton from uric acid gives a resonance-stabilized anion.

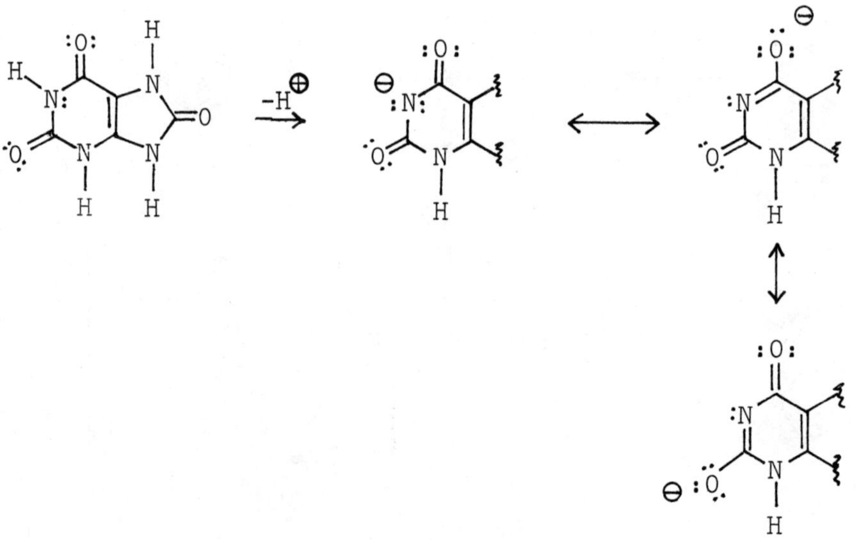

$\xrightarrow{-H^{\oplus}}$ $\longleftrightarrow$ $\longleftrightarrow$

10. A: Ala; S: Ser; G: Gly; V: Val in this answer set.

(a) A-S-G-V; A-S-V-G; A-G-S-V; A-G-V-S; A-V-S-G; A-V-G-S;

S-G-V-A; S-V-G-A; G-S-V-A; G-V-S-A; V-S-G-A; V-G-S-A;

S-A-G-V; S-A-V-G; S-G-A-V; S-V-A-G; G-S-A-V; G-V-A-S;

G-A-S-V; G-A-V-S; V-S-A-G; V-G-A-S; V-A-G-S; V-A-S-G.

In each, the N-terminal amino acid is on the extreme left and the C-terminal amino acid is on the right.

(b) A-V-S-G; A-V-G-S; S-V-G-A; S-V-A-G; G-V-A-S; G-V-S-A.

(c) A-G-S-V (others are possible)

$$H_2N{-}\underset{\underset{CH_3}{|}}{CH}{-}\overset{\overset{O}{\|}}{C}{-}NH{-}\underset{\underset{H}{|}}{CH}{-}\overset{\overset{O}{\|}}{C}{-}NH{-}\underset{\underset{CH_2OH}{|}}{CH}{-}\overset{\overset{O}{\|}}{C}{-}NH{-}\underset{\underset{CH(CH_3)_2}{|}}{CH}{-}CO_2H$$

ala (1) gly (2) ser (3) val (4)

11. (a) 5040 (b) 40320 (c) 362880

12. 21,400

13. Phe-Val-Asp-Glu-His

14. (a) $C_6H_5CH_2OCOCl \xrightarrow{H_2NCH(CH_3)CO_2H} C_6H_5CH_2OCONHCH(CH_3)COOH$

$\xrightarrow{ClCOOEt}$ $\sim\sim NHCH(CH_3)COOCOOEt \xrightarrow{H_2NCH_2COOEt}$

$\sim\sim NHCH(CH_3)CONHCH_2COOEt \xrightarrow{OH^{\ominus}/H_2O} \xrightarrow{H_3O^{\oplus}}$

$C_6H_5CH_2OCONHCH(CH_3)CONHCH_2COOH \xrightarrow{H_2/Pd}$ ala-gly

$+ CO_2 + C_6H_5CH_3$

(continued next page)

(b) $C_6H_5CH_2OCOCl \xrightarrow{H_2NCH_2COOH} \xrightarrow{ClCOOEt} \xrightarrow{H_2NCH(CH_3)COOEt}$

$\xrightarrow[OH^{\ominus}]{H_2O} \xrightarrow{H_3O^{\oplus}} C_6H_5CH_2OCONHCH_2CONHCH(CH_3)COOH$

$\xrightarrow{H_2/Pd}$ gly-ala-gly + CO_2 + $C_6H_5CH_3$

15. (a) $CH_3\text{-}\underset{\underset{CH_3}{|}}{C}\text{=CH-}CO_2H \xrightarrow{H_2/Pt} CH_3\underset{\underset{CH_3}{|}}{C}H\text{-}CH_2\text{-}CO_2H \xrightarrow[PBr_3]{Br_2}$

$(CH_3)_2CHCH(NH_2)CO_2H \xleftarrow{NH_3} (CH_3)_2CHCHBrCO_2H$

(b) $CH_3\underset{\underset{Br}{|}}{C}H\text{-}CO_2H + OH^{\ominus} \xrightarrow{H_2O} CH_3\text{-}\underset{\underset{Br}{|}}{C}H\text{-}CO_2^{\ominus} \xrightarrow{\text{potassium phthalimide (N-K)}} \xrightarrow[H_2O/heat]{OH^{\ominus}} CH_3\underset{\underset{NH_2}{|}}{C}H\text{-}CO_2H$

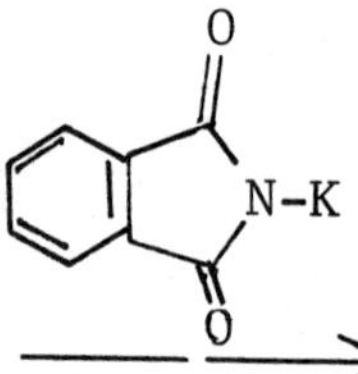

(c) $C_6H_5CH_2CH_2CO_2H \xrightarrow[PBr_3]{Br_2} C_6H_5CH_2CHBrCO_2H \xrightarrow{NH_3} C_6H_5CH_2CH(NH_2)CO_2H$

(d) $CH_2\text{=CH-}CH_2CH_2\overset{\overset{O}{\|}}{C}CH_3 \xrightarrow[OH^{\ominus}]{Cl_2} \xrightarrow{H_3O^{\oplus}} CH_2\text{=CH-}CH_2CH_2CO_2H \xrightarrow{HBr/R\cdot}$

$BrCH_2CH_2CH_2CH_2CO_2H \xrightarrow{SOCl_2} \xrightarrow{CH_2N_2} \xrightarrow[H_2O]{Ag^{\oplus}} Br(CH_2)_5CO_2H$

$\xrightarrow[PBr_3]{Br_2} Br(CH_2)_4\underset{\underset{Br}{|}}{C}HCO_2H \xrightarrow{NH_3} H_2N(CH_2)_4\underset{\underset{NH_2}{|}}{C}HCO_2H$

(e) $HOCH_2\underset{NH_2}{CH}CO_2H$

$BrCH_2CO_2C_2H_5 \xrightarrow{Zn} BrZnCH_2CO_2C_2H_5 \xrightarrow{H_2C=O} \xrightarrow{H_3O^{\oplus}}$

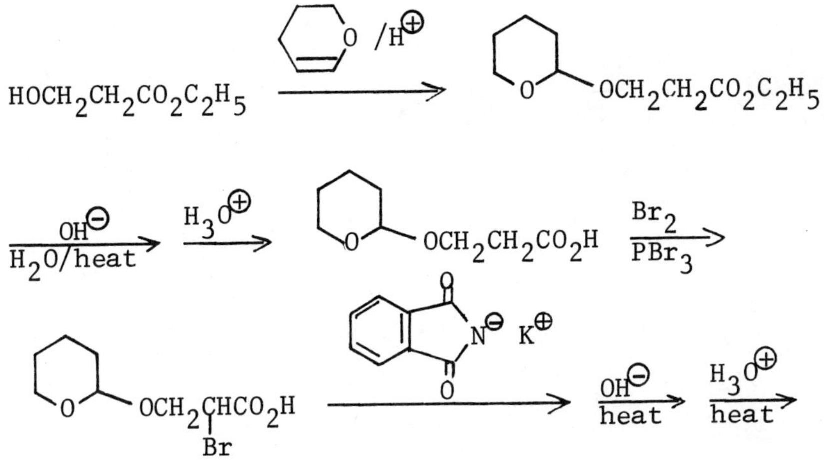

HOCH2CH2CO2C2H5
O /H⊕
O
OCH2CH2CO2C2H5
OH⊖
H2O/heat
H3O⊕
OCH2CH2CO2H
Br2
PBr3
OCH2CHCO2H
Br
N⊖ K⊕
OH⊖
heat
H3O⊕
heat

$HO-CH_2-\underset{NH_2}{CH}-CO_2H$

(f)

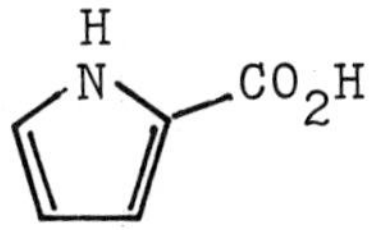

H
N
CO2H

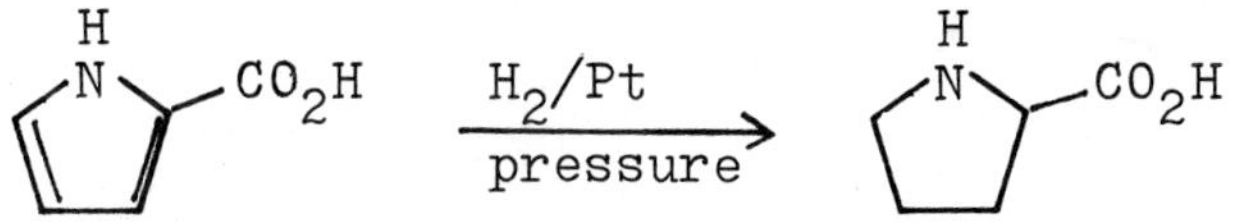

H
N
CO2H
H2/Pt
pressure

(g) $HO_2CCH_2\underset{NH_2}{CH}CO_2H$

$HO_2C-CH_2CH_2-CO_2H \xrightarrow[PBr_3]{Br_2} HO_2CCH_2\underset{Br}{CH}CO_2H \;\frac{OH^{\ominus}}{H_2O}$

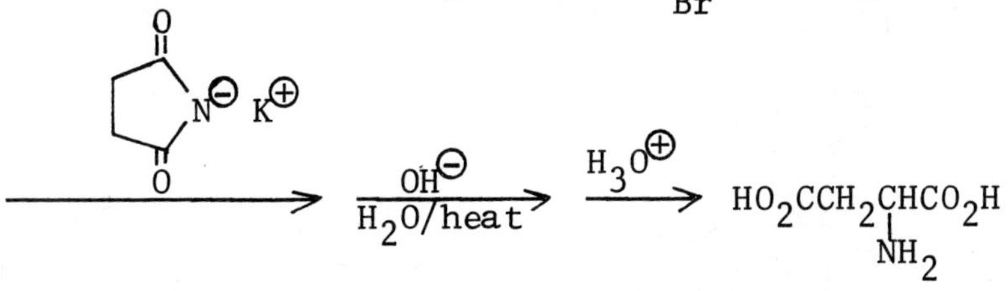

O
N⊖ K⊕
O
OH⊖
H2O/heat
H3O⊕
HO2CCH2CHCO2H
NH2

16. (a) $\frac{pK_1 + pK_2}{2} = PI$

(b) (i) $(CH_3)_2CHCH_2CH(\overset{\oplus}{N}H_3)CO_2H$ (ii) $H_3\overset{\oplus}{N}(CH_2)_4CH(\overset{\oplus}{N}H_3)CO_2H$

(iii) $H_3\overset{\oplus}{N}CH_2CO_2H$ (iv) $H_3\overset{\oplus}{N}CH_2CONHCH_2COOH$

17. pro-gly-phe, gly-phe-ser, phe-ser-pro, ser-pro-phe, pro-phe-arg

18. (a) $(\text{glu-gly-pro-trp})^{1 \ldots 2}\,(\text{leu-glu-glu-glu-glu-ala-ala-tyr})^{3 \ldots 4}\,(\text{gly-trp})^{5 \ldots 6}\,(\text{met-asp-phe})^{7 \ldots 8}$

(b) Other possibility is:

1 ∼∼∼∼ 2-5 ∼∼∼∼ 6-3 ∼∼∼∼ 4-7 ∼∼∼∼ 8

19. 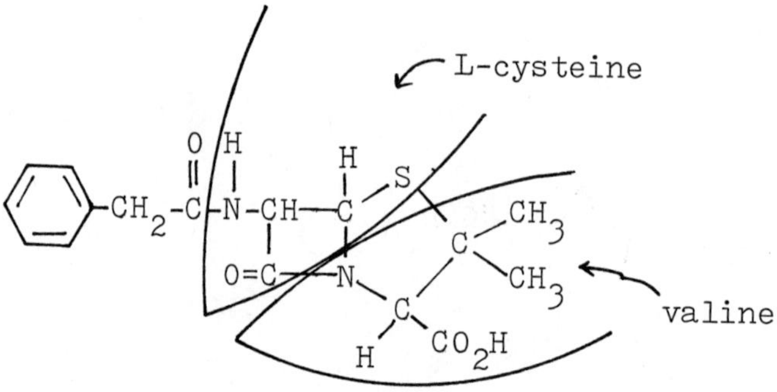
L-cysteine
valine

20. (a) No. The minimum composition of a peptide is two amino acids joined by an amide bond.

(b) $CH_2NH_2CH_2CO_2H$

(c) $(CH_3)_2CHCHO \xrightarrow[\text{base}]{HCHO} HOCH_2C(CH_3)_2CHO \xrightarrow{NaCN} \xrightarrow{H_3O^{\oplus}}$

(continued next page)

$$HOCH_2C(CH_3)_2\underset{\substack{|\\OH}}{C}HCN \xrightarrow{H_2O/OH^{\ominus}} HOCH_2C(CH_3)_2\underset{\substack{|\\OH}}{C}HCOOH \xrightarrow[(-H_2O)]{H^{\oplus}}$$

$$\text{lactone: } CH_2—C(CH_3)_2,\ O,\ CHOH,\ C{=}O \xrightarrow[\text{base}]{H_2NCH_2CH_2COOH} \text{product}$$

21. (a) A: $CH_3CHClCH_2O\overset{O}{\overset{\|}{C}}Cl$ B: $CH_3CHClCH_2O\overset{O}{\overset{\|}{C}}NH_2$

(b) the acid chloride reacts by an addition-elimination reaction. The alkyl chloride reacts with $(CH_3)_3N$ by S_N2.

(c) $CH_3\underset{\substack{|\\ \oplus N(CH_3)_3}}{C}HCH_2O\overset{O}{\overset{\|}{C}}NH_2 \quad Cl^{\ominus}$

not an amino acid

22. (a) $(CH_3)_2CHCH_2CH_2CO_2H \xrightarrow[PBr_3]{Br_2} (CH_3)_2CHCH_2\underset{\substack{|\\Br}}{C}HCO_2H$

$$\xrightarrow{\text{potassium phthalimide } (:N:^{\ominus}\ K^{\oplus})} \xrightarrow{OH^{\ominus}/H_2O} (CH_3)_2CHCH_2CH(NH_2)CO_2H$$

(b) $CH_3\text{-}\underset{\substack{|\\CH_3}}{C}H\text{-}CHO \xrightarrow[NaCN/C_2H_5OH]{(NH_4)_2CO_3/H_2O} \xrightarrow[\text{heat}]{OH^{\ominus}/H_2O} \xrightarrow{H_3O^{\oplus}}$

$CH_3\text{-}\underset{\substack{|\\CH_3}}{C}H—\underset{\substack{|\\NH_2}}{C}H\text{-}CO_2H$

(continued next page)

(c) $CH_3CH_2NO_2 \xrightarrow[OH^{\ominus}/H_2O]{H_2C=O} \underset{\displaystyle NO_2}{CH_3\overset{}{C}H}\text{-}CH_2OH \xrightarrow{CrO_3/H^{\oplus}} \underset{\displaystyle NO_2}{CH_3CH}COOH$

$\xrightarrow{H_2/Pt} \underset{\displaystyle NH_2}{CH_3CH}COOH$

(d) $CH_3\text{-}CH_2\text{-}\overset{\displaystyle H}{\underset{\displaystyle CH_3}{C}}\text{-}Br \xrightarrow[ether]{Mg} CH_3CH_2\underset{\displaystyle CH_3}{CH}\text{-}MgBr \xrightarrow{CH_2O} \xrightarrow{H_3O^{\oplus}}$

$CH_3CH_2\underset{\displaystyle CH_3}{CH}\text{-}CH_2OH \xrightarrow{CrO_3/Ac_2O} \xrightarrow{H_3O^{\oplus}} CH_3CH_2\underset{\displaystyle CH_3}{CH}CHO$

$\xrightarrow[NaCN/C_2H_5OH]{(NH_4)_2CO_3/H_2O} \xrightarrow[heat]{OH^{\ominus}/H_2O} \xrightarrow{H_3O^{\oplus}} CH_3CH_2\underset{\displaystyle CH_3}{CH}\text{—}\underset{\displaystyle NH_2}{CH}\text{-}CO_2H$

23\. (a) $HS\text{-}CH_2\underset{\displaystyle NH_2}{CH}CO_2H + (C_6H_5)_3C^{\oplus} \xrightarrow{-H^{\oplus}} (C_6H_5)_3CSCH_2\underset{\displaystyle NH_2}{CH}CO_2H$

$(C_6H_5)_3CCl \xrightarrow{-Cl^{\ominus}} (C_6H_5)_3C^{\oplus}$

(b) nitrogen is not as nucleophilic as is sulfur

(c) $(C_6H_5)_3CSCH_2\underset{\displaystyle NH_2}{CH}CO_2H \xrightarrow{AgNO_3} (C_6H_5)_3\overset{\oplus}{C}\overset{\ominus}{NO_3} + AgSCH_2\underset{\displaystyle NH_2}{CH}CO_2H$

$\xrightarrow{HCl}$ $AgCl + HSCH_2\underset{\displaystyle NH_2}{CH}CO_2H$

(as salt)

24. (a)

$$\text{dihydropyran} \xrightarrow{H^{\oplus}} \left[\text{oxocarbenium ion} \leftrightarrow \text{oxonium ion} \right] \xrightarrow{R\ddot{O}H} \text{protonated acetal} \xrightarrow{-H^{\oplus}} \text{2-alkoxytetrahydropyran (}\ddot{O}R\text{)}$$

(b) These derivatives are acetals and are stable in basic solution.

25. (a) They form resonance-stabilized cations, e.g.,

$$R\text{-}C(=\ddot{N}\text{-}H)\ddot{N}H_2 \xrightarrow{H^{\oplus}} \left[R\text{-}C(=\overset{\oplus}{N}H_2)\ddot{N}H_2 \leftrightarrow R\text{-}C(\ddot{N}H_2)=\overset{\oplus}{N}H_2 \right]$$

(b)

$$R\text{-}C(=\ddot{N}H)\ddot{N}H^{\ominus} \longleftrightarrow R\text{-}C(\ddot{N}H^{\ominus})=NH$$

(c)

$$H_3\overset{\oplus}{N}\text{-}C(H)(CO_2^{\ominus})\text{-}(CH_2)_3\text{-}NH\text{-}C(=\overset{\oplus}{N}H_2)\text{-}NH_2$$

(continued next page)

(d) Resonance structures of the nitroguanidine unit: $H_2N-C(=\overset{\oplus}{N}H-NO_2)-NH-$ ⟷ $\overset{\oplus}{H_2N}=C(-NH-NO_2)-NH-$ ⟷ $H_2N-C(-NH-NO_2)=\overset{\oplus}{N}H-$

26. (a) $C_6H_{11}-N=C=N-C_6H_{11}$ (DCC) + H_2O ⟶ $C_6H_{11}-NH-C(O)-NH-C_6H_{11}$

DCC

$C_6H_{11}-N=C=N-CH_2CH_2-N(morpholine)$ + H_2O ⟶ $C_6H_{11}-NH-C(O)-NH-CH_2CH_2-N(morpholine)$

(b) The morpholine fragment contains a basic nitrogen and this fragment will form a water-soluble salt in hydrochloric acid. The remaining nitrogens (those which are part of the urea fragment) are not sufficiently basic to form a salt.

(c) Yes. A and its mirror image are not superimposable. The carbodiimide unit has the same geometry as an allene unit.

$C_6H_{11}-N=C=N-CH_2CH_2N(morpholine)$ and its mirror image $(morpholine)NCH_2CH_2-N=C=N-C_6H_{11}$

27. $R-CH(CO_2H)-NH-C(O)-CH(CH_2CH_2SCH_3)-NHC(O)R'$ + $:N\equiv C-Br$ ⟶ $CH_3-\overset{\oplus}{S}(C(Br)=\overset{\ominus}{N}:)-CH_2CH_2-CH(\sim)-C(O)-NH\sim$ ⟶ $CH_3-\overset{\oplus}{S}(C\equiv N)-CH_2CH_2-CH(\sim)-C(O)-NH\sim$ ↓ $(-CH_3SCN)$

(continued next page)

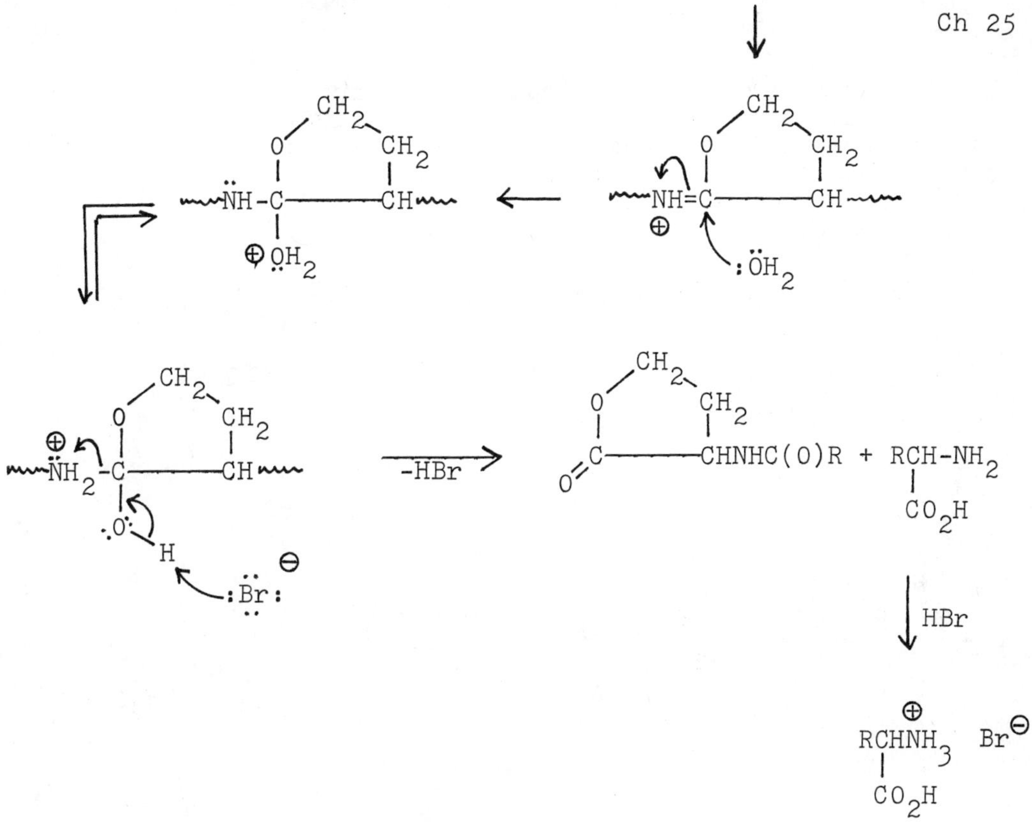

28. (a) O_2N-C$_6$H$_3$(CO$_2$H)-S-S-C$_6$H$_3$(CO$_2$H)-NO_2 + $HSCH_2$-CH(NH_2)-CO_2H $\xrightarrow{S_N2}$ O_2N-C$_6$H$_3$(CO$_2$H)-$SSCH_2CH(NH_2)CO_2H$

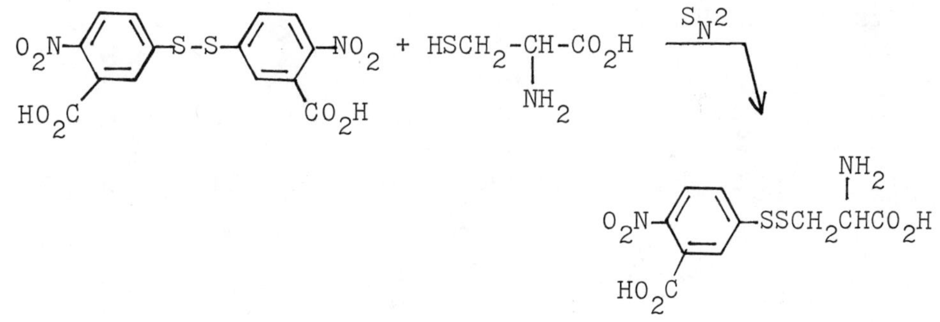

(b) determine the number of SH groups present before reduction with $LiAlH_4$ (i.e., the number of free -SH groups in the protein).

(c) carry out the sequence as outlined in the problem but delete the first stage.

29. The protein in its native state has an overall structure, complete with helical segments, which is chiral. Some of this is

(continued next page)

lost when the protein's secondary structure is weakened. However, the chiral centers still give some ord curve. This means that the native protein is the sum of the contributors due to the chiral centers and the chiral conformations.

30. (a) $CH_3C(=O)NHC(CO_2C_2H_5)_2CH_2CH_2CN \rightarrow CH_3C(=O)NHC(CO_2C_2H_5)_2CH_2CH_2CH_2NH_2$

A B

B $\downarrow$ C: cyclic lactam — ring C(=O)–NH–CH_2–CH_2–CH_2–C, with the ring carbon bearing $C_2H_5O_2C$ and CH_3-C(=O)-NH

C $\xrightarrow{HCl/H_2O,\ heat}$ D

$Cl^{\ominus}\ H_3\overset{\oplus}{N}CH(CO_2H)CH_2CH_2CH_2\overset{\oplus}{N}H_3\ Cl^{\ominus}$

D C

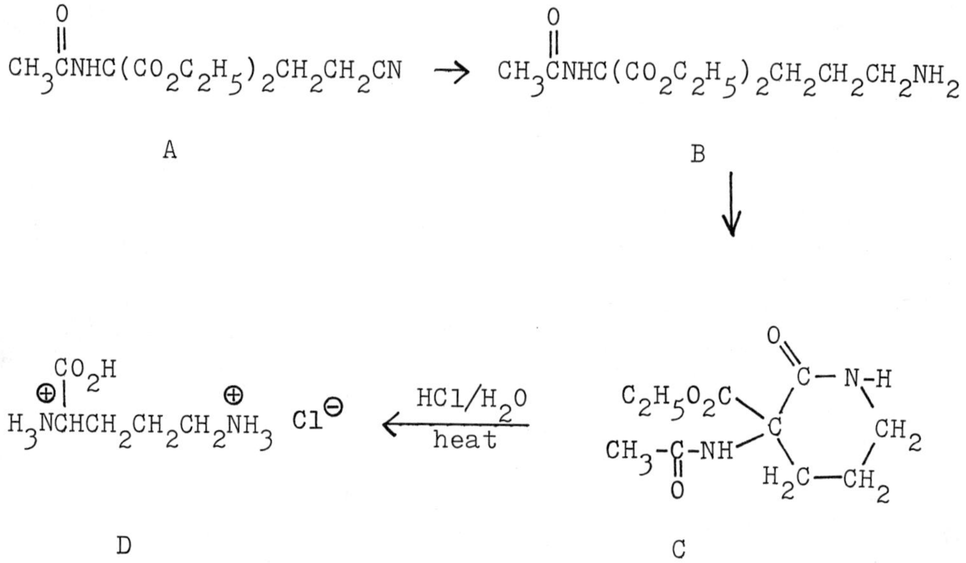

Ornithine: $H_2NCH(CO_2H)CH_2CH_2CH_2NH_2$

(b) $H_2N\text{-}CH(COO^{\ominus})CH_2CH_2CH_2\text{-}NH_2\text{:}$ $\xrightarrow{H_2N\text{-}C\equiv N\text{:}}$ $H_2N\text{-}CH(COO^{\ominus})CH_2CH_2CH_2\text{-}\overset{\oplus}{N}H_2\text{-}C(NH_2)\text{=}\ddot{N}\text{:}^{\ominus}$

ornithine in basic solution

$\downarrow$

$H_2N\text{-}CH(COO^{\ominus})CH_2CH_2CH_2\text{-}NH\text{-}C(NH_2)\text{=}NH$

arginine as anion

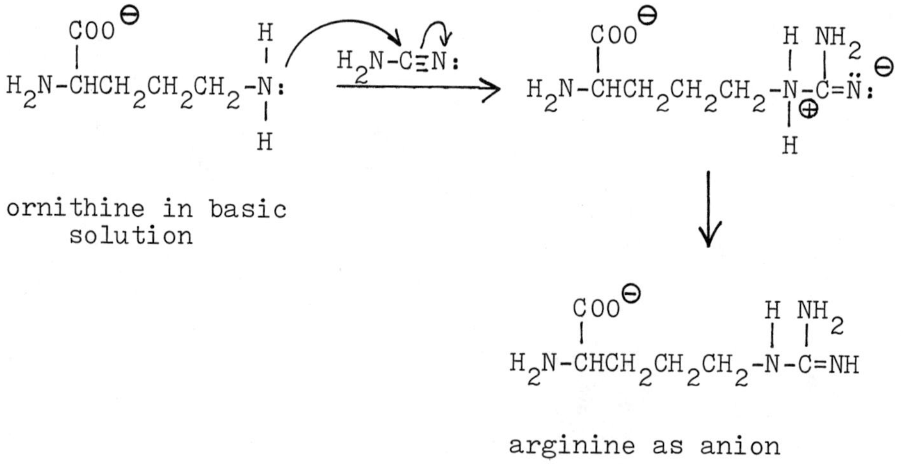

31. 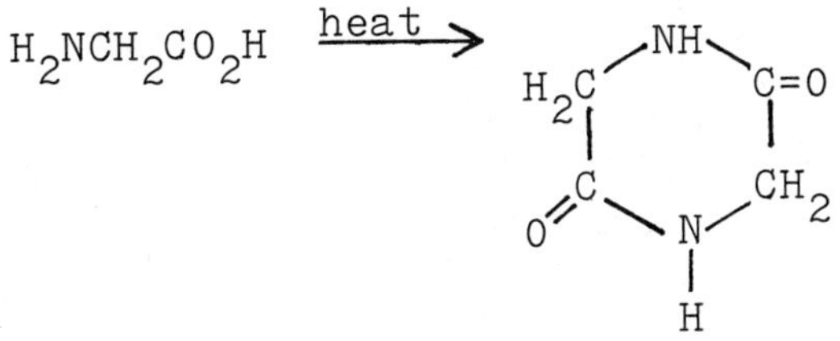 $H_2NCH_2CO_2H \xrightarrow{heat}$ [cyclic product: ring of H_2C, NH, C=O, CH_2, N–H, C=O]

$CH_3CH(NH_2)-CH_2CO_2H \xrightarrow[heat]{-NH_3} CH_3-CH=CHCO_2H$

A

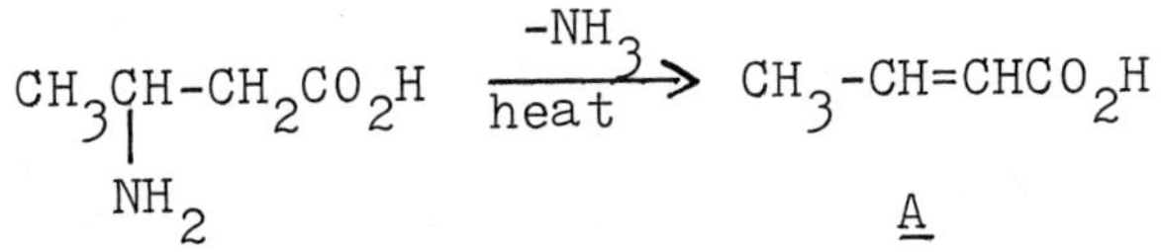

3-aminobutanoic acid

$CH_3CH(NH_2)CH_2CH_2CO_2H \longrightarrow$ [ring: CH_3-CH, CH_2, CH_2, C=O, HN] + H_2O

B

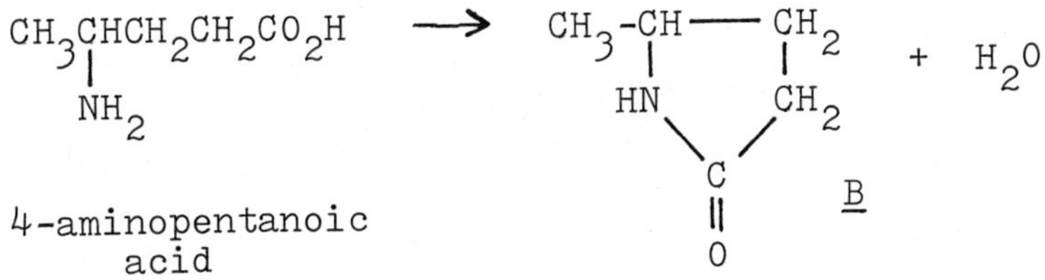

4-aminopentanoic acid

A lactam is a cyclic amide.

32. 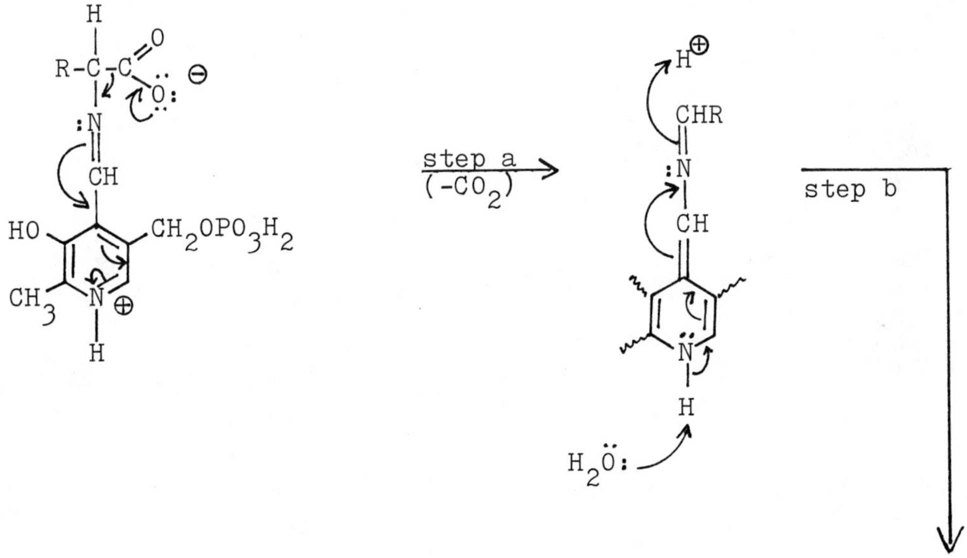 [structure: R-C(H)(-C(=O)O:⊖)-N=CH on pyridinium ring with HO, $CH_2OPO_3H_2$, CH_3, N⊕-H] $\xrightarrow[(-CO_2)]{step\ a}$ [structure: $H^{\oplus}$, CHR=N-CH= ring, N-H, $H_2\ddot{O}$:] step b

(continued next page)

$\text{RCH}_2\text{N}{=}\text{CH}$–(pyridoxal ring) $\xrightarrow[\text{step c}]{H^{\oplus}/H_2O}$ $\text{RCH}_2\text{NH}{-}\overset{\oplus}{\text{C}}\text{H}$–(ring) $\xrightarrow{H_2\ddot{O}:}$ $\text{RCH}_2\text{NH}{-}\text{CH}(\overset{\oplus}{O}H_2)$–(ring) $\rightleftharpoons$ $\text{RCH}_2\overset{\oplus}{\text{N}}\text{H}_2{-}\text{CH}(\ddot{O}H)$–(ring) $\longrightarrow$ pyridoxal phosphate (HO, CH_3, CHO, $CH_2OPO_3H_2$ on pyridine ring) + $RCH_2\ddot{N}H_2$

33. $N(CH_3)_2$ / naphthalene / $SO_2NHCH_2CO_2H$ dansyl derivative of glycine

The dansyl derivative is a sulfonamide which is not readily hydrolyzed.

34. A left-handed helix of L-amino acids is a diastereomer of the right-handed helix of the same amino acids.

35. By considering the R of $R\text{-}\underset{NH_2}{\underset{|}{CH}}\text{-}CO_2H$ to be arranged on a polymer backbone consisting of peptide linkages.

36. (a) The methionine is oxidized to a sulfoxide which is chiral. Creation of the chiral center generates two diastereomers

(continued next page)

which differ in configuration at the sulfinyl group.

(b) Since oxidation of cys might produce a disulfide (but would not create a new chiral center) only one product would be formed.

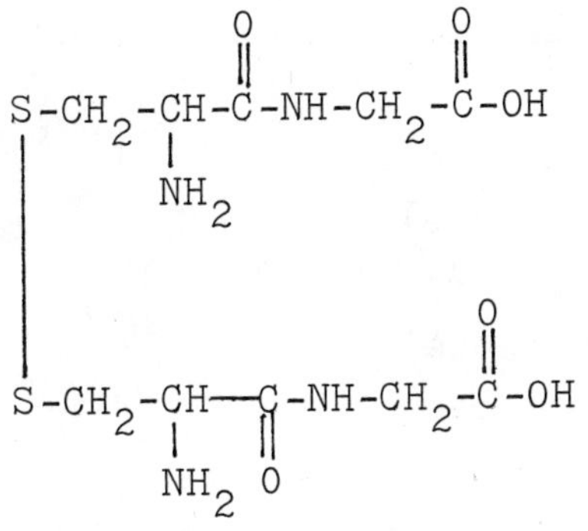

```
                O          O
                ||         ||
S-CH2-CH-C-NH-CH2-C-OH
|          |
|          NH2
|
|                              O
|                              ||
S-CH2-CH--C-NH-CH2-C-OH
           |   ||
           NH2 O
```

37. $\left[\alpha\right]_D^{25} = \frac{100\alpha}{\ell c}$; $\alpha = \frac{\ell \times c \times \left[\alpha\right]_D^{25}}{100}$

ℓ = 1 dm, c = 2

(a) (i) + 0.036°

(ii) - 0.22°

(iii) + 0.25°

(iv) - 1.72°

(b) $= \frac{2\ (+1.8)}{100} + \frac{2\ (-11.0)}{100} = -0.18^{\circ}$

(c) Zero. They are enantiomers, so the solution is racemic.

CARBOHYDRATES - MONOSACCHARIDES

ANSWERS TO QUESTIONS

1. Since it lacks a hydroxy group, formaldehyde should not be considered the simplest carbohydrate.

2. (a) aldose, hexose (aldohexose)
   (b) aldose, hexose (aldohexose)
   (c) ketose, tetrose (ketotetrose)
   (d) aldose, hexose (aldohexose)
   (e) aldose, hexose (aldohexose)
   (f) aldose, pentose (aldopentose)
   (g) ketose, pentose (ketopentose)

3.

(a)
CHO
H—OH
HO—H
H—OH
H—OH
CH_2OH

(b)
CHO
HO—H
H—OH
HO—H
HO—H
CH_2OH

(c)
HO—H
HO—H
O H—OH
HO—H
—H
CH_2OH

(d)
H—OH
HO—H
O H—OH
HO—H
—H
CH_2OH

Examine Problem 9 for a comparison with part b.

4.

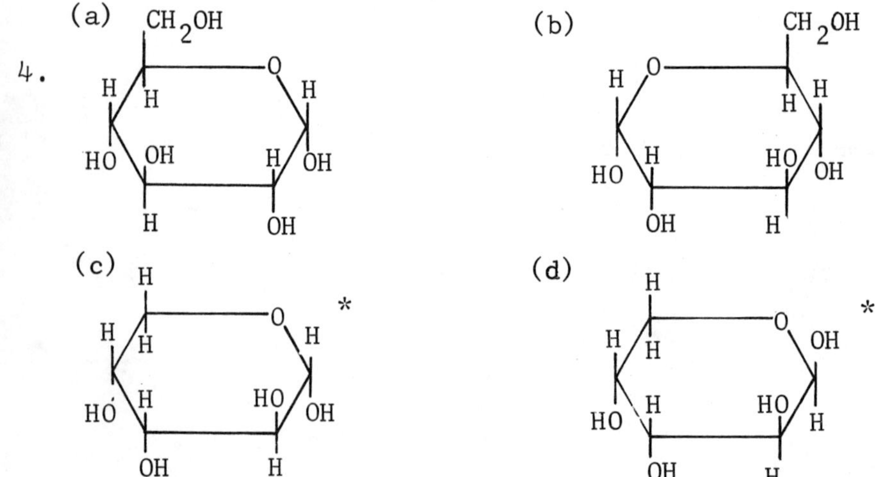

*These can be constructed easily by beginning with sequence A in Figure 26A-3 and forming the pyranose ring immediately after turning the structure 90°.

5. The α and β forms of any sugar are diastereomers, not enantiomers. The only pair of enantiomers in the set are α-L-glucopyranose and α-D-glucopyranose.

6.

α-D-glucopyranose

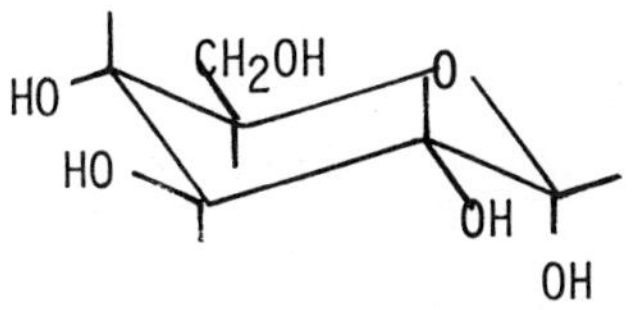

HO, CH_2OH, O, HO, OH, OH

β-D-glucopyranose

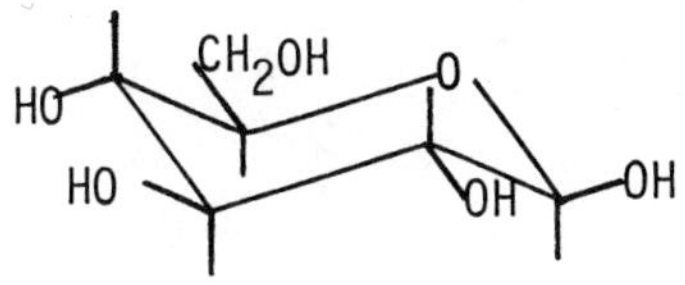

HO, CH_2OH, O, HO, OH, OH

α-L-glucopyranose

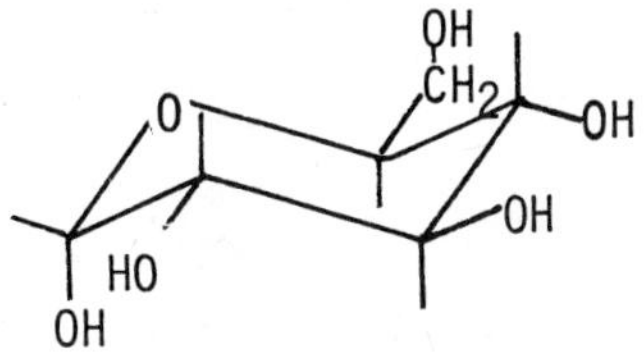

OH, CH_2, O, OH, OH, HO, OH

β-L-glucopyranose

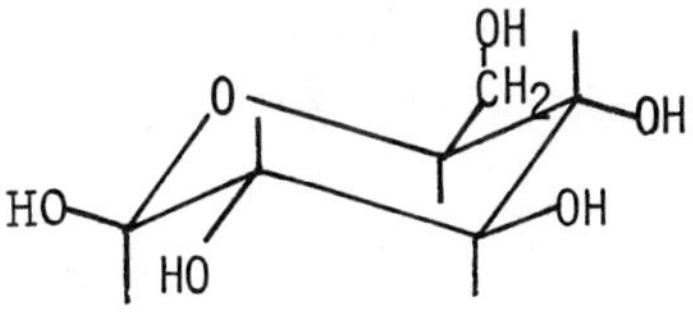

OH, CH_2, O, OH, HO, OH, HO

7. (a) They are diastereomers and of unequal energy. Since $\Delta G = RT\ln K$, K can equal 1 only when $\Delta G = 0$.

(b) For the same reason. They are diastereomers, not enantiomers.

8. The carbon at which D-idose and D-talose differ (C3) is not involved in hemiacetal formation. Mutarotation signals a change in configuration at the anomeric carbon, C1 in these compounds.

9. (a) Inversion of configuration at C5 of D-glucose produces an L-aldohexose.

   (b) Since the configuration at C5 has been changed relative to all of the other chiral centers, we have produced L-idose and not L-glucose.

10. (a)

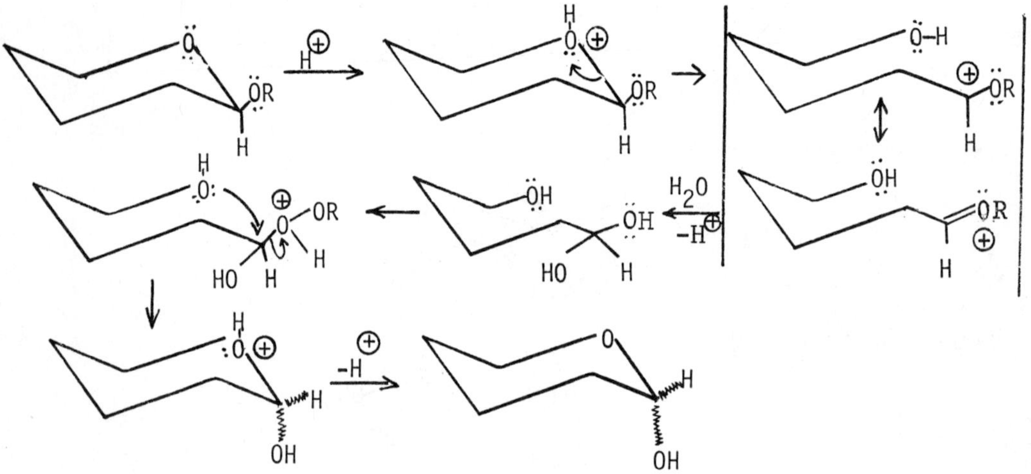

    (b) Both involve a resonance-stabilized carbocation intermediate.

11. (a) D-Mannitol and D-glucitol are epimers at C-2. Both are formed since hydride reduction occurs at both faces of the C=O of D-fructose.

    (b) The two products are diastereomers. They have different energies and the formation of each requires a different activation energy.

12. (a)

CH_2OH
H — OH
H — OH
HO — H
H — OH
CH_2OH

(b)

CH_2OH
HO — H
HO — H
H — OH
HO — H
CH_2OH

(c)

CH_2OH
H — OH
H — OH
HO — H
HO — H
CH_2OH

and

CH_2OH
HO — H
H — OH
HO — H
HO — H
CH_2OH

(d)

CH_2OH
HO — H
H — OH
HO — H
HO — H
CH_2OH

13. Reduction of D-erythrose leads to an optically <u>inactive</u> product.

14. allose and galactose

15. Yes. The glycaric acid derived from D-glucose is shown below.

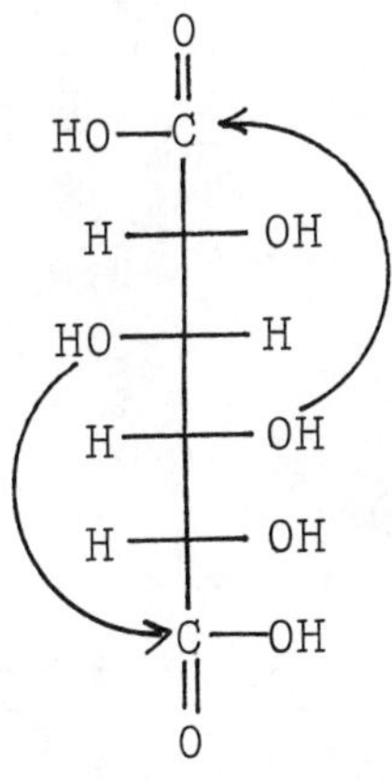

Cyclization involving the OH and C=O groups indicated by arrows give the dilactone.

16. Two (E,Z isomers)

17. Fructose is converted to glucose and mannose (aldoses) because of tautomerism.

18. (a)

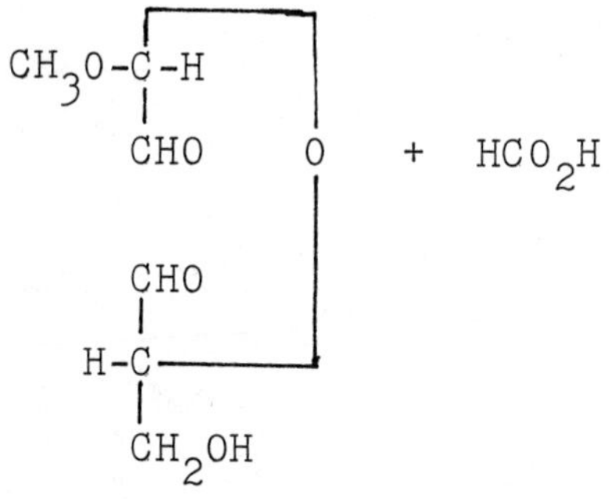

```
        ┌──────────┐
CH3O-C-H           │
     |             │
     CHO           O    +  HCO2H
                   │
     CHO           │
     |             │
   H-C─────────────┘
     |
     CH2OH
```

(b) Two

(c) Yes. It has two chiral centers, and it is derived from an optically active material by a process which does not change the configurations at the chiral centers.

19. L-glucose and L-fructose

20.

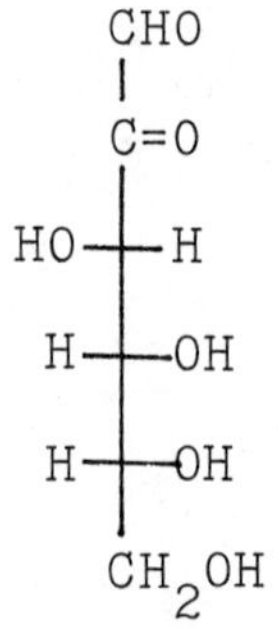

```
     CHO
     |
     C=O
     |
HO───┼──H
H────┼──OH
H────┼──OH
     |
     CH2OH
```

21.

```
        H    ↷:NH2NHC6H5                        ⊕
        |    ↓                                  NH2NHC6H5
C6H5-C——C-C6H5        1           H          |                  2
        |    ‖↷       ——→     C6H5-C——C-C6H5         ——→
        OH   :O:                    |          |                  ↓ H⊕
                                    OH        :O:
                                               ⊖
```

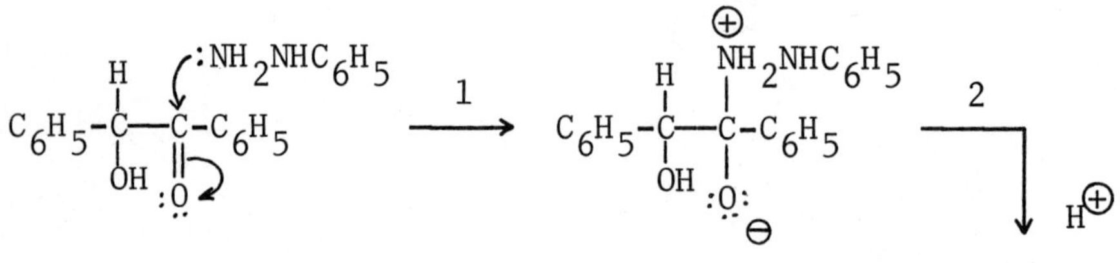

(continued next page)

C_6H_5-C(H)(OH)-C(C_6H_5)($\overset{\oplus}{N}H_2NHC_6H_5$)(OH) $\overset{3}{\rightleftharpoons}$ C_6H_5-C(H)(OH)-C(C_6H_5)(NH-NHC$_6$H$_5$)($\overset{\oplus}{O}H_2$) → C_6H_5-C(H)(OH)-C(C_6H_5)=N-NHC$_6$H$_5$

→ C_6H_5-C(OH)=C(C_6H_5)-NH-NHC$_6$H$_5$ ⇌ C_6H_5-C(=O)-CH(C_6H_5)-NHNHC$_6$H$_5$ →

C_6H_5-C($\overset{\oplus}{N}H_2NHC_6H_5$)($O^{\ominus}$)-CH($C_6H_5$)-NH-NHC$_6H_5$ → (repeat step 2) → (repeat step 3) → C_6H_5-C(=N-NHC$_6$H$_5$)-CH(C_6H_5)-NH-NHC$_6$H$_5$

↓ $H^{\oplus}$

C_6H_5-C(=N-NHC$_6$H$_5$)-CH(C_6H_5)-NH-$\overset{\oplus}{N}H_2C_6H_5$ (with $C_6H_5NHNH_2$) → C_6H_5-C(=N-NHC$_6$H$_5$)-C(C_6H_5)=NH (with :$NH_2NHC_6H_5$)

→ C_6H_5-C(=N-NHC$_6$H$_5$)-C(C_6H_5)($NH^{\ominus}$)($\overset{\oplus}{N}H_2NHC_6H_5$) $\xrightarrow{H^{\oplus}}$ C_6H_5-C(=N-NHC$_6$H$_5$)-C(C_6H_5)($\overset{\oplus}{N}H_3$)(NH-NHC$_6$H$_5$) (with $C_6H_5NHNH_2$)

↓ $-NH_3$, $-H^{\oplus}$

C_6H_5-C(=N-NHC$_6$H$_5$)-C(=N-NHC$_6$H$_5$)-C_6H_5

the osazone of benzoin

22. Tautomerism plays a critical role in both. In osazone formation the mechanism requires both an imine-enamine isomerization and a keto-enol isomerization. Only keto-enol isomerization is required in the Lobry de Bruyn-Alberda van Ekenstein rearrangement.

23. (a) D-mannose

(b) D-erythrose

(c) L-erythrose

24. (a) $\xrightarrow[H_2O]{OH^{\ominus}}$ (D-glucose and D-fructose also formed)

(b) $\xrightarrow{H_2NOH}$ $\xrightarrow{Ac_2O}$ $\xrightarrow[CH_3OH]{CH_3O^{\ominus}}$ $\xrightarrow{-HCN}$

(c) $\xrightarrow{CH_3OH/H^{\oplus}}$ (α also formed)

(d). $\xrightarrow[H_2O/heat]{HI/P}$

(e) $\frac{\text{Ruff degradation}}{\text{(Figure 26A-8)}}$ D-arabinose $\xrightarrow{repeat}$ D-erythrose

$\xrightarrow[heat]{HNO_3}$

(f) $\xrightarrow[heat]{Ac_2O/HOAc}$

25. Because of the hydrogen bonding possible between water molecules and the large number of OH groups in sugars.

26. Let x = fraction of α-D-mannose at equilibrium; 1-x is the fraction of β-D-mannose.

$$x(29.3) + (1-x)(-17.0) = 14.2$$

$$29.3x-17 + 17x = 14.2$$

$$x = .67$$

$$1-x = .33$$

(continued next page)

Thus at equilibrium the mixture contains 67% α and 33% β -D-mannose.

27. (a) 32 (extend Figure 26A-1 one more level for D and L forms)

(b) 32 (open chain form); 64 (α and β-pyranose forms)

28.

(a) 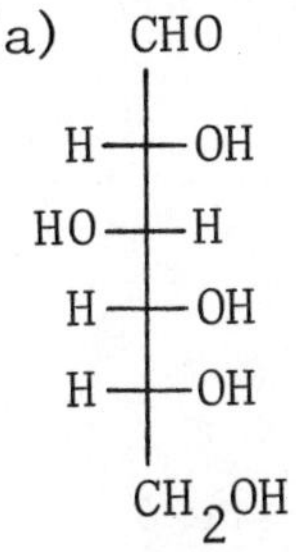

D-glucose

(b) 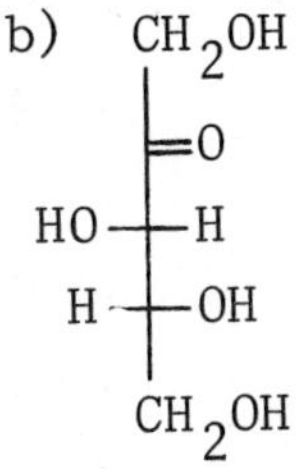

D-xylulose

(c) 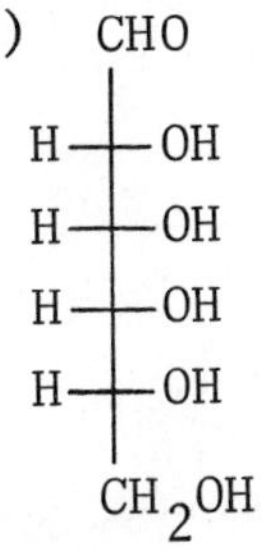

D-allose

(d) 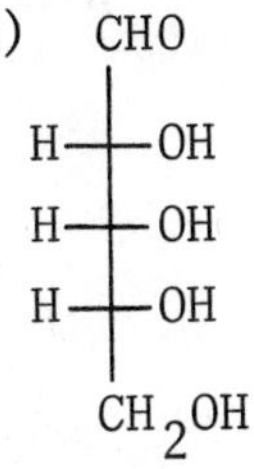

D-ribose

(e) 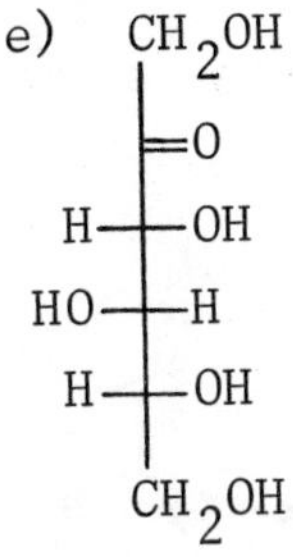

D-sorbose

(f) 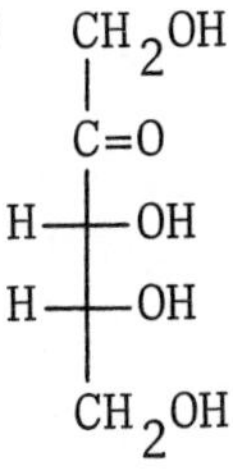

D-ribulose

(g) 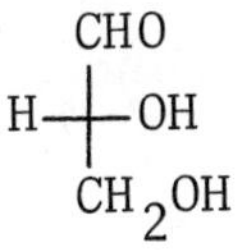

D-glyceraldehyde

(h) 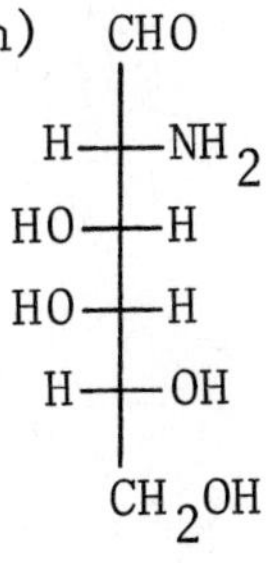

D-galactosamine

(i) 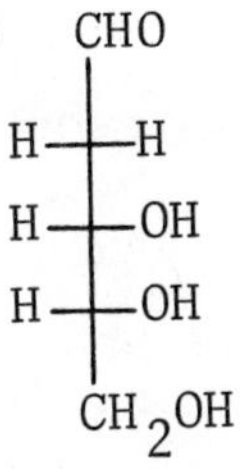

2-deoxy-D-ribose

29.

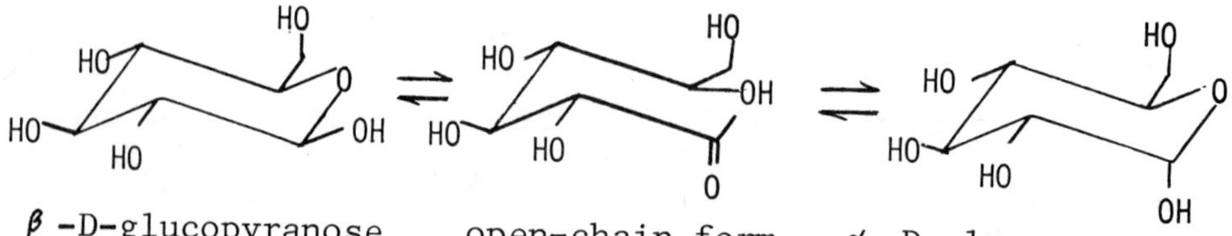

β -D-glucopyranose   open-chain form   α -D-glucopyranose

(Note that C-H fragments have been deleted)
The β form is more stable since all substituents are equatorial.

30. (a)

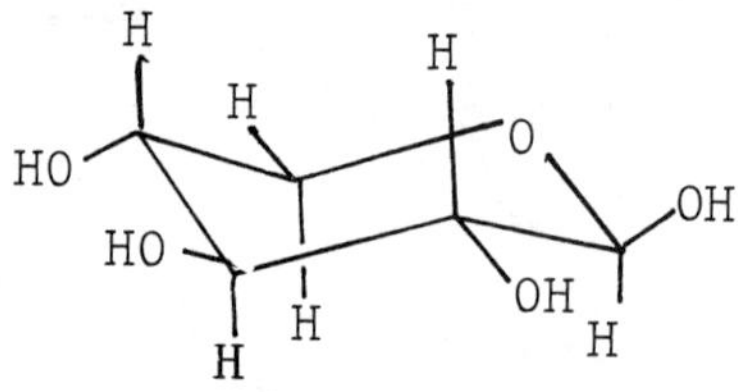

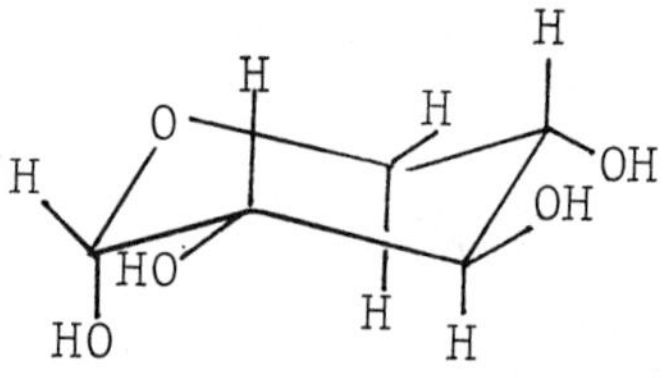

β -D-xylopyranose   α -D-xylopyranose

(b)

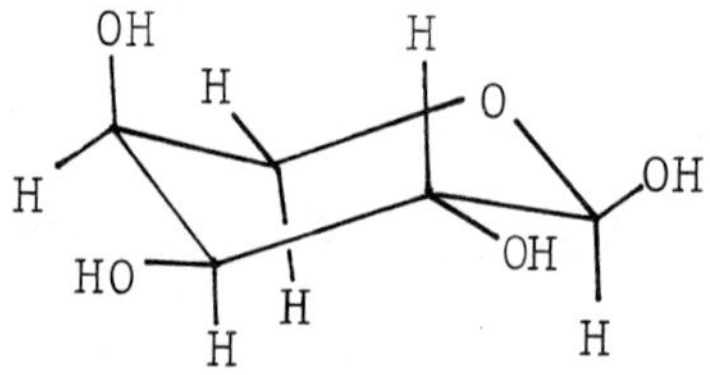

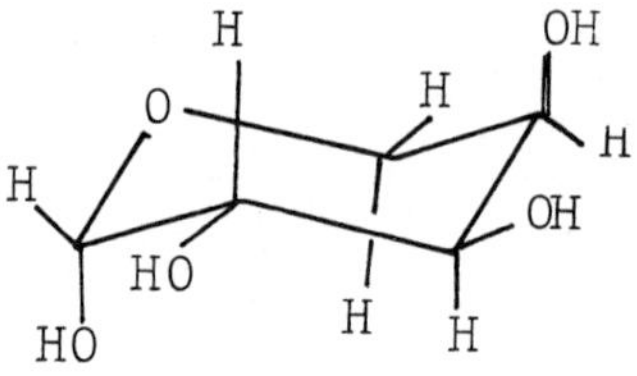

β-L-xylopyranose   α -D-xylopyranose

31. (a) Mutarotation is the experimentally determined change in optical rotation brought about by epimerization. Epimerization reflects the change in configuration at one chiral center in a molecule which contains more than one such center. Anomerization is a specific type of epimerization at the hemiacetal (hemiketal) carbon.

(b) If the molecule contains many chiral centers, a configurational change at one may not be detectable.

(continued next page)

(c) The pyranoside, an acetal, will not spontaneously open to the free carbonyl compound.

32.

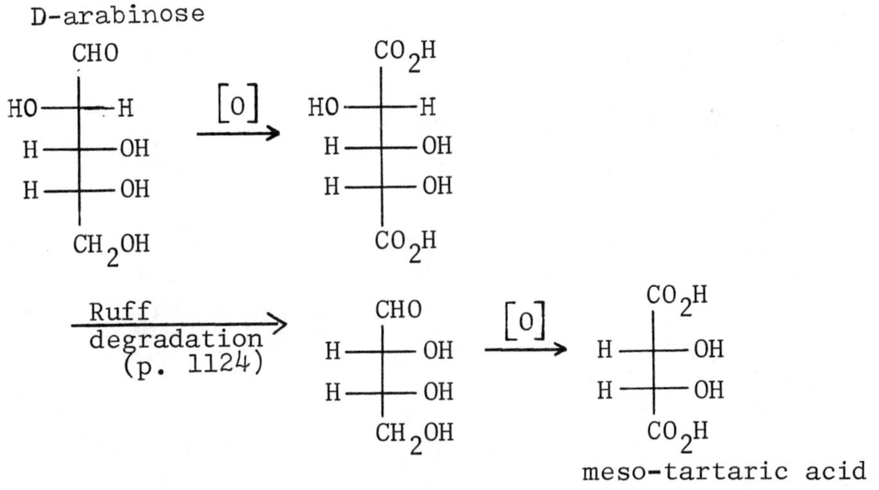

D-arabinose

CHO		CO2H
HO—H	[O] →	HO—H
H—OH		H—OH
H—OH		H—OH
CH_2OH		CO_2H

Ruff degradation (p. 1124) → CHO; H—OH; H—OH; CH_2OH [O] → CO_2H; H—OH; H—OH; CO_2H

meso-tartaric acid

33. L-gulose, L-idose and L-xylose

34. D-galactose, D-talose and D-lyxose

35.

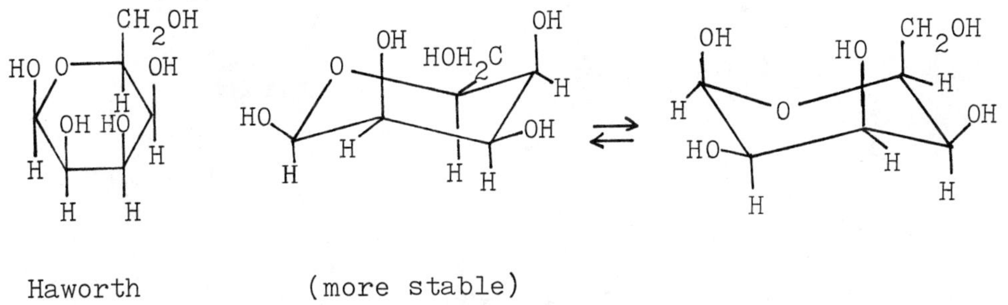

Haworth

(more stable)

36. (a) The aldehyde group of D-lyxose becomes a new chiral center. The epimers D-talose and D-galactose are formed.

(continued next page)

CHO / HO—H / HO—H / H—OH / CH_2OH —Kiliani-Fischer→ CHO / HO-C-H + CHO / H-C-OH

(b) Since talose and galactose are diastereomers and are produced at different rates through diastereomeric activated complexes, they are produced in unequal amounts.

(c) L-galactose has the opposite configuration from D-galactose at all carbons and is not formed.

37.

(a) CHO / H-C-OH (R) / H-C-OH (R) / H-C-OH (R) / CH_2OH
D-ribose

CHO / HO-C-H (S) / H-C-OH (R) / H-C-OH (R) / CH_2OH
D-arabinose

(b) D-ribose (see a)

CHO / HO-C-H (S) / HO-C-H (S) / HO-C-H (S) / CH_2OH
L-ribose

(c) CHO / H-C-OH (R) / HO-C-H (S) / H-C-OH (R) / H-C-OH (R) / CH_2OH
D-glucose

CHO / HO-C-H (S) / HO-C-H (S) / H-C-OH (R) / H-C-OH (R) / CH_2OH
D-mannose

38. Erythrose has the erythro configuration; threose has the threo configuration.

39. (a) see p. 1124.

(b) Carried to the tetrose stage, D-glucose gives D-erythrose and D-galactose gives D-threose. Sodium borohydride reduction of each produces a tetraol, and that from erythrose will be optically active.

(c) no

(d) no

40. A:

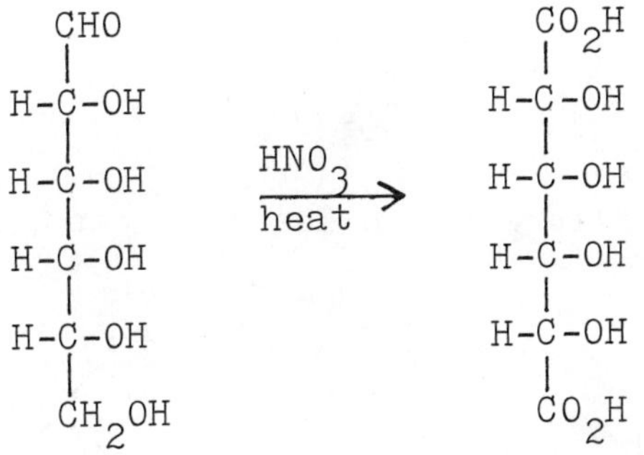

CHO
H-C-OH
H-C-OH
H-C-OH
H-C-OH
CH_2OH

D-allose

$\xrightarrow[\text{heat}]{HNO_3}$

CO_2H
H-C-OH
H-C-OH
H-C-OH
H-C-OH
CO_2H

B

41. (a) CH_2OH ring: O, =O, H, H, HO, OH, H, H, OH

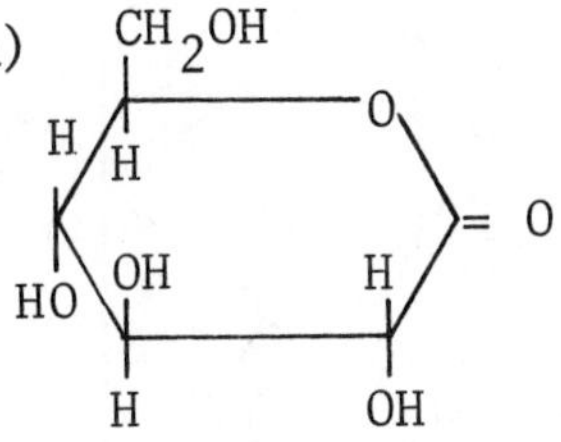

(b) $CH=N-NHC_6H_5$
$CH=N-NHC_6H_5$
HO-C-H
H-C-OH
H-C-OH
CH_2OH

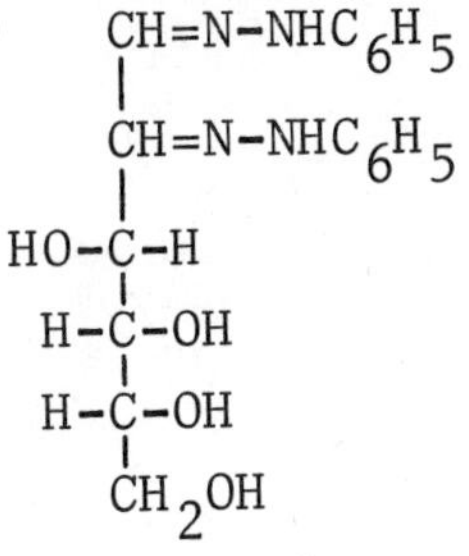

(c) CO_2H
H-C-OH
HO-C-H
H-C-OH
H-C-OH
COOH

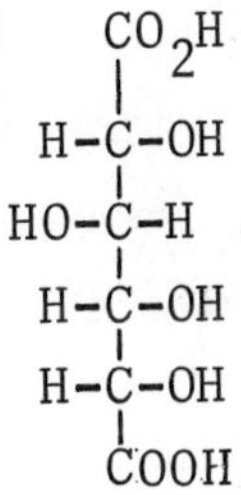

(d) H-C-OAc
H-C-OAc
AcO-C-H
H-C-OAc
H-C-O- (ring to top carbon)
CH_2OAc

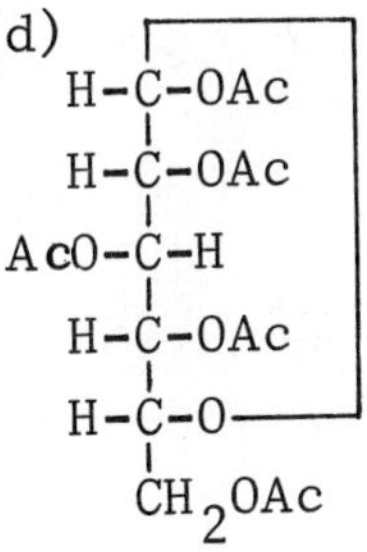

(e) CH_2OH
H-C-OH
HO-C-H
H-C-OH
H-C-OH
CH_2OH

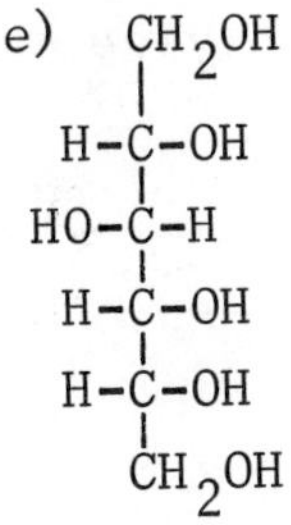

(f) $CH_2O + 5\,HCO_2H$

(g) n-hexane

(h) HC=NOH
H-C-OH
HO-C-H
H-C-OH
H-C-OH
CH_2OH

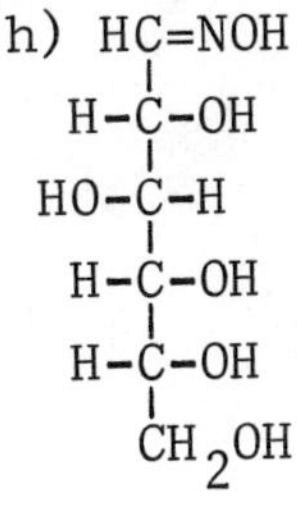

(i) $H-C-OCH_3$
$H-C-OCH_3$
CH_3O-C-H
$H-C-OCH_3$
H-C—O— (ring to top carbon)
CH_2OCH_3

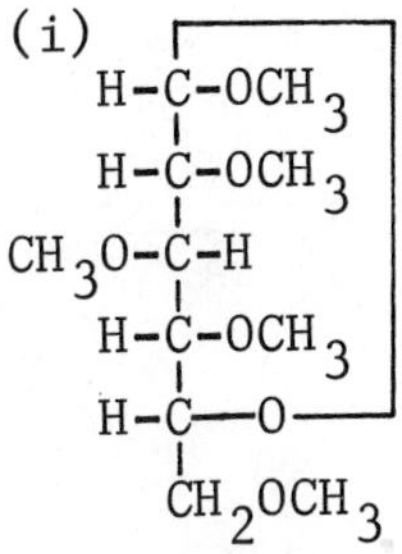

(continued next page)

(j) H-C-OC_2H_5 / H-C-OH / HO-C-H / H-C-OH / H-C-O / CH_2OH (ring)

(k) H-C-$O_2CC_6H_5$ / H-C-$O_2CC_6H_5$ / $C_6H_5C(=O)$-O-C-H / H-C-$O_2CC_6H_5$ / H-C / $CH_2O_2CC_6H_5$ (ring O)

(l) H-C-OCH_3 / CHO / O / CHO / H-C / CH_2OH

(+ HCO_2H)

(m) D-arabinose

42. They are both free radical reactions which lead to loss of CO_2.

43.

acid protons

$\xrightarrow{-H^{\oplus}}$

etc.

44.

Either D-glyceraldehyde or dihydroxyacetone in base produces a mixture of the two.

$$\begin{array}{c}CHO\\ H-C-OH\\ CH_2OH\end{array} \underset{H_2O}{\overset{B^{\ominus}}{\rightleftharpoons}} \left[\begin{array}{c}CHO\\ {}^{\ominus}:C-OH\\ CH_2OH\end{array} \longleftrightarrow \begin{array}{c}H-C-O^{\ominus}\\ \| \\ C-OH\\ CH_2OH\end{array}\right] \rightleftharpoons \begin{array}{c}CH_2-O^{\ominus}\\ C=O\\ CH_2OH\end{array} \underset{B^{\ominus}}{\overset{H_2O}{\rightleftharpoons}} \begin{array}{c}CH_2-OH\\ C=O\\ CH_2OH\end{array}$$

(continued next page)

Condensation between one molecule of each gives D-fructose and D-sorbose (other isomers are also possible).

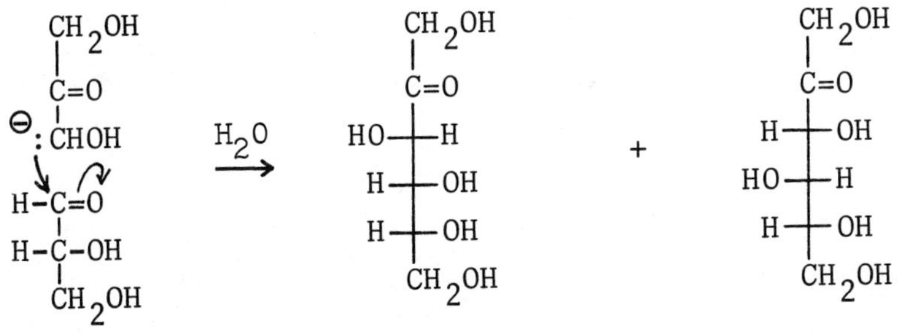

CH_2OH–C=O–$^{\ominus}$:CHOH + H–C=O / H–C–OH / CH_2OH $\xrightarrow{H_2O}$ D-fructose + D-sorbose

D-fructose

D-sorbose

D,L-Dendroketose is formed by an aldol condensation between two molecules of dihydroxyacetone.

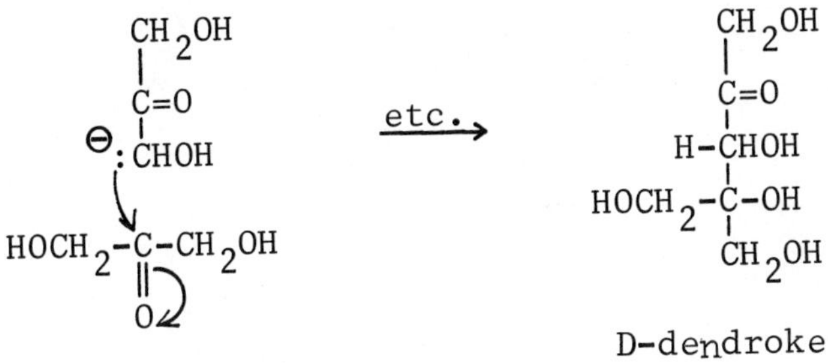

CH_2OH–C=O–$^{\ominus}$:CHOH + $HOCH_2$–C(=O)–CH_2OH $\xrightarrow{\text{etc.}}$ CH_2OH–C=O–H–CHOH–$HOCH_2$–C–OH–CH_2OH

D-dendroketose

If we assume that racemization of D- and L-glyceraldehyde occurs rapidly (via the carbanion), then either gives the same products.

45\.

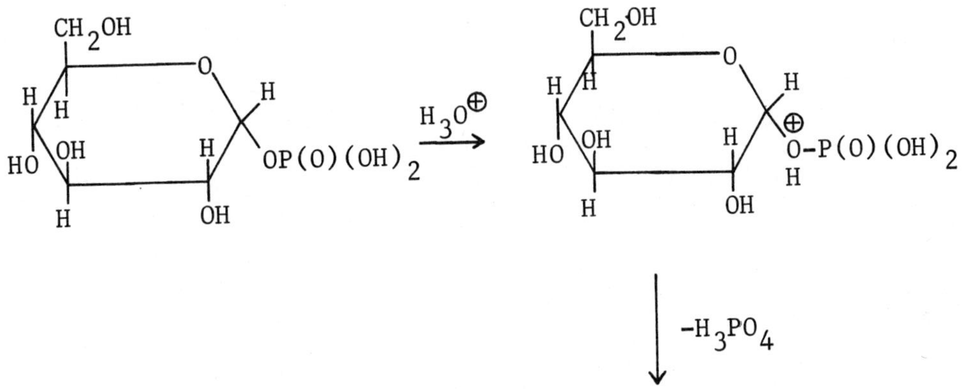

$\xrightarrow{H_3O^{\oplus}}$ ... $\overset{\oplus}{O}$–P(O)(OH)$_2$ ... OP(O)(OH)$_2$

$\downarrow -H_3PO_4$

(continued next page)

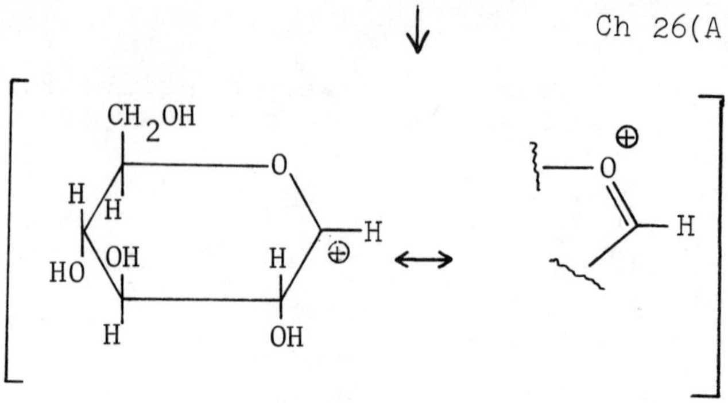

Hydrolysis which cleaves the C-O bond produces a resonance-stabilized cation.

CHAPTER 26 (B)
OLIGOSACCHARIDES AND POLYSACCHARIDES

ANSWERS TO QUESTIONS

1. (a) α-form.

(b)

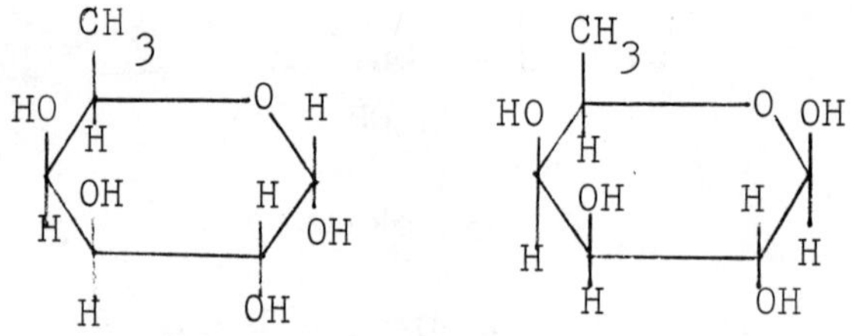

α-D-fucose $\qquad$ β-D-fucose

(c)

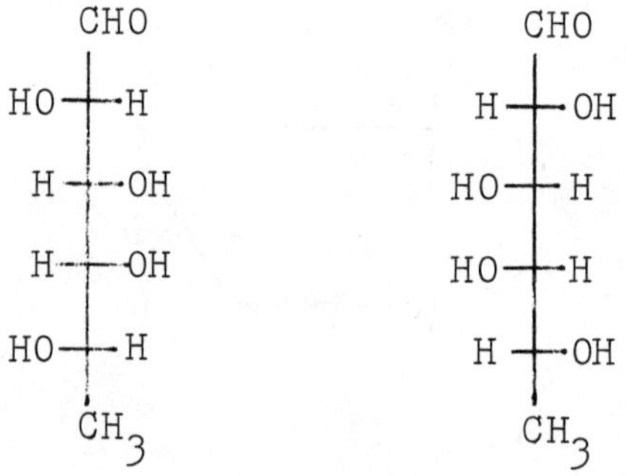

L-fucose $\qquad$ D-fucose

(continued next page)

2. Mutarotation which gives an equilibrium mixture of α- and β-forms.

3. The reaction occurs in solution where the acyclic form is present.

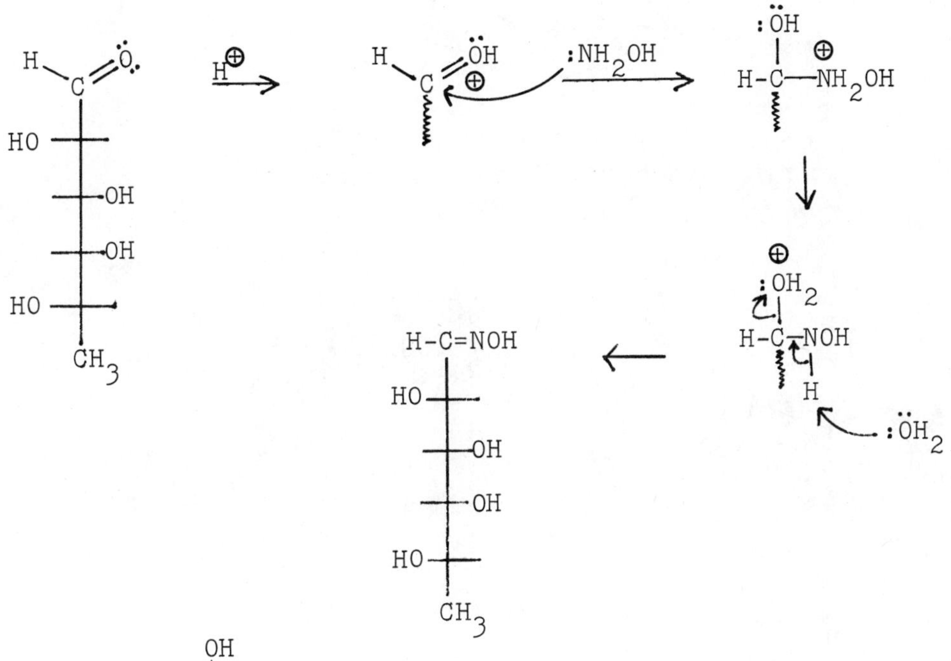

H–C=O (HO, OH, OH, HO, CH_3) $\xrightarrow{H^{\oplus}}$ H–C=$\ddot{O}H^{\oplus}$ $\xrightarrow{:NH_2OH}$ H–C(:$\ddot{O}$H)–$NH_2^{\oplus}OH$

↓

H–C=NOH (HO, OH, OH, HO, CH_3) ⟵ H–C($:OH_2^{\oplus}$)–NOH(H) ← $:\ddot{O}H_2$

4.

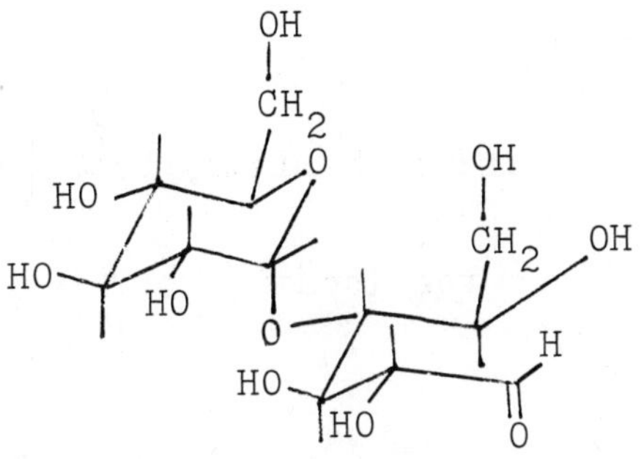

5. O-α-D-glucopyranosyl-(1,4)-α-D-glucopyranose

6.

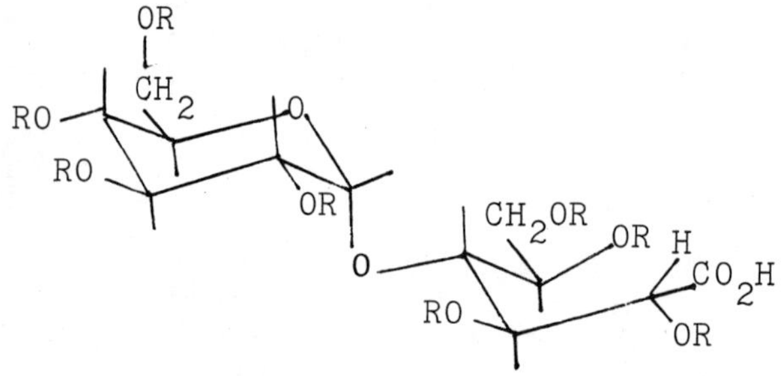

A: R = H

B: $R = CH_3$

(Note that ring hydrogens have been deleted)

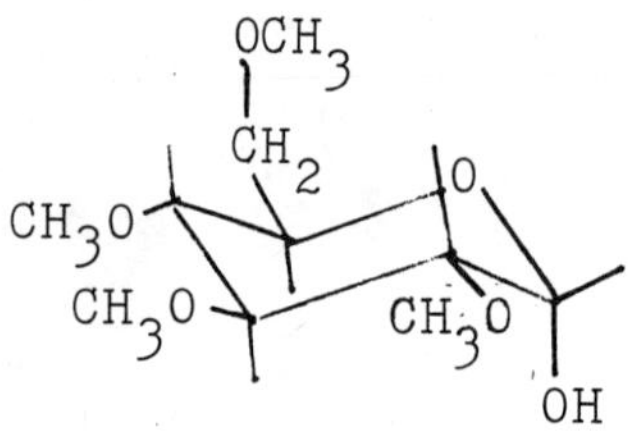

C

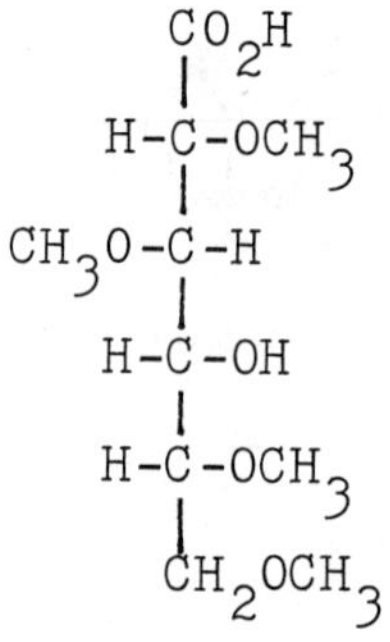

D

(i) the α anomer of C establishes the stereochemistry between both rings;

(ii) the free -OH at C4 of D establishes the location of ring attachment.

7.

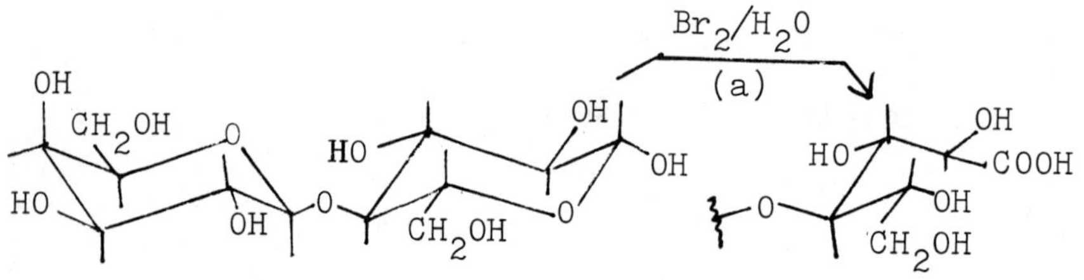

Br_2/H_2O (a)

lactobionic acid

(b) hydrolysis

D(+)-galactose + D(-)gluconic acid

8. (a) D-(+)-glucose and D-(+)-galactose

(b) The reaction is hydrolysis of an acetal.

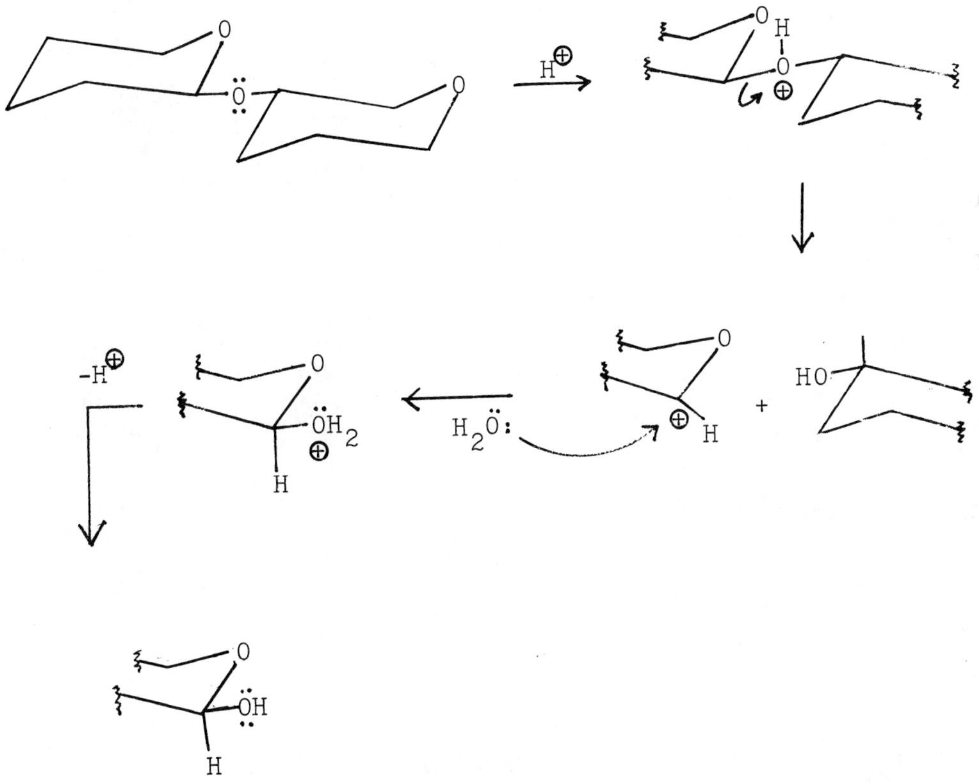

9. Yes. Lactose undergoes mutarotation in solution producing the acyclic form which contains a carbonyl group.

10. Yes. α-Lactose also contains the β-galactoside linkage (α- and β-lactose differ only in the orientation of the free OH at the anomeric center of the glucose ring).

11. Yes. Maltose contains an α-glucoside linkage and cellobiose contains a β-glucoside linkage.

12. 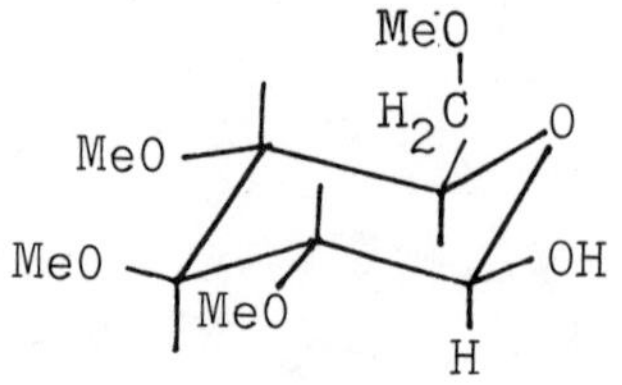 [chair structure: pyranose ring with MeO, MeO, MeO, H_2C–OMe, O, OH, H] and

CO_2H
H–C–OCH_3
CH_3O–C–H
H–C–OH
H–C–OCH_3
CH_2OCH_3

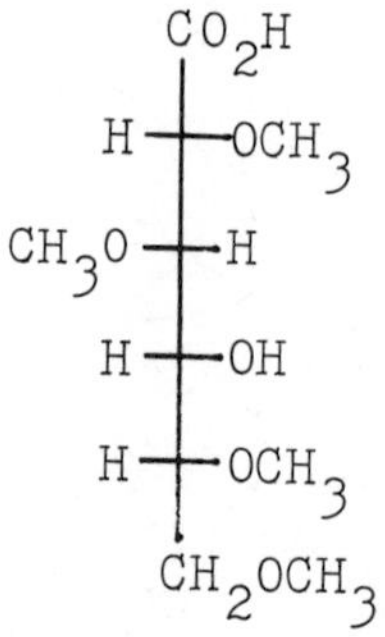

13. -19.9°. Invert sugar, the major constituent of honey, is about as sweet as sucrose. Because of its greater tendency to not crystallize, invert sugar is preferred over sucrose in the preparation of some syrups and confections. "Inversion" in bees is accomplished using the enzyme invertase.

14. Yes. It contains the hemiacetal structure in one ring, i.e.,

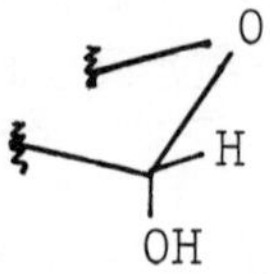

[structure: O, H, OH] and can undergo ring opening in solution.

15. (a) Reduce with HI/P. Arabinose gives pentane; glucose gives hexane (Ch 26A). Or, treat with acid. Arabinose gives furfural, glucose gives 5-hydroxymethyl furfural (Ch 26A).

(b) Ribose is oxidized by HNO_3 to an inactive dicarboxylic acid.

(c) A Kiliani synthesis using glucose gives epimers. When oxidized with HNO_3, one gives an active dicarboxylic acid, the other gives an inactive dicarboxylic acid. A Kiliani synthesis using mannose gives epimers. When oxidized with HNO_3 both give active dicarboxylic acids.

(d) Determine the molecular weights (freezing point depression method, see Chapter 3). The osazone of maltose is soluble in hot water, that from glucose is not. Oxidize with

(continued next page)

Br_2/H_2O to the glyconic acids and determine the equivalent weights.

(e) Cellobiose will reduce Tollen's and Fehling's solutions.

(f) Treat with β-galactosidase which calalyzes only the hydrolysis of lactose.

16. (a) An α-D-glucopyranose unit (six-membered ring with OH on C-1 down in Haworth structure) is bonded through an oxygen ("O") to a β-D-fructofuranoside unit (five-membered ring structure of the acetal of β-D-fructofuranose with OH at C-2 up). C-1 of the first ring is linked to C-2 of the second ("1,2").

(b) A β-D-glucopyranose unit (six-membered ring with OH on C-1 up in Haworth structure) is bonded through an oxygen ("O") to an α-D-glucopyranose (same as first unit except OH on C-1 down). C-1 of the first ring is linked to C-4 of the second ("1,4").

(c) A D-glucose structure in which H-C-OH at C-2 is replaced by H-C-NHC(O)CH_3. Acetamido group is -NHC(O)CH_3. 2-Deoxy sugars have a -CH_2- in place of -CHOH- at C-2.

17. (a) anomers

(b) A glucoside is a specific example of a glycoside.

(c) Cellobiose has a glycosidic bond between C-1 and C-4 of two simple sugars; gentiobiose has a glycosidic bond between C-1 and C-6 of two simple sugars.

(d) none

(e) A glyconic acid is a monocarboxylic acid resulting from oxidation of the aldehyde function of an aldose; a glycaric acid is a dicarboxylic acid from oxidation of aldoses and ketoses.

(f) Reducing sugars contain a hemiacetal linkage (reduce Tollen's, Fehling's or Benedict's solutions); nonreducing sugars do not.

(g) Mannose is an aldohexose. Cellobiose is a disaccharide consisting of two aldohexoses (glucose) units.

(continued next page)

(h) Epimers are two stereoisomers of a structure containing several chiral centers that differ in configuration at only one chiral center. Anomers are specific examples of epimers which differ in configuration only at the hemiacetal or hemiketal carbon.

(i) Both are acetals. A furanoside has a five-membered ring; a pyranoside a six-membered ring.

(j) N-glycoside-the atom at C-1 is nitrogen; O-glycoside-the atom at C-1 is oxygen.

18. (a) see p. 1143 for cellulose structure

(b) hydrolysis of cellulose to monosaccharide (glucose) units

19. Since D-raffinose is nonreducing, all anomeric carbons are involved in glycosidic linkages. Possible structures are:

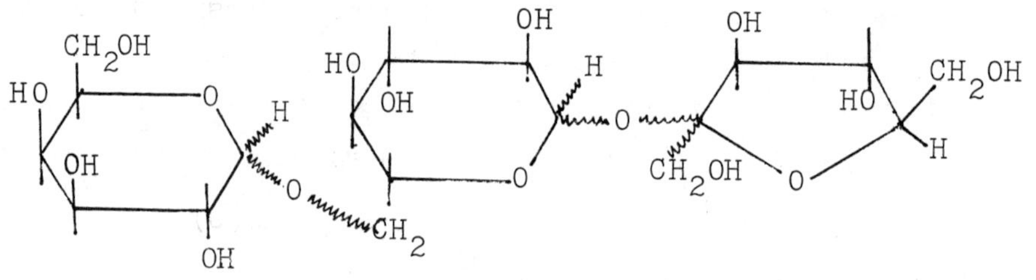

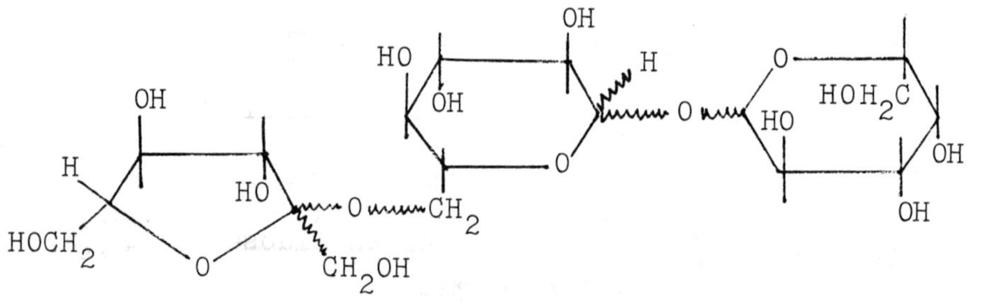

The stereochemistry of the glycosidic linkages and the arrangement of monosaccharide units must be established.

20.

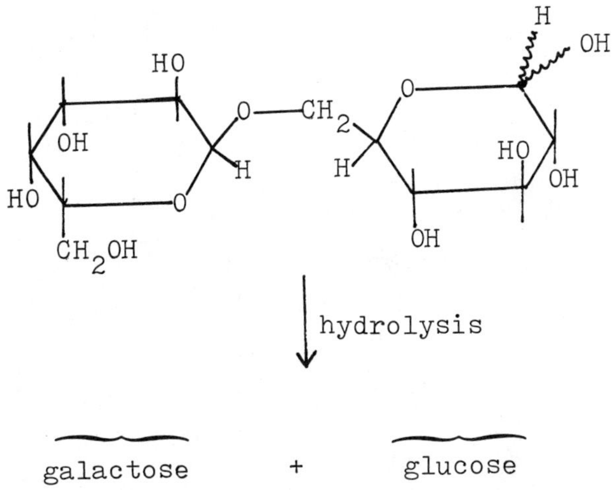

hydrolysis

galactose + glucose

21. In the mechanism of mutarotation, a proton is removed by a base from the anomeric hydroxyl group and a proton is donated to the ring oxygen. 2-Hydroxypyridine is an example of a bifunctional catalyst which can accomplish both of these steps.

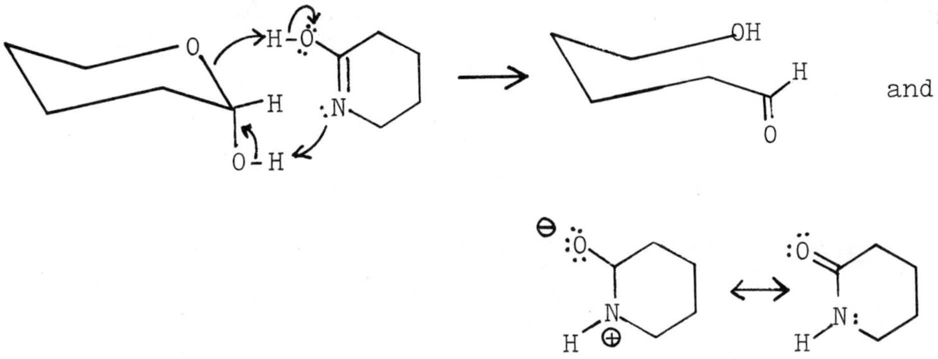

and

22. (a) (i) indicate an α-configuration at the glucoside portion of the structure

(ii) contains two D-glucose units

(iii) neither glucose portion has a hemiacetal group

(iv) shows that the glucose portions are pyranosides

(continued next page)

The structure consistent with these data is:

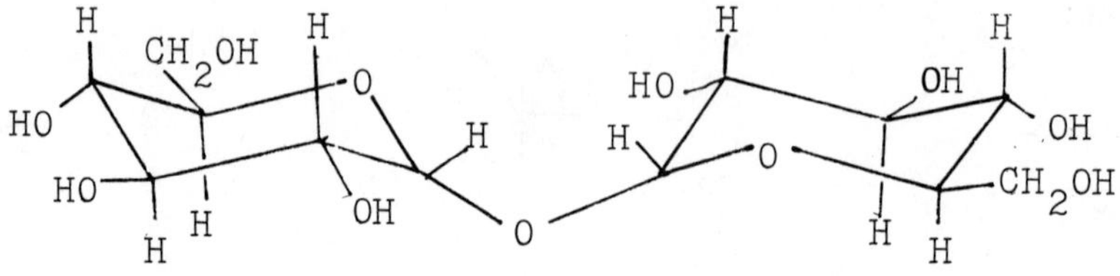

(b) No. It is an acetal, but a hemiacetal is necessary for mutarotation.

23.

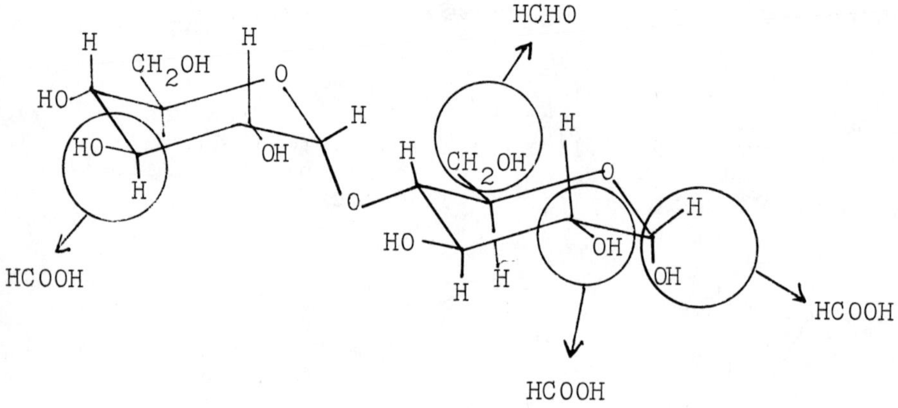

The large organic fragment is OHCCHOCHOCHCHO (with CHO on the central carbon, and CH_2OH and CHO substituents)

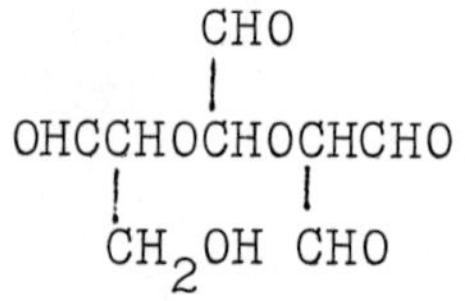

24. amylose $\xrightarrow{HIO_4}$

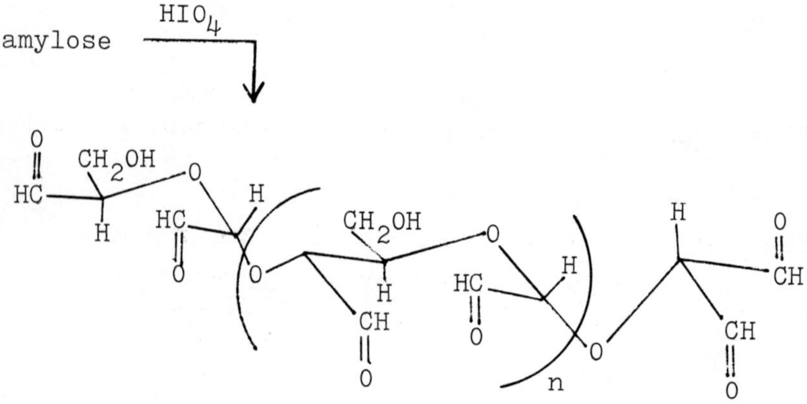

A

(continued next page)

B: all aldehyde groups are oxidized to acid groups, -CHO ⟶ -COOH

C: n molecules of HOOCCH(OH)-CH(OH)CH$_2$OH

D: n molecules of HC(=O)COOH

25.

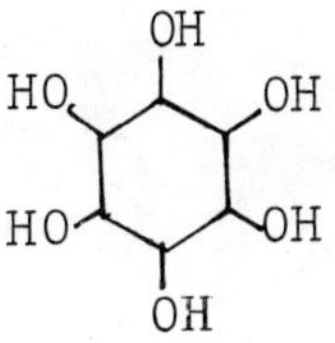

inactive 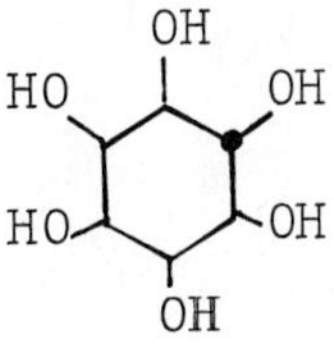 inactive 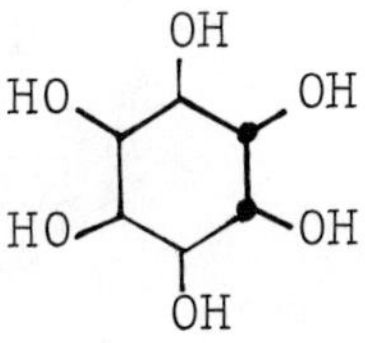 inactive 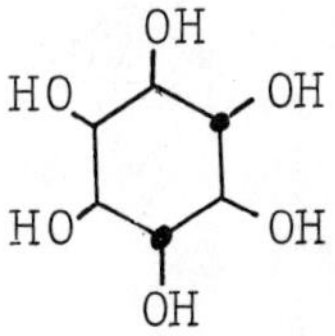 inactive

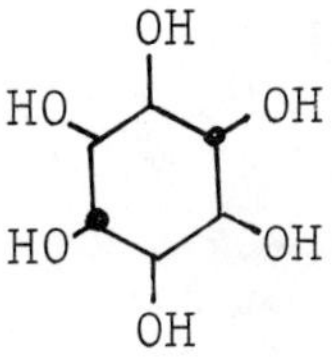

inactive 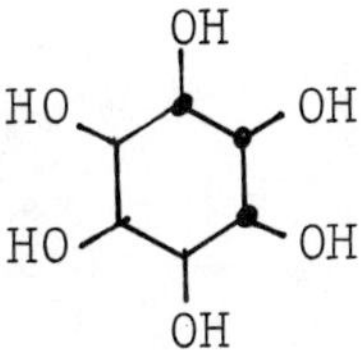 inactive 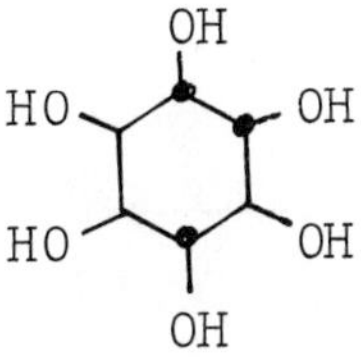 (d,ℓ) 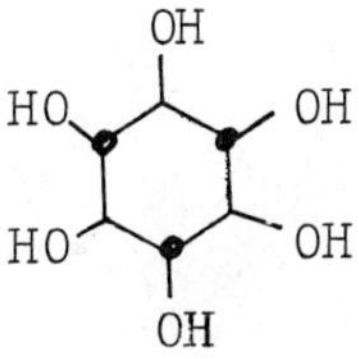 inactive

CHAPTER 27, NUCLEIC ACIDS

ANSWERS TO QUESTIONS

1. (a) [pyrimidine structure] (b) [purine structure]

2. No. Atom locations differ. They are tautomers rather than resonance structures.

3. (a) [structure with D labels] (b) [structure with O^{18} labels]

4. $H_2NC(=O)NH_2 \xrightarrow{EtO^{\ominus}} H_2NC(=O)\ddot{N}H^{\ominus}$

[mechanism: $H_2N-C(=O)-\ddot{N}H^{\ominus}$ attacks C=O of $(Et)(C_6H_5)C(CO_2Et)C(=O)OEt$ → tetrahedral intermediate $(Et)(C_6H_5)(EtO_2C)C-C(O^{\ominus})(OEt)-NH-C(=O)-NH_2$]

$\downarrow (-EtO^{\ominus})$

(continued next page)

$\xleftarrow{EtO^{\ominus}}$

$\xrightarrow{(-EtO^{\ominus})}$

5. (a) $C_6H_6 \xrightarrow[Fe]{Br_2} C_6H_5Br \xrightarrow[ether]{Mg} C_6H_5MgBr + \xrightarrow{\text{ethylene oxide}} \xrightarrow{H_3O^{\oplus}}$

$\xrightarrow[heat]{MnO_4^{\ominus}/H^{\oplus}} \xrightarrow{Cl_2/PCl_3} C_6H_5CHClCO_2H \xrightarrow{OH^{\ominus}} \xrightarrow{CN^{\ominus}}$

$\xrightarrow[heat]{EtOH/H^{\oplus}} C_6H_5CH(CO_2C_2H_5)_2 \xrightarrow[(b)\,C_2H_5Br]{(a)\,C_2H_5O^{\ominus}} C_6H_5C(CH_2CH_3)(CO_2C_2H_5)_2$

(b) $CH_3CH_2MgBr + \xrightarrow{\text{ethylene oxide}} \xrightarrow{H_3O^{\oplus}} CH_3CH_2CH_2CH_2OH \xrightarrow[pyridine]{CrO_3}$

$\xrightarrow{CH_3MgCl} \xrightarrow{H_3O^{\oplus}} \xrightarrow{SOCl_2} \xrightarrow[ether]{Mg} CH_3CH_2CH_2CH(CH_3)MgCl$

now repeat synthesis in part <u>a</u> using $CH_3CH_2CH_2CH(CH_3)MgCl$ instead of C_6H_5MgBr

6. Two

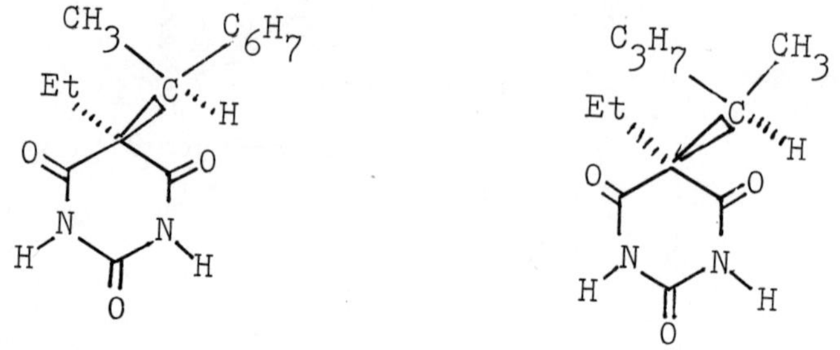

7.

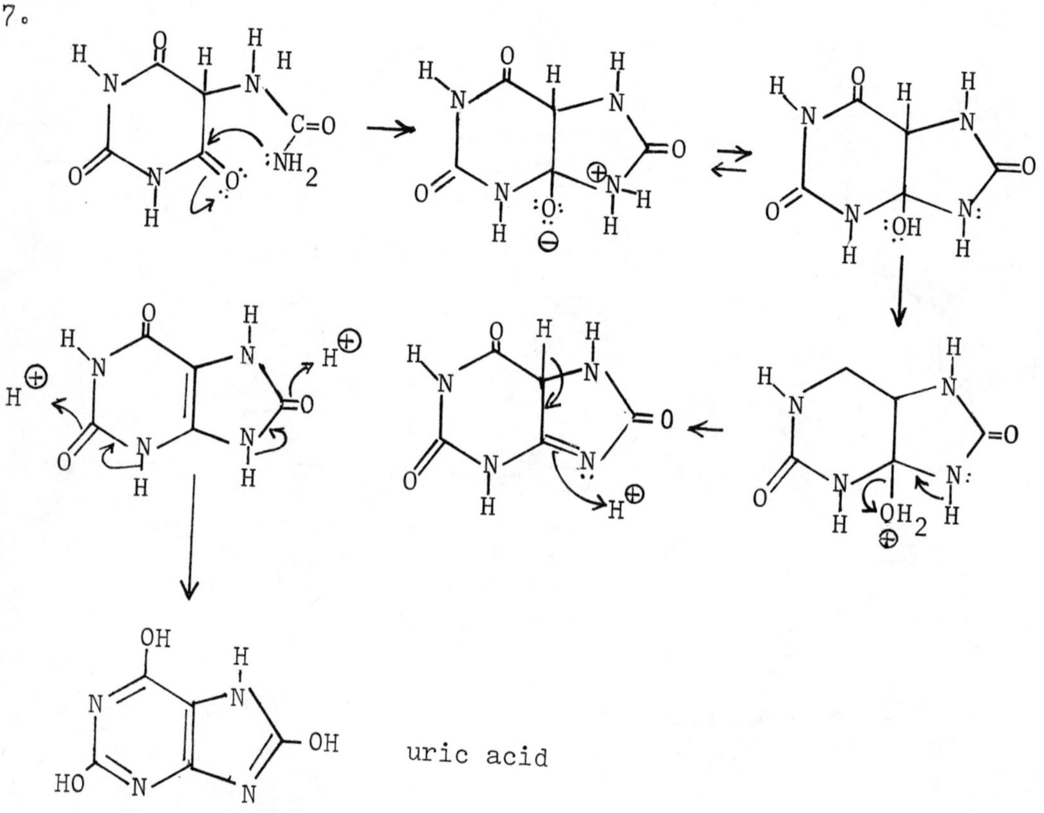

uric acid

8.

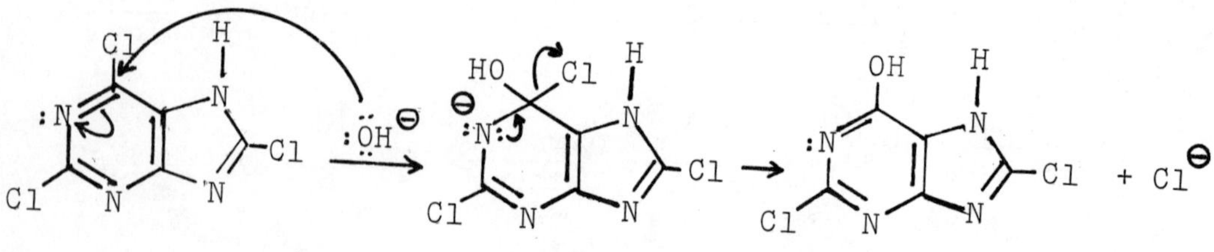

9. (a) HN=C(NH⁻)(NH$_2$) + OC_2H_5–C(=O)–CH_2–CN ⟶ HN=C(NH$_2$)–NH–C(O⁻)(OC_2H_5)–CH_2–CN ⟶ ($-C_2H_5O^{\ominus}$)

⟶ HN=C(H$_2$N)–NH–C(=O)–CH_2–C≡N ⟵ base

H–N–C=O, HN=C, CH_2, N⁻ : C≡N ⟵

HN=C, N–C=O, CH_2, N–C=N⁻

H_2O

HN=C, N–C=O, CH_2, N–C=NH ⇌ H_2N–C, N=C–OH, CH, N=C–NH_2

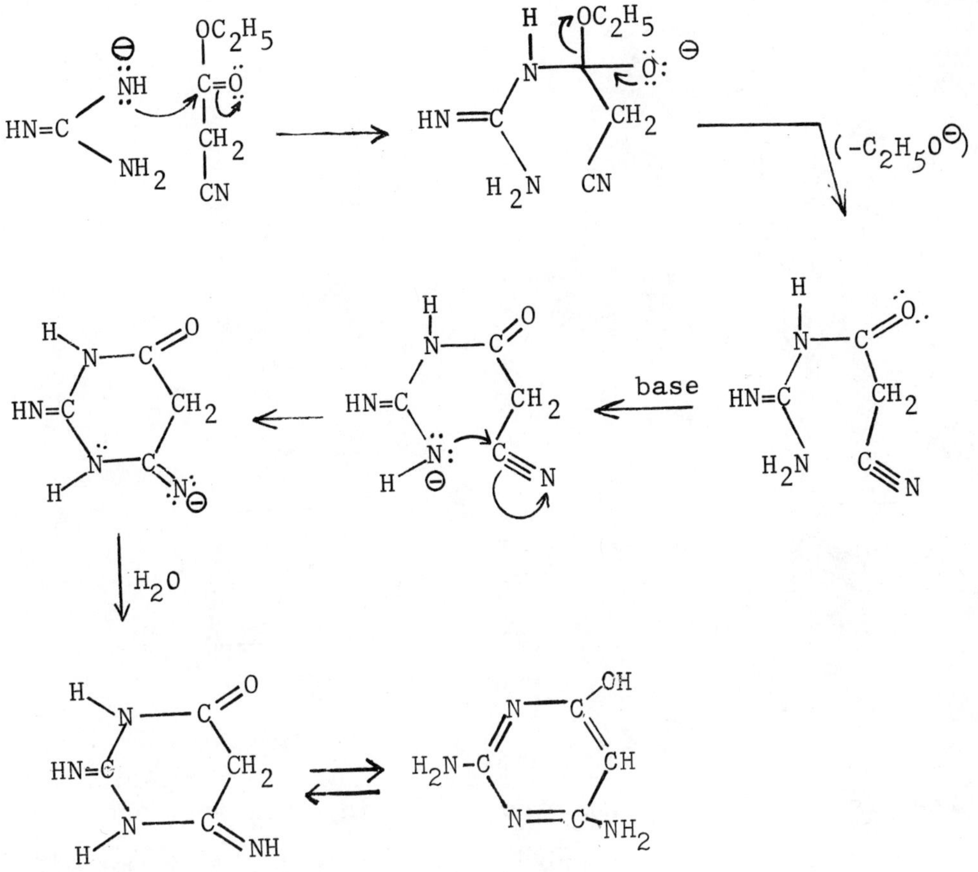

(b), (c) All protons are exchangeable.

10. OEt, N, N⊕, OEt, R + :$\ddot{O}H_2$ ⟶ OEt, N, N, ⊕OH$_2$, OEt, R ⇌ OEt, N, N, O–H, Et, O⊕H, R

(-EtOH, -H$^{\oplus}$)

OEt, N, N, O, R

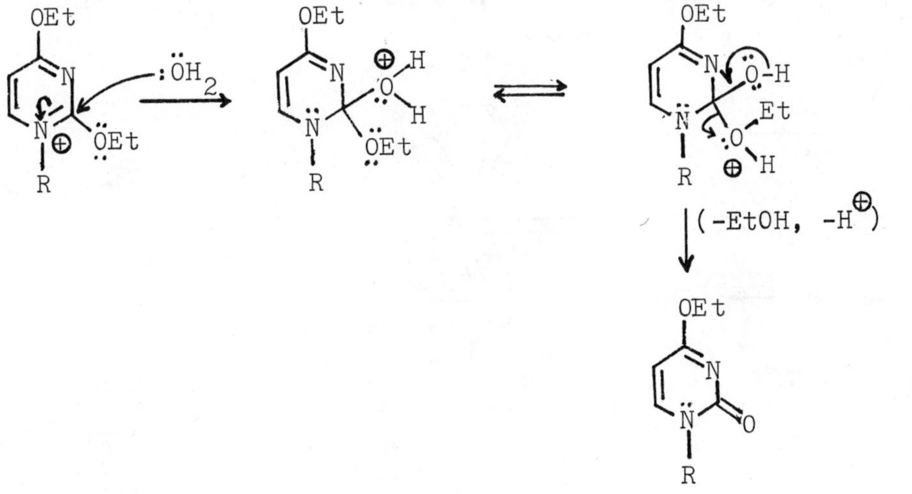

11. 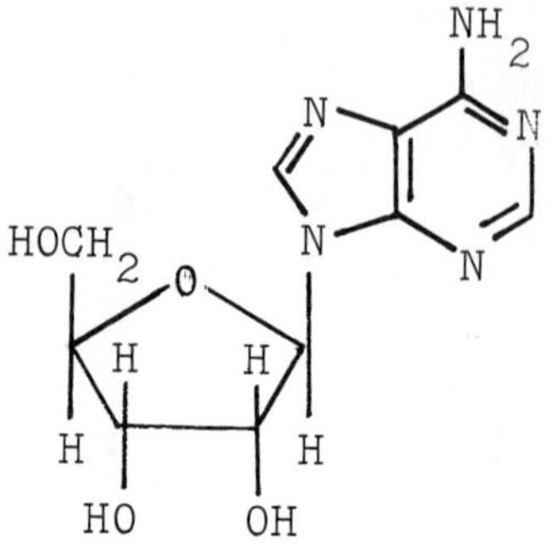 [structure: adenosine — $HOCH_2$ furanose ring with O, H, H, HO, OH, attached at N-9 to adenine ring bearing NH_2]

12. 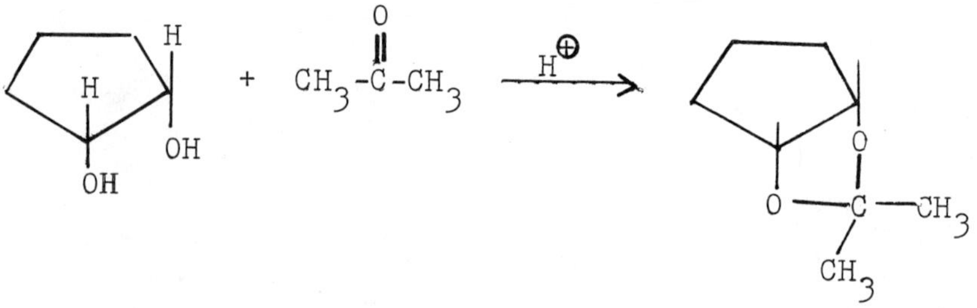 [cyclopentane-1,2-diol (cis, H, H, OH, OH)] + $CH_3-\overset{O}{\overset{\|}{C}}-CH_3$ $\xrightarrow{H^{\oplus}}$ [cyclopentane fused acetonide: O—C(CH_3)(CH_3)—O]

13. See Figure 27-1 for adenosine, guanosine, uridine, and thymidine. Cytidine is similar to deoxycytidine except C-2' is H-C-OH rather than H-C-H.

14.

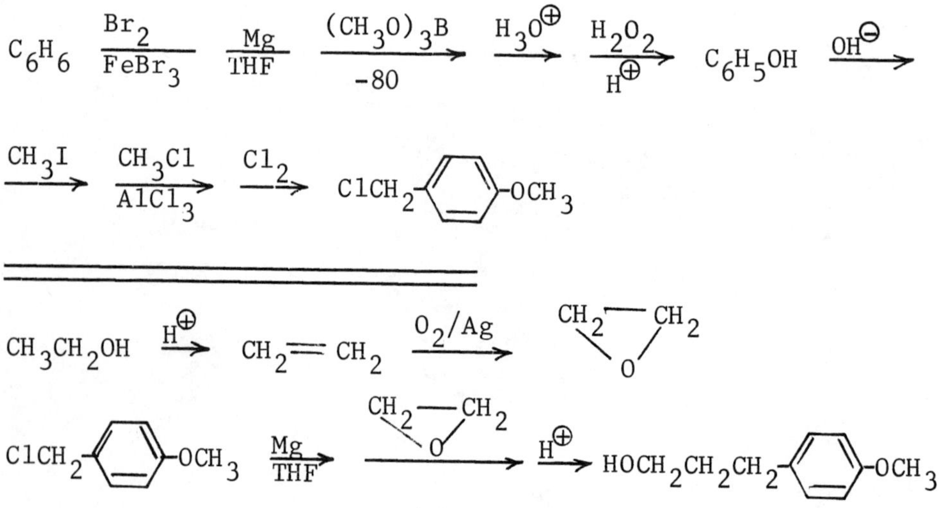

$C_6H_6 \xrightarrow[FeBr_3]{Br_2} \xrightarrow[THF]{Mg} \xrightarrow[-80]{(CH_3O)_3B} \xrightarrow{H_3O^{\oplus}} \xrightarrow[H^{\oplus}]{H_2O_2} C_6H_5OH \xrightarrow{OH^{\ominus}}$

$\xrightarrow{CH_3I} \xrightarrow[AlCl_3]{CH_3Cl} \xrightarrow{Cl_2} ClCH_2-C_6H_4-OCH_3$

$CH_3CH_2OH \xrightarrow{H^{\oplus}} CH_2{=}CH_2 \xrightarrow{O_2/Ag}$ CH_2-CH_2 (epoxide, O)

$ClCH_2-C_6H_4-OCH_3 \xrightarrow[THF]{Mg} \xrightarrow{\text{ethylene oxide}} \xrightarrow{H^{\oplus}} HOCH_2CH_2CH_2-C_6H_4-OCH_3$

(continued next page)

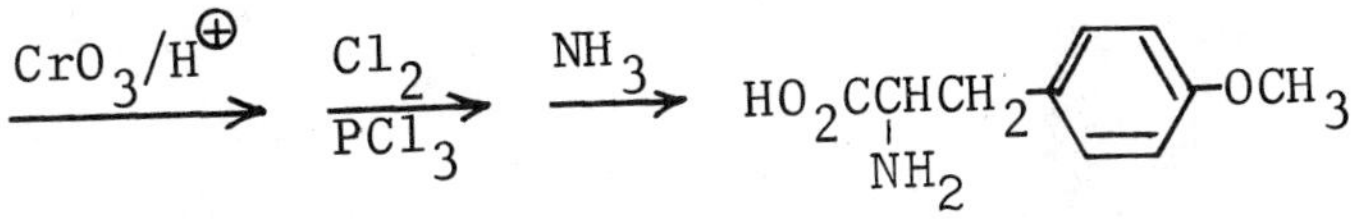

$\xrightarrow{CrO_3/H^{\oplus}}$ $\xrightarrow[PCl_3]{Cl_2}$ $\xrightarrow{NH_3}$ $HO_2C\underset{NH_2}{C}HCH_2$–$C_6H_4$–$OCH_3$

15. (a) Any of several polynucleotides that control the synthesis of proteins. These are grouped according to function: messenger RNA, ribosomal RNA, and transfer RNA. All are characterized by the presence of ribose in the nucleotide units of the polymer.

(b) A polynucleotide containing genetic information. The sugar that is part of the polymer backbone is 2-deoxyribose.

(c) A nucleoside (see e) in which the sugar is ribose.

(d) A nucleoside (see e) in which the sugar is deoxyribose.

(e) An N-glycoside in which the sugar is either ribose or deoxyribose, and the aglycone is one of several derivations of either pyrimidine or purine.

(f) A phosphoric ester of a nucleoside. The nucleotide is the fundamental repeating unit in nucleic acids (polynucleotides).

(g) A class of drugs that act as general depressants of the central nervous system. Derivatives of barbituric acid.

(h) A glycoside in which the aglycone is attached to C-1 by nitrogen instead of oxygen.

(i) Two intertwined helical coils.

(j) A region in the DNA molecule that is coded for the synthesis of a particular protein.

(k) A combination of three nucleotides that, when read in the proper direction, specifies a specific amino acid. Codon and anticodon for a specific amino acid are complementary.

(l) A combination of three nucleotides that, when read in the proper direction, specifies a particular amino acid.

(m) A disease caused by any change in a codon which results in a mutation.

16. (a) 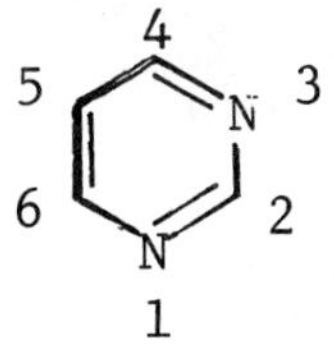 [pyrimidine ring, positions numbered 1–6, N at 1 and 3] (b) 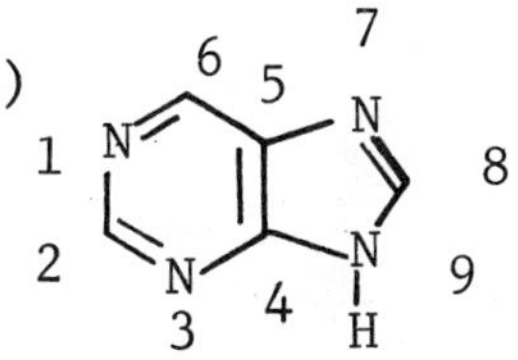 [purine ring, positions numbered 1–9, N at 1, 3, 7, 9; N-H at 9] (c)

CHO
|
CH_2
|
H-C-OH
|
H-C-OH
|
CH_2OH

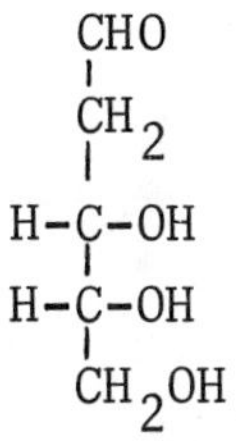

(continued next page)

(d) 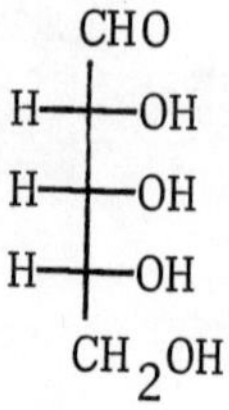 CHO / H—C—OH / H—C—OH / H—C—OH / CH_2OH

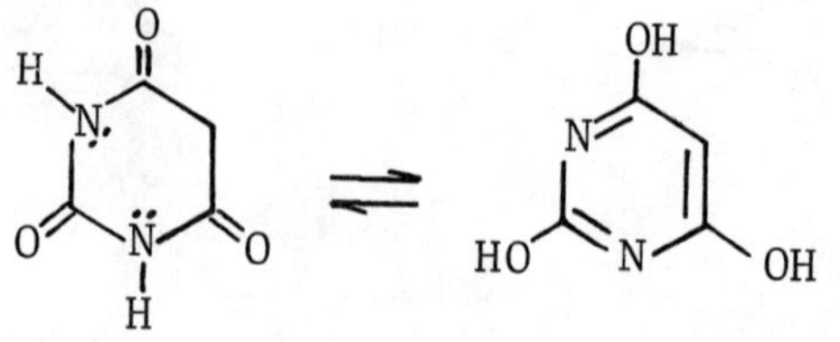

(e) [structure: barbituric acid (keto form) ⇌ 2,4,6-trihydroxypyrimidine; labels: O, H, N, N, O, O, H ⇌ OH, N, HO, N, OH]

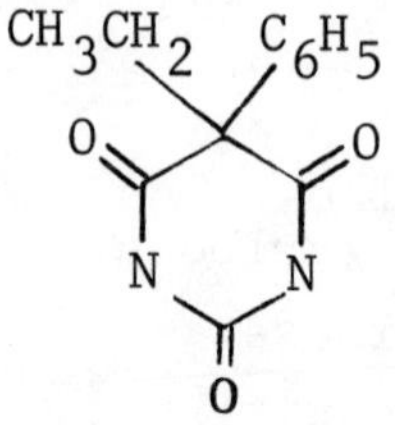

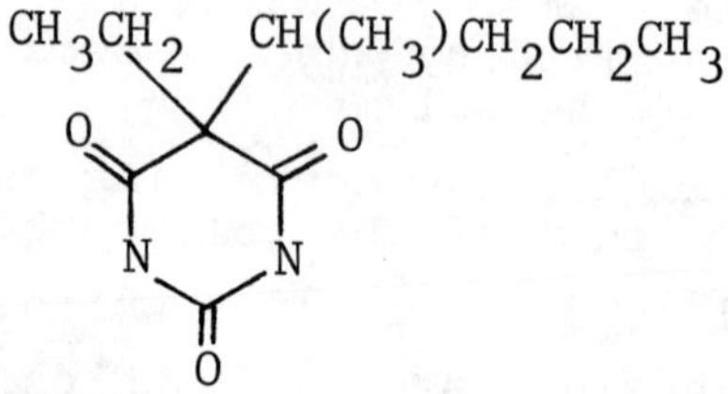

(f) [structure: labels CH_3CH_2, C_6H_5, O, O, N, N, O]

(g) [structure: labels CH_3CH_2, $CH(CH_3)CH_2CH_2CH_3$, O, O, N, N, O]

(h) [structure: labels H, N, O, O=, N, $HOCH_2$, O, H, H, H, H, OH, OH]

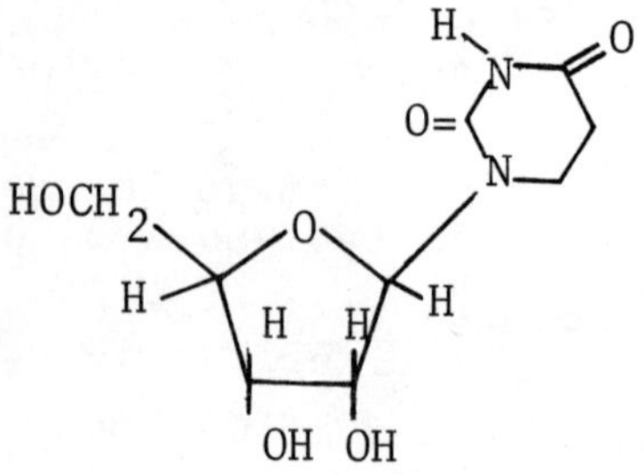

(i) 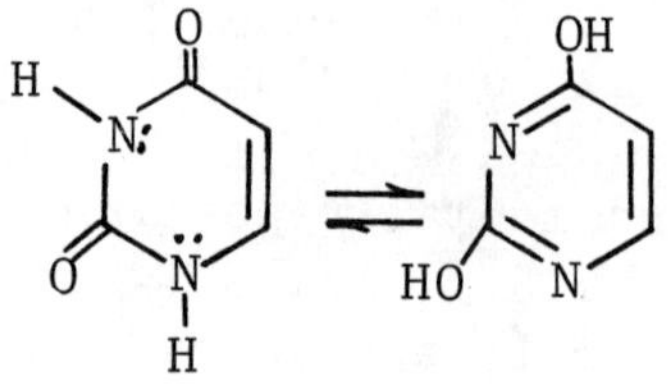 [structure: labels O, H, N, O, N, H ⇌ OH, N, HO, N]

(j) 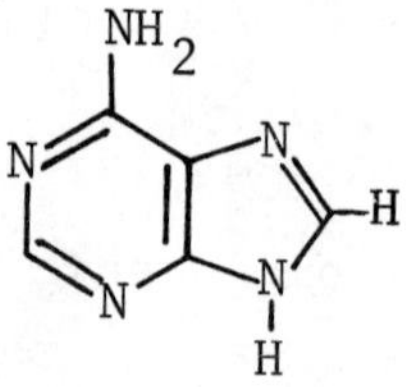 [structure: labels NH_2, N, N, N, H, N, H]

(k) 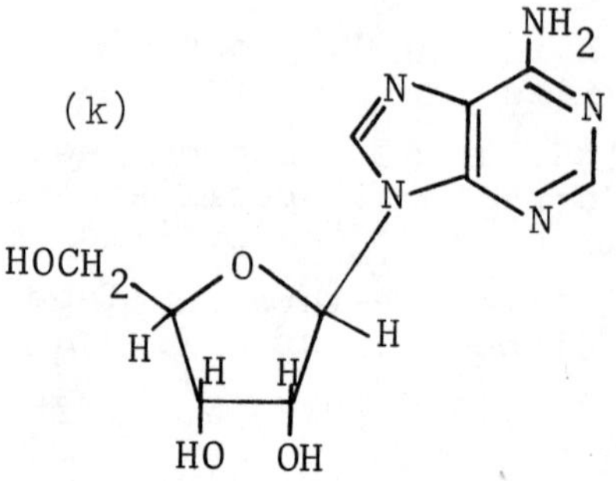 [structure: labels NH_2, N, N, N, N, $HOCH_2$, O, H, H, H, H, HO, OH]

(l) (see k for adenosine) [structure: labels ⊖, O, ⊕, H-O-P—O-CH_2, OH, O, H, H, HO]

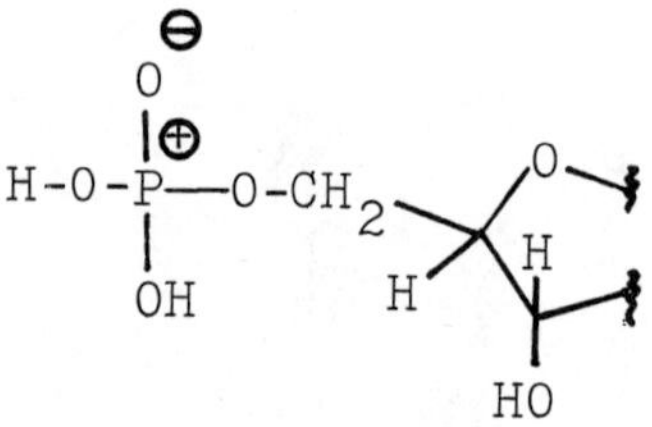

(continued next page)

(m)

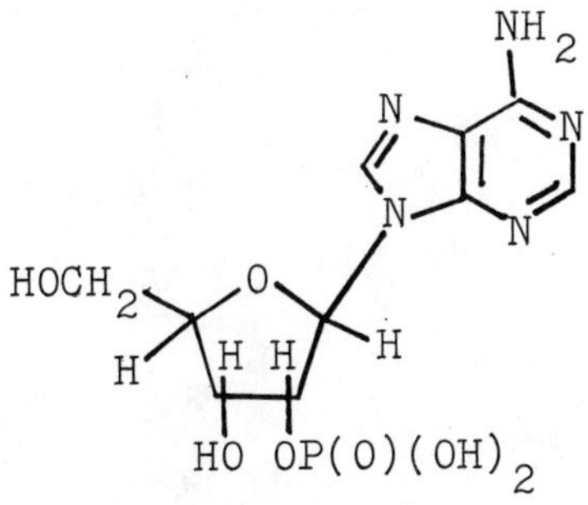

NH$_2$, N, N, N, N, HOCH$_2$, O, H, H H, H, HO OP(O)(OH)$_2$

(n)

O, CH$_3$, H$_3$C, N, N, H, O, N, N, CH$_3$

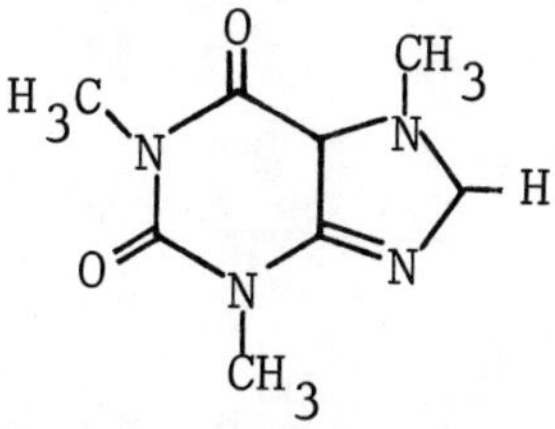

(o) (see k for adenosine)

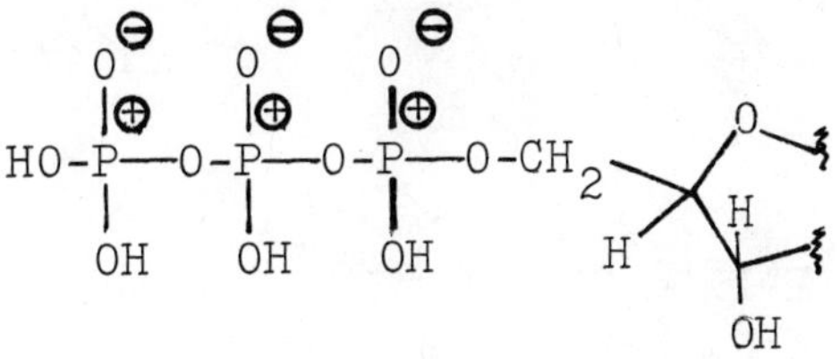

$$HO\text{-}\overset{O^{\ominus}}{\underset{OH}{\overset{|}{\underset{|}{P^{\oplus}}}}}\text{—}O\text{-}\overset{O^{\ominus}}{\underset{OH}{\overset{|}{\underset{|}{P^{\oplus}}}}}\text{—}O\text{-}\overset{O^{\ominus}}{\underset{OH}{\overset{|}{\underset{|}{P^{\oplus}}}}}\text{—}O\text{-}CH_2\text{—}$$

O, H, H, OH

17. (a) $$(C_2H_5O\overset{O}{\overset{\|}{C}})_2\underset{C_2H_5}{\underset{|}{C}}\text{-}C_6H_5 + NH_2\overset{O}{\overset{\|}{C}}NH_2 \xrightarrow[C_2H_5OH]{Na^{\oplus}\,{}^{\ominus}OC_2H_5} \xrightarrow{H_2O} \text{phenobarbitol}$$

(b) $$(C_2H_5O\overset{O}{\overset{\|}{C}})_2\underset{C_2H_5}{\underset{|}{C}}\text{-}\overset{CH_3}{\overset{|}{C}}CH_2CH_2CH_3 + NH_2\overset{O}{\overset{\|}{C}}NH_2 \xrightarrow[C_2H_5OH]{Na^{\oplus}\,{}^{\ominus}OC_2H_5} \xrightarrow{H_2O} \text{Nembutal}$$

(c) $$(C_2H_5O\overset{O}{\overset{\|}{C}})_2CH_2 + NH_2\overset{O}{\overset{\|}{C}}NH_2 \xrightarrow[C_2H_5OH]{Na^{\oplus}\,{}^{\ominus}OC_2H_5} \xrightarrow{H_2O} \text{barbituric acid}$$

(continued next page)

(d) barbituric acid (from e)

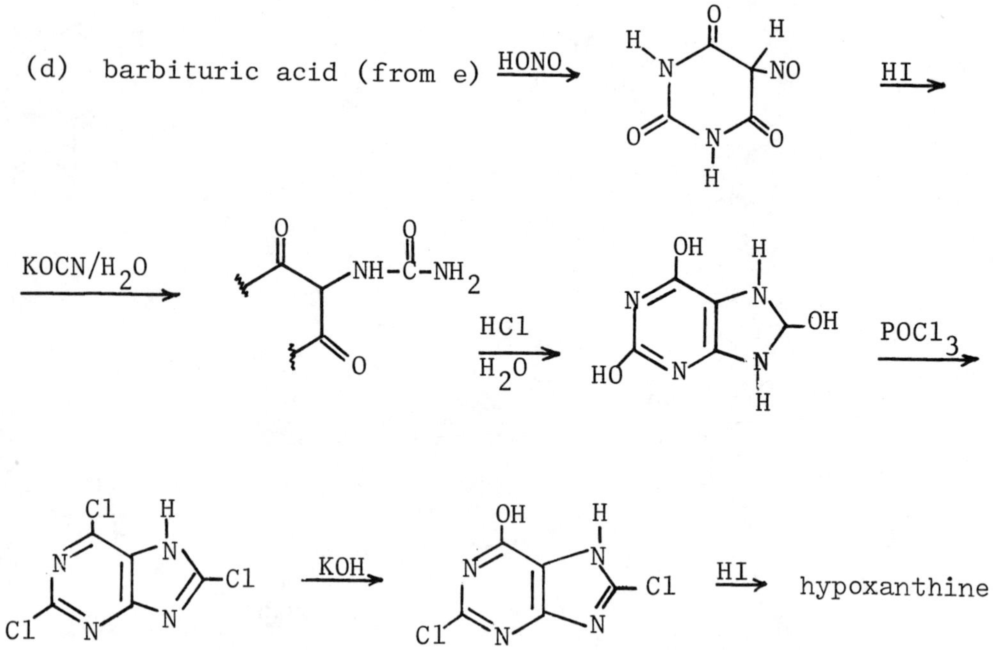

18. Both pCpCpA and pApCpC have the backbone structure below. For pCpCpA, x and y are cytidine, and z is adenosine. For pApCpC x is adenosine, y and z are cytidine.

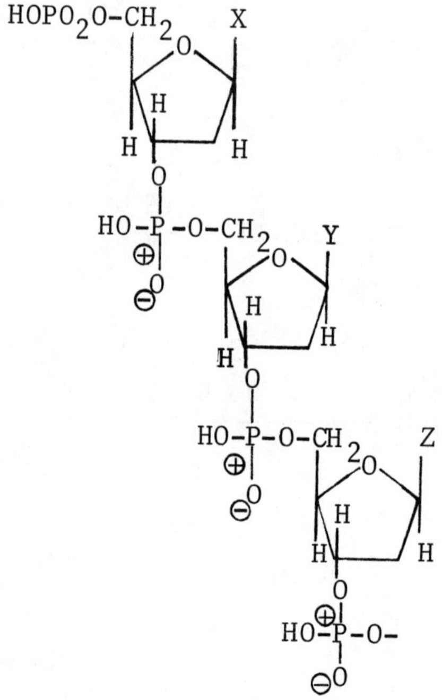

19.

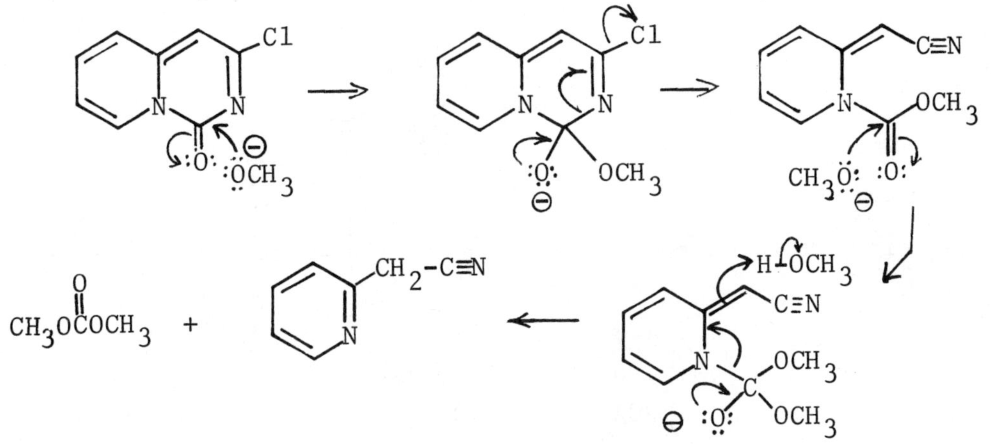

$CH_3O\overset{O}{\overset{\|}{C}}OCH_3$ + 2-pyridyl-$CH_2-C{\equiv}N$

20. The CH_3: ring-H ratio is 9:1 in caffeine, 3:1 in both theophylline and theobromine. Theophylline will have a CH_3 absorption (that on N flanked by two carbonyl groups) further downfield than any CH_3 absorption in the nmr spectrum of theobromine.

21. They can act as hydrogen bond acceptors in the keto form but not in the enol form.

22. The presence of deoxyribonucleic acid in DNA in the place of ribonucleic acid in RNA.

23. Messenger RNA: A polynucleotide whose function is to transmit information from within the nucleus (on DNA) to ribosomes outside the nucleus.

Ribosomal RNA: The RNA found in ribosomes. Its function in peptide synthesis is less clearly understood than is that of either messenger RNA or transfer RNA.

Transfer RNA: A ribonucleotide whose function is to transfer a particular amino acid to the ribosome at the proper time for a peptide synthesis. Refer to p. 1173 for an analysis of the role of each type of RNA in protein biosynthesis.

24. (a) Nucleotides are hydrolyzed to nucleosides and phosphoric acid.

(b) A ribotide has a phosphoric acid moiety, a riboside does not.

(c) They both contain ribose.

25. The others exist in the keto form, e.g., 2-hydroxypyridine is:

3-Hydroxypyridine cannot undergo this tautomeric shift.

26. (a) valine: CAC, TAC, AAC, GAC
    (b) leucine: TAA, CAA, AAG, GAG, TAG, CAG
    (c) glycine: ACC, GCC, TCC, CCC

27. The dotted lines indicate the number of hydrogen bonds.

28. One. Sickle cell hemoglobin, which has valine for glutamic acid in the sixth amino acid from the N-terminus, is formed by substituting U for A as the middle base of the mRNA codon for glutamic acid.

29. UAA, UAG and UGA. These have been designated "ochre", "amber" and "umber" respectively.

30. The third letter is the least important. For example, valine, is coded by GUG, GUA, GUU, and GUC.

31. It is caused by the synthesis of hemoglobin which contains a valine in place of glutamic acid. The disease gets its name because the red cell's misshapen "sickle" appearance.

32. (a) see the example in answer 18. -CpApTp- has the same basic structure with x, y and z cytidine, adenosine, and thymidine respectively.
    (b) -pApTpG-
    (c) No. T does not occur in RNA; it is replaced by uracil in RNA (see Table 27-3, p. 1163).

33. The cation is bridged, and the nucleophile may attack only from the top.

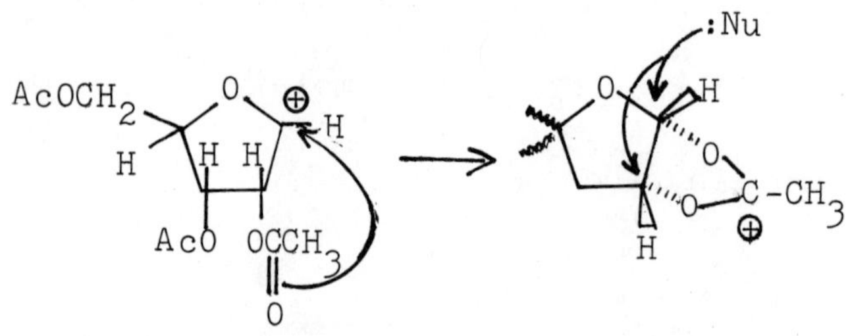

34. Ring opening occurs primarily by attack of $OH^{\ominus}$ at the less sterically hindered 5' position which gives the 3'-phosphate.

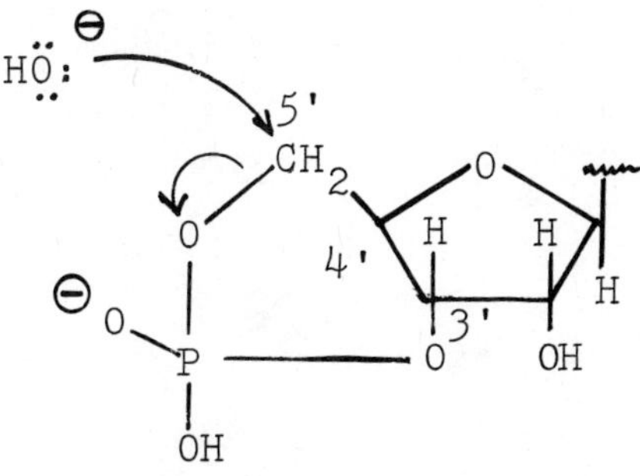

CHAPTER 28

INFRARED SPECTROSCOPY, ULTRAVIOLET SPECTROSCOPY AND MASS SPECTROMETRY

ANSWERS TO QUESTIONS

1. (a) $v = \nu\lambda$

$$3 \times 10^{10} \text{ cm sec}^{-1} = \nu \times 10 \text{ cm}$$

$$3 \times 10^{9} \text{ sec}^{-1} = \nu$$

$$E = h\nu$$

$$E = (6.62 \times 10^{-27} \text{ erg-sec})(3 \times 10^{-9} \text{ sec}^{-1}) = 19.86 \times 10^{-18} \text{ erg}$$

$$\frac{19.86 \times 10^{-18} \text{ erg}}{4.184 \times 10^{7} \text{ erg cal}^{-1}} = 4.75 \times 10^{-25} \text{ cal}$$

(b) 4.75×10^{-26} cal

(c) 4.75×10^{-27} cal

2. 1 nanometer = 10^{-9} meter

(a) 10^{9} nm

(b) 10^{12} nm

(c) 2.5×10^{7} nm

(d) 3.67×10^{7} nm

(e) 1.3×10^{5} nm

3. (a) ultraviolet
(b) ultraviolet
(c) visible (blue)
(d) visible (green)
(e) infrared

4. (a) wave number (cm^{-1}) = $\frac{1}{\lambda}$ and $\lambda = c/\nu$

$E = Nh\nu = Nh\ c/\lambda = Nhc$ (wave number)

$$E = (6.02 \times 10^{23})(6.62 \times 10^{-27} \text{ erg-sec})(3 \times 10^{10} \text{ cm/sec})(4000 \text{ cm}^{-1})$$

$E = 4.78 \times 10^{11}$ erg/mole <u>and</u>

$$4.78\ 10^{11} \text{ erg/mole } \left(\frac{1 \text{ cal}}{4.184 \times 10^{7} \text{ erg}}\right)\left(\frac{1 \text{ kcal}}{1000 \text{ cal.}}\right) = 11.4 \text{ kcal/mole}$$

(b) 7.47×10^{10} erg/mole or 1.78 kcal/mole

The end toward higher wave number is richer in energy.

5. (a) The Cl_2 vibration requires more energy.

(b) I_2: $214 = \frac{1}{2 \times 3.14 \times 3 \times 10^{10}} (f/\mu)^{\frac{1}{2}}$

$4.03 \times 10^{13} = (f/\mu)^{\frac{1}{2}}$

(continued next page)

$\mu = \dfrac{M_I M_I}{M_I + M_I}$; $M_I = 127/6.02 \times 10^{23} = 2.11 \times 10^{-22}$

$\mu = 1.05 \times 10^{-22}$

$4.03 \times 10^{13} = (f/1.05 \times 10^{-22})^{\frac{1}{2}}$

$f = 1.7 \times 10^{5}$ dynes/cm

(c) Cl_2: $M_{Cl} = 5.90 \times 10^{-23}$

$f = 3.4 \times 10^{5}$ dynes/cm

(d) bond stretching does not result in a dipole moment change.

6. See page 1187 for derivation of formula. ^{13}C is represented as C^*.

(a) $$\left(\frac{\nu_{C\text{-}H}}{\nu_{^{13}C\text{-}H}}\right)^2 = \frac{\dfrac{M_C^* M_H}{M_C^* + M_H}}{\dfrac{M_C M_H}{M_C + M_H}} = \frac{M_C^* M_C + M_C^* M_H}{M_C M_C^* + M_H M_C} \cong 1$$

$$\frac{\nu_{C\text{-}H}}{\nu_{C^*\text{-}H}} = \sqrt{1} = 1 \quad ; \quad \nu_{^{13}C\text{-}H} = 3050 \text{ cm}^{-1}$$

(b) $$\frac{\nu_{C\text{-}H}}{\nu_{C^*\text{-}^2H}} = \sqrt{1.87} = 1.37 \quad ;$$

$$\nu_{C^*\text{-}^2H} = \frac{3050}{1.37} = 2232 \text{ cm}^{-1}$$

(continued next page)

(c) see page 1187 for calculation of $\frac{\text{C-H}}{\text{C-D}}$

$$\frac{\nu_{\text{C-H}}}{\nu_{\text{C-D}}} = \sqrt{1.86} = 1.36 \; ; \; \nu_{\text{C-D}} = \frac{3050}{1.36} = 2238 \text{ cm}^{-1}$$

$^{13}C-^{2}H$ shows the greatest shift due to isotope effect.

7. 0.25 Å to 0.001 Å = 2.5×10^{-9} to 10^{-11} cm

$$E = h\nu = \frac{hc}{\lambda}$$

For 2.5×10^{-9} cm wavelength:

$$E = \frac{(6.62 \times 10^{-27} \text{ erg-sec})(3.0 \times 10^{10} \text{ cm/sec})}{2.5 \times 10^{-9} \text{ cm}}$$

$E = 7.9 \times 10^{-8}$ erg (or 7.9×10^{-15} J or 1.9×10^{-15} cal.)

For 10^{-11} cm

$$E = \frac{(6.62 \times 10^{-27} \text{ erg-sec})(3.0 \times 10^{10} \text{ cm/sec})}{10^{-11}}$$

$E = 2.0 \times 10^{-5}$ erg (or 2.0×10^{-12} J or 4.8×10^{-13} cal.)

8. The infrared spectrum of dimethyl sulfone (SO_2 group) shows both symmetric and asymmetric stretching vibrations.

9. Set 1: n-heptane: C-H bending (~ 1380 cm^{-1}) singlet

2,4-dimethylpentane: C-H bending (~ 1380 cm^{-1}) doublet

(continued next page)

Set 2: 1-hexyne: ≡C-H stretch and C≡C stretch

acetylene(symmetric): ≡C-H stretch

2-butyne: neither ≡C-H nor C≡C stretches

Set 3: propylene: =C-H (~ 3050 cm^{-1}) and C=C (~ 1650 cm^{-1}) stretches (also C-H bending ~ 910 and ~ 990 cm^{-1} for $RCH=CH_2$).

ethylene(symmetric): =C-H stretch, no C=C stretch

tetramethylethylene (symmetric): lacks =CH and C=C stretches

Set 4: 1-hexene: =CH (~ 3050 cm^{-1}) and C=C (~ 1650 cm^{-1}) stretches.

cyclohexene: =CH stretch but no C=C stretch

cyclohexane: lacks =CH and C=C stretches

Set 5: all can be distinguished by out-of-plane C-H bending vibrations:

cis-2-butene: 735-670 cm^{-1}

trans-2-butene: 965 cm^{-1}

1-butene: 910 and 990 cm^{-1}

Set 6: chlorobenzene: =C-H (~ 3050 cm^{-1}) and C=C (~ 1600 and 1500 cm^{-1}) stretches (the others lack these absorptions)

2-chlorobutane: C-H bending ~ 1380 cm^{-1} for CH_3.

chlorocyclohexane: lacks C-H bending vibration ~ 1380 cm^{-1} for CH_3.

10. The four C-Cl stretching frequencies are from the four isomers:

$(CH_3)_2C(CH_2Cl)$–cyclohexyl $(CH_3)_3C$–C(Cl) cyclohexane $(CH_3)_3C$–cyclohexyl–Cl^*(axial)

$(CH_3)_3C$–cyclohexyl–Cl^*(equatorial)

*(may be substituted at C-2, C-3 or C-4 ring carbons)

11.

CH_3 CH_3 >CHCH$_2$C(=O)-H

C-H stretch (alkane)   C-H stretch (aldehyde)   C=O stretch   C-H bend (isopropyl)

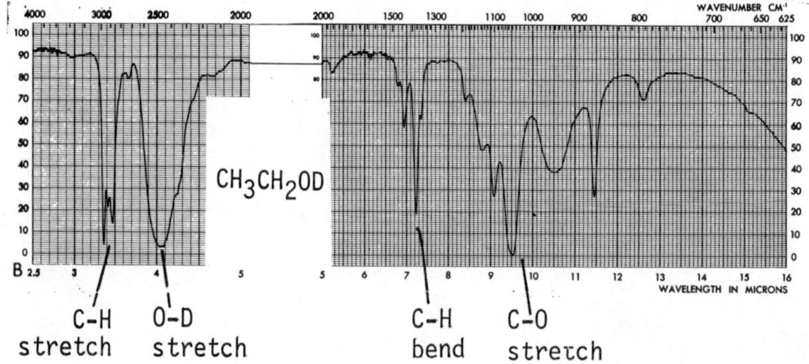

(continued next page)

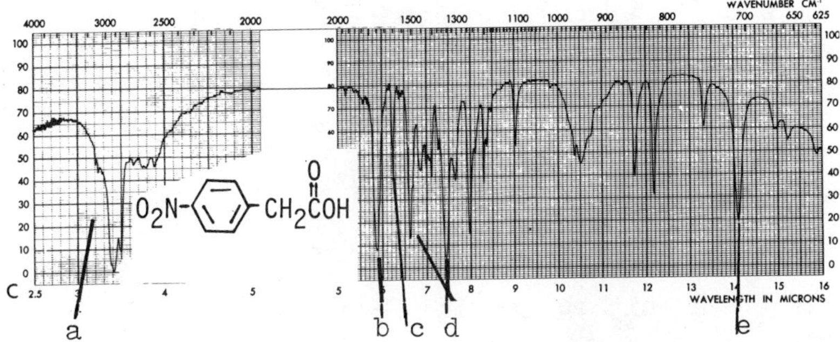

a: Acids typically show an "envelope" of absorptions from 3300 to 2500 cm^{-1} which contains =C-H, C-H (alkyl) and O-H stretches in this example.

b: C=O stretch, c: C=C stretch, d: N=O stretches

e: =C-H out-of-plane bending (1,4-disubstituted benzene)

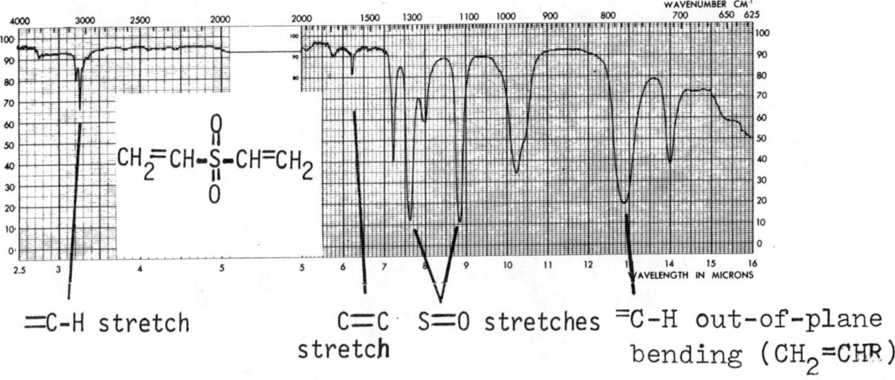

=C-H stretch C=C stretch S=O stretches =C-H out-of-plane bending (CH_2=CHR)

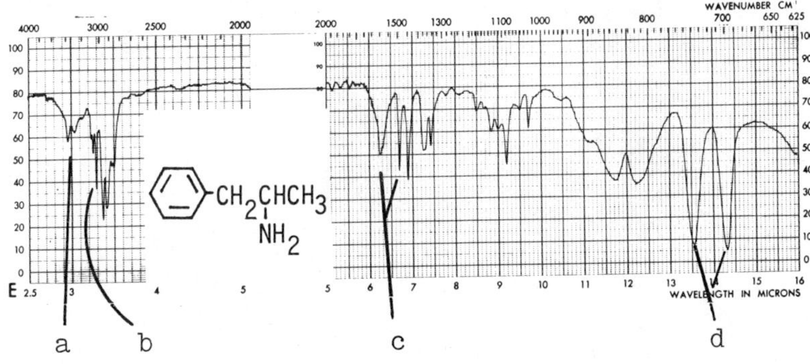

a: N-H stretches (symmetric and asymmetric)

b: C-H stretches (benzene >3000 and alkyl < 3000 cm^{-1})

c: C=C stretches (benzene), d: =C-H out-of- plane bending (monosubstituted benzene)

12. A: The spectrum shows a predominance of axial C-OH ($\sim$980 cm^{-1})

B: The spectrum shows a predominance of equatorial C-OH ($\sim$1020 cm^{-1})

C: The spectrum shows only equatorial C-OH ($\sim$ 1020). No absorption 900-1000 cm^{-1} for axial C-OH.

13. Check your equations in Section 28.10 of the text.

14. blue (the color observed is that which is complementary of the color absorbed)

15. There is less conjugation in the all-<u>cis</u> isomer due to greater steric hindrance (increasing conjugation moves absorption to longer wavelength). Conjugation is the most extensive in the all-<u>trans</u> isomer; intermediate in the <u>trans</u>-<u>cis</u> isomer.

16. (a) 214nm + 2(5nm) = 224nm

(b) 214nm + 4(5nm) = 234nm

(c) 214nm + 4(5nm) + 5nm (exocyclic) = 239nm

(d) 214nm + 2(5nm) = 224nm

(e) 214nm + 2(5nm) = 224nm

(f) 214nm + 4(5nm) + 5nm (exocyclic) = 239nm

17.

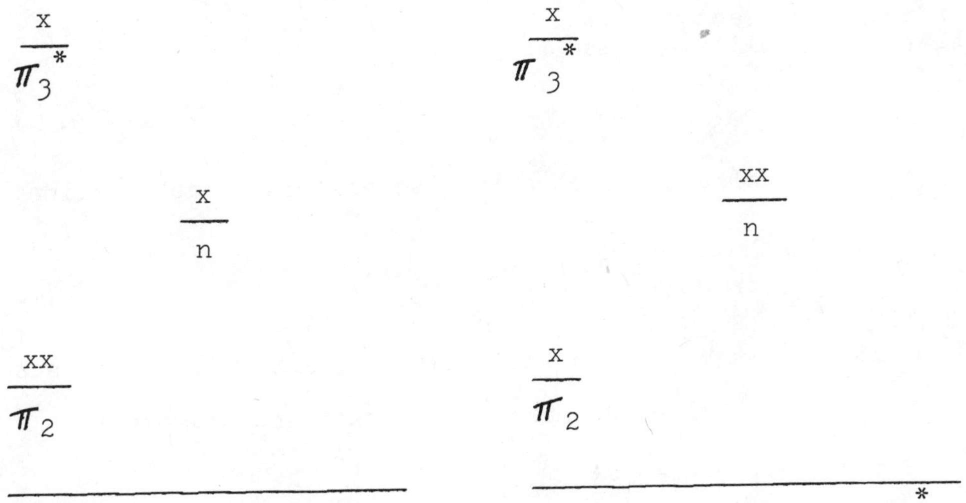

x	x
π_3^*	π_3^*
x	xx
n	n
xx	x
π_2	π_2
appearance of $n \longrightarrow \pi^*$ excited state	appearance of $\pi \longrightarrow \pi^*$ excited state

18. All groups in Portion A are not held planar as they are in thalidomide and Portion B. Since the degree of planarity influences the extent of delocalization (and ΔE in the appropriate transition) this structural feature is important in the model.

19. (a) 98 (the most intense peak)

(b) and (c) are the same, 156.

20. Since the compound has a molecular ion with an even mass number, it contains an even number of nitrogens (see the nitrogen rule, p. 1212).

21. m/e values

OH	17	CH_3OH	32
NH_3	17	HCN	27
H_2O	18	HCO_2H	46
H_2CO	30	HNCO	43
CO	28		

Only OH and NH_3 would not have been resolved.

22. In order to react they must collide in space. They are so far apart that collisions are very rare.

23. 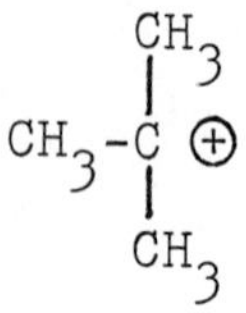 $CH_3-\overset{CH_3}{\underset{CH_3}{|}}{C}^{\oplus}$ cleavage occurs to give the most stable cation.

24. (a) and (b) As branching increases, fragmentation increases to give the most stable cationic fragments (stability: $3^o > 2^o > 1^o$)

(c) Both F and P are isotopically homogeneous and their presence should not contribute to an M+1 peak.

25. (a) A m/e = 32

B m/e = 15

C m/e = 29

D m/e = 31

(b) The base peak, with RA = 100, is m/e = 31

(c) The parent peak is m/e = 32

(d) $CHO^{\oplus}$ is protonated carbon monoxide ($H-C\equiv\overset{\oplus}{O}:$)

$\overset{\oplus}{CH_2}OH$ is protonated formaldehyde ($H_2\overset{\oplus}{C}-\ddot{O}-H \leftrightarrow H_2C=\overset{\oplus}{\ddot{O}}H$)

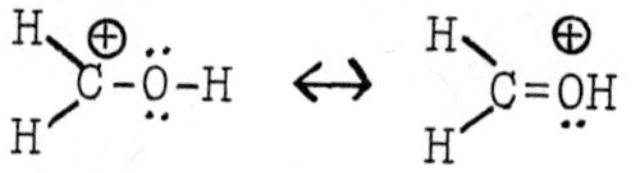

CHAPTER 29, NUCLEAR MAGNETIC RESONANCE SPECTROSCOPY

ANSWERS TO QUESTIONS

1. $e^{-\Delta E/RT} = e^{-0.0057/592} = 0.999991$

   nuclei in excited state = 2,000,010 - 1,000,010 = 1,000,000

   nuclei in ground state = 1,000,010

   ratio: $\frac{1,000,000}{1,000,010} = 0.99999$

2. (a) $\delta = \frac{\Delta\nu \times 10^6}{60 \times 10^6} = \frac{100 \times 10^6}{60 \times 10^6} = 1.67$

   (b) 1.00 (c) 0.75 (d) 1.50 (e) -0.14

3. (i) (a) 54 Hz
   (b) 78 Hz
   (c) 432 Hz
   (d) 438 Hz

   (ii) (a) 90 Hz
   (b) 130 Hz
   (c) 720 Hz
   (d) 730 Hz

4. $\tau = 10 - \delta$

   (a) τ 9.1
   (b) τ 8.7
   (c) τ 5.0
   (d) τ 0.0
   (e) τ -1.0

5. This difference is accounted for by comparing the anisotropic effects of the two groups. Compare the orientation of the protons relative to the induced magnetic fields created by the π electrons. See Figure 29-6.

6. (a) Each shows a single absorption: methyl ether near 3.5 δ and acetone near 2.0 δ .

   (b) $(CH_3)_3CC(CH_3)_3$ shows only a single line in the nmr spectrum. The spectrum of $(CH_3)_2C{=}CH_2$ is more complex.

   (c) The nmr spectrum of toluene shows a methyl singlet near 2.0 δ .

   (d) Since chlorine is more electronegative than bromine, the $-CH_2Cl$ absorption (quartet) will appear further downfield than $-CH_2Br$.

   (e) The $-OCH_3$ singlet will appear further downfield than the $-\overset{O}{\overset{\|}{C}}-CH_3$ singlet.

7. See the shielding constants, Table 29-2.

(a) $0.23 + 0.47 + 0.47 = 1.17\ \delta$

(b) $0.23 + 0.47 + 2.56 = 3.26\ \delta$

(c) $0.23 + 1.14 + 2.56 = 3.93\ \delta$

(d) $0.23 + 1.70 + 1.70 = 3.63\ \delta$

(e) $0.23 + 0.47 + 1.85 = 2.55\ \delta$

8. (a) 4 (b) 6

9. The boldface protons are equivalent in structures (a), (g) and (h).

10. The boldface protons will have the same chemical shift in structures (a), (b), (d), (e) and (g).

11. (a) 1
(b) 1
(c) 7 (triplet + quartet)
(d) 10 (septet + triplet)
(e) 7 (triplet + quartet)
(f) 1
(g) 8 (singlet + triplet + quartet)

12. No. For this to be a doublet there must be another proton responsible for the splitting; thus, there would have to be another signal. The two lines are two singlets.

13. Fluorine causes splitting of the proton signal.

14. $(Cl)(H_c)C{=}C(H_a)(H_b)$ All protons nonequivalent

Coupling constants: J_{H_a,H_b} J_{H_a,H_c} J_{H_b,H_c}

15. (a) quartet (1H) and doublet (3H); quartet further down-field.

(b) a singlet

(c) triplet (1H) and doublet (2H); triplet further down-field.

(d) singlet (1H) and singlet (3H).

16. Slow rotation about the carbon-nitrogen bond of $-\overset{O}{\overset{\|}{C}}-N<$ makes the two ethyl groups nonequivalent.

17. Check *your* definitions on the following pages:

(a) 1235 and 1256
(b) 1222 and 1256
(c) 1223
(d) see Fig. 29-3 and 1258 (upfield)
(e) 1223
(f) 1222 and 1257
(g) 1225 and 1256
(h) 1238 and 1258
(i) 1238 and 1256
(j) 1229 and 1257
(k) 1231 and 1256
(l) 1239

18. In the diagrams below each division represents 0.5 Hz

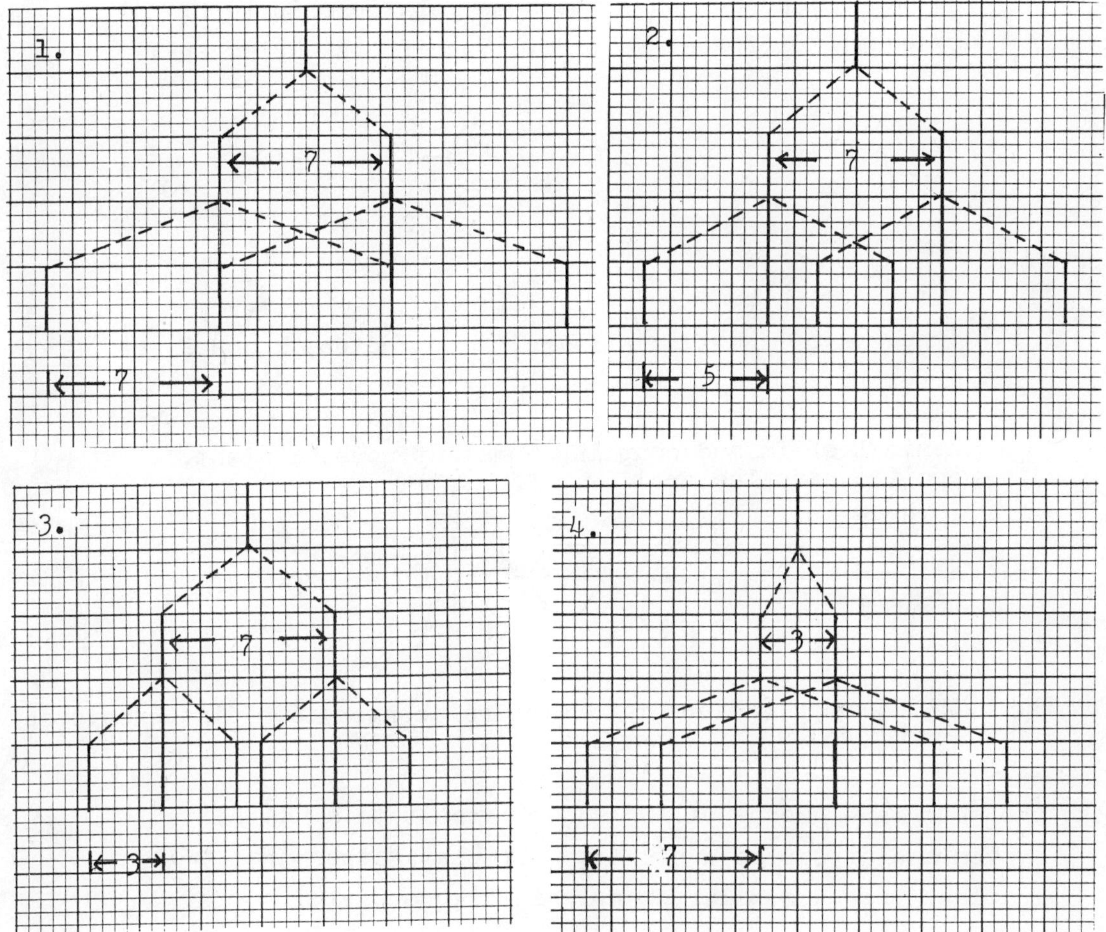

If the resolution is 2 Hz, the number of lines appearing in the spectra are: (1) 4, (2) 6, (3) 6, (4) 5.

If the resolution is 1 Hz, the number of lines appearing in the spectra are: (1) 4, (2) 6, (3) 6, (4) 6

19. (a) There are many examples including the protons of:

(i) CH_3CH_3 (ii) $ClCH_2CH_2Cl$, etc.

(b) For example, compounds of the type:

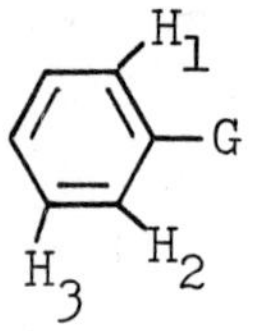

[structure: benzene ring bearing G, with H_1 and H_2 ortho to G and H_3 meta]

where protons H_1 and H_2 (chemical shift equivalent) do not have the same relationship to H_3 and couple differently to H_3. Specific examples are:

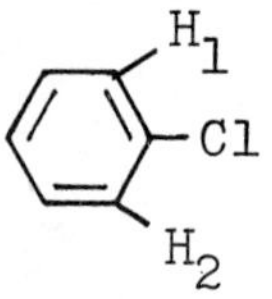

[structures: chlorobenzene with H_1, H_2 ortho to Cl] and [4-methylanisole (CH_3O–C$_6$H$_4$–CH_3) with H_1, H_2 ortho to CH_3] etc.

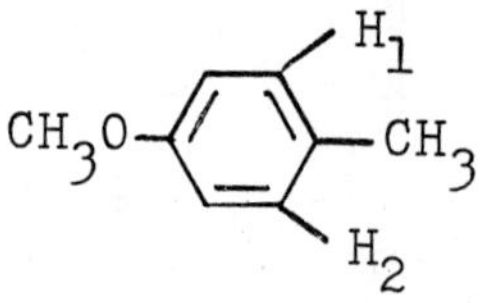

(c) Compounds of the type Y–C(H_3)(Z)–C(H_1)(H_2)–R Protons 1 and 2 are chemical shift and coupling constant nonequivalent. Specific examples are:

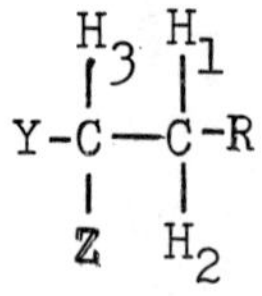

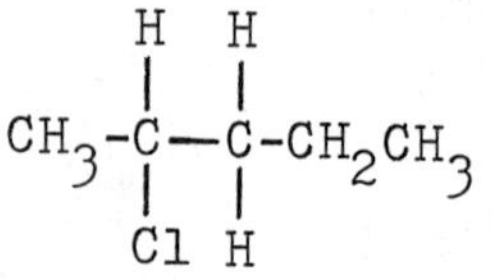

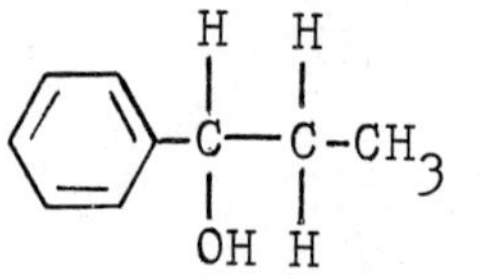

CH_3–CH(Cl)–CH_2–CH_2CH_3 , C_6H_5–CH(OH)–CH_2–CH_3 , etc.

20. The methyl groups (adjacent to a chiral center) are diastereotopic and are chemical shift nonequivalent.

21. The methylene groups of ICH_2CH_2Cl constitute an AA'BB' system and appear as a complex multiplet rather than two triplets (see a similar example in Figure 29-18).

22. (a) identical
(b) enantiotopic
(c) enantiotopic
(d) enantiotopic
(e) diastereotopic
(f) identical
(g) diastereotopic
(h) none of these
(i) enantiotopic
(j) diastereotopic
(k) identical
(l) diastereotopic
(m) diastereotopic
(n) identical
(o) diastereotopic
(p) enantiotopic
(q) diastereotopic
(r) diastereotopic
(s) enantiotopic
(t) diastereotopic
(u) identical
(v) identical

23. Yes. A careful analysis of such features as the N-H absorption (amine NH vs. amide NH) and the coupling between the methine hydrogens (in a four-membered ring vs. acyclic) could provide the information necessary to distinguish these.

24. (a) In D_2O, hydrogen-deuterium exchange occurs giving CH_3OD, and only the protons of the methyl group appear in the spectrum.
(b) In the presence of a trace of HCl or H_2O, exchange of the OH protons is rapid, and the NMR spectrometer sees only an average enviornment. Thus, the OH signal is not split by hydrogens on the adjacent carbon, and the signal for the protons of the methylene are not split by the proton of the OH group.

25. Two singlets. A doublet (a signal split by one adjacent proton) would require that another signal be present.

26.

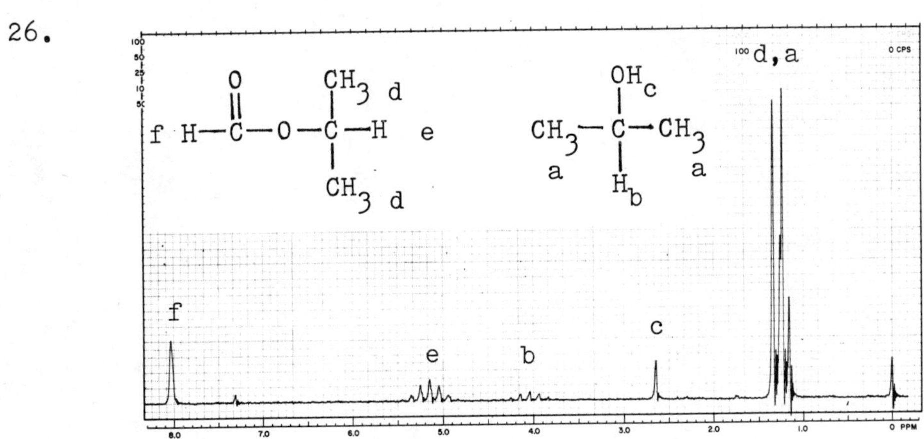

27. (a) $CH_2{=}C(CH_3){-}\overset{O}{\overset{\|}{C}}{-}OH$ or $CH_3{-}CH{=}CH{-}\overset{O}{\overset{\|}{C}}{-}OH$

(in either, a multiplet for CH_3 must be due to long-range coupling)

(b) $CH_3CH_2CH(Br)\overset{O}{\overset{\|}{C}}OH$

(c) $CH_3CH_2O\overset{O}{\overset{\|}{C}}CH_3$

(d) $C_6H_5{-}CH(NH_2)CH_3$

(e) CH_3OH

(f) $CH_2ClCHCl_2$

(g) CF_3CH_2OH

(h) $ClCH_2CH_2CH_2Cl$

(i) CH_2 — CH_2
| |
CH_2 — O

28. A mixture of $\underset{CH_3}{\overset{H}{\diagdown}}C{=}N{\diagup}OH$ and $\underset{CH_3}{\overset{H}{\diagdown}}C{=}N{\diagdown}OH$

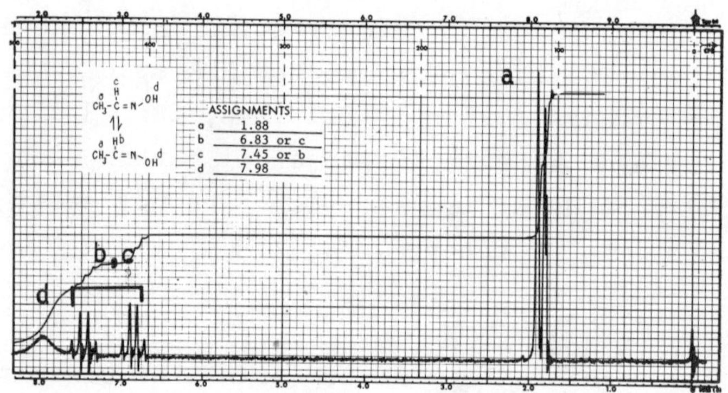

29. At -130° interconversion of A and B is slow enough that each can be detected in a mixture. A should be more stable (maximum separation of fluorine dipoles). At room temperature, interconversion is so rapid that the NMR spectrometer "sees" only an average fluorine enviornment.

$Cl_2FC{-}CFCl_2$ (A) $\rightleftharpoons$ $Cl_2FC{-}CFCl_2$ (B)

A B

30. A

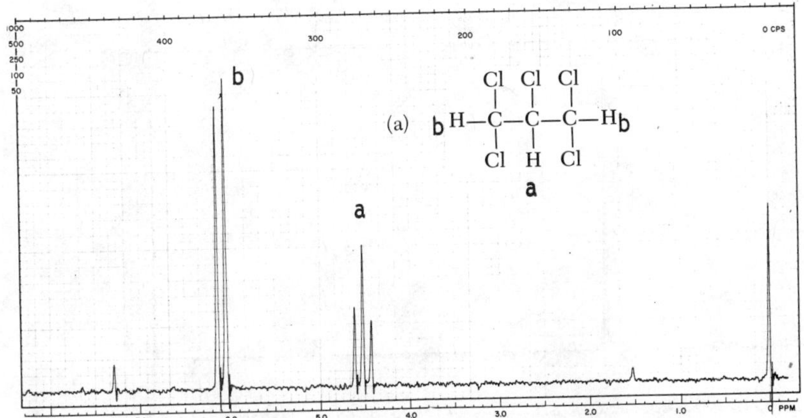

The signal for the "b" protons is further downfield because of the deshielding effect of the two geminal chlorines.

B

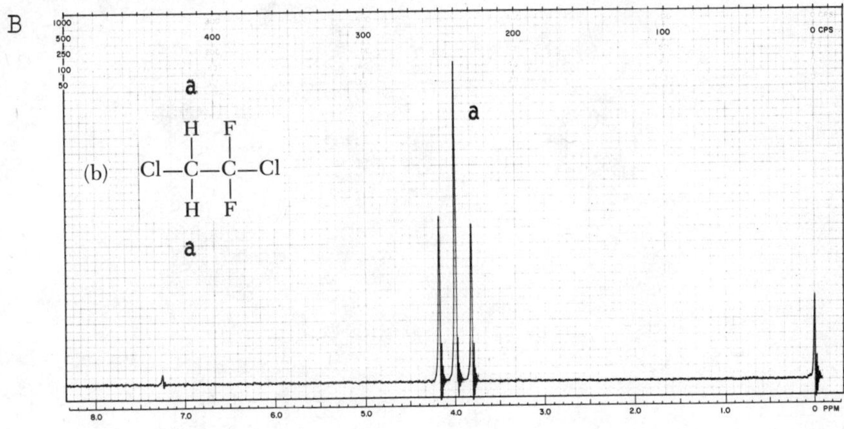

The signal for the protons is split into a triplet by the two fluorine atoms. J is larger than usually observed for proton-proton splittings.

C

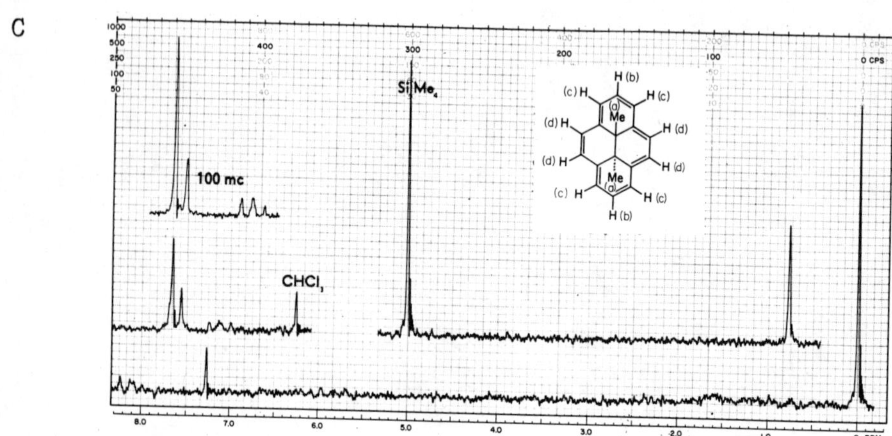

The absorption which appears upfield of TMS at -4.2 δ is for the methyl groups which experience unusual shielding because of their orientation relative to the ring current.

D

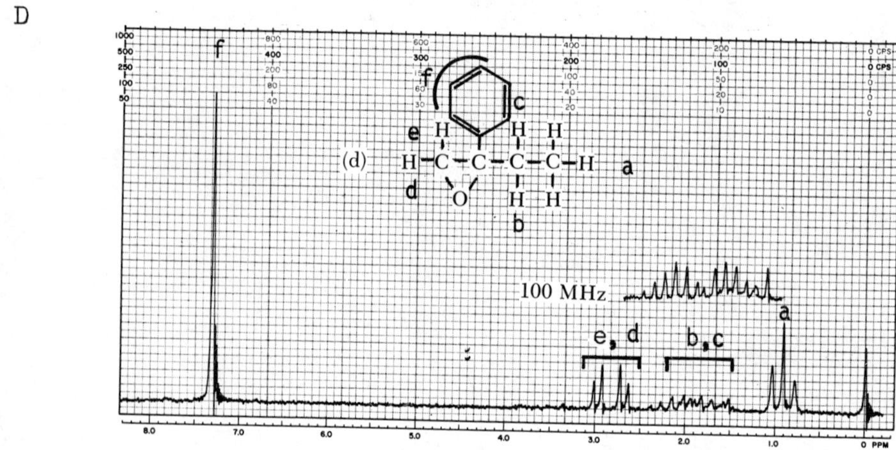

Note the increased resolution of the b,c protons at 100 MHz.

31. At high temperature the methyl groups are equivalent because of the rapid interconversion:

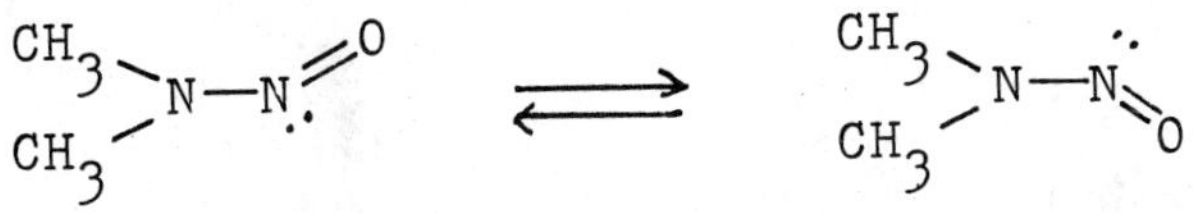

$$(CH_3)_2N{-}N{=}O \rightleftharpoons (CH_3)_2N{-}N{=}O$$

32.

H_B |← 7.2 →| H_A

|← 15 Hz →|← x →|← 15 Hz →|

1 2 3 4

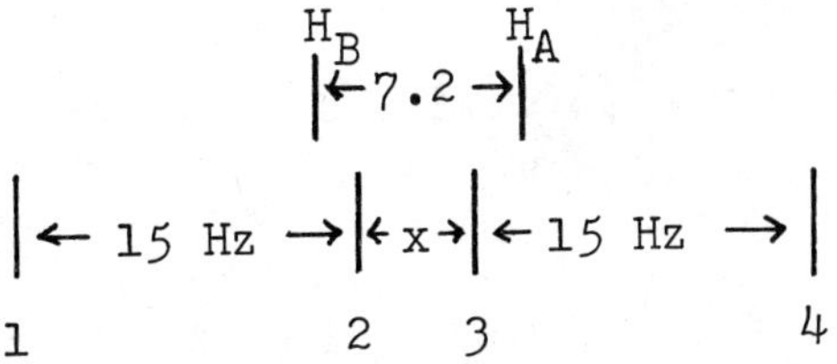

$\delta H_A - \delta H_B = 1.00 - 0.88 = 0.12\,\delta = 7.2$ Hz

$$7.2 = \sqrt{(H_1 - H_4)(H_2 - H_3)} = \sqrt{(30 + x)(x)}$$

$$(7.2)^2 = (30 + x)(x)$$

$$51.8 = 30x + x^2 \quad \text{and} \quad x^2 + 30x - 51.8 = 0$$

solve for x with quadratic formula

$$x = \frac{-b \pm \sqrt{b^2 - 4ac}}{2a} = \frac{-30 \pm \sqrt{1107}}{2}$$

x = 1.5 Hz

The midpoint of the AB pattern is 0.94 δ or 56.4 Hz

Thus:

$$H_1 = 56.4 + 0.75 + 15 = 72.15 \text{ Hz} = 1.20\,\delta$$

$$H_2 = 56.4 + 0.75 = 57.15 \text{ Hz} = 0.95\,\delta$$

$$H_3 = 56.4 - 0.75 = 55.65 \text{ Hz} = 0.93\,\delta$$

$$H_4 = 56.4 - 0.75 - 15 = 40.65 \text{ Hz} = 0.68\,\delta$$

33. (a) Integration of peak areas will distinguish 1-butene (3 vinyl protons) from the others (2 vinyl protons in each).

(b) Isobutylene, <u>cis</u>-2-butene, and <u>trans</u>-2-butene each have two equivalent vinyl protons. The NMR spectrum of each has two singlets; therefore these are not readily distinguished by NMR (they can be distinguished by IR).

34. The spectra should show the following features:

(a) singlet near 1 δ (CH_3)

(b) singlet near 3 δ ($-CH_2I$)

(c) triplet near 1 δ (CD_2H-)

doublet near 3 δ ($-CH_2I$)

(d) doublet near 1 δ ($-CHD_2$)

doublet near 3 δ ($-CHDI$)

(e) pair of doublets near 1 δ (CH_2D, diastereotopic protons)

4 line pattern near 3 δ (CHDI), (this proton is split by two diastereotopic protons)

35. All are calculated from the expression:

$$= \frac{\Delta\nu \times 10^6}{\text{oscillator frequency (Hz)}}$$ (see p. 1223)

Example: CH_3R (0.8-1.2)

$0.8 = \frac{\Delta\nu}{60}$; $\Delta\nu = 48$ Hz

$0.8 = \frac{\Delta\nu}{100}$; $\Delta\nu = 80$ Hz

$0.8 = \frac{\Delta\nu}{300}$; $\Delta\nu = 240$ Hz

$1.2 = \frac{\Delta\nu}{60}$; $\Delta\nu = 72$ Hz

$1.2 = \frac{\Delta\nu}{100}$; $\Delta\nu = 120$ Hz

$1.2 = \frac{\Delta\nu}{300}$; $\Delta\nu = 360$ Hz

36. 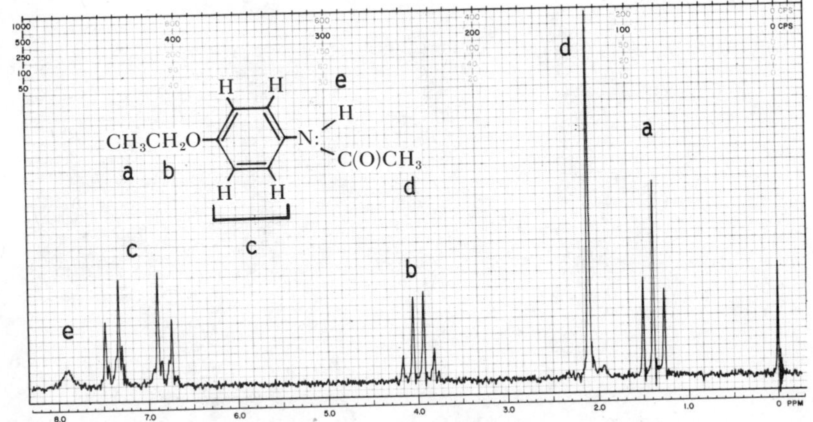

37. The hydrogens are chemical shift and coupling constant equivalent.

38. Yes. H_a and H_e do not have a symmetric relationship with bonds "a" but they do with bonds "b" (or "c").

39. (a) $CH_3\overset{O}{\overset{\|}{C}}NH_2$

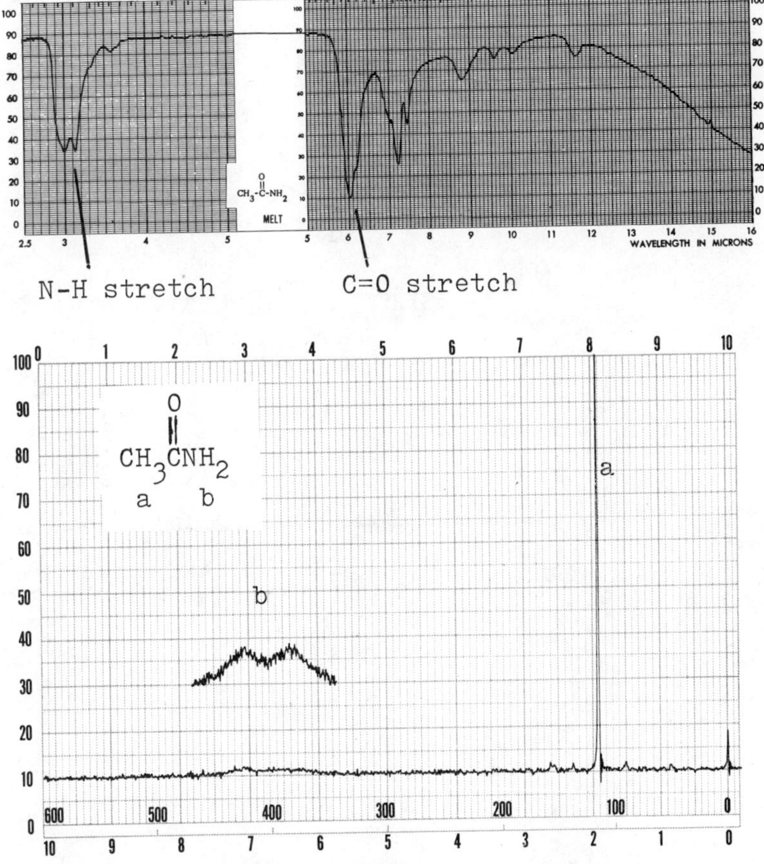

(b) $H_2N-C_6H_4-CH_2-C\equiv N$

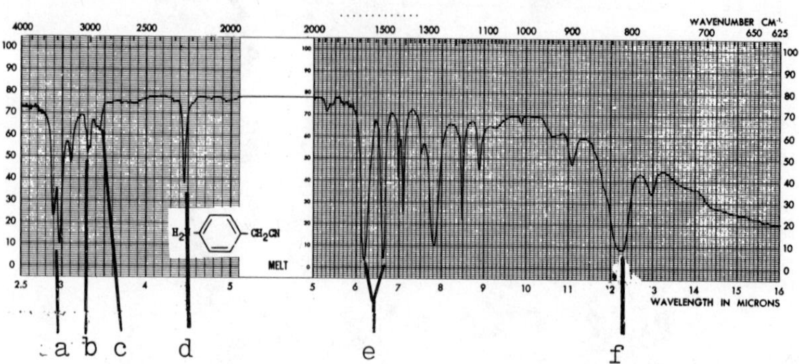

a: N-H stretch (symmetric and asymmetric), 1° amine

b: =C-H stretch (benzene)

c: C-H stretch (alkyl)

d: C≡N stretch

e: C=C stretch (benzene)

f: =C-H out-of-plane bending (p-disubstituted benzene)

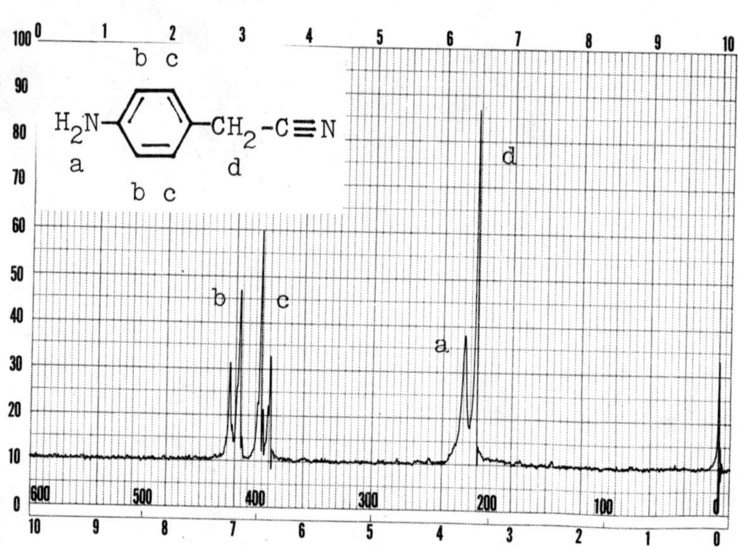

(c) $CH_3-C_6H_4-Br$

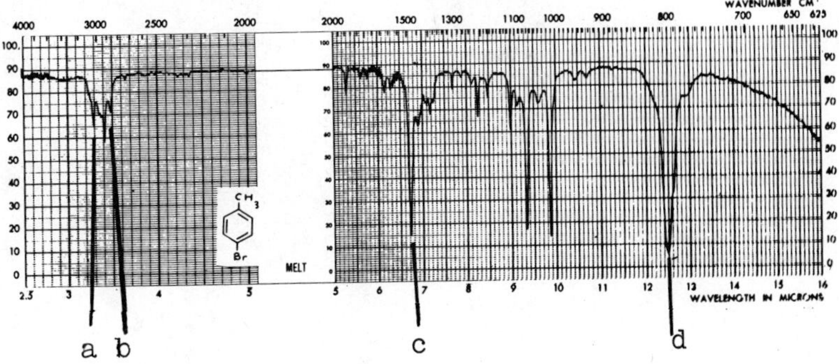

a: =C-H stretch (benzene)

b: C-H stretch (alkyl)

c: C=C stretch (benzene)

d: =C-H out-of-plane bending (p-disubstituted benzene)

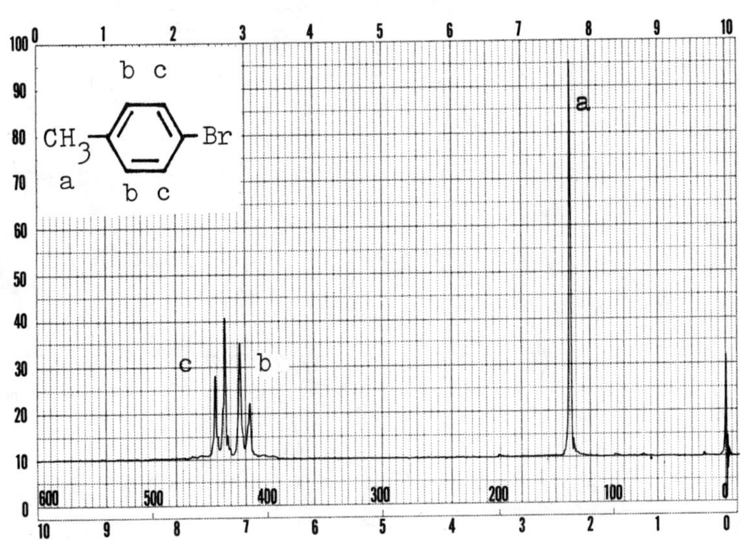

(d) $(CH_3CH_2)_3COH$

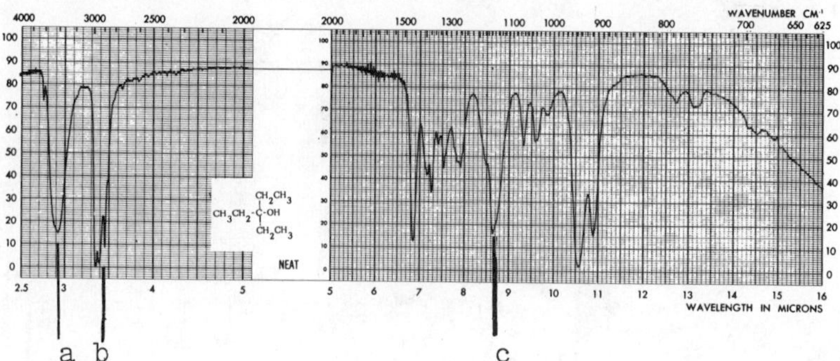

a: O-H stretch (alcohol)

b: C-H stretch (alkane)

c: C-O bending (3° alcohol)

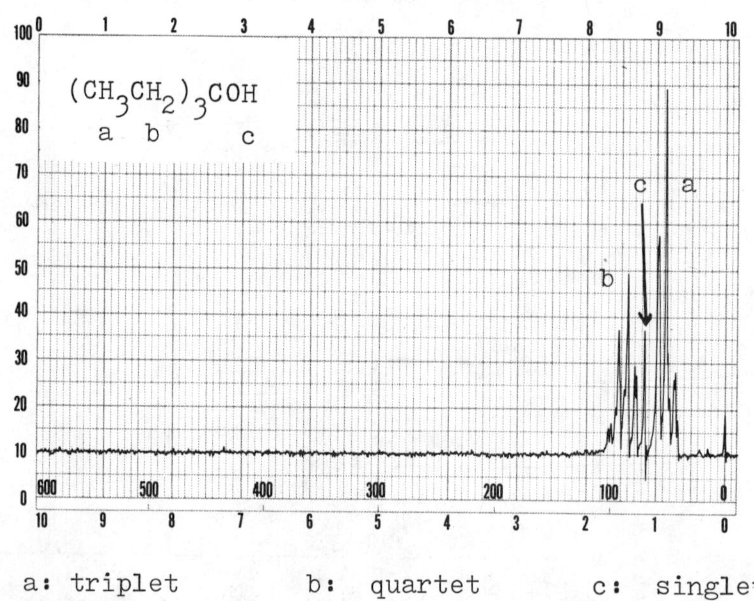

a: triplet b: quartet c: singlet

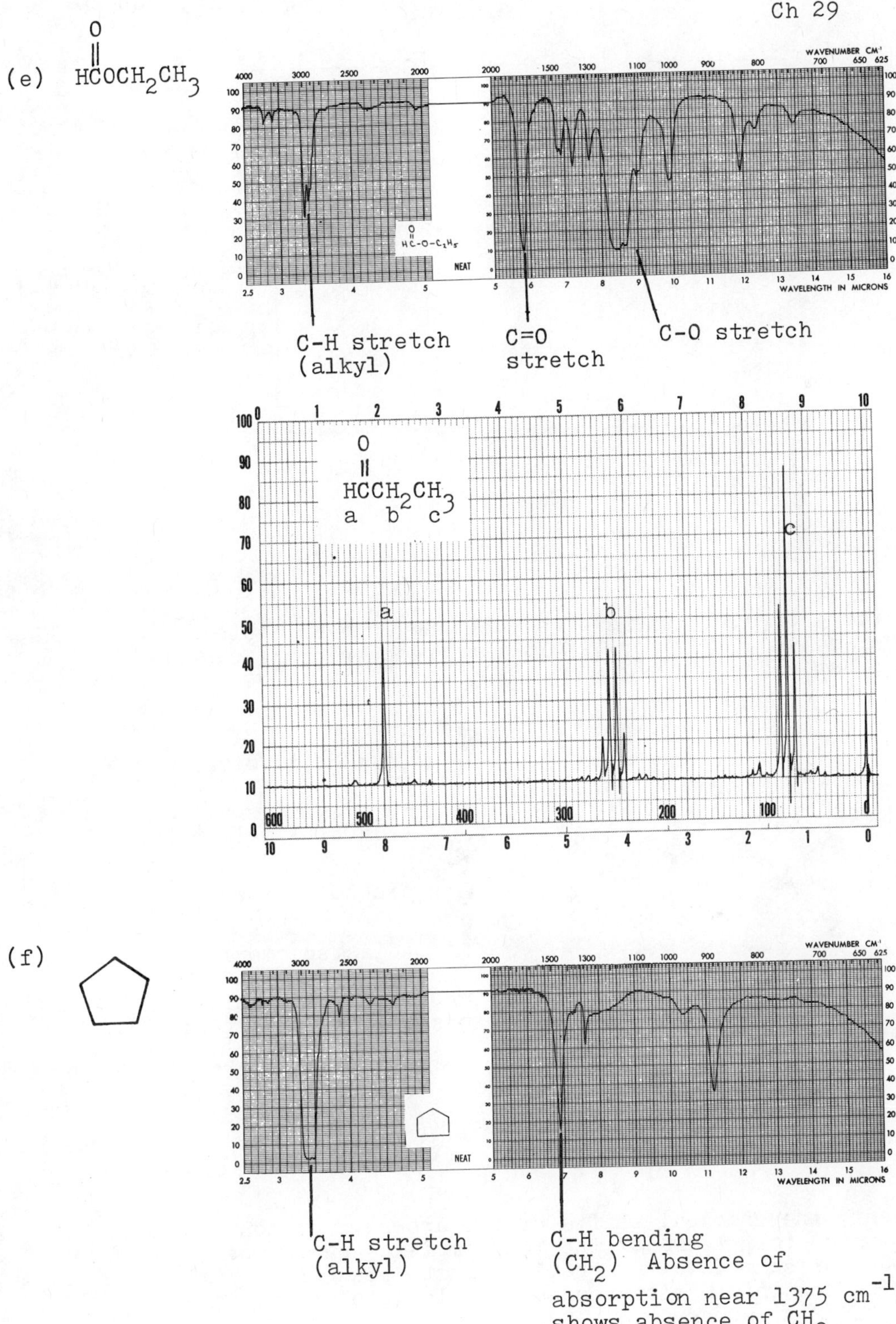
(e) HCOCH2CH3
C-H stretch (alkyl)
C=O stretch
C-O stretch
HCCH2CH3
a b c
a
b
c
(f)
C-H stretch (alkyl)
C-H bending (CH2) Absence of absorption near 1375 cm-1 shows absence of CH3.

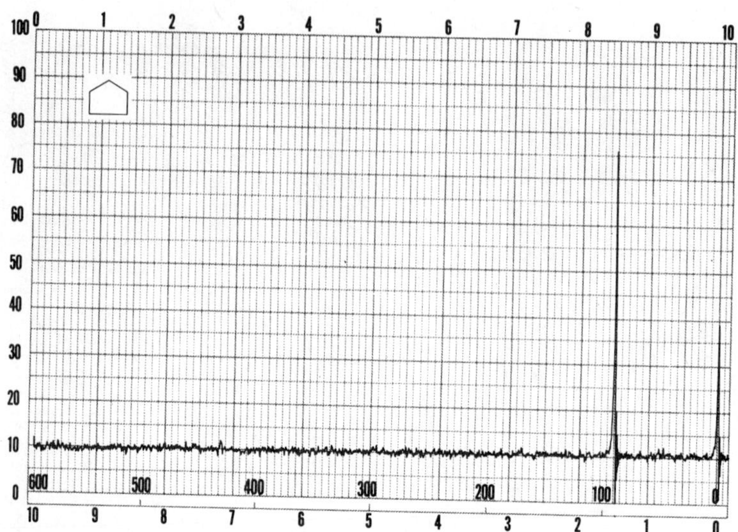

40. A mixture of enantiomers of a compound in a suitable chiral solvent [(+) or (-) isomer] produces complexes which are diastereomeric, e.g.,

$$(\pm)\text{-ROH} + (+)\text{-solvent} \longrightarrow \begin{cases} [(+)\text{ROH } (+)\text{solvent}] \\ [(-)\text{ROH } (+)\text{solvent}] \end{cases}$$

alcohol-solvent complexes (diastereomeric)

Thus, the protons of a particular group in one complex will not be equivalent to the protons of the same group in the other complex.

In a mixture containing (±) $CH_3C(OH)(C_6H_5)CH_2CH_3$ and a chiral solvent, the chemical shifts of the aromatic proton resonances (singlets) of the two complexes can be compared, for example. Integration of the peak areas will provide the information needed to calculate the percentage of each isomer.

(continued next page)

Matching the two aromatic proton resonances with the correct isomers can be accomplished by obtaining the NMR of each pure enantiomer in the chiral solvent.

41. Yes. They are equivalent. This can be determined by replacing each hydrogen in turn by some other group such as deuterium. In this case all structures are identical.

$$\mathrm{(D)(H)C{=}C(H)(H)} \qquad \mathrm{(H)(D)C{=}C(H)(H)} \qquad \mathrm{(H)(H)C{=}C(H)(D)}$$

$$\mathrm{(H)(H)C{=}C(D)(H)}$$

all identical

42. None of these. Replacing the hydrogens in turn by another atom or group such as deuterium gives structural isomers rather than stereoisomers.

$$\mathrm{(Cl)(H)C{=}C(Br)(D)} \quad \text{and} \quad \mathrm{(Cl)(D)C{=}C(Br)(H)}$$

structural isomers

PERIODIC CHART OF THE ELEMENTS

1A	2A	3B	4B	5B	6B	7B	8B	8B	8B	1B	2B	3A	4A	5A	6A	7A	8A
																1 **H** 1.00797 ±0.00001	2 **He** 4.0026 ±0.00005
3 **Li** 6.939 ±0.0005	4 **Be** 9.0122 ±0.00005											5 **B** 10.811 ±0.003	6 **C** 12.01115 ±0.00005	7 **N** 14.0067 ±0.00005	8 **O** 15.9994 ±0.0001	9 **F** 18.9984 ±0.00005	10 **Ne** 20.183 ±0.0005
11 **Na** 22.9898 ±0.00005	12 **Mg** 24.312 ±0.0005											13 **Al** 26.9815 ±0.00005	14 **Si** 28.086 ±0.001	15 **P** 30.9738 ±0.00005	16 **S** 32.064 ±0.003	17 **Cl** 35.453 ±0.001	18 **Ar** 39.948 ±0.0005
19 **K** 39.102 ±0.0005	20 **Ca** 40.08 ±0.005	21 **Sc** 44.956 ±0.0005	22 **Ti** 47.90 ±0.005	23 **V** 50.942 ±0.0005	24 **Cr** 51.996 ±0.001	25 **Mn** 54.9380 ±0.00005	26 **Fe** 55.847 ±0.003	27 **Co** 58.9332 ±0.00005	28 **Ni** 58.71 ±0.005	29 **Cu** 63.54 ±0.005	30 **Zn** 65.37 ±0.005	31 **Ga** 69.72 ±0.005	32 **Ge** 72.59 ±0.005	33 **As** 74.9216 ±0.00005	34 **Se** 78.96 ±0.005	35 **Br** 79.909 ±0.002	36 **Kr** 83.80 ±0.005
37 **Rb** 85.47 ±0.005	38 **Sr** 87.62 ±0.005	39 **Y** 88.905 ±0.0005	40 **Zr** 91.22 ±0.005	41 **Nb** 92.906 ±0.0005	42 **Mo** 95.94 ±0.005	43 **Tc** (99)	44 **Ru** 101.07 ±0.005	45 **Rh** 102.905 ±0.0005	46 **Pd** 106.4 ±0.05	47 **Ag** 107.870 ±0.003	48 **Cd** 112.40 ±0.005	49 **In** 114.82 ±0.005	50 **Sn** 118.69 ±0.005	51 **Sb** 121.75 ±0.005	52 **Te** 127.60 ±0.005	53 **I** 126.9044 ±0.00005	54 **Xe** 131.30 ±0.005
55 **Cs** 132.905 ±0.0005	56 **Ba** 137.34 ±0.005	57 ***La** 138.91 ±0.005	72 **Hf** 178.49 ±0.005	73 **Ta** 180.948 ±0.0005	74 **W** 183.85 ±0.005	75 **Re** 186.2 ±0.05	76 **Os** 190.2 ±0.05	77 **Ir** 192.2 ±0.05	78 **Pt** 195.09 ±0.005	79 **Au** 196.967 ±0.0005	80 **Hg** 200.59 ±0.005	81 **Tl** 204.37 ±0.005	82 **Pb** 207.19 ±0.005	83 **Bi** 208.980 ±0.0005	84 **Po** (210)	85 **At** (210)	86 **Rn** (222)
87 **Fr** (223)	88 **Ra** 226.05	89 **†Ac** (227)	104 (257)	105													

*Lanthanum Series

58 **Ce** 140.12 ±0.005	59 **Pr** 140.907 ±0.0005	60 **Nd** 144.24 ±0.005	61 **Pm** (147)	62 **Sm** 150.35 ±0.005	63 **Eu** 151.96 ±0.005	64 **Gd** 157.25 ±0.005	65 **Tb** 158.924 ±0.0005	66 **Dy** 162.50 ±0.005	67 **Ho** 164.930 ±0.0005	68 **Er** 167.26 ±0.005	69 **Tm** 168.934 ±0.0005	70 **Yb** 173.04 ±0.005	71 **Lu** 174.97 ±0.005

†Actinium Series

90 **Th** 232.038 ±0.0005	91 **Pa** (231)	92 **U** 238.03 ±0.005	93 **Np** (237)	94 **Pu** (242)	95 **Am** (243)	96 **Cm** (247)	97 **Bk** (249)	98 **Cf** (249)	99 **Es** (254)	100 **Fm** (253)	101 **Md** (256)	102 **No** (253)	103 **Lr** (257)

Printed in U. S. A.

Atomic Weights are based on C^{12}—12.0000 and Conform to the 1961 Values

Infrared Absorption Frequencies of Common Functional Groups. (N.B. Colthup; Courtesy of American Cyanamide Company, Stamford, Conn. Used with permission of the Journal of the Optical Society of America.)

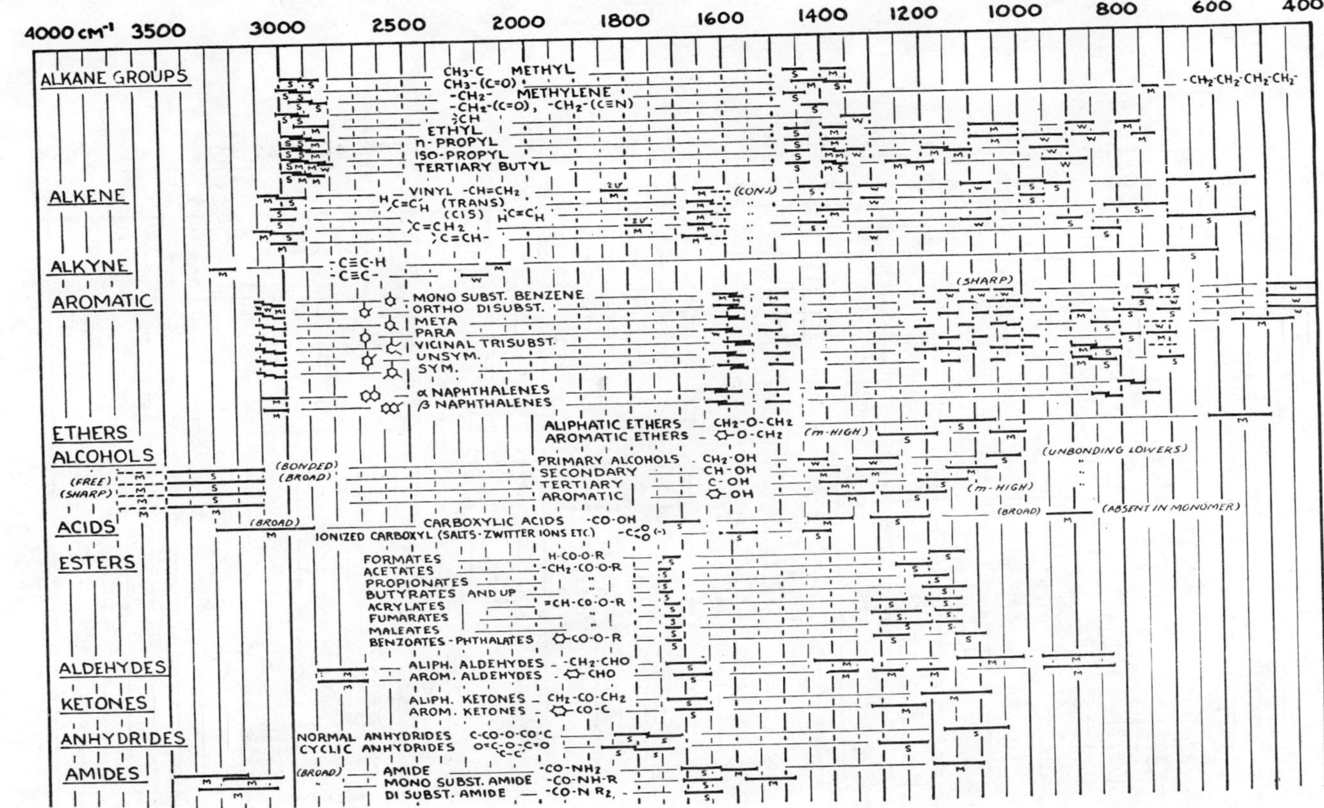

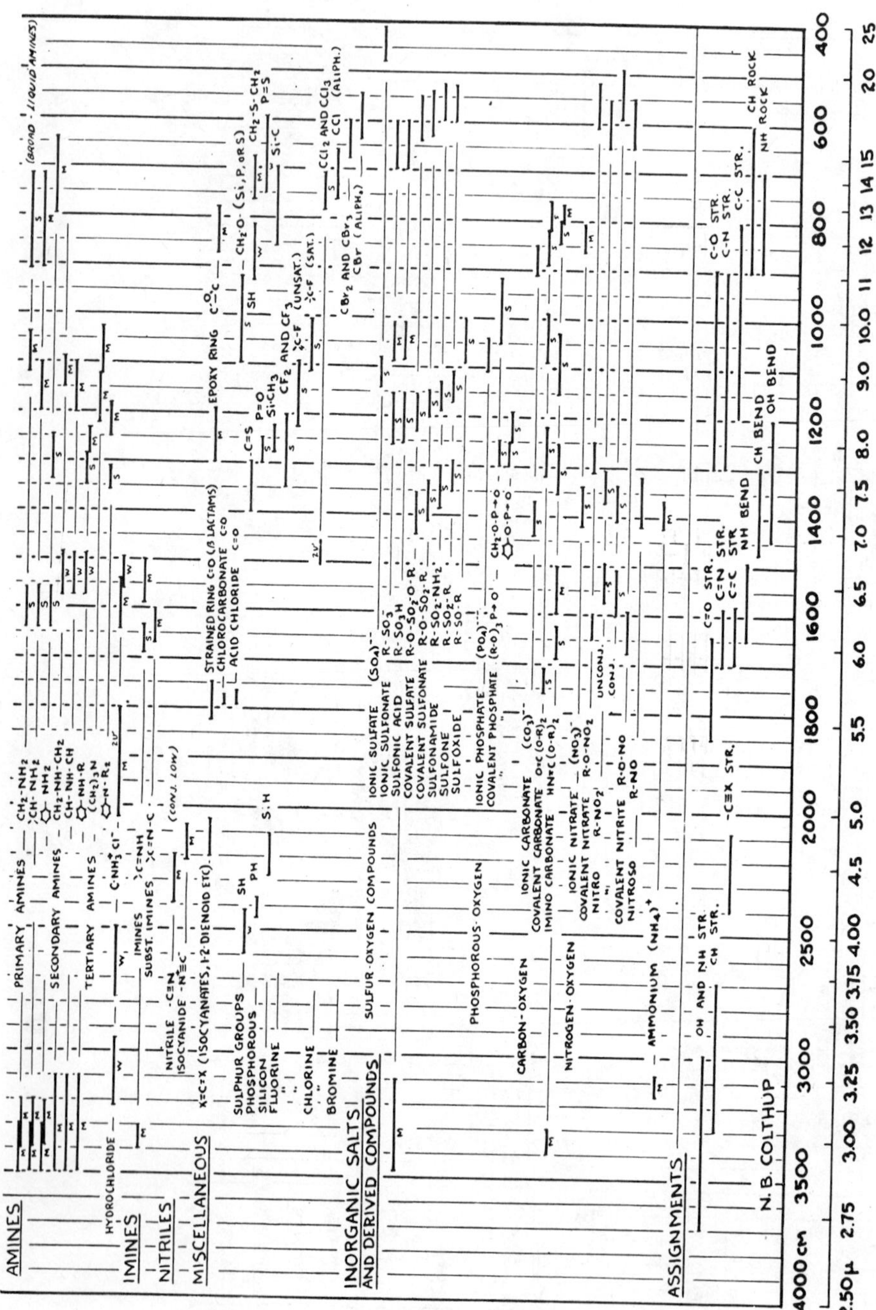
AMINES
PRIMARY AMINES
SECONDARY AMINES
TERTIARY AMINES
HYDROCHLORIDE
IMINES
NITRILES
MISCELLANEOUS
SULPHUR GROUPS
PHOSPHOROUS
SILICON
FLUORINE
CHLORINE
BROMINE
INORGANIC SALTS AND DERIVED COMPOUNDS
SULFUR-OXYGEN COMPOUNDS
PHOSPHOROUS-OXYGEN
CARBON-OXYGEN
NITROGEN-OXYGEN
AMMONIUM
ASSIGNMENTS
OH AND NH STR.
CH STR.
N. B. COLTHUP
4000 CM 3500 3000 2500 2000 1800 1600 1400 1200 1000 800 600 400
2.50μ 2.75 3.00 3.25 3.50 3.75 4.00 4.5 5.0 5.5 6.0 6.5 7.0 7.5 8.0 9.0 10.0 11 12 13 14 15 20 25

CHEMICAL SHIFTS OF PROTON GROUPINGS*

Methyl Groups CH_3—		*Methylene Groups* RCH_2—	
Group	**Shift, δ ppm**	**Group**	**Shift, δ ppm**
CH_3—R	0.8–1.2	R—CH_2—R	1.1–1.5
CH_3—CR=C<	1.6–1.9	R—CH_2—Ar	2.5–2.9
CH_3—Ar	2.2–2.5	R—CH_2—C(=O)—R	2.5–2.9
CH_3—C(=O)—R	2.1–2.4	R—CH_2OH	3.2–3.5
CH_3—C(=O)—Ar	2.4–2.6	R—CH_2OAr	3.9–4.3
CH_3—C(=O)—OR	1.9–2.2	R—CH_2—OC(=O)R	3.7–4.1
CH_3—C(=O)—OAr	2.0–2.5	R—CH_2—Cl	3.5–3.7
CH_3—N<	2.2–2.6	*Methine Groups* R_2CH—	
CH_3—OR	3.2–3.5	R_3CH	1.4–1.6
CH_3—OAr	3.7–4.0	R_2CHOH	3.5–3.8
CH_3—OC(=O)R	3.6–3.9	Ar_2CHOH	5.7–5.8
		Other Groups	
Unsaturated Groups		ROH	3–6
RCH=C<	5.0–5.7	ArOH	6–8
Ar—H	6.0–7.5	RCO_2H	10–12
R—C(=O)—H	9.4–10.4	RNH—	2–4

*In this table, R = saturated carbon (CH_3—, —CH_2—, —CH, —C(—)(—)—;) Ar = aromatic ring.